Introduction to Engineering Materials:
The Bicycle and the Walkman

C.J. McMahon, Jr.
and
C.D. Graham, Jr.

*Professors of Materials Science and Engineering
University of Pennsylvania*

The trademark "Walkman" is the property of the Sony Corporation.

Copyright 1992 C.J. McMahon, Jr

All rights reserved. This book, or parts thereof, may not be reproduced in any form without written permission of the publisher.

ISBN 0-9646598-0-8

Merion Books
Philadelphia, PA

Distributed by:
  Whitman Distribution Center
  10 Water Street
  Lebanon, NH 03766
  Fax (603) 448-2576

Library of Congress Catalog Card Number 95-77924

Preface

This book was developed for a one-semester introductory course in materials science and engineering designed to serve both as a first course for majors in the field and a complete introduction for non-majors. The course has been taken successfully by students with only advanced secondary-school preparation in physics, chemistry, and calculus.

In a departure from the conventional one, the book employs what are, in effect, two large case studies: the bicycle and the Walkman, to provide paradigms for structural materials and device materials. This method has been found to provide a framework that enables students to organize and retain the diverse subject matter of a first course in MSE.

Our objective has been to provide the student with a coherent knowledge of the subject with enough depth that no further formal instruction is needed for an engineer outside the field of MSE. This requires that breadth of coverage be restricted. The topics have been selected so that the needs of students from other engineering fields are satisfied and no major topic covered in a conventional introductory course is omitted.

This book has benefitted greatly from the advice of many students in our course at Penn from 1987 to 1995 and particularly from the execution of illustrations by Jodi Forlizzi, editing by Edward Hewett, the photography of Felice Macera, the metallography of Elmer Anderson, the design and layout by Kathy McLaughlin and the various contributions of Helen McMahon, Ransom Weaver, Alex Radin, Marybeth Meyer, Sue Flom, Pat Overend and Cliff Warner. Gwyn Jones of Merlin Metalworks and Harry Havnoonian of Cycle Sport have provided invaluable advice and assistance.

The authors would be pleased to receive comments and suggestions. Contact can be made by FAX at 215-573-2128 or by e-mail at cmcmahon@lrsm.upenn.edu. Current information on the availability of supplementary multimedia material on CD-ROM can be obtained on the world-wide web at the following address: http://www.lrsm.upenn.edu/bw/BW.html

Dedicated to the memory of our friend and colleague, Professor Peter Haasen.

CONTENTS

1 THE BICYCLE IN HISTORY 1
1.1 The Significance of the Bicycle 1
1.2 The Development of the Bicycle 4
References 9

2 WHEELS AND SPOKES; CORROSION RESISTANCE 10
2.1 The Bicycle Wheel 10
2.2 Carbon Steel *vs.* Stainless Steel Spokes 13
2.3 Corrosion of Steels 13
2.4 Corrosion Protection of Carbon Steel 14
2.5 Constitution of Stainless Steel Spokes 17
Summary 19
Exercises 20

3 MECHANICAL BEHAVIOR OF SPOKES 22
3.1 Types of Stress and Strain 22
3.2 Small-Scale *vs.* Large-Scale Deformation 25
3.3 Measurement of Stress *vs.* Strain — The Tensile Test ... 27
3.4 Elastic Behavior 29
3.5 Plastic Behavior 32
3.6 Large-Scale Plastic Flow — The Stress-Strain Curve 37
Summary 41
Appendix 3.1 - The Load Cell 42
Appendix 3.2 - Criterion for Necking in a Tensile Test ... 42
Exercises 43

4 MICROSTRUCTURE AND CRYSTAL STRUCTURE OF STAINLESS STEEL SPOKES 45
4.1 Metallographic Sample Preparation 45
4.2 Optical Microscopy 45
4.3 Imaging of Grain Boundaries by Reflection Contrast ... 46
4.4 The Structure of a Perfect FCC Crystal 48
4.5 Crystallographic Indices of Planes 49
4.6 Direction Indices 52
Summary 53
Appendix 4.1 - The Scalar Product of Two Vectors 54
Exercises 54

5 SLIP, DISLOCATIONS, AND STRAIN HARDENING IN STAINLESS STEEL SPOKES 56

5.1 Slip in FCC Crystals. 57
5.2 Types of Dislocation. 59
   5.2.1 The Edge Dislocation. 59
   5.2.2 The Screw Dislocation 60
   5.2.3 Mixed Dislocations 62
   5.2.4 General Characteristics of Dislocations 65
5.3 Imaging of Dislocations 66
5.4 The Stress Field of a Dislocation 67
   5.4.1 General Aspects of Stress Fields 67
   5.4.2 The Stress Field of a Screw Dislocation. 68
   5.4.3 The Stress Field of an Edge Dislocation. 70
   5.4.4 Mixed Dislocations 71
   5.4.5 Energy of a Dislocation 71
5.5 Mechanism of Strain Hardening 72
   5.5.1 Plastic Strain and Dislocation Density 73
   5.5.2 Yielding and Dislocation Multiplication. 74
   5.5.3 The Physical Basis of Strain Hardening 78
      Summary. 81
      Appendix 5.1 - Bragg's Law of Diffraction 82
      Appendix 5.2 - Stress Field of an Edge Dislocation. . . . 83
      Exercises 85

6 ANNEALING OF COLD-WORKED SPOKES 87

6.1 Thermodynamic Principles 87
6.2 Stages of Annealing 90
6.3 Recrystallization. 90
   6.3.1 Mechanism of Recrystallization 90
   6.3.2 Kinetics of Recrystallization 91
6.4 Recovery 94
6.5 Grain Growth 97
   6.5.1 Mechanism of Grain Growth 97
   6.5.2 Kinetics of Grain Growth. 99
   6.5.3 Softening During Grain Growth 102
6.6 Annealing of Carbon-Steel Spokes 103
   6.6.1 The As-Received Condition 103
   6.6.2 The Annealed Condition 104
   6.6.3 The "Fully-Annealed" Condition 106
      Summary. 107
      Appendix 6.1 - Measurement of Hardness 108
      Appendix 6.2 - The Equilibrium Concentration of Lattice Vacancies. 109
      Exercises 110

7 CARBON STEEL SPOKES; PLASTIC DEFORMATION AND FATIGUE RESISTANCE 112
7.1 Slip in BCC Iron 112
7.2 Interaction of Solute Atoms with Dislocations 113
7.3 Segregation of Solutes to Dislocations 115
7.3.1 Substitutional Diffusion 115
7.3.2 Interstitial Diffusion 116
7.4 The Nature of Metal Fatigue 118
7.5 The S-N Curve 120
7.5.1 Analysis of the Failure of a Stainless Steel Spoke 122
7.6 The Fatigue Limit of Carbon Steel 124
Summary 125
Appendix 7.1 Brittle Fracture in Carbon Steel 126
Appendix 7.2 Crystal Symmetry 128
Exercises 130

8 ALLOYS FOR SOLDERING AND BRAZING; PHASE DIAGRAMS 131
8.1 Soldering and Brazing 131
8.2 Low-Melting Alloys for Soldering; Phase Diagrams 133
8.2.1 Lowering of the Melting Temperature; Salt-Water Mixtures 134
8.2.2 A Hypothetical Simple-Eutectic Phase Diagram 135
8.2.3 The Lead-Tin System (Soft Solder) 140
8.2.4 Microstructures of Hypoeutectic Lead-Tin Alloys 146
8.2.5 The Microstructure of a Hypereutectic Pb-Sn Alloy ... 150
8.3 The Phase Rule 151
8.4 Brazing Alloys 152
Summary 156
Appendix 8.1 The Allotropic Transformation of Tin 156
Appendix 8.2 The Phase Rule 157
Exercises 159

9 PHASE TRANSFORMATIONS IN CARBON STEEL SPOKES .. 161
9.1 The Iron-Carbon Phase Diagram 162
9.2 Eutectoid Decomposition 164
9.2.1 Isothermal Transformation 164
9.2.2 Transformation during Cooling 168
9.3 Transformations in a Carbon Steel Spoke 171
9.3.1 The "Equilibrium" Condition 171
9.3.2 Isothermal Transformation 171
9.3.3 Transformations during Cooling of a 1040 Steel Spoke 173
9.3.4 Processing of Spokes 174
9.4 The Martensite Transformation 175
9.4.1 Hardening of Steel 175
Summary 179
Exercises 180

10 THE CHAIN AND BEARINGS; RESISTANCE TO FRICTION AND WEAR; HARDENING OF STEEL; CERAMIC BEARINGS 182

- 10.1 Friction 182
- 10.2 Wear 184
- 10.3 Lubrication 185
- 10.4 The Use of Hardened-Steel Components 186
  - 10.4.1 The Hardness of Martensite 187
  - 10.4.2 Hardenability of Steel 188
  - 10.4.3 Tempering of Steel 189
  - 10.4.4 Heat Treatment of Steels 191
  - 10.4.5 Surface Hardening 193
- 10.5 Cracks in Solids 197
- 10.6 The Use of Ceramics for Bearings 199
  - 10.6.1 Crystal Structures 200
  - 10.6.2 Dislocation Motion and Ductility 201
  - 10.6.3 Forming of Ceramics, Sintering 202
  - 10.6.4 Silicon Nitride Bearings 205
    - Summary 208
    - Appendix 10.1 Twinning in Crystals 209
    - Appendix 10.2 Linear-Elastic Fracture Mechanics 211
    - Exercises 214

11 WHEEL RIMS; PRECIPITATION-HARDENED ALUMINUM ALLOYS 216

- 11.1 Mechanism of Precipitation Hardening 217
- 11.2 Requirements for Precipitation Hardening 219
- 11.3 Principles of Precipitation 220
- 11.4 Factors that Control γ 222
- 11.5 Stages of Precipitation Hardening 226
- 11.6 Fabrication of Wheel Rims 227
  - Summary 230
  - Appendix 11.1 Strengthening Effects of Precipitate Particles 230
  - Exercises 232

12 THE BICYCLE FRAME 233

- 12.1 The Frame as an Engineered Structure 233
  - 12.1.1 Forces in a Bicycle Frame; Basic Definitions and Rules 234
  - 12.1.2 Approximation of the Rear Portion as a Pin-Jointed Frame 235
  - 12.1.3 Calculation of Reaction Forces 236
  - 12.1.4 Calculation of Forces in the Links 236
  - 12.1.5 Forces and Moments on the Front Section 238
- 12.2 The Ideal Case: Pure Bending 240
  - 12.2.1 Moments of Inertia 244
- 12.3 Application of Beam Theory to the Front Section 246
- 12.4 Bending of the Frame Due to Out-of-Plane Forces 247
- 12.5. Alloys for Bicycle Frames; General Considerations 250

12.6 Comparisons of Steel, Aluminum-Alloy, and
     Titanium-Alloy Frames........................254
    12.6.1 Road Bikes............................254
    12.6.2 Mountain Bikes 254
12.7 Steel Frames................................255
    12.7.1 Effects of Brazing on Steel Tubing 255
    12.7.2 Lugs; the Fabrication of Complex Shapes........258
    12.7.3 Fabrication of Steel Frames by Welding 260
    12.7.4 Butted Tubing..........................263
12.8 Joining of Light-Metal Frames 263
    12.8.1 Al-Alloy Frames........................263
    12.8.2 Titanium Alloy Frames 266
    12.8.3 Hybrid Frames 270
        Summary.................................271
        Appendix 12.1 - Bending *vs.* Axial Loading 272
        Appendix 12.2 - Buckling of a Thin Strut under Axial
        Compression 273
        Exercises.................................275

13 POLYMERIC MATERIALS; TIRES 277
13.1 Characteristics of Polymers.....................277
13.2 Polyethylene...............................281
    13.2.1 Molecular Structure 281
    13.2.2 Crystallization 282
    13.2.3 The Glass Transition....................284
13.3 Viscoelasticity..............................285
13.4 Rubber 289
        Summary.................................293
        Appendix 13.1 Secondary Bonds 294
        Appendix 13.2 Architecture of Some Simple Polymers 296
        Exercises.................................297

14 COMPOSITE MATERIALS; TIRES, FRAMES, AND WHEELS . 299
14.1 Principles of Fiber Reinforcement...................300
    14.1.1 Longitudinally Stressed Fibers in a Continuous-Fiber
        Composite 301
    14.1.2 Transversely Stressed Fibers.................303
    14.1.3 Discontinuous Fibers 304
    14.1.4 Composite Behavior 305
14.2 Fibers For Reinforcement 306
    14.2.1 Organically Based Fibers 306
    14.2.2 Inorganically Based Fibers 311
14.3 Network Polymers for Composites and Adhesives 314
    14.3.1 Formation of Network Polymers................314
    14.3.2 Adhesive Bonding 316
14.4 Carbon-Fiber Reinforced Composite Frames.............317
    14.4.1 Case Study of a Hybrid Composite Frame 317
    14.4.2 Monocoque Frames 322

14.4.3 Composite Wheels 324
Summary. 327
Exercises 328

15 THE WALKMAN; MAGNETIC RECORDING AND PLAYBACK 330
15.1 Historical Background 330
15.2 General Operation of the Walkman. 331
15.3 Magnetic Materials. 333
15.4 Magnetic Quantities 334
15.5 Magnetic cgs Units. 335
15.6 Types of Magnetic Behavior 335
15.7 Magnetic Domains 337
15.7.1 Magnetic Anisotropy 338
15.7.2 Domain Walls. 338
15.7.3 Magnetization Processes 339
15.8 The Curie temperature 340
15.9 Magnetic Materials in the Walkman™ 341
15.9.1 The Tape Head 341
15.9.2 Eddy Currents. 342
15.9.3 Ceramic Tape Heads 342
15.10 Materials for Magnetic Recording Tape. 344
15.11 Magnetic Materials for Motors and Headphones 346
Summary. 349
Appendix 15.1 Some Units Employed for Magnetism . 349
Exercises 350

16 ELECTRONIC MATERIALS IN THE WALKMAN 352
16.1 Introduction 352
16.2 Electrical Conductivity 353
16.3 Electron Energy Bands and Conductivity 355
16.4 Intrinsic Semiconductors. 356
16.5 Extrinsic Semiconductors 359
16.6 Purification and Crystal Growth of Silicon for
Semiconductors 362
16.6.1 Zone Refining. 362
16.6.2 Growth of Single Crystals. 364
16.7 Semiconducting Devices. 364
16.7.1 Junction Devices; The Rectifier 365
16.7.2 The Junction Transistor. 366
16.7.3 The Field-Effect Transistor 367
16.8 Integrated Circuits 368
16.9 Other Semiconductor Materials. 372
16.10 Examples of Related Devices 372
16.10.1 The Light-Emitting Diode. 372
16.10.2 The Solar Cell. 373
16.10.3 The Laser 374
16.10.4 The Piezoelectric Phono Pickup 376

16.11 Batteries 378
  16.11.1 The Zinc-Carbon Cell 380
  16.11.2 The Alkaline-Manganese Oxide Cell 381
  16.11.3 Sealed Nickel-Cadmium Batteries 382
   Summary 383
   Exercises 385

Appendix 1 Conversion of Units i

Appendix 2 Physical Constants ii

Appendix 3 SI Prefixes ii

Appendix 4 Physical Properties of Common Metals iii

Appendix 5 Elastic Constants of Common Metals iii

Appendix 6 Ionic Radii (Å) iv

Appendix 7 Specific Modulus Values v

Appendix 8 Properties of Some Engineering Materials vi

Index . . . . vii–xviii

1
THE BICYCLE IN HISTORY

"Let me tell you what I think of bicycling," said suffragette Susan B. Anthony in an interview for New York World in 1896. "I think it has done more to emancipate women than anything else in the world. I stand and rejoice every time I see a woman ride on a wheel. It gives women a feeling of freedom and self-reliance."

1.1 The Significance of the Bicycle

The bicycle was the first widely available means of individual transportation, and it began the era of high-speed, long-range personal transport. It has had enormous social and technological impact, providing freedom of travel to ordinary people and contributing to a host of social transformations. It led directly to the automobile beginning in 1885 when Gottlieb Daimler produced a motorized bicycle, shown in Fig. 1.1. Karl Benz independently unveiled a motorized tricycle the next year, shown in Fig. 1.2. The U. S. auto industry began in 1896 with Henry Ford's bicycle-derived vehicle, illustrated in Fig. 1.3.

Fig. 1.1 Daimler's first vehicle, produced in 1885, a wooden bicycle powered by a 1-cylinder gasoline engine. (Smithsonian Institution)

Fig. 1.2 Karl Benz's first vehicle, produced in 1886, a tricycle with a 1-cylinder engine. (Smithsonian Institution)

CHAPTER 1 THE BICYCLE IN HISTORY

Fig. 1.3 Henry Ford's first automobile, a 2-cylinder machine produced in 1896. (Smithsonian Institution)

The preceding fifty years of bicycle development yielded a number of inventions that made these automotive precursors possible, many of which are still found in motorcars today. These include the pneumatic tire, the differential gear (which allows two side-by-side wheels to turn at slightly different speeds when a vehicle is rounding a corner), the tangent-spoked wheel (in which the spokes brace the rim against the torque applied during acceleration and deceleration), the perfection of ball bearings and the bush-roller chain for power transmission, the concept of gearing, gear ratios, and free-wheeling (in which the driving wheels are allowed to rotate free of the driving mechanism), and, of course, various braking systems, which were made necessary by the introduction of free-wheeling.

Many features of today's automobiles are direct descendants of bicycle technology. The free-wheeling concept is used in clutch assemblies, and the derailleur idea is used in transmissions. The timing-chain that turns the camshaft is a sometimes bicycle-type bush-roller chain, and both drum and caliper-disc brakes were bicycle developments. It is no mystery how this technology was transferred so rapidly, since a great number of the early auto manufacturers got their start making and repairing bicycles. Of course, one of the most important contributions of the bicycle pioneers was the development of methods for mass production of intricate, highly reliable and easily repairable machines.

A number of advances in materials technology stem from the development of the bicycle. These include the processing of thin-walled, seamless drawn-steel tubing, brazing, electric welding, heat treatment and case-hardening of steel, and the use of fibers for reinforcement (which was necessary for the pneumatic tire).

The bicycle literally paved the way for the automobile. Cyclist associations organized public support for paving roads in Britain, continental Europe and the U.S. In the U.S., the League of American Wheelmen, founded in 1880, was the leading force in promoting the formation of state highway departments and laws granting access to the roads to cyclists (to the dismay of the horse-and-carriage set). The later transition to automobiles followed naturally. It is ironic that the demand for private means of transportation, established by the bicycle and subsequently filled mainly by the automobile as infrastructure and the national economy developed, ultimately relegated the bicycle to a recreational status.

It may be, however, that the wheel has not yet turned completely. The automobile is now facing pressures from the forces of energy conservation and pollution control, to say nothing of the fitness movement. We may not yet have seen the end of the bicycle as a means of widespread "serious" transport in developed countries, possibly in conjunction with mass transit.

It may be less obvious that the airplane is the other major offspring of the bicycle. However, a careful look at the early flying machines of the Wright brothers and Glenn Curtiss shows the genealogy quite clearly. The Wrights were in the business of bicycle manufacturing, and Curtiss was a bicycle racer. The Wrights were familiar with the technology needed for minimizing weight in a high-strength, stiff structure, and they employed a framework of drawn-steel tubing, braced like a bicycle frame, along with bicycle wheels and the cabling used on bicycles for brakes and gears to manipulate their control panels. With this they used the lightweight motors that came from motorized bicycles. An example of an early airplane is shown in Fig.1.4.

Fig. 1.4 Glenn Curtiss's 1908 "June Bug," which utilized bicycle wheels and a front fork, giving it the first steerable tricycle landing gear. (Glenn H. Curtiss Museum, Hammondsport, N. Y.)

Most of the early pilots got started with bicycles. Curtiss edged out the Wrights in getting the first U.S. pilot license. Orville Wright, who along with Wilbur invented the flying machine, ironically received the second license. Both Curtiss and Orville Wright had also raced bicycles. In France, the Farman brothers, also bicycle racers, took up flying and became airplane manufacturers. The same thing happened in Germany, where August Euler (German pilot license no. 1) established the first airplane factory there. German license no. 2 went to Hans Grade, also a bicycle racer. Hélène Dutrieu of Belgium was a bicycle racer and daredevil stunt rider who became one of the first women to fly in Europe. Alessandro Anzani, a professional Italian bicycle sprint champion, became a pioneer in airplane-engine manufacture. As Kyle and Gronen* point out, none of this was accidental; the same skills needed for balancing a bicycle and banking it on turns, practiced to the point where they become instinctive, could be transferred directly to flying. For the leading cyclists of the early 20th century, flying was a logical extension of cycling.

In retrospect it is apparent that the development of the bicycle was instrumental in ushering in the modern age. Accordingly, it is instructive to examine that development in a bit more detail, as a kind of case-study of the growth of a technology.

1.2 The Development of the Bicycle

The idea of a human-powered wheeled vehicle is apparently rooted deep in antiquity. The concept of the bicycle can be traced back at least to the 1690 invention by the Frenchman Comte de Sivrac of what came to be called a *célérifère*. An apparently similar machine was described by Blanchard and Magurier in the *Journal de Paris* in July, 1779, and was called a *velocipede*. It consisted of two wheels, one behind the other, connected by a wooden bar on which sat the rider, who propelled the device by pushing against the ground with his feet. The velocipede was used for recreation and sport, and it had limited popularity for a few decades, during which it was made lighter and a bit more comfortable to ride.

In an 1817 publication, Freiherr Karl Drais von Sauerbronn described a breakthrough steerable velocipede. The front wheel pivoted on the frame so that it could be turned by means of handlebars. This improved velocipede evolved in small matters of convenience over the next several years; an example is the "Gentleman's Hobby Horse" shown in Fig. 1.5. It had wooden spoked wheels and iron strips for tires, the shocks from which were dampened somewhat by a padded seat and chest-bar.

The next set of developments came in the propulsion system. A major advance was made by Kirkpatrick Macmillan, a Scottish blacksmith, in 1834-40 when he devised a set of pedals connected by cranks and levers to the rear axle to drive the rear wheel while keeping the feet clear of the ground. He also improved the "gear ratio" by making the rear wheel larger than the front, thus making the velocipede travel farther for each full cycle of the pedals.

Around 1863, the period of explosive development and popularity of what was now called the "bicycle" was ignited by the attachment of cranks and pedals to the front wheel by Pierre Lallement in Paris. After unsuccessful attempts to excite French public interest in an improved model in 1865, Lallement emigrated to Ansonia, Connecticut, where he patented his machine in 1866, giving birth to the bicycle industry in the U.S. In this still-wooden bicycle, facetiously referred to as the "boneshaker", the front wheel was now enlarged to improve the gear ratio. A somewhat later version with an iron frame is shown in Fig. 1.6. Ironically, the first commercially successful manufacture of bicycles (according

---

* See references at end of chapter.

to David Herlihy, a bicycle historian in Boston) began in Paris around 1867 by the firm of Pierre Michaux, which utilized Lallement's invention.

The Hanlon brothers of New York City patented a bicycle in 1868 with solid rubber tires, a front mudguard, and a primitive set of brakes, and Reynolds and Mays exhibited a machine in London in 1869 which had suspension wheels, supported by wire spokes in tension.

Fig. 1.5 Gentleman's Hobby Horse, c. 1818. (Smithsonian Institution)

Fig. 1.6 A velocipede, colloquially known as a "boneshaker," made in Boston, Massachusetts, around 1868. It has wood-spoked wheels with iron tires and bronze pedals. The frame is made of a heavy forged iron bar with iron forks. (Smithsonian Institution)

In England the Coventry Sewing Machine Company began to manufacture bicycles in 1869 under the leadership of James Starley, who carried the gear-ratio idea to what now appears to have been radical extremes. Using his new concept of tangent spoking to improve stability, he enlarged the front wheel still further and shrank the rear wheel in his "Ariel," introduced in 1871. By 1874 he was producing a 50-to-60 pound machine with a 54-inch front wheel. Subsequently, the weight was reduced by the use of steel tubing for the frame and hollow rims for the wheels, and friction was reduced by the use of cone bearings and later ball bearings. This high-wheeler, which became known as the "Ordinary," set the standard for the next decade or so.

A racing version weighing only 24 pounds, made in 1886, is shown in Fig. 1.7(a). Its strengths were speed and ease of handling, but it was not comfortable to ride, and the rider was constantly exposed to the risk of "taking a header," in which one is pitched over the handlebars when the front wheel strikes an obstacle. If the speed were high enough, even a minor obstacle would be sufficient to throw the rider. An American product, also from 1886, the Columbia Light Roadster Ordinary, built by the Pope Manufacturing Company of Hartford, Connecticut, and weighing 36 pounds fully equipped, is illustrated in Fig. 1.7(b).

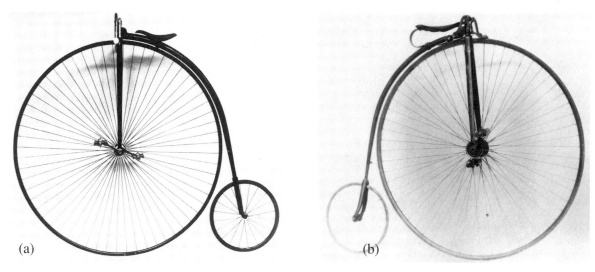

Fig. 1.7 (a) The Humber "Genuine Beeston" Racing Ordinary of 1886. It has a 52-inch front wheel and 18-inch rear wheel, with solid-rubber tires and no brakes (to reduce weight). (b) The Columbia Light Roadster Ordinary of 1886. It has wheels similar to the Humber (with ball bearings), but it has a friction brake acting on the front tire. The wheel rims are rolled from seamless steel tubing, and the frame is cold-drawn seamless steel tubing. (Smithsonian Institution)

The deficiencies of the Ordinary led to a decline in its popularity and the demand for the "safety" bicycle, which was introduced by Starley in 1885; he called it the "Rover." As shown in Fig. 1.8, it had a diamond frame, a chain-and-sprocket drive to the rear wheel, wheels of almost equal size, and a seat for the rider that was so far to the rear that the risk of a "header" was all but eliminated.

The Rover utilized the 1879 advances of Lawson, who had also equalized the wheel sizes and introduced rear-wheel chain drive, but apparently prematurely for the market. Starley also benefitted from the bush-roller chain, introduced by Renold in 1880, which greatly reduced the friction and wear that plagued earlier chain designs.

Fig. 1.8 The Rover safety bicycle, introduced in 1885 by J. K. Starley of England, widely regarded as marking the final development of the bicycle form (From Archibald Sharp: see References at end of chapter.)

Fig. 1.9 The Humber, 1890. (From A. Sharp, loc. cit.)

The final major advance in this "golden age of the bicycle" came in 1889, when John Dunlop, a veterinary surgeon from Belfast, patented the pneumatic tire. The present-day configuration of the bicycle was set by 1890 with the

Humber, with its straight-tube diamond frame, shown in Fig. 1.9. Both the pneumatic tire and the safety bicycle displaced their predecessors entirely in a rather short period of time. The pneumatic tire went from zero to essentially 100% of the market in England between 1889 and 1896. The substitution of the safety model for the Ordinary was complete within less than ten years.

These are good examples of *substitution transformations,* which occur continually in society and nature. They follow a characteristic S-shaped curve in a plot of the fractional extent of the transformation *vs.* time, as shown schematically in Fig. 1.10. Materials scientists refer to such a transformation as a *nucleation and growth* phenomenon; something new is nucleated in many scattered locations, and these pockets of newness grow at an increasing rate (e.g., as the news of the improvement spreads). As the transformation proceeds, the untransformed portion diminishes, and the rate of transformation must therefore drop. One would find the same kind of curve for the introduction of hybrid corn in agriculture, or of coal, and later oil and gas, for home heating. In the present study of bicycle materials, this kind of curve will be encountered in two types of solid-state transformations.

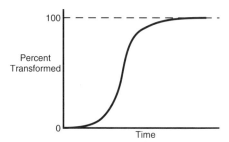

Fig. 1.10 Schematic illustration of the kinetics of a substitution transformation.

By 1910 most present-day bicycle equipment had been introduced, including the freewheel mechanism, caliper and drum brakes, derailleur-type gears, and the hub gear. With the introduction of the safety, the popularity of the bicycle grew explosively. By 1899 there were several hundred factories in the U.S. producing close to one million bicycles per year.

The success of the bicycle in creating widespread demand for a private mode of transportation and stimulating the paving of roads and highways finally led to the shift in public attention to motorcycles and automobiles. Thus, the evolution of the bicycle came to a standstill soon after the turn of the century. It was not until the relatively recent spin-off of materials developed for aerospace applications, including high-strength aluminum and titanium alloys and fiber-reinforced composites, that the excitement returned to bicycle technology. Since the story of the bicycle parallels that of the development of structural materials over the last century and a half, it is in many ways the ideal vehicle for the study of that subject.

References

Archibald Sharp, *Bicycles and Tricycles, An Elementary Treatise on Their Design and Construction,* Longmans, Green, London, 1896. (Reprinted by MIT Press, Cambridge Mass., 1979.)

Encyclopaedia Britannica, 11th Edition, Vol. VII, Cambridge University Press, 1911.

S. S. Wilson, "Bicycle Technology," *Scientific American,* March 1973, p. 81.

C. Kyle and W. Gronen "The Bicycle-Airplane Connection," *Air & Space,* Feb/Mar 1990, p. 88.

© King Features Syndicate

2
WHEELS AND SPOKES; CORROSION RESISTANCE

2.1 The Bicycle Wheel

The modern bicycle wheel functions in a fundamentally different way from the traditional wagon wheel. As depicted in Fig. 2.1(a), the load on a wagon wheel is supported by the compression of the bottom spoke, with the rim transmitting lesser compressive loads to the adjacent spokes. The spokes in the upper part of the wheel carry none of the load. The spokes must be thick enough to resist failure by buckling, which occurs when the compressive load on a thin rod, or column, reaches a critical value such that even a slight deviation from pure axial alignment of the load causes instability and collapse of the column.

In comparing the bicycle wheel to the wagon wheel, an illustrative analogy would be that the former is to the latter as a suspension bridge is to a cathedral. The wagon wheel and the cathedral are both restricted to compressive loading of their components, whereas the suspension bridge and the bicycle wheel both employ stretched wires to support their loads. The analogy is imperfect, however, in that the loaded hub of the bicycle wheel is not simply suspended from the upper spokes, like a roadway is hung from its suspension cables. Rather, the spokes in a bicycle wheel are pre-tensioned during construction of the wheel, like the strings on a musical instrument.

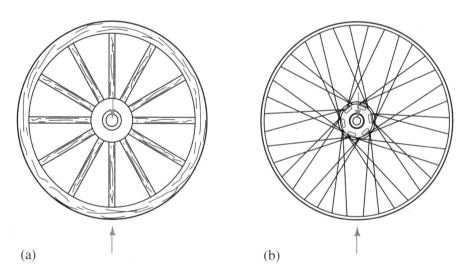

Fig. 2.1 (a) A wagon wheel, showing that the load is carried by the compression of the bottom spoke. (b) A modern bicycle wheel. (From J. Brandt, *The Bicycle Wheel*, Avocet, Menlo Park, CA, 1981, pp. 11 and 12.)

Figure 2.2 illustrates the geometry of a spoke and a nipple and the method of their attachment to the hub and rim of a wheel. The tensioning of each spoke is carried out by turning the nipple with a spoke wrench. When a wheel is properly tensioned, the rim is loaded in uniform circumferential compression, such that it is not distorted out of its plane. In this case the wheel is said to be "true."

In contrast to the wagon wheel, when a bicycle wheel is loaded, the tension in the bottom few spokes is decreased significantly, and the tension in all the rest of the spokes is increased by a small amount. The important point is that all the spokes remain stressed in tension; no spoke is ever in compression. In fact, it is physically impossible for a spoke to be loaded in compression, due to the method of attachment to the rim (cf. Fig. 2.2). This is essential, of course, because the spokes are thin wires, and the critical load for buckling would be very small.

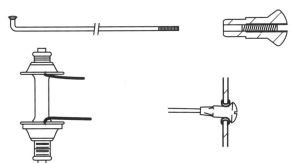

Fig. 2.2 Schematic illustration of a spoke and a nipple and of the method of their attachment to the hub and rim of a wheel. (Adapted from J. Brandt, loc. cit.)

The tensile stress in each spoke is considerable. A typical tensioning load is 90 kg, and a typical spoke diameter is 1.6 mm. This gives a tensile stress of about 440 MPa, which is around one-half the value needed for permanent deformation (i.e., the elastic limit). (The primary strengthening mechanism for spokes is the subject of Chapter 5.)

It must also be recognized that the stress on a given spoke in a rotating wheel varies with time and also varies with the actions of the rider and the conditions of the road surface. For example, if the rider is coasting (i.e., not pedaling) on a smooth surface, a given spoke would be partially unloaded every time this spoke became the bottom spoke. The effect of a pot-hole in the road would be to accentuate this unloading. This cyclic behavior is illustrated schematically in Fig. 2.3. Cyclic loading can lead to failure by fatigue; the mechanism of fatigue failure and how it can be resisted are dealt with in Chapter 7.

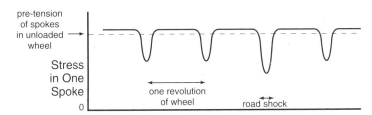

Fig. 2.3 Schematic illustration of the variation of the stress on a particular spoke with time as a wheel rotates, including the effect of a discontinuity in the road surface. (After C.J. Burgoyne and R. Dilmaghanian, *J. Eng. Mech.*, ASCE, Vol. 119, March 1993, p. 439.)

Finally, with regard to wheel design, one should consider why the spokes are laid up in the pattern depicted in Fig. 2.1(b), rather than in the radial pattern found in early bicycles, as illustrated in Fig. 1.7, for example. The reason has to do with the torsional stiffness of the wheel, or the resistance of the wheel to a relative rotation between the hub and the rim. For example, when the torque of pedaling is applied to the radially spoked wheel shown in Fig. 1.7 (cf. Fig. 2.4a), the hub has a tendency to "wind-up," or to rotate more than the rim. In other words, the rotation of the rim tends to lag behind that of the hub.

The tangent-spoked wheel was invented to offset this tendency. In such a wheel the spokes are laid-up in pairs, with each spoke lying tangentially to the circumference of the hub, rather than radially. The effect can be illustrated by Fig. 2.4(b), which shows that the winding-up of the hub would cause one spoke of each pair to lose tension and for its mate to gain tension. Thus, half the spokes would lengthen and the other half would shorten, by a small amount. This allows some wind-up, but, as Sharp has shown, the amount is only about one percent of the amount one would find in a radially spoked wheel.

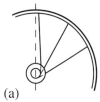
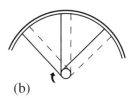

(a)   (b)

Fig. 2.4 (a) Schematic illustration of the tendency for the hub to "wind up" relative to the rim in a radially spoked wheel; (b) shows why this tendency is greatly diminished in a tangentially spoked wheel, in which the spokes act in pairs. (From Archibald Sharp, *Bicycles and Tricycles, An Elementary Treatise on Their Design and Construction,* Longmans, Green, London, 1896; reprinted by MIT Press, Cambridge Mass. 1979, pp. 340 and 342.)

Simple though they are in appearance, an understanding of the structure, properties, behavior, and manufacture of bicycle spokes requires one to delve rather deeply into the subject of materials science. There are two kinds of spokes available: stainless steel and coated carbon steel. They have different basic crystal structures; their microstructures (i.e., the structure one perceives with a microscope) are different, and they respond to corrosive environments and to cyclic loading in different ways. The manufacture of both types of spoke, however, involves wire-drawing, and this process is what imparts most of the strength to the spokes. The study of these topics will occupy the next five chapters of this book. Of course, the topics have much wider applicability, and one can build upon them to study other components of the bicycle.

2.2 Carbon Steel *vs.* Stainless Steel Spokes

The choice of carbon *vs.* stainless steel spokes hinges on which factor is considered more important: resistance to corrosion or resistance to fatigue failure. In the corrosion process, solid material is lost by the conversion of metal atoms to metal ions. Iron or carbon steel corrodes rapidly when exposed to moisture and oxygen. The available means of corrosion protection for carbon steel are only transitory, as shall be seen. However, carbon steel has an advantage over stainless steel in that it is more resistant to the other common mode of spoke failure, fatigue.

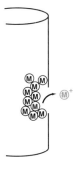

Fatigue is a process of gradual fracture of a component subjected to a cyclic, or repeated, load. In this process a fine crack nucleates at the surface (usually) and grows to a size that produces sudden failure. (This is what happens when one breaks a short piece of wire from a longer piece by repeated bending back and forth.) Obviously, as a bicycle wheel turns, the spokes go through cycles of greater and lesser tension (cf. Fig. 2.3); hence, they are subject to fatigue failure. Such a failure in the front wheel can cause the rider to be pitched over the handlebars, if the broken spoke falls across the front fork and causes the wheel to jam. Cyclists who ride thousands of miles per year are more concerned about fatigue failure than about corrosion, so they sometimes choose carbon steel spokes. They re-lace their wheels with new spokes so frequently that there is not enough time for significant corrosion to develop in the carbon steel spokes (which are temporarily protected by coatings of cadmium or zinc).

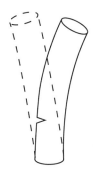

The questions to be addressed in this chapter are the following: Why and how do carbon steel spokes corrode, and how does one protect them? Why is this protection transitory? What are other possible means of protection and why do we reject them? By what mechanism does stainless steel resist corrosion?

Fatigue will be dealt with in Chapter 7; here, it suffices to say that detection of a fatigue crack is difficult, since it usually remains a hairline crack almost to the point of rapid failure. Rusting, on the other hand, is much more obvious, and the failure of a rusted spoke can be readily anticipated.

2.3 Corrosion of Steels

Corrosion involves the ionization of metal atoms and the loss of these ions into solution or into a corrosion product. Since the ionization reaction means giving up electrons, a flow of electrons away from the site of this reaction must occur to avoid a build-up of negative charge. Thus, corrosion is an electrochemical reaction.

The site where the loss of metal occurs is called the anode, or anodic region, and the electrons flow through the metal to a site, called a cathode, where they are consumed in a cathodic reaction. In the case of iron, the anodic reaction is usually

$$Fe \longrightarrow Fe^{++} + 2e^-$$

and the cathodic reaction, in the presence of water and sufficient oxygen, is usually

$$H_2O + 1/2\ O_2 + 2e^- \longrightarrow 2\ OH^-$$

The corrosion product, rust, forms from

$$Fe^{++} + 2\ OH^- \longrightarrow Fe(OH)_2$$

The actual electrochemical mechanism can be appreciated if one considers how a rust pit forms. A pit begins at some inhomogeneity on the surface, such as an impurity particle, and the above reactions occur. The pit-type geometry forms because the anodic reaction continues to occur underneath the rust cover as shown in Fig. 2.5.

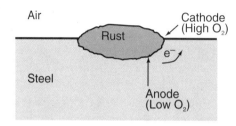

Fig. 2.5 Corrosion occurs under the rust, where the oxygen content is lower. The result is the formation of a pit.

It is useful to consider the formation of a rust pit in some detail, because it helps one understand the electrochemical nature of corrosion more clearly. The important questions are: where is the anode and where is the cathode? Since the cathodic reaction employs water and oxygen to use up electrons, it must occur where the water and oxygen are available. This locates the cathode at the surface of the steel at the periphery of the rust. The anode is then at a location where water and oxygen are less concentrated, which is underneath the rust. Thus, iron is ionized under the rust, and the electrons flow to the surface alongside the rust, to be consumed in the cathodic reaction. Hence, material loss proceeds underneath the rust, and the result is a pit.

A rust pit on a spoke would be dangerous. Not only would it reduce an already-small cross section, but it would act as a notch where stress would be concentrated. This would be a likely spot for the initiation of a fatigue crack.

2.4 Corrosion Protection of Carbon Steel

Carbon steel spokes are protected from corrosion by coating them with a more reactive "sacrificial" metal. If the coating is zinc, the product is referred to as *galvanized* steel; it can be made, for example, by dipping the steel into a bath of molten zinc. A sacrificial metal is one which undergoes the anodic reaction in preference to another, more "noble" metal. That is, the sacrificial metal has a greater tendency to lose electrons. This tendency can most usefully be expressed in what is called a "Galvanic series." Table 2.1 gives such a series for some metals and alloys in sea water. This table gives only relative positions, rather than quantitative differences, because it is not based upon a standardized testing condition.

SECTION 2.4 CORROSION PROTECTION OF CARBON STEEL

Table 2.1 Galvanic Series in Sea Water

Platinum	Cathodic - noble
Gold	
Titanium	
18-8 Stainless steel	
Copper	
Brasses (Cu-Zn)	
Nickel	
Lead-tin solders	
Iron or carbon steel	
Aluminum	
Cadmium	
Zinc	
Magnesium	Anodic - active

The meaning of Table 2.1 is that zinc or cadmium, being more active than iron or carbon steel in this series, will, if electrically connected to iron, act as the anode, and the electrons released will then flow to the iron, which will be forced to serve as the cathode, as shown in Fig. 2.6. The cathode does not dissolve, so the iron remains intact as long as any of the more active metal remains close by.

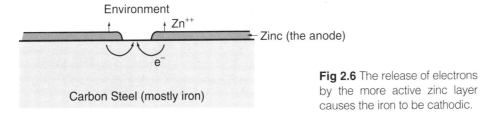

Fig 2.6 The release of electrons by the more active zinc layer causes the iron to be cathodic.

One might think to protect the carbon steel spokes by chromium plating, which gives a brighter surface than does zinc. This would be a mistake, because the chromium plating is not sacrificial. Chromium plating is actually mostly nickel (plated over a thin layer of copper, used to make a better bond with the steel), as shown in Fig. 2.7. The relatively thick nickel layer is then covered by a very thin layer of chromium to keep the surface bright.

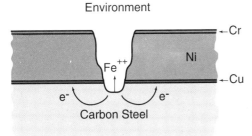

Fig. 2.7 If the coating on a chromium-plated steel is breached, the exposed carbon steel can become a very active anode, and a pit can form.

If this coating is breached locally (e.g., by mechanical damage), the iron is exposed in a small area which is surrounded mainly by nickel. Since nickel is noble with respect to iron (Table 2.1), it acts as a cathode and the iron as an anode, a rust pit forms. It tends to form quickly, since the cathode:anode area

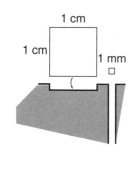

ratio is large; this gives a high *current density* at the anode.

The electrons that leave the anode area represent the corrosion current. In the case where iron atoms become ferrous ions, two electrons leave for every ion created. One could calculate the weight loss by measuring the corrosion current and knowing the atomic weight of iron. For example, a current of about 3×10^{15} electrons would remove one atom layer from an anode area of $1 cm^2$. However, if the anode area were confined to $1 mm^2$, the same current would make a pit 100 atom layers deep. Thus, for a given area of cathode, the smaller the anode area is, the deeper the pit.

A rust pit is undesirable for cosmetic reasons, but it can also serve as a stress concentrator, leading to early fatigue failure.

Chromium resists corrosion by a process called *passivation*. Passivation involves the rapid formation of a thin oxide layer by the chemical combination of oxygen atoms with surface atoms of the metal; this isolates the metal from the environment and prevents further oxidation. The oxide, which is so thin that it is transparent, must be highly stable, well-bonded to the metal, and free of pores and other defects. Once formed, it must have an extremely slow rate of thickening. With chromium, the protecting oxide is Cr_2O_3.

Passivation is also responsible for the corrosion resistance of metals like aluminum and titanium, which also form very stable oxides. That is, the driving force to form the oxide is large. Once formed, the oxide is very difficult to reduce. (That is why these metals are relatively expensive to produce; it requires large amounts of energy to reduce the oxide ores to produce the free metal.)

A solid piece of chromium would retain its passivated character even if the surface oxide were scratched away. This is because there is always more of the chromium below to form oxide and re-establish the protection. Only factors which inhibit the oxide formation would cause problems. Chloride ions act this way, so salt water environments are dangerous to metals protected by passivating oxides.

Stainless steel works on the same principle as a piece of solid chromium. There are enough chromium atoms dispersed in the crystal lattice of stainless steel to provide an oxide sufficiently rich in chromium to passivate the surface.

In contrast to stainless steel, the problem with so-called chromium plating is that, once the underlying carbon steel is exposed, a protective oxide does not re-form over it; therefore, any defect in the plating will ultimately lead to a tiny rust pit. (The rust can be cleaned off by abrasion, but the reaction will continue, and more rust will eventually erupt from the pit. The process can be retarded somewhat by applying a coating, such as a wax, which tends to seal off the pits.) Clearly, stainless steel is far superior to chromium-plated carbon steel.

2.5 Constitution of Stainless Steel Spokes

In stainless steel, which is about 70% iron, the chromium exists in *solid solution* in the iron, meaning that the chromium atoms simply substitute for iron atoms in the crystal structure, as indicated in Fig. 2.8. This is called a *substitutional* solid solution. The solid solubility can be large when the solute atom is nearly the same size as the host atom and not too different chemically. Another kind of solid solution occurs with solute atoms which are very much smaller than the host atom and are thus able to fit into the interstices between the host atoms in its crystal (cf. Fig. 2.8). This is called an *interstitial* solid solution. The classical example is carbon in iron, which will be discussed in great detail later.

A solid solution is an example of a solid *phase*. A phase can be defined as a homogeneous body of matter having a distinct structure (i.e., atomic arrangement) and which can, at least in principle, be mechanically separated from a surrounding phase (or phases).

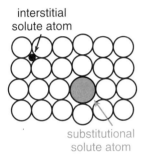

Fig. 2.8 Stainless steel is a solid solution, having both substitutional solutes (Cr and Ni) and interstitial solutes (C and N). The latter are usually treated as impurities in stainless steels.

Example 2.1:

Q. How many phases are present in a glass of ice water?

A. Three: liquid, solid, and vapor.

Q. Is the number of phases changed if something is dissolved in the water?

A. No, but the water becomes a liquid solution, and the ice would become a solid solution, if the substance dissolves in it.

It has been found empirically that at least 12wt% chromium in solid solution is necessary to make stainless steel. However, it turns out that an Fe-12 % Cr alloy tends to have low *ductility* and *toughness,* so it would not be easy to make a wire spoke out of it. Ductility refers to the ability of a piece of metal to be stretched (permanently) by plastic deformation. (It is related to malleability, the ability to be processed (by rolling) into thin sheets.) Toughness refers to the resistance of a material to the propagation of a crack; it is the opposite of brittleness.

The low ductility and toughness of Fe-Cr alloys at ordinary temperatures is related to their crystal structure, which is the same as in pure iron: *body-centered cubic,* or BCC, as shown in Fig. 2.9. To get around the low-ductility problem, it was found possible to change the crystal structure of a Fe-Cr alloy by adding a sufficient amount of nickel. Alloys will be discussed later, but that is why the nickel is present; it does not play an important role in the corrosion protection.

The most common stainless steel is the "18-8" austenitic stainless steel, meaning it contains 18wt% chromium and 8wt% nickel and has the *face-centered cubic,* or FCC, crystal structure of the phase called *austenite,* named in honor of W. C. Roberts-Austen, an early pioneer in the study of steels. It is illustrated in Fig. 2.10. One attractive feature of this crystal structure is that it is almost always ductile and can be readily cold-drawn into wire, ideal for making spokes.

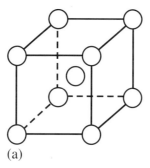
(a)

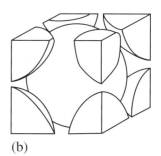
(b)

Fig. 2.9. The BCC unit cell, expressed as (a) a ball-and-stick model, showing the locations of the atom centers, and (b) a hard-sphere model, showing that atoms touch along the diagonals of the cubic unit cell (i.e., the body diagonals).

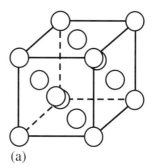
(a)

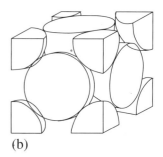
(b)

Fig. 2.10 The FCC unit cell. The atoms touch along each of the face diagonals.

Example 2.2:

Q. In a cubic crystal (e.g., BCC or FCC) the lattice parameter a is the length of an edge of the unit cell. Calculate the length of the diagonal of a face and of the cube (the body diagonal).

A. Face diagonal: $\sqrt{a^2 + a^2} = a\sqrt{2}$

Body diagonal: $\sqrt{a^2 + a^2 + a^2} = a\sqrt{3}$

Example 2.3

Q. Calculate the packing density of a BCC crystal; i.e., the percentage of the unit cell occupied by atoms, assuming the hard-sphere model.

A. In a BCC unit cell the atoms touch along the body diagonal. Let the atomic radius be r. Then $4r = \sqrt{3}a$, and the volume of the unit cell is

$$a^3 = \frac{64r^3}{3\sqrt{3}}$$

There are two atoms per unit cell:

1/8 per corner x 8 corners = 1, plus the center atom = 2

Therefore, the total volume of atoms is $2 \times 4/3\,\pi\,r^3$, and

$$\text{the packing density} = \frac{\text{volume of atoms}}{\text{volume of unit cell}} = \frac{8/3\,\pi\,r^3}{64/3\sqrt{3}r^3} = 0.68$$

Thus, 68% of the BCC unit cell is occupied by solid matter, according to the hard-sphere model.

Example 2.4

Q. In an 18-8 stainless steel, what is the atomic % chromium?

A. 18-8 stainless steel contains 18wt% chromium and 8wt% nickel.

Approximate atomic weights (from the Periodic Table, inside back cover):

Fe 56

Cr 52

Ni 59

Consider 100g of the stainless steel; it has 18g Cr, 8g Ni, and 74g Fe.

$$\text{Approximate atomic \% Cr} = \frac{18/52}{74/56 + 18/52 + 8/59} = 19$$

Summary

The study of the bicycle begins with the tensioned-spoke wheel. The spokes are arrayed tangentially with respect to the hub to minimize "windup" in the rear wheel during pedaling. The spokes are never fully unloaded as the wheel turns, but the "down" spokes are partly unloaded, and this is countered by a small increase in tension in the rest of the spokes. Thus, the spokes are subjected to cyclic load variations as the wheel rotates. This raises the possibility of fatigue failure, a subject to be covered in Chapter 7.

Another type of failure, corrosion, is treated here. Corrosion involves loss of metal atoms at an anode and consumption of the resulting electrons at a cathode. Carbon-steel spokes are temporarily protected from corrosion by coatings of cadmium or zinc, which act as sacrificial anodes. Stainless steel spokes are self-protecting by the formation of a passive surface layer comprising a very thin mixed-Fe-Cr oxide. This requires the presence of chromium atoms in solid solution in the iron; a minimum of 12wt% chromium in solid solution is required for this protection; bicycle spokes normally contain about 18wt%. About 8wt% nickel is also added (making "18-8" stainless steel, which is the most common variety) for the purpose of converting the crystal structure from BCC to FCC. The latter is much more amenable to mechanical processing by wire drawing. The permanent corrosion resistance of the stainless steel makes it preferable to the more fatigue-resistant carbon steel for most riders.

Exercises - Chapter 2

Terms to Understand

- Anodic Reaction
- Cathodic Reaction
- Consumption of Electrons
- Pitting Corrosion
- Noble *vs.* Active
- Sacrificial Anode
- Passivated Surface
- Stainless Steel
- Solid Solution - substitutional, interstitial
- Phase
- Hard-sphere model
- Face-centered Cubic, FCC
- Body-centered Cubic, BCC

Problems

2.1 Aside from cost considerations, is gold plating a good way to protect iron against corrosion? Explain your answer.

2.2 How would this differ from zinc coating (galvanizing) the iron?

2.3 What is the risk of connecting a new copper water pipe to an existing iron pipe? How can this risk be reduced, while continuing to use the same two dissimilar pipes?

2.4 Copper and brass are often plated with nickel, as are the brass nipples used to attach spokes to wheel rims. (Further examples are plumbing fixtures such as faucets.) Explain the purpose of the nickel and describe how it functions.

2.5 When a piece of carbon steel is exposed to humid air, rust forms all over the surface. Therefore, localized anodes and cathodic regions must exist on the surface of the piece. Speculate on what might cause this.

2.6 The decorative steel straps on the steel plates bordering the South Street bridge near the University of Pennsylvania campus (Fig. 2.11) have been buckled out as a result of crevice corrosion. From the discussion of pitting corrosion, explain in terms of oxygen concentrations why the rust formed under the straps. (This represents a direct conversion of chemical energy into mechanical work.)

Fig. 2.11. Decorative steel straps on steel plates which border the South Street bridge leading to the Penn campus. The straps have buckled out from the rust which formed beneath them owing to crevice corrosion.

2.7[1] Zinc anodes weighing 100g are attached to the steel hull of a ship to protect it from corrosion. If an average corrosion current of 0.01A is measured through each anode, how long will they last? Suggest a reason why zinc anodes are normally used for this purpose, instead of, say, magnesium or cadmium; give a concrete basis for your answer. (See Appendix 2 at back of book for useful data.)

2.8[1] Explain why using copper rivets in a steel plate would cause less of a corrosion problem than steel rivets in a copper plate.

2.9 Write the anode reactions and use simple sketches to show the path of flow of electrons for corrosion under the following three conditions:

(a) A zinc-coated carbon-steel spoke with a gap in the coating.

(b) A nickel-coated carbon-steel spoke with a gap in the coating.

(b) Two overlapping carbon-steel plates held together by a carbon-steel bolt and nut. (No coatings on any of them.)

2.10 Evaluate the feasibility of making stainless-type steel spokes using an Fe-Al alloy or an Fe-Ti alloy, in the event of a scarcity of chromium.

2.11 What is the total number of atoms contained within an FCC unit cell? What is the packing density (i.e., the fraction of the unit-cell volume occupied by atoms, using the hard-sphere model)? Note: The atoms touch along face diagonals.

2.12 A common carbon steel used for spokes contains 0.4wt% carbon. Convert this to atomic % carbon. (Use the Periodic Table for atomic weights.)

2.13 Aluminum costs more per ton than does steel, but aluminum has virtually replaced steel for beverage containers. Give as many factors as you can think of that could explain this replacement.

[1] Contributed by Professor D. L. Callahan, Rice University.

3
MECHANICAL BEHAVIOR OF SPOKES

3.1 Types of Stress and Strain

In order to analyze mechanical deformation, it is necessary to determine the relevant loads and to convert each load into a *stress,* meaning load per unit area. The various types of stress, and the corresponding strains, are defined in this section.

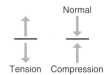

The basic types of stress are normal and shear. A normal stress may be either tensile or compressive.

A spoke in a tensioned wheel is loaded axially (i.e. along its length) by a tensile force and is said to be loaded in simple tension. The tensile stress is defined as the force, F, divided by the area, A, or

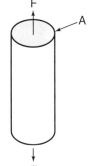

$$\sigma = \frac{F}{A}$$

where the lower-case Greek letter sigma, σ, is used to designate a normal stress, that is, a stress which acts perpendicular, or normal, to the plane of interest. This tensile stress is considered to act uniformly across the cross-section. Since the force in the spoke is constant along its length, so is the tensile stress. As noted in Chapter 2, a typical tensioning load of 90 kg applied to a spoke of 1.6mm diameter would produce a tensile stress of about 440 MPa.

The primary design concern would be that the stress in a spoke be kept below the level that would stretch it permanently. That is, the spoke should behave elastically. This means that the distortion caused by the load would be completely reversed when the load is removed. A body which behaves elastically follows *Hooke's law.*

The original formulation of Hooke's law in 1679 stated that the distortion of an elastic body is proportional to the force applied. This is the behavior of a spring, which follows the law (or equation)

$$F = k\,x$$

where F is the applied force, x is the extension of the spring, and k is called the *spring constant.*

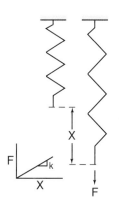

This law applies equally well to a spoke, but here the extension would be much smaller than for a spring; i.e., the spring constant for the spoke would be very much larger. One would say that the spoke is much more elastically stiff than a spring. The form of Hooke's law applied here uses stress instead of force, and, instead of extension, uses the extension per unit length, or strain, of the spoke. To denote strain the lower-case Greek letter epsilon, ε, is used, and Hooke's law is written as

$$\sigma = E \varepsilon$$

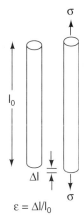

where E is known as the *elastic modulus,* or *Young's modulus,* in honor of Thomas Young, who first recognized this relationship more than one hundred years after Robert Hooke's original observation. This equation applies to any normal stress, either tensile or compressive. In general, tension and compression can be treated as differing only in the sign of the stress; tensile stresses are conventionally taken to be positive, and compressive stresses are, therefore, taken as negative.

Stress and strain, instead of force and displacement, are used to take account of the dimensions of a body. In Section 3.4 the physical origins of ε and E will be considered. Irreversible distortion, or plastic deformation, which occurs when the elastic limit is exceeded, is discussed in Section 3.5 and in Chapter 5.

If a spoke is loaded in bending, then both tensile and compressive stresses are present. For example, in Fig. 3.1 the top of the spoke is in compression, and the bottom is in tension. (Obviously, somewhere in between the stress must pass through zero.) Bending loads are extremely important in any structure, because they produce much higher tensile (and compressive) stresses than are usually found in axial loading (cf. Appendix 12.1). This is a matter of common experience. If one wishes to break a stick, one loads the stick in bending, e.g., across the knee, rather than by pulling on it axially, because less force is needed in bending to reach the fracture stress of the stick.

Fig. 3.1 Illustration of a spoke loaded by a bending moment M.

The bending of the spoke in Fig. 3.1 was accomplished by a *bending moment,* M, which is equal to a force multiplied by the length of the *moment arm* through which the force acts. The moment arm is the perpendicular distance from the line of action of the force to the point about which the force acts to rotate the body. A moment is analogous to the torque exerted on a bolt by a wrench.

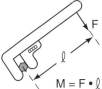

In addition to tensile and compressive stresses, there may also be shear stresses in a bent structural member, such as a beam. This is illustrated in the cantilever beam in Fig. 3.2(a) and by the division of the beam into two *free-bodies* in Fig. 3.2(b). A free-body is a portion of a structure that is isolated so that one can denote all the forces that act on that part of the structure. Since the downward force, F, must be balanced by an equal and opposite force, there must exist a force F acting upward in the plane of the joint between the beam and the wall. Similarly, the requirement for a balance of vertical forces means that there must be the forces shown acting on the faces of the cut between the two free bodies in Fig. 3.2(b). (The cut could have been made on any transverse plane.) The right-hand free body exerts a downward force on the left-hand one, and the left-hand free body exerts an upward force on the right-hand one. These forces acting in a plane (instead of perpendicular to it as in the case of a normal stress) are known as shearing forces; they act on every transverse plane in this beam.

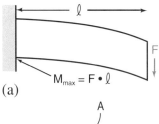

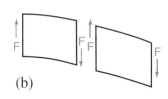

Fig. 3.2 (a) A cantilever beam loaded by the force F at a distance l from the built-in end. (b) Division of the beam into two free bodies by cutting along an arbitrarily selected transverse plane.

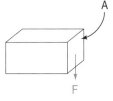

A shear stress is denoted by the Greek letter tau, τ, and is defined as the shearing force divided by the area of the plane on which it acts: $\tau = F/A$, analogous to a normal stress.

The shear strain, denoted by the Greek letter gamma, γ, is defined as the tangent of the angle of shear, which is equal to the displacement in the direction of the stress per unit length normal to the stress.

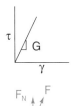

Hooke's law for elastic deformation in shear is

$$\tau = G\gamma$$

where G is the shear modulus.

Any force acting on a plane can be resolved into its normal and shearing components to give a normal stress and a shear stress.

Before proceeding further, one should reflect upon the physical significance of the various types of stress. A tensile stress can lead to fracture in a body that has a tendency to be brittle, i.e., one which can be broken with very little plastic distortion. This does not happen in a well-behaved metallic alloy, but it can, for example, in a carbon-fiber-reinforced plastic (CFRP), as will be seen in Chapter 14. In other words, a tensile stress tends to cause a crack to propagate, if the plane of the crack lies nearly normal to the stress. Conversely, a compressive stress would tend to close a crack; therefore, one does not have to worry nearly so much about compressive stresses.

To illustrate the significance of this point, consider the construction methods of masonry structures like cathedrals and aqueducts, for example. Stone and mortar are brittle materials and are likely to contain crack-like flaws; they must not be loaded in tension. Our predecessors learned this the hard way, and they developed design features, like arches, which ensured that all joints would be in compression. The only ancient materials that could be loaded safely in tension were bone, wood, rope, and leather, all of which depend on the presence of strong fibers. Even here, tension can be applied only in the fiber direction; these materials are weak in the direction perpendicular to the fibers.

A material in which the properties vary with direction is said to be *anisotropic*, while one in which the properties are the same for all directions is called *isotropic*. In present-day structures like bicycles, airplanes, bridges, etc., one relies mainly on metallic materials to carry tensile stresses safely, and these are usually approximately isotropic. However, fiber-reinforced composites are being applied increasingly where minimizing weight is important, and these materials are often anisotropic.

Shear stresses of sufficient magnitude tend to deform materials plastically. Plastic deformation is an alternative to fracture; that is, an applied force can do work on a solid by either fracture or plastic shear. Many metallic materials tend to be *ductile,* as opposed to brittle, because of their ability to deform plastically and, thus, to resist fracture. One challenge in developing better metallic materials is to devise methods for strengthening them against plastic deformation in order to carry higher applied stresses, without restricting the plasticity so much that they become brittle. The approaches used to strengthen metallic materials will be dealt with in much of this book.

3.2 Small-Scale *vs.* Large-Scale Deformation

Materials engineers tend to approach the subject of mechanical behavior from the perspective of either large-scale or small-scale deformation. Spokes can serve to illustrate this very well. When a spoke is in use on a bicycle, one is concerned with small-scale deformation, which extends from the elastic range through the early stages of plastic deformation. When the bicycle is in use, a large-scale plastic strain would result only from some traumatic event. On the other hand, during manufacture of spokes, in which steel rods are drawn into wires by pulling them through dies, the imposed plastic strains are obviously quite large, and one is not at all concerned with elastic behavior or with the early stages of plastic flow. This dual perspective is reflected in the two ways used to define stress and strain.

So-called *engineering* stress and strain are used to characterize small-scale-deformation. They are defined with respect to the initial dimensions of the object being considered, in this case a bicycle spoke. Thus, engineering stress (tensile or compressive) is given by

$$\sigma = \frac{F}{A_0}$$

and engineering strain by

$$\varepsilon = \frac{\Delta l}{l_0}$$

(If the loading were compressive, Δl, ε, F, and σ would all be given a negative sign.)

These definitions are applicable when one is interested in elastic behavior; i.e., in the application of Hooke's law: $\sigma = E\varepsilon$. They also are used when referring to the *yield* stress, which is the stress used to characterize the onset of plastic flow. (See Section 3.3.) Note that, while stress has dimensions of force per

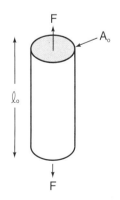

unit area (N/m² or pascals, Pa), strain is dimensionless. Thus, Young's modulus has the same units as stress.

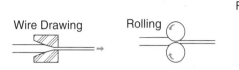

Fig. 3.3 Examples of metal-forming operations which entail large plastic strains.

At the other extreme, manufacturing operations like wire-drawing, rolling, or forging, employ large plastic strains, as illustrated in Fig. 3.3. Here, the changes in dimensions are so large that the initial dimensions lose their relevance. Therefore, *true* stress and strain should be used; these employ the instantaneous dimensions of the object:

$$\text{true stress} = \frac{F}{A}$$

and

$$\text{true strain} = \sum_{l_0}^{l_f} \frac{\Delta l}{l}$$

where A is the instantaneous area, l is the instantaneous length, and the summation is made over all $\Delta l/l$ from the initial length (l_0) to the final length (l_f). Of course, the summation is evaluated as an integral:

$$\text{true strain} = \int_{l_0}^{l_f} \frac{dl}{l} = \ln \frac{l_f}{l_0}$$

Since the focus in this book is on small-scale deformation, rather than on deformation processing, engineering stress and strain can be used throughout.

With regard to elastic behavior, it should be noted that Young's modulus, E, defined by Hooke's law, is a property of the material; it depends on the interatomic bonding forces, as will be seen below. Materials that are elastically stiff have a high E. For example, as seen in Table 3.1, steel is three times as stiff as aluminum. Therefore, bicycle spokes are made of steel, but the rims are made from aluminum to minimize weight.

Example 3.1

Q. What is the elastic strain in a carbon-steel spoke stressed to 350 MPa ?

A. From Table 3.1, E for carbon steel is 210 GPa. Therefore,

$\varepsilon = \sigma/E = 350$ MPa/210GPa $= 1.67 \times 10^{-3}$ or 0.167%

Q. Does it matter whether some plastic yielding has occurred during the loading?

A. No. The elastic strain depends only on the stress and Young's modulus.

A note on units is in order here. As in most present-day texts, the units employed in this book are mainly the Système Internationale (SI) units (kilograms, meters, newtons, etc.). However, an occasional exception is made in which English units (pounds, inches, etc.) are used. These are still the units in which bicycles are measured in the U. S., and many test results are reported in these units; hence, it is necessary to be bilingual in this sense. One can translate from one set of units to the other using the conversion tables in the back of the book. For example, since one pound is equal to 4.448 newtons, and one inch equals 25.4 millimeters, or 0.0254 meters, one pound per square inch (psi), converted to SI units, is

$$1 \text{ psi} = \frac{4.448}{(0.0254)^2} = 6894 \text{ N/m}^2$$

In the context of materials deformation, one normally deals with thousands of psi, which is denoted ksi. The counterpart in SI units is the mega-newton per square meter, MN/m^2, or mega-pascal, MPa. The important conversion factor is, therefore,

$$1 \text{ ksi} = 6.9 \text{ MPa}$$

Table 3.1 Young's modulus of materials used in bicycles.

Material	Young's Modulus	
	GPa	psi
Iron	210	~30x10^6
Carbon and low-alloy steels	210	~30x10^6
18-8 stainless steel	210	~30x10^6
Aluminum	70	~10x10^6
Titanium	116	~17x10^6
Glass (for fibers)	76	~11x10^6
Carbon fibers Type 1	390	~56x10^6
Type 2	250	~36x10^6
Kevlar	130	~18x10^6
Rubbers	0.01 - 0.1	~10^3 - 10^4
Epoxies	2 - 5.5	~3 - 8x10^5

3.3 Measurement of Stress *vs.* Strain — The Tensile Test

The tensile test gives important information about the mechanical properties of a material. In the case of a bicycle spoke, it is necessary to know Young's modulus in order to know how much tensile stress is generated from a given strain imposed by tightening a nipple (cf. Fig. 2.2). It is also important to avoid permanent deformation of the spoke during tensioning of the wheel. Thus, the applied tensile stress should not be more than about half the yield stress. The yield stress is defined as the stress at which the deformation becomes inelastic (i.e., partly permanent) by some specified amount.

A tensile test is carried out on a machine that can exert sufficient force to stretch the specimen of interest; for a spoke a capacity of about one thousand pounds is needed. One end of the specimen is attached through a suitable linkage to a load cell, and the other is attached to a movable cross-head, which is driven by either large screws or by a hydraulic piston. A screw-driven machine is illustrated schematically in Fig.3.4(a). The elongation of the specimen can be measured by an extensometer, which is a displacement transducer, or a device that converts a displacement to an electrical signal. The voltage outputs from the load cell and the extensometer are sent to the Y and X terminals, respectively, of an X-Y recorder and the small-scale-deformation portion of the stress-strain curve is displayed, as illustrated in Fig. 3.4(b). Most load cells and extensometers employ electrical-resistance strain gauges, the operation of which is described in Appendix 3.1.

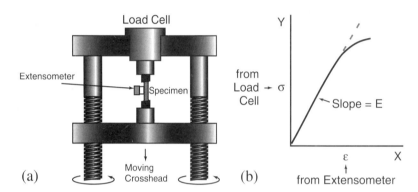

Fig. 3.4 (a) Schematic representation of tensile-testing apparatus. (From H. W. Hayden, W. G. Moffat, and J. Wulff, *The Structure and Properties of Materials Vol. III*, John Wiley & Sons NY, 1965, p.2.) (b) Initial portion of σ-ε curve plotted by an X-Y recorder, showing plastic yielding, causing the departure from linear elastic behavior.

In the stress-strain curve shown in Fig. 3.4(b), Young's modulus, E, is simply the slope of the initial, linear portion of the curve. The yield stress is less-obviously characterized, since the transition from elastic to inelastic behavior occurs gradually. In order to have a material parameter that can be defined unambiguously, it is convention to select some small amount of permanent (i.e., plastic) deformation as the indicator that the purely elastic behavior has come to an end. For example, a plastic offset (from the elastic line) of 0.002 inch in a one-inch gauge section (i.e., a plastic strain of 0.2 percent) could be selected; this gives the 0.2% yield stress. Obviously, the definition of the yield stress is arbitrary in the sense that the plastic strain must be specified.

The initial parts of the stress-strain curves of the two common types of bicycle spoke are shown in Fig.3.5; the yield stress (arrows) is defined by the 0.2% plastic offset criterion. The differences between carbon steel and stainless steel spokes will be discussed in detail later. For the present, note that the yield strength and the elastic modulus are similar for the two types. Therefore, one can substitute one for the other in a bicycle wheel without compromising the mechanical integrity of the wheel.

In the design of a bicycle wheel it is necessary to use Young's modulus to calculate the amount of distortion that would occur when the wheel is loaded. Also, the yield stress must be known to ensure that the spokes are not deformed plastically by any reasonable loading condition. Thus, the tensile test has played an important role in the development of bicycle spokes and wheels.

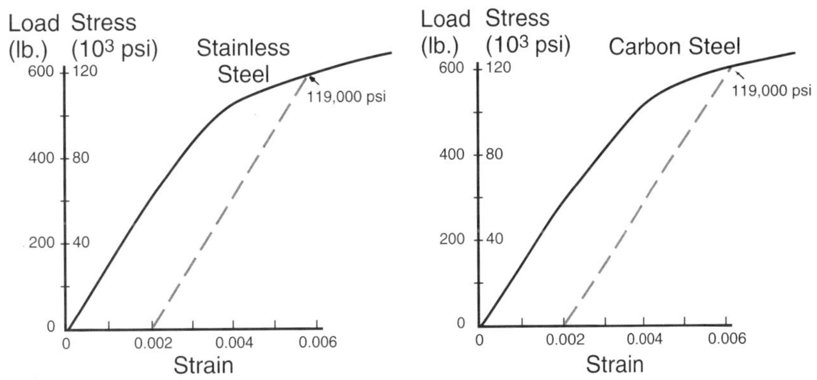

Fig. 3.5 The initial stress-strain curves of a carbon-steel and a stainless-steel spoke; the strain was measured with a strain-gauge extensometer and displayed on an XY recorder.

Example 3.2

Q. For the stress-strain curve of the stainless-steel spoke shown in Fig. 3.5, show how the 0.2% yield stress and the total strain at this stress are calculated. The spoke diameter is 2mm (0.08in).

A. The load at which the stress-strain curve crosses the line representing a plastic strain of 0.002 is 2578 N (600 lb). (Note that this line has a slope equal to E, 210 GPa (30×10^6 psi). Since the cross-sectional area is 3.14mm² (0.005in²), the yield stress is:

$$\frac{\text{load at 0.2\% plastic offset}}{\text{original cross-sectional area}} = \frac{2578 \text{N}}{3.14 \times 10^{-6} \text{m}^2} = 821 \text{MPa (119,000psi)}$$

The total strain (ε_t) = elastic strain (ε_e) + plastic strain (ε_p). At yielding, $\varepsilon_p = 0.002$ and $\varepsilon_e = 821$ MPa/210 GPa = 0.004. Therefore, $\varepsilon_t = 0.006$.

3.4 Elastic Behavior

Table 3.1 indicates that carbon steel, which is usually more than 90 volume-percent iron (the rest being iron carbide), has essentially the same Young's modulus as stainless steel, which is about 70 volume-percent iron (the rest being mainly chromium and nickel dispersed in solid solution in the iron). Thus, the modulus is determined by the majority (or solvent) element in these iron-based alloys. On the other hand, Table 3.1 shows that aluminum has a Young's modulus only one-third that of iron. To understand the physical basis for differences in E from one material to another, elastic strain must be examined on the atomic scale.

Elastic deformation is easiest to envision in terms of a two-dimensional array of atoms, as shown in Fig. 3.6. Note that this deformation may be viewed equivalently either as a vertical separation of atoms or as shearing of the rows of atoms oriented at ±30° to the horizontal. This shearing occurs because a solid reacts to a stretching force so as to minimize the increase in volume which accompanies a tensile strain. Thus, the solid contracts laterally as it stretches, and this contraction is accomplished by the shears depicted in Fig. 3.6

Fig. 3.6 (a) A portion of a close-packed plane of atoms; each atom touches six neighbors. This is the densest possible packing of spheres of equal size. (b) An (exaggerated) elastic strain is produced in this plane by the application of a tensile stress.

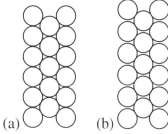

The amount of contraction per unit tensile strain is a property of the material and is called *Poisson's ratio*. If the tensile strain occurs along the y axis and the lateral contraction along the x axis, the Poisson's ratio, usually denoted by the Greek letter nu, ν, is expressed as follows:

$$\nu = \frac{-\varepsilon_x}{\varepsilon_y}$$

Values of ν are generally in the range 0.3 to 0.35.

Since the property of a material which expresses the resistance to elastic shear is the shear modulus, and since a tensile strain with a Poisson contraction necessitates elastic shearing, there must be a relationship connecting the three material constants E, G, and ν. For an isotropic solid this relationship is

$$G = \frac{E}{2(1+\nu)}$$

Crystals are not isotropic, so the relationship is more complicated and varies with the crystallographic directions involved, but solids made up of a large number of small crystals having random orientations can be approximated as isotropic. This is the case for metallic materials used in bicycles, so the above expressions can be applied here.

Elastic strain is produced when the applied stress works against the interatomic bonding forces; i.e., the forces that caused the atoms to condense into a crystal. Interatomic bonding is actually the resultant of two opposing forces: an attractive force that works over a long range (relative to the atom diameter) and a repulsive, short-range force which comes into play when the atoms are close together. The corresponding attractive, repulsive, and resultant energies, from which the forces are derived, are shown schematically in Fig. 3.7(a).

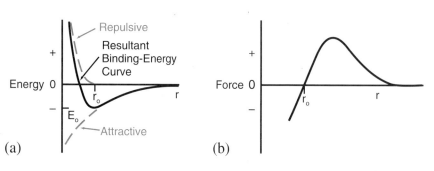

Fig. 3.7 (a) An interatomic bonding energy curve as the resultant of an attractive and a repulsive energy. The equilibrium separation of atoms is r_o, and the binding energy is E_o. (b) The force vs. displacement curve (i.e., the "stress-strain" curve for the pair of atoms), from the first derivative of the energy curve in (a).

The nature of the attractive energy depends on the type of bond being formed. For example, in an ionic crystal like NaCl, it is simple Coulombic attraction between the ions of opposite charge. The simplest metals are, essentially, positive ions surrounded by a sea of negative charge; i.e., the valence, or "free," electrons. Therefore, metallic bonding is more complicated, but the shape of the attractive-energy curve is roughly similar to the simple ionic case. The steeply rising repulsive energy comes into play when the ion cores are pressed so closely together that they begin to overlap; this drastically raises the energies of the electrons in the overlap regions. It is because of this steep repulsive energy that ionic or metallic crystals can be modeled as arrays of hard spheres.

The binding energy is the sum of the attractive and repulsive energies. The interatomic force is just the derivative of that energy with respect to the distance of atomic separation, as shown in Fig. 3.7(b). Since Young's modulus is proportional to the slope of the force vs. displacement in a tensile test, it must be proportional to the slope of the interatomic force vs. displacement curve shown in Fig. 3.7(b) at the equilibrium atomic separation (where the force passes through zero). (Note that elastic strains involve very small interatomic displacements; they are on the order of 10^{-3} times the atomic diameter or less. Thus, the displacements in Fig. 3.6 are highly exaggerated.)

Three points about elastic behavior should be emphasized:

1. Hooke's Law ($\sigma = E\varepsilon$) is only true for small displacements; at very large elastic strains the σ–ε behavior would be nonlinear. That is, the force vs. displacement curve shown in Fig. 3.7(b) is approximately linear only in the vicinity of $r = r_0$.

2. E is the same in tension and compression for small strains.

3. E is related to the second derivative, i.e., the curvature, of the binding-energy curve at its minimum. That is, it is not only a function of the attractive component, as might be imagined, but it is also influenced by the steepness of the repulsive component.

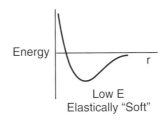
Low E
Elastically "Soft"

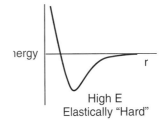
High E
Elastically "Hard"

As shown in Fig. 3.6, an elastic strain decreases the packing density of the atomic array. That is, the volume per atom is increased, as is the average interatomic spacing (r in Fig. 3.7). This must involve an increase in energy of the crystal, because it was originally in a state of minimum energy. This is equivalent to saying that the crystal is no longer at the minimum in the binding energy curve in Fig. 3.7(a). Thus, there is stored elastic energy in an elastically deformed solid, just as there is stored energy in a stretched spring. The amount of stored energy is simply the product of the average force times the displacement, or the area under the force vs. displacement curve, as shown in Fig. 3.8.

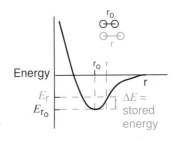

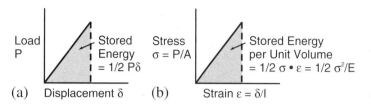

Fig. 3.8 (a) The stored elastic energy is the area under an elastic load-displacement curve. (b) The stored energy per unit volume is the area under the σ–ε curve.

Since stress is force/area and strain is displacement/length, the area under the σ-e curve is stored-energy/volume. (Force x displacement = energy, and length x area = volume.) Strain energy is of great importance in some areas of materials science, but the stored energy in the spokes in a bicycle wheel plays no direct role in the functioning of the bicycle. It is just a consequence of the fact that the spokes are tensioned.

Example 3.3

Q. Calculate the stored elastic energy per unit volume in the spoke loaded to the 0.2% yield stress in Example 3.2.

A. Stored energy/volume = $1/2\ \sigma^2/E$ = $1/2(821\text{MPa})^2/210\text{GPa}$ = 1600×10^3 Pa

Therefore, the stored energy/volume = 1600 kJ/m³.

$$1\text{Pa} = \frac{1\text{N}}{\text{m}^2} = \frac{1\text{ N-m}}{\text{m}^3} = \frac{1\text{J}}{\text{m}^3}$$

3.5 Plastic Behavior

As already noted, when the applied stress becomes sufficiently high, a metallic material, like a steel, departs from linear elastic behavior, because plastic deformation begins. The stress-strain relationship is no longer described by Hooke's Law. (However, elastic strain at a given stress, even in the non-linear region, is still obtained from Hooke's law.) Because the accumulation of plastic strain is so gradual, the yield stress is arbitrarily defined at a certain amount of plastic strain, as shown in Fig. 3.5.

Plasticity in a crystalline solid is fundamentally different from elastic deformation. As indicated schematically in Fig. 3.9, plastic flow occurs by a shearing of a crystal along a certain crystallographic plane in a certain crystallographic direction. This process is called *slip*. The planes and directions will be discussed later, but the important points for the moment are the following:

1. Plastic displacement occurs by the shearing of one part of the crystal relative to another in units of the interatomic spacing along the direction of slip.

2. Plastic strain is permanent; it does not reverse itself when the stress is removed. That is, the block of crystal in Fig. 3.9(b) is in a new position of equilibrium; there is no restoring force tending to return it to the configuration of 3.9(a).

3. Slip does not change the symmetry of the crystal; the crystal structure is the same after slip as before.

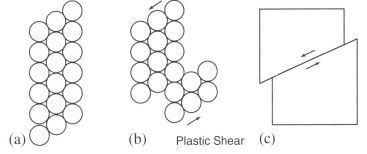

Fig. 3.9 Schematic representation of plastic shear in a crystal. (a) is the original block of atoms. (b) is the block after plastic shear. (c) represents shear strain in a continuum block.

One can observe the effects of slip by examining under a microscope the surface of a deformed piece of polycrystalline brass, shown in Fig. 3.10. The piece was heated to a high temperature to produce coarsening of the crystallites, called *grains*, and then polished to a mirror-like finish before being squeezed in a vise. The explanation of how this microscopic image is formed is given in the next chapter; here, we are only concerned with the slip lines that appear in individual grains. These lines are manifestations of the steps formed in the surface of these grains due to the intense plastic offset which has occurred there. Obviously, in this material the plastic strain is concentrated in well-defined, narrow bands.

If the surface of a pre-polished tensile specimen of the coarse-grained brass were to be observed with a microscope during a tensile test, one would find that the process of yielding takes place in the following way: First, very faint slip lines (meaning small surface steps) occur in a few isolated grains. At this stress, the deviation from linear elasticity is hardly detectable with an ordinary displacement transducer, since these only detect displacements equivalent to strains of 10^{-5} or greater. As the stress is further increased, the slip lines in the isolated grains become more apparent. That is, the slip steps become larger, and more slip lines occur in these and other grains. At some point the deviation from linear behavior appears on the X-Y recorder. When the plastic strain reaches 10^{-3} (0.1%), most of the grains will contain observable slip lines.

Fig. 3.10 The surface of a pre-polished, coarse-grained piece of brass after being squeezed in a vise to produce abundant slip.

It might be imagined that slip occurs by the sliding of one part of a crystal over another in the manner of rigid blocks, the way a pack of cards can be sheared. However, it can be shown that this would require an extraordinarily high stress, and it is now known that slip actually occurs progressively, rather than all at once. A useful analogy is the displacement of a rug on a floor when one creates a ripple in the rug and then pushes the ripple across the rug to the far side, as shown in Fig. 3.11. The amount of displacement achieved in this manner is equal to the amount of rug in the ripple (b in Fig. 3.11). The displacement is achieved with far less force than would be required to move the whole rug at once. The ripple in the rug is analogous to a linear defect in a crystal called a *dislocation*. An illustration of one of the two basic kinds of dislocation is given in Fig. 3.12.

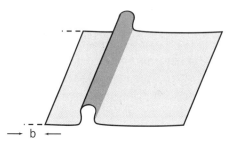

Fig. 3.11 A rug may be given a shear displacement with respect to the floor by propagating a ripple across it.

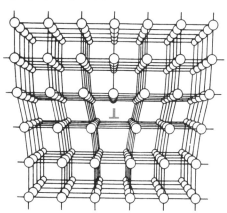

Fig. 3.12 An edge dislocation in a crystal. The dislocation comprises the bottom edge of the extra plane of atoms contained in the upper half of the crystal. (From A. G. Guy, *Elements of Physical Metallurgy*, Addison-Wesley, Reading, MA, 1959, p. 110.)

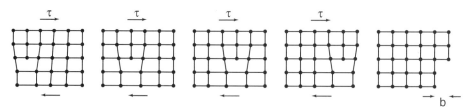

Fig. 3.13. The glide of an edge dislocation under the influence of a shear stress τ.

The plastic shear in crystalline solids occurs by the motion of dislocations, as illustrated by Fig. 3.13. One must understand a certain amount about dislocations to comprehend how crystalline solids deform and how they can be strengthened (i.e., made more resistant to deformation). This is particularly important in bicycle spokes, since they are manufactured by plastic deformation (i.e., wire drawing), and this large-scale plastic deformation gives them most of their strength, a process known as *strain hardening*. Strain hardening is the increase in yield stress imparted to a material by plastic deformation; it results from an increase in the density of dislocations.

Also, the superior fatigue resistance, or resistance to fracture from repeated stressing, of carbon steel spokes can only be understood in terms of the special interaction between dislocations and interstitial atoms, like carbon, in carbon steel. This will be made clear in subsequent chapters. For now, the following points about dislocations need to be made.

1. The component of stress that causes dislocations to "glide" along a particular plane in a crystal, producing slip, is a shear stress. This shear stress lies in the plane of glide and acts in the direction along which slip occurs, as indicated in Fig. 3.13. The magnitude of the shear stress is the shearing force divided by the area of the plane on which it acts.

2. After a metallic material is heated to a high temperature (but not so high as to melt it) and cooled to room temperature, it contains a relatively small number of dislocations. When plastic deformation occurs by the motion (glide) of these dislocations, the dislocations multiply (by a process to be described later), and, therefore, the separation distance between dislocations becomes progressively smaller.

3. The region in a crystal around the dislocation line is obviously distorted; i.e., the atoms are displaced from their equilibrium positions. Thus, there must be a stress field and an associated stored elastic energy associated with the dislocation. It will be shown later that, if similar dislocations approach each other, their stress fields will interact to cause mutual repulsion. The repulsive force will be shown to be inversely proportional to the dislocation spacing, just as with two electrically charged particles of like sign. Since dislocation multiplication decreases their average spacing, it must increase their average mutual repulsion. This makes glide more difficult, and more stress must be applied to cause a continuation of dislocation motion. Thus, plastic deformation results in strain hardening.

Dislocation Repulsion

The shear stress that causes dislocation motion can be related to an applied tensile stress in the following way: Consider a cylindrical rod, like a spoke, having a cross-sectional area A_0, subjected to a tensile force P. This tensile force can be resolved into components that lie in, and perpendicular to, any plane. These are the shear and normal forces and can be designated S and N, respectively. The area of this arbitrary plane is denoted as A.

The orientation of the plane is specified by the angle, θ, between the tensile axis and the plane normal. The complementary angle, λ, is between the tensile axis and the vector S.

The area A is given by:

$$A = A_0/\cos\theta$$

The force vector, S, which is the shearing force on the arbitrary plane, is given by:

$$S = P\cos\lambda$$

The shear stress, τ, on that plane is defined as S/A and is given by:

$$\tau = P \cos \lambda / A_0 / \cos \theta = (P/A_0) \cos \lambda \cos \theta$$

Since P/A_0 is by definition σ, the resolved shear stress can be written:

$$\tau = \sigma \cos \lambda \cos \theta$$

This expression is known as *Schmid's law*. It can be shown (cf. Problem 3.8) that the maximum value of τ, i.e., of $\cos \lambda \cos \theta$, occurs when $\lambda = \theta = 45°$, or when $\cos \lambda = \cos \theta = \sqrt{2}/2$. This maximum value is, therefore, $\tau = \sigma/2$.

Figure. 3.14 shows three orientations of the arbitrary plane. Of the three, (a) would have the maximum shear stress, because λ and θ are both close to 45°. In (b), where θ is small, the resolved shearing force S is small; hence, so is τ. In (c), where θ is large, S is large, but the plane would have a very large area; hence τ is small. The latter result may appear counterintuitive, until it is remembered that stress is defined as force per unit area.

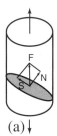
(a)↓

(b)↓

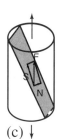
(c)↓

Fig. 3.14 Resolution of a tensile force (F) into shear (S) and normal (N) force components on three arbitrary planes.

Returning to the description of the yielding process, it should now be apparent why a few isolated grains in a polycrystalline tensile specimen would yield before the others. These are the grains with the highest shear stress on the particular crystallographic plane on which dislocation motion is easiest. (The crystallography of slip will be discussed in Chapter 5.)

Example 3.4

Q. Calculate the shear stress on the planes and directions defined by the following λ and θ values for an applied tensile stress of 100 MPa. (Assume that the shear direction, tensile axis, and shear plane normal are coplanar, as would be the case in a continuum solid.)

λ	θ
15°	75°
40	50
75	15

A. Use Schmid's law: $\tau = \sigma \cos \lambda \cos \theta$
For $\sigma = 100$ MPa,

λ	θ	τ (MPa)
15°	75°	25
40	50	49
75	15	25

3.6 Large-Scale Plastic Flow — The Stress-Strain Curve

One often wants to know more about the mechanical behavior of a material than just the Young's modulus and the yield stress. Much useful information is contained in the total stress-strain curve, obtained by continuing the tensile test until the specimen fails. The strain involved is usually well beyond the range of most extensometers, so in the complete tensile test the extension of the specimen is measured in a comparatively crude manner. This is generally adequate, since there is no need for high precision when the strains are very large.

The total engineering stress-strain curve is actually plotted in the form of a load-elongation curve. The curves have the same shape because of the definition of engineering stress and strain (i.e., load and elongation divided by constants, A_0 and l_0, respectively). The curve is often plotted on a strip-chart recorder, which can record only one input signal. This comes from the load cell of the machine, and it goes to the Y axis. The chart is driven at a constant rate, so the measure along the X axis is simply time. However, if the movable cross-head on the testing machine is also driven at a constant rate, then a distance along the X axis is proportional to the displacement of this cross-head. The rates of travel of the chart and the cross-head can each be set as one wishes. Examples of the stress-strain curve from each kind of spoke are shown in Fig. 3.15.

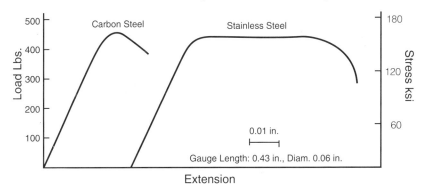

Fig. 3.15 Comparison of the tensile behavior of carbon steel and stainless steel spokes, showing the load plotted against the displacement of the movable cross-head of the machine. The load is proportional to the engineering stress (load/original area) and the displacement is proportional to the engineering strain (extension/the nominal gauge length of the specimen). Virtually all of the plastic extension occurs in this gauge section.

Several aspects of the engineering stress-strain curves are worth noting:
1. The initial linear portion represents elastic deformation, but the slope is not Young's modulus, because the displacement of the whole load train is being recorded, not just that of the gauge section of the specimen.
2. After yielding, a large amount of strain hardening can occur. It is this behavior that makes metals so useful; plastic flow leads not to immediate collapse or fracture, but to a stronger material.
3. Ultimately, the curve passes through a maximum. This maximum represents the ultimate load-bearing capacity of the specimen. It is called the *ultimate tensile strength,* UTS, and is the maximum load divided by the original area of the gauge section.

The maximum in the stress-strain curve occurs because the specimen gets thinner as it is plastically stretched; in this sense it becomes weaker. The strain hardening is initially more than sufficient to compensate for the thinning of the specimen, but ultimately it is not, because the rate of strain hardening decreases with strain. (The reason for this will be shown in Chapter 5.) At the point where the strain hardening is just balanced by the thinning, the specimen becomes "plastically unstable," and at some point along the gauge section the specimen begins to neck down. (The criterion for necking is given mathematically in Appendix 3.2.) Essentially all subsequent plastic flow is then concentrated in this necked region, and failure occurs there also, since this is now the region of maximum stress.

The neck is also a region in which damage accumulates in the form of internal cavities. These generally nucleate at impurity particles and grow with continued plastic flow. They finally link up to form a large internal crack-like flaw, which leads to final failure. The process of necking and failure is illustrated schematically in Fig. 3.16, and an example of failure by rupture is shown.

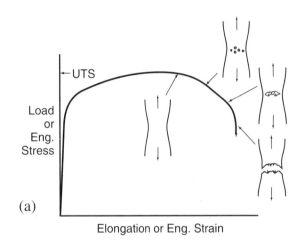
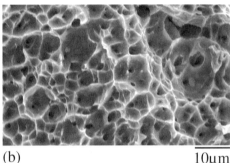

Fig. 3.16 (a) Schematic illustration of necking, cavity formation and coalescence, and rupture. (b) Fracture surface of a stainless steel spoke. (Scanning electron micrograph by Cliff Warner, Univ. of Penn.)

The reason the cavities grow in the center of the necked region is that the triaxial tension is greatest there. In a tensile specimen the state of stress in the center of a neck is a combination of simple, or uniaxial, tension (i.e., the tension applied by the machine), and transverse tension. The latter arises from the fact that the necked region tends to continue to thin down with the ongoing plastic flow, but the thicker regions alongside the neck do not, because very little further plastic flow occurs there. Thus, the thicker regions exert a constraint on the material in the neck. This constraint manifests itself as a transverse tensile stress, which is equal in all radial directions.

The stress state in the center of the neck is illustrated schematically in Fig. 3.17. The transverse tensile stress reaches a maximum in the center of the neck and falls to zero at the free surface, since a free surface is, by definition, free of normal components of stress (i.e., components of stress perpendicular to the surface). If the tensile test is interrupted just before failure and the specimen is then sectioned longitudinally and examined under a microscope, it is found that cavities have formed internally in the region of the neck, as shown schematically in Fig. 3.16. The cavities are large near the central axis and diminish in size toward the surface of the neck.

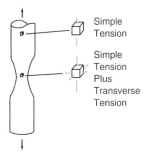

Fig. 3.17 The state of stress in the center of a neck contains a transverse component, caused by plastic constraint.

The type of failure involving hole growth is called *rupture*. It is a displacement-controlled (as opposed to a stress-controlled) mode of failure, because it results from the damage that accumulates with strain. That is how ductile materials fail.

Ductility is normally expressed in one of two ways: The preferred way is the *percent reduction in area*, %RA, which is simply the total reduction of area divided by the original area times 100. That is,

$$\%\text{RA} = \frac{A_0 - A_f}{A_0} \times 100$$

where A_0 is the original cross-sectional area of the gauge section and A_f is the final area in the neck. This parameter best expresses the resistance of the material to rupture. For example, the cleaner the material, i.e., the lower the concentration of impurity particles (which serve as nuclei for cavities), the greater is the %RA.

The less-preferred measure of ductility is the *percent elongation*, %elong. which represents the total plastic extension of the specimen (measured from the stress-strain curve) divided by the original nominal gauge length. That is,

$$\%\text{elong.} = \frac{l_f - l_0}{l_0} \times 100$$

where l_0 is the original length of the gauge section, and l_f is the final length. Percent elongation is easier to determine, but it is less precise, and its physical meaning is less clear, since it combines the uniform elongation before necking with the elongation which occurs only in the neck. This makes the percent elongation a function of the original gauge length, which is not the case with the %RA.

In the operation of a bicycle, the ductility of the spokes is not usually a consideration, except that some ductility is obviously required. However, during the wire-drawing operation used to make spokes, the ductility is clearly very important. If the wire cannot be greatly elongated and thinned, then extensive wire drawing cannot be carried out. The drawing process relies heavily on the fact that the wire-drawing dies exert a transverse compression on the wire, as illustrated in Fig. 3.18.

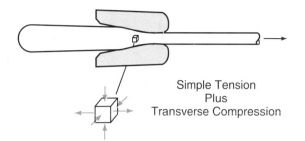

Fig. 3.18 Schematic representation of a wire being drawn through a die. The die exerts transverse compression and allows large tensile elongation without rupture.

The transverse compression plays two important roles. First, it inhibits the growth of any cavities which may form in the wire. That is, it has an effect opposite to that of the transverse tension from plastic constraint in the neck of a tensile specimen. Secondly, the transverse compression acts to elongate the wire by squeezing it, just as one can elongate a roll of clay by making a fist around it and squeezing. The effect can be understood mathematically by considering the two-dimensional cases shown in Fig 3.19.

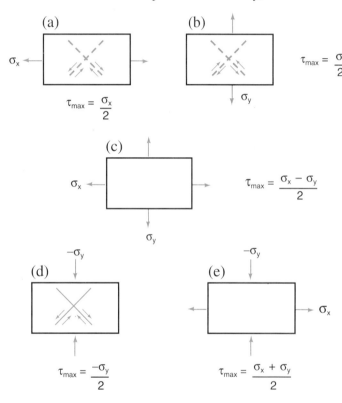

Fig. 3.19 Illustrations of the maximum shear stresses which arise from various combinations of tensile and compressive stresses. (a) Longitudinal tension only. (b) Transverse tension only; note that the sign of the shear stress is reversed from (a). (c) Combination of (a) and (b); note that the maximum shear stress is reduced by the transverse tension. (d) Transverse compression only; note that the sign of the shear stress is the same as in (a). (e) Combination of (a) and (d); note that the shear stress is greatest in this case, which is analogous to the case of wire drawing.

Because the transverse compression imposed by the wire-drawing die increases the maximum shear stress in the wire, plastic flow is enhanced; hence elongation and thinning of the wire proceed more readily. Of course, the frictional force exerted by the dies must be overcome by the drawing force, and the drawing force must also overcome the strain hardening effect. After a certain amount of drawing, the wire must be softened by *annealing*, meaning heating so as to soften it, before more drawing can be done. Otherwise, the drawing force would become so large that the wire would neck while it is being drawn. (Annealing will be discussed in Chapter 6.)

The strain hardening produced by wire drawing is the primary means of strengthening bicycle spokes. The effect of the wire drawing on the tensile strength of a stainless steel spoke is illustrated in Fig.3.20; here, the stress-strain curve of an as-drawn stainless-steel spoke is compared to that of an annealed spoke. Note that the wire-drawing process produces a level of strain hardening much greater than can be achieved in a tensile test. This is a direct result of the enhancement of plastic flow caused by the transverse compression from the drawing die.

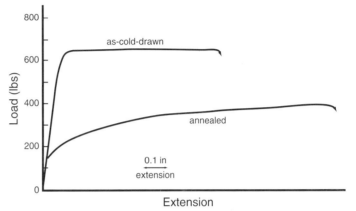

Fig. 3.20 Stress-strain curves of an as-cold-drawn stainless steel spoke and the same spoke after being annealed.

Summary

Solids can be loaded by normal stresses (either tension or compression) or by shear stresses, or they can be subjected to a bending moment, which produces a non-uniform stress field comprising a mixture of tension, compression, and sometimes shear. The tensile stress in a spoke must be kept well below the yield stress; i.e., the loading must be in the elastic range. The stress created by a given amount of strain (imposed by turning the threaded nipple) can be calculated by means of Hooke's law, with knowledge of Young's modulus. Both the yield stress and Young's modulus can be measured in the tensile test by employing an extensometer, which is used in the study of small-scale deformation. Here, engineering stress and strain are always used.

When the consideration is production of spokes, rather than their use in a wheel, large-scale-deformation behavior is the relevant issue. This is exemplified by the tensile test carried out beyond the point of necking to fracture to measure the UTS and the % R.A. or the % elongation. The transverse tension in the neck of the specimen, imposed by the constraint of the un-necked regions, leads to void formation in the interior of the neck. The growth and coalescence of the voids produces failure by rupture. By contrast, a wire-drawing die imposes transverse compression, which has two effects: Void formation is suppressed, and the shear stress is increased. The latter produces enhanced plastic flow in the die, which in turn has two effects: It facilitates the reduction in wire diameter at a drawing force below that corresponding to the UTS of the wire, and it produces much more strain hardening than does deformation by simple tension. Thus, wire drawing serves both to fabricate the spokes from rod stock and to strengthen it for use in a wheel.

Appendix 3.1 - The Load Cell

Load cells have wide applications; one is the common bathroom scale. Most operate on the principle of the resistance strain gauge.

The electrical resistance of a wire, R, is given by the product of the resistivity of the material, ρ, times the length of the wire, divided by its cross-sectional area: $R = \rho l/A$. When the wire is stretched, l increases and A decreases; this means R must increase. Thus, if a wire carrying a constant current, i, is stretched, the voltage drop along the wire must change, according to Ohm's law:

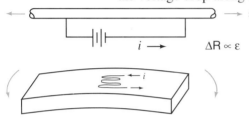

$$V = iR$$

In a load cell a strain gauge comprising a long, folded wire is bonded to a beam which is bent as the load is applied.

As the beam bends, the wire of the strain gauge is stretched. The effect of the change in its resistance is measured by the use of a Wheatstone-bridge circuit in which the strain gauge is one of the legs. With the appropriate electronics, the effect is indicated as a voltage in the millivolt range. This is, in turn, calibrated (by application of known weights to the beam to which the gauge is bonded) to read out as a load. Obviously, the load cell must be powered by a constant-current source.

The allowable load range for a given load cell must be such that the beam deforms only elastically.

Appendix 3.2 - Criterion for Necking in a Tensile Test

The mathematical criterion for reaching a maximum in the load *vs.* elongation curve is that the variation in load becomes zero. Load, P, is stress, σ, times area, A. Therefore, at maximum P:

$$dP = 0 = \sigma dA + Ad\sigma$$

or

$$\frac{d\sigma}{\sigma} = -\frac{dA}{A}$$

Physically, dσ/σ can be interpreted as the fractional increase in load due to strain hardening, and -dA/A as the fractional decrease in cross-section as the specimen is plastically elongated. Before necking, the former exceeds the latter:

$$\frac{d\sigma}{\sigma} > -\frac{dA}{A}$$

and the specimen is plastically stable. After necking, strain hardening continues in the necked region, but it is insufficient to compensate for the thinning in the neck. That is,

$$\frac{d\sigma}{\sigma} < -\frac{dA}{A}$$

This is known as the *Considère criterion*.

Exercises - Chapter 3

Terms to Understand

Engineering stress, strain
True stress, strain
Tensile test
Extensometer, strain gauge
Stress-strain curve
Yield stress
Stored elastic energy
Slip, slip lines
Shear strain
Dislocation
Dislocation glide

Schmid's law
Strain hardening
Ultimate tensile strength (UTS)
Necking
Necking criterion
Plastic constraint
Rupture
Ductility
%RA, %elong.
Wire drawing

Problems

3.1 For small-scale deformation, engineering stress and strain are good approximations for true stress and strain. Explain why.

3.2 A carbon-steel spoke with an elastic modulus of 210 GPa is loaded to a tensile stress of 240 MPa. The total strain is found by means of an extensometer to be 0.3%. Calculate how much plastic strain has occurred.

3.3 Calculate the stored elastic energy per unit volume in the spoke in Problem 3.2. How can this be related to the binding-energy curve (cf. Fig. 3.7.)?

3.4 If the UTS of a stainless steel spoke is 1725 MPa (250ksi), what is the *elastic* strain at this UTS? (Use the engineering stress approximation.)

3.5 What is the average (plastic) shear strain in a crystal in which one out of every twenty atom planes (on average) is displaced in shear by 4 atom spacings? (The type of shear is shown in Fig. 3.9 and at right.)

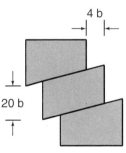

3.6 To the extent that the elastic strain is negligibly small, there is no volume change associated with plastic strain. Explain. (cf. Fig. 3.9).

3.7 What is the maximum shear stress on a plane the normal of which lies at 35° to the tensile axis, when a force of 4450N (1000lb) is applied to a bicycle spoke with a diameter of 2mm?

3.8 Prove mathematically that the maximum shear stress in a tensile specimen is half the applied tensile stress. (Hint: express θ in terms of λ, since their sum is 90°, and then maximize τ with respect to λ by differentiation.)

3.9 Write an expression for the maximum shear stress in a spoke as it is being drawn through a die when the longitudinal stress is σ_l and the radial stress from the die is σ_r.

3.10 Explain why wire drawing produces more strain hardening in a spoke than can be produced in a tensile test; cf. Fig. 3.20. (Assume the starting condition in both cases is an annealed wire.)

3.11 It can be seen from Fig. 3.20 that the load needed to produce plastic flow in the cold-drawn spoke in a tensile test is above 600lb. However, the load required for reduction of the wire diameter during cold drawing of the spoke was significantly lower. Explain why.

3.12 A tensile sample has an initial diameter of 11.3mm, and length of 50mm. After the sample fractures, the diameter at the neck is 9.6mm.

   a. Express the ductility as percent reduction in area.

   b. What can you say about the ductility expressed as percent elongation?

3.13 A stainless steel bicycle spoke 1.8mm in diameter is loaded in service with a tensile force of 1300N.

   a. What is the tensile force in pounds?
   b. What is the tensile stress in the spoke?
   c. Compare the tensile stress to the expected yield strength.
   d. What is the elastic strain in the spoke?
   e. If the spoke is 25.30 cm long when unstressed, how long is it when the maximum load is applied?

3.14 In producing wire for bicycle spokes, fairly long lengths of wire are pulled through a wire-drawing die as in Fig. 3.18. If the wire emerging from the die has a diameter of 0.125 inch, and the force required to pull the wire through the die at a constant rate is 1200 lb,

   a. What is the tensile stress in the wire?
   b. If the length of the wire (measured from the die to the point where the force is applied) is 35.2 ft with the force applied; by how much will the wire shorten if the force is removed?

4
MICROSTRUCTURE AND CRYSTAL STRUCTURE OF STAINLESS STEEL SPOKES

4.1 Metallographic Sample Preparation

As shown in Fig. 3.20, there is a dramatic difference in the tensile behavior between an annealed and a cold-drawn stainless steel spoke. To begin to explore the reasons for this difference, one can examine the microstructure of the spokes in an optical microscope; this process is called *optical metallography.* In order to do a metallographic examination, it is necessary to cut, mount, polish, and etch the specimen to be examined. This process is outlined in Fig. 4.1.

1. Piece to be analyzed is cut to reveal the desired section.

2. Cut piece is mounted in plastic (e.g., Bakelite® or Lucite®).

Longitudinal   Transverse

3. Specimen surface is ground flat on a series of abrasive papers having ever-finer sizes of silicon carbide (SiC) particles.
4. The ground specimen is polished on a series of polishing wheels using ever-finer grits of diamond dust or Al_2O_3 suspended in oil or water, respectively, to produce a mirror-like surface.
5. The sample is etched in an appropriate reagent, often a dilute acid.

Fig. 4.1 Process for preparing a metallographic specimen; for example, a longitudinal and a transverse cross-section of a bicycle spoke.

4.2 Optical Microscopy

When the preparation is completed, the specimen may be examined in a metallurgical microscope, a schematic example of which is shown in Fig. 4.2. The special characteristic of this microscope is that it is configured for the observation of opaque specimens. (Geologists use the same kind of microscope for rocks and minerals; they call it a petrographic microscope.) One can see that the illumination is directed from the light source down through the objective lens to the specimen. The light that is reflected directly back from the specimen passes again through the objective lens and then up through the half-silvered reflector and the eyepiece. (Obviously, there is considerable loss of intensity from the light's having to pass twice through the reflector, but this is unavoidable and is compensated for by the use of a bright, well-focussed lamp as the light source.)

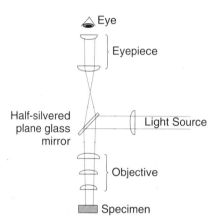

Fig. 4.2 Schematic representation of a metallurgical microscope.

One can observe the microstructure through the eyepiece, or, by mounting a camera over the eyepiece, one can make a photomicrograph (not a microphotograph) of the specimen. This was done for the specimen of a cold-drawn stainless-steel spoke; the results are shown in Fig. 4.3.

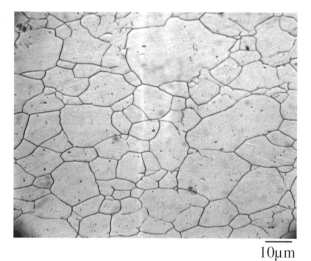

Fig. 4.3 Microstructure of a stainless steel bicycle spoke in the cold-drawn (i.e., as-received) condition.

4.3 Imaging of Grain Boundaries by Reflection Contrast

The microstructure of the stainless steel spoke in Fig. 4.3 shows a polycrystalline aggregate, the nature of which can be understood by using a two-dimensional analog of a polycrystal, which can be produced by a bubble raft, as shown in Fig. 4.4. This is made by blowing bubbles of equal size in a soap solution held in a broad, flat container. The bubbles coalesce into two-dimensional arrays floating on the surface of the solution, and they tend to form ordered regions in which each bubble has six neighbors. Such a region is analogous to a close-packed plane of atoms in a crystal; this configuration is the densest possible packing of spheres of equal size. With continued blowing of bubbles, the regions grow until they merge. This process is analogous to the solidification of a polycrystalline material (like stainless steel).

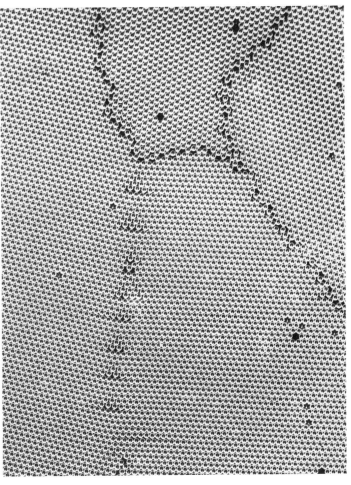

Fig. 4.4 A two-dimensional analog of a polycrystal, made with a bubble raft, showing the varying degrees of atomic disorder at grain boundaries. (From P. G. Shewmon, *Transformations in Metals*, McGraw-Hill, N.Y. 1969, p. 26.)

The arrangement of the "atoms" in the boundaries of the two-dimensional "grains" of the bubble raft depends on the degree of misorientation of the two contiguous grains and the orientation of the boundary. However, two things are clear:

1. In a two-dimensional grain, each atom touches 6 neighbors; however, the atoms along the grain boundaries have, on the average, fewer than 6 touching neighbors. Since the arrangement of atoms inside the grain is the minimum-energy configuration, atoms along a grain boundary (or on a surface) must be in a higher energy state than atoms fully within a grain. Thus, there is associated with grain boundaries in a solid an extra energy per unit area (in three dimensions), which is analogous to the surface energy of a solid or liquid.

2. Some boundaries are more disordered and loosely packed than others. Therefore, the grain boundary energy varies from one boundary to another.

Since atoms along grain boundaries are in a relatively high energy state, they go into solution more readily than atoms fully within grains when the specimen is etched in a chemical reagent. For this reason, grooves tend to develop along grain boundaries, the depth of which depends on the duration of etching attack and the atomic arrangements in the boundary. This grooving causes light to be scattered, rather than reflected back through the objective lens, when the etched sample is viewed in the microscope. This scattering delineates the grain boundaries, as illustrated by Fig. 4.5.

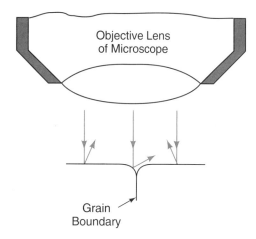

Fig. 4.5 The groove formed by etching makes a grain boundary appear as a dark line. (From W. D. Callister, Jr., *Materials Science and Engineering,* John Wiley, NY, 1985, p. 61. Reprinted by permission of John Wiley & Sons, Inc.)

4.4 The Structure of a Perfect FCC Crystal

As discussed in Chapter 2, the austenitic stainless steel used for bicycle spokes has an FCC crystal structure, the unit cell of which has already been illustrated in Fig. 2.10. A hard-sphere model extending over a few dozen of these unit cells is shown in Fig. 4.6(a). This kind of model is quite reasonable for many metallic crystals, because the attractive (bonding) forces are very nearly non-directional (being provided by the interaction between the positive ion cores and the "sea" of negative charge of the delocalized, or "free," valence electrons). Also, the steeply-rising repulsive force is felt only at close atomic spacings. Thus, a metal crystal can be approximated as a densely packed array of mutually attracting hard spheres.

The FCC structure represents the highest possible packing density for spheres of equal size, because it comprises close-packed planes stacked one on top of the other. One of the close-packed planes is illustrated in Fig. 4.6(c); this is revealed by removing a few of the corner atoms. An FCC crystal is constructed by stacking close-packed planes in a particular sequence, which can be described by considering the three sets of atom positions shown in Fig. 4.7. If one places the first close-packed plane in the A sites, the second in the B sites, and the third in the C sites, and repeats this ABC stacking indefinitely, the result is the FCC crystal shown in Fig. 4.6.

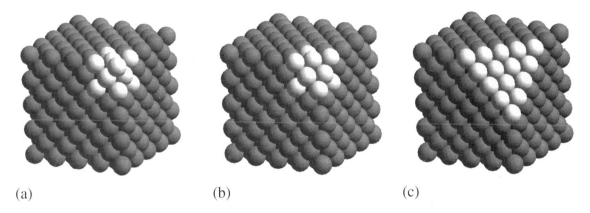

Fig.4.6 Hard-sphere model of an FCC crystal, showing (a) a unit cell, (b) one sphere removed to show a close-packed plane, and (c) more spheres removed to enlarge the close-packed plane.

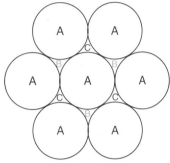

Fig. 4.7 A close-packed plane has three sets of sites, A, B, and C, as shown.

It can be seen that each atom has 12 neighbors in a close-packed crystal: six around it in a close-packed plane and three touching it in the adjacent planes above and below. This is the crystal structure one would expect in an ideal metal, which reaches its minimum energy condition by achieving the maximum density.

The FCC crystal structure is characteristic of platinum, gold, silver, copper, aluminum, and nickel, for example. The reason the stainless steel has this crystal structure is that it contains at least 8wt% nickel. This allows the high-temperature form of iron, which is also FCC, to be retained at low temperatures. If it were not for this alloying, the steel would be body-centered-cubic, BCC (cf. Fig. 2.9), at lower temperatures. Chromium is also BCC, so it is completely compatible with the normal low-temperature form of iron. Note that the BCC crystal structure has no close-packed planes; thus, it is less densely-packed than the FCC structure. (When all atoms in a crystal are the same size, the packing density depends only on the crystal structure.) In the case of the transition metals, like iron and chromium, the BCC crystal structure implies that the bonding has a directional component.

4.5 Crystallographic Indices of Planes

Close-packed planes are important in an FCC crystal, not only because of the way they are stacked, but also because slip occurs preferentially on close-packed planes. Therefore, these planes play an important role in plastic deformation of stainless steel spokes. In order to be able to discuss slip planes and slip directions in metals, it is necessary to learn the system of notation used to designate particular planes and directions. Here, the discussion is confined to cubic crystals; that is, the system of notation to be described applies equally well to FCC or BCC crystals.

Returning to Fig. 4.6(b), it is apparent that the choice of corners from which to remove atoms was arbitrary; any one of eight corners could have been chosen. However, if the corner diagonally across the cubic array from the top-right-front corner had been chosen, the exposed close-packed plane would be parallel to that seen in Fig. 4.6(b). Crystallographically, these planes are considered equivalent. Therefore, one can say that there are only four distinct close-packed planes in the FCC crystal lattice: the ones revealed by removing atoms from any of the four top corners of the cubic block shown in Fig. 4.6(a).

The notation developed to identify planes in cubic crystals is the system of *Miller indices*, which is described by the following procedure. For simplicity one can think of a simple cubic lattice, made of the three perpendicular sets of parallel lines which define the unit cell, as shown in Fig. 4.8.

> 1. Select any lattice point as the origin of coordinates; use a right-handed set of Cartesian coordinates x,y,z, as shown in Fig. 4.8.
>
> 2. As an example, consider the shaded face of the unit cell in Fig. 4.8.
>
> 3. Write down the *intercepts* of the plane in units of the *lattice parameter*. (The lattice parameter is the length of an edge of the unit cell.) In this case the intercepts are ∞, 1, ∞. Because the plane is parallel to the x and z axes, it is considered to intercept them at infinity.
>
> 4. Take the *reciprocals* of this set of intercepts. These are the Miller indices of the plane. Thus, $1/\infty$, $1/1$, $1/\infty$ becomes 010.
>
> 5. To indicate the specific plane, enclose the set of indices in parentheses: (010). Note that the indices are <u>not</u> separated by commas.

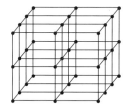

intercepts: ∞, 1, ∞

Fig. 4.8 A simple cubic lattice and the unit cell, showing the plane to be identified and the origin of coordinates chosen for this purpose.

Any face of a cubic unit cell can be represented by some permutation of 0, 1, and 0. The top plane, using the axes shown in Fig. 4.8, is (001); the front plane is (100). To index the rear plane, simply move the origin to a front corner; the indices of the plane are then ($\bar{1}$00). This, of course, is parallel to the front plane and is crystallographically equivalent to (100). (Any two planes whose Miller indices are related by multiplying through by -1 are equivalent.)

To refer to cube planes in general, curly brackets are used. Thus, {100} means any cube face, and it includes all the specific cube-plane indices. Note that, by convention, minus signs are written above a negative index, and the index is called *bar* one, *bar* two, etc.

Note also that Miller indices are always expressed as sets of integers. Hence, fractional reciprocal intercepts are always cleared by multiplication through by the lowest common denominator. For example, consider the plane shown shaded in Fig. 4.9(a), the intercepts of which are 1, 2, and 1. The reciprocals are 1, $^1/_2$, and 1, and these are multiplied through by 2 to get the Miller indices: (212).

Note further that, when any set of Miller indices is multiplied through by any number, the result is a plane parallel to the original; an example is shown in Fig. 4.9(b).

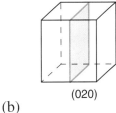

Fig. 4.9 (a) The (212) plane. (b) The (020) plane is parallel to the (010) plane.

The above procedure is valid for all cubic lattices. The fact that the BCC and FCC lattices have lattice points at the body center and at the face centers, respectively, of the unit cells is irrelevant.

Returning to the FCC crystal depicted in Fig. 4.6(b), the close-packed plane shown there can be represented in either of the ways depicted in Fig. 4.10. One would normally give the Miller indices as (111).

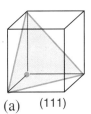 

(a) (111)   (b) ($\bar{1}\bar{1}\bar{1}$)

Fig. 4.10 (a) The (111) plane. (b) The ($\bar{1}\bar{1}\bar{1}$) plane. Each is referred to the origin indicated by the filled dot. These planes are equivalent, since they are related by a -1 multiplication, and the choice of the origin is arbitrary.

Example 4.1

Q. Sketch the other three close-packed planes in the FCC crystal shown in Fig. 4.6 and give their Miller indices.

A. The three planes can be revealed by removing atoms from the other three corners of the block in Fig. 4.6. If the (111) plane, which is already indexed, is considered as the southeast (SE) corner, then the others are as shown below:

SW corner   NW corner   NE corner

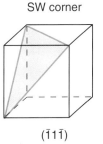 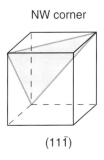 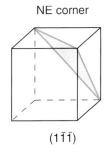

($\bar{1}11$)   ($11\bar{1}$)   ($1\bar{1}1$)

The origin is always chosen for convenience; obviously, the plane must not pass through the origin. The Miller indices are given for each plane.

Note that none of the four sets of indices: (111), (1–11–), (111–) and (11–1–) can be obtained by a -1 multiplication of any of the others. Therefore, they are distinct planes. These are the only {111}-type planes in the FCC crystal lattice; the ones revealed by removing atoms from the bottom corners of the block in Fig. 4.6 are parallel to, and therefore equivalent to, the ones shown above.

Another important plane in cubic crystals is parallel to a cube axis and passes through a cube face. This family of planes has the indices {110}. Two examples are shown in Fig. 4.11.

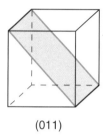
(011)

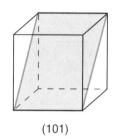
(101)

Fig. 4.11 Two examples of {110} planes.

Example 4.2

Q. Referring to Fig. 4.11, sketch the other four {110}-type planes in a cubic lattice and give their indices.

A. The four planes are shown below, with their indices:

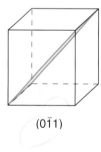
($0\bar{1}1$)

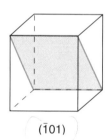
($\bar{1}01$)

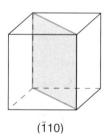
($\bar{1}10$)

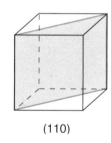
(110)

4.6 Direction Indices

It is also important to have a system for denoting directions in crystals. For example, it is necessary to describe slip in terms of a slip plane and a slip direction. Direction indices are nothing more than the smallest set of indices that describes a vector in units of the lattice parameter, using the cube axes as the coordinate system. For specific directions square brackets are used: [ ]. For example, the cubic axes are [100], [010], and [001]. To refer to these axes generically, angled brackets are used: < >. For example, [100] means the direction along the x axis, and <100> means any of the three axial directions.

A great convenience in the cubic lattice is the fact that the [hkl] direction is perpendicular to the (hkl) plane. Examples are shown in Fig. 4.12. (This is only true for cubic crystals.) Note that multiplying direction indices through by -1 simply gives the reverse direction; it is crystallographically equivalent to the original. The body diagonals of a cube have <111> indices, and the face diagonals have <110> indices. Table 4.1 summarizes the convention regarding brackets around indices.

Fig. 4.12 Examples of [hkl] directions lying perpendicular to (hkl) planes.

Table 4.1 Convention for brackets around indices.

	Planes (Miller indices)	**Directions**
Specific	( )	[ ]
Generic	{ }	< >

Example 4.3

Q. Show that three <110> directions lie in the (111) plane and give their indices.

A. The directions are sketched in the unit cell shown here; they could also have been represented by their opposites: [$\bar{1}$01], [1$\bar{1}$0], and [01$\bar{1}$].

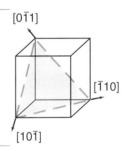

Summary

Properties of materials depend on structure on several dimensional scales. What is commonly referred to as the microstructure is the arrangement of phases and grains as perceived with the optical (metallographic) microscope. Specimen preparation usually involves grinding, polishing, and etching, and the microstructure is imaged by means of reflection contrast.

On the atomic scale, the crystal structure of solids can be analyzed by means of x-ray diffraction (cf. Appendix 5.1), utilizing crystallographic notation to specify various planes and directions. This notation is introduced here for the cubic lattices, and the FCC structure characteristic of stainless-steel spokes is used as the working example.

Appendix 4.1 - The Scalar Product of Two Vectors

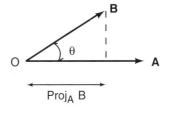

The *scalar*, or *dot*, product of two vectors **A** and **B** is defined by the equation

$$\mathbf{A} \cdot \mathbf{B} = |\mathbf{A}| \, |\mathbf{B}| \cos \theta$$

where θ is the angle between **A** and **B**.

Thus, the dot product may be interpreted geometrically as the length of **A** times the length of the projection of **B** onto **A**. In the present context, the usefulness of the dot product lies in the fact that, if **A** and **B** are perpendicular, $\cos \theta = 0$, and the dot product is zero.

The dot product can be calculated by writing the vectors in terms of their components,

$$\mathbf{A} = a_1 \mathbf{i} + a_2 \mathbf{j} + a_3 \mathbf{k}$$

$$\mathbf{B} = b_1 \mathbf{i} + b_2 \mathbf{j} + b_3 \mathbf{k}$$

and multiplying the components, as follows:

$$\mathbf{A} \cdot \mathbf{B} = a_1 b_1 + a_2 b_2 + a_3 b_3$$

Thus, the $[1\bar{1}0]$ direction lies in the (111) plane, because

$$1 \times 1 + \bar{1} \times 1 + 0 \times 1 = 0$$

Exercises - Chapter 4

Terms to Understand

Metallography
Photomicrograph
Polycrystalline material
Grain boundary
Grain boundary energy

Close-packed plane
Miller indices
Crystal lattice
Scalar (dot) product

Problems

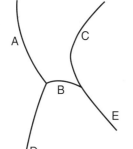

4.1 Referring to Fig. 4.4, how would you expect the grain boundaries (denoted A, B, C, D, and E in the sketch to the left) to rank in order of grain boundary energy; use three groups: high, medium, and low energy, and explain your rankings.

4.2 A two-dimensional analog of the atomic packing density is the number of atoms per unit area on a crystallographic plane. (Only atoms whose centers lie on the plane are included in the number.) Calculate the areal packing density on the (001), (011), and (111) planes in the FCC and in the BCC structures.

4.3 Given that the density of tungsten is 19.25 g/cm³, the atomic weight is 183.85, and the crystal structure is BCC, calculate the size of the tungsten unit cell (i.e., a) and the radius of a tungsten atom.

4.4 Given the fact that platinum is FCC with atomic weight 195, calculate the mass of one unit cell. Then, given the density = 21.5 g/cm³, calculate the size of the unit cell and the diameter of a Pt atom. Finally, calculate the areal density of packing of Pt atoms on a close-packed (111) plane. (Consider the plane to pass through the centers of the atoms.)

4.5 A simple cubic unit cell has an atom of radius r at each corner of the cube. Each edge of the cube has length 2r. Calculate the atomic packing density of this structure.

4.6 Sketch a unit cell, and show the following:

$$[100]\ [110]\ [120]\ [0\bar{1}0]\ [01\bar{1}]\ [012]$$

and

$$(001)\ (1\bar{1}\bar{2})\ (1\bar{1}1)$$

4.7 Show by means of a sketch that the (112) and (111) planes both contain the $[\bar{1}10]$ direction.

4.8 Using the fact that the [hkl] direction is perpendicular to the (hkl) plane in a cubic lattice and the concept of a scalar product of two vectors (i.e., a dot product; cf. Appendix 4.1), show that the $[\bar{1}10]$ direction must lie in both the (112) and (111) planes. (Pay attention to the difference between plane and direction indices; i.e. ( ) vs. [ ].)

4.9 Find the three {110} planes which contain the $[\bar{1}\bar{1}1]$ direction. Illustrate this with a sketch and prove your answer by means of dot products.

4.10 Draw a unit cell and show a (115) and a (112) plane.

5

SLIP, DISLOCATIONS, AND STRAIN HARDENING IN STAINLESS STEEL SPOKES

The stress-strain curves of the stainless-steel spokes shown in Fig. 3.20 exemplify the unique behavior of metallic structural materials. This is the only class of materials which exhibits both large-scale plastic flow and strain hardening. Strain hardening is not only a primary strengthening mechanism in metals, but it is the reason why stable plastic flow is possible (cf. Appendix 3.2). This behavior of metals can be contrasted with that of other crystalline materials, like oxides, carbides, nitrides, etc., which generally exhibit negligible plasticity. These are brittle materials because they are likely to contain crack-like defects, and these cracks cannot be rendered harmless by plasticity (which blunts the tip of a sharp crack). In viscous amorphous materials, plastic flow can occur, but strain hardening cannot. In order to understand these differences, a knowledge of the properties and behavior of crystal dislocations is required.

Some of the specific questions to be addressed in this and later chapters are the following:

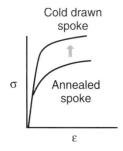

1. How does strain hardening occur, and why does it saturate; i.e., why does the rate of strain hardening decrease with strain?

2. Besides strain hardening, how else can one make a high-strength metallic alloy? That is, how can one inhibit the motion of dislocations, so that a high applied stress is required for plastic flow? Applications are high-strength, thin-walled steel tubing or high-strength aluminum alloys for lightweight bicycle frames, for example.

3. Why do carbon-steel spokes have a fatigue limit, whereas stainless-steel spokes do not? (At stresses below the fatigue limit, fatigue failure does not occur, regardless of the number of stress cycles applied.)

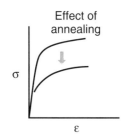

4. Why do strain-hardened materials like spokes soften when heated to around half the melting temperature (on the absolute scale, degrees Kelvin)?

5. Why do high-strength (precipitation-hardened) aluminum alloys soften at even lower temperatures?

5.1 Slip in FCC Crystals

In metallic crystals, slip tends to take place on planes with the closest packing and in directions of closest packing on those planes. The simplest to understand are FCC metals, of which stainless steel is a good example. Face-centered cubic crystals are made of stacked close-packed planes (cf. Figs. 4.6 and 4.7), and each close-packed plane contains three directions, called *close-packed directions,* along which atoms touch. These directions are illustrated in Fig. 5.1(a).

When slip occurs on a close-packed plane, it means that the adjacent (parallel) plane is translated along one of the close-packed directions. The slip displacements occur in units of the interatomic spacing. For example, one unit of slip is depicted by the vector marked "b" in Fig. 5.1(a). If the plane shown were designated as an A plane (cf. Fig. 4.7) and the next plane up is called a B plane, then one would say that the slip vector b represents the translation of an atom from one B site to the next. In an FCC crystal the close-packed planes are {111} planes, as seen earlier. One of these, the (111) plane, is shown in Fig. 5.1(b), along with the three close-packed directions, i.e., the <110> directions, in that plane.

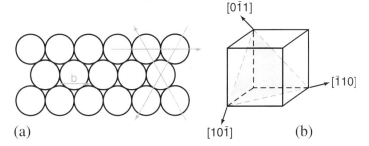

Fig. 5.1 (a) A close-packed plane of hard spheres, showing three close-packed directions and a slip vector. (b) An FCC unit cell showing a slip plane and three slip directions in that plane.

The reason close-packed directions are preferred for slip is that this requires the least work. One could say, rather loosely, that the "bumps" are smaller in such a direction. A bit more scientifically, one would say that shear of one plane of atoms over another periodically decreases the packing density of the crystal in that region, as the atoms move from one set of low-energy sites to the next. Any decrease in density requires an input of energy, i.e., work. The decrease in density during slip is obviously smallest along a close-packed direction; hence, work is minimized. (This simple explanation is appropriate for close-packed metals, but other factors become important for more complex crystal structures.)

Example 5.1

Q. Show that the three slip vectors illustrated in Fig. 5.2(a) can be expressed as $\frac{a}{2}[\bar{1}10]$, $\frac{a}{2}[10\bar{1}]$, and $\frac{a}{2}[0\bar{1}1]$, where a is the lattice parameter.

A. The vectors can be drawn to span atom centers or to span interstices (the spaces between atoms), as indicated in Fig. 5.2(a), and they can also be depicted in a unit cell, as shown in Fig. 5.2(b). The vector in the $[\bar{1}10]$ direction has the components $-\frac{a}{2}, \frac{a}{2}, 0$. Factoring out the $\frac{a}{2}$, one writes $\frac{a}{2}[\bar{1}10]$. The other two vectors are treated similarly.

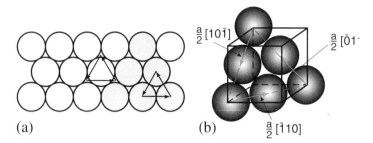

Fig. 5.2 (a) The slip vectors in a close-packed plane may be drawn spanning either interstices or atom centers. (b) The same vectors are shown in a unit cell.

The reason a close-packed plane is usually chosen for slip, if one is available in the crystal structure, is essentially the same as for the choice of a close-packed direction: The "bumps" are smaller. This can be illustrated by comparing the packing of a {111} and a {100} plane in an FCC crystal, as shown in Fig. 5.3. Possible slip vectors are shown in each case. Not only are the slip vectors longer for the {100} plane, but the vertical displacements that occur when an adjacent plane is translated along this vector are larger than for the {111} plane. When two adjacent planes are in their normal, equilibrium relationship, they are in a minimum-energy configuration. When they are translated past each other, they pass through states of increased energy, as indicated schematically in Fig. 5.3(b). The difference between a minimum and a maximum in the plot of energy *vs.* displacement represents the work done during slip of one interatomic distance. The minimum work is done when slip occurs on a close-packed plane in a close-packed direction. The force necessary for slip would be the maximum slope of the energy *vs.* displacement curve. This is obviously greater for the non-close-packed plane in Fig. 5.3.

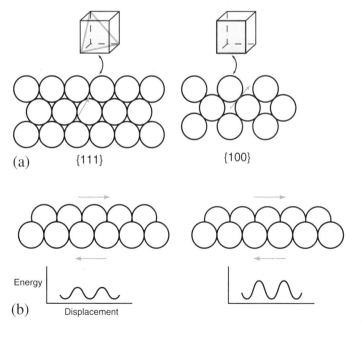

Fig. 5.3 Comparison of the atomic packing on {111} *vs.* {100} planes in an FCC crystal, showing (a) plan views of these planes and their locations in a unit cell, (b) two adjacent planes in an elevation view, and the variation in energy as the upper plane is translated over the lower plane in the direction of the vectors shown.

A combination of a slip plane and slip direction lying in that plane is called a *slip system;* the slip systems in an FCC metal are referred to generically as {111}<110>. As already noted in Chapter 3, slip occurs not by whole blocks of crystal sliding past one another, but by the passage of dislocations. In other words, slip is a consecutive, rather than a simultaneous, process. In the rest of this chapter, the properties of dislocations that are important for the understanding of plastic yielding, strain hardening, and other strengthening mechanisms, as well as the fatigue limit in carbon-steel spokes, are described. Some details about dislocation properties and behavior are contained in Appendix 5.2 and the worked examples throughout the chapter.

5.2 Types of Dislocation

There are two basic types of dislocation: *edge* and *screw.* They can each produce slip on any slip system, but they have very different geometries and behaviors. Dislocations can also be of mixed type (edge and screw combined).

5.2.1 The Edge Dislocation

The edge dislocation has already been illustrated in Figs. 3.12 and 3.13. One can imagine the creation of an edge dislocation by the process shown in Fig. 5.4. The top half of the block of crystal is sheared to the right with respect to the bottom half, but the displacement does not occur all at once over the whole plane of shear. Rather, the sheared region spreads from left to right, leaving a step on the left-hand face of the block. The boundary between the sheared (i.e., slipped) and unsheared portions of the plane of shear (i.e., slip plane) is defined as the dislocation line.

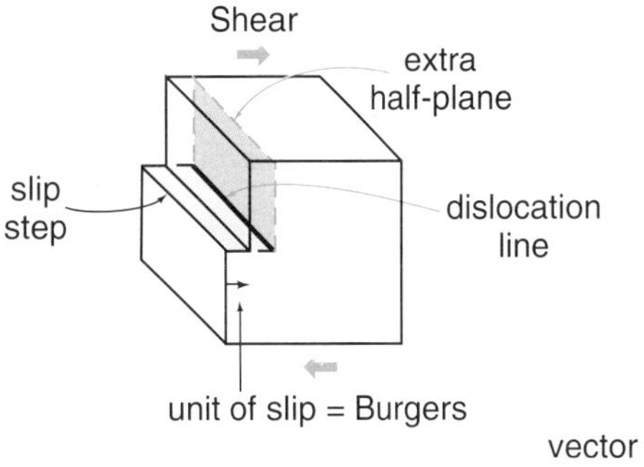

Fig. 5.4 Slip has occurred in the darkly shaded region of this block of crystal. The boundary between the slipped and unslipped region is called a dislocation; it is a linear crystal defect. In this case an edge dislocation has been created.

The partial shearing of the block in Fig. 5.4 has left N vertical planes of atoms in the upper half of the block squeezed into the space occupied by N-1 planes in the lower half of the block. The extra half-plane in the upper part is found at the end of the sheared region. As the sheared region continues to spread, the location of the extra half-plane moves to the right, as seen in Fig. 5.5*. When the shear is complete, the extra half-plane appears as a step on the right-hand side of the block (cf. Fig. 5.5c). The bottom edge of the extra half-plane locates the line of the edge dislocation.

---

*Note that the extra half-plane is a configuration, like the ripple in the rug in Fig. 3.11. When we say that the dislocation moves, we mean that the configuration moves; we do not mean to imply that a particular half-plane of atoms is transported from one side of the crystal block towards the other.

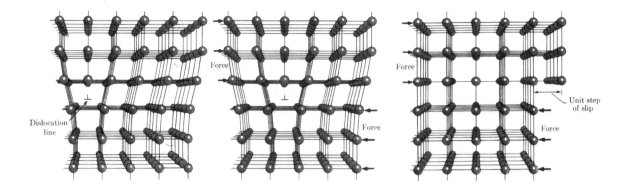

Fig. 5.5 An atomistic representation of an edge dislocation at several stages of the glide process, which produces slip. In the final panel the dislocation has emerged at the surface, leaving a slip step, and the crystal is left dislocation-free. (From A. G. Guy, loc. cit., p. 109.)

An important property of a dislocation is the slip vector, or *Burgers vector*. The Burgers vector denotes the magnitude and direction of slip produced when the dislocation glides. It also indicates the size of the slip step produced when an edge dislocation exits the crystal face, as in Figs. 3.13 and 5.5. An edge dislocation has the following important characteristics:

1. The Burgers vector is perpendicular to the dislocation line.

2. Glide occurs in the direction of the Burgers vector.

3. The glide plane of an edge dislocation is fixed. It is defined by the dislocation line and the Burgers vector. Because the edge dislocation involves extra material, it can only move off its glide plane if atoms are added to, or subtracted from, the extra half-plane. (This would be a non-conservative process, and it must involve diffusion of atoms ((i.e., an elevated temperature)) as will be seen in Chapter 6.)

It is important to reiterate that slip does not change the geometry (i.e., the symmetry) of a crystal. A crystal is the same after the passage of a dislocation as it was before. This means that the Burgers vector must produce an "identity translation." Thus, the Burgers vector of a slip dislocation in an FCC crystal must connect one atomic site with the next along the slip direction, as illustrated in Fig. 5.2(a). The Burgers vector of a slip dislocation in an FCC crystal is designated as $a/2 <110>$; this nomenclature is clarified by Example 5.1. Three examples of Burgers vectors are shown in Fig. 5.2(b).

5.2.2 The Screw Dislocation

The other basic type of dislocation, the screw dislocation, can be formed by shearing the kind of block shown in Fig. 5.4, but in a direction 90° from that used to form the edge dislocation. This is illustrated in Fig. 5.6. In this case the slip step is on the side face of the block.

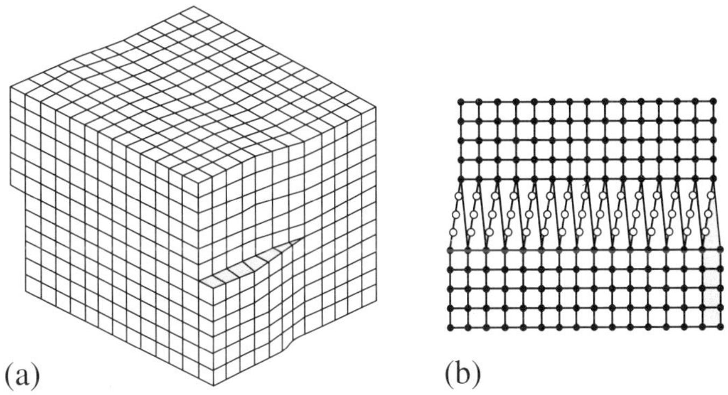

Fig. 5.6 (a) Illustrates the type of shear that produces a screw dislocation. (b) Shows the displacement of atoms in the core of the screw dislocation. (Adapted from W.T Read, Jr., *Dislocations in Crystals,* McGraw-Hill, N.Y., 1953, p. 17.)

There is no extra half-plane of atoms associated with a screw dislocation. Rather, the atoms in the zone along the dislocation line (i.e., in the core of the dislocation) are displaced into a helical configuration. The atomic planes perpendicular to the dislocation line are distorted from a stack of parallel planes into a spiral ramp (like the threads of a screw) in the dislocation core. Looking vertically downward on the slip plane, the core would appear as shown in Fig. 5.6 (b).

The special characteristics of a screw dislocation are as follows:

1. The Burgers vector is parallel to the dislocation line.

2. Glide occurs in the direction perpendicular to the Burgers vector, as shown in Fig. 5.7, but the Burgers vector is still, by definition, in the direction of shear.

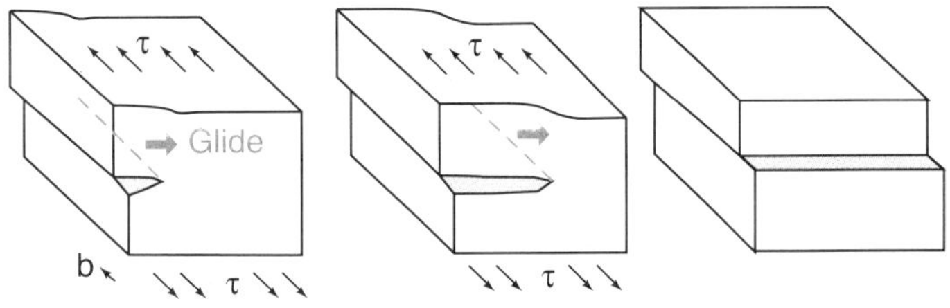

Fig. 5.7 Slip by glide of a screw dislocation. Note: τ still lies along b, but the direction of glide is now perpendicular to τ and b.

3. Since the Burgers vector and the dislocation line are parallel, they do not define a glide plane. Therefore, from a purely geometrical standpoint, a screw dislocation can glide on any of the infinite number of planes that contain the Burgers vector. Thus, if a crystal had no preferred slip planes (e.g., close-packed planes), the slip step could be quite wavy, as shown in Fig. 5.8(a). In most materials, however, there are preferred slip planes (like {111} planes in FCC stainless steel), and the slip steps from screw dislocations are made up of straight segments, as shown in Fig. 5.8 (b).

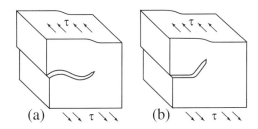

Fig. 5.8 Schematic illustration of slip steps that would occur when a screw dislocation glides on more than one plane. In (a), no particular slip plane is preferred. (b) represents the case of FCC crystals in which close-packed planes are preferred.

5.2.3 Mixed Dislocations

A mixed dislocation is simply one for which the Burgers vector is neither parallel nor perpendicular to the dislocation line, but somewhere in between. One can be constructed by shearing a block of crystal starting at a corner, as shown in Fig. 5.9. The shear along one edge of the block produces a dislocation which varies continuously from pure edge to pure screw. That is, the angle between the Burgers vector and the dislocation line varies from 90° at point A to 0° at point B. Figure 5.10 shows that the glide of this dislocation (i.e., the spreading of the slipped region) produces the same kind of shear of the block as would occur with a pure edge or pure screw dislocation.

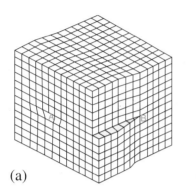
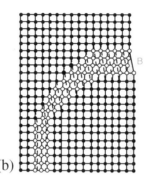

Fig. 5.9 (a) Illustrates a crystal sheared to produce a mixed dislocation, which runs from A to B. (b) shows the displacements of atoms in the core of the dislocation, which is pure edge at A and pure screw at B. (Adapted from W. T. Read, Jr., loc. cit., p. 18.)

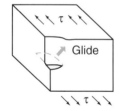
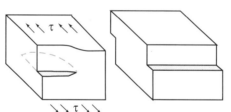

Fig. 5.10 Slip by glide of a mixed dislocation.

The important characteristics of the three dislocation types can be summarized as follows:

Type	Angle between b and dislocation line	Direction of shear	Direction of glide	b in FCC crystal
Edge	90°	along b	perpendicular to the line	$a/2<110>$
Screw	0°	"	"	"
Mixed	varied	"	"	"

62

Example 5.2

Q. What happens in a crystal in which the extra half-plane of an edge dislocation lies below the glide plane, as in Fig. 5.11 (using a ⊤ sign to indicate the dislocation, as shown).

A. The shear in the crystal occurs in the opposite sense to the case in which the half-plane lies above the glide plane. One can assign a positive sense to an edge dislocation with half-plane up and a negative sense to one with half-plane down. Note that a dislocation of negative sign glides in a direction opposite to that for a positive dislocation upon the application of a shear stress. This is illustrated in Fig. 5.11.

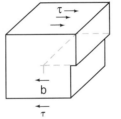
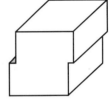

Fig. 5.11 A negative edge dislocation subjected to the same positive shear stress as applied in Figs. 5.4 and 5.5. Note that the negative dislocation glides in the opposite direction to the ones shown in those figures.

Example 5.3

Q. What happens in a crystal with two edge dislocations of opposite sign on the same glide plane? of the same sign?

A. If the two are of opposite sign, the application of a shear stress of one sense will drive them apart, as in Fig. 5.12(a), and a shear stress of the reverse sense will bring them together, whereupon they will annihilate each other, as in Fig. 5.12(b). If they are of the same sign, they will both move in the same direction upon the application of a shear stress, but there will be a mutual repulsion between them. (This will be elaborated upon in Section 5.5.) The dislocations of opposite sign experience a mutual attraction.

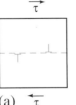

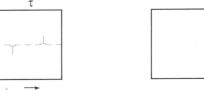

Fig. 5.12 (a) A pair of edge dislocations of opposite sign, showing the result of the application of a positive shear stress to this pair of dislocations; slip steps are formed on opposite side of the crystal block as each dislocation emerges at the surface. (b) shows the result of the application of a negative shear stress.

Example 5.4

Q. Can one speak of positive and negative screw dislocations?

A. Yes. Fig. 5.13 shows what this means. Note that the assignment of which is called positive and which negative is completely arbitrary. It is only important that the assignment be consistent.

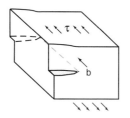
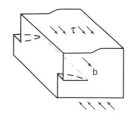

Fig. 5.13 Screw dislocations of opposite sign.

Example 5.5

Q. What would happen if, when a shear stress is applied to a crystal block, slip started in the center of the block, instead of at an edge?

A. By definition, the boundary between slipped and unslipped regions of a slip plane is a dislocation line. If the slipped region is contained entirely inside the crystal, the dislocation line must form some kind of loop that closes on itself. Fig. 5.14(a) gives an example, drawn under the approximation that all parts of the dislocation line have the same elastic-strain energy; this means that the loop would be circular. Note that all parts of the loop have the same b, since the same kind of slip has occurred throughout the slipped region. Note, also, that diametrically opposed portions of the loop are dislocations of the same type, but of opposite sign. In Fig. 5.14(a), the loop is + edge at 12:00 o'clock and - edge at 6:00 o'clock, + screw at 3:00 o'clock and - screw at 9:00 o'clock. At all other positions the dislocation line is of mixed character.

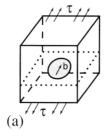
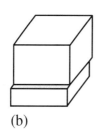

(a)   (b)

Fig. 5.14 (a) A dislocation loop created by shear starting inside a crystal. (b) The sheared block after the loop has exited at the surfaces.

Example 5.6

Q. What would happen if the shear stress on the crystal block in Fig. 5.14(a) were continuously increased?

A. The loop would continue to expand and would finally emerge from the side surfaces, leaving slip steps on the front and back sides, as shown in Fig. 5.14(b).

Example 5.7

Q. What would happen if the shear stress on the block in Fig. 5.14(a) were suddenly removed, assuming that the intrinsic resistance to dislocation glide in that crystal is very small?

A. The loop would shrink by inward glide, owing to the mutual attraction of diametrically opposed segments (parallel dislocations of opposite sign). Finally, the loop would self-annihilate.

5.2.4 General Characteristics of Dislocations

It is important to maintain perspective about dislocations and not get bogged down in details. They are analogous to ripples in rugs (cf. Fig. 3.11). When a ripple moves (i.e., "glides"), the rug is translated over the floor, like one part of a crystal block over another. Moving a rug like this is much easier than moving it all at once, because the resistance to ripple glide is much smaller than the frictional force opposing the motion of the whole rug. The same is true for a dislocation in metals. Considering the atomistic representation of an edge dislocation (Fig. 5.5), one can see that the crystal planes are elastically warped in opposite directions on either side of the dislocation line. The restoring forces arising from the distorted interatomic bonds would tend to make these planes straighten up. Thus, the dislocation "feels" two equal and opposite forces. An applied shear stress causes an imbalance of these forces, and in a soft FCC crystal (e.g., pure copper) even a small imbalance is sufficient to cause glide.

The ease of dislocation glide in any metallic crystal depends on two factors:

1. The nature of the intrinsic interatomic bonding forces, and

2. The types and densities of barriers, such as foreign atoms, particles having a different crystal structure, other dislocations, etc. (These are factors extrinsic to the crystal *per se*.)

The bonding force in a simple FCC metal is essentially non-directional. Thus, dislocation glide, which involves changing the angular relationships between rows of atoms, is relatively easy. Therefore, pure copper, silver, gold, platinum, and aluminum, for example, are plastically "soft." On the other hand, BCC transition metals, like iron, chromium, molybdenum, and tungsten, have more complicated bonding. In these metals, edge dislocations still glide fairly easily, but the cores of screw dislocations are highly distorted. The small displacements of atoms in the direction of b along the screw dislocation line are more like a spiral staircase in BCC crystals than like the spiral-ramp configuration in FCC crystals. This makes screws relatively immobile in BCC metals (like iron) at low temperatures. At higher temperatures, thermal vibrations of atoms "smear out" the steps in the screw dislocation core, and glide becomes easier; i.e., BCC metals become more deformable as temperature is increased, and this causes their tendency toward brittleness to decrease.

In a covalently bonded crystal like silicon, the bonding is completely directional, and even edge dislocation glide is impossible below temperatures of around 800°C (silicon melts at 1410°C). Thus, silicon is brittle, except at very high temperatures.

A new factor comes into play in ionic crystals. Not only are there size differences between positive and negative ions, but the ions are constrained to lie in an ordered array, with negative ions surrounded by positive ions, and vice versa. Since dislocation glide would break up the order on many of the potential glide planes, slip occurs on few systems, and polycrystalline ionic solids are generally brittle.

The extrinsic barriers to dislocation glide are the means whereby crystalline solids, particularly metals, are strengthened. These must be employed carefully, since, when plasticity is inhibited, ductility and toughness usually suffer. Bicycle technology employs a variety of strengthening mechanisms. In spokes the most important one is dislocation density; i.e., strain hardening. To understand strain hardening one needs to learn how dislocations interact. They do so by virtue of their stress fields. Before moving to this topic, it is useful to consider how dislocations in crystals can be observed directly.

5.3 Imaging of Dislocations

The true role of dislocations in crystals was not widely recognized until the application of the electron microscope to solids in the 1950s. To prepare a sample for transmission electron microscopy (TEM), one first cuts a thin slice of a metal or alloy; the slice is then thinned further by electrolytic dissolution to a fraction of a micrometer. This is thin enough that an electron beam accelerated by a voltage of around 100kV can pass through it. The essential features of a standard electron microscope are illustrated in Fig. 5.15(a), and some dislocations in a thin foil of stainless steel are shown in Fig. 5.15(b).

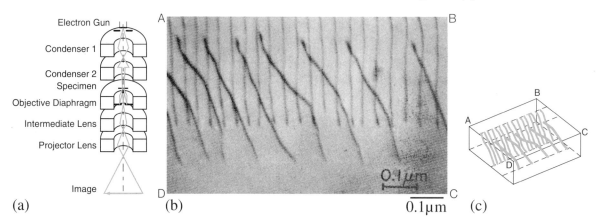

Fig. 5.15 (a) Schematic illustration of essential features of a transmission electron microscope. (b) Dislocations in a thin foil of stainless steel. (c) Illustration of the foil which produced the image shown in (b). ((b) and (c) from D. Hull and D. J. Bacon, *Introduction to Dislocations,* 3rd ed., Pergamon Press, Oxford, 1984, p. 34. Reproduced by permission of Butterworth-Heinemann Ltd.)

The dislocations in the slice of stainless steel have been revealed by a process called *diffraction contrast* (as opposed to the process of reflection contrast through surface roughening in optical microscopy). Since the electron beam has a wavelength smaller than the interatomic spacing in a crystal, the beam can be diffracted by the crystal lattice. (The law of diffraction was first worked out by Sir Lawrence Bragg in 1912, and it is explained in Appendix 5.1.) Because some atomic planes are bent in the vicinity of a dislocation (cf. Fig.5.5), the electron beam is diffracted at a different angle than in the undistorted crystal remote from the dislocation line. Thus, this part of the diffracted beam is not focused on the photographic plate, and the dislocation line appears dark.

The arrays of parallel dislocations in Fig. 5.15(b) are actually segments of dislocations which have been captured by the sectioning of the specimen, and the segments run from the bottom surface of the foil to the top, in the manner shown in Fig. 5.15(c).

5.4 The Stress Field of a Dislocation

The fact that metals deform plastically by way of dislocation motion was deduced long before there was direct evidence that dislocations actually exist in crystals. One of the originators of the dislocation concept, G. I. Taylor, also conceived the first theory of strain hardening. In it he recognized that plastic deformation involves the multiplication of dislocations (which means that their average spacing must decrease), and that there is a mutual repulsion between parallel dislocations of like sign. This repulsion exists because the displacements of the atoms in the crystal in the vicinity of the dislocation line give rise to a stress field. The stress can be calculated through Hooke's law, if the displacements are known.

5.4.1 General Aspects of Stress Fields

The stress field of a dislocation is, in a certain sense, analogous to the electric field around a charged particle. Just as the strength of the electric field depends upon the distance from the charged particle, the stress field of a dislocation depends on the distance from the dislocation. Therefore, one must be able to deal with the value of a stress field at some point in a solid. The way to do this for an edge dislocation is to imagine an infinitesmal cube centered on the point of interest, as shown in Fig. 5.16. The z coordinate of the cube is aligned with the dislocation line.

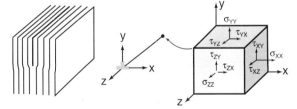

Fig. 5.16 Illustration of the components of stress at some point remote from a dislocation line by consideration of the stresses acting on the surface of an infinitesmal cube centered at that point.

The components of the stress field are defined as the stresses that act on the surface of the cube; all the possible positive components of stress are shown on the cube in Fig 5.16. Each of these has an equal and opposite component acting on the respective opposite faces.

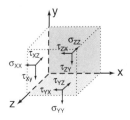

The scheme used to designate the various stress components involves a double subscript, the first to indicate the plane on which the stress acts, and the second to indicate the direction in which it acts. Thus, σ_{xx} represents a normal stress on a plane perpendicular to the *x* axis and acting in the positive *x* direction. ($-\sigma_{xx}$ would act in the negative *x* direction.) Similarly, τ_{yz} is a shear stress on a plane perpendicular to the *y* axis and acting in the *z* direction. A bit of thought will quickly lead to the conclusion that Fig 5.16 does, in fact, show all of the possible (positive) components of stress on the surface of the cube. These stress components can be represented in the following matrix:

$$\begin{matrix} \sigma_{xx} & \tau_{xy} & \tau_{xz} \\ \tau_{yx} & \sigma_{yy} & \tau_{yz} \\ \tau_{zx} & \tau_{zy} & \sigma_{zz} \end{matrix}$$

It is easy to show that there are only six independent components in this matrix, since:

$$\tau_{xy} = \tau_{yx} \qquad \tau_{xz} = \tau_{zx} \qquad \tau_{yz} = \tau_{zy}$$

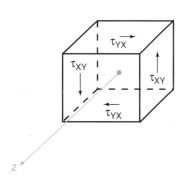

These relationships stem from the fact that the cube is stationary and at equilibrium. For example, let the origin of coordinates be in the center of the cube, let the faces be of unit area (so that stresses are numerically equal to forces), and consider the moments which tend to spin the cube about the *z* axis. The forces τ_{xy} and $-\tau_{xy}$ would tend to spin it counter-clockwise, and the forces $\pm\tau_{yx}$ would tend to spin it clockwise. Since the cube does not spin, and the moment arms of these forces are all equal to half the length of the cube edge, the respective force components must be equal; thus, $\tau_{xy} = \tau_{yx}$. The other relationships can be proved by considering moments about the other two axes.

It is important to point out that, in some cases, some of the terms in the matrix are equal to zero. An example is the stress field around a substitutional solute atom in an alloy like stainless steel. If the solute atom is too large for its lattice site (i.e., larger than the host atoms), then the solute atom is compressed symmetrically in three dimensions by the surrounding host atoms.

This is exactly the kind of stress field (or pressure field) that would be experienced by a body submerged in a fluid, like a ball below the surface of the ocean. For this reason, this stress state is called *hydrostatic compression*. In this case all the shear-stress components are zero (i.e., the ball remains spherical), and the normal (i.e., σ) terms are all equal.

Even before the existence of dislocations in crystals was first postulated, the stress fields in distorted blocks (as in Figs. 5.4 and 5.6) had been determined in the context of continuum elasticity theory. The case of the screw dislocation is particularly straightforward, and the stress field can be calculated easily.

5.4.2 The Stress Field of a Screw Dislocation

The stress field of a screw dislocation can be demonstrated by considering a cylindrical block of radius r and length L, as shown in Fig. 5.17. In a thought experiment, the block is slit longitudinally through to its central axis and displaced axially by an amount b before re-joining the faces of the slit. The block is now elastically strained, and the strain is obviously greatest along the central axis where the shearing was terminated. The strain field in the block gives rise to a stress field (through Hooke's law).

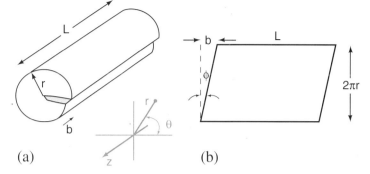

Fig. 5.17 (a) Creation of a screw dislocation along the axis of a cylindrical block. (b) The shear strain produced in the cylindrical surface at a distance r from the core of the dislocation.

The stress field in the dislocated cylinder approximates that of a screw dislocation in a crystal. The approximation arises from the use of an isotropic elastic continuum to represent a crystal, which, of course, is made up of discrete atoms. This kind of approximation is made routinely in materials science, because continuum elastic theory is very well developed, and atomistic calculations are still rather new and complex.

Because the screw-dislocated cylinder has (obviously) cylindrical symmetry, it is more convenient to use cylindrical coordinates instead of Cartesian (x,y,z) coordinates to express the components of the stress field. In cylindrical coordinates, the axis of symmetry is designated as the z axis, and any point is given by its z coordinate along with its radial coordinate, r, and its angular coordinate, θ.

Planes are denoted in the same way as in Cartesian coordinates: The z plane is perpendicular to the z axis; the "r plane" is the cylindrical surface perpendicular to any radius. The θ plane contains the z axis and is perpendicular to the angular coordinate, θ.

Applying this scheme to the dislocated cylinder in Fig. 5.17, it is clear that a θ plane has been sheared in the z direction. Therefore, the necessary shear stress is $\tau_{\theta z}$.

To determine the stress field in the cylinder, it is necessary first to calculate the strain and then to use the version of Hooke's law that applies to shear deformation:

$$\tau = G\gamma$$

where G is the shear modulus, or the slope of the plot of the shear stress, τ, vs. the elastic shear strain, γ. The strain is found by mentally unwrapping the outer surface of the cylinder so it becomes a flat sheet. One can see in Fig. 5.17 (b) that what was originally a rectangle of dimensions L x $2\pi r$ has been sheared by an amount b into a parallellogram.

Applying this to a screw dislocation in a crystal, b would be the Burgers vector of the dislocation. The shear strain is defined by $\tan \phi$, where ϕ is the angle of shear. Thus,

$$\gamma = \tan \phi = \frac{b}{2\pi r}$$

and Hooke's law gives

$$\tau = \frac{Gb}{2\pi r}$$

69

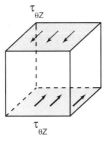

This shear stress $\tau_{\theta z}$ is the same on all radial (θ) planes. That is, the stress field would be the same regardless of which θ plane was picked to be slit and sheared in the z direction. Note, also, that the shear stress varies as $1/r$. That is, the stress in the body is large near the dislocation line, and it falls off with the radial distance from the dislocation.

The stress field of a screw dislocation is very simple; it is pure shear and does not involve any dilatational terms. That is, to a first approximation, there is only a shape change in the block; there is no volume change. In the matrix of stress components, all terms would be zero except for $\tau_{\theta z}$ and $\tau_{z\theta}$.

This stress state is in contrast, for example, to the stress state in a spoke loaded in uniaxial tension in which there are shear stresses (as calculated using Schmid's law), along with the tensile stress (which causes a distortion of interatomic bond lengths and is, therefore, dilatational).

It is clear that the stress field of a screw dislocation would go to infinity as r goes to zero; this is physically unrealistic. Continuum elasticity theory is not capable of giving the actual stress level in the dislocation core. This problem is circumvented by using the continuum theory to give the stress field outside the core only. The core is defined by a cut-off radius, r_0, which is a few atomic diameters. Problems involving the core region must be handled by atomistic calculations, but these are not necessary for most problems in dislocation mechanics.

5.4.3 The Stress Field of an Edge Dislocation

Since an edge dislocation involves an extra half-plane of material and is, therefore, not axi-symmetric (i.e., the same for all values of θ), its stress field is more complicated than that of a screw. It is obvious from inspection of Fig. 5.5 that the material above the glide plane is in a state of compression and that below in a state of expansion. Thus, for an edge dislocation there must be dilatational stress terms as well as shear stresses. However, all these terms contain the factor Gb/r, so the stress field of an edge dislocation shares certain features with that of a screw, including the $1/r$ falloff of stress with distance from the dislocation line.

The expressions for the various terms of the stress field of an edge dislocation are given in Appendix 5.2. These expressions employ Cartesian coordinates because of the symmetry involved. For the present purpose it is not necessary to examine the details of these equations, except for two important points: The first is that the stress field of the edge dislocation has normal-stress, or dilatational, components as well as a shear stress component. This means that edge dislocations (and mixed dislocations) interact with other defects that have a dilatational component to their stress fields. By "interact" is meant either to be attracted to, in cases where the stress fields have terms of opposite sign and thereby tend to cancel each other (as with charged particles of opposite sign), or to be repelled by, in cases where the stress field terms have the same sign and thereby reinforce each other.

A classic example is the interaction of an edge dislocation and a misfitting substitutional solute atom. An oversize solute atom would exert a repulsive force on an edge dislocation if the solute were on the same side of the glide plane as the extra half-plane of the dislocation. This can be understood by consideration of the model of the edge dislocation shown in Fig. 5.5, in which one can see that the atoms in the vicinity of the extra half-plane are squeezed together. This creates a compressive stress field, which interacts with the compressive stress field around the oversized solute in a repulsive manner in the same way that two negatively charged particles would repel each other. This kind of interaction is obviously important in the context of strengthening mechanisms.

It is important to emphasize that two defects can interact elastically only if their respective stress fields contain matching non-zero components. This can be made clear by considering the cases of an edge dislocation, a misfitting substitutional solute atom, and a screw dislocation. The edge will interact with the solute, but the screw will not, as seen by the matchup of the stress matrices:

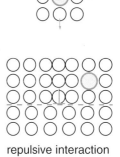
repulsive interaction

$$\begin{matrix} \sigma_{xx} & \tau_{xy} & 0 \\ \tau_{yx} & \sigma_{yy} & 0 \\ 0 & 0 & \sigma_{zz} \end{matrix} \qquad \begin{matrix} \sigma_{xx} & 0 & 0 \\ 0 & \sigma_{yy} & 0 \\ 0 & 0 & \sigma_{zz} \end{matrix} \qquad \begin{matrix} 0 & 0 & 0 \\ 0 & 0 & \tau_{yz} \\ 0 & \tau_{zy} & 0 \end{matrix}$$

EDGE   SOLUTE   SCREW

|—— Interaction ——| |—— No Interaction ——|

The second important point is that the shear component of the stress field of an edge dislocation, although not axisymmetric, does depend on $1/r$, as in the case of a screw dislocation. Thus, the general feature of strain hardening, which is the interaction between the shear-stress fields of parallel dislocations of like sign, is essentially the same for all kinds of dislocations.

5.4.4 Mixed Dislocations

The stress field of a mixed dislocation can be estimated by treating it as a superposition of stress fields of two dislocations having Burgers vectors b_E and b_S which would add up to the resultant b of the mixed dislocation, as shown in Fig. 5.18.

Fig. 5.18 The Burger's vector of a mixed dislocation can be resolved into screw and edge components.

5.4.5 Energy of a Dislocation

As shown in Section 3.4, any solid under stress contains strain energy, which is the energy stored in the distorted atomic bonds. Since the crystal lattice around a dislocation line is distorted, there must be energy associated with any dislocation. It is possible to derive an expression for the strain energy per unit length of a dislocation outside the core region, because the stress field is known. The screw dislocation serves as the simplest case.

Consider the cylinder in Fig. 5.17 which contains a screw dislocation. The stress field is given by $\tau = Gb/2\pi r$, and the strain field is given by $\gamma = b/2\pi r$. From Section 3.4 the strain energy per unit volume is given by the area under the stress strain curve for a specimen in simple tension (in which the stress is uniform all the way through the specimen). This area is σ·ε/2. For a solid in pure shear the analogous expression is τ·γ/2. Since the stress in the cylinder in Fig. 5.17 is not uniform (i.e., it varies with 1/r), the term τ·γ/2 must be integrated over the volume of the cylinder. Thus,

$$\text{Strain energy} = 1/2 \int_V \tau\gamma \, dV$$

One can substitute for τ by using Hooke's law, τ = Gγ, and use the fact that, for a cylinder, dV = 2πrdr times L, the length of the cylinder. Hence,

$$\text{Strain energy} = 1/2 \int_V G\gamma^2 \, dV = 1/2 \, L \int_{r_0}^{r} G\left(\frac{b}{2\pi r}\right)^2 2\pi r \, dr$$

When the integral is evaluated, the result is

$$\text{Strain energy/unit length of dislocation} = \frac{Gb^2}{4\pi} \ln \frac{r}{r_0}$$

The result of such a calculation for an edge dislocation turns out to be almost the same, in spite of the more complicated stress field. Hence, for any dislocation, edge, screw, or mixed, it is sufficient to make the following approximation: Since the energy increases with ln r, most of the energy is in the region near the dislocation, and one can account for almost all the energy if r is set equal to $10^4 \, r_0$. Then,

$$\ln r/r_0 = \ln 10^4 = 2.3 \log 10^4 = 4(2.3), \text{ and } \frac{4(2.3)}{4\pi} \text{ is approximately one.}$$

Therefore,

$$\text{Strain energy/unit length} \approx Gb^2$$

The relatively small contribution to the total energy from the region $r < r_0$ does not affect the approximation greatly.

Note that more strain energy is stored in a stiffer material (higher G), as one might expect, and that the energy of a dislocation increases with the square of the Burgers vector. This is why the dislocations observed in crystals tend to be those with the minimum Burgers vector.

5.5 Mechanism of Strain Hardening

The essence of strain hardening can be summarized as follows: As plastic strain proceeds, the dislocation density increases, and as the separation between dislocations decreases, the force of mutual repulsion between neighboring dislocations gets larger. This means that more stress must be applied to the material for plastic flow to continue.

To understand this phenomenon thoroughly, it is necessary to examine first the relationship between plastic strain and dislocation density, then the process of yielding and dislocation multiplication, and finally the nature of the force between neighboring dislocations. The result will be that the rate of strain hardening decreases as strain increases in a material. Thus, the flow stress, or the stress needed to maintain plastic flow, will be found to vary parabolically with plastic strain:

$$\sigma_{fl} \propto \varepsilon_{pl}^{1/2}$$

5.5.1 Plastic Strain and Dislocation Density

Here it will be shown that, in a metal undergoing plastic strain, the dislocation density increases with strain in an approximately linear fashion. First, it is necessary to show the relationship between plastic strain and the motion of a dislocation. Then, a simple expression can be written to relate plastic strain to the motion of a number of dislocations over some average distance.

Consider first a cubic unit volume of crystal and allow one dislocation to glide completely through this volume. The plastic offset would be equal to the Burgers vector of the dislocation, b. If this happened in every unit of volume in the crystal, the average shear strain would be the displacement, b, divided by the spacing of the glide planes, assuming one glide plane per unit volume, or

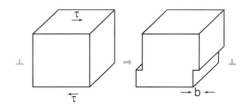

$$\gamma_p = b/1$$

Now consider such a unit volume containing one dislocation, and let the dislocation glide by a small amount dx. The plastic offset is now only a fraction of b, and this fraction is $dx/1$, which is the portion of the length of the glide plane traversed by the dislocation when it glides by the amount dx. If this happened in every unit of volume in the crystal, the average shear strain would be the displacement, $bdx/1$, divided by the spacing of the glide planes (i.e., unity).

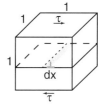

$$d\gamma_p = b dx$$

The tensile strain would be approximately given by

$$d\varepsilon_p = 0.7 \, b \, dx$$

if it is assumed that the glide planes lie, on the average, at about 45° to the tensile axis.

Dislocation density, ρ, is defined as the total dislocation length per unit volume, or, equivalently for a small volume in which all dislocations can be considered to be parallel, the number of dislocations passing through a unit of area.

Now consider a unit volume of material having a dislocation density ρ, and during an increment of plastic strain let each dislocation move by dx. The plastic shear strain, $d\gamma_p$, would be $\rho b dx$. Integrating from zero plastic tensile strain, the result would be

$$\varepsilon_p = 0.7 \, \rho \, b \, \bar{x}$$

where $\bar{x}$ is the average distance moved by the dislocations. In a polycrystalline metal, $\bar{x}$ can be set approximately equal to some fraction of the grain size. (It decreases during plastic strain, but to a much smaller degree than the increase in ρ.) Therefore, the dislocation density must increase approximately linearly with plastic strain. That is, dislocations must multiply during plastic strain.

The need for dislocation multiplication can be illustrated by a simple example. A typical ductile metal can elongate plastically on the order of 50% in a normal tensile test. An annealed metal, after having been made into a tensile specimen, would have a dislocation density of less than 10^6 cm of dislocation per cm^3 (or 10^6 dislocations per cm^2 of area). Considering a tensile specimen with a two-millimeter-diameter gauge section (roughly, like a spoke) in which the average slip plane is oriented at 45° and in which the grain size is 0.1mm (which is fairly large for a typical structural alloy), assume that the average distance of dislocation glide is one-tenth of the grain size. The amount of plastic tensile strain that would be possible from motion of the pre-existing population of dislocations, assuming that they were all mobile and all contributed to the plastic strain, would be given approximately by:

$$\varepsilon_p = 0.7 \rho b \bar{x}$$

Let b be 2.5×10^{-8}cm, which is about right for a common metal, like iron or copper. Thus,

$$\varepsilon_p = 0.7 \times 10^6/cm^2 \times 2.5 \times 10^{-8} cm \times 0.001 cm$$

$$= 0.0018\%$$

Therefore, the initial dislocation population could only account for a tiny fraction of the amount of plastic strain actually observed. The conclusion must be that, to achieve the amounts of plastic strains observed in tensile tests, ρ must increase by at least several orders of magnitude during the test. In the above example, ρ would have to increase by more than a factor of 3×10^4 to account for the 50% strain that would occur in the test.

5.5.2 Yielding and Dislocation Multiplication

To understand these processes, it is first necessary to have an accurate idea of what the initial dislocation structure in an annealed crystalline solid would look like.

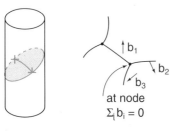

A dislocation is a linear defect with a beginning and an end. Because a dislocation represents the boundary between slipped and not-yet-slipped portions of a crystal, it cannot end inside the crystal except at some other defect. This other defect could be, for example, a surface (either the external surface or that of an internal void), a grain boundary, the interface of a particle of a second phase, or another dislocation line. In the latter case, a three-dislocation node is formed, and Burgers vectors of the three must sum to zero.

The proper mental picture of a dislocation, therefore, is a linear defect that may be pinned at each end, like a stretched rubber band. As seen in Section 5.4.5, the dislocation has an energy per unit length, or a line tension, of approximately Gb^2. Thus, as with the rubber band, a force must be applied to the dislocation to make it bow out into a loop, thereby increasing its length (and total energy).

Such a force is exerted by a shear stress, which has a vector component parallel to the Burgers vector of the dislocation. That is, dislocation glide is caused only by that component of a shear stress which lies along the Burgers vector of the dislocation.

Because a dislocation is not a physical entity, but is, rather, a configuration of matter, the idea of a force on a dislocation needs to be defined carefully. It is useful to speak of a force acting on a dislocation, because a dislocation can move under the influence of a shear stress, and work is done by this motion. However, the magnitude of the force for a given shear stress is derived in a somewhat

SECTION 5.5.2 YIELDING AND DISLOCATION MULTIPLICATION

roundabout way. The method is simply to calculate the work done on a crystal when a dislocation moves, and then to take the derivative of that work with respect to the distance moved and call that the force. (Work = force x displacement.) This is done as follows:

Consider a block of crystal containing a dislocation to which a shear stress, τ, is applied, as shown in Fig. 5.19. The area of the glide plane is A. If the dislocation were to glide across this entire glide plane, from one side to the other, the two parts of the crystal would be displaced with respect to each other by b, the Burgers vector of the dislocation, Fig. 5.19(b). The work done by the shear stress, which lies along the direction of b, would then be the force times the displacement, or

$$\text{WORK} = \tau A b$$

If, on the other hand, the dislocation were to glide over only a small portion of the glide plane, Fig. 5.19(c), then only a fraction of this work would be done. Let the distance of glide be dx. The area swept out is then Ldx. The fractional amount of work done, dW, should then be Ldx/A times the total possible work, τAb. Thus,

$$dW = (Ldx/A)(\tau Ab) = \tau Lbdx$$

If the force on the dislocation is defined to be dW/dx, then this force is $L\tau b$, and one can write the force per unit length of dislocation as

$$\boxed{\text{Force per unit length} = \tau b}$$

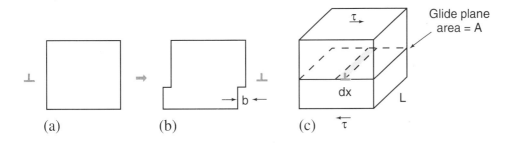

Fig. 5.19 (a) and (b) A dislocation before and after passing through a volume of crystal; the passage causes a shear displacement of b. (c) A dislocation in that volume which glides only by an amount dx; the shear displacement is now only a fraction of b.

Note that this force acts normal to the dislocation line at every point along the line (i.e., in the direction of glide). In this sense it is analogous to the force on a bubble due to the internal pressure of the air trapped inside the bubble. (Note that, strictly speaking, the force per unit length should be written as the dot-product of two vectors: $\boldsymbol{\tau} \cdot \mathbf{b}$.)

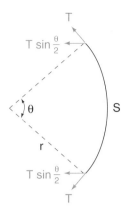

Returning to the case of a dislocation segment pinned at both ends, the shear stress necessary to bow-out a dislocation segment of length S can now be derived by considering the equilibrium between the restoring force, acting to the left, which results from the line tension (T = Gb² per unit length)* and the force exerted by the shear stress τ, acting to the right, given by τbS.

The segment has a radius of curvature equal to r, and it subtends an arc θ, expressed in radians. Therefore,

$$\theta = \frac{S}{r}$$

The total force acting to the left is the sum of the horizontally resolved components of the line tension at each end of the segment, given by

$$T \sin \frac{\theta}{2}$$

The approximation that $\sin \theta/2 \approx \theta/2$, when θ is given in radians, can be used, and the force balance can then be written

$$\tau b S = 2 T \sin \frac{\theta}{2} = T\theta = \frac{TS}{r} = \frac{Gb^2 S}{r}$$

The final result is that the shear stress needed to bow out a dislocation segment to a radius r is, approximately,

$$\tau = \frac{Gb}{r}$$

Obviously, the shear stress necessary to bow out a loop increases as the radius of the loop decreases. Conversely, if the shear stress is removed, the loop should become straight again, owing to the restoring force.

With these basic ideas about dislocation behavior in hand, it is now possible to understand the mechanism of dislocation multiplication, which was conceived independently by two theorists, F.C. Frank and W.T. Read, around 1950. A useful mental image of this simple but brilliant idea can be gotten by visualizing the ripples generated when a pebble is dropped into a quiet pond. The appearance of dislocation loops emanating from a dislocation source on a glide plane would be similar to the circular ripples which radiate from the point of impact of the pebble on the surface of the water.

Dislocation multiplication starts with a segment of dislocation of length L which can bow-out between pinning points, and it follows a process illustrated schematically in Fig. 5.20. In order for the segment to become a loop, the stress must increase as the radius of curvature decreases, according to $\tau \approx Gb/r$. When the loop reaches a semi-circular shape, further bowing-out proceeds with an increasing radius. Therefore, past this point, the applied stress of ~Gb/L/2 is more than necessary to hold the loop in place, and the loop expands in an unstable manner. The unstably expanding loop folds back on itself and transforms into a full circle when the two parts touch, as shown, because these two parts annihilate each other, being of opposite sign. The complete loop encloses a segment like the original one, lying between the two pinning points. A further increase in the stress causes the complete loop to expand and the residual segment to bow out as before. The process then repeats over and over, giving a set of concentric loops expanding away from the source. A similar process occurs at many locations in a metallic material, gradually filling the material with new dislocations.

* Note: Gb² = energy per unit length = force

SECTION 5.5.2 YIELDING AND DISLOCATION MULTIPLICATION

Fig. 5.20 Stages of bowing out of a segment of dislocation into a loop which then becomes unstable and produces a full loop and the starting configuration over again. This is known as the Frank-Read process.

The early stages of plastic flow in an annealed metallic material can now be described. It is now known that dislocations are emitted from grain boundaries during the process of yielding. This is illustrated schematically in Fig. 5.21(a), in which the dislocation is emitted as an elongated loop which, here, is mainly of screw type. The region of slip plane swept out by the loop is represented by shading. The unique feature of a screw dislocation, as noted earlier, is that it is not constrained to glide on any one plane. Rather, it can change planes, as shown previously in Fig. 5.8, by a process known as *cross slip*. This can happen along a portion of the emitted screw dislocation, as shown in Fig. 5.21(b), and it can happen a second time, returning the dislocation to the original slip plane, as shown in Fig. 5.21(c). The total process shown here is known as *double cross slip*.

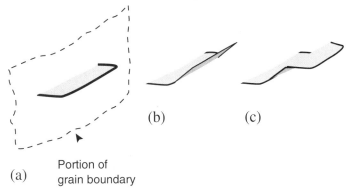

Fig. 5.21 (a) A dislocation loop, mainly screw, emitted from a grain boundary in the early stages of yielding. (b) A segment of the emitted screw dislocation has cross slipped onto another {111} plane. (c) This segment cross slips again back to the original {111} plane; the segment is now pinned at its end points and is constrained to bow out as a loop.

The dislocation configuration formed by this process is of particular importance, because it creates a segment of dislocation that is pinned at its end points. This is the starting configuration for dislocation multiplication by the Frank-Read mechanism. The reason the segment is pinned is that the parts of the dislocation left on the first cross-slip plane are edge in character (because they lie perpendicular to the Burgers vector) and, therefore, cannot glide in any direction except parallel to b. This means that the segment which has undergone double cross slip is constrained to bow out in a new loop, with its end points anchored in place.

5.5.3 The Physical Basis of Strain Hardening

Since any dislocation has a stress field with a shear stress component, one dislocation may exert a force on another dislocation. The most relevant situation is that of dislocations that lie more or less parallel to one another. If two such dislocations are of the same sign, they have similar stress fields, and the mutual interaction force is repulsive. That is, the stress field of one dislocation exerts a force per unit length ($=\tau b$) on the other. Since the shear stress component for any dislocation is approximately Gb/r, the interaction force varies as $1/r$. That is, the force gets stronger as the dislocation spacing decreases. Thus, the flow stress increases with $1/\bar{r}$ where $\bar{r}$ is taken to be the average dislocation spacing.

The average spacing is given by

$$\bar{r} = \frac{1}{\sqrt{\rho}}$$

where ρ is the dislocation density, because ρ is defined as the total length of dislocation line per unit volume, which is equivalent to the number of dislocations passing through a unit of area for the case of parallel dislocations. Thus, $1/\rho$ is area per dislocation, and the square root of the area per dislocation is the dislocation spacing.

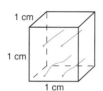

The relationship between the average spacing and the dislocation density can be illustrated by considering a unit volume of material (e.g., a 1-cm cube) that contains four more or less parallel dislocations. The dislocation density is

$$\rho = 4\,\text{cm/cm}^3 = 4 \text{ dislocations/cm}^2$$

and the average spacing is

$$\bar{r} = 1/2 \text{ cm/dislocation} = \frac{1}{\rho^{1/2}}$$

It was demonstrated earlier that ρ varies approximately linearly with plastic strain, ε_{pl}. Since flow stress varies with $1/\bar{r}$, it must vary as $\rho^{1/2}$. Therefore, the conclusion is that in a material with uniformly spaced parallel dislocations,

$$\sigma_{fl} \propto \varepsilon_{pl}^{1/2}$$

which means that the plastic part of the stress-strain curve is parabolic. This is equivalent to saying that strain hardening proceeds at a decreasing rate as plastic strain increases, or the strain-hardening effect tends to saturate, as shown schematically in Fig. 5.22. This effect is observed in Fig. 3.20 for the stainless-steel spokes.

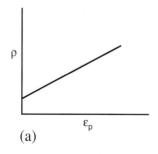
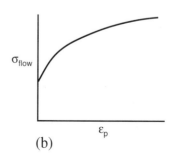

Fig. 5.22 The linear variation of dislocation density with plastic strain (a) leads to (b) a parabolic stress-strain curve.

Example 5.8

Q. Write each individual slip system in the set {111}<110> for an FCC crystal.

A. For the (111) plane (cf. Fig. 5.1(b)) the result is

$(111)[\bar{1}10]$
$(111)[10\bar{1}]$
$(111)[0\bar{1}1]$

For the rest the result is

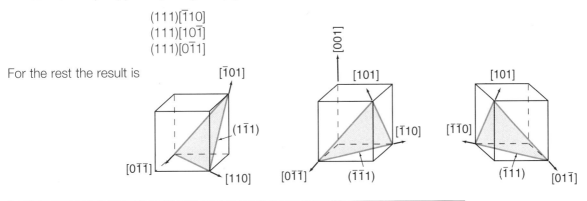

Example 5.9

Q. If the slip step left by a cross-slipping screw dislocation in an FCC crystal is made up of segments on the (111) and $(\bar{1}\bar{1}1)$ planes, as shown below, what must be the Burgers vector of the dislocation?

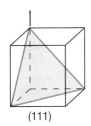 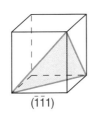 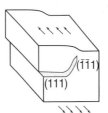

(111) $(\bar{1}\bar{1}1)$

A. The Burgers vector must be of the type $a/2$ <110>, since it is an FCC crystal. The only <110> direction which lies in both of the observed glide planes is the $[\bar{1}10]$ direction (or, equivalently, $[1\bar{1}0]$.) Therefore, the Burgers vector must be $a/2\,[\bar{1}10]$ or $a/2\,[1\bar{1}0]$.

Example 5.10

Q. Prove that the $[\bar{1}10]$ direction lies in both the (111) and $[\bar{1}\bar{1}1]$ planes.

A. In a cubic crystal the [hkl] direction is normal to the (hkl) plane. Therefore, the normals to the two {111} planes are in the [111] and $[\bar{1}\bar{1}1]$ directions, respectively. By definition, a plane normal is perpendicular to any vector in the plane. Therefore, the dot-product of a plane normal and any vector in the plane must be zero. Taking the dot-product, it is found that

$[\bar{1}10]\,[111] = \bar{1}\times1 + 1\times1 + 0\times1 = 0$, and
$[\bar{1}10]\,[\bar{1}\bar{1}0] = \bar{1}\times\bar{1} + 1\times\bar{1} + 0\times1 = 0$

Therefore, the $[\bar{1}10]$ direction lies in both planes.

Example 5.11

Q. Consider an interaction between an oversize substitutional solute atom and an edge dislocation in which the solute atom could lie either above or below the glide plane, as indicated below. Which of the positions would produce a repulsive interaction and which would produce an attractive interaction?

A. The oversize solute atom puts the surrounding lattice into a compressive stress state that is spherically symmetrical, assuming the atom acts like a hard sphere and the crystal is elastically isotropic. This means that it has a stress field of purely hydrostatic compression. Therefore, if the solute atom lies above the glide plane (position A), the stress fields reinforce, and the interaction is repulsive. Conversely, in position B the interaction is attractive. Note that, in either position, glide of the dislocation past the solute would be inhibited.

Example 5.12

Q. What would the answer have been if the dislocation in Example 5.11 had been a screw instead of an edge?

A. There is no geometrically defined glide plane for a screw dislocation; also, the stress field is axi-symmetric. Therefore, the interaction depends only on r, not on θ. Also, the stress field of the screw has no normal components, while that of the substitutional solute has only normal components. Therefore, there is no interaction resulting from the solute-size effect.

Example 5.13

Q. Suppose that a substitutional solute atom alters the atomic bonding locally so as to reduce the shear modulus G. How would that affect the interactions between the solute and any dislocation?

A. Since the dislocation energy per unit length is proportional to G, this energy would be lower in the vicinity of the solute atom. Thus, the solute atom would attract the dislocation and could act as a local trap for the dislocation. This effect alone would tend to inhibit glide. (Obviously, the interaction would be repulsive if G were locally increased by the solute atom.) These effects can play a role in solid-solution hardening.

Example 5.14

Q. Calculate the force per unit length on an edge dislocation with Burgers vector $b = 2.5 \times 10^{-10}$ m on the indicated slip plane in the mono-crystalline tensile specimen subjected to a tensile stress, $\sigma = 100$ MPa, shown below. (Note that the tensile axis (TA), the slip-plane normal (N), and the slip direction (b) are not co-planar.)

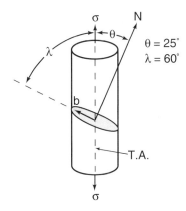

A. Since the force per unit length is given by τb, one simply needs to calculate τ in the direction of b. This is given by the Schmid law:

$$\tau = \sigma \cos \lambda \cos \theta$$
$$= 100\text{MPa} \cos 25 \cos 60 = 45 \text{ MPa}$$

Therefore, the force per unit length is

$$\tau b = 45\text{MPa} \cdot 2.5 \times 10^{-10}\text{m}$$
$$= 113 \times 10^{-4} \text{N/m}$$

Summary

The process of plastic deformation can be summarized as follows: As the stress on a material is increased, dislocation loops are emitted from grain boundaries. Double cross-slip creates dislocation segments pinned at their end points, and the longest segments bow out first, followed by shorter and shorter segments as the stress rises. The response of the material, therefore, becomes gradually less elastic. (However, the plastic displacement in this early stage of plasticity is too small to be detected by the transducers generally used in tensile testing.) At a sufficiently high stress, dislocation multiplication takes place, and the plastic displacement increases rapidly; this constitutes yielding. The dislocation density continues to increase, and the dislocation interactions become more intense; this is strain hardening. The flow stress continues to rise until necking occurs; past this point, the load-elongation curve falls until rupture occurs.

Some important features of dislocations are the following:

- They create a stress field that is inversely proportional to the distance from the dislocation. That is, the stress field has a 1/r dependence.

- One dislocation, A, exerts a force (per unit length) on another dislocation, B, equal to the dot product of the shear stress of A acting at B and the Burgers vector of B. Thus, the force between parallel dislocations is inversely proportional to their spacing.

- Dislocations have an energy per unit length equal approximately to Gb^2. Thus, a strain-hardened material is in a high-energy state.

Appendix 5.1 - Bragg's Law of Diffraction

Because x-ray or electron beams can have wave lengths comparable to the atomic-spacing in crystals, these beams can be diffracted by crystals. This was demonstrated with x-rays by von Laue in 1912, and the geometrical law governing diffraction by crystals was worked out by W.L. Bragg in the same year.

To understand this, recall that an x-ray or electron beam can be considered as an electric field which varies sinusoidally with time. When an atom in a crystal is encountered by an incident beam, the electrons are set into oscillation, and they, in turn, radiate energy in all directions. (Any accelerating electric charge radiates electromagnetic energy, as in a radio antenna, for example.)

Incident beam; the electric field, E, varies with time.

The electrons in the atom are oscillated by E and energy radiates over 360°.

When an x-ray or electron beam hits a crystal and penetrates a micrometer or so, all atoms in this region are made to radiate energy in all directions. In certain directions this "scattered" radiation is in phase, and a "diffracted" beam results. In all other directions the scattered radiation is out of phase, and the beams cancel each other.

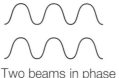

Two beams in phase

Two beams 180° out of phase

To derive Bragg's law, consider the top few atomic planes of a crystal, as in Fig. A5.1.1. A diffracted beam will result if the incident beam strikes the top plane at an angle θ, such that scattered beams from atom planes parallel to the top plane, with planar spacing d, are in phase with the incident beam. That is, the phase difference between the two must be an integral number of wave lengths. In Fig. A5.1.1, the incoming beams, with wave length λ and incident angle θ, are in phase along the plane front xx'. Consider beam 1 scattered from atom K and beam 2 scattered from atom L, giving scattered beams 1' and 2'. If the path difference MLN is nλ, where n is an integer, then beams 1' and 2' are in phase along the plane front yy' and they reinforce one another.

Fig. A5.1.1 Diffraction of an x-ray or electron beam from a crystal. Adapted from B.D. Cullity, *Elements of X-ray Diffraction*, Addision-Wesley, Reading, MA, 1956, p.120.)

From geometry,

$$ML = LN = d \sin \theta$$

Therefore, the diffraction condition is

$$n\lambda = 2d \sin \theta$$

Thus, for this combination of λ, d, and θ, Bragg's law is said to be satisfied. Whenever this occurs, a diffracted beam results. The intensity of the diffracted beam depends on the kinds of atoms present and their density (number per unit area); the intensity is not predicted by Bragg's Law.

Appendix 5.2 - Stress Field of an Edge Dislocation

The edge dislocation has a stress field that is more complicated than that of a screw dislocation. To derive the expressions for the stress components of an edge dislocation requires a knowledge of the theory of elasticity and of stress analysis. For present purposes it is sufficient merely to quote the results:

Normal Stresses:

$$\sigma_{xx} = -Dy \frac{(3x^2 + y^2)}{(x^2 + y^2)^2} \qquad \sigma_{yy} = Dy \frac{(x^2 - y^2)}{(x^2 + y^2)^2}$$

$$\sigma_{zz} = \nu(\sigma_{xx} + \sigma_{yy})$$

Shear Stress:

$$\tau_{xy} = \tau_{yx} = Dx \frac{(x^2 - y^2)}{(x^2 + y^2)^2}$$

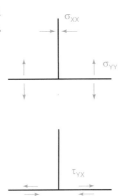

The remaining components, $\tau_{xz} = \tau_{zx}$ and $\tau_{yz} = \tau_{zy}$, are zero. That is, there is no component of shear stress in the direction along the line of an edge dislocation.

The parameter ν is a material constant, called *Poisson's ratio* and is defined by:

$$\nu = \frac{-\varepsilon_{zz}}{\varepsilon_{xx}} = \frac{-\varepsilon_{yy}}{\varepsilon_{xx}}$$

for an isotropic material. For example, it characterizes the contraction of a solid along the two transverse axes, y and z, when the solid is stretched along the longitudinal axis, x. The value of ν for most interesting materials is in the range 0.3 to 0.35.

The parameter, D, is defined by:

$$D = \frac{Gb}{2\pi(1-\nu)}$$

where G is the shear modulus and b is the Burgers vector of the dislocation.

Although the expressions for these stress components look quite different from the $\tau = Gb/2\pi r$ derived for the screw dislocation, there is an important similarity. First, the term $(1 - \nu)$ is about equal to 2/3. Thus, $D \approx Gb/4$. Secondly, the radial distance in the x,y plane from the dislocation line is

$$r = (x^2 + y^2)^{1/2}$$

and inspection of the functions of x and y in the three stress components shows that they have a 1/r character. Of course, they depend on whether y is positive or negative and whether y is greater than, less than, or equal to x at the point x, y where the stress state is to be calculated. Hence, the stress components change sign in various segments of the x, y plane; but, the magnitude of the stress at any point still varies approximately as $Gb/2\pi r$.

It was noted earlier that, for a screw dislocation, it is more convenient to use cylindrical coordinates r, θ, z instead of Cartesian coordinates x, y, z, as illustrated in Fig. A5.2.1. In this case the stress field of the screw dislocation is characterized by:

$$\tau_{\theta z} = Gb/2\pi r$$

(This is equal to $\tau_{z\theta}$, which exerts a twist to the end of the cylinder shown in Fig. 5.17; this is irrelevant for our purposes.)

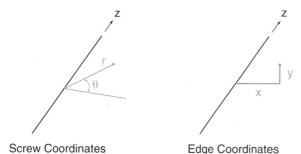

Fig. A5.2.1 Coordinate systems used for screw and edge dislocations

Finally, it should be emphasized that shear stress components produce shape changes, and normal stress components produce volume changes, or dilatations.

Exercises Chapter 5

Terms to Understand

Close-packed plane
Close-packed direction
Burgers vector
Slip system
Glide
Edge *vs.* screw dislocation
Mixed dislocation
Dislocation loop
Dislocation interactions
Diffraction contrast

Stress field of dislocation
Dilatational strain
Dislocation-solute interactions
Dislocation energy, line tension
Force on a dislocation
Dislocation density
Dislocation multiplication
Stress to bow a loop
Parabolic hardening
Cross slip, double cross slip

Problems

5.1 When a screw dislocation changes glide planes, the process is called cross slip. Determine the direction of the Burgers vector of the screw dislocations that can undergo the following types of cross slip:

From $(1\bar{1}1)$ to (111)

From $(1\bar{1}1)$ to $(\bar{1}11)$

5.2 It is physically possible to have dislocations with the Burgers vector $a\langle100\rangle$ in an FCC crystal. Compare the energy of such a dislocation with that of one with the common Burgers vector $a/2\langle110\rangle$, using the approximation that energy per unit length is equal to Gb^2, and comment on why the former is rarely observed. Remember, for the vector $\mathbf{a} = a_1\mathbf{i} + a_2\mathbf{j} + a_3\mathbf{k}$,

$$|a| = \sqrt{(a_1^2 + a_2^2 + a_3^2)}$$

5.3 Consider a segment of dislocation pinned securely at each end as shown, with a Burgers vector as indicated. If the mobility of an edge dislocation in this material is <u>much</u> larger than the mobility of a screw dislocation (as in the case of iron at temperatures around 77K), the segment would not form a symmetrical loop when subjected to a shear stress in the direction of the Burgers vector. With the aid of a sketch, show the expected response of the segment to such a stress.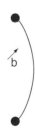

5.4 Show the connection between the dislocation loop pictured in Fig. 5.14 and the operation of a Frank-Read source.

5.5 List the strengthening mechanisms (i.e., ways to impede dislocation glide) that you have learned about so far, and explain briefly how they operate.

5.6 Consider the edge dislocation shown below, in which the extra half-plane is a (010) plane of the crystal. If the presence of this dislocation were to be revealed by diffraction contrast in the TEM, it would be a good idea to use the beam diffracted from (010) planes; explain why. What would be the result if the beam diffracted from (100) planes were used to form the image?

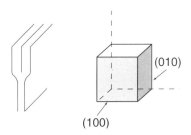

5.7 The energy per unit length of a screw dislocation = $Gb^2/4\pi(\ln r/r_0)$ obviously depends on the value chosen for r (i.e., the extent of the crystal around the dislocation within which the energy is stored). Make a plot of energy/length vs. r (not ln r). (Let $r_0 = 2b$, and express r in terms of b; i.e., $r = 10b$, $10^2 b$, etc.) Plot the energy in units of Gb^2, and extend the plot to $r=10^3 b$.

5.8 An FCC crystal is subjected to a tensile stress, σ, of 140 MPa along the $[\bar{1}\bar{1}2]$ direction. Calculate the force per unit length on dislocations that lie on the $(\bar{1}\bar{1}1)$ plane and have the following Burgers vectors: $a/2[\bar{1}10]$ and $a/2[101]$. Let $a = 2.5 \times 10^{-10}$ m.

5.9 Imagine an isolated substitutional solute atom which causes a local reduction in the shear modulus of the crystal. By means of schematic sketch showing the solute atom and a screw dislocation, show the nature of the expected interaction between them as the screw attempts to glide past the solute atom. Do the same for the case where the local shear modulus is increased.

5.10 Iron is BCC at room temperature, with $a = 0.28$ nm.

    a. Calculate the length of the Burgers vector in iron.
    b. Calculate the shear modulus in iron, given $\nu = 0.33$ and $E = 210$ GPa
    c. Calculate the approximate energy per unit length of a dislocation line in iron.

5.11 Consider a reaction in which two dislocations combine to make a third. Written in terms of the Burgers vectors, a potential reaction is

$$1/2\,[111] + 1/2\,[1\bar{1}\bar{1}] \longrightarrow [100]$$

Is this physically possible from the standpoint of vector addition? Is it energetically favorable or unfavorable? Explain your answer

6
ANNEALING OF COLD-WORKED SPOKES

Spoke manufacture involves drawing rods through a series of dies of decreasing hole size until wire of the desired diameter is produced. This cannot be done in one continuous series of draws, starting with thick rods, because the accumulation of strain hardening would become too great. The force needed to pull the wire through a die cannot exceed that corresponding to the UTS, or the wire will neck and rupture. Therefore, the wire must be softened periodically during the drawing process. This is accomplished by heating the wire above some threshold temperature in a process known as *annealing*. The annealing process involves some important internal changes in the wire which permit it to transform from a high-energy, *cold-worked* state to a lower energy state. (Cold-working a metallic material means deforming it plastically at a temperature low enough that the strain-hardened condition is retained indefinitely.) To understand the annealing process, it is necessary to think in terms of thermodynamics, the basic principles of which will be described before proceeding further.

6.1 Thermodynamic Principles

The cold-worked state is thermodynamically unstable because of the internal energy stored in the deformed grains, mainly in the form of dislocations (Gb^2 per unit length), but also as *point defects,* mainly excess lattice vacancies. A lattice vacancy is a lattice site from which an atom is missing. As will be shown later, this is a "natural" kind of defect in a crystalline solid. That is, at any temperature above absolute zero, a crystal at equilibrium will have a certain fraction of vacant lattice sites. The process of plastic flow results in a vacancy concentration higher than the equilibrium value. The return to equilibrium entails, in part, the "annealing out" of the excess vacancy concentration.

The central concept of thermodynamics is that of equilibrium. The equilibrium state of a system is the state which persists interminably, without change, regardless of attempts to perturb the system by small amounts. For example, water represents an equilibrium configuration of oxygen and hydrogen (at room temperature and one atmosphere of pressure). Shaking water in a container does not cause it to separate into its elements.

It is common experience that the attainment of equilibrium in any system involves a reduction of energy to a minimum value. Thus, water runs downhill to minimize its potential energy (in the gravitational field of the earth). The cold-worked condition is not at equilibrium, because it is not in a state of minimum energy. Although the cold-worked condition is unstable, it persists at low temperatures, because the processes that produce softening, which are lumped together under the term "annealing," must be *thermally activated*. That is, thermal energy is required for the several kinds of atomic rearrangements which must take place. This can be envisioned schematically with the aid of Fig. 6.1; the system can pass from the high-energy state to the more stable, low-energy state only if an activation energy is supplied. The net energy difference between the two states is the *driving force* for the reaction.

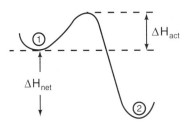

Fig. 6.1 Schematic illustration of the requirement of an activation energy to allow a system to pass from high-energy state 1 to a lower-energy state 2.

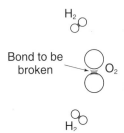

An analogy can be made to a container of oxygen and hydrogen gas at room temperature and pressure. Although water would be the equilibrium condition, some energy must be supplied to the gas mixture to break the bonds of the O_2 molecules so the oxygen atoms can react with the H_2 molecules. The energy released by this reaction is much larger than the activation energy needed to get it started. Thus, the invested activation energy is "paid back" once the reaction starts.

A minimum of energy is not the only factor which determines the equilibrium state. This can be illustrated by considering a box large enough to contain many gas molecules, but which contains only a few, as illustrated schematically in Fig. 6.2, which shows four possible configurations of the molecules.

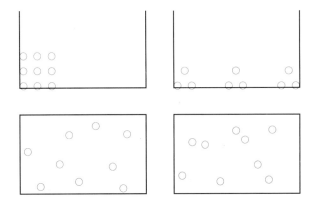

Fig. 6.2 Four possible arrangements of n molecules in a box large enough to contain N molecules, where N is much greater than n.

Assume that the temperature and pressure are constant. Ignore, for the moment, the influence of energy on the molecular arrangement, and let all arrangements have equal probability. It is intuitively obvious that the molecules are more likely to be found at random locations than in either of the two orderly arrangements shown. Thus, one would say that, if the molecules were stacked in an orderly array in one corner of the box, the system would not be at equilibrium, because there would be a tendency for it to transform spontaneously into a more probable configuration. This means that the criterion for equilibrium must somehow take probability into account.

Imagine that the box can be divided up into cells, each of which can contain one gas molecule, and the molecules are permitted to occupy any of the cells. There is a very large number of possible random, disorderly configurations and only a relatively small number of possible orderly arrangements. Therefore, one would conclude that, at equilibrium, the molecules will be arranged in some disorderly array. It does not matter which disorderly array, since they are all equivalent in the sense that they all lack order. (An orderly array is one for which there is some scheme whereby, if the position of one molecule is known, then the positions of all others can be calculated by some formula.) This does not mean that the molecules are never arranged in an orderly array. Ordering may indeed occur at some instant in time, but this arrangement does not persist. The equilibrium condition is the one that would be found at virtually any instant in time, and this would be characterized by a minimum of order. The thermodynamic variable that characterizes the state of disorder is *entropy,* denoted by S. Thus, all other things being equal, the equilibrium state is characterized by a maximum of entropy.

Suppose now that the temperature of the system is reduced to the point where the molecules condense from a gaseous to a crystalline state. The entropy has now been reduced to a very low value, because the energy part of the criterion for equilibrium has taken control. That is, because of the existence of the kind of binding energy illustrated in Fig. 3.7(a), the crystalline state is the most stable at a sufficiently low temperature. Therefore, the equilibrium criterion must be mediated somehow by temperature.

The universal criterion for equilibrium was deduced by J. Willard Gibbs and involves the minimization of a *free energy,* which is an energy modified to take probability into account. It is written

$$G = H - TS$$

where G is called the *Gibbs free energy,* T is the absolute temperature (degrees Kelvin), and H is called the *enthalpy.* The latter is equal to E + pV, where p means pressure, V means volume, and E is the *internal energy,* e.g., the binding energy of the molecules (cf. Fig. 3.7). In solids and liquids the volume is negligible compared to that in the gaseous state, and at one atmosphere the pV term is also negligible. That is, one can usually ignore changes in energy arising from changes in volume at one atmosphere pressure. Therefore, H can normally be considered equivalent to E for condensed systems.

Thus, equilibrium is characterized by a minimum value of G, and any spontaneous change in nature must involve a decrease in G. That is, a stable system is one for which G is a minimum. At zero Kelvin, this is equivalent to a minimum in energy. However, as the temperature is increased, entropy becomes more and more important. For example, at a certain temperature, the TS term becomes large enough to overcome the binding energy of the molecules in the box, and the gaseous state takes over.

These principles will now be applied to the process of annealing a cold worked bicycle spoke. The heat supplies the activation energy needed for various stages to occur so the spoke can find its way to a minimum free energy.

6.2 Stages of Annealing

The three stages of annealing are *recovery, recrystallization,* and *grain growth.* The latter two stages can be followed by measurement of *hardness,* which can be defined as the resistance of a material to indentation by a loaded indenter. Several common hardness tests are described in Appendix 6.1; their utility lies in their simplicity and speed and the fact that specimen preparation is minimal. Physically, hardness represents a generalized resistance of a material to plastic deformation under a rather complicated, but reproducible, state of stress. Therefore, it is a useful parameter for comparative purposes or for following trends of hardening or softening, but not for understanding phenomena at a fundamental level. The way hardness varies during cold working and annealing is shown schematically in Fig. 6.3.

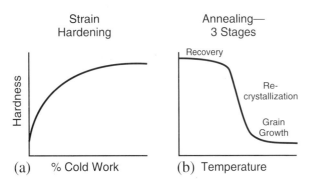

Fig. 6.3 Variation of hardness during (a) cold working and (b) annealing. (During annealing a fixed time at each temperature is assumed.)

6.3 Recrystallization

6.3.1 Mechanism of Recrystallization

A dramatic change in a cold-worked material occurs in the recrystallization stage of the annealing process. Here the microstructure is transformed from deformed (i.e., hard) grains with a high dislocation density to equi-axed grains with a very low dislocation density. The new grains are soft, because the strain-hardening has been eliminated. This transformation is an example of a class of transformations that occur by nucleation and growth.

In recrystallization, new grains are nucleated in regions of particularly high dislocation density, and these grains grow by consuming the surrounding deformed, high-energy grains. The growth process involves simply the transfer of atoms across the boundaries of the new grains from the high-energy to the low-energy side.

Thus, the driving force is the difference in energy per atom between the old grain with many dislocations and the new grain with almost none. That is, the driving force is the stored energy of cold work. The recrystallization process can be observed by means of transmission electron microscopy, as illustrated by Fig. 6.4.

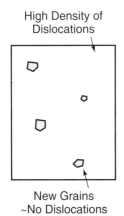

 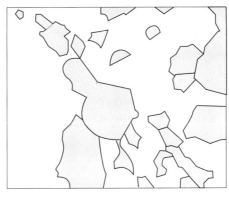

0.6μm

Fig. 6.4 (a) Recrystallized grains in cold-worked iron annealed 5 min at 500°C (From W. C. Leslie et al. in *Iron and Its Dilute Solid Solutions,* Interscience, NY, 1963, p. 164.) (b) Tracing of (a) showing new, dislocation-free grains and the surrounding matrix.

6.3.2 Kinetics of Recrystallization

Since the growth process requires atomic migration, the atoms near the boundary must acquire some kinetic energy. This is where the activation energy Δh_{ACT} comes in. (Lower-case h is used here to refer to energy per atom, whereas upper-case H will refer to energy per mole of atoms, meaning Avogadro's number of atoms.) The bonds holding the atom in the deformed grain must be broken before the atom can jump across the gap to the new, low-energy grain. The rate of such a process is equal to the number of attempts per second to cross the energy barrier times the probability that any given attempt will be successful. The number of attempts per second can be taken to be equal to the frequency of atomic vibration, ν, in the direction of the jump. The probability of success is essentially equal to the probability that the atom has the necessary kinetic energy, Δh_{ACT}.

The most probable distribution of energy among a collection of N particles, where N is large, can be calculated by the methods of statistical mechanics. A useful result is the Maxwell-Boltzmann distribution, depicted in Fig. 6.5, showing the probability that a particle has a given energy.

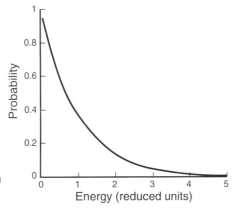

Fig. 6.5 The Maxwell-Boltzmann distribution of energy.

The important feature of this distribution is that the probability that a particle has a given energy decreases exponentially as the value of this energy increases. One is interested only in atoms with a kinetic energy toward the high end of the distribution, because Δh_{ACT} is not small. The kinetic energy of any particle depends on temperature. For example, the kinetic theory of gases shows that the average gas particle has a kinetic energy of 1/2 kT for each of the three orthogonal directions of motion of the particle, where k is Boltzmann's constant. In a collection of particles in which the energy is distributed according to the Maxwell-Boltzmann function, it can be shown that, at a temperature T (in degrees Kelvin), the probability that a particle has an energy Δh (e.g., the increase in energy per atom needed for a process) is given by

$$\text{Probability} = \exp\left(-\frac{\Delta h}{kT}\right)$$

Thus, the rate at which atoms cross the barrier to the new grain is given by:

$$\text{Rate} = \nu \exp\left(-\frac{\Delta h}{kT}\right)$$

which is the same as:

$$\text{Rate} = \nu \exp\left(-\frac{\Delta H}{RT}\right)$$

for one mole of jumping atoms, R being the gas constant = $N_o k$, where N_o is Avogadro's number. A process that follows this rate equation, known as the *Arrhenius equation,* is termed a thermally-activated process.

The exponential dependence on temperature results in a steeply rising rate of jumping of atoms from old to new grains over a small temperature range. This is reflected in Fig. 6.3(b), which indicates that recrystallization occurs within a fairly well-defined range of temperature.

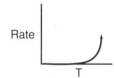

Recrystallization is a substitution reaction (like the one that occurred with the introduction of pneumatic tires in 1889; cf. Fig. 1.10). It occurs by nucleation and growth, following a classical pattern when the fraction transformed is plotted against time. This is shown schematically in Fig. 6.6, along with an example of interrupted recrystallization in cold-worked iron.

Recrystallization occurs sooner as the annealing temperature is raised in the temperature range of this process (because the rate varies exponentially with temperature). The physical explanation of the classical S-shaped curve is the following: The transformation occurs at the boundaries of the growing clusters of new grains (cf. Fig. 6.6b), which are scattered about the volume of the cold-worked material. As the clusters grow larger, so does the total area along their outer boundaries. Because the reaction occurs by atoms jumping across boundaries, the increase in boundary area causes the transformation to go faster. Also, additional new grains are nucleated as time passes. However, as time passes the clusters of new grains begin to impinge upon one another. Wherever this happens, the transformation must cease in the regions where overlap would otherwise occur. From this time on, the volume transformed per unit time must decrease. The fraction transformed goes asymptotically to unity as the last bit transforms.

SECTION 6.3.2 KINETICS OF RECRYSTALLIZATION

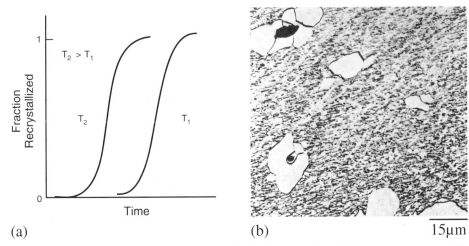

Fig. 6.6 (a) Kinetics of recrystallization, a substitution reaction which occurs more rapidly as temperature is increased. (b) Early stage of recrystallization in cold-worked iron, initiated at oxide inclusions; optical photomicrograph. (From W. C. Leslie et al. in *Iron and Its Dilute Solid Solutions,* Interscience, NY, 1963, p. 169.)

It should be pointed out that the apparent "recrystallization temperature" depends on the following factors:

1. The annealing time. The longer the time, the lower the temperature at which dramatic softening occurs (within a limited temperature range).

2. The amount of prior cold work. The more driving force available, the lower the recrystallization temperature. (Again, within limits.)

3. The purity of the material. The growth stage occurs more easily in a pure, single-phase material. For example, high-purity aluminum can recrystallize at room temperature, whereas commercial-purity aluminum must be heated above 200°C.

Example 6.1

Q. In Fig. 6.6(a), it is indicated that recrystallization occurs in a shorter time if the annealing temperature is raised. Give a physical explanation for this.

A. The rate-controlling step in recrystallization is the transfer of atoms across the interface between the cold-worked matrix and the new, dislocation-free growing grain. Thus, the rate of growth of the new grains increases exponentially with temperature according to:

$$\text{rate} = \text{constant} \times \exp(-\Delta H_{act}/RT)$$

where ΔH_{act} is the activation energy (cf. Fig. 6.1 and Appendix 6.2) needed to transfer one mole of atoms across the interface, and T is the absolute temperature (degrees Kelvin). R = 8.31 J/mole-K.

Example 6.2

Q. Suppose that ΔH_{act} for the growth stage of recrystallization of a stainless-steel spoke were 170kJ/mole. Calculate the annealing temperature needed to double the rate of growth found at 1073K (=800°C). (This would approximately halve the time needed for recrystallization.)

A. The following relationship must be solved for T_2, given that $T_1 = 1073K$.

$$\frac{\text{Rate 2}}{\text{Rate 1}} = 2 = \frac{\exp-\left(\frac{170 \times 10^3}{8.31 \cdot T_2}\right)}{\exp-\left(\frac{10 \times 10^3}{8.31 \cdot T_1}\right)}$$

The constant in the rate equation cancels out. Rearranging,

$$2 \exp-\left(\frac{170 \times 10^3}{8.31 \cdot 1073}\right) = 1.05 \times 10^{-8} = \exp-\left(\frac{170 \times 10^3}{8.31 \cdot T_2}\right)$$

Taking the logarithm of both sides

$$-18.37 = -\frac{170 \times 10^3}{8.31 \cdot T_2}$$

giving

$$T_2 = 1113K = 840°C$$

Thus, the rate of recrystallization would double with an increase in annealing temperature from 800 to 840°C.

6.4 Recovery

It should not be supposed that nothing happens in advance of recrystallization in a cold-worked material. Prior to recrystallization, the cold-worked state undergoes a kind of relaxation process, called *recovery*. During recovery the excess concentration of point defects, mainly lattice vacancies, decays toward the equilibrium level. (The excess point defects are generated as a result of the intersection of dislocations gliding on different slip systems.) The equilibrium level of vacancies in a crystal at any temperature can be derived mathematically by means of elementary statistical mechanics, as shown in Appendix 6.2. Examples of lattice vacancies in a bubble raft are shown in Fig. 4.4.

The physical explanation of this derivation is the following: When a lattice vacancy is formed, the internal energy of the crystal is increased, because the atoms around the vacancy are displaced from their equilibrium lattice positions. That is, there is a stored elastic energy associated with a vacancy. This means that at zero Kelvin the free energy is minimized only if there are no vacancies. However, at a higher temperature, entropy must play a role. A perfect crystal can be envisioned as having lattice vacancies only on its outer surface. This is a highly ordered configuration of the vacancies. The entropy of the crystal would be increased if some of those vacancies were distributed randomly on lattice sites.

Therefore, to minimize G=H-TS at some temperature T, the vacancy concentration inside the crystal would rise until the increase in TS (decrease in -TS) begins to be offset by the increase in H. This is the equilibrium concentration, because it minimizes G.

In actuality, the vacancy concentration can rise only if the temperature is high enough to provide thermal activation (in the form of kT) to permit atoms to change places with vacancies. This process is called *self-diffusion*. During self-diffusion, vacancies jump randomly about in a crystal, which is equivalent to saying that atoms jump randomly about through interchange with vacancies.

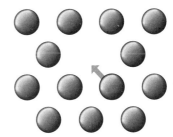

The result of the derivation in Appendix 6.2 is that the equilibrium vacancy concentration is

$$\text{vacancy concentration} = \exp-\left(\frac{\Delta H_f}{RT}\right)$$

where ΔH_f is the increase in the enthalpy (internal energy) of the crystal when one mole of vacancies is formed.

The recovery process may be studied in several ways. One utilizes the fact that vacancies contribute to electrical resistivity, because they interrupt the periodic variation of electron density in a metallic crystal lattice. Since they act as isolated scatterers, the increase in resistivity is essentially proportional to the vacancy concentration. Thus, one can follow the "annealing out" of vacancies by monitoring the drop in resistivity with time at a fixed temperature, as in Fig. 6.7.

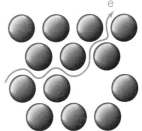

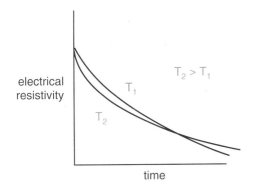

Fig. 6.7 Schematic representation of the kinetics of recovery. Note that both T_1 and T_2 are below the recrystallization temperature.

The question arises: Where do the vacancies go as they anneal out? Most go to dislocations with some edge component, because they can eliminate part of the extra material associated with an edge dislocation. This causes the dislocation to climb, a process illustrated in Fig. 6.8. One can think of the extra half-plane as being compressed along its sides by the surrounding crystal into which it is jammed (like a wedge into a piece of wood). When the dislocation climbs, the compressive stress σ_{xx} (cf. Appendix 5.2) does work, and it relaxes in the same way that the stress in a person's hand is relaxed when toothpaste squirts out of a tube. Thus, the driving force for recovery is, again, the stored energy of cold work.

CHAPTER 6 ANNEALING OF COLD-WORKED SPOKES

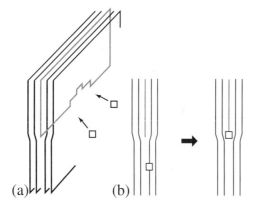

Fig. 6.8 Illustration of climb of an edge dislocation due to the absorption of vacancies, represented (a) in three dimensions and (b) in two dimensions.

The dislocation climb that occurs during recovery allows the tangled networks of dislocations resulting from cold work to rearrange into lower-energy configurations. An example is shown in Fig. 6.9. This rearrangement usually does not lower the dislocation density enough to produce a significant amount of softening, as reflected in the hardness plot of Fig. 6.3.

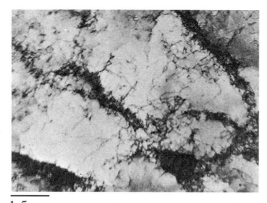
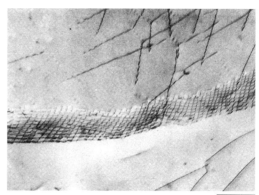

1.5μm   (a) Iron, cold-rolled 16% at room temperature   (b) Same, after 16h at 550°C   0.5μm

Fig. 6.9 Effect of recovery on dislocation cell structure in cold-rolled iron (A.S. Keh in *Direct Observation of Imperfections in Crystals,* J.B. Newkirk and J.H. Wernick, Eds., Interscience, NY, 1962, p. 213.)

Example 6.3

Q. In Fig. 6.7 it is indicated that recovery becomes more rapid if the temperature is raised. Why should this be so?

A. The annihilation of vacancies occurs by diffusion of vacancies to dislocations with some edge component. This diffusion occurs by successive interchanges between vacancies and atoms in the crystal lattice. In each interchange, a neighboring atom "jumps into" a vacant site. Since this requires thermal activation, the rate of jumping must increase exponentially with temperature.

Example 6.4

Q. In Fig. 6.7 the schematic curves for the two temperatures are drawn to cross at long annealing times. Explain why this should be expected.

A. The electrical resistivity is proportional to the vacancy concentration, and the equilibrium vacancy concentration increases exponentially with temperature (cf. Appendix 6.2). Therefore, the ultimate level of resistivity should be higher for the higher annealing temperature.

6.5 Grain Growth

6.5.1 Mechanism of Grain Growth

Just after the completion of recrystallization, a material usually has a fine grain size, especially if the amount of prior cold work was large (meaning that there were very many sites for nucleation of new grains). Since atoms in a grain boundary have higher energy than those in the crystal lattice, the total internal energy of a fine-grained material is larger than that of a coarse-grained material. This provides a driving force for grain growth.

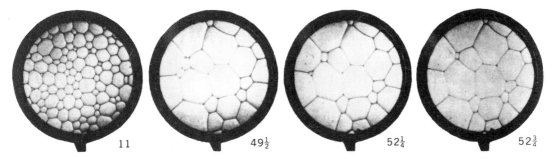

Fig. 6.10 Stages of bubble growth in a two-dimensional cell at the indicated times, in minutes. (C. S. Smith in *Metal Interfaces,* ASM, Metals Park, OH, 1952, p. 81.)

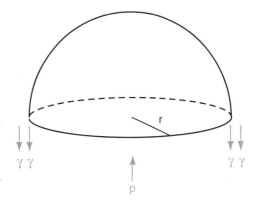

One can observe the analog of grain growth in a network of bubbles (e.g., of a soap solution), an example of which is shown in Fig. 6.10. The driving force here is the surface energy associated with the bubble walls. The mechanism of bubble growth depends on the fact that the pressure inside a small bubble is larger than that inside a large bubble, as demonstrated by the following argument.

Consider the equilibrium of forces on the upper half of a soap bubble. The force pulling the bubble down is due to the surface tension, γ, and is equal to $2\gamma 2\pi r$, since the bubble has 2 surfaces, and the length over which γ operates is $2\pi r$, the bubble circumference. The upward force due to the internal pressure, p, is $p\pi r^2$, since the pressure acts on the cross-sectional area.

At equilibrium

$$p\pi r^2 = 2\gamma 2\pi r$$

or

$$p = \frac{4\gamma}{r}$$

Because of the pressure gradient between a small bubble and a large one, air diffuses through the connecting wall, and the larger one grows at the expense of the smaller, as illustrated in Fig. 6.11(a).

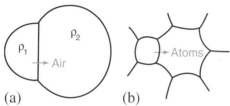

Fig. 6.11 (a) In a pair of soap bubbles the larger one grows at the expense of the smaller by diffusion of air through the connecting bubble wall, driven by the pressure difference. (b) In a polycrystal a large grain grows at the expense of its smaller neighbors by the jump of atoms from the smaller grains to the larger one.

In the case of a polycrystalline solid, the analog of the diffusion of air between bubbles is the transport of atoms across the boundaries from small grains to large grains, as indicated schematically in Fig. 6.11(b). Atoms along a grain boundary on the convex side have, on the average, fewer neighbors than those on the concave side, as indicated in Fig. 6.12(a). Thus, the average energy per atom of the former is slightly higher. (The minimum energy state is that in which an atom is completely surrounded by a crystal with atoms in their correct positions; in an FCC crystal each would have 12 neighbors.)

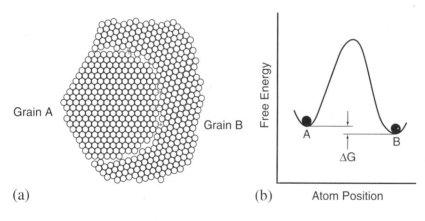

Fig. 6.12 Atoms at the boundary of the left-hand (convex) grain in (a) have, on average, fewer neighbors than those across the boundary on the concave side, causing an energy gradient across the boundary, as shown in (b).

At an elevated temperature the atoms along a grain boundary hop back and forth from one side to the other and tend to stick longer to the side on which they have more neighbors (i.e., lower energy). Hence, over time there is a net motion from the smaller grain to the larger. That is, the grain boundary tends to migrate towards its center of curvature, and the larger grain grows at the expense of the smaller. One can imagine the surface tension "squeezing" the high-energy atoms from the smaller to the larger grain, just as the high-pressure gas is squeezed from the smaller soap bubble into the larger.

The "grain boundaries" of a bubble network meet at dihedral angles of 120°. This is due to the necessity for equilibrium among surface tensions, which is

given by
$$\gamma_{12} = \gamma_{23} \cos \theta/2 + \gamma_{13} \cos \theta/2$$
or, since in a soap froth,
$$\gamma_{12} = \gamma_{23} = \gamma_{13} = \gamma,$$
$$\gamma = 2\gamma \cos \theta/2$$

Thus,
$$\cos \theta/2 = 1/2, \therefore \theta = 120°$$

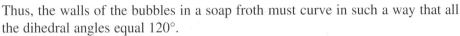

Thus, the walls of the bubbles in a soap froth must curve in such a way that all the dihedral angles equal 120°.

If the grain boundaries of a polycrystalline solid all had equal energy, as do the walls of a bubble network, they would also meet at 120° angles. In reality, the energies vary with the atomic structure of the boundary (cf. Fig. 4.4). However, for most boundaries in a polycrystalline aggregate of randomly oriented grains the variation is not large, and the bubble network can still be used as a model.

In a two-dimensional representation of a three-dimensional network, as shown in Fig. 6.13, the requirement that all triple points have 120° angles between the constituent boundaries, means that for grains with six sides, the boundaries are straight. For grains with three, four or five sides, the sides are concave inward; i.e., the center of curvature lies toward the grain center. However, for grains with more than six sides, the sides are concave outward. Thus, according to Fig. 6.12, the larger grains should grow at the expense of the smaller grains, as long as there is enough thermal energy to activate the hopping of atoms across the boundary.

Fig. 6.13 Schematic representation of grain growth. Large grains (>6 sides) tend to grow, and small grains (< 6 sides) tend to shrink. (after J.E. Burke, from P.G. Shewmon, *Transformations in Metals*, McGraw-Hill, NY, 1969, p.120.)

6.5.2 Kinetics of Grain Growth

The kinetics of grain growth can be expressed by plotting the average grain diameter *vs.* time, as in Fig. 6.14.

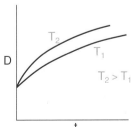

Fig. 6.14 Schematic illustration of the parabolic variation of average grain diameter D with time t during grain growth at different temperatures T_1 and T_2.

The shape of the curves can be explained by the following argument:

1. The driving force for growth is given by :

$$\frac{\gamma \times \text{grain-boundary area}}{\text{total grain volume}}$$

   where γ is the average grain-boundary energy per unit area.

2. If the average grain diameter is D, then the average grain-boundary area is proportional to D^2 and the average grain volume is proportional to D^3. Thus, the driving force is proportional to γ/D.

3. It is reasonable to set the rate of grain growth, dD/dt, proportional to the driving force (at a fixed temperature). Thus:

$$\frac{dD}{dt} = \frac{k\gamma}{D}$$

   where k is some constant depending on grain geometry, temperature (exponentially), and the material.

This equation, when integrated, yields a parabolic growth law:

$$D = k\gamma\, t^{1/2}$$

if the approximation is made that the grain size is very small (≈ 0) at time zero.

This analysis is idealized, because it ignores the effects of impurity atoms and particles of second phase, both of which cause grain growth to be slower than would be predicted otherwise. The impurity atoms are, in effect, unwanted elements in solid solution. The second-phase particles can be unwanted dirt, or they can be intentionally present, e.g., for purposes of inhibiting grain growth.

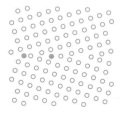

Solute elements tend to segregate (i.e., by diffusion) to grain boundaries at low temperatures, because the total internal energy of the polycrystal is thereby reduced. For example, an oversize solute often can fit into a grain boundary with less local distortion than within a grain, because the grain boundary can have an excess of free volume, because of looser atomic packing in the boundary (cf. Fig. 4.4). Alternatively, solutes that tend to develop directional bonds with the host atoms may be able to do that in a grain boundary, where the packing arrangement permits it. If the jumping rate of the segregated solute atom is slower than the host atoms, then grain boundary migration is retarded.

This drag on grain boundary mobility diminishes as temperature increases, because segregated solutes tend to leave the grain boundary and return to their random arrangements in solid solution at higher temperatures. The explanation for this is that solute segregation reduces the entropy of a polycrystal (i.e., makes it less randomly organized). This has an effect opposite to the reduction of internal energy which promoted the segregation in the first place. In other words, at low temperatures, H dominates and segregation occurs, but, as T increases, segregation diminishes, because -TS becomes more important. Thus, the minimization of G=H-TS requires that desegregation take place at elevated temperatures. This would have the effect of making the grain boundaries more mobile, and the parabolic growth kinetics would be more likely to be observed.

The effect of second-phase particles can be analyzed from the standpoint of interfacial energy. A particle has an interfacial area between itself and the matrix, whether it is within a grain or on a grain boundary. Since a grain boundary also has an energy per unit area associated with it, anything that reduces the total grain boundary area must reduce the energy of the system. If a particle resides on a grain boundary, part of the area of the boundary is replaced by the cross-sectional area of the particle, as shown in Fig. 6.15.

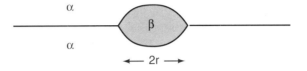

Fig. 6.15 A particle of the β phase on a grain boundary of the α phase reduces the total α grain boundary area.

Thus, the total interfacial energy is reduced by:

$$\gamma_{\alpha\alpha}\pi r^2$$

where $\gamma_{\alpha\alpha}$ is the average grain boundary energy of the matrix (α) phase. This amount of energy must be supplied to pull a grain boundary away from the particle.

The regions of grain boundary remote from the particle would tend to migrate as they would normally. The local pinning by the particle causes the boundary to become curved into a cusp, as shown in Fig. 6.16, thus causing the total grain boundary area to increase. When the size of the cusp gets large enough, it becomes energetically favorable for the boundary to separate from the particle, and a local spurt of grain boundary migration then occurs.

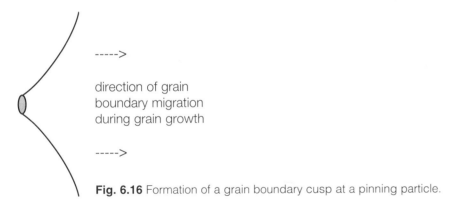

Fig. 6.16 Formation of a grain boundary cusp at a pinning particle.

Clearly, the more densely arrayed the particles are in the solid, the less bowing-out there will be between particles. (Note the analogy with dislocations.) Thus, an array of many fine particles will retard grain growth more effectively than fewer large ones.

The particle-pinning effect also tends to diminish as temperature is increased, owing to two effects. First, the solubility of a second phase tends to increase with temperature (the entropy effect again). As the volume fraction of particles is thereby reduced, there is less pinning. Second, the interfacial energy acts as a driving force to make particles coarsen. This occurs by the diffusion of the constituent solute element through the matrix from smaller particles to larger particles (cf. Section 6.6). With coarsening, the particle spacing increases, and again the pinning effect is reduced.

6.5.3 Softening During Grain Growth

Grain growth produces gradual softening for two reasons. The more important reason is that irregularities on grain boundaries are prime sources of dislocations during deformation. Therefore, as the grain boundary area per unit volume decreases during grain growth, so does the density of dislocation sources per unit volume. Hence, the dislocation density at any stage in the deformation tends to be lower. (That is, the plastic strain is accomplished by fewer dislocations moving greater distances, on the average.) Thus, grain coarsening during annealing weakens the strain-hardening effect during subsequent deformation.

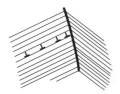

The other aspect of grain growth that produces softening is that grain boundaries act as barriers to slip bands because of the mismatch of slip systems from grain to grain. This effect tends to diminish in importance as plastic strain increases and the dislocation arrangements become more complicated, because slip bands are less well defined at large plastic strains.

Applying an understanding of annealing to bicycle spokes, one can appreciate that, since strain hardening is used to strengthen spokes, the wire drawing must be carried out well below the recrystallization temperature range. (Deformation in or above the temperature at which recrystallization occurs is called *hot working*.) Furthermore, one must not heat a spoke into the recrystallization range; otherwise, much of its strength would be lost, and premature failure could result.

Example 6.5

Q. How is grain growth different from the growth stage in recrystallization, since both involve growth of grains?

A. The driving force is different. In recrystallization, atoms hop from highly dislocated grains to undislocated grains. That is, they flow down a gradient of internal energy. In grain growth the atoms hop back and forth between undislocated grains. They simply tend to stick longer to the side with more nearest neighbors.

Example 6.6

Q. How does surface energy, γ, act to drive grain growth? What is the physical mechanism?

A. Surface energy (J/m^2) also has the units of surface tension (N/m), and the dihedral angles at three-grain junctions can be thought of as being governed by the equilibrium of surface tension forces. This is what controls the curvature of a grain boundary, depending on the number of sides of the two neighboring grains. The sign of the curvature determines the direction of net drift of the hopping atoms.

6.6 Annealing of Carbon-Steel Spokes

The microstructure of a carbon steel spoke comprises two phases: about 95% of the steel is almost-pure BCC iron (cf. Fig. 2.9), and the balance is iron carbide, Fe_3C, called *cementite,* which is a hard, brittle phase. Since the plastic deformation of carbon steel spokes during wire drawing occurs almost entirely in the BCC-iron phase, known as *ferrite,* the presence of the cementite will be ignored for now. The ferrite is not quite pure iron; it contains carbon in addition to some other, less important, elements in solid solution. The carbon plays several important roles, as will be seen, but first the microstructure in the cold-drawn condition and after two kinds of annealing treatments will be examined.

6.6.1 The As-Received Condition

Photomicrographs of a longitudinal and a transverse section of a carbon-steel spoke in the as-received (i.e., cold-drawn) condition are shown in Fig. 6.17. The microstructure obviously consists of light-etching and dark-etching regions strung out in the direction of wire drawing, as seen in the longitudinal section. The distribution of light and dark constituents is non-directional when viewed in the transverse section, as might be expected from the cylindrical symmetry of the wire-drawing die (cf. Fig. 3.18).

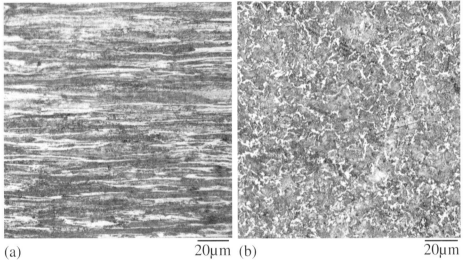

Fig. 6.17 Microstructure of an as-received carbon-steel spoke at 500X magnification. (a) longitudinal section; (b) transverse section. (Metallography by E. Anderson, Univ. of Penn.)

Recalling that the contrast in the optical microscope comes from surface roughness, as illustrated schematically in Fig. 6.18, one can interpret Fig. 6.17 as comprising one constituent which is essentially un-attacked by the etchant (i.e., is flat, reflects the incident light, and appears white) and another constituent which is heavily roughened by the etchant; therefore, it scatters the incident light and appears dark.

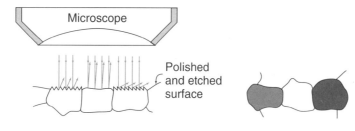

Fig. 6.18 Schematic representation of three regions in which etching produces different degrees of roughening. This gives variations from light to dark contrast in the image viewed in the microscope (From W.D. Callister, Jr., loc. cit., p. 60. Reprinted by permission of John Wiley & Sons, Inc.)

6.6.2 The Annealed Condition

The microstructure of the carbon steel spoke can be simplified by annealing. In order to recrystallize the steel, 700°C was chosen, since this temperature is greater than 0.5 T_{MP} in degrees Kelvin, and self-diffusion occurs rapidly above this temperature. Two annealing times were used: 3h and 24h. After the shorter annealing time, the microstructure, shown in Fig. 6.19(a) and (b), consisted of two phases. One is the fine-grained polycrystalline matrix comprising ferrite (BCC iron), and the other consists of particles of iron carbide, or cementite, Fe_3C, of less than 1µm diameter, which are strung out in the wire-drawing direction. It is apparent that the ferrite grain size in the transverse section of the recrystallized spoke, shown in Fig. 6.19(b), ranges from about 1 to 7µm, even though the annealing temperature is quite high. This is a good example of how a dispersed second phase can inhibit grain growth.

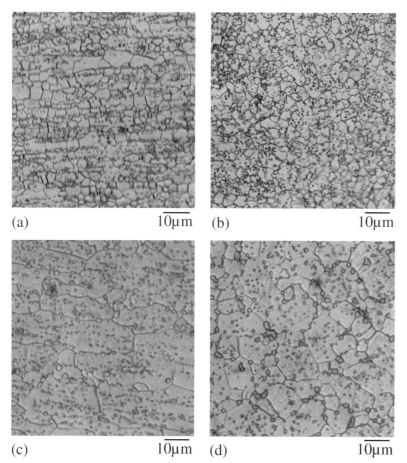

Fig. 6.19 Microstructure of a carbon steel spoke after annealing at 700°C for the indicated times, shown at 1000x magnification; (a) 3h, longitudinal section, (b) 3h, transverse section, (c) 24h, longitudinal section, (d) 24h, transverse section. (Metallography by E. Anderson, Univ. of Penn.)

The effect of prolonging the annealing time to 24h is shown in Fig. 6.19(c) and (d). Now the ferrite grain size in the transverse section ranges from about 5 to 25μm, and the cementite particles have coarsened substantially. Fig. 6.19(c) shows that the grain growth in the longitudinal direction during the extended anneal has been even larger.

Two distinct but interrelated phenomena are illustrated in Fig. 6.19. The first is that the cementite particles coarsen with time during annealing. That is, the average particle size increases. The number of carbides per unit volume decreases, and their spacing increases. (Of course, the total volume of carbide remains constant during coarsening.) The second phenomenon is grain growth in the ferrite phase. However, this growth is constrained to occur mainly in the longitudinal direction of the wire, owing to the presence of the longitudinal strings of cementite particles. As the spacing between the particles increases during coarsening, the constraint on grain growth in the ferrite is gradually reduced. Both phenomena are driven by interfacial energy: carbide coarsening by the ferrite/cementite interfacial energy, and ferrite grain growth by the energy of the ferrite/ferrite boundaries.

The evolution of the microstructures shown in Fig. 6.19 can be summarized as follows: During the anneal at 700°C, the cold-worked ferrite recrystallizes rapidly, but grain growth is inhibited by the fine distribution of cementite particles. This microstructure, however, contains a large amount of interfacial energy, which provides a driving force for the coarsening of the cementite and grain growth in the ferrite. The coarsening reaction proceeds by the diffusion of carbon atoms from small cementite particles to larger ones.

The reason for this flux of carbon is the curvature of the interface, as is the case with grain growth (cf. Fig. 6.12). The carbon atoms in solid solution in the ferrite next to a cementite particle can be thought of as analogous to atoms in the gaseous state above a solid. Then, the "partial pressure" of the carbon next to a small particle would be greater than that next to a larger one.

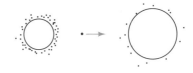

The reason for the greater carbon concentration around small particles is simply that the carbon at the surface of a smaller particle is less-completely surrounded by the cementite crystal structure and is, therefore, less-tightly bound. The difference in "partial pressure" translates to a difference in carbon concentration in the ferrite between small and large cementite particles, and this concentration gradient drives the diffusion of carbon from the smaller to the larger particles.

As the cementite particles coarsen they become fewer in number and, therefore, more widely spaced; the ferrite grain boundaries then have fewer barriers to pin them, and grain growth becomes increasingly less inhibited. This is especially true in the longitudinal direction, but the strings of carbide particles continue to inhibit grain growth in the lateral direction. Thus, the "memory" of the prior deformation processing persists in the annealed microstructure.

6.6.3 The "Fully-Annealed" Condition

When a carbon steel spoke is heated to 950°C and slowly cooled (e.g., by switching off the furnace), the result, shown in Fig. 6.20, is completely unlike that found in Fig. 6.19. In the steel industry, this condition is called the "fully-annealed" condition, but in truth it should be called the "fully transformed" condition. Not only is there no memory of the elongated cold-drawn condition, but the steel is now divided into roughly equal volumes of light and dark constituents. The former is ferrite, and the latter is called *pearlite* and consists of alternate layers of ferrite and cementite. The layers are so finely spaced that the etched pearlite is dark. This is close to the condition that existed prior to the final cold-drawing operation, which produced the microstructure shown in Fig. 6.17. That is, during the cold drawing, the ferrite and pearlite constituents were deformed into cigar-shaped regions mixed on a fine scale. The explanation for the radical change in microstructure shown in Fig. 6.20 will be given in Chapter 9.

Fig. 6.20 Microstructure of a carbon steel spoke that has been "austenitized" at 950°C and furnace cooled. The light regions are ferrite, and the dark are pearlite, comprising lamellar ferrite and cementite. (Metallography by E. Anderson, Univ. of Penn.)

20μm

Summary

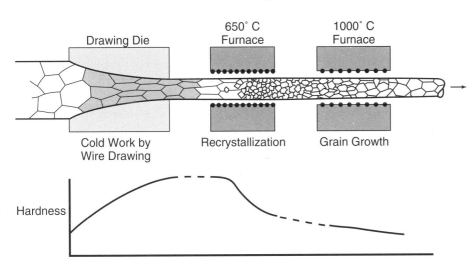

At certain stages of processing rod into wire for spokes it is necessary to soften the metal by annealing, because the wire can become so strain-hardened that there is danger of breakage during wire drawing. Most of the softening occurs in a rather narrow temperature range by the process of recrystallization, in which new dislocation-free grains nucleate and grow into the cold-worked matrix. This is a substitution reaction, and it follows the classical S-curve behavior (fraction transformed plotted *vs.* time for a fixed annealing temperature).

At temperatures below the recrystallization range, the cold-worked metal undergoes the process of recovery in which excess lattice vacancies anneal out, mainly by annihilation at dislocations, leading to dislocation climb and rearrangement. This process can be monitored by measurements of electrical resistivity (because vacancies scatter conduction electrons in a crystal lattice). It exhibits the isothermal kinetics of a decay reaction that occurs homogeneously throughout the solid, in contrast to the heterogeneous nature of recrystallization. The driving force for both recovery and recrystallization is the stored energy of cold work, which is essentially equal to the total length of dislocations times Gb^2.

At high annealing temperatures, after recrystallization is completed, the process of grain growth can occur. This is driven by the total interfacial energy associated with grain boundaries, and in the ideal case the grain size would increase with the square root of time at a constant annealing temperature. Of course, the rate would increase exponentially with temperature. The rate of grain growth can be decreased by the segregation of solute atoms to grain boundaries and by the presence of second-phase particles in grain boundaries. Both effects tend to diminish as the annealing temperature is raised. Concepts introduced in this chapter include self-diffusion and the equilibrium concentration of lattice vacancies as a function of temperature. The connection between grain boundary energy and the growth of large grains at the expense of small ones is also explained.

In a carbon-steel spoke, the presence of a second phase, cementite (Fe_3C) inhibits grain growth in the ferrite matrix after the latter recrystallizes. Extended annealing, however, produces coarsening of the cementite, and grain growth in the ferrite can then proceed. The starting condition before wire drawing was one which involved austenitization (cf. Chap. 9) and transformation during cooling to ferrite and pearlite, the latter being a mixture of ferrite and cementite. In steel-industry terminology this is called "full-annealing" if the cooling is slow and "normalizing" if air-cooling is used.

Appendix 6.1 - Measurement of Hardness

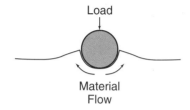

The term *hardness* is generally used to connote resistance to plastic deformation. (It is sometimes also used in the context of elastic behavior, as the opposite to elastically soft, or springy.) There exist a number of simple, common hardness tests, all of which characterize the resistance of a material to penetration by an indenter of a particular size and geometry under a known load.

As one can visualize, this is a rather complicated mode of plastic deformation, so the result of such a test cannot be interpreted in the same way as a uniaxial tensile test. Rather, it measures the combined effects of yielding and strain hardening in a multiaxial stress state. Thus, hardness data are used for purposes of comparison, rather than to probe a fundamental material property. The features of the various common hardness tests are summarized in Table A6.1.1.

Table A6.1.1 Features of Common Hardness Tests

Test Name	Indenter	Load(s)	Parameter Measured	Units of Hardness	Comments
Brinell	10 mm sphere	500, 1000 or 3,000 kg	Diameter of indentation; then calculate projected area*	kg/mm^2	Large indentation; a large volume of material is sampled.
Vickers, or Diamond Pyramid	Symmetrical diamond pyramid	1 g to 120 kg	two axes of diamond-shaped indentation; then calculate projected area*	kg/mm^2	One of the most precise methods; can be used with very low loads to give microhardness.
Knoop	Elongated diamond pyramid	1 g to 120 kg	The major axis of the elongated indentation; then calculate projected area*	kg/mm^2	Measurement precision higher than Vickers, but subject to effects of anisotropy of material.
Rockwell	diamond cone or 1/16 in. sphere	60 to 150 kg	Machine measures depth of penetration and gives output on a dial	none	Quick and easy, but less precise.

*The calculations are tabulated and are available in handbooks; one can, therefore, read out the hardness number, given the measured diameter (for a given load).

Mineralogists use the Mohs scale, which ranks minerals in a relative way such that any mineral scratches the one below and is scratched by the one above. Diamond is at the top, and talc is at the bottom; silica glass ranks below hardened steel, but above gold, for example. This is still a measure of resistance to plastic deformation, but on a more extreme level, since a scratch is formed by a kind of plowing action in which a material is deformed to the point of failure in shear.

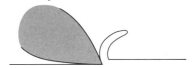

Fig. A6.1.1 Schematic illustration of deformation involved in a scratch-hardness test. It is akin to machining on a lathe.

Appendix 6.2 - The Equilibrium Concentration of Lattice Vacancies

Consider a crystal containing N atoms and n vacancies. An expression for the free energy associated with vacancies can be written as follows

$$G = H - TS = nh - TS$$

where h is the enthalpy ($\approx$ local crystal-distortion energy) associated with each vacancy. (The perfect crystal is the zero-energy reference state.) It is desired to calculate the value of n for which G is a minimum; i.e., for which

$$\frac{dG}{dn} = 0$$

To do this, Boltzmann's statistical definition of entropy is employed,

$$S = k \ln W$$

where W is the number of "microstates per macrostate," or the number of distinguishable ways of arranging the N atoms and n vacancies on the N + n lattice sites. If the sites are filled one by one, starting with an empty lattice, there are N + n choices for the first move, N + n - 1 for the second, N + n - 2 for the third, and so on. For the first two, there are (N + n)(N + n - 1) combinations of arrangements. The total number of combinations for all arrangements is, then, (N + n)! However, since all atoms are indistinguishable, as are all vacancies, the number of distinguishable combinations is much less and can be written as:

$$W = \frac{(N + n)!}{n! \, N!}$$

because any of the n vacancies or N atoms can be interchanged with another vacancy or atom, respectively, without creating a new atom-vacancy arrangement. Therefore, the total number of combinations must be reduced by dividing by n! to account for the number of ways of arranging n vacancies on a given set of n lattice sites, and similarly for the N atoms on a given set of N lattice sites.

Stirling's approximation: $\ln x! \approx x\ln(x) - x$ can be used to write:

$$\ln W \approx (N + n)\ln(N + n) - (N + n) - n \ln(n) + n - N \ln(N) + N$$

Hence, the entropy associated with the presence of vacancies is

$$S = k\,[(N + n)\ln(N + n) - n\ln(n) - N\ln(N)]$$

Therefore,

$$G = nh - kT\,[(N+n)\ln(N + n) - n\ln(n) - N\ln(N)]$$

and for equilibrium G must be a minimum; therefore,

$$dG/dn = 0 = h - kT\,[\ln(N + n) + 1 - \ln(n) - 1]$$

or

$$h = kT\ln[(N+n)/n]$$

and, transposing, the vacancy concentration at equilibrium at temperature T is

$$\frac{n}{N + n} = \exp\left(-\frac{h}{kT}\right)$$

On a per-mole basis, one uses $R = kN_0$, where R is the gas constant and N_0 is Avogadro's number, and $H = N_0 h$ is used to express the enthalpy of formation of one mole of vacancies. Since $n \ll N$, one can write

$$\frac{n}{N} \approx \exp\left(-\frac{H}{RT}\right)$$

As an example, $H \cong 23\,\text{kcal/mole}$ for copper, and the gas constant R has a value of approximately 2 cal/mol-K. At 1000K (727°C), therefore,

$$\frac{n}{N} \approx 10^{-5}$$

That is, one lattice site in 100,000 in copper is vacant at equilibrium at 1000K. (Copper melts at 1083°C.)

Exercises - Chapter 6

Terms to Understand

Cold work	Electrical resistivity
Annealing	Grain growth
Recrystallization	Grain boundary energy
Driving force	Grain boundary curvature
Hardness test	Effect of grain size on strength
Nucleation and growth	Vacancy concentration at equilibrium
Thermal activation	Entropy
Activation energy	Equilibrium in a cellular network
Kinetics	Enthalpy
Stored energy of cold work	Gibbs free energy
Recovery	Grain boundary pinning

Problems

6.1 Explain the difference between activation energy and driving force.

6.2 Why does hardness decrease during recrystallization much more than during grain growth?

6.3 Explain the driving forces for recrystallization and grain growth.

6.4 Explain why high-purity aluminum recrystallizes at a much lower temperature than commercial purity aluminum.

6.5 Explain from the standpoint of kinetics why recrystallization and recovery are fundamentally different kinds of transformations. Find an analog of each in nature or society.

6.6 With regard to Example 6.2, calculate the further increase in temperature needed to double the rate of recrystallization again.

6.7 Determine the geometry of a grain-boundary triple junction in which one of the three grain boundaries has an energy of one-half the average, while the other two are average.

6.8 Explain why the kinetics of grain growth become much more nearly parabolic (rate $\propto \tau^{1/2}$) as the annealing temperature is increased, whereas the time exponent is usually less than 1/2 at lower temperatures.

6.9 If grain boundaries are assumed to be the major sources of dislocations in a material, and if dislocation sources are uniformly distributed on grain boundaries, then the dislocation density, ρ, must be proportional to the grain boundary area per unit volume. Use this idea to show that the flow stress should vary with grain size according to

$$\sigma_{flow} = \sigma_0 + k\, D^{-1/2}$$

where σ_0 is the stress needed to overcome the effects of lattice friction, solute atoms, etc.; d is the average grain diameter, and k is a constant.

6.10 Figure 6.19 shows the evolution of the microstructure of a carbon steel spoke during annealing for times up to 24h. What would be the ultimate microstructure if the annealing were to be carried out *ad infinitum* (assuming cementite would not transform to graphite, which is the ultimately stable phase)?

6.11 What would you expect to happen if the transformed microstructure shown in Fig. 6.20 were to be similarly annealed at 700°C *ad infinitum*?

7

CARBON STEEL SPOKES; PLASTIC DEFORMATION AND FATIGUE RESISTANCE

7.1 Slip in BCC Iron

Referring to the microstructure of a fully annealed carbon-steel spoke, shown in Fig. 6.21, it will be shown in Chapter 9 that the pearlite constituent is about 90% BCC ferrite and 10% carbide, so the steel as a whole is about 95% ferrite. Therefore, the mechanical behavior of the spoke depends mainly on the deformation behavior of the ferrite. The deformation properties of BCC metals differ markedly from those of FCC metals. The difference in both the crystal structure and deformation behavior is the result of a directional component of interatomic bonding, which is absent in the FCC metals. Another important factor is the unique behavior of interstitial solutes in BCC metals, particularly with regard to the interaction between interstitial atoms and dislocations. The classic example is that of carbon in iron, and this behavior plays a central role in the special fatigue behavior of carbon steel, which depends on the interaction between carbon atoms and dislocations. This topic will be covered in the present chapter. Another manifestation of this interaction is the dependence of hardness on carbon content in quenched (i.e., "hardened") steels, which will be described in Chapter 10.

As one can see from Fig. 2.9, the atoms in a BCC crystal touch along the body diagonals of the unit cell. Thus, the <111> directions are close-packed directions, and the Burgers vectors of the slip dislocations lie along these directions. However, the slip plane in a BCC crystal is less well defined. There is no close-packed plane; the {110} planes are the most nearly close packed, and these tend to be the slip plane of choice. However, other planes which contain a <111> direction, such as {112}, can also act as slip planes. For our purposes, the slip systems will be taken to be {110}<111>; an example is shown in Fig.7.1.

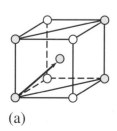
(a)

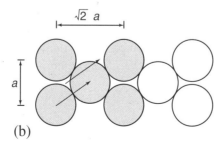
(b)

Fig. 7.1 Two representations of a BCC slip system: (a) in a unit cell, and (b) on a hard-sphere model of a {110} plane.

Except for the differences in the crystallography of slip, the previous discussion of dislocation behavior in FCC stainless steel applies to BCC iron. An additional characteristic of BCC transition metals like iron, however, is that, as mentioned in Section 5.2.4, the displacements of atoms along the core of a screw dislocation do not take the smooth form of a helical ramp, but tend to be concentrated in certain radial planes (i.e., planes which contain the Burgers vector and intersect along the core of the dislocation). This is not important at higher temperatures, because thermal vibration of the atoms tends to wash out this kind of irregularity. However, at low temperatures this distortion of the core of screw dislocations restricts their mobility. As a result, plasticity is made more difficult (e.g., Frank-Read-loop expansion requires a much higher stress), and the yield stress increases steeply as the temperature is reduced. This is the reason for the ductile-brittle transition found in carbon steel at low temperatures. (cf. Appendix 7.1.)

Example 7.1

Q. Sketch each individual slip system in the set {110} <111> for a BCC crystal.

A.

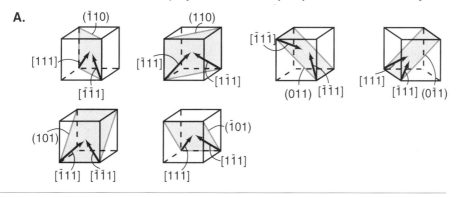

7.2 Interaction of Solute Atoms with Dislocations

As discussed earlier in Section 5.4.3, a dislocation can interact with a misfitting solute atom, because they both have elastic stress fields. If the stress fields have the same sign, the interaction is repulsive, and if they are of opposite sign, the interaction is attractive. This is easy to understand in the case of an edge dislocation. Figure 7.2 gives the four possible interactions. All of them decrease the mobility of an edge dislocation or a mixed dislocation. (A mixed dislocation has some edge character, meaning that there is some part of an extra half-plane associated with the dislocation.) This reduction in mobility means that additional stress must be applied to force a dislocation to glide through a crystal containing such solutes. This phenomenon is known as *solid-solution hardening*.

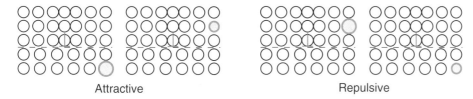

Attractive   Repulsive

Fig. 7.2 The four possible interactions between misfitting substitutional solute atoms and dislocations with edge, or partly edge, character.

A pure screw dislocation has no extra material associated with it. Its stress field is one of pure shear; there are no dilatational terms. Therefore, there is, to a first approximation, no interaction between a screw dislocation and a substitutional solute, since the solute has only a hydrostatic stress field (i.e., equal in all directions) but no shear component (cf. Section 5.4.3). In general, the same considerations are valid for an interstitial solute in an FCC lattice, since here again the distortion around the solute atom is spherically symmetrical. However, this is not the case for an interstitial solute in a BCC crystal, and this is of critical importance in carbon steel.

A special feature of an interstitial solute in a BCC crystal, like carbon or nitrogen in iron, is that the distortion is not purely dilatational. There is a component of shape distortion of the site in which carbon atoms reside. That is, the site which produces the minimum overall strain energy in the lattice is located in the center of a cube face, as shown in Fig. 7.3. (The center of a cube edge is an equivalent site.) This site is called an octahedral site, because it has six surrounding iron atoms which lie at the apices of an octahedron (cf. Fig. 7.3). The site is not symmetrical, since the distance between one pair of iron atoms is a, the lattice parameter, whereas the distance between the other two pairs is $\sqrt{2}a$. When a carbon atom occupies this site, the latter two pairs are attracted inward, but the third pair is pushed outward. This converts the symmetry of the crystal from cubic to tetragonal in the vicinity of the carbon atom. (A tetragonal unit cell can be obtained by stretching or compressing a cube along one axis; see Appendix 7.2 for more on crystal symmetry.) Because of the localized shape change of the unit cell, the stress field of an interstitial atom in a BCC lattice contains non-zero shear-stress terms.

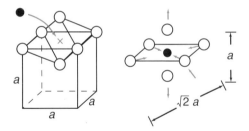

Fig. 7.3 An interstitial atom in the octahedral site of a BCC lattice produces a tetragonal distortion.

The result of this distortion is that carbon atoms in BCC iron are attracted even to pure screw dislocations. This has profound effects, which will be described after considering how the solutes get to the dislocations.

Example 7.2

Q. Show that the octahedral site in the center of the face of the BCC unit cell in Fig. 7.3 is equivalent to the center of an edge of the unit cell.

A. Since all the atoms in the crystal lattice are equivalent, the unit cell could have been centered on any atom. Two alternatives are shown. It can be seen that the center of the top face in one unit cell is the center of an edge in the other.

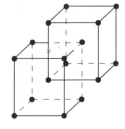

7.3 Segregation of Solutes to Dislocations

The driving force for the segregation of solutes to dislocations is the reduction of the overall elastic strain energy in the crystal lattice. That is, segregation produces a decrease in enthalpy, which in turn produces a decrease in free energy, G=H-TS, if temperature is low enough. This caveat is necessary, because segregation of randomly distributed solute atoms to dislocations increases the orderliness of the solute-atom arrangements and, therefore, decreases the entropy of the system. This would contribute to an increase in free energy, and it can only occur at low temperatures, where the TS term is small. The conclusion from thermodynamics, therefore, is that the influence of entropy tends to drive the solutes back into random solid solution as the temperature is increased. (This is true of any segregation process, including segregation of solutes to grain boundaries or the free surface.) Of course, if the temperature is so low that diffusion cannot occur in a reasonable time, then segregation is precluded, and equilibrium cannot be reached.

7.3.1 Substitutional Diffusion

In discussing solute segregation, one must distinguish between diffusion of substitutional *vs.* interstitial solutes. For substitutional solutes to diffuse, lattice vacancies are required. The process is illustrated in Fig. 7.4, and can be viewed either as vacancy migration or as the interchange of position between an atom and a vacancy. (If there is no solute involved, only host atoms, the process is called *self-diffusion*.)

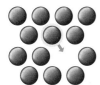

Fig. 7.4 Substitutional or self-diffusion occurs by vacancy migration.

In order for one such jump to occur, there are two requirements. First, a site next to the solute atom must be vacant. Second, there must be enough kinetic energy in the cluster of atoms around the vacancy for the solute to leave its site and jump into the vacant site, overcoming the constraint of its neighbors. Thus, the probability of a successful jump is the product of two probabilities: the probability that one adjacent site is vacant (which is proportional to the concentration of vacancies in the lattice) and the probability that sufficient kinetic energy can be localized at the jumping atom. Both probabilities are exponential functions of temperature. That is why significant diffusion of substitutional solutes occurs only at elevated temperatures (meaning temperatures on the order of half the melting temperature in degrees Kelvin or higher).

The important features of substitutional diffusion can be summarized as follows:

1. The mechanism involves the interchange between atoms and vacancies. This occurs in a random manner, except that it can be biased by a gradient in strain energy (i.e., by a strain field, as near a dislocation).

2. The probability of any particular interchange depends on two factors: the probability that a site next to a particular atom is vacant (exp-(ΔH_f/RT), where ΔH_f is the energy of vacancy formation), and the probability that the atom has enough thermal energy to make the jump into the vacancy (exp-ΔH_m/RT), where ΔH_m is the activation energy for vacancy motion.

3. Thus, the rate of substitutional diffusion is proportional to

$$\exp-\left(\frac{\Delta H_f}{RT}\right)\exp-\left(\frac{\Delta H_m}{RT}\right) = \exp-\left[\frac{(\Delta H_f + \Delta H_m)}{RT}\right] = \exp-\left(\frac{Q}{RT}\right)$$

The term $Q = \Delta H_f + \Delta H_m$ is the activation energy for substitutional diffusion, or for self-diffusion.

7.3.2 Interstitial Diffusion

The situation for diffusion of interstitial solutes is different than for substitutional solutes, since no vacancy is required, as indicated by Fig. 7.5. The concentration of interstitial solutes is always so small that there is little likelihood that two adjacent interstitial sites are occupied. Therefore, the activation energy, Q, for interstitial diffusion contains only one term: the enthalpy of motion, or the kinetic energy required for the interstitial atoms to jump through the small "cage" formed by the host atoms that restrict the path to a neighboring interstitial site.

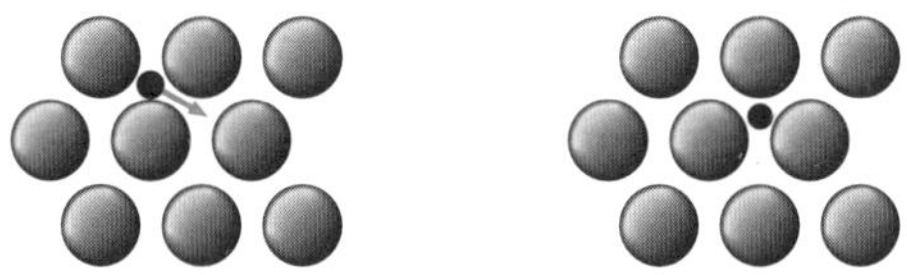

Fig. 7.5 Diffusion of interstitial solute atoms does not require the presence of lattice vacancies.

Since the vacancy concentration is irrelevant here, interstitial diffusion can occur at temperatures far below those needed for substitutional diffusion. For example, a carbon atom in iron at room temperature changes its interstitial site once per second, on the average. This means that all dislocations in iron will gradually become "decorated" with segregated carbon (or nitrogen) atoms.

The site for interstitial segregation to an edge dislocation is illustrated in Fig. 7.6. Since this is a minimum-energy configuration, the dislocations can become immobilized, or "pinned," if enough of this segregation occurs. That is, extra work must be done to free a pinned dislocation from an "atmosphere" of segregated carbon. This has a profound influence on the deformation behavior of iron and, in particular, on its resistance to fatigue failure, which is the topic of the following sections.

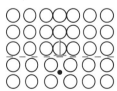

Fig. 7.6 An interstitial solute atom is attracted to the expanded lattice in the core of an edge dislocation.

Example 7.3

Q. There are three sets of octahedral sites in a BCC lattice, one set along each of the three coordinate axes of the unit cell, as shown below. These sets can be called x, y, and z sites, respectively. Deduce the effect of the application of a tensile stress along the [100] axis of a crystal of BCC iron which contains interstitial carbon in solid solution. Assume that the temperature is high enough for interstitial diffusion.

A. Any octahedral site occupied by a carbon atom is distorted by elongation along one of the three axes and contraction along the other two. A carbon atom in an x site causes the unit cell to elongate along the [100] axis. Therefore, the application of a tensile stress along the x axis would favor the x sites as sites for carbon atoms; the distortion of the unit cell would then be similar to the elastic strain caused by the applied stress. The carbon atoms, initially deployed at random among the three sets of sites (to maximize entropy), would then diffuse into x sites. As this occurred, the crystal would continue to extend in the [100] direction until all of the carbon atoms had reached x sites. If the stress were removed, the process would reverse until the carbon distribution again became random. The result would be a time-dependent strain, as shown at right. The stress can be applied or removed quickly, but the attainment of the extra strain, or its decay to zero, would take time, since diffusion is required. This kind of strain is called *anelastic* strain, because it is not plastic and is not governed by Hooke's law.

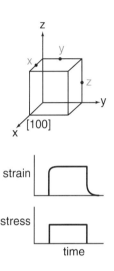

Example 7.4

Q. Lattice vacancies are equilibrium defects in crystals, meaning that, at a finite temperature, a certain fraction of lattice sites will be vacant, even though vacancies increase the internal energy of the crystal lattice. Why is this true for vacancies (and other point defects) but not for defects like dislocations or grain boundaries?

A. The entropy associated with point defects is very large, especially compared with that associated with line defects or planar defects. The reason is that the number of choices for the position of a vacancy is essentially equal to the number of atoms in the crystal. Because a very large number of atoms is associated with a dislocation or grain boundary, the choice of locations is much more limited. Therefore, the entropy effect associated with dislocations and grain boundaries is too low to offset the increase in enthalpy associated with them. Hence, the presence of dislocations or grain boundaries always raises the free energy of a material above its minimum value.

7.4 The Nature of Metal Fatigue

Fatigue is a common mode of failure in metallic materials that are subjected to repetitive loading. This is obviously the case with spokes, and fatigue is usually the life-limiting factor for spokes given high-mileage usage. For such usage, cyclists often select carbon steel spokes. The reason will be examined in some detail here, after it is established just what is meant by fatigue failure.

Anyone who has broken a piece of wire by bending it back and forth repeatedly has performed a rudimentary fatigue experiment. It is common experience that the greater the bending deflection on each cycle, the smaller the number of cycles needed to break the wire. It would take very many cycles, indeed, if the bending were so slight as to be almost entirely elastic. Such is the case for most cases of fatigue of mechanical components.

Research has shown that fatigue failure occurs by the formation of a crack, usually at the surface of a component but sometimes at an internal defect, and the growth of this crack by a small increment during each loading cycle. The crack begins in a region of concentrated plastic flow, as in an intense slip band, which gradually creates a microscopic fissure where it meets the surface. This process can be accelerated by corrosion. The number of cycles needed for crack nucleation can be greatly reduced by the presence of stress concentrations, such as sudden variations in cross-section, deep scratches, corrosion pits, etc., as indicated schematically in Fig. 7.7.

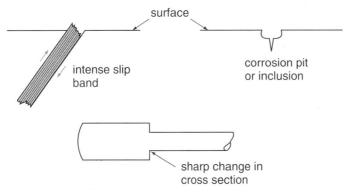

Fig. 7.7 Common sites of fatigue-crack initiation.

The most common mode of crack advance is the plastic opening, or blunting, which occurs on each loading cycle; this is illustrated by Fig. 7.8. This blunting simply increases the total surface area of the crack, and, when the crack closes during the unloading or compression part of the cycle, it is necessarily longer than before. All of this occurs on the microscopic scale, and it is usually not until late in the specimen or component life that the crack is large enough to detect visually. Since inspection procedures are seldom good enough to reveal growing fatigue cracks, there are many sudden, surprising failures of mechanical components annually. These occur when cracks become so large that the applied load is sufficient to cause rapid propagation during the final loading cycle.

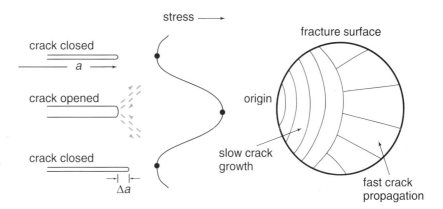

Fig. 7.8 Fatigue crack of length *a* grows by a plastic-blunting mechanism.

Since bicycle components are loaded cyclically, they are subject to fatigue failure. One common site of such failure is the connection of the pedal to the crank that turns the chainwheel. The pedal rotates on a stem which is normally screwed into the end of the crank. When the rider drives the pedal downward with his foot, the stem is loaded like a cantilever beam, and the maximum stress is found at the juncture of the stem and the crank. The stress is further concentrated here by the notch provided by the last screw thread, as shown schematically in Fig. 7.9.

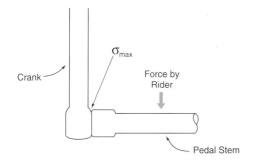

Fig. 7.9 Location of maximum tensile stress due to bending of the pedal stem under the downward force of the rider's foot.

An analogous configuration can be found in the oarlock of a racing shell. Here, every rowing stroke loads the pin on which the oarlock rotates in the same manner as a pedal stem. An example of a fatigue failure of such a stem is shown in Fig. 7.10.

Fig. 7.10 Fracture surface of an oarlock pin that failed in fatigue.

The fatigue crack in Fig. 7.10 started in a thread root at the point indicated by the arrow, and propagated initially in microscopic steps. The crack front had the classical thumbnail shape, and the slight rubbing of the two faces of the crack produced the smooth appearance. Variations in the oxide film (owing to variations in temperature and relative humidity) provide a record of the crack front shape at several locations. As the crack progressed, the steps (one per rowing stroke) grew larger, and the fracture surface became more ragged. When the fatigue crack had progressed slightly more than half-way through the pin, the stem snapped off during the final stroke.

This kind of failure can be dangerous, because it usually comes without warning. Cracks are often obscured by other parts and are, therefore, difficult to inspect. This is the case with pedal stems and oarlock pins. With the sudden fracture of a heavily loaded part, injuries are likely to occur. Bicycle riders have been severely injured by falls after failure of a pedal stem. This kind of problem is often the fault of a manufacturer, on the grounds that the stress analysis during the design of the component was inadequate (i.e., not up to the state of the art). However, users of devices with heavily stressed components that are cyclically loaded can help protect themselves by being aware of the possibility of fatigue and by periodically inspecting, or even replacing, critical components.

7.5 The S-N Curve

The study of fatigue behavior was first systematized by A. Wöhler, a German railway engineer, in the 19th century. He showed that the fatigue life of any component could be characterized by plotting the number of cycles to failure as a function of the maximum tensile stress per cycle, assuming all cycles to be identical. (The latter is usually true only in the laboratory, but that is the source of most fatigue data.) Common lab tests include a rotating cylindrical beam subjected to a side-wise force, which produces a sinusoidal variation in tensile stress at any point on the surface, as shown in Fig. 7.11.

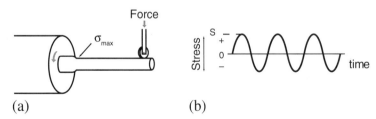

Fig. 7.11 (a) Schematic representation of the rotating-beam fatigue test. (b) Sinusoidal variation of stress with time at some point on the surface, when the rotation rate is constant.

The Wöhler curve, or S-N curve, conventionally shown with the independent variable, stress (S), along the ordinate (as in Fig. 7.12), has been a common basis for design against fatigue for many decades. One can test a given material in a particular condition of heat treatment, grain size, yield strength, etc. and determine the S-N curve. The intended life, or "design life," of a particular component, in terms of the number of cycles one can anticipate before the component is replaced or discarded, is then established (often assumed) by the design engineer. From the S-N curve, the maximum allowable tensile stress can then be specified, under the assumption that all loading cycles will be identical.

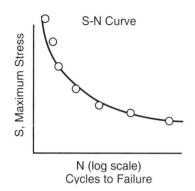

Fig. 7.12 The S-N, or Wöhler, curve; this is the most common representation of fatigue behavior.

The initiation and growth of a fatigue crack involves localized plastic deformation. Therefore, increasing the resistance of a material to plastic deformation, i.e., strengthening it, has the effect of raising the S-N curve to higher levels of stress. That is, it would require more cycles at a given stress to produce fatigue failure. Conversely, anything that reduces the yield strength would also reduce the fatigue resistance. This is particularly true for a reduction in the strength of surface regions, since that is where fatigue cracks typically start. In fact, surface regions are often specifically strengthened to improve fatigue resistance. A common way of doing this is "shot peening" which involves blasting a surface with small steel balls to strengthen it by strain hardening.

It should be noted that a large number of supposedly identical specimens run at the same maximum stress will usually show a distribution of fatigue life scattered about a mean value, as shown schematically in Fig. 7.13. In principle, the mean value should be used for the S-N curve shown in Fig. 7.12. This means that the earliest fatigue failures will occur at a life somewhat shorter than indicated by the curve.

Fig. 7.13 There will be a distribution of fatigue lives at a given stress level; hence, an S-N curve should show a scatter band.

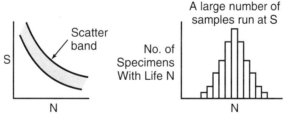

The problems encountered in designing against fatigue failure on the basis of S-N data are as follows:

1. The scatter in the fatigue life for a given stress (cf. Fig. 7.13) can be large. This may force the designer to choose between an unreasonably low design stress (e.g., an unreasonably large and heavy component) or some finite probability of fatigue during the life of the component. The remedy for this problem is to control the manufacturing process so that microstructure and properties vary as little as possible from one specimen to the next. This reduces the scatter of the S-N data and makes it easier to design for almost-zero failures.

2. It is often unreasonable to expect that all loading cycles will be identical. In this case fatigue data should be obtained under conditions of varying maximum stress to mimic the expected service conditions. This kind of testing can be complicated and expensive and carries with it a set of assumptions that may be inadequate. However, in some critical applications, like airplane wings, for example, there is no alternative.

3. In components of varying cross section, like a shaft with varying diameter or with screw threads or other stress concentrations (cf. Fig. 7.7), a careful stress analysis is required to estimate the stress-concentration factor at each discontinuity. It must be emphasized that the maximum tensile stress on the component, i.e., at the most severe stress concentration, is the stress that must be used in the S-N diagram to estimate the number of cycles to failure.

4. The effects of corrosion, superimposed on those of cyclic loading, are often extremely difficult to predict. Even if relevant laboratory data are available, the service conditions could vary unpredictably.

All of these problems necessitate periodic inspection of critical components or replacement of components that cannot be adequately inspected. Applying this to the bicycle would mean replacement of spokes after a certain number of miles ridden and periodic removal of the pedals for inspection of the roots of the screw threads where they attach to the cranks.

7.5.1 Analysis of the Failure of a Stainless Steel Spoke

A road bike with about 9000 miles on its wheels since the spokes were replaced was ridden over a bump in the pavement, and one of the spokes broke with a loud snapping noise. Inspection showed that the fracture took place at a thread root inside the nipple. Fig. 7.14(a) shows the nipple after it had been sectioned longitudinally to reveal the remaining piece of the spoke, and the fracture surface of the mating piece is shown in Fig. 7.14(b). The failure was obviously caused by a fatigue crack that began at the point indicated by the arrow and propagated more than halfway through the cross section before rapid failure occurred. On the fatigue portion of the fracture surface (lighter portion), coarse "thumbnail" marks denoting locations of crack arrest can be seen.

The fatigue portion was examined at higher magnifications in a scanning electron microscope (cf. Section 14.4.1), and the point of initiation of the fatigue crack is shown in Fig. 7.14(c). Crack-arrest striations on a finer scale can be seen here. Finally, at still higher magnification (Fig. 7.14d), striations demarking crack arrest with a spacing of 2 to 3 micrometers can be seen. The latter striations probably mark the progress of the fatigue crack during each revolution of the wheel. The larger striations probably correspond to road bumps of various magnitudes. The larger the bump, the greater the crack opening and the more pronounced the striation. The presence of the thumbnail pattern and the striations at high magnification are the classic hallmarks of a fatigue failure.

SECTION 7.5.1 ANALYSIS OF THE FAILURE OF A STAINLESS STEEL SPOKE

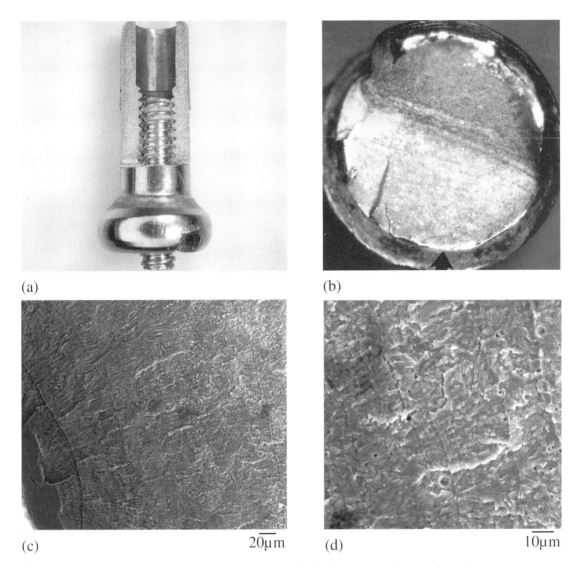

Fig. 7.14 (a) Sectioned nipple containing the end of a fractured stainless steel spoke. (b) The fracture surface of the mating piece; the arrow denotes the origin of the fatigue crack; the lighter portion is that over which the fatigue crack grew, and the darker portion is that which failed rapidly at the final bump in the road. (c) The origin of the fatigue crack at high magnification, showing crack-arrest striations. (d) Part of (c) at still higher magnification, showing striations with a 2-to-3 micrometer spacing. (Contributed by Cliff Warner, Univ. of Penn.)

7.6 The Fatigue Limit of Carbon Steel

There are two general classes of S-N behavior. In one class, which includes carbon steel, there is a threshold value of S below which the fatigue life becomes essentially infinite. This value is called the *fatigue limit*. This is shown schematically in Fig. 7.15. Thus, in principle, one can design against fatigue in a material with a fatigue limit simply by keeping the maximum stress below the fatigue limit. In the other class of behavior, which includes austenitic stainless steels, there is no threshold of any practical significance, which means that fatigue failure is an ultimate certainty once the requisite number of cycles is reached. The question to be addressed is: what is the reason for the fatigue limit in ferritic steels? It has to do with the interaction between carbon atoms and dislocations.

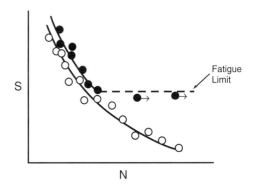

Fig. 7.15 Schematic illustration of S-N curves for materials with and without a fatigue limit.

It was explained in Section 7.2 that carbon atoms in solution in BCC metals occupy octahedral sites around which the lattice is distorted locally into a tetrahedral symmetry. Since this local shape change involves shear strains in addition to the dilatational strains associated with the interstitial solute, the carbon atoms interact with all dislocations in the BCC lattice. The attraction between carbon atoms and dislocations in BCC iron is strong, and because diffusion of this interstitial atom is rapid at room temperature, carbon can migrate readily to dislocations. This causes the dislocations to become pinned in place, so that they do not glide when a shear stress is applied.

It has been shown by experiment that the fatigue limit can be removed from a low-carbon steel by a heat treatment in hydrogen, which removes carbon. (The carbon diffuses to the surface and reacts with the hydrogen to form methane, CH_4.) It is believed that the carbon produces the fatigue limit by pinning dislocations; this can prevent concentrated slip bands from forming, or it can stop the development of early-stage fatigue cracks. The pinning is present before the steel is cyclically loaded, and more pinning may occur by the diffusion of carbon to new dislocations formed after cyclic loading has started. Above the fatigue limit the stress is high enough to produce fresh (i.e., not pinned) dislocations and to cause them to multiply at a rate too rapid for them to be pinned by diffusion of carbon during the cyclic loading.

S < Fatigue Limit

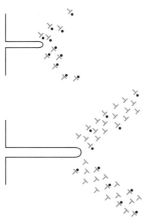

S > Fatigue Limit

The fatigue behavior of carbon steel *vs.* stainless steel spokes is shown in Fig. 7.16. These results are for segments of spokes loaded in cantilever bending. This is not the same as in-service loading, but it permits a comparison of the material behavior. It can be seen that, for a given maximum load, the life of the carbon steel exceeded that of the stainless steel by a substantial amount. This difference increased as the load was reduced. At the lowest stress employed, the carbon steel appears to be approaching its fatigue limit; the testing could not be carried out long enough to establish what that limit is.

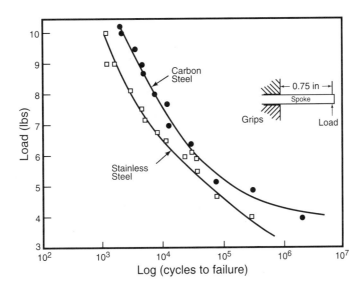

Fig. 7.16 Fatigue behavior of carbon steel and stainless steel spokes loaded in repeated cantilever bending.

The importance of a fatigue limit for design purposes is that, as long as the maximum cyclic stress is kept below this value, the life should be essentially infinite with respect to fatigue. Of course, any surface damage to a spoke that produces a stress concentrator, such as a deep scratch or a corrosion pit, can cause the local stress to exceed the fatigue limit. Thus, even with carbon steel, spokes must be inspected periodically for surface damage. Fatigue failures are to be avoided, especially in the front wheel, since the rider may be pitched over the handlebars if a broken spoke falls across the front fork and jams the wheel.

Summary

Carbon-steel spokes are 95% BCC iron, or ferrite, and as such their deformation behavior differs markedly from that of the FCC stainless steel spokes. Carbon-steel spokes would become brittle at very low temperatures because of the inherent low mobility of screw dislocations in the BCC lattice; they are ductile at room temperature owing to the presence of sufficient thermal energy to "smear out" the complicated core structure of the screw dislocations, which enables them to glide more readily. In addition, interstitial solutes occupy octahedral sites in the BCC lattice, and these are not spherically symmetrical. Therefore, the lattice is locally distorted into a BCT, or body-centered tetragonal, symmetry around the interstitial atom, such as with carbon in ferrite. The resulting stress field around the interstitial atom contains shear-stress terms, as well as the dilational terms found in the FCC lattice. The presence of the shear-stress terms means that interstitial atoms in a BCC lattice would interact with all dislocations, including pure screws. This effect leads to dislocation pinning and, therefore, to the fatigue limit observed in carbon steel at low cyclic stresses.

The process of fatigue was shown to be one of nucleation of cracks, usually at surfaces, as a result of plastic deformation, even on a very small and localized scale, and the growth of these cracks by way of the plastic blunting mechanism. This growth can be arrested at low cyclic stresses by the diffusion of carbon atoms to the plastic zone at the tips of small cracks; this produces the fatigue limit. Interstitial diffusion was shown to be much faster than substitutional diffusion, owing to the fact that vacant lattice sites are not necessary for the former.

Appendix 7.1 Brittle Fracture in Carbon Steel

Many metals and alloys undergo a ductile-to-brittle transition when cooled to relatively low temperatures. The underlying reason for this, in general, is that certain kinds of dislocation motion become difficult at low temperatures. In iron and other BCC metals, screw dislocation mobility decreases at low temperatures. This has to do with atomic displacements in the cores of screw dislocations in BCC metals, as discussed in Section 7.1. The result is a steeply rising yield stress with decreasing temperature and a corresponding decrease in ductility; the behavior of BCC iron is contrasted with that of FCC copper in Fig. A7.1.1.

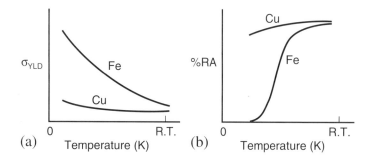

Fig. A7.1.1 Schematic representation of variation of (a) yield stress and (b) ductility with temperature in BCC iron and FCC copper.

The high yield stress at low temperatures in BCC metals makes it possible to attain levels of tensile stress at which the cohesive forces between atoms can be overcome, so that brittle fracture can occur. In a typical case, a few dislocations move and become blocked by some strong obstacle, such as a hard particle. This concentrates stress at the particle, and a microcrack can form, starting with cracking of the particle. This microcrack can propagate through the metal if the applied stress is high enough and if plastic flow at the crack tip is difficult; both of these conditions apply in iron, and therefore carbon steel, at low temperatures. In iron and carbon steels the brittle fracture usually occurs by cleavage along {100} planes, giving a fracture appearance as shown in Fig. A7.1.2.

APPENDIX 7.1 BRITTLE FRACTURE IN CARBON STEEL

The tendency of carbon steel to behave in a brittle manner at low temperatures has been responsible for many catastrophic failures, even the fracture of entire ships. A well known example is that of the S.S. Schenectady, which split in two shortly after outfitting. As shown in Fig. A7.1.3, it fractured on a cold day while sitting alongside a pier in quiet water.

Fig. A7.1.2 An example of cleavage fracture in iron. The {100} cleavage cracks must change planes at every grain boundary. (This makes fine-grained steels tougher than coarse-grained steels.)

Fig. A7.1.3 The S.S. Schenectady after it fractured in half on a cold winter day while tied up at a fitting-out pier in calm weather. This is an example of brittle fracture of carbon steel presumably triggered by residual stresses from construction. (Final report of a board of investigation convened by order of the Secretary of the Navy to inquire into the design and methods of construction of welded steel merchant vessels, U.S. Gov't Printing Office, 1947, Fig. 11.)

Appendix 7.2 Crystal Symmetry

Most metals and alloys crystallize in simple structures, because the constituent atoms are the same or nearly the same, in size, and the bonding is essentially metallic: Positive ions share a delocalized pool of "free" electrons. Thus, the constraints of ionic compounds, in which negative ions must surround positive ions, and covalent solids, in which bond lengths and bond angles are important, are absent. The most common metallic structure is cubic, which has the highest symmetry.

For example, a cubic crystal has three 4-fold axes of rotational symmetry, which are the three cube axes. That is, one can rotate a cube by 90°, 180°, 270°, or 360° around any <100> axis and find that the atom positions are indistinguishable before and after the rotation. By the same token a <111> direction is a 3-fold rotation axis, and a <110> direction is a 2-fold rotation axis, as shown in Fig. A7.2.1.

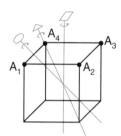

Fig. A7.2.1 Axes of rotational symmetry in a cubic crystal: □4-fold axis: A_1 becomes A_2; △3-fold axis: A_1 becomes A_3; ○2-fold axis: A_1 becomes A_4. (From B. D. Cullity, *Elements of X-Ray Diffraction*, Addison-Wesley, Reading, MA, 1956, p.34.)

These statements are true for both FCC and BCC crystals, which correspond to two of the fourteen Bravais lattices. Figure A7.2.2 shows the unit cells of the lattices representing the fourteen possible ways of arranging points in space in a "space lattice," in which the unit cell is repeated in three dimensions. There are no others!

When a cubic unit cell is distorted along one axis, the result is a tetragonal lattice. It has lower symmetry than a cubic lattice: one 4-fold and two 2-fold axes of rotation.

Crystal symmetry has a profound effect on physical properties. For example, cubic crystals are isotropic with regard to heat flow, thermal expansion, and diffusion. All others are anisotropic: e.g., the rates of heat and mass flow vary with the crystallographic direction. In cubic crystals only three elastic constants are needed to describe all possible stress-strain relationships. In the triclinic lattice, which has the least symmetry, twenty-one are needed.

APPENDIX 7.2 CRYSTAL SYMMETRY

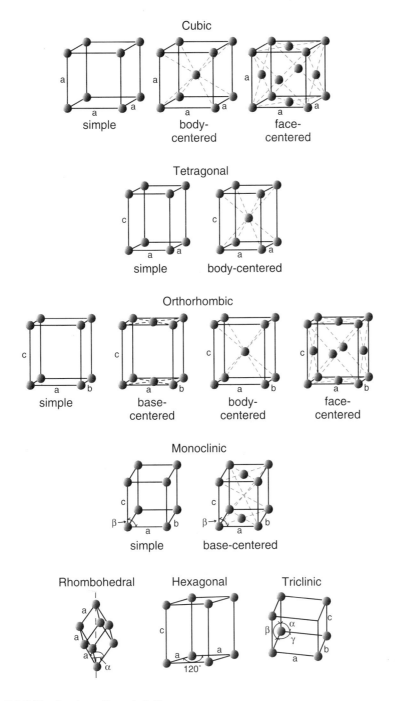

Fig. A 7.2.2 The fourteen Bravais lattices.

Exercises - Chapter 7

Terms to Understand

Ferrite	Self-diffusion
Cementite	Fatigue crack nucleation
Pearlite	Fatigue crack growth
Austenite	S-N diagram
Solute segregation	Fatigue limit
Octahedral site in BCC	Ductile-brittle transition
Tetragonal distortion	Cleavage fracture
Substitutional solute diffusion	Crystal symmetry
Interstitial solute diffusion	Shot peening

Problems

7.1 The lattice parameter (i.e., the unit-cell dimension) of BCC iron at room temperature is 2.87Å (0.287nm). Calculate the diameter of the largest interstitial atom that could fit into an octahedral site without distorting the lattice.

7.2 If the interstitial atom in Problem 7.1 could take the shape of an oblate spheroid, calculate the major and minor diameters of such a spheroid that would just fill the octahedral site in iron without distorting the lattice.

7.3 The diameter of a carbon atom in diamond or graphite is about 1.5Å. Assuming this value for a carbon atom in iron, calculate the two kinds of displacement of iron atoms that would occur when a hard-sphere carbon atom occupies an octahedral site such that all six iron atoms touch the carbon atom, which is assumed to be spherical.

7.4 Fig. 7.3 shows an interstitial carbon atom located in the center of the face of a BCC unit cell. Show that the position at the center of an edge of the unit cell is crystallographically equivalent. That is, the surrounding atoms are at the same distances and in equivalent directions.

7.5 Where in a bicycle spoke is a fatigue failure most likely to occur? Why?

7.6 For a 26-inch diameter bicycle wheel, the predicted fatigue life of a spoke is 500,000 cycles; how far can it travel before fatigue failure is likely?

7.7 Sketch two stress-strain curves, one for a ductile material and one for a brittle material. Label the yield stress, the UTS, and the elongation at fracture.

7.8 Although fatigue is a fracture process, the rate of fatigue-crack propagation depends on the resistance to plastic deformation. That is, as the yield stress is decreased, fatigue cracks grow farther per cycle. Considering the mechanism of fatigue crack growth, explain this.

8

ALLOYS FOR SOLDERING AND BRAZING; PHASE DIAGRAMS

8.1 Soldering and Brazing

Throughout most of the history of bicycles, frames have been made of steel tubes held together by brazed joints. Brazing differs from soldering only in the temperature range in which it is carried out. The term soldering is used for joints made below about 450°C and brazing for those made above that temperature. In both processes the parts to be joined remain solid, and only the filler metal is melted. The molten filler metal is drawn into the crevice of the joint by capillarity before it solidifies. (Capillarity is a phenomenon that depends on differences in interfacial energies; it is why liquids are absorbed by paper tissues and is described in more detail below.) In household plumbing systems made with copper tubing, joints are soldered with a Pb-Sn alloy so-called "soft solder." The joints are made with copper fittings that form a sleeve around the tubing and provide a large amount of joint surface along which the molten solder can be drawn, as shown in Fig. 8.1.

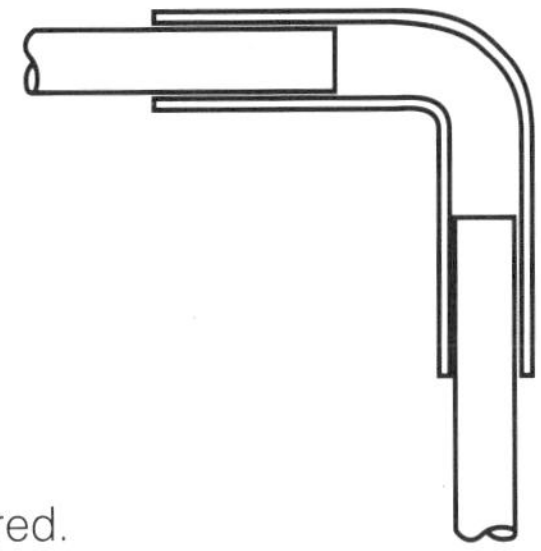

Fig. 8.1 Right-angle connection in a plumbing line to be soldered.

Capillary action can draw solder up into the joint if the solder "wets," or spreads out on, the surface of the copper. Figure 8.2 shows schematically two extremes of wetting and non-wetting. The extent of wetting can be characterized by the contact angle, θ. When wetting is nearly perfect, θ approaches zero, and the liquid spreads out on the solid, as is needed in soldering or brazing.

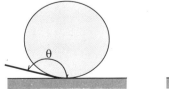 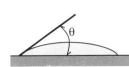

Fig. 8.2 A liquid droplet on a solid plate, showing (a) a small degree of wetting; the contact angle θ is large. (b) A large degree of wetting.

The value of θ depends on the energies of the various surfaces or interfaces involved. Atoms at the interface between two phases are in a higher energy state than atoms in the interior of either phase, because they are partly surrounded by the "wrong" structure and/or composition. The energy of an interface depends on the degree of "wrongness." For example, a metal/metal interface would have a lower energy than a metal/metal-oxide interface, because in the latter case one of the phases would have non-metallic bonding. A liquid-metal/solid-metal interface would be higher in energy than one between two solid metals because of the greater structural dissimilarity in the former.

An equilibrium of forces, similar to that discussed in Section 6.5, must exist along the triple junction where the three phases meet: the solid, liquid, and vapor (air). This equilibrium is depicted in Fig. 8.3.*

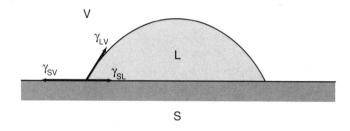

Fig. 8.3 Schematic representation of equilibrium of surface tensions at a solid-liquid-vapor triple junction.

The horizontal equilibrium condition is given by:

$$\gamma_{SV} = \gamma_{SL} + \gamma_{LV} \cos \theta$$

Hence, θ is given by:

$$\cos \theta = (\gamma_{SV} - \gamma_{SL})/\gamma_{LV}$$

For good wetting, θ should approach zero; therefore, cos θ should be as large as possible. That means it is necessary to maximize γ_{SV} and minimize γ_{SL}. (Normally, one cannot do much about γ_{LV} in soldering.**) It can be seen from Fig. 8.3 that γ_{SV} is the force tending to spread the liquid, and γ_{SL} is the force tending to pull it into a ball. In other words, wetting involves the substitution of a low-energy (solid/liquid) interface for a high-energy (solid/vapor) interface.

In order to achieve wetting, it is necessary to raise the energy of the solid surface. Oxidation of a copper surface reduces its γ_{SV}. Therefore, to raise the surface energy, it is necessary to remove the oxide and expose bare metal.

---

* In this figure only the horizontal components of the surface tension are considered; the vertical components are irrelevant, and to deal with the equilibrium of these components would require one to know how the solid is distorted on the atomic scale at the triple junction.

** This is not the case with water. Here one can use a wetting agent, like a soap, which lowers γ_{LV} and thereby promotes spreading of the water on a solid. Sometimes one wants to retard wetting; waxing an automobile reduces γ_{SV} and increases γ_{SL}, thereby causing water droplets to "ball-up," like solder on an untreated copper surface.

An interface between the copper oxide and molten solder would have a high energy (γ_{SL}), because the liquid alloy and the oxide are so different chemically. This is the opposite of what is desired for wetting, but this problem is rectified when the oxide is removed. For copper plumbing, one applies hydrochloric acid to the tube and fitting in order to remove most of the oxide. The rest of the oxide is removed by applying a flux, which, when heated by a torch, can entrain the residual oxide in a liquid. This liquid does not mix with the liquid solder, and, being lighter in mass, it is swept away ahead of the solder as it spreads over the clean metal surface. The metal/metal interface formed when a liquid metal spreads on a clean solid metal has a low energy, owing to the ease of metallic-bond formation. Thus, the term ($\gamma_{SV} - \gamma_{SL}$) is maximized by cleaning the oxide off the copper and allowing the molten solder to spread on the cleaned surface.

The other factor in making a successful joint is the clearance between the two metal pieces. The requirement for a small clearance can be explained by a simple exercise in physics, as in Fig. 8.4. At equilibrium the upward capillary force, $2\gamma_{NET} \times L$, must be balanced by the downward force due to gravity, mg, where m is the mass of the liquid in the column, and g is the gravitational constant. Since m is given by density, ρ, times the volume of liquid in the column, thL, the force balance can be written:

$$\rho t h L g = 2 \gamma_{NET} L$$

or

$$h = 2 \gamma_{NET} / \rho g t$$

That is, for a given ρ and γ_{NET}, h increases as t decreases. Therefore, one wants the spacing as small as possible while still allowing the liquid to flow into the joint.

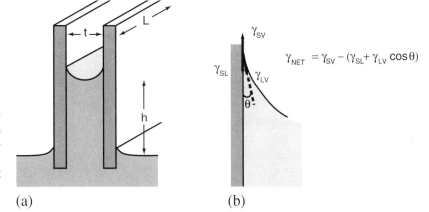

Fig. 8.4 (a) Two parallel solid plates, spaced t apart, are dipped in a liquid, which rises under the influence of the capillary force, γ_{NET}, given in (b), so that t rises to a height above the surrounding liquid.

8.2 Low-Melting Alloys for Soldering; Phase Diagrams

The most obvious characteristic one would desire for a soldering or brazing alloy is that it melt at the lowest possible temperature, consistent with the required mechanical properties of the joint. The reason for using an alloy in the first place, instead of a pure metal, is that the melting point of a substance can often be lowered by mixing in a second substance. This is employed during winter in the use of road salts, which lower the freezing (melting) point of water.

8.2.1 Lowering of the Melting Temperature; Salt-Water Mixtures

A very useful diagram that expresses the lowering of melting points is the *phase diagram*. Parts of the phase diagrams of the salt-water mixtures for the two common road salts are shown in Fig. 8.5. Both these diagrams exhibit what is known as a eutectic reaction, in which a liquid phase freezes at a single temperature to a mixture of two solid phases. This takes place at the *eutectic point*, which is the minimum in the liquid phase field, shown in Fig. 8.5. ("Eutectic" is a Greek word meaning "easy melting.")

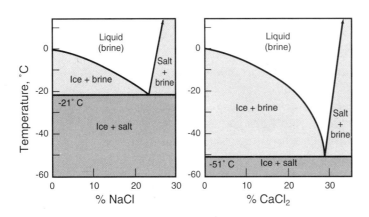

Fig. 8.5 Parts of the phase diagrams for the NaCl-water and $CaCl_2$-water systems.

It is easy to see why calcium chloride has replaced sodium chloride as the road salt of choice wherever its higher cost can be justified. When $CaCl_2$ is applied, slush (i.e., a solid-liquid mixture) would be stable down to a temperature of -60°F (-51°C). That is, the two-phase regions in the phase diagram extend down to that temperature. In contrast, the NaCl-water system becomes completely solid at -6°F (-21°C). No matter how much rocksalt is applied, the mixture would remain solid below that temperature.

The diagrams in Fig. 8.5 are examples of binary phase diagrams. That means they express the equilibrium configuration of two components at different temperatures. (The other relevant thermodynamic variable is pressure, which is taken to be fixed at one atmosphere in most phase diagrams.) The components may be either pure elements or stable compounds, as in Fig. 8.5. (By "stable" it is meant that the compounds do not break down into their constituent elements at the temperatures of interest.) Thus, phase diagrams are simply maps of the temperature and composition regions in which one, or more than one, phase is stable at equilibrium. In such a map, temperature is plotted on the vertical axis, and the composition, in terms of the amount of the second component, is plotted on the horizontal axis. The composition axis is used for two purposes:

1. To express the composition of the alloy, or mixture of components, as a whole. For example, the composition of the eutectic mixture of rocksalt and water is 23.3% NaCl and 76.7% water.

2. To express the composition of the phases present at any given temperature. For example, the NaCl brine that is in equilibrium with ice at -10°C has a composition of 15% NaCl, 85% water.

8.2.2 A Hypothetical Simple-Eutectic Phase Diagram

To illustrate how to "read" a phase diagram and use it to understand the microstructures of binary (i.e., two-component) mixtures, it is useful to consider the hypothetical system depicted in Fig. 8.6. This is a simple eutectic system with three phases: a liquid in which both components A and B are completely miscible, and two solid phases, denoted as α and β. The components could be either pure elements (e.g., metals) or stable compounds (e.g., oxides). The two solid phases are based on the two components. Each solid phase is a solid solution, and each has the crystal structure of the component upon which it is based. Because the discussion here is centered upon metallic alloys, it will be assumed that A and B are both pure metals; however, the principles to be discussed can be applied to any two-component system that forms a simple eutectic.

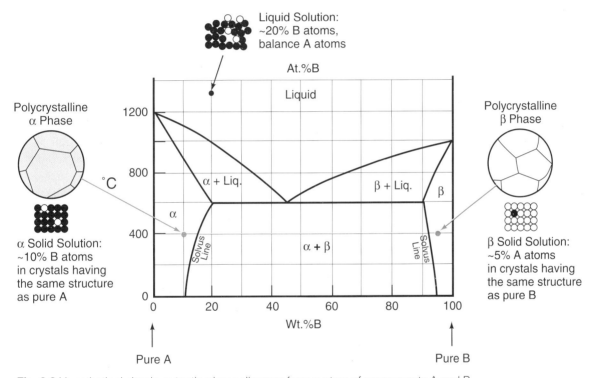

Fig. 8.6 Hypothetical simple-eutectic phase diagram for a system of components A and B.

Figure 8.6 shows that a certain amount of component B can be dissolved in component A to form an α-phase solid solution. Similarly, a β-phase solid solution can be formed by adding A atoms to pure B. This is common in metallic systems because of the flexibility of metallic bonding, in which positive ions are held together by a sea of negative charge comprising the delocalized valence electrons.

It is useful to contrast this with the case of the salt-water mixtures shown in Fig. 8.5. Here, the two solid phases are virtually equivalent to the two components; that is, neither salt nor ice exhibits any appreciable solid solubility for the other. The fact that chloride salts have negligible solubility in ice, and H_2O has negligible solubility in the salts, means simply that the crystal lattice of the salt cannot accommodate water molecules, and that of the ice cannot accomodate the ions of the salt. This should not be surprising. The ionic bonding in salt crystals operates by the electrostatic attraction between positive and negative ions, which are the result of electron transfer from the metal atom to the chlorine atom. In ice, there is no electron transfer, and, therefore, no charged particles. Ice is bonded by a special kind of secondary bond, called a *hydrogen bond*, between water molecules, which are themselves internally bonded by covalent bonding, or the sharing of electrons between hydrogen and oxygen atoms (cf. Appendix 13.1). Because the bonding of these solids is so different, they are incompatible in the solid state and exhibit essentially no mutual solid solubility.

This is in sharp contrast to the case of a metallic alloy, in which both components exhibit similar bonding. In such alloys, the solid solubility of one component for another depends on the size of the constituent atoms and on their degree of chemical similarity. If two metals are sufficiently dissimilar in the configuration of their valence electrons, then they may form *intermetallic compounds.* (See Fig. 8.21 for examples of such compounds in the Cu-Zn alloy system.) In the system depicted in Fig. 8.6, there are no such compounds, and the limitations on the solubility of B in α and A in β would then be mainly the result of the difference in the sizes of the A and B atoms.

It is apparent in Fig. 8.6 that the three single-phase fields are bounded within the diagram by two-phase fields. The boundary line between a one-phase and a two-phase field represents a solubility limit, or the locus of points at which the single phase becomes saturated with solute. For example, the boundary between the α- and α+β-phase fields (called a *solvus line*) represents the saturation of the α phase with component B. Thus, if a crystal of pure A were held at 400°C and B atoms were added to it, this crystal would become saturated if the concentration of B exceeded 15wt%, according to Fig. 8.6. Addition of more B atoms would result in the formation of the β phase. If the overall concentration of B atoms in the alloy were increased, the amount of β phase would increase, and the concentration of B atoms in the α phase would remain the same (i.e., it would be fixed at the saturation value for the temperature 400°C).

Exactly the same kind of argument with regard to the saturation of the β phase at 400°C could be made by starting with pure B at that temperature and continually adding A atoms. The β phase would become saturated with A atoms when the concentration exceeded 8wt%, according to Fig. 8.6. Continuing to add A atoms beyond the saturation point would simply increase the amount of α phase.

The concepts of saturation and of exceeding the solubility limit to form a two-phase mixture are familiar to anyone who cooks. For example, one can make syrup by dissolving sugar in water, and it is well known that more sugar can be dissolved as the water temperature is increased. Thus, the solubility limit of sugar in the "α" phase (i.e., the syrup) increases with temperature.

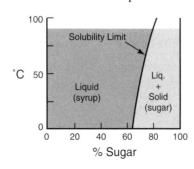

The solubility limit can be exceeded, i.e., the syrup can become supersaturated, by either adding too much sugar at any given temperature or by cooling a syrup of a given sugar concentration. In either case, the excess sugar separates as a second phase; that is, a two-phase region in the sugar-water phase diagram has been entered.

SECTION 8.2.2 A HYPOTHETICAL SIMPLE-EUTECTIC PHASE DIAGRAM

Returning to Fig. 8.6, shown again below as Fig. 8.7, and the discussion of varying the composition of an A-B alloy held at 400°C, it has been established that variations in the B content between 15 and 92wt% affect only the relative amounts of the α and β phases, and not their compositions. That is, for any alloy within this composition range, both phases are saturated with the respective solutes. It should be obvious that, when the alloy composition is just beyond 15% B, it is still almost all α, and, when it has reached, say, 90%B, it must be almost all β. The relative amount of each phase in any alloy containing between 15 and 92%B at equilibrium at 400°C can be found simply by noting the position of the alloy composition along the line in the two-phase region connecting the α and β solubility limits, 15%B and 92%B, respectively. This line is called a *tie line*. The change in the relative amounts of each phase along the tie line at 400°C is illustrated below. (Note that phase diagrams are determined by experimenting with alloys made by weighing out various mixtures of the components. Thus, they are normally expressed in concentrations given in weight percent. This can be converted to atom percent by use of the atomic weights of the components.)

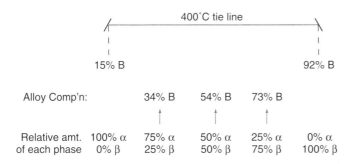

The microstructure, meaning the distribution of phases, of any alloy at any temperature can only be understood if one has reference to the relevant phase diagram. The method of doing this can be illustrated with Fig. 8.7. First, consider the eutectic alloy, A-45%B. This solidifies directly from the liquid state to a two-phase mixture of α and β, according to the phase diagram. The relative amount of each phase can be found by drawing a tie line just below the eutectic temperature, 600°C; i.e., in the two-phase region. The end points of the tie line are at 20 and 90% B, respectively. To make things clear, consider a 100g sample of the eutectic alloy at 599°C. The amount of β is obtained by calculating how far the alloy composition (45%B) is along the line from 20 to 90%. Thus,

$$\text{Amount of } \beta = \frac{45-20}{90-20} \text{ times } 100g = 35.7g$$

The amount of α must, therefore, be 64.3g. This can be checked by seeing if it is in agreement with the compositions of the alloy and of each phase. The alloy must have 45g of B in it, and

Amount of B in β phase = 0.90 x 35.7g = 32.1g,
Amount of B in α phase = 0.20 x 64.3g = 12.9g

137

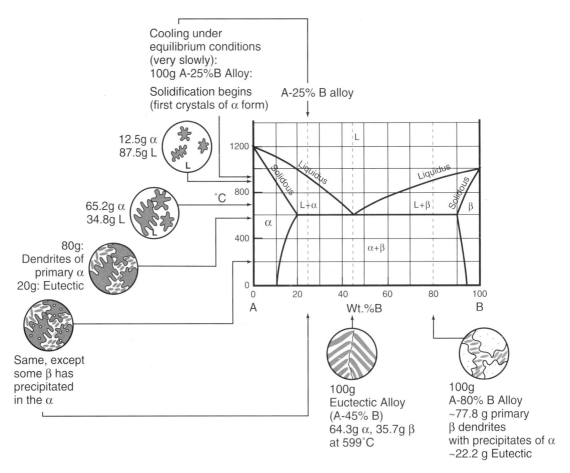

Fig. 8.7 Hypothetical A-B phase diagram, illustrating microstructures of various alloys.

Since the sum of the amounts of B in each phase equals 45g, the calculation is correct. This calculation follows the *lever rule*, so called because it is reminiscent of a lever balanced on a fulcrum (represented by the alloy composition):

```
_____25_____45_____
   |               |                |
   α             Alloy              β
 64.3g           100g             35.7g
```

That is,

$$\alpha \text{ fraction} = \frac{45}{25 + 45} = 45/70 = 0.643$$

$$\beta \text{ fraction} = \frac{25}{25 + 45} = 25/70 = 0.357$$

SECTION 8.2.2 A HYPOTHETICAL SIMPLE-EUTECTIC PHASE DIAGRAM

This procedure can be used to analyze the evolution of microstructure during the cooling of an alloy of A-25%B from the liquid state, as shown schematically in Fig. 8.7. Again, assume 100g of alloy. The alloy first enters the two-phase region when it cools below 920°C. Under equilibrium conditions (i.e., very slow cooling), crystals of α would nucleate in the liquid and would grow in the form of *dendrites,* meaning a branched structure. (Dendritic is from the Greek and means "tree-like.") The composition of this first-formed α would be 10%B, as given by drawing a tie line at 920°C.

The lever rule can be applied at lower temperatures, say 900 and 700°C, to follow the course of solidification.

At 900°C,

Composition of α = 11%B

Composition of liquid = 27%B

Amount of α = $\frac{27-25}{27-11}$ × 100g = 12.5g

Amount of liquid = 87.5g

At 700°C,

Composition of α = 17%B

Composition of liquid = 40%B

Amount of α = $\frac{40-25}{40-17}$ × 100g = 65.2g

Amount of liquid = 34.8g

It is apparent that the concentration of B in both the solid and liquid increases as solidification proceeds. This does not violate the conservation of matter, because the solid remains much lower in B than the initial liquid (i.e., the alloy as a whole). The solid and liquid compositions are seen to follow the lateral boundaries of the two-phase region, denoted as the *liquidus* and *solidus* lines, respectively, in Fig. 8.7. When the eutectic temperature, 600°C, is reached, the amount of liquid remaining has fallen to

$$\frac{25-20}{45-20}(100) = 20g$$

and the composition of this liquid has reached 45%B, which is the eutectic composition. This eutectic liquid at 600°C must then undergo the eutectic transformation into the two solid phases, α and β, simultaneously. Afterwards, the microstructure comprises 80g of primary α dendrites, formed above 600°C, and 20g of the eutectic mixture.

The total amount of α in the A-25%B alloy at this point includes the 80g of primary α and the α in the eutectic. The latter was previously calculated to be 64.3% of the total eutectic, which would mean that there is 0.643 × 20g = 12.9g of α in the eutectic, for a total of 92.9g of α in the alloy.

This calculation can be checked using a tie line in the two-phase region just below the eutectic temperature:

$$\text{Amount of } \alpha \text{ in the alloy} = \frac{90-25}{90-20} \times 100g = 92.9g$$

Because the solvus lines slope outward below the eutectic temperature, the relative amounts of α and β at equilibrium must change as the temperature is further reduced. The primary α, which contained 20%B at the eutectic temperature, can hold only 12%B at 200°C, for example. Therefore, some β phase must precipitate in the dendrites of primary α, indicated schematically in Fig. 8.7.

The eutectic must also adjust itself somewhat. A eutectic alloy at 200°C would comprise

$$\text{Amount of } \alpha = \frac{94\text{-}45}{94\text{-}12} = 60\%$$

$$\text{Amount of } \beta = 40\%$$

This is in contrast with the 64.3%α, 35.7%β in the eutectic just below the eutectic temperature. Thus, the β regions in the eutectic mixture grow a bit at the expense of the neighboring α regions. This is accomplished by diffusion of A and B atoms so as to adjust the composition and amount of each phase.

So far the discussion has been concerned with a eutectic alloy and a hypoeutectic alloy, "hypo" meaning "less than." It remains only to consider a typical hypereutectic alloy, and the microstructure of one is shown schematically in Fig. 8.7. This A-80%B (or B-20%A) alloy would have

$$\frac{80\text{-}45}{90\text{-}45} = 77.8\% \text{ primary } \beta, \text{ and } 22.2\% \text{ eutectic}$$

just below the eutectic temperature. Upon further cooling, some α would precipitate in the primary β, and the compositions and amounts of α and β in the eutectic would adjust themselves, as in the case of the hypoeutectic alloy.

The principles learned by consideration of the hypothetical eutectic alloy can now be put to use on a common alloy system used for soldering.

8.2.3 The Lead-Tin System (Soft Solder)

Alloys of lead and tin are used as solder for electronic equipment, plumbing, roofing, and other applications. (These alloys are not strong enough for holding the tubes of bicycle frames together; alloys for this purpose will be considered later.) The phase diagram for this system is shown in Fig. 8.8.

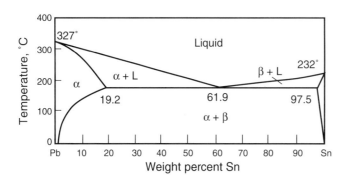

Fig. 8.8 The Pb-Sn phase diagram (From *Metals Handbook,* 8th ed., ASM, 1973, p. 330).

As in Figs. 8.6 and 8.7, this is a simple eutectic system; the addition of tin to lead lowers the melting temperature, as does the addition of lead to tin. The eutectic point is at 61.9% tin. This composition would be selected for soldering electronic equipment or copper tubing, because it requires the minimum temperature for melting, and it freezes at a constant temperature. The latter fact is illustrated schematically in the cooling curves shown in Fig. 8.9. (A cooling curve is determined by placing a thermocouple in a molten alloy and using it, in conjunction with a voltage-measuring device and a strip-chart recorder (cf. Section 3.6), to record the temperature *vs.* time as the alloy is allowed to cool to room temperature.)

SECTION 8.2.3 THE LEAD-TIN SYSTEM (SOFT SOLDER)

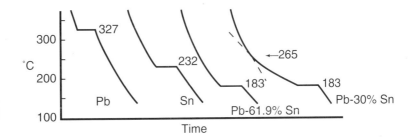

Fig. 8.9 A schematic set of cooling curves for the indicated compositions

Three of the cooling curves in Fig. 8.9 have a similar shape. Each has a horizontal portion, above which one finds a smoothly varying cooling curve of a liquid, and below which is a smoothly varying curve for a solid material. This is the kind of cooling curve found for any pure substance, like water for example.

The solid-liquid equilibrium for water occurs at a fixed temperature: 0°C. When water is cooled to 0°C, even though heat continues to flow out, the temperature must remain constant from the time that ice begins to form until all of the liquid is gone. The cooling curves for lead and tin both follow this behavior, as does the one for the eutectic alloy.

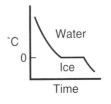

Pb-61.9% Sn

The microstructure of the solidified eutectic alloy is shown in Fig. 8.10. It consists of roughly plate-like, or lamellar, regions of the (dark-etching) Pb-rich phase, here called α, in a matrix of the light-etching, tin-rich β phase. This alloy was 100% liquid at just above 183°C. After the temperature of the liquid fell to 183°C, it remained constant during the liquid-to-solid transformation, even though heat continued to flow out of the container which held the alloy. The reason why the temperature did not fall during freezing is that the release of the latent heat of fusion was just sufficient to balance the heat loss from the container. The thermodynamic basis for this special behavior of pure elements and eutectic alloys will be discussed later, but it should be obvious why this freezing at constant temperature is ideal for electronics and copper tubing. It is necessary for the parts being soldered to be held in a fixed position until the solder is completely solidified, and one would like this to be finished as quickly as possible.

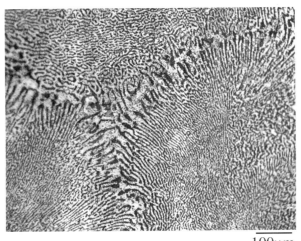

Fig. 8.10 Microstructure of the solidified eutectic alloy Pb-61.9%Sn comprising lamellae of α in a matrix of β.

100μm

Pb-30% Sn

A different kind of behavior is desired for solders used for roofing or filling the large joints in cast-iron waste pipes. Here, an extended range of freezing is useful, because it is necessary to work the solder into large gaps with hand tools, instead of having it flow in by capillary action. The cooling curve for the Pb-30%Sn alloy exhibits this desired behavior. It shows a decrease in slope at about 265°C, which corresponds to the release of the heat of fusion when the first solid begins to form. (This is, in fact, how the point on the liquidus line was determined experimentally.) Thus, the 30% tin alloy starts to freeze at 265°C, but it remains "mushy" down to 183°C, when the remaining liquid freezes at the eutectic temperature.

The cooling behavior and the resulting microstructure of the 30%Sn alloy is correlated with the phase diagram in Fig. 8.11. At 300°C the alloy is completely molten. As noted earlier, solidification starts at 265°C. Since the system has now entered a two-phase region, a tie line can be drawn at 265°C; this reveals that the composition of the first solid (α) that forms has about 10% Sn. When the temperature has fallen to 250°C, the amount of α has increased and the amount of liquid has decreased. The tie line at this temperature shows that the α should now contain about 12% Sn, and the liquid should contain about 37% tin. This means that, as solidification proceeds, the tin is being partitioned between the solid and liquid. That is, the freezing solid has much less tin than the original liquid, and the liquid becomes enriched in tin as the solid forms.

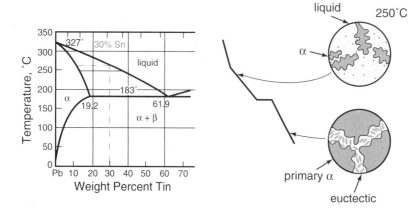

Fig. 8.11 The correlation of the phase diagram, the cooling curve, and the microstructures of the Pb-30%Sn alloy during cooling from the melt.

The relative amounts of solid and liquid in equilibrium at 250°C can be found by applying the lever rule:

$$\text{Relative amount of } \alpha \text{ at } 250°C = \frac{37-30}{37-12} = 28\%$$

$$\text{Relative amount of liquid at } 250°C = \frac{30-12}{37-12} = 72\%$$

When the temperature has fallen to 200°C, the relative amount of α has increased to $(58-30)/(58-18) = 70\%$, because the α now contains 18% Sn and the liquid contains 58% Sn. Finally, when the temperature has reached 183°C, the composition of the liquid has reached 61.9%, which is the eutectic composition. This liquid must now solidify at a constant temperature into the two-phase solid mixture (cf. Fig. 8.10). After this has occurred, the microstructure can be described as having two *constituents*. One is *primary* α, or the α which solidified before the eutectic reaction, and the other is the two-phase eutectic. The lever rule is used to find the relative amounts of each as follows:

$$\text{Relative amount of primary } \alpha = \frac{61.9-30}{61.9-19.2} = 75\%$$

$$\text{Relative amount of eutectic constituent} = \frac{30-19.2}{61.9-19.2} = 25\%$$

The relative amount of α in the eutectic constituent can be found by treating the eutectic as a separate alloy having an overall composition of 61.9% Sn, as follows:

$$\text{Relative amount of } \alpha \text{ in the eutectic} = \frac{97.5-61.9}{97.5-19.2} = 45\%$$

The other 55% of the eutectic is, of course, β.

By making a large enough set of Pb-Sn alloys, one can use the method of cooling curves to trace out the complete liquidus curves for both hypoeutectic and hypereutectic alloys. It would also be found that, if the cooling rate is slow enough to maintain equilibrium conditions at each temperature, the eutectic reaction, signalled by the constant-temperature transformation at 183°C, is found only in the range 19.2% to 97.5%Sn. Therefore, a horizontal line is drawn between these compositions at this temperature (i.e., the eutectic temperature) as the lower boundary of the liquid+solid regions (the upper boundaries of which are the liquidus curves).

So far it has been explained how the two liquidus curves and the eutectic line are determined by the method of cooling curves. The solidus curves can be found in the same way. Figure 8.12 shows schematically how the cooling curve for a Pb-10%Sn alloy should look. Within the two-phase region the cooling rate is retarded by the emission of the heat of fusion by the increasing volume of solid. This produces the decrease in cooling rate at the liquidus temperature. The cooling rate becomes faster again when all the liquid is gone. Thus, the boundaries of both the liquid+α and the liquid+β regions can be determined by means of cooling curves. This method is called *thermal analysis*.

The two solid phases which form in the lead-tin system are called *terminal solid solutions*, which means that they are solid solutions based on each of the two pure components. Thus, the α phase is a solid solution based on FCC lead, and the β phase is a solid solution based on tin, which can have one of two crystal structures, tetragonal or diamond cubic, depending on whether the temperature is above or below 18°C, as explained in Appendix 8.1. The α and β phase fields can be interpreted in the following way: Lead has a solid solubility for tin which varies with temperature; the maximum solid solubility occurs at the eutectic temperature. At any temperature between 183°C and the melting point of Pb, any addition of Sn to the α phase beyond the composition indicated by the solidus line would result in the formation of a liquid phase in the formerly solid alloy. At any temperature below 183°C, excessive addition of tin to the α phase would result in the formation of the Sn-rich β phase, mixed in with the α.

CHAPTER 8   ALLOYS FOR SOLDERING AND BRAZING; PHASE DIAGRAMS

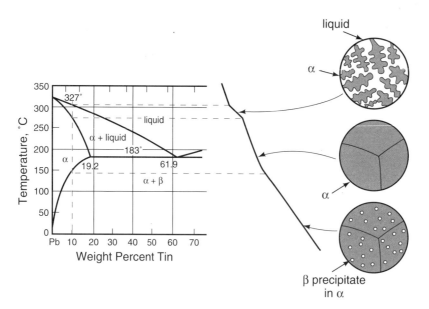

Fig. 8.12 The Pb-rich portion of the Pb-Sn phase diagram, showing the location of the 10%Sn alloy. A cooling curve for this alloy, showing the changes in slope at the liquidus and solidus temperatures, is correlated with representations of microstructures at several temperatures, assuming equilibrium conditions.

The extent of solid solubility of one element in another depends not only on the degree of chemical compatibility, but also on the sizes of the ion cores involved. For a given element, the size of the ion core actually varies, depending on what kinds of atoms surround it and what the crystal lattice is. However, one can get a rough idea from the closest spacing of the atoms in a crystal of the pure element. For lead, this is about 3.5Å, and for tin, it is either 3.0 or 2.8Å, depending on which crystal structure is considered (cf. Appendix 8.1). The point is that there is at least 14% difference in atomic size; this is near the limit beyond which solid solubility would be quite restricted. From the phase diagram, which shows a much larger solubility of tin in lead than *vice versa,* one might surmise that the size difference between the ion cores is less when the lattice is FCC than when the lattice is tetragonal or diamond cubic. However, the solubility difference could also be the result of chemical factors, such as different preferences for metallic *vs.* non-metallic bonding on the part of lead *vs.* tin.

If the 10% Sn alloy were cooled so slowly as to maintain equilibrium, the β phase would begin to precipitate in the solid α once the solvus line was crossed. This is represented schematically in Fig. 8.12 for the temperature of 100°C. In principle, the location of the solvus line could be determined by thermal analysis, because there is a heat of transformation associated with the transition from α to α+β. In practice, this heat effect is too small to be detected. More important, however, is the fact that solid-state transitions involving diffusion of atoms in a crystalline structure are always much slower than the transitions which involve diffusion in liquids. Therefore, the formation of β in α during cooling is likely to be kinetically retarded and thus not likely to occur at the equilibrium solvus temperature of a given alloy, even during very slow cooling. For this reason, the solvus is usually determined by heating specimens of a given composition to a series of temperatures in the vicinity of the expected transition, allowing each specimen to come to equilibrium at the temperature selected for it, and then quenching rapidly (e.g., in cold water) to "freeze in" the microstructure that existed at the selected equilibration temperature. The quenched specimens can then be examined metallographically or by x-ray analysis* to determine the temperature at which the β phase (formerly present) dissolves completely in the α. Obviously, a series of compositions would be needed to make a determination of the solvus line, and this method can only be carried out at temperatures high enough for the Sn atoms to diffuse out of the β phase in reasonable times. This makes the low-temperature portions of a solvus line very difficult to determine. The solvus of the β-phase field is, of course, determined in exactly the same way.

It should be clear that the choice of which component to put on the left in a phase diagram and which to put on the right is completely arbitrary, as is the choice of which phase to call α and which to call β. Thus, one must speak of α and β and hypo- and hyper eutectic only in the context of a particular representation of the phase diagram of an alloy system. One can find lead on the left in some handbooks of phase diagrams and tin on the left in others.

In principle, given the appropriate data, a phase diagram could be calculated using the laws of thermodynamics. One general rule is that between any two neighboring single-phase regions there must be a two-phase region comprising these two phases. The extent of this two-phase region may degenerate to a single point. For example, for a single component this occurs at the melting temperature, where the solid and liquid phases are in equilibrium. At a binary eutectic point, three phases are in equilibrium. More generally, the two-phase regions extend over substantial areas of the phase diagram, such as those which occur above and below the eutectic temperature in Pb-Sn alloys.

Example 8.1

Q. What is the latent heat of fusion, e.g., for the liquid-to-solid transformation of lead, in the context of thermodynamics?

A. The latent heat of fusion is the difference in enthalpy between the liquid and solid states, usually expressed on the basis of one mole of material (i.e., Avogadro's number, N_0, of atoms, or 6.02×10^{23} atoms). From the definition of enthalpy (cf. Section 6.1), $H = E + pV$, the change in enthalpy on freezing at a constant pressure would be given by

$$\Delta H = \Delta E + p\Delta V$$

The volume change would be only a few percent, so the $p\Delta V$ term at one atmosphere of pressure would be negligible. Therefore, the heat of fusion is essentially equal to the total binding energy of N_0 atoms in a crystal of the solid (cf. Fig. 3.7). This heat is released during freezing, and it must be put back into the solid in order to melt it.

---

* Because the β phase has a different crystal structure from the α phase, its x-ray diffraction pattern would show lines (i.e., diffracted beams) at angles different from those for the α phase.

Example 8.2

Q. (a) At what temperature does a Pb-40%Sn alloy begin to solidify upon cooling from the melt?

(b) How much solid is present (under equilibrium conditions) at 200°C?

(c) How much of the alloy is primary α at room temperature?

A. (a) From the phase diagram in Fig. 8.8, it can be seen that the liquidus line is crossed in a 40%Sn alloy at just below 250°C.

(b) The relative amount of α at 200°C is given by the lever rule as

$$\text{Relative amount of } \alpha \text{ at } 200°C = \frac{58-40}{58-18} = 45\%$$

(c) The total amount of primary α just below the eutectic temperature is given by

$$\text{Total primary } \alpha \text{ at } 182°C = \frac{61.9-40}{61.9-19.2} = 51\%$$

However, during cooling to room temperature, some β will precipitate in the α. The amount that should precipitate can be calculated as follows: Consider the primary α as a separate alloy with 19.2% Sn at 180°C. Upon cooling to room temperature, where the solid solubility of tin in lead is taken to be about 1%, and that of lead in tin to be zero, this alloy should have the following amount of β:

$$\text{Relative amount of } \beta = \frac{19.2-1}{100-1} = 18.4\%$$

Therefore, the remaining amount of primary α is given by

$$\text{Amount of primary } \alpha = \frac{100-18.4}{100} \times 51 = 41.6\% \text{ (of the total alloy)}$$

8.2.4 Microstructures of Hypoeutectic Lead-Tin Alloys

As noted earlier, solids tend to crystallize from the melt in the form of branched particles called dendrites. This is observable in transparent crystals, like succronitrile crystallizing in acetone, an example of which is shown in Fig. 8.13(a). Dendritic solidification can also be observed on surfaces of metal castings, as exemplified by Fig. 8.13(b). The primary phases (α or β) that form during solidification of Pb-Sn alloys would also follow this behavior. Therefore, the microstructures of as-cast alloys must be interpreted in this light. That is, the primary phase revealed in a metallographic section is a two-dimensional cut through a number of dendrites.

The reason for the development of this kind of solid/liquid interface in a hypoeutectic Pb-Sn alloy is that the α contains much less tin than the liquid from which it forms. Therefore, in order for the α to grow, tin must be transported away into the liquid by diffusion. If the α/liquid interface remained planar, the tin would tend to pile up along the interface and slow it down. The rate at which tin is transported away from the interface depends on the concentration gradient of the tin in the liquid alongside the solid/liquid interface. An example of such a concentration gradient is shown in Fig. 8.14(a) for the case of a 30%Sn alloy solidifying at 250°C, where the tie line gives the solid and liquid compositions as 12 and 37%Sn, respectively (cf. Fig 8.11). (This fixes the tin concentrations on either side of the solid/liquid interface, where local equilibrium can almost always be assumed.) If a spike were to develop on the interface, the concentration gradient of tin at the tip of the spike would become much steeper, as shown in Fig. 8.14(b), and the diffusion of tin away from the tip would be correspondingly faster than from the planar interface.

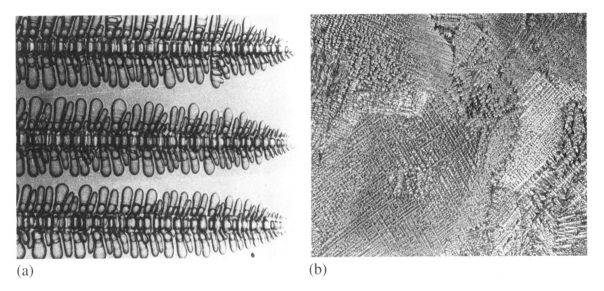

Fig. 8.13 (a) Dendrites formed during precipitation of succronitrile in acetone. (Prof. R. Trivedi, Iowa State Univ.) (b) Dendritic solidification observed on the surface of a Cu-Sn casting. (P. Villanueva, Univ. of Penn.)

In effect, at the tip of a spike, the tin can be rejected into a larger volume of liquid. Therefore, a spike can grow faster than the rate of advance of the planar interface, because the rate of advance is controlled by the rate of diffusion of tin in the liquid. As the spike thickens, it develops smaller spikes along its length for the same reason. (When a pure substance, like water, is supercooled, an analogous argument can be made in terms of the removal of the heat of fusion; here, the tip of the spike projects into a liquid that is cooler than along a planar solid/liquid interface. Therefore, heat flow out of the tip would be faster, and the tip would grow out ahead of the planar interface.)

Fig. 8.14 (a) Schematic representation of the concentration gradient at a planar interface between solid α and the liquid in a Pb-30%Sn alloy freezing at 250°C. The dashed lines are lines of constant concentration and are an analog of a contour map. (b) The steeper concentration gradient at the tip of a spike (representing the beginning of a dendrite).

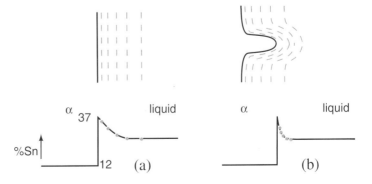

The microstructure of a slowly solidified Pb-30%Sn alloy is shown in Fig. 8.15. It shows sections of dendrites of primary α surrounded by the eutectic constituent. The primary α contains small particles of β that precipitated during cooling between 183°C and room temperature. The eutectic appears different from that formed in the eutectic alloy (cf. Fig. 8.10), although it is clearly a two-phase constituent. The eutectic in Fig. 8.15 appears to contain less α and more β than it should. The reason for this is the influence of the primary α. It can only be understood by first considering why the microstructure in the eutectic alloy in Fig. 8.10 is lamellar.

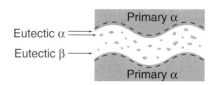

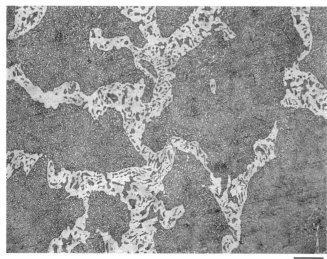

Fig. 8.15 Microstructure of a slowly solidified Pb-30%Sn alloy. The dark areas are primary α dendrites, which contain small particles of precipitated β. The dendrites are surrounded by the two-phase eutectic.

The key factor is that, in the eutectic reaction: liquid—>α+β, both solid phases must form simultaneously. This involves large mass transport in the liquid, because the α phase can hold only 19.2%Sn, and the β phase can hold only 2.5%Pb, whereas the liquid from which they form contains 61.9%Sn and 38.1%Pb. Thus, the simultaneous formation of α and β requires an enormous redistribution of Pb and Sn by diffusion in the liquid. If the alloy is losing heat at all rapidly, then the diffusion distance must be short. This is accomplished by the lamellar formation of α and β depicted schematically in Fig. 8.16, in which the redistribution of the Sn from the regions becoming α to those becoming β at the solid/liquid interface is indicated by the arrows. The flux of Pb atoms goes in the opposite direction. The same kind of short diffusion distance could be accomplished if the α phase took the form of needles instead of plates (lamellae). In the present case, plates are necessary owing to the large volume fraction of α in the eutectic.

Fig. 8.16 Schematic representation of freezing of a lamellar Pb-Sn eutectic; the flux of Sn in the liquid is indicated by arrows; the flux of Pb is in the opposite directions.

Another important factor in the morphology of a eutectic microstructure is the direction of heat flow. Figure 8.16 is drawn with the assumption that the solid/liquid interface moves left to right, which means that heat flow is assumed to be unidirectional toward the left. The eutectic microstructure shown in Fig. 8.10 can be interpreted as the result of heat flow in the direction normal to the plane of the photomicrograph. That is, the solid/liquid interface can be imagined as moving toward the viewer and to be divided into cells, within each of which the α/β lamellae are aligned. Thus, the region of the microstructure illustrated in Fig. 8.10 shows the juncture of three of these cells.

Returning to Fig. 8.15, it is seen that the presence of primary α dendrites perturbs the normal eutectic solidification. The reason for this is two-fold. First, the advance of a solid/liquid interface during solidification of the interdendritic liquid cannot be unidirectional because of the convoluted shape of this liquid region. Second, because there is α already present, the liquid tends to phase-separate by "plating-out" some of the eutectic α as a thin layer on the primary α dendrites, while the remaining eutectic α and the eutectic β solidify in a somewhat disorganized manner farther away from the surfaces of the dendrites. (See inset, Fig. 8.15.) The plating-out of one phase of a eutectic on a pre-existing primary phase is known as *eutectic divorcement*.

The foregoing discussion of eutectic morphology should illustrate the fact that the phase diagram cannot be used to predict morphology. It only deals with the amounts and compositions of phases at equilibrium at a given temperature in a given alloy. This caveat cannot be stressed too strongly. It is easy to depart from equilibrium during solidification, because solid-state diffusion is necessary to maintain the equilibrium composition of a solid phase as temperature falls. This fact is illustrated very well in an alloy of Pb-10%Sn, in which one would not expect to find any eutectic microstructure, judging from the phase diagram (cf. Fig. 8.12). The microstructure of a cast sample of this alloy is shown in Fig. 8.17; there is a small amount of eutectic present, even though the last liquid should have disappeared well above 200°C. The explanation of this non-equilibrium solidification can be made with the aid of Fig. 8.18.

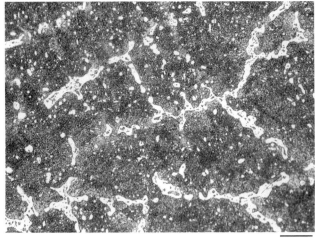

Fig. 8.17 The microstructure of a cast Pb-10%Sn alloy.

50μm

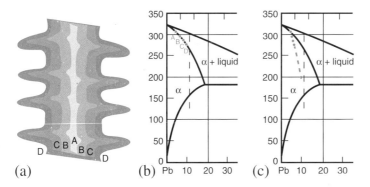

Fig. 8.18 (a) A portion of an α dendrite showing the positions of the solid/liquid interface at the temperatures indicated on the solidus curve in (b). (c) The non-equilibrium solidus, which represents the average composition of the α, resulting from the insufficiency in the time for diffusion in the solid α.

A portion of an α dendrite is shown schematically in Fig. 8.18(a), in which the shaded regions are meant to indicate the positions of the solid/liquid interface at four different temperatures. These temperatures are shown by the points on the solidus curve of the phase diagram shown in Fig. 8.18(b), which also correspond to the equilibrium α composition at each of these temperatures. It is obvious that, in order for the α dendrite to maintain the equilibrium composition as it cools, there must be time allowed for Sn to diffuse in from the surface to raise the Sn level in the core of the dendrite. If the time allowed during solidification is insufficient, then the average composition of the dendrite must be lower than the phase diagram would indicate at any given temperature below point A in Fig. 8.18(b). This would result in the *non-equilibrium solidus*, which is a plot of the average dendrite composition *vs.* temperature, shown in Fig. 8.18(c). The effect is that some liquid still remained at 183°C, and this underwent the eutectic reaction.

Essentially what has occurred is that the liquid remained too rich in Sn, because the dendrite grew without absorbing its full share of the Sn. This excess Sn concentration in the liquid continued to increase as more and more α formed; finally the temperature reached 183°C and there was still some liquid present, the Sn concentration of which had reached 61.9%Sn. The eutectic reaction was the inevitable result.

8.2.5 The Microstructure of a Hypereutectic Pb-Sn Alloy

All of the foregoing regarding a hypoeutectic alloy applies to a hypereutectic alloy. This follows from the fact that the designations are arbitrary; the phase diagram can be switched to its mirror image without loss of validity. Thus, little need be said further about the Sn-rich side of the Pb-Sn system. Most of it can be expressed by the microstructure of a cast Sn-10%Pb alloy, shown in Fig. 8.19.

The large white areas are the arms of primary β dendrites. Between the dendrite arms is a divorced eutectic in which the eutectic β has largely plated out on the β dendrites, leaving the α to form a coarse lamellar structure with the remaining β. This alloy would not be used for soldering, because tin is more expensive than lead, and because the volume change associated with the allotropic transformation in tin (cf. Appendix 8.1) could compromise the mechanical stability of an alloy with such large amount of tin.

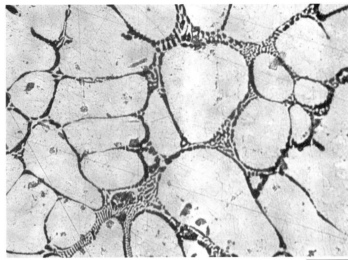

Fig. 8.19 Microstructure of a cast alloy of Pb-90%Sn showing dendrites of primary β and a divorced eutectic in which the eutectic β has separated from the eutectic α.

8.3 The Phase Rule

A proper phase diagram must be consistent with the laws of thermodynamics, as derived by J. Willard Gibbs. In particular it must obey the Gibbs *phase rule*, applicable to any system at equilibrium. The phase rule gives the number of thermodynamic variables (like temperature, pressure, and composition of each phase) which are not fixed in a system of C components and P phases existing together at equilibrium. This number of free variables is called *the number of degrees of freedom*, F, and the phase rule states that:

$$F = C - P + 2$$

For example, consider the two-phase equilibrium of ice and liquid water at 0°C, one atmosphere pressure. For this one-component system, the phase rule states that F = 1-2+2 = 1. This means that one variable, either temperature or pressure, can be varied while still maintaining the two-phase equilibrium. Thus, the freezing point is different from 0°C if the pressure is varied, but, for any given pressure, the freezing temperature is fixed. Both variables cannot be changed arbitrarily at the same time; otherwise, the system would depart from the two-phase equilibrium. (It would revert to a one-phase state.)

Since it is generally stipulated that pressure on an alloy will always be fixed at one atmosphere, one degree of freedom is used up *a priori*. Therefore, for most cases the phase rule can be recast to reflect the isobaric (constant pressure) case:

$$F = C - P + 1$$

As shown in more detail in Appendix 8.2, the phase rule comes directly from the definition of chemical equilibrium in a multi-phase system: that the composition of each phase remains constant. That is, there is no driving force for the transfer of any component from one phase to another. Such a driving force would be a difference in *chemical potential* between any two phases. This is analogous to a difference in electrical potential, which leads to the transfer of electrical charge in a conductor. Thus, the microstructure in Fig. 8.19, for example, shows

F=Degrees of freedom
C=No. of components
P=No. of phases

the tin-rich β phase in equilibrium with the lead-rich α phase. "Equilibrium" means that there is no net flow of either element between the two phases, in spite of the large compositional difference between them. Therefore, the chemical potential of lead must be the same in each phase, and similarly for tin; this fact leads directly to the phase rule, as shown in Appendix 8.2. The F degrees of freedom in an alloy is nothing more than the number of free variables in the set of equations that express the fact that, at equilibrium, the chemical potential of each of the C components is equal in each of the P phases.

The constant-temperature portion of the cooling curves for the pure lead and pure tin, shown in Fig. 8.9, can be understood in exactly the same terms as would hold for the case of freezing of water: For the isobaric two-phase equilibrium at freezing in a one-component system, $F = 1-2+1 = 0$. Therefore, all variables are fixed, so the freezing of a pure substance must occur at a constant temperature.

The freezing of the eutectic alloy at the constant temperature 183°C occurs by the following reaction:

$$\text{Liquid} \longrightarrow \alpha + \beta$$

Therefore, at this temperature there is a three-phase equilibrium in this two-component system, and the phase rule states that $F = 2-3+1 = 0$. Again, freezing must occur at a constant temperature at the eutectic composition. Thus, the freezing of a eutectic at a constant temperature is a thermodynamic necessity, as long as the system is at equilibrium. It is possible to supercool a eutectic liquid and to make it freeze at some lower temperature, but the system is then far from equilibrium, and the phase rule does not apply.

8.4 Brazing Alloys

A common method for joining steel bicycle tubing employs *lugs*, illustrated in Fig. 8.20, to serve the same purpose as the copper fitting shown in Fig. 8.1. The stresses on a bicycle frame can be large, and Pb-Sn solder is too weak for this application, since room temperature is about two-thirds of the Pb-Sn eutectic temperature on the absolute scale. Hence, diffusion in Pb-Sn solder is rapid, and the solder would deform by *creep*, which is a time-dependent process of plastic flow involving self-diffusion (dislocation climb; cf. Fig. 6.8) occurring in metals at high temperatures (around 0.5 T_{MP} and above). Therefore, it is necessary to find an alloy that liquifies at a temperature much higher than 183°C.

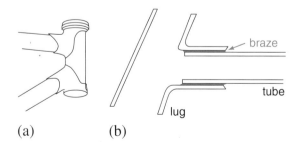

Fig. 8.20 (a) Lugged joints connecting the top tube and down tube with the head tube of a bicycle. (b) Schematic representation of the brazed joint between a lug and a tube.

One looks for an alloy for which room temperature is no higher than one-third the liquation temperature (in degrees Kelvin). Thus, the liquation temperature should be at least 900K, or about 625°C. The alloy should also be readily available, preferably inexpensive, and able to form a low γ_{LS} with clean iron. There are essentially two families of alloys that are used for brazing steel frames; one is based on the silver-copper system and the other on copper-zinc. For hand-made lugged frames one can use either. The silver-rich alloys are based on the silver-copper eutectic system, the phase diagram of which is shown in Fig. 8.21(a). It is a simple eutectic and is essentially similar to the Pb-Sn system. The eutectic temperature of the binary alloy is 780°C, and this can be lowered to around 670°C by additions of zinc and tin, thereby creating a quaternary, instead of a binary, alloy.

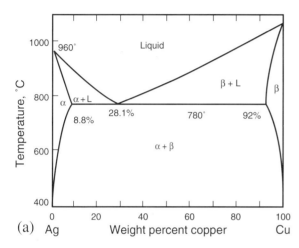 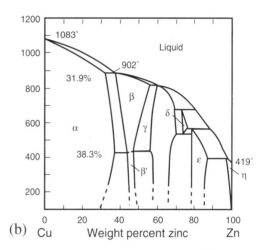

Fig. 8.21 (a) The Ag-Cu phase diagram. (b) The Cu-Zn phase diagram. (From M. Hansen and K. Anderko. *Constitution of Binary Alloys,* McGraw-Hill, N.Y. 1958, pp. 18 and 649, resp.)

The copper-zinc alloys are based on the alloy 60%Cu-40%Zn, sometimes called "Muntz" metal. The relevant part of the binary Cu-Zn phase diagram is shown in Fig. 8.21(b). This portion of the diagram shows what is called a peritectic reaction, in which there is a three-phase equilibrium at around 900°C. The reaction upon cooling is

$$\text{Liquid} + \alpha \longrightarrow \beta$$

The rest of the diagram is complicated and, fortunately, irrelevant. The important feature is that the 60-40 alloy must be heated above 900°C in order to liquify it. This liquation temperature is not changed much by the common additions of about one percent tin, iron, or nickel, so the steel tubes must be heated to a much higher temperature for brazing with the copper-zinc alloys than for the silver-based alloys.

An example of a brazed joint between a lug and a tube (both carbon steel) is shown in Fig. 8.22(a). Here, the steel is attacked by the etchant of nitric acid in alcohol, but the Cu-Zn brazing alloy is not. The width of this joint is 0.02mm, or 0.0008in.

Fig. 8.22 (a) Brazed joint between a lug and a tube, both made of carbon steel; etched to reveal the microstructure of the carbon steel. (b) Another brazed joint etched to reveal the microstructure of the 60Cu-40Zn brazing alloy.

Another example, in which the Cu-Zn alloy has been etched (using a $NH_4OH-H_2O_2$ mixture), is shown in Fig. 8.22(b). The brazing alloy obviously comprises two phases in roughly equal proportions; the matrix is the β phase, and the elongated second phase is α. This microstructure can be rationalized with the aid of the phase diagram. It can be seen that the 60-40 alloy would first form the β phase upon solidifying and that, when cooled below about 800°C, the α phase would precipitate in the β. It appears that the α forms as plates having a crystallographic relationship with the β matrix.

Hand brazing is done by heating the joint area with an oxygen-acetylene torch. The alternative procedure for lugged frames, which are produced in higher volume, is to use *furnace brazing*. Here, the copper-zinc-based alloys are always used, and the frame is assembled and held together in a heat-resistant framework, with the brazing alloy in the form of powdered metal or thin sheets inserted into the joints. When the framework is loaded into the brazing furnace and allowed to reach temperature, the brazing alloy liquifies and spreads along the joint. In contrast with the localized heating during hand brazing, the heating during furnace brazing is obviously uniform over the entire frame.

Lugged construction can be used only when the tube diameters and the angles between the tubes correspond to a commercially available lug design. This is often not the case for frame-builders, and the traditional alternative has been to employ a process called *fillet brazing* or *braze welding*. For this, the copper-zinc alloys are always used, and this method is capable of producing strong and attractive joints. An example is shown in Fig. 8.23(a). In fillet brazing by hand, a torch is used to deposit the molten metal in the joint and to shape the fillet (pronounced *fill-it*) so that it has a smooth contour with large radii of curvature. The tubing is heated only locally, but for a much longer time than with the brazing of lugged joints.

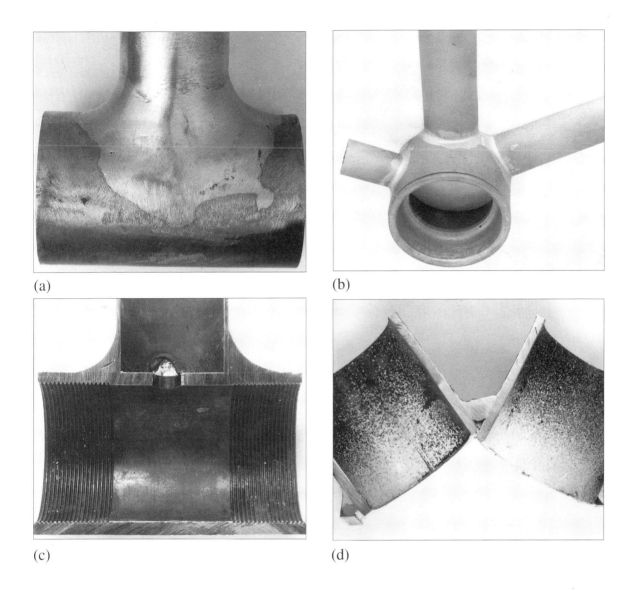

Fig. 8.23 Examples of joints made by fillet brazing. (a) A strong, hand-brazed joint with a large fillet. (b) A fillet-brazed bottom bracket with minimal fillets, resulting in a weak joint. (c) and (d) The cross-sections of these two joints.

There are commercially produced bicycles with fillet-brazed joints in which relatively small amounts of filler metal are applied, such as shown in Fig. 8.23(b). When a frame is painted, the nature of the joints is often not apparent. For the case shown in Fig. 8.23(b), when the paint was removed by sand-blasting, it was revealed that the bottom-bracket joints would have questionable resistance to fatigue failure, given the lack of cross-section in the joint.

Summary

Steel bicycle frames have traditionally been constructed from tubes joined by brazing, which is the high-temperature version of soldering, usually employing capillary action to draw a molten alloy into the crevice between the tube and a sleeve called a lug. This requires good wetting of the solid pieces by the liquid alloy, and for this the surfaces of the solid must be cleaned of oxide films. In addition, the process is made easier if the melting temperature of the liquid metal can be reduced significantly by alloying. The effects of alloying elements in this regard can be represented by the use of phase diagrams, which are essentially maps showing the equilibrium configurations of alloys as a function of composition and temperature. Simple eutectic phase diagrams are discussed in terms of (a) the experimental methods used to determine them, (b) the use of tie lines and the lever rule in determining the amounts and compositions of phases in two-phase regions, and (c) the microstructures that develop as a result of cooling from the liquid phase.

The reason for the formation of lamellar eutectics is given in terms of the need for a short diffusion distance during solute partitioning between the two phases. The reason for dendritic solidification is given in terms of the instability of a planar interface from which solute is being rejected during crystal growth. Solders made from lead-tin alloys are used as examples, and the silver-based and copper-zinc alloys used in bicycle construction are discussed.

The idea of a chemical potential as the driving force for mass flow is introduced, and the Gibbs phase rule is then derived in Appendix 8.2 from the condition that, at equilibrium, no net mass flow between coexisting phases can occur, meaning that the chemical potential of each component of an alloy must be the same in each of these coexisting phases.

Appendix 8.1 The Allotropic Transformation of Tin

Many pure elements undergo solid-state phase transformations, called *allotropic* transformations. Iron changes from BCC to FCC upon heating past 910°C. Tin undergoes a transformation from *diamond cubic* to *body-centered tetragonal* upon heating through 13°C; these structures are depicted in Fig. A8.1.1. Since tin is a Group IV element in the Periodic Table, like carbon, silicon, and germanium, it is not surprising that it can take on the crystal structure of diamond, as do silicon and germanium.

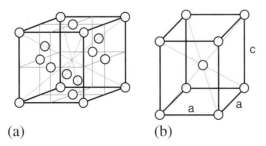

Fig. A8.1.1 The crystal structures of tin: (a) diamond cubic, (b) body-centered tetragonal.

(a)   (b)

In the diamond-cubic structure each atom has four neighbors, and each pair shares two electrons in covalent bonding, as in Fig. A8.1.2. In this way each atom can have a complete 8-electron outer shell, like an inert gas. This is a good example of how a bonding type can dictate the crystal structure. The Bravais lattice of the diamond cubic structure is FCC; however, there are two atoms associated with each lattice point in the diamond cubic structure.

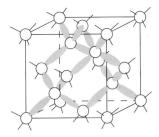

Fig. A8.1.2 Covalent bonds in a diamond cubic crystal. (From L. H. Van Vlack, *Elements of Materials Science and Engineering*, 4th Ed., Addison-Wesley, Reading, MA, 1975, p. 37.)

Appendix 8.2 The Phase Rule

The phase rule is a simple equation that gives the number of fundamental variables that can be altered in value while maintaining the particular phase state, meaning an equilibrium of one phase, two phases, three phases, etc., in a system containing a known number of components. It is based on the general principle that the condition for equilibrium is the minimization of the Gibbs free energy, G, with respect to all relevant variables. It can be written

$$F = C - P + 2$$

where C is the number of components, P is the number of phases existing together in equilibrium, and F is the number of "degrees of freedom," meaning the number of unfixed variables.

Equilibrium means that no spontaneous changes occur in the system; the system is essentially changeless with time. A simple example is an ideal gas, which follows the law

$$pV = nRT$$

given in any elementary chemistry course, where p is the pressure, T the absolute temperature, n is the number of moles of the gas, V is the volume, and R is the gas constant. The *intensive* variables that determine the state of the gas are p and T; n and V are *extensive* parameters which depend on the amount of gas. Since in an ideal gas, C = 1 and P = 1, the number of degrees of freedom is 2. That means both temperature and pressure may be varied while still maintaining a one-phase equilibrium. It does not mean that p and T can be varied without limit. If T were reduced enough, or if p were increased enough, the gas would presumably condense. The phase rule only says that there can be some change in the two variables while still maintaining the gaseous state.

In an alloy with C components and P phases there is an additional factor relevant to the concept of equilibrium, and that concerns the possibility of net mass flow among the phases. By definition, equilibrium requires that all phases attain a steady state with regard to temperature, pressure, and composition; therefore, there can be no net mass flow between phases. In a system at equilibrium at constant p and T, a change in free energy in phase α resulting from the transfer of *dn* moles of component *i* from phase α to phase β would be offset exactly by an equal and opposite change in the free energy of phase β. The variation of free energy with n_i at constant p, T, and n_j, where n_j refers to all of the other components beside *i*, is defined as the *chemical potential*, μ_i

$$\mu_i = \left[\frac{\partial G}{\partial n_i}\right]_{p,T,n_j}$$

The chemical potential can be thought of as the driving force for mass flow, in the same way that a temperature gradient drives heat flow, and a gradient in electrical potential drives current flow. Thus, equilibrium requires that the chemical potential of each component be the same in all of the P phases. Only then will there be no mass flow. This means that for the ith component

P - 1 equations

$$\mu_i^\alpha = \mu_i^\beta = \mu_i^\gamma = \ldots = \mu_i^P$$

There is a similar set of P-1 independent equations for each of the C components; giving a total of C(P-1) equations.

C(P-1) equations:

$$\mu_1^\alpha = \mu_2^\beta = \cdots = \mu_1^P$$
$$\mu_2^\alpha = \mu_2^\beta = \cdots = \mu_2^P$$
$$\vdots$$
$$\mu_c^\alpha = \mu_c^\beta = \cdots = \mu_c^P$$

Now the composition of each phase can be specified by C-1 composition terms, since if the concentrations of all but one of the components is known, then that of the remaining one is given by subtracting the sum of the known concentrations from 100%. Therefore, in a system with P phases, there are P(C-1) independent concentration variables. If we ignore electrical, magnetic, surface, or other such influences, the only other relevant variables are temperature and pressure. Hence, the total number of variables that remains undetermined is

$$[P(C-1) + 2] - [C(P-1)] = C - P + 2$$

This is what is meant by the degrees of freedom, F. It is the number of variables which remain to be specified in order to define the state of a system completely. There is some freedom in the specification of these remaining variables. As noted in Section 8.3, it is common to give up a degree of freedom by fixing the pressure at one atmosphere, in which case the phase rule is written

$$F = C - P + 1$$

Exercises Chapter 8

Terms to Understand

Miscible
Capillarity
Soldering
Brazing
Lever rule
Lamellar
Eutectic divorcement
Phase rule
Cooling curve
Eutectic
Liquidus

Solidus
Solvus
Tie line
Primary phase
Dendrite
Non-equilibrium solidus
Hypoeutectic
Hypereutectic
Chemical potential
Allotropic transformation

Problems

8.1 Apply the phase rule to the Pb-30%Sn alloy in the two-phase region between 265°C and 183°C; explain the meaning of F=1. (Remember, pressure is fixed at 1atm.)

8.2 Calculate the amount of primary α phase in alloys of Pb-40%Sn and 60%Sn cooled to 182°C under equilibrium conditions.

8.3 What would happen if the microstructure shown in Fig. 8.10 were heated to 175°C and held there for a long time? Explain your answer.

8.4 Given below is the microstructure of a Ag-Cu alloy of unknown composition. Deduce what the composition is and explain your reasoning. (Hint: Focus first on the eutectic constituent, and decide which phase is which, based on the phase diagram.)

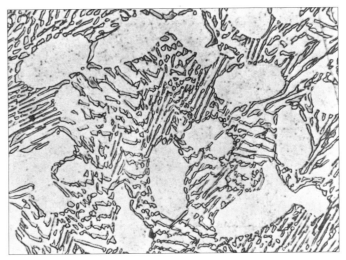

(Photomicrograph from R. E. Reed-Hill, *Physical Metallurgy Principles*, Van Nostrand, N. Y., 1973, p. 544.)

8.5 Sterling silver is Ag-7.5%Cu. Using the phase diagram in Fig. 8.21(a), tell what the microstructure would look like if the alloy were melted and cooled slowly under equilibrium conditions to room temperature, and what it would look like if it were cooled too rapidly for equilibrium to be achieved. Make sketches to illustrate your answer. (Review Fig. 8.18.)

8.6[1] The copper in sterling silver can provide solid-solution strengthening. Sterling silver equilibrated at 785°C and cooled rapidly is stronger and can be polished to a higher luster than if cooled slowly. Explain why.

8.7 Describe the changes that would occur in an alloy of Ag-70% Cu cooled from the liquid state to room temperature. Sketch the resulting microstructure and give the relative amount of each phase and each microstructural constituent (i.e., primary phase and eutectic). (Refer to Fig. 8.21 and review Fig. 8.7 and relevant text.)

8.8[1] Use the phase rule to calculate the number of degrees of freedom, F, for an alloy of Pb-20wt%Sn in the vicinity of 250°C and at 183°C. Explain the physical meaning of F in each case.

8.9[1] Consider a solid composed of one mole of A atoms and one mole of B atoms. What was the change in configurational entropy, compared with the pure A and B, that resulted from mixing the A and B? (Use Stirling's approximation.) Did the effect of the entropy *per se* cause an increase or decrease in the Gibbs free energy, G?

8.10[1] An alloy of Ge-50%Si is cooled very slowly from 1500°C. Using the phase diagram given below, please answer the following questions:

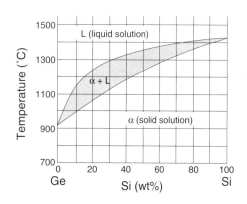

(a) At what temperature does freezing begin?
(b) At what temperature is freezing complete?
(c) At what temperature is the mixture half solid and half liquid?
(d) What is the composition of the first solid to form?
(e) What is the composition of the last liquid to freeze?
(f) What are the weight fractions of the two phases at 1250°C?
(g) What are the compositions of the two phases at 1250°C?
(h) What are the atomic fractions of Si and Ge in the alloy?

8.11 The crystal structure of silicon is diamond cubic (cf. Fig. A8.1.2). Deduce the crystal structure of germanium from the phase diagram above, and explain your answer.

[1]Contributed by Professor D. L. Callahan, Rice University.

9
PHASE TRANSFORMATIONS IN CARBON STEEL SPOKES

In Fig. 6.20, the microstructure of a carbon steel spoke that was heated to 950°C and slowly cooled was shown. In Fig. 9.1, this microstructure is compared with that of the same kind of spoke cooled more rapidly from 950°C. The starting condition of the spoke before being wire drawn was somewhere in the range between these two conditions. The main objective of the present chapter is to understand the kinds of phase transformations that produce these microstructures, as well as the one which results when the steel is quenched rapidly, as into cold water, from such a temperature.

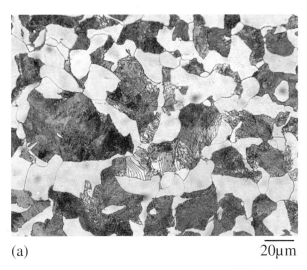

(a)

Fig. 9.1 Comparison of microstructures of a carbon steel spoke heated to 950°C and (a) furnace cooled or (b) air cooled.

(b)

9.1 The Iron-Carbon Phase Diagram

The microstructures comprise white-etching regions of a single phase, called *ferrite*, and grey-to-black-etching regions of a two-phase constituent, called *pearlite*. The pearlite will be shown to be analogous to the eutectic constituent found in the Pb-Sn system (cf. Fig. 8.10), and the ferrite is the analog of the primary phase which forms during cooling of a hypoeutectic alloy (cf. Fig. 8.15). To understand these microstructures, one needs not only the relevant phase diagram, but also another type of diagram that portrays kinetic information about the phase transformations. Kinetic diagrams are needed here, whereas they were not needed when discussing the liquid-to-solid transformation, because the transformations in steel occur in the solid state, in which diffusion rates are very slow compared with diffusion in the liquid state. As a result of the role of diffusion-controlled kinetics, it is possible to achieve a great variety of microstructures in steels. These microstructures have different properties. Therefore, to control the properties of a steel, one must have a clear understanding of how to control the microstructure. This starts with the phase diagram, shown in Fig. 9.2.

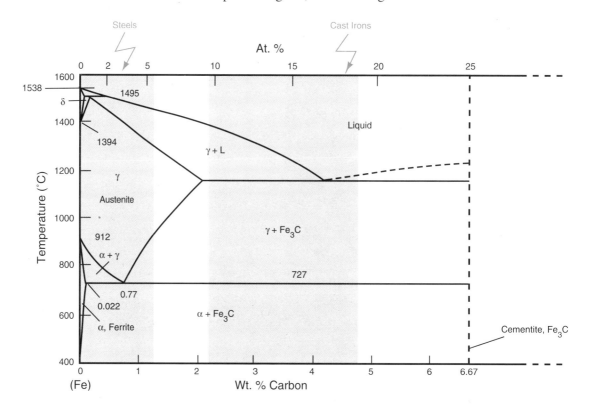

Fig. 9.2 The Fe-Fe$_3$C phase diagram. This is a metastable portion of the Fe-graphite system. The portions of the diagram relevant to steels and cast irons are indicated.

The iron-carbon phase diagram is of central importance to metallurgists, but only two regions of the diagram are widely used. One is relevant only to cast irons; this is the region which contains the eutectic reaction at 1148°C, where a relatively low-melting liquid freezes to the FCC γ phase plus graphite in the true equilibrium diagram.

The crystal structure and bonding of graphite is very far from metallic. It consists of plates of connected, covalently bonded, six-carbon rings, with Van der Waals bonding between the plates. Because of its non-metallic nature, graphite is rather difficult to nucleate in a solid metallic matrix. This is true in steels even when the carbon content is nearly 2%.

The difficulty of graphite nucleation results in the formation of the metastable Fe$_3$C phase *(cementite)* instead, especially in solid-state transformations. Although graphite can be made to form in high-carbon cast irons, it can only form in steels (<2%C) after heating for years at temperatures above 600°C. Thus, when dealing with most applications of steels, it is appropriate to work with the metastable Fe-Fe$_3$C diagram, shown in Fig. 9.2; this can be treated as an essentially stable system.

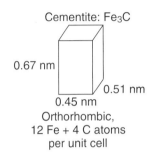

Cementite is an example of a metal carbide. As such, it is properly considered a ceramic, in which the bonding is only partly metallic, the rest being mainly covalent. That is, some of the valence electrons are delocalized, as in metallic bonding, and the rest are shared between iron and carbon atoms. The crystal structure is dictated by the geometry that suits this kind of bond formation and by the necessity to fit together atoms of very different size.

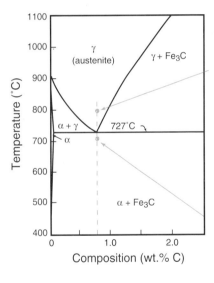
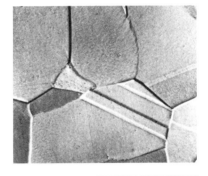
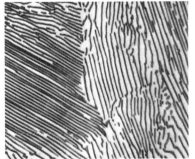

Fig. 9.3 The eutectoid region of the Fe-Fe$_3$C diagram, showing the transformation of austenite (γ) to pearlite (α + Fe$_3$C) in a eutectoid alloy. (From E.C. Bain and H.W. Paxton, *Alloying Elements in Steel,* ASM, Metals Park, OH, 1961, pp. 21, 27)

The part of the Fe-Fe$_3$C diagram of most interest is that containing the *eutectoid* reaction, shown isolated in Fig. 9.3. This is similar to a eutectic reaction, except that the single phase at the high temperature is a solid instead of a liquid. Thus, at 0.77%C the FCC γ phase, called *austenite,* transforms on cooling to BCC α, or ferrite, plus cementite. As in a eutectic alloy, the latter two phases are arranged in an intimate mixture, comprising alternating lamellae of ferrite and cementite. This is the constituent called pearlite.

9.2 Eutectoid Decomposition

The reason for the lamellar morphology of pearlite is exactly the same as explained in Fig. 8.16 for the formation of the Pb-Sn eutectic. During the decomposition of austenite, a large-scale redistribution of carbon must occur, because the ferrite can contain, at most, 0.02wt%C, and the cementite needs 6.7wt% (25at%)C. In order for these two phases to form in times on the scale of seconds or minutes, the diffusion distances must be short. This is achieved in a lamellar microstructure, as shown in Fig. 9.4, and the partitioning of the carbon can be carried out in the interface between the advancing pearlite colony and the parent austenite.

It must be emphasized that the phase diagram does not indicate that the austenite should decompose into pearlite. It only indicates that ferrite and cementite should form. Thus, the lamellar morphology is dictated by kinetic, rather than thermodynamic, considerations. Moreover, the only way to produce pearlite is to cause austenite to decompose in a certain range of rates.

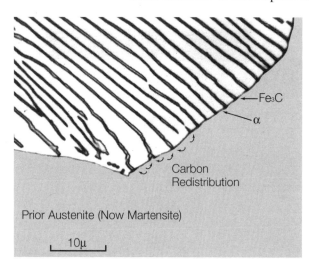

Fig. 9.4 Part of a pearlite colony growing into austenite. The edge-wise growth of the alternating plates of α and Fe_3C provides a mechanism of rapid transformation, since only short-range diffusion of carbon is required. (From P.G. Shewmon, loc. cit., p. 227.)

9.2.1 Isothermal Transformation

The decomposition of austenite was first studied in a systematic way by isothermal transformation of specimens that were quenched from the austenite region to some temperature below the eutectoid temperature, held for a time, and then quenched rapidly to room temperature. The process is illustrated in Fig. 9.5.

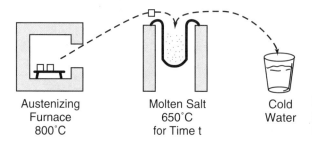

Fig. 9.5 Schematic representation of the experimental procedure used to carry out the partial transformation illustrated in Fig. 9.6.

The microstructure formed by the partial isothermal transformation of a eutectoid steel (~0.8%C) is shown in Fig. 9.6. The pearlite nucleated at the austenite grain boundaries and then grew in nodules into the austenite grains. When the transformation was interrupted, the remaining austenite transformed to martensite. (This involves a diffusionless process which is so rapid it cannot be suppressed.)

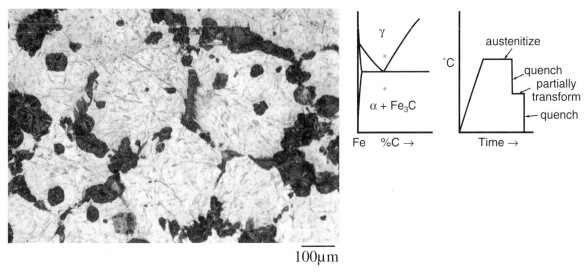

100μm

Fig. 9.6 The formation of nodules of pearlite at austenite grain boundaries during cooling of a eutectoid steel; the transformation was interrupted by quenching. The heat treatment employed is shown schematically, in conjunction with the phase diagram. (Photomicrograph by E. Anderson, Univ. of Penn.)

When the eutectoid transformation is compared with the freezing of a eutectic alloy, it is apparent that equilibrium conditions are not maintained in the steel. Otherwise, there would be a cooling curve with an isothermal portion, reflecting the three-phase equilibrium at the eutectoid temperature, as shown for the Pb-Sn eutectic in Fig. 8.9. The reason the eutectoid decomposition occurs far from equilibrium is that the driving force for this transformation is smaller than that driving the eutectic solidification.

The driving force for the eutectoid decomposition is simply the difference between the free energy of the austenite phase and that of the α+Fe$_3$C mixture. This free energy difference can be written

$$\Delta G = G_{\alpha+Fe_3C} - G_\gamma$$

and can be understood from a schematic plot of free energy *vs.* temperature, which can be assumed to be linear over a restricted range of temperature. Since G=H-TS, the free energy of both the austenite and the α+Fe$_3$C mixture must decrease with increasing temperature, as in Fig. 9.7. The curves must cross at 727°C, where all three phases are in equilibrium. Above 727°C, austenite is stable; therefore, the austenite free-energy curve lies below the one for α+Fe$_3$C. The reverse is true below 727°C. The vertical distance between the two curves at any temperature is the driving force for the transformation from one condition to the other. Thus, it is apparent that the driving force for the γ → α+Fe$_3$C transformation is zero at 727°C, and it increases continuously as the temperature falls below 727°C.

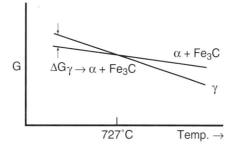

Fig. 9.7 Schematic representation of the free energy vs. temperature for austenite and α+Fe$_3$C, showing the equilibrium at 727°C and the increasing driving force (ΔG) for austenite decomposition below that temperature.

Because the initial and final states are both solids, the scale of the free-energy change for a eutectoid decomposition is much smaller than for eutectic solidification in which the initial state is a liquid with weak bonding between atoms, and the final state is solid with strong bonding between atoms. Thus, the austenite must be undercooled in order to produce pearlite, because the pearlite is a rather high-energy configuration of α+Fe$_3$C. This is a result of the large amount of area of α/Fe$_3$C interface, and a concomitant large interfacial energy. That is, the ΔG must be increased by undercooling to "pay for" the energy needed to form the large amount of α/Fe$_3$C interface in the pearlite. (The ΔG is energy released in the transformation, and the interfacial energy is consumed.) The fact that the transformation product is not in its minimum-energy configuration attests to the importance of the kinetic factors in the formation of pearlite.

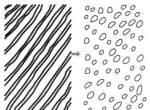

If the transformation depicted in Fig. 9.6 had been allowed to go to completion, and if the steel had then been held at this high temperature for a long time, the lamellar pearlite would have gradually broken up into globules of Fe$_3$C in a matrix of α by a process called *spheroidization*. The total amount of interfacial energy would continue to fall as coarsening of the particles of cementite occurred.

The decomposition of austenite in a eutectoid steel (i.e., 0.77%C) is a classical substitution reaction in which one constituent substitutes completely for another. In that respect, it is similar to the recrystallization of a cold-worked metal; both cases involve only short-range diffusion at the interface between the old and the new constituents. As in any substitution reaction involving nucleation and growth of the new constituent, the kinetics can be described by an S-shaped curve, in which, after some incubation time, the rate of transformation goes through a maximum. After that, the rate decreases, because the untransformed volume is depleted, and the transformed regions impinge on one another.

As already seen in Fig. 6.6, the kinetics of this kind of transformation can be expressed as a series of "S-curves" at different transformation temperatures. The S-curve portrays the fraction transformed as a function of time at a constant transformation temperature. To determine such a set of curves for a eutectoid steel, one follows the procedure illustrated in Fig. 9.5, austenitizing a series of specimens and transferring them, one at a time, into a hot-salt bath at the selected transformation temperature. Each specimen is then held for a different time at the transformation temperature before the transformation is arrested by quenching the specimen to room temperature. The amount of transformation can then be assessed by metallographic examination, as done for Fig. 9.6. This process is repeated for different transformation temperatures. The results would appear as shown in Fig. 9.8.

SECTION 9.2.1 ISOTHERMAL TRANSFORMATION

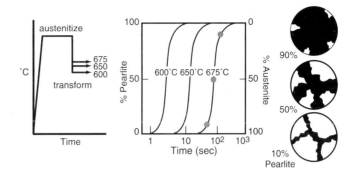

Fig. 9.8 S-curves showing how, in a eutectoid steel, the volume transformed from austenite to pearlite changes with time at different temperatures. The corresponding heat treatments and microstructural changes are also illustrated.

It would be found that the rate of transformation reaches a maximum at a particular temperature. The overall transformation kinetics can be expressed as a set of "C-curves," constructed from a series of S-curves, as shown in Fig. 9.9. The S-curve for a given transformation temperature provides, for example, the times for the beginning, the mid-point, and the end-point of pearlite formation.

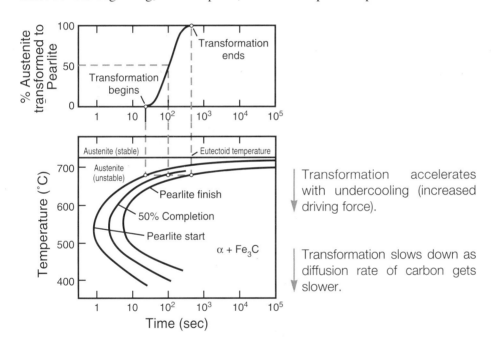

Fig. 9.9 The S-curve is used to construct the C-curve, as shown. A specimen quenched to 680°C and held for about 100 seconds would show 50% pearlite when quenched to room temperature. (Adapted from *Atlas of Isothermal Transformation and Continuous Transformation Diagrams*, Amer. Soc. for Metals, Metals Park, Ohio, 1977, p. 28.)

It can be seen from Fig. 9.9 that the rate of transformation goes through a maximum at around 550°C, referred to as the "nose" of the C-curve. The rate decreases as the transformation temperature is raised above the nose, because the thermodynamic driving force for the γ-to-pearlite transformation falls to zero as the temperature approaches 727°C. The rate also decreases as the temperature is lowered below the nose of the C-curve, owing to the temperature dependence of the rate of diffusion of carbon (cf. Fig. 9.4). That is, at low temperatures there is plenty of driving force, but here the transformation rate is controlled by the rate of carbon diffusion. Above the nose, the diffusion rate is rapid enough, and the driving force is rate-controlling.

When a eutectoid steel is transformed below the nose of the C-curve, the transformation product is still α+Fe$_3$C, but it is not called pearlite, because the lamellar microstructure cannot form at such low temperatures. Instead, a very fine microstructure is formed which comprises plate-like ferrite with cementite particles either surrounding or embedded in the ferrite plates, depending on whether it forms just below the nose or at even lower temperatures. This two-phase constituent is known as *bainite,* and its microstructure is so fine that an electron microscope is required to analyze it.

upper bainite

lower bainite

The C-curve is part of an *isothermal transformation* diagram, or IT diagram. It is useful for the study of transformation kinetics aimed at a fundamental understanding of the process. However, such a diagram should not be used directly for the heat treatment of steel. For a heat treatment consisting of austenitizing and cooling, one needs information about the kinetics of transformations occuring over a range of temperatures during continuous cooling.

9.2.2 Transformation during Cooling

A *continuous-cooling-transformation* (CT) diagram can be constructed from the results of several kinds of experiments. One involves the metallographic examination of specimens cooled for specific times at known rates before being quenched to halt the transformation. A useful method for this employs specimens, called *Jominy bars,* cooled one-dimensionally in a special fixture, which is illustrated in Fig. 9.10(a). In this procedure, a set of bars is austenitized, and then one bar is quickly transferred from the austenitizing furnace to a quenching fixture. Here, a stream of water of a specified temperature flows from a nozzle of specified diameter so that it reaches a specified height above the nozzle. Most of the heat flow out of the bar occurs at the bottom face, against which the water is directed. Comparatively little heat loss occurs by radiation from the sides of the bar. Under these conditions, the cooling rate at each position up the bar (in the interior, where the surface effects are absent) can be calculated. The cooling curves can be plotted on a diagram such as Fig. 9.10(b).

After a chosen time, the bar is quickly removed from the end-quench fixture and immediately plunged into cold water. The intention is to freeze-in the microstructure which had evolved up to this chosen quenching time. (The untransformed austenite remaining at that time transforms during the final quench in a diffusionless manner to *martensite,* which does not etch and therefore appears white.) An example of an interrupted transformation is seen in Fig. 9.6 in the regions that did not transform to pearlite. (The martensitic transformation is discussed in Section 9.4.)

After the transformation is interrupted by quenching, a flat face is formed by grinding along the long axis of the bar. This face is then polished and etched and examined metallographically, and the amounts of ferrite, pearlite, and bainite observed at each position are recorded. This process is repeated with the remaining bars for different choices of quenching times. Each bar is allowed to cool one-dimensionally for the chosen time before being completely quenched.

A diagram of the type shown in Fig. 9.10(b), containing cooling curves for closely spaced points along the bar, is used to plot the points at which each transformation begins and ends.

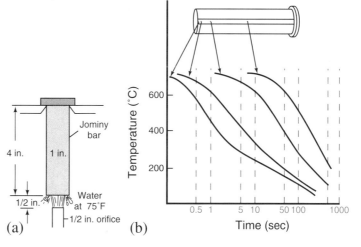

Fig. 9.10 (a) Schematic illustration of Jominy end-quench test used to produce essentially one-dimensional cooling of an austenitized steel bar. (b) Cooling curves for various positions along the bar; the vertical lines indicate the total time allowed for one-dimensional cooling in any bar before the transformation is interrupted by quenching the whole bar.

The important feature of this technique is that each position along the length of the bar corresponds to a given, known cooling curve. Thus, the position where the initial stage of ferrite formation, for example, is detected (by observation of a small amount of ferrite in a matrix of martensite) moves away from the water-cooled end as the time allowed for transformation is increased. The curves for the various stages of transformation can be constructed from the metallographic data, because the vertical lines representing the transformation time and the cooling curves for each position in the bar can give as many time-temperature data points as desired.

The results of such an experiment on a eutectoid steel are shown in Fig. 9.11, in which the CT diagram is compared with the IT diagram. This illustrates the general fact that the nose of the CT diagram occurs at lower temperatures and longer times than the nose of the IT diagram.

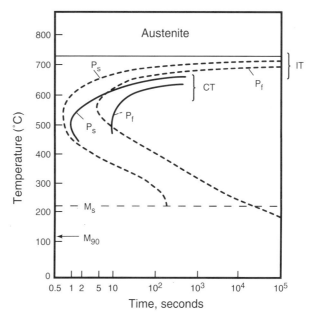

Fig. 9.11 Comparison of the isothermal and continuous-cooling-transformation diagrams for a eutectoid steel. (From *Atlas of Isothermal Transformation and Transformation Diagrams*, ASM, 1977, p. 28)

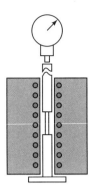

An alternative way to make a CT diagram is to use a *dilatometer*. This is a setup that measures length changes in a rod-like specimen as it is either heated or cooled at a controlled rate. Since the transformation γ-to-α involves a volume expansion (BCC α is not close packed), the dilatometer can detect the beginning of the formation of ferrite and of pearlite (which is mostly ferrite). For the highest precision, the two methods can be used to supplement each other.

Example 9.1

Q. What are the relative amounts of ferrite and cementite in the pearlite of a slowly cooled eutectoid steel?

A. This can be found by use of Fig. 9.2 and the lever rule, as follows:

$$\text{Relative amount of } \alpha = \frac{6.67-0.77}{6.67-0.02} = 88.7\%$$

$$\text{Relative amount of Fe}_3\text{C} = \frac{0.77-0.02}{6.67-0.02} = 11.3\%$$

Example 9.2

Q. Using the IT diagram in Fig. 9.9, find the time necessary for the formation of 50% pearlite in a eutectoid steel by isothermal transformation at 600°C, and compare the expected appearance of that pearlite with pearlite formed in a similar way at 650°C. Consult Fig. 9.4 to visualize the growth of pearlite into austenite.

A. The time to reach 50% pearlite at 600°C is found from the time axis of Fig. 9.9 to be just under 5 seconds, compared with about 12 seconds at 650°C.

650°C pearlite

600°C pearlite

Since the growth of pearlite requires diffusive partitioning of carbon at the advancing lamellar front, and since the transformation proceeds more rapidly at 600 than at 650°C (even though the diffusivity of carbon is lower at 600°C) the diffusion distance at 600°C must be shorter than at 650°C. The conclusion is that the pearlite formed at the lower temperature must be finer; i.e., the lamellae must be closer together. (The finer pearlite would appear darker when etched, because the grooves formed at the α/Fe$_3$C interfaces would tend to overlap to a greater degree.)

9.3 Transformations in a Carbon Steel Spoke

9.3.1 The "Equilibrium" Condition

The microstructure of the furnace-cooled carbon-steel spoke in Fig. 9.1(a) can now be evaluated. The sample was austenitized at 950°C and slowly cooled. The microstructure comprises roughly equal amounts of pearlite and ferrite. The first question is: What is the carbon content? This can be answered by first considering how the microstructure was formed and then using the lever rule. The ferrite is the *proeutectoid* phase. That is, it formed before the eutectoid transformation, and is analogous to the primary constituent which forms before the eutectic in the solidification of an alloy like a hypoeutectic Pb-Sn solder (cf. Fig. 8.15). The ferrite is stable as soon as the temperature falls below the boundary between the γ and $\gamma+\alpha$ phase fields; this is analogous to the liquidus line in a eutectic diagram. As the temperature falls, more ferrite forms, and the carbon content of the remaining austenite increases toward 0.77%. That is, as ferrite forms, carbon must diffuse away from it into the surrounding austenite.

After the maximum amount of proeutectoid ferrite has formed, the remaining austenite decomposes eutectoidally. If the microstructure is roughly 50% proeutectoid ferrite, then the carbon content of the steel must be about halfway between 0.02 and 0.77%. That means the spokes must have been made from steel with about 0.4%C. A metallurgist would then guess that a 1040 steel had been used. The 10 refers to a "plain carbon" steel, meaning nominally unalloyed, and the 40 means 0.4%C. (Any low-alloy steel is designated by the first two digits, which refer to the specific alloy elements present; the last two digits refer to the carbon content in this AISI/SAE system of steel designation.*)

In actual fact, a plain carbon steel does have elements other than iron and carbon which are added on purpose. For example, manganese is added, at levels usually below 1%, to scavenge sulfur, which is a deleterious impurity. In this way the sulfur is collected into manganese-sulfide inclusions, which are less deleterious, depending on their size, shape, and quantity. A bit of silicon or aluminum may also be added to the molten steel for purposes of tying up oxygen and nitrogen. Again, this would produce less-deleterious "dirt" in the form of oxides and nitrides. However, the amounts of these other elements are low enough to ignore when using the phase diagram; hence the designation "plain-carbon" steel.

9.3.2 Isothermal Transformation

The isothermal transformation kinetics of a 1040 steel can be represented by the IT diagram in Fig. 9.12, which also indicates the relationship to the phase diagram and the evolution of microstructures at temperatures above and below the eutectoid temperature. The application of this diagram can be described as follows: Let three specimens of a carbon-steel spoke be austenitized at 850°C and rapidly cooled to 750°C, then held for 2, 10, and 300 seconds, respectively. These isothermal holds are terminated by a rapid quench to room temperature. The schematic microstructures indicate that the steel was still fully austenitic after two seconds, but that after 10 seconds ferrite had begun to form at the austenite grain boundaries. The fact that carbon is rejected from the transformed regions is represented in the figure by speckles along the ferrite/austenite

---

* AISI = American Iron and Steel Institute; SAE = Society of Automotive Engineers

boundaries. After 300 seconds, the steel had essentially come to equilibrium, with about 40% ferrite and 60% austenite, as can be checked by using the lever rule in the two-phase region of the phase diagram. (A tie line can be drawn at 750°C, and the end points would be found at about 0.01 and 0.65%C; the overall alloy composition is, of course, 0.4%C.)

Alternatively, a transformation temperature of 675°C could be chosen. Let five specimens be austenitized, cooled rapidly to 675°C, and held for the following times, respectively: 0.5, 1, 2, 6, and 60 seconds. The microstructures at these times would be the following:

0.5 sec   The specimen is still fully austenitic.

1 sec   Ferrite has begun to form at austenite grain boundaries; carbon is being expelled into the surrounding austenite.

2 sec   The formation of proeutectoid ferrite is almost complete; the carbon content of the remaining austenite is approaching the eutectoid concentration (0.77%).

6 sec   Pearlite begins to form where the carbon concentration is high.

60 sec   Transformation is complete; microstructure is like Fig 9.1(a).

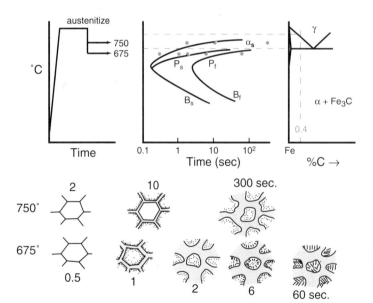

Fig. 9.12 The IT diagram for a 1040 steel, shown in relation to the phase diagram. The evolution of microstructure during isothermal holds at 750°C and 675°C is shown schematically.

The major difference from the IT diagram for a eutectoid steel (Fig. 9.9) is that the formation of proeutectoid ferrite must be accounted for. The ferrite nucleates on the austenite grain boundaries and grows into the austenite grains at a rate determined by the rate of diffusion of carbon away from the interface. That is, the carbon "piles up" in the austenite ahead of the advancing ferrite, and the ferrite cannot advance at a rate faster than that allowed by the diffusion of carbon down the resulting concentration gradient. This carbon pileup is represented by the speckles alongside the ferrite in the microstructures shown in Fig. 9.12 and is depicted schematically in Fig. 9.13 for ferrite forming at temperature T_1.

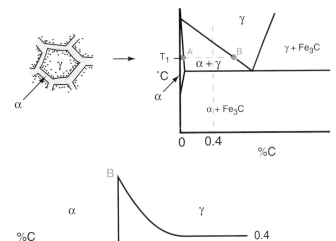

Fig. 9.13 Schematic representation of the piling-up of carbon in the austenite ahead of an advancing ferrite region at temperature T_1. The carbon must diffuse away (down the concentration gradient) in order that the ferrite can grow, since the ferrite can hold no more carbon than A%. The austenite in equilibrium with ferrite (i.e., at the interface) must contain B%, according to the phase diagram.

Note that the relevant diffusion distance for the formation of proeutectoid ferrite is much larger than for the formation of pearlite (cf. 9.4), since all the diffusion during pearlite growth takes place at the pearlite/austenite interface. Thus, pearlite can grow much more rapidly than large, "blocky" regions of ferrite.

The formation of proeutectoid ferrite is governed by its own C-curve. Since this ferrite forms prior to the pearlite, its C-curve lies at shorter times. Also, instead of having an asymptotic relationship with the eutectoid temperature, the ferrite-start curve has an asymptotic relationship with the temperature above which ferrite is unstable, i.e., about 800°C for a 1040 steel.

9.3.3 Transformations during Cooling of a 1040 Steel Spike

The CT diagram for a 1040 steel is shown in Fig. 9.14. Again, it differs from the CT diagram for the eutectoid (1080) steel in that the curve for the start of the formation of proeutectoid ferrite must be shown. Note, also, that the noses of both the IT and CT diagrams for the 1040 steel occur at shorter times than for the 1080 steel (Figs. 9.9 and 9.11). This is a manifestation of the fact that the transformation from γ to $\alpha+Fe_3C$ is retarded by an increase in the carbon content.

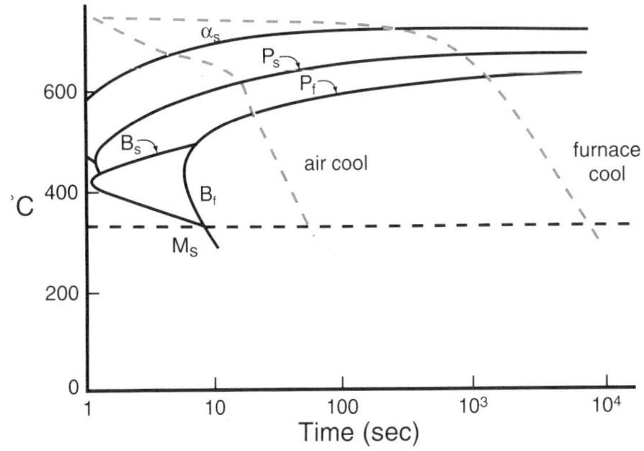

Fig. 9.14 The continuous-cooling-transformation diagram for a 1040 steel. (Adapted from C. F. zurLippe and J. D. Grozier, in *Atlas of Isothermal Transformation and Continuous Transformation Diagrams*, ASM, 1977, p. 418.)

The reason for the different appearance of the furnace-cooled and air-cooled spokes (Fig. 9.1) can be explained with the aid of the CT diagram in Fig. 9.14. Two cooling curves, representing air cooling and furnace cooling, respectively, are plotted on the CT diagram. For the case of air cooling, the time period available for the formation of proeutectoid ferrite (before the pearlite reaction starts) is only on the order of 10 seconds long. However, for the case of furnace cooling, the time for this transformation extends over several hundred seconds. Therefore, it is quite reasonable that in the furnace-cooled spoke the formation of proeutectoid ferrite has gone to completion, whereas, in the air-cooled spoke, the transformation was obviously interrupted soon after it had begun.

In addition to the fact that the formation of proeutectoid ferrite did not go to completion in the air-cooled spoke, one other feature of this microstructure deserves special mention. That is the extensions of ferrite plates jutting out from the blocky ferrite which formed along the austenite grain boundaries.

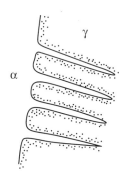

These ferrite side-plates are called *Widmanstätten* ferrite, after the man who first reported such a structure in meteorites. They form for the same reason that dendritic arms form at an advancing planar interface during solidification (cf. Fig. 8.14). An advancing plate of ferrite can reject carbon from its tip much more rapidly than from a flat interface, because the concentration gradient at the tip is much steeper. (The diffusion distance at the tip is about equal to the radius of curvature of the tip.) Thus, when ferrite is made to form quickly and at relatively low temperatures (cf. Fig. 9.1b), the planar interface becomes unstable, and what starts out as blocky ferrite switches to a plate-like growth mode.

If the cooling rate is such that the temperature gets too low for pearlite to form, then, as mentioned earlier, the two-phase decomposition product takes on the bainitic morphology. This can be viewed as the natural extension of Widmanstätten-ferrite formation. The only real difference between the two is that bainite has fine particles of cementite embedded in, or along, the ferrite plates.

Finally, it should be noted that, since the proper amount of proeutectoid ferrite did not form during air cooling of the 1040 steel spoke, the austenite that finally transformed to pearlite, or pearlite plus bainite, could not have had the proper amount of carbon (0.77%). Obviously, this was no impediment to the formation of the two-phase constituents. This emphasizes the point that, under non-equilibrium conditions, the phase diagram cannot be relied upon to give the compositions of the various constituents.

9.3.4 Processing of Spokes

In the manufacture of carbon steel spokes the starting condition was probably not the furnace-cooled, or "annealed," condition, because furnace cooling after austenitization is slow and, therefore, expensive. Rather, the air-cooled or "normalized" condition was probably used. The 1040 steel wire, which contained almost 7% cementite in the form of fine pearlite, was cold-drawn to the desired diameter. Cementite is normally a non-plastic, brittle material; however, it can be deformed plastically during cold drawing, because the drawing die imposes large radial pressures on the wire, opposing the tendency to brittle fracture or hole formation. (Decades ago Bridgman showed that one could even deform marble plastically, if sufficient hydrostatic pressure were imposed along with an applied tensile stress.)

Thus, during cold drawing, both the ferrite and cementite are plastically deformed to very large strains so that both phases become strung-out in the drawing direction (cf. Fig. 6.17). In the cold-drawn condition, in addition to being strengthened by the presence of a very high dislocation density, the 1040 steel has a component of strengthening by what might be called the "composite" effect. Here, a softer material (e.g., ferrite) is intimately bonded to a dispersed material (e.g., pearlite) that is harder and stiffer. The cementite carries more of the load than would an equivalent volume fraction of the ferrite. This composite effect is employed in many new "engineered" structural materials, including fiber-strengthened polymeric materials, which will be discussed in Chapter 14 in the context of tires and bicycle frames.

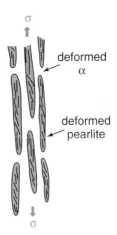

Example 9.3

Q. In Fig. 9.14 the cooling curve for the air cooling exhibits a decrease in slope following the beginning of the formation of ferrite. Explain.

A. The decrease in slope arises from the release of the heat of transformation ($\gamma \rightarrow \alpha$), and the phenomenon is the same as the change in slope that occurs when a liquid alloy cools past the liquidus temperature (cf. Fig. 8.9). In the case of the transformation in steel, the effect is called *recalescence*. (Heat effects are used in *thermal analysis*, which is an alternative method for determination of CT diagrams; it requires small specimens, sensitive measurement of temperature *vs.* time, and carefully controlled cooling rates.)

9.4 The Martensite Transformation

It has been pointed out that austenite that has not transformed *via* a diffusion-controlled transformation (e.g., to ferrite or pearlite) will transform in a diffusionless manner to martensite when quenched to room temperature. This new phase is not shown on the phase diagram, because it is not an equilibrium phase; it is only metastable. However, it is the desired phase for hardening of steel, because the hardness of martensite can be very high, depending on carbon content. This property is used to great advantage in components subject to wear, which is dealt with in the next chapter. For the present, only the nature of the martensite transformation will be described.

9.4.1 Hardening of Steel

Steel is hardened by heating it into the austenite region of the phase diagram and quenching, usually in water or oil, as shown schematically in Fig. 9.15. The purpose of the rapid cooling is to avoid the formation of ferrite or pearlite, which are relatively soft constituents. If the steel is cooled rapidly enough to temperatures where diffusion is slow, then the unstable austenite must transform martensitically.

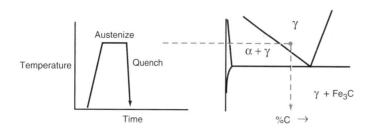

Fig. 9.15 Heat treatment required to harden a 1060 steel, i.e., to transform it to martensite.

In a martensitic transformation, no diffusion occurs. That is, the compositions of the parent phase and the product are identical; there is no partitioning of solute during the transformation. A martensitic transformation takes place by rapid shearing of the parent lattice into the lattice of the new phase. This occurs by the propagation of special kinds of dislocations along the advancing interface of the new phase, which usually forms as plates having a special crystallographic relationship (called a *habit* relationship) to the parent phase. The shearing from the parent FCC crystal to the new martensite crystal is complicated and can be viewed as comprising a homogeneous primary shear, similar to that in crystallographic *twinning* (see Appendix 10.1), and a nonhomogeneous secondary shear caused by dislocations and twins on the submicroscopic scale. Figure 9.16 shows schematically what this would look like in a grain of austenite. The reason for the lens-like shape of the martensite plate is that this minimizes the strain energy induced in both the martensite and in the parent phase by virtue of the difference in specific volume between the two phases. (Austenite is close-packed; martensite is not and therefore has a greater specific volume.)

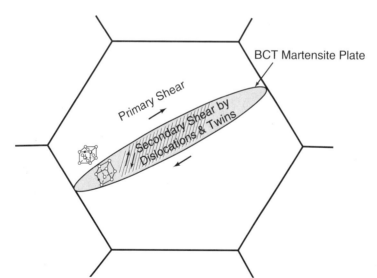

Fig. 9.16 Schematic representation of a martensite plate formed in a grain of austenite.

Note that the high density of defects in the martensite plate (from the secondary shear) means that the martensite comes with some strain hardening already built in. The interaction between these defects and carbon atoms (trapped in solid solution during the quench) is of major importance in the high hardness possible in iron-carbon martensite.

The special feature of Fe-C alloys that makes the martensite transformation a hardening process is that austenite can dissolve rather large amounts of carbon, but ferrite dissolves only a very small amount. The difference in the maximum carbon solubility in the two phases is two orders of magnitude (0.02 vs. 2.0%). This means that, when austenite containing a fair amount of carbon is quenched, the carbon is trapped in a crystal structure that is trying to shear into BCC ferrite. The result is a *body-centered tetragonal* (BCT) crystal structure, which can be viewed as a BCC crystal distorted along one axis, as shown in Fig. 9.17.

The tetragonal distortion is the result of the trapped carbon, which, as described in Section 7.2, occupies octahedral sites. A dilute solution of carbon in α iron causes only local tetragonal distortion (cf. Fig. 7.3), but in a concentrated solution there is an overall distortion of the lattice into a tetragonal crystal structure. This has two lattice parameters, instead of one, and the volume of the unit cell is $a \cdot a \cdot c$, rather than a^3. X-ray measurements of Fe-C martensites of varying carbon contents have shown that, as carbon content is decreased, the c parameter decreases and the a parameter increases, so that they converge to the lattice parameter of BCC iron at zero carbon content. That is, when carbon-free γ-iron is quenched and made to transform martensitically, the product is BCC.

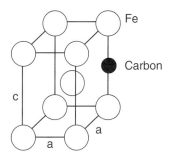

Fig. 9.17 (a) Unit cell of BCT iron-carbon martensite and (b) the effect of carbon concentration on the lattice parameters. Note that at 0% carbon $c = a = 2.86$ Å, the lattice constant for BCC α-iron. (From E.C. Bain and H. W. Paxton, loc. cit. p. 36.)

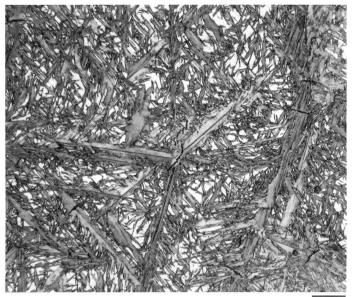

Fig. 9.18 Martensite plates in a partly transformed 1.4% steel. The white background is retained austenite. The martensite is slightly tempered at 160°C. (From A. R. Marder and G. W. Ruyak, Bethlehem Steel Corp.)

In a partially transformed, high-carbon steel the martensite plates can be made to stand out against the white background of retained austenite, as in Fig. 9.18, by *tempering,* which involves the precipitation of Fe_3C from the supersaturated (i.e., unstable) martensite. When tempering is carried to completion, the product is BCC ferrite with finely distributed carbide particles. In the early stages of tempering, the carbides are submicroscopic, but they cause a polished surface to become roughened by the etching process. Therefore, the incident light is scattered randomly, and the martensite plates appear dark.

The reason this steel transformed only partially when quenched to room temperature can be explained as follows: An Fe-C martensite is not only a metastable phase; it is also highly distorted by the supersaturation of carbon. Therefore, a large driving force is necessary for its formation. This driving force is provided by undercooling the austenite to a very large degree (quickly enough to avoid ferrite and pearlite, of course). The amount of martensite that forms depends on the degree of undercooling. The first martensite plates form at the *martensite-start* temperature, or M_s. Progressively more plates form as the temperature is reduced below the M_s. The reason still more driving force is needed is that there is a spectrum of potential nucleation sites for the plates, and only those requiring the least driving force are activated at the M_s.

The temperature at which the martensitic transformation is essentially complete is called the *martensite-finish* temperature, or M_f. (The transformation is never totally completed by a cooling treatment, because it is impossible to form new plates once the untransformed volume falls to a very low level; thus, any quenched steel would have at least a few volume percent of retained austenite, which could be seen only with the aid of an electron microscope.)

Because the martensitic transformation is diffusionless, it cannot be represented by a C-curve on a transformation diagram, in the manner of ferrite, pearlite, and bainite. That is, the amount of martensite formed depends on the temperature reached during cooling, not on the amount of time spent at any given temperature. Thus, the existence of the martensite transformation can be indicated on a transformation diagram only by the temperature at which it begins and the fractions of martensite that form at temperatures below the M_s.

It has been seen from the phase diagram in Fig. 9.2 that carbon is an austenite stabilizer; that is, an increase in carbon content expands the temperature range over which austenite is stable. Also, the amount of distortion in a martensite plate is a function of the amount of carbon trapped in it. This increases the driving force (i.e., undercooling) needed to drive the transformation. Therefore, the M_s temperature decreases as the carbon content increases, as shown in Fig. 9.19. So does the M_f temperature. When the carbon content is higher than about 1%, the M_f is far enough below room temperature that the retained austenite (which etches white) is easily seen in an optical microscope (Fig. 9.18).

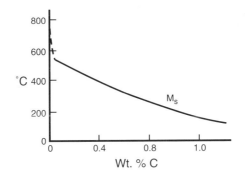

Fig. 9.19 The decrease in the M_S temperature with increasing carbon content in Fe-C alloys. (From *Principles of Heat Treatment of Steel,* G. Krauss, ASM, 1980, p. 52)

Example 9.4

Q. In Fe-C martensite the BCT crystal structure is the result of the entrapment of carbon atoms in one of the three sets of octahedral sites depicted in Example 7.2. Let the z sites be the ones occupied in a martensite plate in a 0.43%C steel; calculate the fraction of possible sites occupied by carbon.

A. Converting 0.43%C to atomic percent (approximately $56/12 \cdot 0.43 = 2$ at%), we find that there is about one carbon atom for every 50 iron atoms in the steel. Like a BCC crystal, a BCT crystal has two iron atoms and one z-type octahedral site per unit cell (4 z edges, each shared by 4 unit cells). Therefore, there are 25 possible z-type octahedral sites for every 50 iron atoms. Thus, in the martensite plate in the 0.43%C steel, only one octahedral site in 25 is occupied by a carbon atom. Obviously, the influence of carbon in iron must be very large for such a low site occupancy to change the crystal structure.

Summary

The various microstructures developed by heat treatment of a carbon steel spoke can be understood by use of a portion of the Fe-Fe$_3$C phase diagram. This portion contains the solid-state analog of a eutectic reaction, called a eutectoid reaction, in which a high-temperature solid phase decomposes upon cooling into two other solid phases. In this case the FCC (austenite) γ-phase transforms to pearlite, comprising a lamellar mixture of BCC (ferrite) α-phase and Fe$_3$C (cementite). Because the driving force for this reaction is smaller than for the eutectic reaction (i.e., the ΔG is smaller), and because solid-state diffusion is much slower than diffusion in a liquid, the eutectoid decomposition is comparatively sluggish. For this reason it is necessary to use transformation diagrams (or time-temperature diagrams) to display the kinetics of transformation of steels of various compositions. The isothermal-transformation diagram is useful to gain an understanding of a transformation, but a continuous-transformation diagram is more useful for purposes of heat treatment. With the CT diagram for the 1040 carbon steel spoke one can readily understand the difference in the microstructures of a slowly cooled *vs.* an air-cooled spoke (after austenitization).

The C-curve shape of a transformation diagram can be understood by considering the effect of undercooling on the thermodynamic driving force and the countervailing effect on the rate of diffusion. Finally, the diffusionless transformation of austenite to martensite upon quenching so rapidly as to suppress the diffusion-controlled formation of ferrite and cementite was considered, along with the effect of the carbon content on the degree of tetragonality of martensite and the M_s temperature.

Exercises Chapter 9

Terms to Understand

Austenite	C-curve
Cementite	Isothermal-transformation (IT) diagram
Ferrite	Continuous-transformation (CT) diagram
Pearlite	Plain-carbon steel
Eutectoid	Normalized steel
Proeutectoid ferrite	Martensitic transformation
Bainite	Widmanstätten ferrite
Martensite	M_s temperature
Jominy bar	M_f temperature
Dilatometer	Retained austenite
Thermal analysis	Undercooling
Recalescence	Body-centered tetragonal, BCT

Problems

9.1 Explain why the proeutectoid ferrite in Fig. 9.1(b) appears to lie in a cellular network.

9.2 Given a 1030 steel being cooled very slowly from the austenite region:

(a) At what temperature can ferrite first start to form?

(b) Give the relative amount and the composition of each phase present at 750°C, assuming equilibrium conditions.

(c) Give the relative amount of each phase and of each microstructural constituent present at 725°C, again assuming equilibrium.

9.3 Using Fig. 9.4, estimate the volume fraction of Fe_3C in pearlite. Using the approximation that ferrite and Fe_3C have the same density (i.e., volume fraction equals weight fraction), compare your estimate with the value calculated from the phase diagram with the lever rule.

9.4 A eutectoid steel is austenitized at 800°C and quenched rapidly to 625°C, where it is held for isothermal transformation. Using Fig. 9.9, give the time (approximately) to form 75% pearlite. What would the other 25% be? What would the latter be if the steel were quenched rapidly to room temperature at this time?

9.5 A one-inch diameter bar of 1080 steel is austenitized and quenched so that the microstructure at room temperature contains martensite, bainite, and pearlite.

   (a) Sketch the cross-section of the bar and indicate on it where you would expect to find the largest fraction of each of these three constituents.

   (b) Name the *phases* present in the quenched bar.

9.6 A 1090 steel is austenitized at 850°C and very slowly cooled.

   (a) At what temperature could a new phase first appear? What is this phase?

   (b) Sketch the microstructure you would expect to find at room temperature, and give the relative amount of each constituent.

9.7 If a 1080 steel were quenched to martensite and then tempered for a long time at, say, 650°C, the microstructure would comprise spheroids of Fe_3C in a matrix of ferrite. How could this microstructure be converted to pearlite?

9.8 What experiments would you use to develop a heat treatment for a 1040 carbon steel spoke that would result in a microstructure resembling that of a slowly cooled 1060 steel. (An approximation giving the principles employed is sufficient.)

9.9 Devise a heat treatment of a 1040 carbon-steel spoke which would produce about 50% ferrite and 50% martensite.

9.10 The growth rate of pearlite remains essentially constant as the growth proceeds, whereas the growth rate of proeutectoid ferrite slows down as growth proceeds. Explain.

10

THE CHAIN AND BEARINGS; RESISTANCE TO FRICTION AND WEAR; HARDENING OF STEEL; CERAMIC BEARINGS

The efficient use of human power is obviously crucial to the success of the bicycle. Minimization of energy wasted on friction is just as important as the minimization of weight. Smoothness of operation and sureness of mechanical connections are part of this picture, and these require the minimization of wear. Friction and adhesive wear are closely related, and the remedies for friction and wear are similar, so the two phenomena can be discussed in the same context. From the materials standpoint, the ideal approach would be two smooth, hard components separated by a film that can withstand a large pressure but has almost no resistance to shear.

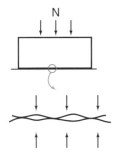

10.1 Friction

The root cause of friction and adhesive wear of metallic components, like bicycle chains and bearings, is the fact that metal surfaces are never smooth and inert; rather, they are rough and chemically active. This means that mating surfaces make contact at their respective asperities; it also means that chemical bonding can occur if there is metal-to-metal contact, because of the nature of metallic bonding.

When metal atoms are isolated from each other, as in a vapor, their valence electrons occupy discrete energy levels, each of which can hold only two electrons, according to the Pauli Exclusion Principle. When these electrons become constrained in a small volume in the crystalline state, a large number of energy levels is required, and these levels are squeezed so closely together that they form a quasi continuum, referred to as an *energy band*. The formation of such a band is illustrated schematically in Fig. 10.1.

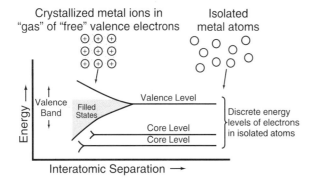

Fig. 10.1 Schematic illustration of the formation of an energy band from a discrete electron-energy level of isolated metal atoms when these atoms are condensed into a crystal. A valence band is shown here to be partly filled; i.e., only a fraction of the available energy states is occupied with valence electrons.

The configuration of an energy band is such that it is asymmetric with respect to the original valence level of isolated atoms. That is, the bottom of the energy band extends farther below the original valence level than the top of the band does above it. This means that, even if the valence band were filled in the crystalline state, the average level of the valence electrons would be below that in the isolated-atom state. This reduction in the average energy of the valence electrons resulting from band formation is the source of the binding energy which causes the crystallization to occur (below the freezing point).

When clean metal surfaces are brought into close proximity, the electrons in the valence band on one side of the interface co-mingle with those on the other side, forming a metallic bond in the same way as occurs during crystallization. This "cold welding" process occurs easily when metal surfaces are rubbed together in a vacuum, where the oxide which covers most metallic surfaces is unable to re-form. This problem can also occur at normal atmospheric pressure, albeit less readily. It is known as "seizing" when it occurs between the components of a machine. On the microscopic scale, it is the predominant mechanism behind friction and adhesive wear of metallic components.

When two metal surfaces are pressed together by a large force normal to the boundary plane, flattening of the surface asperities can occur by means of plastic deformation. The flattening increases the contact area in the interface and, thus, the area over which cold welding can occur. This is why the frictional force between two sliding objects increases with the normal force acting across the contact surface. The frictional force is defined as the force that opposes the relative sliding of two solids subjected to both a normal force, N, and a shearing force, S. Experiments have shown that the frictional force, f, is proportional to N through the coefficient of friction, μ.

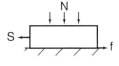

The μ for the beginning of motion, static friction, is generally greater than for sliding friction, because there is more cold welding at the beginning of sliding.

This relationship can be understood in a general way if strain hardening is ignored and the solids are approximated as "perfectly plastic," meaning that the stress-strain curve is horizontal after yielding occurs (in compression) at $N/A = \sigma_y$. Thus, the contact area, A, increases with N according to

$$N = \sigma_y A$$

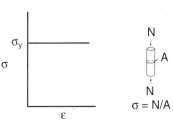

where σ_y is the tensile yield stress.

If bonding occurs across the boundary, then sliding requires the shearing of the metallic bridges. This requires a shear stress of at least $\sigma_y/2$, which, from Schmid's law (cf. Section 3.5), would be the yield stress in pure shear. Thus, the frictional force can be approximated by:

$$f = (\sigma_y/2)A$$

Substituting $\sigma_y = N/A$ gives $f = \mu N$, with $\mu = 0.5$.

This model is too simplistic; it assumes that cold welding occurs along the whole contact area. However, it is reasonable to expect that the amount of cold welding would increase with deformation of the asperities (i.e., with N), since this deformation would help to break up the oxide layers and other barriers that inhibit cold welding. Thus, in general, in order to reduce μ, one or more of the following must be done:

1. Reduce the deformation of the asperities; i.e., harden the mating solids.

2. Reduce the cold welding; i.e., reduce metal-to-metal contact.

3. Create a boundary region that is weak in shear.

10.2 Wear

The shearing of metallic bridges just described produces *adhesive wear*. When the cold-welded asperities are subjected to large local plastic deformation, they are strain hardened for some distance back from the original interface, as shown schematically in Fig. 10.2(a). Thus, when plastic shearing of this metallic bridge takes place, it tends to occur not on the original interface but in the softer metal some distance away, as in Fig. 10.2(b). This results in transfer of metal from one side of the interface to the other, producing adhesive wear of the side that lost the metal. Further shear of the type shown in Fig. 10.2(c) produces debris in the form of bits of hardened metal sheared from rubbing surfaces. Subsequent oxidation can lead to hard metal-oxide particles. When such particles are pushed along the surfaces ahead of other sliding asperities, microscopic grooves are cut in the surface. This is the mechanism of *abrasive wear*.

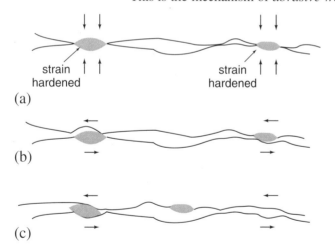

Fig. 10.2 Mechanism of adhesive wear, involving cold welding followed by shearing of the metal bridges.

Abrasive wear is akin to metal removal in a machining process, but on the microscale. It involves plowing furrows in the surface of a metal when a particle is pushed along the surface. It occurs by means of plastic deformation so large that the material is completely sheared away.

To reduce abrasive wear one must:

1. Eliminate hard particles (e.g., dirt) as much as possible.

2. Separate the surfaces by creating an interfacial layer that is thicker than the size of the hard particles and is weak in shear.

Applying these principles to the chain and bearings of a bicycle, one would want to make the mating surfaces as hard as possible, thereby reducing the flattening of the asperities, and to use lubricants to create an interfacial layer that would prevent cold welding and would be weak in shear. Another possible alternative is to consider the use of ceramic components. Not only are ceramics much more difficult to deform plastically than metals, but their surfaces are less reactive. Thus, cold welding is less likely.

10.3 Lubrication

Regardless of the material used, a primary resource to combat friction and wear is lubrication. The functions of a good lubricant should already be clear from the foregoing discussion of friction and wear. There are essentially two types of lubrication: *boundary lubrication,* which operates at low sliding velocities to provide a layer of material weak in shear that prevents metallic contact, and *hydrodynamic lubrication,* which serves the same function and gives a much lower coefficient of friction at high velocities.

A common boundary lubricant is an oxide film which prevents cold welding between metallic components. The absence of such a film on surfaces cleaned (e.g., by ion bombardment) in ultra-high vacuum (e.g., outer space) leads to cold welding; this is a problem encountered in space technology.

An oxide film can be removed by mechanical abrasion. When this happens the result is the production of debris that can lead to rapid abrasive wear. Therefore, oxide films are not generally considered appropriate for lubrication. Sulfides would be much better, because they tend to be soft, particularly in shear, and sulfur adsorbs strongly on most metallic surfaces. Molybdenum disulfide is a well-known solid lubricant, because it forms platelike crystals that slide past each other easily. In this respect it is similar to graphite, having only van der Waals forces between the crystal planes (cf. Section 9.1).

A more common type of boundary lubricant employs chain-like organic molecules, one end of which has a chemical group that bonds strongly to a metal surface. The molecular coatings act somewhat like the bristles on a brush on each opposing surface and thereby tend to prevent metal-to-metal contact. Although strong (covalent) bonds exist within the molecules, the intermolecular forces are of the weak van der Waals type. Hence, the molecular films are weak in shear.

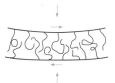

When the relative motion between metallic components occurs at high speeds, hydrodynamic lubrication can occur. This relies on the principle that it is difficult to squeeze out all of a fluid trapped between two solids under pressure. A classical example is a shaft turning in a bearing ring containing oil.* Generally, some force presses the shaft against one side of the ring. As the shaft turns, oil is dragged into the wedge-like space between the shaft and ring, as shown in Fig. 10.3(a). Ideally, the shaft is supported fully on this film, which may be under very high pressure. In this case, all the shear occurs in the oil film, and the value of μ may be as low as 0.001. The actual value depends on the viscosity, η, of the oil, the relative velocity, v, of the components, and the pressure, p, as shown in Fig. 10.3(b). In special cases of extremely high speeds, pressurized air can be used; examples are high-speed dental drills and gyroscopes.

---

* Another example is the hydroplaning of an automobile on a wet road.

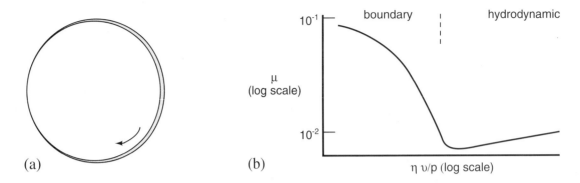

Fig. 10.3 (a) Schematic illustration of trapping of a viscous oil between a rotating shaft and the support ring; the shaft rides on a pressurized film of oil. (b) Dependance of μ on the viscosity of, and pressure on, the oil film and the relative velocity of the moving parts in the boundary and hydrodynamic regions.

For the wheel bearings of a bicycle, the lubricant should provide adequate boundary lubrication at low speeds and good hydrodynamic lubrication at high speeds. Thus, the chain-like organic molecules must bond at one end to the metal, and the chains must be long enough to provide sufficient viscosity to resist the high pressures and form a thick-enough boundary film. However, the viscosity must not be raised so much as to increase the coefficient of friction to unacceptable levels. The lubricant should be resistant to oxidation and breakdown at high temperatures and, of course, be noncorrosive. The art and science of formulating lubricants include not only the configuration of the organic molecules, but also the additives which impart these other characteristics.

A much more demanding set of requirements is found in modern automobile engines. Here, the lubricant must be formulated to withstand the extreme pressures of high-compression engines, to provide a detergent action to remove combustion products, and to have a relatively constant viscosity over a fairly large temperature range. Thus, an engine oil must not be so viscous as to prevent starting in the dead of winter, but it must still protect the engine from wear at high speeds in the summer. Compared to this, lubrication of a bicycle is fairly straightforward.

10.4 The Use of Hardened-Steel Components

In present day technology, the "materials fix" used to combat friction and wear in the bearings and chain of a bicycle is to make them of hardened steel. A steel commonly used in bicycle bearings is designated 52100, which means it contains chromium and 1% carbon. The high carbon content is used to provide a microstructure of hard plate-like martensite in which are embedded small particles of carbide, as illustrated by Fig. 10.4. This microstructure can be obtained by austenitizing a hypereutectoid steel somewhat above the eutectoid temperature (where high-carbon austenite and residual (spheroidal) carbides would be formed) and quenching to room temperature. Several precautions are needed in the heat treatment of this kind of steel; they are discussed in Section 10.4.4.

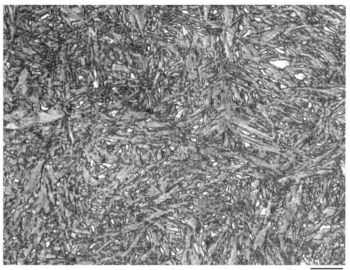

Fig. 10.4 Microstructure of a 52100 steel bearing.

10.4.1 The Hardness of Martensite

The shearing process that transforms austenite to martensite in Fe-C alloys is complicated on the atomic scale. Part of the shearing occurs by the creation and glide of slip dislocations or by the formation of twins on a very fine scale. (Twinning is explained in Appendix 10.1.) Accomplishment of the martensitic transformation partly by slip predominates in low-carbon martensites, whereas this occurs mainly by twinning in high-carbon steels. In either case, the martensite has a very high defect concentration as soon as it is formed, and the interaction of the trapped carbon with these defects is the primary source of the hardening phenomenon. The hardness of martensite increases sharply with the carbon concentration, as shown in Fig. 10.5. The effect of carbon is so great that the hardness of martensite can be considered to be a function of only the carbon content.

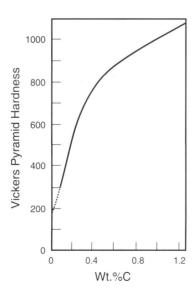

Fig. 10.5 The hardness of martensite depends only on the carbon content. (From E. C. Bain and H. W. Paxton, loc. cit., p. 3.)

10.4.2 Hardenability of Steel

The hardening of small components, such as bicycle chains and bearings, is fairly straightforward, because, for a carbon content above about 0.4%, it is easy to cool a small component quickly enough to form martensite. However, a component with a thick cross-section would present a problem, because the cooling rate in the interior would be considerably slower than at the surface. Hence, the interior could transform to the unwanted soft products, ferrite and pearlite. For such cases the diffusion-controlled decomposition of γ must be retarded by the addition of alloy elements. Some elements, like manganese and nickel, stabilize the austenite phase and thereby lower the driving force for the $\gamma \rightarrow \alpha$ transformation, allowing it to occur at lower temperatures where martensite or bainite is favored. Other elements, like chromium, molybdenum, and vanadium, are strongly attracted to carbon atoms and interfere with the diffusion of carbon in austenite. This would be expected to retard a diffusion-controlled reaction. These alloying elements are said to impart *hardenability* to a steel. This does not mean that they increase the hardness of the martensite, but rather that they enhance the tendency to form martensite (or bainite) during quenching. A comparison of the CT diagrams of a 1040 steel and a 4340 steel (which contains about 1.7%Ni, 0.7%%Cr, and 0.3%Mo) is shown in Fig. 10.6. The latter can be transformed to martensite or bainite with a rather slow cooling rate, compared to the 1040 steel.

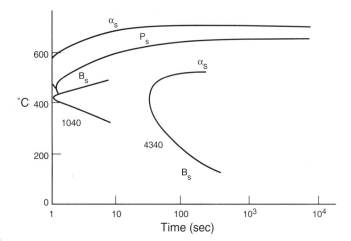

Fig. 10.6 Comparison of the CT diagrams for a 1040 and a 4340 steel.

The standard way of measuring the hardenability of a steel is the Jominy end-quench method, illustrated in Fig. 9.10. In this case, the one-dimensional cooling is allowed to continue until the whole bar is cooled essentially to room temperature. A flat surface is created by grinding the bar longitudinally, and hardness measurements are made at intervals along the bar starting from the quenched end. A comparison between a 1040 steel and a 4340 steel is made in terms of the Jominy hardness curves in Fig. 10.7. The alloy elements obviously make a very large difference in the hardenabiltiy of a 0.4%C steel.

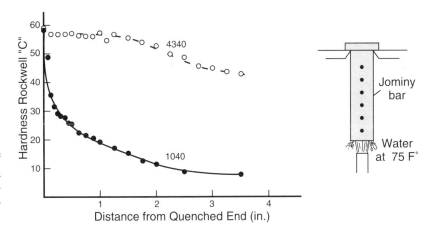

Fig. 10.7 Comparison of Jominy hardness curves of a 1040 and a 4340 steel, showing the much greater hardenability of the latter.

10.4.3 Tempering of Steel

The hardness of a quenched steel brings with it a tendency toward brittleness. Therefore, one usually sacrifices some of the hardness to gain toughness, which is the reason for tempering a hardened steel. This is illustrated schematically in Fig. 10.8. The extent of tempering is selected to fit the application. For example, knife blades are made from high carbon steel and are only lightly tempered (i.e., tempered at a relatively low temperature), since hardness is much more important than toughness in this application. However, bicycle chains are complex in geometry and are subjected to complicated stresses, so toughness is important. Therefore, steels with lower carbon content are used, and tempering is carried out at somewhat higher temperatures.

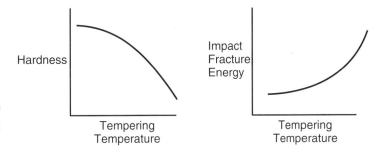

Fig. 10.8 Idealized representation of variation of hardness and toughness during tempering.

The martensite in lower-carbon steels is lath-like, rather than plate-like, and this makes the microstructure fine and complex. Observation by transmission electron microscopy is necessary to clarify it. An example is shown in Fig. 10.9.

Tempering occurs in stages. First, a transition carbide having a hexagonal crystal structure forms, and the martensite becomes somewhat less tetragonal. Then, cementite forms and the martensite becomes still less tetragonal. At higher temperatures or longer times, the cementite particles coarsen (driven by the minimization of interfacial energy) and the matrix becomes BCC.

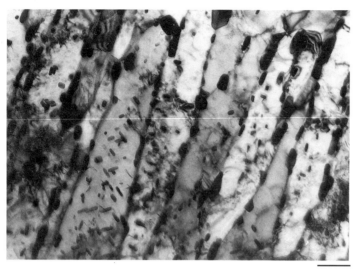

Fig. 10.9 A 4130 steel water quenched from 900°C and tempered for 1h at 650°C. (Courtesy F. Woldow and Prof. George Krauss, Colorado School of Mines.)

3μm

In an alloy steel, when the temperature is high enough for the diffusion of substitutional solutes, alloy carbides can form. In the case of chromium, these can be M_7C_3 or $M_{23}C_6$, where M is either chromium or iron. These form initially in a fine distribution. Since they act as effective barriers to dislocation motion, they provide some additional hardening, superimposed on the softening due to coarsening of cementite. This is generally called *precipitation hardening*, but in alloy steels it is called *secondary hardening*. The secondary hardening produced by chromium and molybdenum is illustrated in Fig. 10.10. It is important not only in bearings, but also in cutting tools such as drill bits. In the latter case, the secondary hardening (utilizing Cr, Mo, W, and V) retards softening, and therefore wear, during high-speed metal cutting.

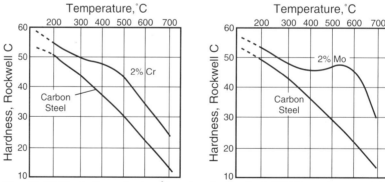

Fig. 10.10 Illustration of secondary hardening in 0.35%C steel caused by formation of chromium and molybdenum carbides. (From E.C. Bain and H. W. Paxton, loc. cit., pp. 197 and 200)

The fact that chromium and molybdenum form fine-scale, stable carbides is why Cr-Mo steels are used for lightweight bicycle tubing, as will be discussed in the next chapter. The steel in Fig. 10.9 is a bicycle-tubing alloy. The coarse, inter-lath carbides are probably M_3C, and the fine intra-lath ones are probably alloy carbides.

10.4.4 Heat Treatment of Steels

When applied to steels, the term "heat treatment" generally means "hardening," which involves austenitizing, quenching, and tempering. In addition to achieving the desired final hardness, one wants to avoid two potential pitfalls: Cracking or distortion can result from quenching, and dimensional instability can result from retained austenite. Cracking or distortion is caused by the large temperature gradients between the inside and outside of an object, which accompany rapid cooling. These thermal gradients mean that the volume changes from thermal contraction or a phase transformation occur nonuniformly, and the inevitable result is the buildup of stresses, called *thermal stresses,* between the inside and the outside of a rapidly cooled object.

Consider the effect of thermal contraction alone. When a hot cylinder is quenched, the outside contracts and thus squeezes the inside, which responds by deforming plastically (because its strength is relatively low at the high temperature). By the time the inside starts its thermal contraction, the outside is already cold and hard. Hence, the inside is constrained from contracting fully. The residual stress pattern, therefore, is tension on the inside and compression on the outside.

If there is a defect of sufficient magnitude on the inside, the tensile stress could cause the object to crack. If the geometry of the object were more complicated, i.e., if the cross-section were not uniform and the object lacked simple symmetry, then the thermal stresses could cause distortion by plastic deformation.

Alternatively, consider the effect of a dilatational phase transformation, like that of close-packed austenite to non-close-packed martensite. The same reasoning would lead to the conclusion that the residual stress pattern would be compression on the inside and tension on the outside. Here, a defect in the outside region could grow into a crack. This becomes more probable as carbon content is increased, because the toughness of martensite decreases as hardness (i.e., carbon content) increases.

Now, consider the effects of increasing the carbon content of a steel. As pointed out in Section 9.4, the M_s and M_f temperatures fall as the carbon content increases. This means that, in a high-carbon steel, the volume changes due to the martensitic transformation take place at low temperatures. For this reason, the associated stresses are less likely to be relieved by plastic flow. The combination of the low transformation temperatures and the high hardness of the martensite makes cracking during a water quench a distinct possibility when the carbon content is greater than about 0.3%, depending on the size of the object.

The way to prevent quench cracking and distortion is to use a slower quench. For example, oil, or even hot oil, is often used instead of water. This usually calls for more hardenability than is found in carbon steels, so alloy steels must be used, except with thin cross-sections. This is one reason a chromium steel like 52100 (1.45%Cr, 1.00%C) is used for bicycle bearings.

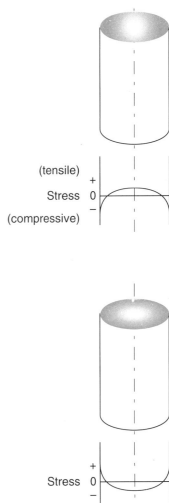

The problem of dimensional stability is related to the amount of retained austenite (e.g., cf. Fig. 9.18), which is increased by high carbon contents (i.e., low M_s and M_f). If a steel with retained austenite is allowed to sit at or above room temperature for a long time, some of the austenite could begin to transform isothermally to bainite. This could occur because the lower part of the C-curve for the bainite transformation (cf. Fig. 9.11) continues to have meaning, regardless of the fact that most of the steel has transformed martensitically. Since the γ-to-bainite transformation is dilatational, such a transformation could produce dimensional distortion (or even cracking if the hardness is high enough) after a component, like a bearing, is machined to high-precision tolerances.

To prevent this distortion, two measures can be employed. One is to cool a quenched high-carbon steel to cryogenic temperatures (like 77K, using liquid nitrogen) immediately after the quench, in order to approach or to get below the M_f. The other is to temper the steel at temperatures high enough to transform the retained austenite to bainite during the tempering treatment. In this case it is desirable to have chromium or other carbide-formers in the steel, to retard softening during tempering (cf. Fig. 10.10). This is another reason for using 52100 steel for bearings.

Bicycle chains are made of thin-walled parts. Since thin high-carbon components have sufficient hardenability to be oil-quenched, alloy steels are not needed here. The normal approach is to use the minimum carbon content needed to give the desired hardness, since the toughness of steels of equivalent hardness is inversely related to carbon content. (Remember that hardness is controlled by the tempering treatment, as well as by the carbon content of martensite.) Figure 10.11 shows the microstructure of a pin from a chain. It comprises a mixture of bainite and tempered martensite. From the standpoint of mechanical behavior, these two constituents can be considered more or less equivalent.

Fig. 10.11 Microstructure of the pin from a bicycle chain, consisting of bainite and tempered martensite.

Example 10.1

Q. Assume that a highly alloyed steel, when hardened by quenching to room temperature, has 10% retained austenite. Calculate the volume expansion that would occur if this austenite later transformed to a ferrite-like phase (i.e., martensite or bainite).

A. FCC austenite has a packing density of 0.74, while that of ferrite is 0.68 (cf. Example 2.3). Therefore, a $\gamma \to \alpha$ transition would involve a volume expansion, $\Delta V/V = (0.74-0.68)/0.74 = 8\%$. In a steel with 10% retained austenite, the expansion would be $0.1 \times 8 = 0.8\%$, which would be intolerable in a precisely manufactured component, like a bearing set. A $\Delta V/V$ of 0.008 would translate approximately to a linear strain, $\Delta l/l$, of 0.003 (since $\Delta V/V \approx 3\Delta l/l$, if second-order terms and higher are ignored in the expression $\Delta V/V = [(1+\Delta l)^3 - 1]/1$). Such a transformation would obviously produce a very large residual stress in the component.

10.4.5 Surface Hardening

The simplest and cheapest method to improve wear resistance would be to use a high-carbon steel in the quenched-and-lightly-tempered condition. This is appropriate for the bearings, which are loaded mainly in compression. Here, the brittleness of high-carbon martensite is not a problem. However, the chain obviously cannot be made from a brittle material.

One possible approach for the chain would be to use a low-carbon steel for the components and then to create a high-carbon surface by *case hardening*. This can be done by *carburization*, which involves heating the steel in an atmosphere that deposits carbon on the surface in the temperature range where austenite is stable (because austenite has a high solubility for carbon). For example, one could use either a $CH_4 + H_2$ or a $CO + CO_2$ atmosphere. These produce carbon according to:

$$CH_4 \to C + 2H_2 \quad \text{or} \quad 2CO \to C + CO_2$$

The carbon then diffuses into the steel, and the high-carbon surface region gives a hard, wear-resistant case supported by a lower-carbon, and therefore tough, core. The process and its effects are illustrated in Fig. 10.12.

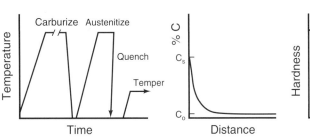
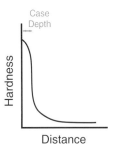

Fig. 10.12 Thermal cycle and resulting carbon and hardness profiles in case hardening by carburizing.

The case thickness should be sufficient to give the desired wear resistance without being so great as to allow large cracks to form, since the latter could lead to fracture of the component. Since hardness and toughness are inversely related (cf. Fig. 10.8), a hard case would tend to be brittle and the softer core would tend to be tough. Surface hardening allows the surface to be hard and wear-resistant without having low toughness throughout the component. That is, the tough core can act as a crack-arrestor, if a crack should happen to occur in the case. The subject of cracks will be addressed in the next section.

The case depth would obviously increase with the temperature and time of carburization, and it is necessary to be able to predict and control this process. For this one must understand something about diffusion in solids.

The analysis of diffusion is exactly analogous to the analysis of heat flow; the same equations are used. It is also analogous to the flow of electric current in a conductor. In all three cases the flow of something is proportional to a driving force. The latter can be a gradient in mass concentration, or temperature, or electrical potential. For mass flow, the constant of proportionality is the *diffusion coefficient* (also called the *diffusivity*). This is analogous to the thermal conductivity or electrical conductivity. (As discussed in Section 8.3, the fundamental driving force for mass flow is actually the gradient in chemical potential; however, in many cases this is approximately equivalent to the concentration gradient.)

Thus, for diffusion one can write:

$$\text{Mass flux} = J \text{ (mass/unit area/unit time)} = \text{constant} \cdot dC/dx,$$

where C is the concentration (mass/unit volume) and x is the distance along which diffusion occurs. For simplicity, flow in only one dimension is assumed, but the extension to three dimensions is straightforward.

The preceding expression is known as *Fick's first law* and is normally written as:

$$J = -D \, dC/dx$$

where D is the diffusion coefficient. Note that J>0 when dC/dx<0.

To predict the case depth in surface hardening of steel by carburization, it is necessary to determine how much carbon accumulates in a given region during some diffusion time, Δt. That is, the rate of accumulation of carbon, $\Delta C/\Delta t$, in some region of thickness, Δx, must be calculated. For this a simple mass balance is used: The accumulation in the region Δx in time Δt equals the flux in, J_1, minus the flux out, J_2. That is, $\Delta C \times \text{volume} = (J_1 - J_2) \times \text{area} \times \text{time}$

$$\Delta C \, A \, \Delta x = (J_1 - J_2) \, A \, \Delta t$$

or

$$\Delta C \, \Delta x = -\Delta J \, \Delta t$$

and

$$\frac{\Delta C}{\Delta t} = -\frac{\Delta J}{\Delta x}$$

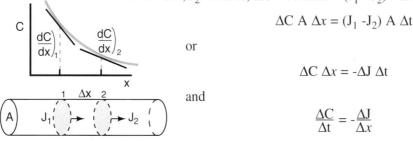

If the Δs become infinitesimal, this can be written in differential form:

$$dC/dt = -dJ/dx \quad \text{or} \quad dC/dt = d/dx(D \, dC/dx)$$

This becomes

$$dC/dt = D \, d^2C/dx^2$$

for the case where D does not depend on C, which is an approximation often made for ease of calculation. This diffusion equation is known as *Fick's second law* and is applied whenever one wishes to calculate C as a function of t and x.

To solve this second-order differential equation, the boundary conditions must be established. For example, in carburization, at time t = 0, the surface concentration is C_s and the bulk concentration is C_o, while at t = ∞ the bulk concentration becomes C_s. Solutions for the diffusion equation for various sets of boundary conditions are available, mainly from studies of heat flow. For one-dimensional carburization when C_s is fixed, the solution for the concentration at any value of x, C_x, at time t is

$$\frac{C_x - C_o}{C_s - C_o} = 1 - \text{erf}\left(\frac{x}{2\sqrt{Dt}}\right)$$

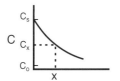

The error function, erf, is used in the statistics of random processes, and its appearance in this solution to the diffusion equation emphasizes the fact that the dispersion of carbon in the solid comes by way of random jumping of carbon atoms among interstices in the crystal lattice. Values of the error function are given in Table 10.1.

Table 10.1 erf z for various values of z

z	erf (z)	z	erf (z)	z	erf (z)
0	0	0.55	0.5633	1.3	0.9340
0.025	0.0282	0.60	0.6039	1.4	0.9523
0.05	0.0564	0.65	0.6420	1.5	0.9661
0.10	0.1125	0.70	0.6778	1.6	0.9763
0.15	0.1680	0.75	0.7112	1.7	0.9838
0.20	0.2227	0.80	0.7421	1.8	0.9891
0.25	0.2763	0.85	0.7707	1.9	0.9928
0.30	0.3286	0.90	0.7970	2.0	0.9953
0.35	0.3794	0.95	0.8209	2.2	0.9981
0.40	0.4284	1.0	0.8427	2.4	0.9993
0.45	0.4755	1.1	0.8802	2.6	0.9998
0.50	0.5205	1.2	0.9103	2.8	0.9999

Note that z, the argument of erf z, is a dimensionless quantity. Thus, $\sqrt{Dt}$ has the dimensions of distance. (D has the dimensions m²/sec or cm²/sec in compilations of data.) The quantity $\sqrt{Dt}$ is, therefore, a characteristic diffusion length, and it is often used as a rough estimate of the length scale over which diffusion has occurred during time t for a given value of D.

In order to calculate the depth of carburization, one must know the value of D at the temperature of interest. Values of D have been determined experimentally for various systems. They are tabulated according to the known exponential dependence of D on temperature, T, expressed in degrees Kelvin:

$$D = D_o \exp{-(Q/RT)}$$

where Q is the *activation energy* and R is the gas constant (cf. Section 7.3.1). When Q is given in cal/mole, then R is $\approx$ 2cal/mole-K. Values of D_0 and Q for several systems are given in Table 10.2.

Table 10.2 Selected Values of D_0 and Q

System	D_0(cm²/sec)	Q(kcal/mole)
C in α-Fe	6.2×10^{-3}	19.2
C in γ-Fe	0.1	32.4
Ni in Ni	1.3	66.8
Nb in Nb	12	105

The carburization process is considerably more expensive than ordinary heat treatment because of the specialized equipment and skills as well as the time and energy required. If case hardening were used for the components of bicycle chains, the cost of the chains would be beyond the reach of most customers. The alternative is simply to use a medium-carbon steel and to temper (after austenitizing and quenching) to a hardness that corresponds to a toughness sufficient for the application. This requires that some wear resistance be sacrificed. The economic question is whether it is cheaper to replace periodically the worn chains made of conventionally heat-treated steel or to pay the initially higher cost of case-hardened chain, which would have much longer life. The nature of the product and the market have decided in favor of the former. This means that lubrication and cleanliness are doubly important in the lives of chains.

Although carburization is not used for chains, it is routinely used for the freewheel cogs. Surface hardening is needed here to provide reasonable wear life for the gear teeth.

Example 10.2

Q. Referring to Fig. 10.12, it is desired to carburize a 1010 steel to a case depth of 0.2mm (0.008in), defined as the distance below the surface at which the hardness (and, therefore, the carbon content in a quenched steel) is halfway between the value at the surface and the value in the relatively soft core. Calculate the time needed for this at 925°C in an H_2/CH_4 mixture, which can saturate the surface with carbon.

A. Assume that the surface is flat enough so a one-dimensional diffusion analysis is sufficient. A solution is required to Fick's second law:

$$dC/dt = D\, d^2C/dx^2$$

where C is the carbon concentration (which is a function of distance from the surface, x) t is time, and D is the diffusivity of carbon in iron at the appropriate temperature. The solution, given in Section 10.4.5, is

$$\frac{C_x - C_0}{C_s - C_0} = 1 - \text{erf}\left(\frac{x}{2\sqrt{Dt}}\right)$$

where C_x is the carbon concentration at any point x (here taken to be 0.2mm); C_0 is the initial carbon concentration in the steel (0.1%); C_s is the carbon concentration at the surface, and erf is the error function (Table 10.1).

If the surface is saturated at 925°C, then from the phase diagram (cf. Fig. 9.2) the carbon content of the austenite at the surface, C_s, is 1.25%. The desired carbon concentration at $x = 0.2$mm is then $(1.25 - 0.1)/2 + 0.1 = 0.675$%. That is,

$$\frac{C_x - C_0}{C_s - C_0} = 0.5$$

and the error function can be written

$$\text{erf } z = 1 - 0.5 = 0.5$$

Entering this value in Table 10.1 and interpolating, z is found to be 0.477. Thus,

$$\frac{x}{2\sqrt{Dt}} = 0.477$$

For $x = 0.02$cm,

$$\sqrt{Dt} = \frac{0.02}{2(0.477)}$$

A value is now required for D. Consulting the data in Table 10.2, $D = D_0 \exp(-Q/RT)$ for $T = 925 + 273 = 1198$K can be written

$$D = 0.1 \exp{-\frac{32400}{2 \cdot 1198}} = 1.3 \times 10^{-7} \text{cm}^2/\text{sec}$$

The required time is then obtained from

$$1.3 \times 10^{-7} \, t = \left(\frac{0.01}{0.477}\right)^2$$

giving $t = 3380$ sec $= 56$ minutes.

10.5 Cracks in Solids

Case hardening is used when it is desirable to have a very hard, wear resistant surface without the danger of brittle fracture of the whole component. Whether or not a crack will propagate in a given material is the subject of fracture mechanics. The tendency for crack propagation depends on four factors:

1. The component of stress acting normal to the plane of the crack, σ.
2. The square root of the crack length, $\sqrt{2a}$.
3. The sharpness of the crack, or the radius of curvature at the crack tip, ρ.
4. The inherent toughness of the material, which means the ability of the material to deform plastically so as to cause the crack tip to become more blunt.

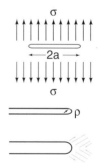

The question of whether or not a crack propagates must depend on the value of the stress at the crack tip. It can be shown from the theory of elasticity that, for an elliptical crack in a plate subjected to a tensile stress, σ, the maximum stress (at the crack tip) is given by

$$\sigma_{max} = \sigma \left(1 + 2\sqrt{\frac{a}{\rho}}\right)$$

For example, for a circular hole, $a=\rho$, and the stress-concentration factor is 3. That is,

$$\sigma_{max} = 3\sigma$$

Thus, the stress-concentration factor increases directly with the length of the crack and inversely with the radius at its tip. Since plastic flow would make the radius larger, it must decrease the stress-concentration factor.

The science of fracture mechanics was developed from the theory and experiments of Griffith, who studied the fracture of glass, choosing it as an ideal model material. In linear elastic fracture mechanics, the stress at a crack tip is characterized by a parameter called the *stress-intensity factor*, K, given by

$$K = Y\sigma\sqrt{a}$$

where Y is a geometrical constant. The basic premise is that a crack will begin to extend in an unstable manner when the stress-intensity factor reaches a critical value, K_c, which is a property of the material. A tough material is one with a high K_c.

In an elastic material (like glass) in which plastic flow does not occur, K_c can be related to Young's modulus and to the energy to create new surface in the material. This was done by Griffith, and the theory is outlined in approximate fashion in Appendix 10.2.

This approach applies to metallic materials under conditions in which plasticity is confined to a small region around the crack tip. In this case the *small-scale-yielding approximation* is used. Here, it is assumed that whatever occurs at the crack tip depends on the elastic stress field outside the plastic region, and that the latter is very small, compared to the length of the crack.

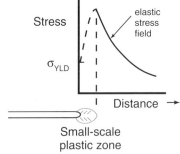

Since K is a parameter which describes the elastic stress field, it must also control the events in the plastic region. Therefore, although K_c for a metallic material cannot be related directly to known material properties, it can be measured by experiment and then used as an empirical material parameter. The small-scale-yielding approximation can be used either when the yield stress of the material is high, or when the applied stress is low (i.e., K is small). The latter case, for example, is used to deal with the propagation of fatigue cracks, even in ductile materials.

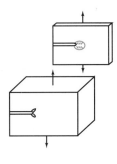

The amount of plasticity at a crack tip depends on the thickness of the cracked body. For a thin sheet, in which stress components normal to the plane of the sheet can be set equal to zero (the condition called *plane stress*), the plastic zone would be larger than in a thick plate, in which strain components normal to the plate can be ignored (the condition of *plane strain*).

The condition of plane strain is used as a reference condition in fracture mechanics. Thus, one can measure the K_c for a material under plane-strain conditions and treat the result as a material constant, to be applied in any case in which the geometrical factor, Y, can be calculated. The critical value of K_c in plane strain is called the *plane-strain fracture toughness*, designated K_{Ic}, and values are tabulated for various high-strength materials, as a function of the yield stress, for example. In a case-hardened steel, the K_{Ic} of the case would be very low, but that of the core would be much higher. If the case thickness is small enough, there is little likelihood that a crack in the case could propagate into the core in an unstable manner and cause the component to fracture.

10.6 The Use of Ceramics for Bearings

Ceramics are generally based on solid oxides, carbides, nitrides, or borides, and they have certain properties which make them attractive for use as bearings for bicycles. These are:

1. High hardness
2. Non-reactive surfaces
3. Low weight

The high hardness means that surface asperities are not easily flattened, and it is difficult to gouge the surface by plastic deformation. Low surface reactivity means that cold welding is difficult. Both these factors favor low friction and long wear. Low weight is a self-evident advantage on a bicycle and results from the fact that ceramics are composed mainly of elements with low atomic weights. Ceramic bearings have been developed for engines designed to run at high temperatures, where the high-temperature strength and chemical inertness of ceramics, compared to metals, are of great advantage. However, now that they are available, their possible use on bicycles may be envisioned for the future.

The main disadvantage of ceramics is their very low toughness, which is a direct result of their great resistance to plastic flow (i.e., high hardness). The high stresses that can occur at the tips of crack-like flaws cannot be relieved readily by dislocation motion. Thus, flaws provide the sites for the initiation of brittle fracture, which occurs when the stress intensity at some flaw reaches the critical value. The energy needed to propagate a crack is small, again because little energy is dissipated by dislocation motion at the crack tip. Thus, the challenge in making structural ceramic components is to eliminate flaws. However, since bearings are loaded in compression, a low toughness is tolerable.

The properties of ceramics stem directly from their structure and bonding. The bonding is usually ionic or covalent. That is, one or more valence electrons are either transferred from one element to another or are shared between the two elements. In many cases, the bonding is actually a combination of the two. That is, a valence electron can be viewed as spending part of its time attached to one element (which is then the anion) and the rest of its time being shared equally between the two reacting elements. In any case, the bonding is very different from that in a metal.

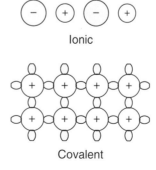

This difference in bonding means that ceramics do not tend to cold weld as metals do when surfaces are rubbed together. There is no sea of free electrons at the surface to merge with the electron sea of a mating surface. The atoms on covalently bonded ceramic surfaces tend to rearrange themselves to minimize the number of "dangling bonds." Ionically bonded ceramics tend to adsorb molecules from the environment to saturate the surface bonds.

10.6.1 Crystal Structures

Ceramics may be crystalline, amorphous (i.e., glassy) or mixtures of the two. The crystal structures are more complex than those found in metals for several reasons, including the following:

1. A ceramic is usually composed of more than one element.

2. The elements are present in a definite ratio, dictated by the necessity for overall charge neutrality in an ionic solid or by the number of possible covalent bonds.

3. In an ionic solid, ions of one sign may be nearest neighbors only of ions of the other sign; the number of nearest neighbors depends on the relative sizes of the ions, because ions are not allowed to overlap. In a covalent solid the bonds must have specific lengths and directions.

Examples of the crystal structures of some important ceramics are shown in Fig. 10.13. Magnesia, MgO, and alumina, Al_2O_3, are mainly ionic. The former has the same crystal structure as rocksalt (NaCl). Silicon carbide is covalent and has a crystal structure similar to diamond or silicon. (Both the rocksalt and diamond cubic structures are FCC, with two atoms (or ions) associated with each lattice site.) In a covalent solid the number of neighbors of any given atom often follow the so-called 8-N rule. The total number of electrons in the shared-electron bonds around any given atom is then eight, the number needed for an inert-gas-type full "outer shell." If an element has a valence of N, it needs 8-N neighbors to complete its outer shell. Since both silicon and carbon have a valence of four, silicon carbide crystallizes in the diamond-cubic structure, as do silicon and carbon themselves. In this structure each atom has four neighbors.

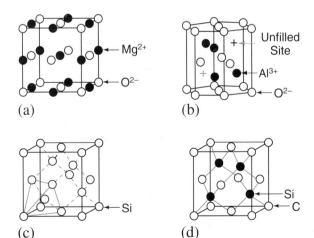

Fig. 10.13 Crystal structures of (a) MgO, the rocksalt structure, (b) Al_2O_3, (c) silicon, the diamond-cubic structure, and (d) silicon carbide. (Adapted from M.F. Ashley and D.R.H. Jones, *Engineering Materials 2*, Pergamon Press, Oxford, 1986, pp. 154, 156)

10.6.2 Dislocation Motion and Ductility

Dislocations exist in crystalline ceramics, as they do in metals. However, the nature of the bonding in ceramics makes dislocation motion difficult, except at high fractions of the melting temperature, where the thermal energy is often sufficient to overcome barriers to dislocation glide. In metals, bonding is more or less non-directional, so the angular relationships between atoms in a crystal may usually be distorted without large expenditures of energy. In addition, all the ions in the valence-electron "gas" are of like sign, so the number of possible slip systems is not constrained in the same way as in ionic crystals. Figure 10.14 shows schematically that shear on certain planes in an ionic crystal would bring ions with like charge into first-neighbor proximity (like the horizontal shear in Fig. 10.14b). This would meet with very high resistance, and slip of this type would not occur except at quite high temperatures.

Glide on the diagonal plane in the ionic crystal would occur with less difficulty, which means that single crystals of the appropriate orientation can deform at reasonably low stresses. However, in polycrystalline materials, in order for the constituent grains to remain bonded together, it is necessary that several slip systems operate in each grain, since each grain of a grain pair tends to deform on its own planes of maximum shear stress. This means that neighboring grains are forced either to take on complex changes in shape or to come apart along the grain boundaries. In the most general case, each grain would need to have five independent slip systems available. For FCC metals, this is not a problem, because there are 12 possible slip systems, and 8 of these are independent (cf. Chap. 4). However, in an FCC ionic crystal with the rocksalt structure, for example, certain of these slip systems are forbidden because of the like-charge-overlap problem depicted in Fig. 10.14(b). This difficulty leads to very limited ductility (and toughness) in ionic polycrystals.

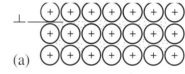

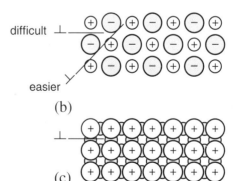

Fig. 10.14 Schematic illustration of dislocation glide in crystals with three types of bonding, showing that glide is relatively unconstrained in (a) metals compared with (b) ionic crystals or (c) covalent crystals. (Adapted from Ashby and Jones, loc. cit., p. 164)

With covalent ceramics, even with single crystals, ductility is limited by the fact that dislocation glide must necessarily break strong localized bonds. This is so difficult that covalently bonded crystals are essentially elastic solids up to about half the absolute melting temperature. That is, dislocation glide requires so much energy at lower temperatures that the necessary applied stresses would be well above those needed to propagate the flaws that inevitably exist (cf. Appendix 10.2). (For example, there are bound to be submicroscopic cracks on the surfaces from contacts with other solids. Even light contact forces generate large stresses, if the contact area is very small.)

10.6.3 Forming of Ceramics, Sintering

Metals are normally produced by processes which begin with melting in a container made of ceramics. This option is not available to most ceramics, because their melting temperatures are so high no material can contain them. In addition, even when a solid ceramic block is produced, it cannot be formed into useful shapes by deformation processing, as metals can, owing to the high hardness and minimal ductility. However, these problems were solved by our ancestors long before the techniques for metal production were worked out. It is known that primitive cultures that had not advanced to the use of metals had nevertheless become quite adept at making a variety of ceramic objects. This technology was also well advanced in western societies long before the industrial revolution, as can be illustrated by consideration of the teacup.

The making of a teacup begins with molding of clay, which is a mix of sheet-like units of a metal silicate and water. The metal silicate may be viewed as an "alloy" of one or more metal oxides, like MgO, CaO, or Al_2O_3, and silicon dioxide, SiO_2.

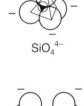

SiO_4^{4-}

$Si_2O_7^{6-}$

The basic building block is the SiO_4^{4-} tetrahedron, which can be thought of as the *monomer* for a variety of structures. (*Mer* is from a Greek word, *meros*, meaning "part.") In this monomer, each oxygen shares one electron with the central silicon atom in four covalent bonds, and each has an additional bonding electron available to attach the monomer to other atoms or groups of atoms. The kind of structure formed depends on the ratio of SiO_2 to metal oxide.

The simplest kinds of silicate structures utilize SiO_4^{4-} monomers or *dimers* of $Si_2O_7^{6-}$ surrounded by metal ions. For example, forsterite, Mg_2SiO_4, is a silicate in which each SiO_4^{4-} tetrahedron is surrounded by four Mg^{2+} ions, which form bridges to adjacent tetrahedra. That is, each Mg^{2+} ion has contributed one electron to an oxygen in each of two SiO_4^{4-} tetrahedra.

If the $MgO:SiO_2$ ratio is a bit higher than 2:1, the structure becomes chainlike, as shown in Fig. 10.15, in which chains of SiO_4^{4-} tetrahedra are held together along their length by metal ions. The oxygens forming the connecting links between tetrahedra are called *bridging oxygens*.

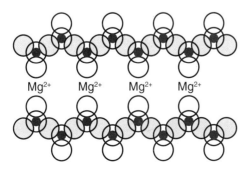

Fig. 10.15 The structure of $MgSiO_3$, a chain-like silicate. The bridging oxygens are shaded blue; the remaining oxygens carry a -1 charge and can bond with metal ions.

At still higher metal-oxide:SiO_2 ratios, the structures become sheet-like. As illustrated in Fig. 10.16, the oxygens at three corners of each tetrahedron are bridging oxygens, and the remaining oxygen is left to form bonds with other ions, particularly metal ions. The bridging oxygens on the lower side of the sheet each have two silicons as neighbors; therefore, they cannot form any more primary bonds. However, the sheets are polarized, with the side having the bridging oxygens appearing negative and the side with the metal ions appearing positive. Hence, adjacent sheets can form secondary bonds, especially with water, which is a polar molecule.

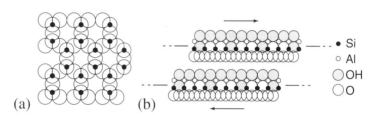

Fig. 10.16 (a) A sheet-like silicate; this structure is found in clay and mica. (b) Schematic illustration of shearing between sheets of clay lubricated with adsorbed water molecules, making the clay highly plastic for easy molding. (From L. H. Van Vlack, *Elements of Materials Science*, 2nd ed., Addison-Wesley, Reading, Mass., 1964, p. 229.)

The secondary bonds between sheets are weak, so the sheets may be easily cleaved apart, as with mica, for example, or sheared, when clay is molded. Thus, in making a teacup, clay comprising sheet-like silicates containing adsorbed water is molded into the desired shape and then heated *(fired)* in a kiln to drive off the excess water. This causes some of the silicate to form a disorderly three-dimensional network, i.e., a glass, in which nearly all the oxygens are bridging oxygens. The glass is fluid at the firing temperature. Upon cooling, the glassy phase becomes rigid and acts to bond the crystalline particles together, as illustrated in Fig. 10.17. This process is an example of *liquid-phase sintering*, the glass phase being a liquid at the firing temperature.

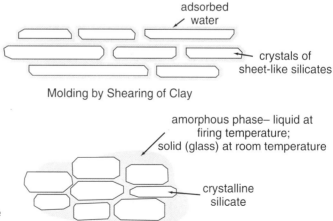

Fig. 10.17 Schematic illustration of liquid-phase sintering during making of pottery.

The sintering of glass spheres by heating them to a temperature at which the glass becomes viscous is illustrated in Fig. 10.18. Capillary action causes the glass at the sphere surfaces to flow to reduce the sharp angles formed along the points of contact.

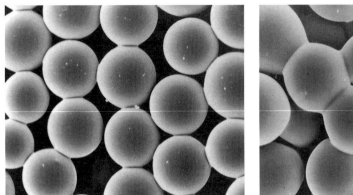
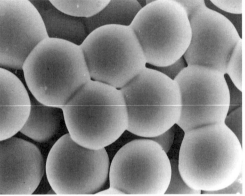

Fig. 10.18 The sintering of glass spheres by heating into the temperature range at which viscous flow can occur. (Courtesy of J. Zhao and M. P. Harmer, Lehigh University.)

Sintering can occur in crystalline solids by diffusion at temperatures above half the absolute melting point; in this case it is called *solid-state sintering*. It can be illustrated by Fig. 10. 19(a) to (c), which shows how two spherical particles can slowly merge by diffusion of atoms along the interface between them into the cusp at the periphery of the contact region. The driving force for this diffusion is, again, capillarity. The solid-vapor surface energy is higher than the grain-boundary energy, and the approach to equilibrium would require that the contact angles increase to the equilibrium value, governed by the balance of interfacial tensions, as shown in Fig. 10.19(d) and (e).

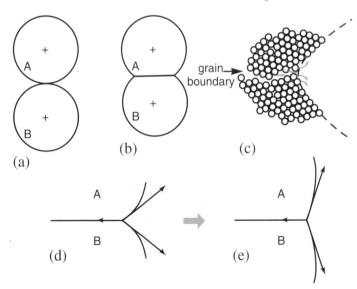

Fig. 10.19. (a) Two spheres in contact before sintering begins. (b) The formation of a solid neck between them as the result of diffusion along the grain boundary which separates them, as shown in (c). (d) The imbalance of interfacial tensions at the periphery of the contact area during the sintering process. (e) The equilibrium balance of interfacial tensions, which would represent the end-point of sintering of this pair of particles.

If sintering were carried out long enough, one could reach 100 percent density. In practice, the process is often stopped short of complete densification, and there is a residue of porosity. As long as the residual pores lie along grain boundaries, as shown in Fig. 10.20(a), they could, in principle, be eliminated, because the grain boundaries are paths of relatively rapid diffusion. However, if grain growth occurs and leaves the pores isolated within grains, as shown in Fig. 10.20(b), then it would be impossible to eliminate them in any reasonable time, owing to the slowness of lattice diffusion.

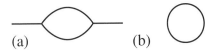

Fig. 10.20 (a) A residual void along a grain boundary, and (b) the same void after grain growth has caused the migrating boundary to leave it behind. Note the equilibrium shape in each case.

10.6.4 Silicon Nitride Bearings

After several decades of development, bearings have been fabricated from silicon nitride, Si_3N_4, a covalently bonded ceramic which occurs in two types, or *polymorphs,* called α and β. The β phase is the stable phase at elevated temperatures. The crystal structures of the two phases are closely related, as shown in Fig. 10.21. Both are made of SiN_4^{8-} tetrahedra arranged in two basic patterns. In this structure, each nitrogen forms bonds with three silicons, in contrast with a silica glass in which each oxygen bonds to two silicons. This is a consequence of the fact that oxygen is in group VI of the periodic table, and nitrogen is in group V.

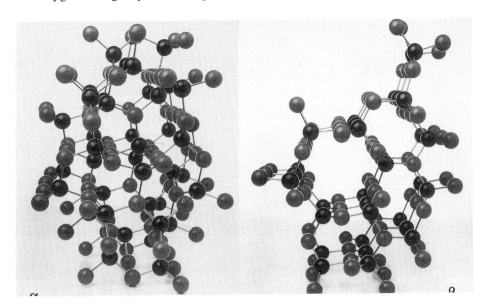

Fig. 10.21 Models of the crystal structures of the two polymorphs of Si_3N_4.

Commercially, Si_3N_4 is made in two forms. One is almost pure silicon nitride, mostly α-phase, but it contains 10 to 30% porosity and is suitable mainly for low-stress applications at high temperatures, where its high creep strength and good resistance to thermal shock can be utilized to advantage. The process by which it is made is called *reaction bonding,* and it starts with fine silicon powder (<5μm) that is then heated in nitrogen gas close to the melting point of silicon (1410°C). The basic reaction is

$$3\ Si + 2\ N_2\ (g) \longrightarrow Si_3N_4$$

The reaction takes only a few hours. This may appear surprising, since the silicon powder starts out with a thin coating of SiO_2, which under certain conditions acts as a reaction barrier. Also, one might expect that a coating of Si_3N_4

would form around the silicon particles and slow the reaction. However, it is now believed that the reaction is not direct, as shown above, but rather involves the transport of silicon through a vapor phase from the solid silicon particles to α-Si_3N_4 nucleated elsewhere. The main carrier gas is probably SiO, although the vapor phase would also contain SiO_2 and Si at this high temperature. The important reactions are the following:

$$Si + SiO_2 \,(g) \longrightarrow 2\, SiO\,(g)$$

$$2Si + O_2 \longrightarrow 2SiO\,(g)$$

$$Si + H_2O\,(g) \longrightarrow SiO\,(g) + H_2$$

The reaction which produces the silicon nitride would then be

$$3SiO\,(g) + 2N_2 \longrightarrow Si_3N_4 + 3/2\, O_2$$

One notable feature of this reaction is that there is almost no overall volume change between the initial silicon powder and the final silicon nitride powder, which has a very fine grain size (<1µm). The volume difference between the two powders is taken up by the porosity. Another remarkable aspect is the rather high strength of the intergranular bonding in the α-Si_3N_4. The fracture stresses measured in four-point bending are in the range 100 to 300 MPa, which is impressive in a brittle solid with such high porosity. It has not yet proved possible to make fully dense, essentially pure α-Si_3N_4. However, if this were accomplished, the material would have a very attractive combination of toughness, high-temperature strength, and thermal-shock resistance.

Reaction-bonded α-Si_3N_4 would not be suitable for bearings, because the high porosity would give inadequate fracture toughness; the bearings would crumble into powder when loaded in compression. However, a fully dense solid based on Si_3N_4 can be made by liquid-phase sintering. In this process, one starts with α-Si_3N_4 powder and adds some mixture of oxides which, when combined with the ever-present SiO_2, gives a eutectic temperature in the desired range. Examples of oxides used for this purpose are Al_2O_3, MgO, and Y_2O_3. The process not only results in a dense product, but also replaces the α-Si_3N_4 with the β phase.

A remarkable feature of β-Si_3N_4 is its capacity to accommodate large amounts of Al_2O_3 and AlN (or Al_3O_3N) in solid solution. Such a solid solution is called a β-SiAlON. In the sintering process, the α-Si_3N_4 dissolves in the eutectic liquid and is transported to particles of β-Si_3N_4, where it reprecipitates as rods of β-SiAlON. The process is illustrated schematically in Fig. 10.22.

The densification process is sometimes assisted by application of a compressive stress during sintering. If the compression is uniaxial, it is called *hot-pressing;* if it is hydrostatic, the process is called *hot-isostatic-pressing,* or *HIP*ing. In the latter case, the powder is encapsulated (for example, in a thin-walled stainless-steel can) and the pressure is applied by a gas, such as argon, in the sintering reaction chamber. Silicon nitride bearings can be machined from pieces cut from hot-pressed plate, but this is very expensive. A more promising approach for commercial applications would be the use of near-net-shape HIPed pre-forms.

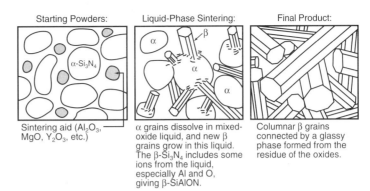

Fig. 10.22 Schematic illustration of the solution-precipitation model for the liquid-phase sintering of Si_3N_4. (After G. Ziegler, *Materials Science Forum*, vol. 47, 1989, p. 162.)

In the ideal process, all of the liquid phase would be taken up in the formation of the SiAlON, and the resulting product would be a fully dense polycrystalline solid. This has not yet been achieved, and all sintered Si_3N_4 is found to have some glassy phase between the β-SiAlON grains. It is found in virtually all grain boundaries. For certain special, low-energy boundaries, and in boundaries along which the oxides are completely consumed in the SiAlON reaction, the glassy phase is confined to three-grain junctions. The two cases are illustrated schematically in Fig. 10.23. For the case of special boundaries, the wetting criterion depends on whether the grain boundary energy γ_{ss} is greater than or less than twice the solid/liquid interfacial energy γ_{sl}. An example of the intergranular glassy phase is shown in Fig. 10.24.

Since the density of Si_3N_4 bearings is less than half that of steel, their use in bicycles would afford a small weight saving, in addition to longer wear-life. Their cost, of course, is much higher, so they could only conceivably be used in the most expensive bicycles.

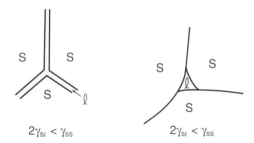

Fig. 10.23 Schematic illustration of two possible morphologies of glassy phase in the β-SiAlON grain boundaries.

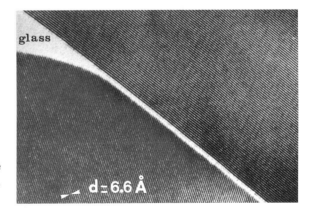

Fig. 10.24 Example of the intergranular glassy phase in a β-SiAlON. (From J. Y. Laval and A. Thorel, *Materials Science Forum*, vol. 47, 1989, p. 143.)

Summary

The nature of friction and wear was discussed in reference to the chain and bearings of a bicycle, as were the principles of lubrication. The need for hard materials for these applications was made obvious. In this context, the hardening, tempering, and hardenabilty of steel were considered, along with the problem of residual stresses from quenching. The subject of diffusion in a concentration gradient was discussed in relation to the case hardening of steel components by carburizing, in which a hard, wear-resistant surface is combined with a softer, tough core. The reason a thin case is desired was explained through the use of fracture mechanics, by which it was shown that the tendency of a crack in the brittle case to propagate through the component depends on the square root of the crack length (i.e., the case thickness). A ceramic is an alternative type of hard material for bearing. After a general discussion of the structures and behavior of ceramics and their production by liquid-phase or solid-state sintering, the case of silicon nitride bearings was considered, including the reaction of nitrogen and silicon to form porous α-Si_3N_4 and then the liquid-phase sintering with added oxides to form β-SiAlON, an example of crystalline grains held together by a glassy phase.

Appendix 10.1 Twinning in Crystals

Twins are a special kind of defect in crystals, in which the lattice points on either side of the "twinning plane" have a mirror-image relationship. They can be the result of growth faults during recrystallization, or they can form as a result of a shear stress during plastic deformation or a martensitic phase transformation. The most commonly observed twins are the growth-fault type, called *annealing twins* in FCC metals, as shown in stainless steel in Fig. A10.1.1. They are characterized by straight, low-energy boundaries which enclose regions showing an etching response different from the surrounding grain. The deformed brass depicted in Fig. 3.10 shows annealing twins being crossed by slip lines; note the orientation of the slip planes within a twin compared to those in the matrix.

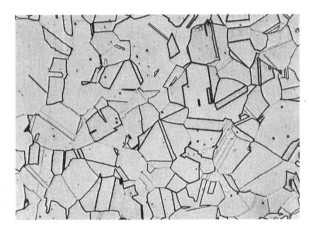

Fig. A10.1.1 Annealing twins in a recrystallized stainless steel spoke.

An annealing twin forms in an FCC crystal when a fault occurs in the stacking of close-packed planes. Referring to Fig. 4.7, a layer of atoms could form on B sites, instead of A sites, during a recrystallization anneal. After that, the normal ABC-type (i.e., CBA) stacking could proceed, and the crystal, except at the stacking fault, would remain FCC. Later on, another stacking fault could reverse the stacking to the original ABC. The region between the two faults is called an annealing twin and is of the type ABCA<u>BCB</u>ACBA.....CB<u>ACA</u>BCABC. Figure A10.1.2 shows the atomic arrangements in a six-layer twin of that type. Within the twinned region, the atoms are in mirror-image positions with respect to those outside, hence the name "twin." The straight boundaries of the annealing twins are thus {111} planes.

Since the stacking at these boundaries is not FCC (i.e., the stacking is ABA-type, instead of ABC-type), the atoms there are in a state of somewhat higher energy, but this *stacking-fault energy* is always much lower than the average grain boundary energy, because the two parts of the crystal are completely coherent. Only the second-neighbor positions are wrong.

The ABABAB... type of close-packed-plane stacking produces a crystal structure called hexagonal close packed (HCP), models of which are shown in Fig. A10.1.3. A number of metals, such as zinc, magnesium, cadmium, and berylium, crystallize in the HCP structure. It is striking that such a seemingly minor difference in stacking can change the symmetry of a crystal from cubic to hexagonal.

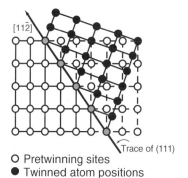

○ Pretwinning sites
● Twinned atom positions

Fig. A10.1.2 Atom positions in a twin in an FCC crystal.

Examination of Fig. A10.1.2 shows that a shear stress in the $[11\bar{2}]$ direction on the (111) plane could produce the kind of shear that could create a twin. This would be a coherent shear, meaning that all the atom movements are coupled and follow a strict pattern, and the atomic displacements in each plane are smaller than would occur in slip. However, a large overall shear strain can be produced, since the atoms move on every plane. Twins that form in response to a shear stress are called *deformation twins*.

The formation of a deformation twin is in some ways similar to the formation of a martensite plate in response to a shear stress, as occurs on a sub-microscopic scale in heavily deformed austenitic (FCC) stainless steel (cf. the strain markings in Fig. 4.3).

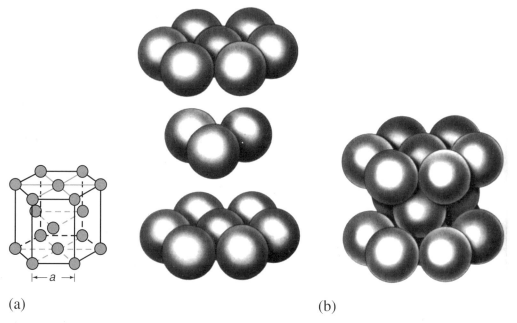

(a)   (b)

Fig. A10.1.3 Models of an HCP crystal: (a) a ball-and-stick model, showing the hexagonal symmetry and (b) a hard-sphere model, showing how the atoms pack together. (From A.G. Guy, *Elements of Physical Metallurgy,* Addison-Wesley, Reading, Mass., 1959, p. 82)

Appendix 10.2 Linear-Elastic Fracture Mechanics

The goal of fracture mechanics is to specify the conditions under which a body of a given geometry containing a crack-like flaw will fracture. It is most easily applicable to bodies that do not undergo large-scale plastic deformation before fracture, and in this case one employs the stress-intensity factor, K, defined by

$$K = Y \sigma \sqrt{c}$$

where Y is a geometry-dependent constant, σ is the stress far from the crack (i.e., the tensile stress normal to the crack plane), and 2c is the crack length (i.e., the dimension along the direction in which crack propagation would occur). The importance of K lies in the fact that the tensile stress in the region of the tip of a crack is proportional to K. The stress-intensity factor can be used as a fracture criterion in any material which can fracture by extension of a crack when a critical value of the tensile stress is reached. This kind of fracture can be called stress-controlled fracture. (Materials with high toughness tend to fail by *rupture,* meaning that they come apart as the result of the growth and coalescence of damage (i.e., voids) produced during plastic deformation. Rupture can be considered strain-controlled fracture. In this case, K is not a good fracture criterion, and another kind of fracture mechanics must be employed.)

If one were to know the critical value of K necessary for crack extension for a given geometry, material, and loading condition (e.g., temperature), one could use this value, K_{CRIT}, in one of two ways:

1. If the size of the worst flaw is known, the critical stress for fracture can be calculated

$$\sigma_{CRIT} = \frac{K_{CRIT}}{Y\sqrt{c}}$$

2. If the service stress, σ, is specified, the size of the largest tolerable flaw can be calculated

$$c_{MAX} = \left(\frac{K_{CRIT}}{Y\sigma}\right)^2$$

The value of K_{CRIT} is always obtained from experiment, and under certain conditions it can be treated as a material constant.

This kind of analysis stems from the work of A. A. Griffith in the years around 1920, culminating in the now-famous Griffith equation, which relates the fracture stress to the crack length in an elastic solid. (In an elastic solid, no plastic deformation occurs; an example is silica-based glass at room temperature, e.g., ordinary window glass. This was the material used by Griffith to check his theory.)

To derive the Griffith equation in an approximate way, consider a semi-infinite plate of unit thickness loaded to a fixed displacement by a stress σ. The strain energy per unit volume is given by

$$SE/vol. = 1/2 \sigma \varepsilon = \sigma^2/2E$$

Imagine that a crack of length 2c forms in the center of the plate shown in Fig. A10.2.1; strain energy is released in the region around the crack.

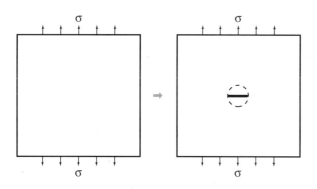

Fig. A10.2.1 Semi-infinite thin plate in which a Griffith crack is formed.

For a first approximation, let the volume in which strain energy is released be πc^2 times unity (the thickness). The total strain energy released is then

$$\text{Total strain energy} = -\left(\frac{\sigma^2}{2E}\right)\pi c^2$$

This energy can offset the energy needed to create the two new surfaces along the crack. This is given by

$$\text{Surface energy} = 2\gamma \, 2c \, 1 = 4c\gamma$$

where γ is this surface energy per unit area.

The total energy change is given by

$$E = \frac{-\pi\sigma^2 c^2}{2E} + 4c\gamma$$

At equilibrium

$$\frac{dE}{dc} = 0 = \frac{-\pi\sigma^2 c}{E} + 4\gamma$$

or

$$\sigma = \sqrt{\frac{4E\gamma}{\pi c}}$$

This is very close to the actual relationship (obtained by a much more complicated derivation). It gives the length of a crack (2c) that is just stable for a given applied stress, σ. At a slightly higher stress the crack would become unstable, as shown schematically in Fig. A10.2.2.

APPENDIX 10.2 LINEAR-ELASTIC FRACTURE MECHANICS

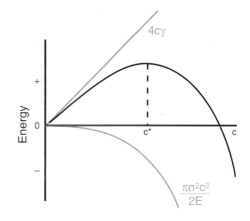

Fig. A10.2.2 For a stress σ, the crack becomes unstable at $c = c^*$, found from maximizing the combination of the energy released and the surface energy required.

The actual Griffith equation for a thin, brittle plate is

$$\sigma = \sqrt{\frac{2E\gamma}{\pi c}}$$

and it can be re-arranged to give K_{CRIT}:

$$K_{CRIT} = \sigma\sqrt{\frac{\pi c}{2}} = \sqrt{E\gamma} \quad \text{(note: here } Y = \sqrt{\frac{\pi}{2}}\text{)}$$

In this case K_{CRIT} is a material constant, since both E and γ are material constants.

In metallic materials in which plastic deformation would occur at the crack tip during crack extension, the actual work of fracture must include the "plastic work," along with γ. Here, K_{CRIT} must be determined by testing of cracked specimens.

This kind of analysis can be used for high-strength materials, in which plastic flow is confined to the crack-tip region. For example, it can be used for hardened steels, and particularly for a carburized surface layer on steel.

Exercises Chapter 10

Terms to Understand

Surface asperities
Cold welding
Perfectly plastic material
Coefficient of friction
Adhesive wear
Abrasive wear
Boundary lubrication
Van der Waals forces
Hydrodynamic lubrication
Hardness of martensite
Twins
Precipitation hardening
Hardening of steel
Hardenability
Thermal stresses
Tempering
Secondary hardening
Widmanstätten ferrite
Carburization
Case hardening

Diffusion coefficient (diffusivity)
Fick's first law
Fick's second law
Stress intensity, stress-intensity factor
Fracture mechanics
Plane stress
Plane strain
Griffith equation
Sintering
Mer
Monomer
Dimer
Bridging oxygens
Reaction bonding
Hot pressing
Hot isostatic pressing (HIP)
Annealing twins
Hexagonal close-packed
Deformation twins

Problems

10.1 Two traditional approaches to minimizing friction and wear are to use hardened-steel components or hardened steel in conjunction with a copper-based alloy containing dispersed particles of a low-melting-point metal, like lead, such that the steel slides relative to the leaded alloy. Explain the principles behind each approach.

10.2 Explain the distinction between the concepts of hardness and hardenability.

10.3 The hardenability curves in Fig. 10.7 both start at essentially the same hardness; explain why. Describe the expected microstructure at a distance of two inches from the quenched end in each steel.

10.4 So-called high-speed steels are used for cutting tools, such as drill bits used to drill holes in metals. The name refers to the fact that the bits resist softening even when driven at high speeds. These steels normally contain chromium and molybdenum; use Fig. 10.10 to explain the principles involved.

10.5 Referring to Example 10.2, calculate the carburizing time for a temperature of 850°C.

10.6 A crystalline powder can be densified by either solid-state sintering, which involves only the powder, or by liquid-phase sintering, which involves the addition of a "sintering agent." The latter is a material that liquifies at the sintering temperature. Describe the mechanism of densification in each case, and describe the expected microstructural evolution as the sintering time approaches infinity.

10.7 Explain why the void in Fig. 10.20(a) is drawn as lenticular, whereas the one in Fig. 10.20(b) is drawn as spherical.

10.8 Referring to Fig. A 10.1.2, draw a cubic unit cell showing the $[11\bar{2}]$ direction lying in the (111) plane.

11

WHEEL RIMS; PRECIPITATION-HARDENED ALUMINUM ALLOYS

Wheel rims must have high strength to support the spoke tension and to resist impacts on rough roads. However, a low mass is essential, because the rims turn at high speeds with a large radius of rotation, and the gyroscopic effect must be minimized, especially in the front wheel, which must be easy to turn while steering.

Most current high-quality wheel rims are made from precipitation-hardened aluminum alloys, which have a high strength-to-mass ratio. These alloys originated from an accidental discovery in 1906 by a German metallurgist, Alfred Wilm, who was trying to find an aluminum alloy which could be hardened by quenching, in the manner of steel. He found that an Al-Cu-Mn-Mg alloy, although not hard immediately after a quench, did harden during aging at room temperature. When Al-based-alloy phase diagrams were investigated, it was realized that the hardening phenomenon was related to the precipitation of a second phase from a supersaturated solid solution. For example, the Al-Cu diagram on the aluminum-rich end, shown in Fig. 11.1, is a simple eutectic between the aluminum-based FCC α phase and the intermetallic compound, $CuAl_2$, denoted as the θ phase. The details of the precipitation reaction, even in this simple system, remained obscure for decades, as did the explanation of its profound strengthening effect.

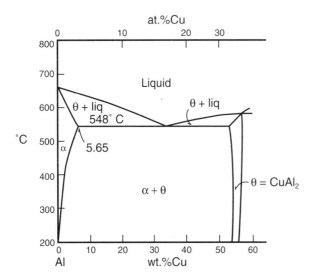

Fig. 11.1 The Al-rich end of the Al-Cu phase diagram, i.e., the Al-$CuAl_2$ diagram. (Adapted from *Metals Handbook,* ASM, 1948, p. 1158.)

11.1 Mechanism of Precipitation Hardening

Precipitation hardening involves the formation of a dense array of strong obstacles to the motion of dislocations. One might think that this is essentially similar to solid solution hardening in a concentrated solid solution; however, this is not the case. In solid solution hardening the resistance to dislocation motion arises only from the elastic interaction between the individual solute atoms and the dislocations. In precipitation hardening the dislocations have to contend with either clusters of solute atoms or fine particles of a second phase with a crystal structure different from the matrix phase.

To achieve the maximum effect, the array of obstacles must be so dense that the dislocations are forced to cut through them in order to glide. The shearing of a particle by dislocation glide is shown schematically in Fig. 11.2, drawn with the assumption that a slip plane in the particle lines up with one in the matrix (which is not necessarily true). Each dislocation shears the particle by an amount equal to the Burgers vector of the dislocation.

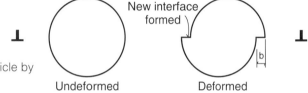

Fig. 11.2 Schematic illustration of the shearing of a particle by a dislocation gliding in the matrix.

An example of particles in an aluminum-lithium alloy that are sheared by the passage of dislocations through them is shown in Fig. 11.3. In this transmission electron micrograph the particles are made to appear white in a black background by using the dark-field technique, in which a diffracted electron beam from a particle is used to form the image on the screen (rather than using a diffracted beam from the matrix, which gives the usual bright field image). In Fig. 11.3 the particles have a crystal structure different from the matrix; therefore, they diffract the electron beam at a Bragg angle different from that of the matrix. Thus, the matrix appears dark. However, the particles are all monocrystalline, and they all bear the same crystallographic orientation relationship with the matrix. Hence, they are all imaged together, i.e., with one diffracted beam.

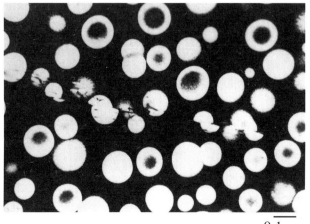

Fig. 11.3 Illustration of particles in an Al-Li alloy which have been sheared by the passage of a number of dislocations. In this case the particles have a crystal structure different from the matrix, and they all "light up" in dark-field illumination, because they all have the same orientation with respect to the matrix. (Dark-field transmission electron micrograph courtesy of Prof. David Williams, Lehigh University.)

0.1μm

Precipitation hardening can be a potent strengthening mechanism in crystalline solids, because particles can present a formidable barrier to the motion of dislocations. The origins of this barrier effect will be described below.

Not all alloys containing an array of dispersed particles are strong. The degree of strengthening depends mainly on the spacing of the particles. The hardening effect is relatively small if the particles are spaced far enough apart that dislocations can bow out in loops between them. If this were to happen, the particles could be bypassed, in the manner depicted schematically in Fig. 11.4, which was originally conceived by Orowan.

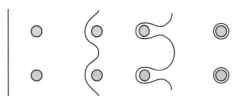

Fig. 11.4 Bypassing of particles by a dislocation that forms a loop by bowing out between the particles. This requires a shear stress of approximately $2Gb/L$, where L is the interparticle spacing; it is known as the *Orowan mechanism*.

As shown in Section 5.6.2, the force to form a loop between pinning points (provided by hard particles) is inversely proportional to the particle spacing. Thus, for maximum hardening the spacing should be such that the shear stress necessary to form a loop is greater than that needed to cut through the particle. The volume fraction of particles should, of course, be as large as possible; that is, at a given (fine) spacing, the particles should be as large as possible.

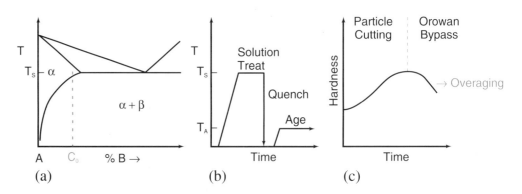

Fig. 11.5 (a) A portion of the phase diagram for a hypothetical alloy that exhibits precipitation hardening. (b) The heat treatment used to produce this hardening. (c) The resulting hardening curve, showing the change in hardness during aging.

The essential features of precipitation hardening can be described with reference to a hypothetical alloy, a partial phase diagram for which is shown in Fig. 11.5(a). The alloy composition is denoted by C_0, and the equilibrium precipitate would be the β phase. The heat treatment for precipitation hardening is shown schematically in Fig. 11.5(b). The alloy is first heated into the single-phase α region (temperature T_S) and quenched rapidly to retain a supersaturated solid solution at room temperature. This is called a *solution treatment*. Next, the solution-treated alloy is aged by heating at some temperature, T_A, in the two-phase region.

A representation of a typical hardening curve is shown in Fig.11.5(c). The solution-treated α is somewhat harder than the annealed condition (i.e., slowly cooled from the α region) because of solid-solution strengthening. The hardness increases further with time at temperature T_A and passes through a maximum, after which *overaging* (caused by coarsening of the precipitate) occurs and the hardness decreases.

Example 11.1

Q. The particles in Fig. 11.3 were obviously too densely spaced for dislocations to bow around them by the Orowan process (shown in Fig. 11.4). Remembering that the particles exist in a thin foil on the order of 0.1μm in thickness, estimate the shear stress needed for the Orowan process to occur. The lattice parameter of aluminum (FCC) is 4.05×10^{-10}m, and the shear modulus is given by $G=E/2(1+\nu)$, where ν, Poisson's ratio, is 0.34.

A. The average particle spacing appears to be approximately 0.2μm. The result of Section 5.6.2 is that the shear stress, τ, to bow out a loop is Gb/r. Taking r to be half the particle spacing, L, τ is $2Gb/L$. Table 3.1 gives E for aluminum as 69GPa; therefore, $G = 69/2(1+0.34) = 25.7$GPa. The Burgers vector, b, is the lattice parameter multiplied by $\sqrt{2}/2$; therefore, the shear stress for the Orowan process is given by

$$\tau = \frac{2 \cdot 25.7 \times 10^9 \text{N/m}^2 \cdot 4.05 \times 10^{-10} \text{m} \sqrt{2}/2}{0.2 \times 10^{-6} \text{m}} = 73.5 \text{MPa}$$

11.2 Requirements for Precipitation Hardening

The primary requirement for precipitation hardening is that the alloy system in question must contain a single-phase field that gets narrower as the temperature is decreased. For example, in Fig. 11.5(a) the α-phase exhibits a decreasing solubility of the solute element B (or, equivalently, of the second phase, β) as the temperature drops. This is obviously necessary for a solution treatment to be carried out.

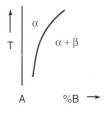

Another requirement is that it must be possible during the aging process to form a precipitate on a very fine scale so that dislocation loops cannot form between the particles, as in Fig. 11.4. The distribution must be so fine that the particles are cut by dislocations gliding in the matrix, as in Fig. 11.2. This is a difficult requirement to meet, because an alloy with a fine-scale precipitate contains a very large total area of particle/matrix interface. That means the content of interfacial energy in the system would be large, unless the particle/matrix interface has an especially low energy. The role of interfacial energy is discussed in more detail in the next section, but here it should be noted that relatively few alloys are known to meet this requirement.

The final requirement is simply that the precipitate that forms must present a significant barrier to the passage of a dislocation. This can occur for any of several reasons, described in Appendix 11.1.

11.3 Principles of Precipitation

The driving force for precipitation is the difference in free energy (per unit volume of alloy) between the supersaturated condition and the condition after precipitation. This free-energy difference is denoted ΔG_v, and it is exactly analogous to the driving force for the decomposition of austenite in steel, as depicted in Fig. 9.7. Thus, ΔG_v times the volume of a precipitate particle is used to express the energy change that drives the formation of a particle. Since this energy is released in the transformation, it is written as a negative term. This is needed to offset the increase in energy that occurs when the interface is formed between the new particle and the matrix. This positive energy term is given by the interfacial energy per unit area of interface, γ, times the area of the interface. Thus, for a spherical particle the total change in free energy is given by

$$\Delta G = - \Delta G_v \, 4/3 \, \pi \, r^3 + \gamma \, 4 \, \pi \, r^2$$

where $4/3 \, \pi \, r^3$ and $4 \, \pi \, r^2$ are the volume and area, respectively, of a sphere. The two opposing energy terms are plotted as a function of particle radius in Fig. 11.6(a), and the resultant, found by summing these two, is plotted in Fig. 11.6(b). Note that when r is small the positive term dominates, but when r is large the negative term dominates. Thus, for small values of r, the formation of a small particle would cause an increase in free energy because of the large surface-to-volume ratio.

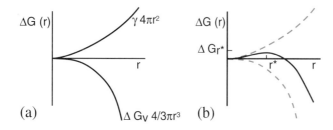

Fig. 11.6 (a) Variation of the volume-energy term and the surface-energy term with the size of a precipitate particle. (b) The resultant total energy change, found by summing the curves in (a).

These ideas can be applied directly to the formation of a water droplet in a supersaturated vapor. The ΔG_v is the difference in free energy per unit volume between the supersaturated vapor and liquid water. One can imagine that, at some instant, a number of water molecules spontaneously collide in the vapor to form a small cluster. The question is whether this cluster will grow into a droplet (by the addition of more molecules) or whether it will re-evaporate. That is, is the cluster stable or unstable? The answer depends on the size of the cluster, because a process will continue spontaneously only if it leads to a reduction in free energy (since equilibrium is characterized by a minimum in the free energy).

The criterion for stability of a particle can be understood by reference to Fig. 11.6(b) and specifically to the particle size r*, at which the curve of total free energy change passes through a maximum. If the cluster is smaller than r*, then it is unstable, because the only way to decrease the free energy is for the cluster to shrink. A cluster with r > r* would be stable and would grow spontaneously into a water droplet. A cluster of the critical size, with r = r*, could go either way. The value of r* is found by taking the first derivative of the expression for the total free-energy change and setting it equal to zero (to locate the maximum of the curve):

$$\left(\frac{dG}{dr}\right)_{r=r^*} = 0 = -\Delta G_v \, 4\pi r^2 + \gamma \, 8\pi r$$

This gives

$$r^* = \frac{2\gamma}{\Delta G_v}$$

From this argument it can be seen that the minimum stable size of a particle can be reduced by having either a small γ or a large ΔG_v.

The same principle applies to precipitation from solid solution. In order to have a fine precipitate, the particle/matrix interfacial energy must be small, and the driving force for precipitation must be large. The driving force ΔG_v increases with undercooling, i.e., cooling below the temperature at which the α phase is no longer stable (which is the $\alpha/\alpha+\beta$ phase boundary) while the interfacial energy γ is approximately independent of temperature. Therefore, fine precipitates are normally characteristic of low aging temperatures.

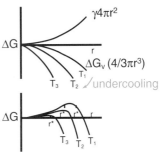

The process just described is called *homogeneous nucleation,* which means that the nucleation of stable particles occurs homogeneously throughout the volume of the supersaturated matrix. This can be difficult in the case of the formation of clouds from supersaturated water vapor. The γ of the liquid/vapor interface is large, because the structure of the two phases is so different, and undercooling may not be sufficient to overcome this. Then, we have to rely on *heterogeneous nucleation* for cloud formation. This involves the formation of water droplets on existing particles, usually some kind of suspended dust. The dust provides a substrate upon which a water droplet can form in the shape of a spherical cap, as shown in Fig. 11.7. The substrate must form a low-energy interface with a water droplet, thereby giving rise to a low contact angle θ between the water and the substrate. It is apparent that such a droplet having the critical radius r* contains far fewer molecules than does an isolated spherical droplet having the same radius (cf. Fig. 11.7). Therefore, the probability that a critical nucleus can form from a simultaneous collision of water molecules is much higher if an appropriate substrate is available for heterogeneous nucleation.

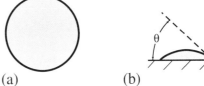

Fig 11.7 (a) A spherical drop having the critical radius r*. (b) A droplet having the same radius of curvature formed on a substrate on which it spreads into a spherical cap having a contact angle θ.

In precipitation from solid solution, the most effective substrate for heterogeneous nucleation is a grain boundary. Thus, a high-γ precipitate is likely to form first along grain boundaries, particularly if the undercooling is not large. This precipitate would be distributed on a scale too coarse to produce significant hardening. This is what occurs in alloy systems in which a low-γ precipitate cannot form. For precipitation hardening to occur, homogeneous nucleation is essential. Therefore, the alloy system must allow the formation of a low-γ precipitate (usually a metastable precursor to the equilibrium second phase).

11.4 Factors that Control γ

The existence of interfacial energy arises from two factors; one is structural and the other is compositional. In solid-state precipitation, if the crystal structure of the new particle is different from that of the matrix phase, as shown schematically in Fig. 11.8, then the atoms along the interface are not completely surrounded by the structure in which their free energy is minimized. For example, the atoms in the β phase (Fig. 11.8) are at their lowest free energy when they are within the β crystal structure. The atoms along the interface are only partly in the β phase; some of their nearest neighbors are in positions characteristic of the α phase. Therefore, they are in a higher energy state than the atoms in the interior of the β phase. This extra energy (per unit area of interface) is the structural component of γ.

Similarly, when the composition of the particle is different from that of the matrix, as it usually is, the atoms bordering each phase are not completely surrounded by the composition that minimizes their free energy. For example, the β phase in Fig. 11.8 is an ordered array in which each A atom is surrounded by a B atom, and *vice versa*. The atoms in the interface do not have nearest neighbors that are all of the opposite type. Therefore, their energy is greater than that of atoms deeper in the β phase. This extra energy (per unit area of interface) is the compositional, or chemical, component of γ. In metallic alloys the structural component of γ tends to be larger than the compositional component.

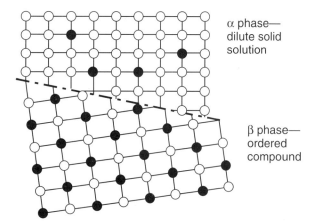

Fig. 11.8 Schematic representation of the interface between a β precipitate and an α matrix.

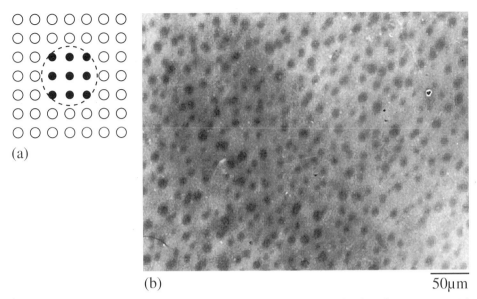

Fig. 11.9 (a) Schematic representation of a spherical particle having the same crystal structure as the matrix. (b) Example of such a particle, called a G-P zone, in an Al-Ag alloy. (From R.B Nicholson, G. Thomas, and J. Nutting, *J. Inst. of Metals,* vol. 87, (1958-59), p. 431.)

The minimum γ in a metallic alloy would be found in the kind of particle represented schematically in Fig. 11.9(a). Here, the two elements are of nearly equal atomic size, and the particle retains the same crystal structure as the matrix. Hence, only a chemical component of γ exists. This occurs in an Al-Ag alloy, as illustrated in Fig. 11.9(b). These particles are not the equilibrium second phase, which is the compound Ag_2Al, indicated by the phase diagram in Fig. 11.10. Thus, when the particles form, the free energy of the system has not reached its minimum value; therefore, these precipitates are metastable. (This means there is still a driving force for the formation of Ag_2Al.) The metastable precipitates shown in Fig. 11.9 are known as Guinier-Preston zones, or G-P zones, after the discoverers of this phenomenon. The G-P zones are responsible for the first stage of precipitation hardening in the Al-Ag alloy.

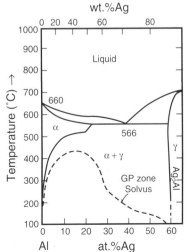

Fig. 11.10 Portion of the Al-Ag phase diagram. (After R. Baur and V. Gerold, *Zeitschrift für Metallkunde,* vol. 52, (1961), p. 671.)

The Al-Cu system also forms G-P zones, but they are plate-like, instead of spherical, as shown schematically in Fig. 11.11, because the copper atoms are substantially smaller than the aluminum atoms of the matrix. This gives rise to the kind of lattice strain shown in Fig. 11.11, and this strain is minimized when the particles form as platelets a few atom layers thick. (The strain-energy effect here overcomes the tendency to form spheres, which minimize the surface-to-volume ratio and, thus, the interfacial area.) The strain energy associated with the Al-Cu G-P zones is an energy-per-unit-volume and is, of course, a positive term. Its effect is to subtract from the ΔG_v and thus to lower the net driving force for precipitation. Therefore, the formation of these G-P zones requires a particularly large degree of undercooling.

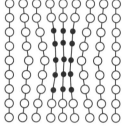

Fig. 11.11 Schematic representation of a G-P zone in an Al-Cu alloy.

Precipitation from solid solution can involve the formation of more than one metastable phase. For example, in the Al-Cu system the precipitation sequence can involve two additional metastable phases after the formation of G-P zones. The next precipitate to form after the G-P zones is called θ". Its crystal structure, as shown in Fig. 11.12, is such that it matches up with the α matrix without strain along two of the crystal axes, and it can be strained to match up along the third axis.

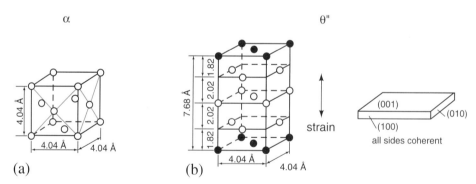

Fig. 11.12 Comparison of the crystal structures of the (a) α and (b) θ" phases in the Al-Cu system. (From D. A. Porter and K. E. Easterling, *Phase Transformations in Metals and Alloys,* Van Nostrand Rheinhold (UK), 1981, p. 297; reproduced by permission.)

When this matching occurs, the particle is said to be *coherent* with the matrix. If the volume of the θ" particle becomes too large, the total strain energy becomes excessive, and the particle then becomes *semi-coherent,* meaning that the mismatch is accommodated by the formation of a periodic array of edge dislocations along the interface. A coherent and a semi-coherent interface are compared schematically in Fig. 11.13. If the crystal structure of the particle is so different from the matrix that no matching is possible, as in Fig. 11.8, it is said to be *incoherent* with the matrix; in this case the structural component of γ is a maximum.

Fig. 11.13 Schematic comparison of (a) coherent and (b) semi-coherent interfaces. (From Porter and Easterling, loc. cit., pp. 144 and 145.)

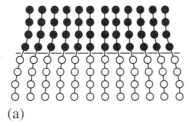
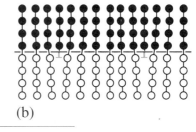

Example 11.2

Q. Where would the solvus line lie for one of the metastable Al-Cu precipitates, compared with the equilibrium solvus shown in the phase diagram in Fig. 11.1?

A. Since the θ phase would eventually form in an alloy quenched from α and aged in the α+θ region, the degree of supersaturation in the quenched α must be larger for the θ phase than for any other precipitate. Therefore, the solvus line for a metastable precipitate must lie to the right of the α/α+θ solvus shown in Fig. 11.1, and, with the sequence of metastable precipitates found in Al-Cu, the less stable the precipitate is, the farther to the right must its solvus be.

Example 11.3

Q. Calculate the maximum possible volume fraction of precipitate in an Al-4%Cu alloy, assuming that the densities of α and all the precipitates are equal.

A. The equilibrium θ precipitate would have the largest volume fraction, because its solvus curve is farthest to the left in the phase diagram of Fig. 11.1. It would contain 52wt%Cu. If the aging temperature were lower than about 250°C, the equilibrium α could be assumed to contain no copper. Therefore, the lever rule would give the weight fraction of θ equal to 4/52 = 0.077. Thus, if the densities of α and θ were the same, the volume fraction of θ would be 7.7%.

Example 11.4

Q. What would be the difference in interfacial energy between the coherent and semi-coherent interfaces shown in Fig.11.13?

A. The difference would lie in the dislocations. It would be a Gb^2-type term per unit length of dislocation for every dislocation in the interface.

11.5 Stages of Precipitation Hardening

The height of the maximum in the ΔG(r) curve in Fig. 11.6(b), denoted ΔG*, represents an energy barrier to the nucleation of a stable particle. The probability that such a particle will nucleate depends exponentially upon ΔG*, since the rate of nucleation is given by an expression of the form

$$\text{Rate} \propto \exp{-(\Delta G^*/RT)}$$

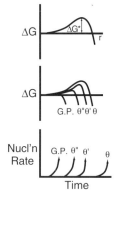

In the aluminum-based alloys discussed here, the G-P zones have the smallest γ, and at large undercooling ΔG$_v$ is large; therefore, ΔG* is small, and the rate of nucleation is large. Hence, at large undercooling, G-P zones form first. In an Al-Cu alloy the θ" phase has the next largest ΔG*, and is the next to form.

When the θ" forms, the G-P zones begin to dissolve in order to supply copper atoms to the new, more stable precipitate. Later, another metastable phase, called θ', forms; it has a semi-coherent relationship with the matrix, which is why its ΔG* is larger than for θ" (i.e., its γ value is larger because of the interface dislocations). When the θ' forms, the θ" begins to dissolve. Finally, the stable θ phase, which has the largest ΔG*, will form if aging is carried out long enough.

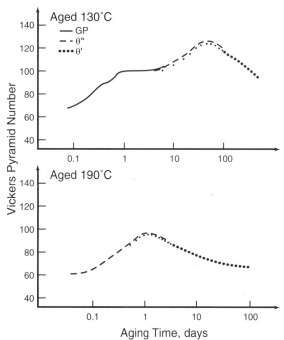

Fig. 11.14 Stages of hardening in Al-4%Cu at two temperatures. (From J. M. Silcock, T. J. Heal, and H. K. Hardy, *J. Inst. of Metals,* vol. 82, 1953-54, p. 239.)

The sequence of precipitation is reflected in the hardening curve (i.e., hardness *vs.* aging time). If precipitation occurs in stages, then hardness increases in corresponding stages. An example from an Al-Cu alloy is shown in Fig. 11.14. From the curve for aging at 130°C one can conclude that the resistance to cutting by dislocations is stronger for the θ" particles than for the G-P zones, since the hardening peak is higher for the θ". Because the θ' has a larger γ than θ", it has a larger r*. That means it forms on a coarser scale, and, after the θ" has dissolved, the hardness starts to decrease, because the dislocations can begin to loop around the θ' particles (cf. Fig. 11.14 for aging at 190°C).

Overaging (meaning a decrease in hardness with continued aging) is caused by the continued coarsening of precipitate particles, making it increasingly easy for dislocations to loop around them. (For a given volume fraction of precipitate, larger particles means fewer particles, therefore a greater spacing between particles.) The coarsening is driven by the tendency to minimize the total amount of particle/matrix interfacial energy in a way analogous to grain growth, except that coarsening requires diffusion of solute from small particles to large particles. This coarsening is exactly analogous to the coarsening of cementite in steel, as discussed in Section 6.6.2. The rate of coarsening depends on the diffusivity of the solute, and therefore on temperature, and on the magnitude of the driving force, which is the particle/matrix interfacial energy.

11.6 Fabrication of Wheel Rims

Wheel rims have quite complicated cross sections, as shown in Fig. 11.15. Manufacturers wish to achieve maximum stiffness with minimum mass, but it is also necessary to provide high strength to hold the bead of a tire that is inflated to a pressure of 100 psi or more. These cross sections are achieved in an economical way by *extrusion,* which is the process by which one gets toothpaste from a tube. This is a highly developed art in the aluminum-fabrication industry. Many high-volume components (e.g., storm-window frames) are made that way.

In the extrusion of an age-hardenable aluminum alloy, a billet of the alloy is heated to a soft condition and then placed in a die held in a large hydraulic press. The die has a hole in one end corresponding to the desired shape. The ram of the press then forces the alloy to flow through the die.

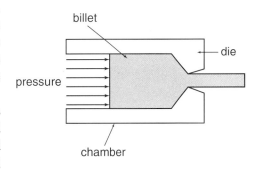

The resulting extrusion is cut to the desired lengths and heat treated by solution treatment, quenching into cold water, and aging as needed. It can be bent into a circle and joined by adhesive bonding, with the aid of thin metal splines that hold the ends together, in order to complete the wheel rim.

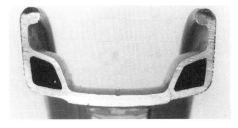 

Fig. 11.15 Example of the cross-section of a wheel rim and the method of closing the circular rim with splines. (Courtesy of Sun Metal Products.)

For components like wheel rims and frame tubes, one uses wrought aluminum alloys, which are designated by a four-digit code followed by a so-called "temper" designation. The latter indicates the state of heat treatment and/or cold work of the material. Some of the commonly used temper designations and their meanings are given in Table 11.1

Table 11.1 Some Temper Designations for Aluminum Alloys

Designation	Meaning
O	Annealed; i.e., slowly cooled from a high temperature.
H	Strain hardened to some degree (indicated by some following digits).
T3	Solution treated (i.e., quenched from the all-α region of the phase diagram), then cold-worked and aged at room temperature.
T4	Solution treated and naturally aged (i.e., at room temperature).
T6	Solution treated and aged at some elevated temperature (so-called "artificially" aged) to maximum hardness.

Commercially-pure aluminum (99.0%Al) is designated 1100. Age-hardenable alloys based on Al-4%Cu are in the 2000 series; those based on Al-Mg are in the 5000 series; those in the 6000 series contain Mg and Si and, sometimes, Cu and/or Cr. The alloys in the 7000 series are based on the Al-Zn-Mg ternary system and often contain small amounts of Mn, Cu, and/or Cr; they are capable of reaching the highest strength.

The current alloy of choice for use in bicycles is 6061 T6; its composition and properties are compared with two other common age-hardenable alloys in Table 11.2. It can be seen that the 6061 T6 does not reach strengths as high as the other two more heavily alloyed compositions. The reason for its selection is that it combines adequate strength with good formability, weldability, and corrosion resistance.

Table 11.2 Compositions and Properties of Some Age-Hardenable Aluminum Alloys

Designation	Composition (wt%)						Mechanical Properties (T6)		
	Cu	Mg	Si	Zn	Mn	Cr	Yld. Str. (MPa)	UTS (MPa)	elong (%)
2014	4.4	0.4	0.8	-	0.8	-	420	490	13
6061	0.25	1.0	0.6	-	-	0.25	280	320	12
7075	1.5	2.5	-	5.5	-	0.3	510	580	11

The 6061 alloy can be approximated as an Al-1.0%Mg-0.6%Si alloy, and the equilibrium condition below about 500°C would consist of the α phase (FCC aluminum with some magnesium and silicon in solid solution) plus the compound Mg_2Si. Normally, for a ternary alloy one must represent the phase diagram by a stack of triangular planes, one for each temperature of interest, of the type shown in Fig. 11.16(a). However, in the Al-Mg-Si case, one can simplify this by using the quasi-binary phase diagram, Al-Mg_2Si, as shown in Fig. 11.16(b).

The commercial alloy 6061 is solution treated at about 530°C and quenched in cold water. It is then aged 6 to 10 hours at 175 to 180°C to give the T6 condition. This treatment will give a very dense, fine, metastable, rod-like precipitate, the composition and crystal structure of which are still a subject of some debate. The precipitate in a model alloy that approximates 6061 is shown in Fig. 11.17.

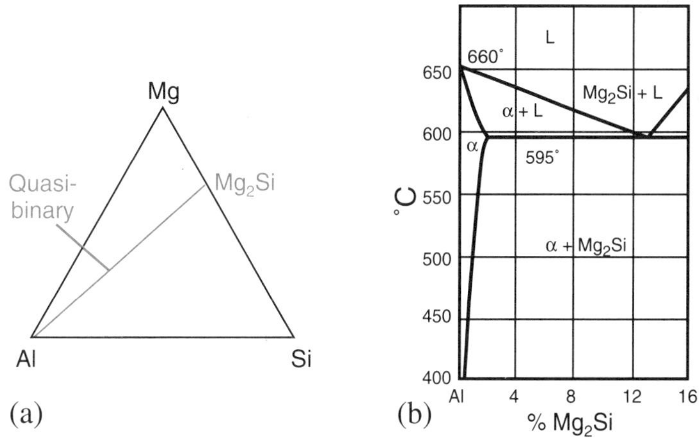

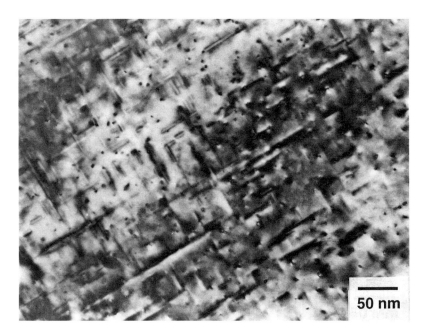

Fig. 11.16 (a) An isothermal plane of the three-dimensional ternary Al-Mg-Si phase diagram. (b) Part of the quasi-binary phase diagram Al-Mg_2Si. (From *ASM Metals Handbook*, 1948 ed., p. 1246.)

Fig. 11.17 Rod-like precipitate of Mg_2Si in Al-Mg-Si alloy aged for 8h at 190°C. Courtesy of Dr. Byung-Ki Cheong and Prof. David Laughlin, Carnegie Mellon Univ.

Summary

Precipitation-hardened aluminum alloys are important for wheel rims as well as for tubing for frames. Through precipitation hardening, the yield strength of aluminum can be raised by more than a factor of ten without a significant change in density. This requires the formation of a finely dispersed precipitate on the size scale of 50 to 100 atom diameters. The spacing of the precipitate must be so small that dislocations are forced to cut through the particles, rather than bypassing them by bowing out in loops. The fineness of the precipitate necessarily means that the total amount of precipitate/matrix interfacial area is very large. The formation of such a precipitate requires a combination of a large thermodynamic driving force (i.e., a large ΔG from undercooling) and a minimal value for the interfacial energy. The latter usually means that the precipitate must be a special metastable phase, rather than the equilibrium phase one would expect from a supersaturated solid solution, since the interfacial energy of the latter is generally large. This requirement puts a severe limitation on the number of alloys systems for which precipitation hardening is possible. The principles discussed here are applied to the 6061 alloy employed for bicycle components, which is strengthened by a metastable phase related to the compound Mg_2Si.

Appendix 11.1 Strengthening Effects of Precipitate Particles

There are several possible reasons why particles would be difficult for a dislocation to cut through. One obvious reason is depicted in Fig. 11.2; it involves the increased amount of particle/matrix interface that results from the shearing of the particle by the dislocation. Since this interface formation requires a certain energy per unit area, the applied stress must do extra work to achieve it.

Another possibility is that the resistance to dislocation motion inside the particle may be greater than that in the surrounding matrix. One would say that the particle presents a greater lattice friction to dislocation motion. Often a precipitate will have a more complicated crystal structure than the simple metal matrix. For example, the particle may be ordered; that is, it may have different types of atoms in specific lattice sites. When one dislocation passes through an ordered structure, it disrupts this order on the slip plane; the order is not restored until a succeeding dislocation comes along. This is depicted in Fig. A11.1.1.

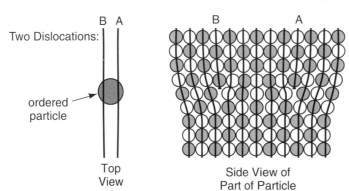

Fig. A11.1.1 Disruption of order by passage of a dislocation through an ordered particle; in the ordered structure, unlike nearest-neighbor atoms are preferred. Dislocation A creates like nearest-neighbors on the glide plane; this planar defect is eliminated by dislocation B, which follows along behind A.

All crystals present some amount of lattice friction to the motion of dislocations. This may be visualized in terms of the variation in the dislocation-core energy as the dislocation moves from one position of minimum energy to the next, as indicated schematically in Fig. A11.1.2. The slope of the energy *vs.* distance curve represents a force (work = force times distance; d(work)/d(distance) = force), and that force must be overcome by the applied shear stress; hence, the lattice-friction stress, sometimes known as the Peierls stress, after the man who first calculated it for a simple crystal structure. In soft FCC metals, like aluminum and copper, this friction stress is small, which is why they are ductile at all temperatures. In BCC transition metals, like iron, it is large for screw dislocations, and screws achieve high mobility only at temperatures well above absolute zero, as discussed in Appendix 7.1. In ordered phases the friction stress can be high as a consequence of the wrong nearest-neighbor pairs created by dislocation motion. This is only one of the possible sources of increased lattice friction that may be found in a precipitate.

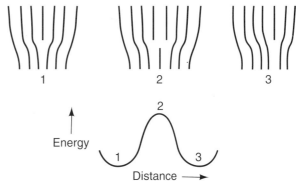

Fig. A11.1.2. Schematic depiction of variation in the energy of a dislocation core as the dislocation moves from one lattice position to the next.

Another possible source of resistance to dislocation motion is a difference in shear modulus between the particle and the matrix. Since the energy of a dislocation is proportional to Gb^2, a dislocation would naturally tend to avoid regions of high G and to be attracted to regions of low G. Thus, a particle with a G different from that of the matrix would interfere with dislocation motion by either repelling the dislocations or by attracting them.

Finally, some precipitate particles or atomic clusters can set up an elastic strain field because they contain solute atoms of a different size than the matrix, or because they are constrained to take on the same lattice spacing as the matrix. One example is the Al-Cu G-P zone shown in Fig. 11.11; another is the precipitate in the Al-Mg-Si alloy shown in Fig. 11.17. The elastic strain field around each particle gives rise to the special diffraction contrast in which the region on either side of the rod-shaped particles appears dark.

Exercises Chapter 11

Terms to Understand

Precipitation hardening
Solid-solution hardening
Orowan mechanism
Solution treatment
Overaging
Critical size
G-P zone
Metastable precipitate
Homogeneous nucleation
Heterogeneous nucleation

Structural component of γ
Chemical component of γ
Coherent precipitate
Semi-coherent precipitate
Incoherent precipitate
ΔG*
T6 temper
6061 aluminum alloy
Ordered precipitate

Problems

11.1 What kind of solvus line is needed for precipitation hardening? Explain your answer.

11.2 What other important characteristics must an alloy system have in order to exhibit precipitation hardening? Explain your answer.

11.3 Explain how the precipitates in a 6061 aluminum might resist being cut by dislocations.

11.4 Show the similarities and differences between the response to heat treatment of a Cr-Mo alloy steel (quenching and tempering) and of an Al-Cu alloy (precipitation hardening). Use schematic plots wherever appropriate and explain what takes place in each alloy.

11.5 The skins of most commercial airplanes are fabricated from panels of precipitation-hardened aluminum-based alloys. However, these alloys cannot be used for many supersonic military aircraft; these require skins of stainless steel or titanium alloys, which results in a weight penalty. Explain why aluminum-based alloys cannot be used, considering the fact that supersonic flight entails aerodynamic heating by collisions of the aircraft with gas molecules in the air.

11.6 Copper with 2% beryllium is a precipitation-hardening alloy; the Cu-rich end of the phase diagram is given here. Describe the experiments you would do to devise the heat treatment that gives the maximum hardness in this alloy. (Keep it as simple as possible.)

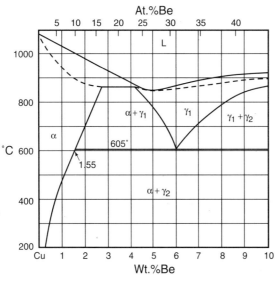

Metals Handbook, 8th ed., ASM, 1973, p. 271.

12
THE BICYCLE FRAME

From the standpoint of structural engineering, the most important component of a bicycle is the frame. It is also the most interesting component with regard to materials engineering. In the present chapter we will consider how the frame responds to applied forces, also called applied loads, and the types of metallic materials and construction used to withstand these loads. Non-metallic composite materials will be treated in Chapter 14. To consider frames properly, it is necessary to know how to analyze stress and deformation in a loaded structural member of a bicycle and then to see how the stresses can be accommodated by intelligent choices of frame geometry, materials, and joining methods.

12.1 The Frame as an Engineered Structure

The evolution of frame design has led to the so-called diamond frame, illustrated in Fig. 1.7 and shown in detail in Fig. 12.1. Stresses in the various parts of such a frame due to the static weight of a rider can be estimated fairly easily, as will be shown below, and they turn out to be rather small. However, stresses arising from some dynamic loads can be much larger and must be given serious attention. The important kinds of loading are indicated schematically in Fig. 12.2.

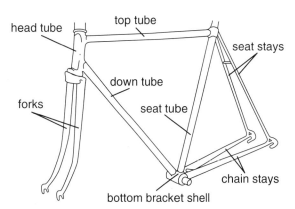

Fig. 12.1 The complete frame of a conventional diamond-frame bicycle.

The potentially large loads depicted in Fig. 12.2 can cause permanent deformation, or even catastrophic fracture, of the bicycle, or they can lead to fatigue cracking (cf. Chap. 7). In addition, the response of the frame to vertical forces from a bumpy road (Fig. 12.2 a) affects riding comfort. A frame that distorts elastically by a relatively large amount (for a given set of forces) is said to be more compliant than one that distorts less. The more compliant the frame, the more comfortable the bicycle is to ride. However, this kind of flexing wastes energy, because the work done by the rider in distorting the frame is not used in forward propulsion. For this reason, a racer opts for a frame that is elastically stiff, rather than compliant.

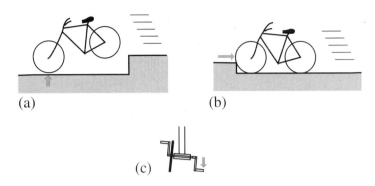

Fig. 12.2 Schematic representation of the types of loading that must be withstood by a bicycle as a result of: (a) a vertical drop after passing over a large bump, (b) an impact from a frontal barrier, and (c) the force of pedaling by a strong rider, e.g., when climbing a steep hill.

12.1.1 Forces in a Bicycle Frame; Basic Definitions and Rules

The analysis of forces employed here will be fairly elementary, but will require some explanation of the methods of engineering statics, which involves the applications of Newton's laws to bodies at rest. That is, only the equilibrium of a stationary bicycle in response to the applied forces will be considered; the dynamics of the moving bicycle will not be treated here. To begin, only the forces that act in the plane of the frame will be considered; later the important out-of-plane forces due to the pedaling action and to the off-center pull of the chain on the rear wheel spindle will be examined briefly. To make an approximate analysis of the forces on the members of a frame, a typical version of a touring bicycle has been selected. The basic geometry is as shown in Fig. 12.3.

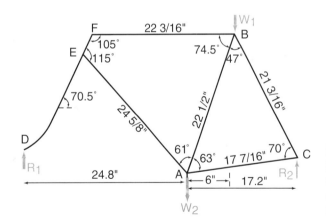

Fig. 12.3 A model of the frame of a common touring bicycle frame, showing the downward forces caused by a rider on the seat, W_1, and pedals, W_2, and the upward (reaction) forces transmitted by the wheel axles, R_1 and R_2.

Before the forces in this frame can be analyzed, one must first know some simple rules of statics:

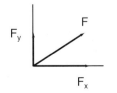

1. A force is a quantity that has both direction and magnitude; therefore, it is a vector quantity. By use of a system of rectangular (i.e., x,y) coordinates, a force vector can be resolved into two components, each of which lies along one of the coordinate axes.

SECTION 12.1.2 APPROXIMATION OF THE REAR PORTION AS A PIN-JOINTED FRAME

2. Newton deduced that any force on a body at rest must be opposed by an equal and opposite force. That is, there can be no net force on a body at rest. The presence of a net force would cause a body to be accelerated, according to the famous Newtonian law: $F = ma$, or force equals mass times acceleration. We will apply this law to the pinned joints on the bicycle frame, resolving forces applied at the joints into x and y components and setting the sums of the vertical and horizontal components equal to zero.

3. This law of Newton applies not only to translational motion, i.e., motion of a body from one place to another, but also to rotational motion of the body about some axis. Thus, Newton's law states that, for a body at rest, there must be no net moment. That is, the sum of all moments must be zero.

12.1.2 Approximation of the Rear Portion as a Pin-Jointed Frame

To estimate the forces in the structural members of the rear portion of the frame shown in Fig. 12.3 which result from the weight of the rider, one can treat this as a truss made of linear members called links held together by pinned joints, such that any pair of links could rotate freely about the pin which joins them. In this way the possibility of bending moments on the links is eliminated, since a bending moment cannot be transmitted through a pinned joint. The advantage of this approximation is that each link can then support only an axial force, either tensile or compressive. That is, the line of force in any link must pass through the pins at each end. With this approximation, it is possible to estimate the forces in the frame using simple engineering statics.

The problem with doing this in the case of the diamond frame in Fig. 12.3 is that it would not be stable with pinned joints, because part of this frame is not perfectly braced, meaning that the two closed sections are not naturally constrained against changing their shape. The rear triangle is so constrained, because the shape of a triangle cannot be changed without lengthening or shortening at least two sides. Thus, the stiffness of the links resists a change in shape of any triangle. This is not true of a four-sided section. For example, the front section in Fig. 12.3 could be distorted by rotations in the joints at E, F, A, and B. (This means that these joints must support bending moments in an actual bicycle.) To analyze the frame using simple statics, it is necessary to turn the front section into two triangles by adding another link between joints A and F, as shown in Fig. 12.4; then the whole frame becomes perfectly braced.

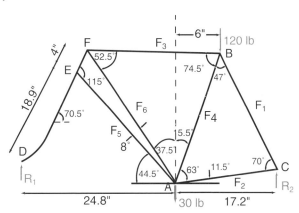

Fig. 12.4 Conversion of the frame shown in Fig. 12.3 into a perfectly braced frame by the addition of a strut to make it a combination of triangular sections.

235

12.1.3 Calculation of Reaction Forces

The first step in the analysis is to determine the magnitude of the upward forces exerted on the frame by the ground through the wheel axles. These are known as reaction forces, because they exist in reaction to the forces created by the mass of the rider and the bicycle. For simplicity, the weight of the bicycle (i.e., its mass times the downward acceleration of gravity), which is much less than the weight of the rider, will be neglected. Assuming the rider to weigh 150 lb. and 1/5 of this weight to be applied to the pedals, the reaction forces can be easily calculated by means of the above rules, as follows:

Select the axis of rotation as the front axle (point D, Fig. 12.4) and set the moments due to W_1, W_2, and R_2 equal to zero. (There is no moment due to R_1, because its moment arm is zero; point D was selected in order to eliminate R_1 as a factor.) Thus, using the dimensions given in Fig. 12.4,

$$120(30.8) + 30(24.8) = R_2(42), \text{ or } R_2 = 105.7 \text{ lb}$$

By setting the sum of the vertical forces equal to zero,

$$R_1 + R_2 = 120 + 30$$

Substituting for R_2,

$$R_1 = 44.3 \text{ lb}$$

Note that most of the weight of the rider is on the rear axle. This is true regardless of the assumed weight distribution between the seat and the pedals, and it makes it difficult for the rider to be pitched over the handle bars. (Of course, it is less difficult if the rider stands on the pedals; then more of the weight would be shifted toward the front axle, because the moment arm from the pedals to the front axle is shorter than from the seat to that axle.) This weight distribution also accounts for the ability of stunt riders to lift the front wheel off the ground by leaning to the rear, thereby increasing the moment arm from their center of mass to the front axle.

12.1.4 Calculation of Forces in the Links

Knowing the reaction force R_2, one can isolate the pinned joint at the rear axle, point C, and calculate the forces on that point due to the two rear links of the frame, remembering that the direction of these forces must be along the links. Because of Newton's law of reaction, these forces must also be the forces in the links, exerted by the pin on the links. The first step in the calculation is to judge the directions of the forces in the links. The frame is loaded by upward forces at each end and downward forces in between. This puts the lower side in tension and the upper side in compression. Thus, one immediately sees that the link AC (representing the chain stays) must be in tension. Therefore, the force, F_2, exerted on the pin at C will be downward and to the left. Because the force R_2 acts upward and the weight of the rider on the seat acts downward, the link BC (representing the seat stays) must be in compression. Therefore, it must exert a force on the pin, F_1, downward and to the right. Since the pin is at rest, the sums of the vertical and horizontal components of the forces, respectively, must be zero, and from trigonometry one can write the following two equations:

Vertical components:
$$F_2 \sin 11.5 + F_1 \sin 58.5 = 105.7$$

Horizontal components:
$$F_1 \cos 58.5 = F_2 \cos 11.5$$

or
$$F_1 = F_2 \frac{\cos 11.5}{\cos 58.5}$$

Substituting this for F_1 in the first equation,
$$F_2 \left(\sin 11.5 + \frac{\cos 11.5 \sin 58.5}{\cos 58.5}\right) = 105.7$$

or $F_2 = 58.8$ lb

Then, from the second equation,
$$F_1 = 110.2 \text{ lb}$$

By isolating the joint at point B, one can use the same method to calculate the force in the top tube, F_3, and in the seat tube, F_4. Finally, the joint at point A is isolated and the forces in the down tube, F_5, and extra link, F_6, are calculated. These calculations are set up as exercises at the end of this chapter, but the results are given in Fig. 12.5.

It can be noted immediately that the force in the added link AF is significant. Thus, this link is important for the stability of a pin-jointed frame. The absence of this link in an actual frame with rigid joints means there is a significant moment at point F. (In the next section the frame design that makes the force on the extra link equal to zero will be shown.)

Note again that the lower links are in tension and the upper ones are in compression. One may be surprised at the small value of the compressive force in the seat tube, but this results from the fact that part of the rider's weight is on the pedals, which has the effect of partly unloading the seat tube. This point illustrates that the calculated values of the forces apply only to the loading conditions assumed at the outset. If, for example, the rider were to shift some weight to the handle bars, the results would be different (See Exercise 12.3).

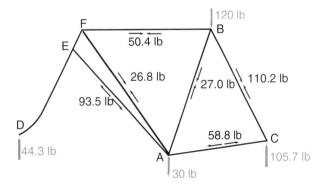

Fig. 12.5 The values of the forces in the members of the rear section of the frame for the loading conditions given in Fig. 12.4.

The value of the tensile stress in a typical steel seat tube (1 1/8 in. diam., 0.056 in. wall thickness, 0.188 in² cross-sectional area) under a load of 27 lb. is given by

$$\sigma = \frac{27}{0.188} = 143 \text{ psi}$$

This is an insignificant stress. Thus, from the standpoint of supporting the static weight of a rider, the rear portion of the frame is much more than adequate. This is not the case for the front section.

12.1.5 Forces and Moments on the Front Section

The front section of the bicycle frame, comprising the front fork and steering head, differs from the rest of the frame in that, even though the joints are assumed to be pinned, it is loaded in bending. The reason is that it has a load applied between its end points. The consequences of the bending are profound, because the tensile stresses developed in bending tend to be much larger than those in axial loading, as shown in Appendix 12.1.

To analyze the front section, it can be isolated and laid down horizontally, as in Fig. 12.6, which shows the forces applied by the top tube and down tube (or, actually, the links used to model these tubes) and by the front-wheel axle. This figure also shows the components of these forces resolved parallel to and perpendicular to the straight part of the front section, which can be characterized as a beam.

The components of force acting parallel to the beam put it in compression. The force balance can be written

$$44.3 \cos 19.5 + 26.8 \sin 33 = 93.5 \cos 65 + 50.4 \sin 19.5 \approx 56.4 \text{ lb}$$

which is accurate to within the error introduced by rounding off to one decimal place. This compressive force is comparable to most of the forces in the rear section and is, therefore, not of any great concern.

Of more importance is the bending of the beam caused by the perpendicular force components, shown in Fig. 12.6(b) and calculated according to the following:

wheel axle: $44.3 \sin 19.5 = 14.8$ lb, acting upward

down tube: $93.5 \sin 65 = 84.8$ lb, acting downward

top tube: $50.4 \cos 19.5 = 47.5$ lb, acting upward

extra link $26.8 \cos 33 = 22.5$ lb, acting upward

SECTION 12.1.5 FORCES AND MOMENTS ON THE FRONT SECTION

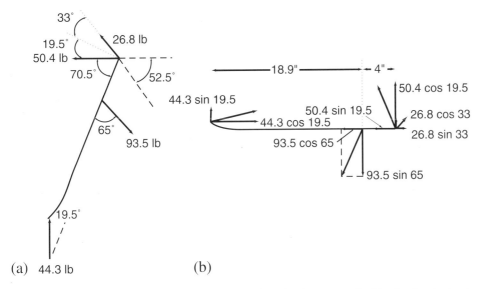

Fig. 12.6 (a) The front section of the frame, showing the forces applied by the front wheel and the rear section; (b) shows the components of these forces acting parallel and perpendicular to the front section, which is laid out horizontally for ease of analysis.

To make this easier to see, the front section can be approximated as a straight beam, as shown in Fig. 12.7(a). It is immediately apparent that the forces sum to zero and that there is a 280 in-lb moment about the inner loading point, clockwise from the left-hand force and counter-clockwise from the right-hand force.

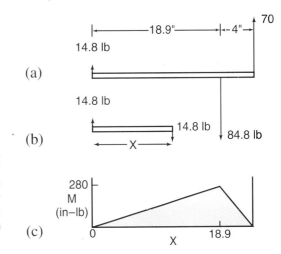

Fig. 12.7 (a) The front section as a beam, showing the perpendicular force components, which are responsible for bending. (b) The left end isolated as a free body, showing the force and moment at the right end (i.e., at the cut) needed for static equilibrium. (c) The bending-moment diagram, showing the value of the bending moment at every point along the length.

The bending of the beam caused by these forces can be analyzed by isolating a section of the beam in a free-body diagram and using it to construct a bending-moment diagram, which is a graphical representation of the bending moment at each point along the beam. The method is illustrated in Fig. 12.7(b), where the left-hand end of the beam is isolated as a free body, and all forces and moments acting on that body are shown. These include the force from the wheel axle and that imposed by the remainder of the beam, which is a downward force of 14.8 lb parallel to the area of the cut, needed to balance the upward force at the left end of the beam. The remainder of the beam also imposes a counter-clockwise moment to balance the clockwise moment imposed by the downward force at the cut. Since the cut was made at a distance x from the left end, the moment at the cut is $14.8x$. With these forces applied to the free body, it is in stable equilibrium; i.e., there is no net force either translating it or causing it to rotate. This is a standard procedure in engineering statics.

The moment at the cut, imposed by the remainder of the beam, obviously increases linearly with x, starting from zero at the left end and increasing to a maximum of $14.8 \times 18.9 = 280$ in-lb at the point of attachment of the down tube. Applying the same procedure to a free body starting from the right end of the beam, the balanced moments start from zero and increase to the same maximum: $70 \times 4 = 280$ in-lb. This is portrayed by the bending-moment diagram in Fig. 2.7(c); it shows a maximum bending moment at the point of attachment of the down tube. The maximum stress due to this bending moment can be calculated by the application of simple beam theory, given in the following section. It should be self-evident that the underside of the beam (i.e., the rear side of the front section) is stressed in tension from the weight of the rider, and the other side is in compression.

Finally, there is a condition, illustrated in Fig. 12.8(a), under which the extra link has zero stress. The requirement is that the point of intersection of the extrapolations of the top tube and down tube be directly over the front axle. In this case, there can be no moment about the point of intersection of the three forces due to any of the three forces shown. The addition of a fourth force not passing through this point (i.e., the force from the extra link) would create an unbalanced moment; therefore, it must be zero. In the frame shown in Fig. 12.3, the extrapolated top and down tubes intersect to the right of the line of action of the upward force at the front axle. Therefore, the extra strut is necessary to cancel the moment about the intersection point due to this upward force, as shown in Fig. 12.8(b). Of course, in the actual bicycle frame, this moment is cancelled by a bending moment in the rigid (i.e., not pinned) joint F.

(a)

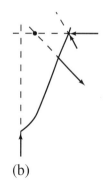

(b)

Fig. 12.8 (a) The frame geometry that obviates the need for the extra link (shown in Fig. 12.4). (b) The usual frame geometry in which the moment due to the force in the extra link would be needed to cancel the moment due to the front reaction force in a pin-jointed frame.

12.2 The Ideal Case: Pure Bending

Stresses in bent beams are usually calculated with reference to the ideal case of pure bending, which means that no shearing forces exist in the beam, only a bending moment. This can be achieved by symmetrical four-point loading, as shown in Fig. 12.9(a). Here, the central portion of the beam (between the inner loading points) is in pure bending, since the shearing forces balance out to zero in that portion. (Prove this for Exercise 12.7 at the end of the chapter.)

It is obvious that the bottom surface of the beam is compressed and the top is extended in the longitudinal direction. Therefore, a plane perpendicular to the longitudinal axis, i.e., the x axis, would experience a tensile stress on its upper part and a compressive stress on its lower part, as indicated schematically by the arrows.

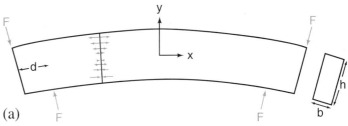

Fig. 12.9 (a) A beam loaded in symmetrical four-point bending. The central portion has no shearing forces and is therefore in pure bending. (b) The bending-moment diagram, showing the constant moment in the central portion.

The resulting deformation is exhibited in the grid inscribed on the rubber beam shown in Fig. 12.10. Clearly, the normal stress on a plane transverse to the long axis of the beam must pass through zero somewhere between the top and bottom surfaces. Since the beam is assumed to be isotropic and homogeneous and the loading symmetrical, the point of zero stress (and strain) must be at the center of the beam. This defines the *neutral axis* of the beam, which is actually a planar surface the area of which remains unchanged when the beam is bent. The neutral axis is the x axis in Fig. 12.9.

An obvious fact about the beam in Figs. 12.9 and 12.10 is that the deformation, and thus the stress, increases with the distance from the neutral axis. In other words, most of the load is carried by the outer portions of the beam. Therefore, if one wishes to design a beam with a given weight per unit length, i.e., with a given amount of material, one should try to concentrate this material as far from the neutral axis as possible. Two beam geometries that accomplish this are the I-beam and the cylindrical tube. The former is used when the beam is to be loaded in only one direction; the latter, being symmetrical, can be loaded in any transverse direction or even twisted about its axis. This is why bicycles are made from tubes.

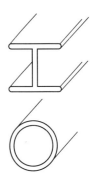

CHAPTER 12 THE BICYCLE FRAME

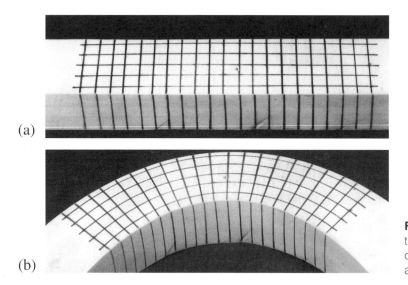

Fig. 12.10 Indication of the nature of the deformation of a grid inscribed on a rubber beam (a) before and (b) after pure bending.

From the appearance of the grid in Fig. 12.10 it is apparent that, in an isotropic, homogeneous beam in pure bending, planes that are perpendicular to the longitudinal axis before bending remain planar and perpendicular to this axis after bending. Moreover, the longitudinal displacements in the beam, indicated by the spreading-apart of the planes in the upper portion and the squeezing-together in the lower portion, are clearly linearly related to y, the distance from the neutral axis. This means that the displacement per unit length, i.e., the strain, can be written as

$$\varepsilon = C\, y$$

where C is a constant. Thus, Hooke's law ($\sigma = E\varepsilon$) can be written as

$$\sigma = E\, C\, y$$

where σ is the longitudinal stress (acting on any transverse plane), and E is Young's modulus.

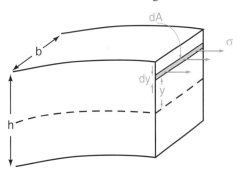

The value of σ at any y can be related to the bending moment, M, (force times moment arm) in the following way. Consider an element of area dA of the cross-section of a bent beam; let this element of area be a distance y from the neutral axis. The longitudinal force on this element is σdA, where dA is $b\,dy$.

Therefore, the corresponding contribution to the bending moment, dM, is

$$dM = \sigma\, b\, y\, dy$$

Remember that the force on this element is due to the stress in the part of the beam to the right of the cut acting on the part of the beam to the left of the cut. Since the stress is tensile in the upper half of the beam, the force acts to the right, and the moment on this element is clockwise. Below the neutral surface, the stress is compressive; therefore, the force is negative (i.e., acts to the left), but, since y is negative, the moment is still clockwise. To sum up the moments on all elements of the area, integrate over all values of y (i.e., from $y = -h/2$ to $y = h/2$). Thus,

$$M = \int_{-h/2}^{h/2} \sigma \, b \, y \, dy$$

and, substituting $\sigma = CEy$,

$$M = EC \int_{-h/2}^{h/2} b \, y^2 \, dy$$

with C and E placed in front of the integral sign, since they are never functions of y. If the beam is not rectangular, b is a function of y.

The integral is known as the *second moment of area* of this plane, or, more commonly, as the *moment of inertia* of this beam. It is designated as I; thus, one can write the total moment as

$$M = C \, E \, I.$$

where

$$I = \int_{-h/2}^{h/2} b \, y^2 \, dy$$

The constant C can now be evaluated with reference to Fig. 12.11, which depicts a beam of original length l_0 bent by an angle θ so that the radius of curvature of the neutral axis is r.

Since ρ is the distance from the center of curvature to the neutral axis (where $y = 0$), one can write

$$l_0 = r \, \theta$$

where θ is the angle (in radians) subtended by the curved beam. (Remember that the length of an arc is given by the radius of curvature times the angle subtended in radians; this is the definition of a radian.)

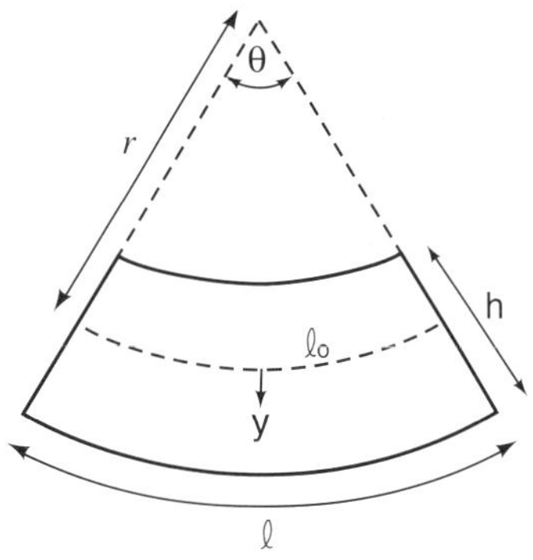

Fig. 12.11 A beam in pure bending; the radius of curvature is r. ($r \gg h$.)

Now, for any other value of y, there will exist a longitudinal strain that is defined as

$$\varepsilon = \frac{\Delta l}{l_0}$$

where $\Delta l = (l - l_0)$ is the change in length of the beam at some value of y.

Therefore, the tensile strain (below the neutral axis) can be written as

$$\varepsilon = \frac{(r+y)\theta - r\theta}{r\theta} = \frac{y}{r}$$

Since the constant C is defined by $\varepsilon = Cy$,

$$C = \frac{1}{r}$$

and this quantity is known as the *curvature* of the beam.

This value for C can now be substituted in the expression for the moment, M=CEI to give:

$$\frac{1}{r} = \frac{M}{EI}$$

Because the stress, σ, is equal to $C E y = \frac{M}{EI} \cdot Ey$, it can be written as

$$\sigma = \frac{My}{I}$$

and the maximum stress in the beam is, therefore,

$$\sigma_{max} = \frac{Mh}{2I}$$

Thus, the maximum stress increases with M and with the height of the beam, h, as one would expect, and it is inversely proportional to I.

12.2.1 Moments of Inertia

The general expression for the moment of area of a beam of width b (the dimension parallel to the axis of bending) is

$$I = \int_{y_{min}}^{y_{max}} b(y)\, y^2\, dy$$

where y is the distance from the neutral axis, as shown in the generalized cross-section in Fig. 12.12(a). Note that, in general, b is a function of y; i.e., $b = b(y)$.

For a rectangular cross-section, Fig. 12.12(b), b is a constant and can be moved in front of the integral sign. The integral can then be evaluated (to within a constant) as

$$\int y^2\, dy = \frac{y^3}{3}$$

Thus, since $y_{min} = -h/2$ and $y_{max} = h/2$, the moment of inertia is

$$I = b\left[\left(\frac{y^3}{3}\right)_{y=h/2} - \left(\frac{y^3}{3}\right)_{y=-h/2}\right]$$

or

$$I = b\left[\frac{h^3}{24} - \frac{-h^3}{24}\right] = \frac{bh^3}{12}$$

This means that I increases very rapidly with h. Therefore, for a beam that contains a given amount of material, one can get the most out of this material by making h as large as possible, as this makes the maximum stress for a given M as small as possible. This is the principle of the I-beam.

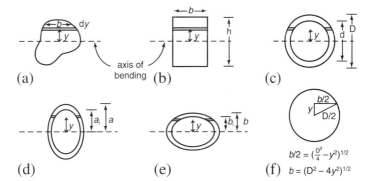

Fig. 12.12 Various cross sections of beams showing an area element in each at a distance y from the neutral axis.

For a cylindrical tube, as shown in Fig. 12.12(c), the moment of inertia is found by first solving for I for a circular cross-section of diameter D equal to the outer diameter of the tube and then subtracting the I for a circular cross-section of diameter d equal to the inner diameter. For the value of b, which obviously varies with y, the relationship shown in Fig. 12.12(f) is used. The I for the circular beam is given by

$$I = \int_{-D/2}^{D/2} (D^2 - 4y^2)^{1/2} y^2 \, dy = \frac{\pi D^4}{64}$$

Therefore, for the circular tube

$$I = \frac{\pi}{64}(D^4 - d^4)$$

A similar procedure is used for the elliptical tube shown in Figs. 12.12(d) and (e). The I for a solid elliptical beam bent around its minor axis (Fig. 12.12d) would be given by

$$I = \frac{\pi}{4} a^3 \cdot b$$

Therefore, for the elliptical tube bent about its minor axis

$$I = \frac{\pi}{4}(a^3 \cdot b - a_i^3 b_i)$$

and about its major axis

$$I = \frac{\pi}{4}(ab^3 - a_i b_i^3)$$

12.3 Application of Beam Theory to the Front Section

In Fig. 12.7 the bending moment on the front section was found to increase from zero at the wheel hub to a maximum at the joint with the down tube. Therefore, the maximum tensile stress is found there as well; that is, on the rear side of the front forks. The way to minimize that stress is to increase the I of the tubular forks by making the tubes elliptical, so that the diameter is increased in the back-to-front direction and decreased in the transverse direction. For example, the touring bicycle from which Fig. 12.3 was drawn has front fork tubes with a front-to-back diameter of 1 1/4 inch and a transverse diameter of 3/4 inch at the top. Because the bending moment decreases toward the bottom at the wheel hub, the tubes can be gradually decreased in size (to reduce weight). On this bicycle the tubes are round at their lower ends, with a 1/2 inch diameter, as shown in Fig. 12.13.

The formula for the moment of inertia of a hollow elliptical tube bent about the minor axis of the ellipse is given in Section 12.2.1 as

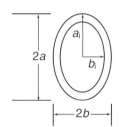

$$I = \frac{\pi}{4}(b \cdot a^3 - b_i \cdot a_i^3)$$

where $2a$ and $2b$ are the major and minor diameters of the ellipse, respectively, and the subscript i refers to the inner dimensions. Using the dimensions given in Fig. 12.13, I is calculated to be 0.0366 in³.

The maximum value of M was found earlier to be 280 in-lb, and the height of the beam above the neutral axis is a, which equals 0.625 in. Therefore, the maximum stress due to bending from the static weight of the rider is

$$\sigma = \frac{Ma}{I} = \frac{1}{2}\frac{280(0.625)}{0.0366} = 2390 \text{ psi}$$

(The 1/2 comes from the fact that the fork is made up of two tubes.) This stress is tensile on the rear side of the tube and compressive on the front side. The highest stress in the rear section of the frame is on the down tube and would be about 500 psi for the standard 0.188in² tube. Thus, the maximum stress due to bending of the front fork is almost five times higher than the stress due to axial loading, which illustrates the importance of bending loads.

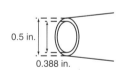

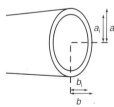

a = 0.625 in. b = 0.375 in.
a_i = 0.52 in. b_i = 0.32 in.

Lower End   Upper End

Fig. 12.13 Dimensions of the cross-section of a front-fork tube of the touring bicycle; the tube varies from elliptical at the top to circular at the bottom.

The maximum tensile stress due to static loading of the front forks is small compared with the stress which could result if the bicycle were ridden against a barrier, like a curb-stone, causing the forks to be bent backward. For example, if the full weight of a 150-lb rider were applied to the moment arm of 18.9 inches, which is the distance from the front-wheel spindle to the point of attachment of the down tube (cf. Fig. 12.4), a tensile stress of

$$1/2 \; \frac{150(18.9)(0.625)}{0.0366} = 24{,}200 \text{ psi}$$

would be generated on the front side of the fork. This stress is high enough to cause some common materials to deform plastically, so the designer would have to decide whether to use a higher-strength material or to increase the moment of inertia of the tubes. This could be done by increasing either the major diameter of the ellipse or the wall thickness of the tube. Obviously, weight is saved by using a material of higher strength.

12.4 Bending of the Frame Due to Out-of-Plane Forces

The two major out-of-plane forces on a bicycle frame are the result of forceful pedaling, which occurs, for example, when a strong rider climbs a steep hill. Both of these forces can produce a large bending moment on a part of the frame. The first is illustrated in Fig. 12.14 and results from the fact that the force on a pedal must lie off the center-line of the frame. In the example used here, the full weight of the rider, 150 lb, is assumed to be applied to one pedal. The moment arm on the touring bicycle was measured to be 5 inches. This moment is supported mainly by the seat tube, and for the purposes of calculating the upper limit of the stress on this tube, it is assumed that the seat tube supports all of the bending moment. The maximum stress is found at the juncture of the seat tube and the bottom bracket. This can be understood by approximating the seat tube as a cantilever projecting from a wall (i.e., the bottom bracket shell). The maximum stress occurs at the point of maximum bending moment, which is at the attachment point (cf. Fig. 3.2).

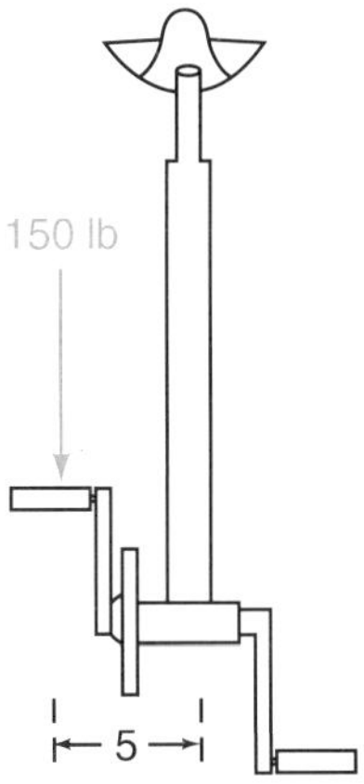

Fig. 12.14 The bending moment applied to the seat tube by the pedaling force.

The bending moment, $M = 150 \times 5 = 750$ in-lb, produces a maximum tensile stress, which, ignoring the stress concentration factor at the joint, is given by

$$\sigma = M\,D/2I$$

where D is the outer diameter of the tube, 1 1/8 inches in this case, since the maximum value of y on this beam is D/2. From Section 2.2.1, I for a circular tube is given by

$$I = \pi/64 \; (D^4 - d^4)$$

where d is the inner diameter of the tube, taken to be 1.025 in in this case, since a typical wall thickness of a steel bicycle tube is 0.05 in. With these dimensions, I is calculated to be 0.0244 in⁴. Thus, the maximum stress is

$$\sigma = \frac{750(1.125)}{2(0.0244)} = 17290 \text{ psi}$$

which is considerably larger than that found previously in the most-highly stressed tube from the forces due to static, in-plane loading. If one takes into account the effect of the small radius at the seat-tube/bottom-bracket joint, this stress could be higher by as much as a factor of two. This is high enough to cause localized plastic deformation in a carbon-steel frame.

The bending moment due to pedaling raises two concerns. The first is that the repeated application of the high stress at the joint could lead ultimately to fatigue cracking, a natural consequence of cyclic plastic deformation. The second concern is of a more immediate nature. If the bottom bracket is rotated sufficiently, then the chain on the chain wheel can rub against the seat tube. This would be rather inconvenient in a bicycle race, so the elastic stiffness of this part of a racing bike is of major importance.

The other important out-of-plane loading results from the fact that the chain is connected to one side of the rear wheel, and the pedaling force in the chain necessarily puts a compressive force and also a bending moment on the chain stay closest to the chain (and to a much smaller extent on the seat stay on the same side; this will be ignored). The configuration is depicted in Fig. 12.15.

The force in the chain can be calculated by taking moments about the crank axis and setting the total moment equal to zero, which gives

$$F_c \, r = P \, R, \text{ or } F_c(2) = 150(7), \text{ or } F_c = 525 \text{ lb}$$

since the crank length on the touring bicycle is 7 inches, and the smallest of the three chain wheels (radius = 2 inches) is used to get the maximum force on the chain. It is again assumed that all of the rider's 150-lb weight is applied to one pedal.

The forces in the chain stays can be calculated with the aid of the diagram in Fig. 12.15(c). The force in the left stay, F_L, is found from the moments around point B. Thus,

$$0.875 F_C = 1.75 F_L \cos 5.3, \text{ or } F_L = 265 \text{ lb}$$

Similarly, the force in the right stay can be obtained from moments around point C:

$$2.625 F_C = 1.75 F_R \cos 5.3, \text{ or } F_R = 790 \text{ lb}$$

One can immediately see that, since the chain tends to twist the bottom bracket in a clockwise sense, the stays must do the reverse; this means that F_L is tensile, and F_R is compressive.

In addition to these forces, the chain tends to twist the quadrilateral ABCD to the side, as shown in Fig 12.16, which puts a bending moment on the stays. If the rear wheel spindle, AD, is isolated, it can be seen that it is acted upon by three non-parallel forces, which, being in static equilibrium, must all pass through one point (about which the total rotational moment would be zero). This point must lie on the line of action of the chain, which exerts one of the forces. This fact can be used to say that the maximum moment arm that F_R could have in bending the right stay would be the distance from point B to the chain, which is 7/8 inch. Therefore, the maximum bending moment, M, would be (7/8 × 790) in-lb, which would produce a maximum stress of $Mb/2I$, where b is the relevant dimension of the elliptical cross-section of the stay.

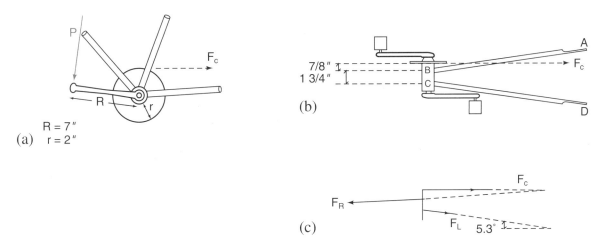

Fig. 12.15 (a) Side view of the area around the bottom bracket, showing the pedaling force and the resultant force on the chain. (b) Overhead view of the chain stays, bottom bracket, and chain. (c) Forces on the bottom bracket exerted by the stays and the chain.

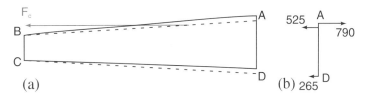

Fig 12.16 (a) The distortion of the quadrilateral formed by the bottom bracket, chain stays, and rear wheel spindle, caused by the force in the chain. (b) The forces on the rear wheel spindle.

The expression for the moment of inertia of a hollow elliptical tube, for bending about the major axis of the ellipse, is given in Section 12.2.1 as

$$I = \pi/4 \, (a \, b^3 - a_i b_i^3)$$

with a and b defined in Fig. 12.13. Using the dimensions found for the elliptical tube in the touring bike, I is calculated to be 0.0125 in³. Therefore, the maximum stress due to bending of the right stay is given by

$$\sigma = \frac{0.875 \;\; 790 \;\; 0.375}{0.0125} = 20{,}740 \text{ psi}$$

One side of the stay is in tension because of this bending, and the other side is in compression. The longitudinal force on this stay, F_R, is compressive, and one can calculate the resulting stress by using the area of the cross-section of the hollow elliptical tube:

$$\text{Area} = \pi \, (a \, b - a_i \, b_i)$$

which, on the touring bike, is 0.2135 in². Therefore, the compressive stress is

$$\sigma = \frac{790}{0.2135} = 3700 \text{ psi}$$

The net result of these two effects is a compressive stress on the outer side of the right stay of 24,440 psi and a tensile stress on the inner side of 17,040 psi. In addition, there is the tensile force of 58.8 lb due to the weight of the rider, resulting in a tensile stress on one side of 29.4/0.2135 = 138 psi, giving a total compressive stress of 24262 psi on the outer side. The designer must ensure that the material employed behaves elastically at this stress level. In actuality, a margin of safety of 50 to 100% would probably be used.

This completes the analysis of the frame of the exemplar bicycle. The methods employed are the same as found in a course in engineering statics or strength of materials. They can be applied to any bicycle frame, given a set of assumed loading conditions. Once the operating loads are known for a particular frame design, the stresses can be dealt with in two ways: one is the choice of tube geometry, particularly with regard to the moment of inertia, and the other is the choice of material. Both of these choices will govern the weight of the frame, which is also an important consideration.

12.5. Alloys for Bicycle Frames; General Considerations

The earliest bicycles were made of wood or wood reinforced with somewhat primitive steel, as illustrated in Fig. 12.17(a). One of the most successful kinds of wood for this purpose was bamboo; an example is shown in Fig. 12.17(b). Aside from the cost of the handwork, the major problem in the development of wood-frame bicycles was joining the individual pieces into a structural network. This has remained a challenge for all subsequent materials. Ironically, bicycle technology has begun to return to fiber-reinforced materials, of which wood was the prototype. However, for most of the past century and still today, the dominant materials have been metallic, primarily steel.

During the century following the 1870s, steel was the dominant material for bicycle frames, and bicycle evolution essentially ceased for the last two-thirds of that period. However, just as aviation technology was an offspring of the bicycle in the early twentieth century, in the 1970s the bicycle became the beneficiary of developments in aerospace technology, particularly with regard to materials. This is nowhere more evident than in frames. Today, there are frames fabricated from high-strength aluminum and titanium alloys and graphite-fiber reinforced polymers. The availability of these materials and the expertise to process them has come directly from the aerospace field, which, in a sense, is now repaying its technological debt. In the rest of this chapter the application of metallic alloys to frame building is examined.

SECTION 12.5 ALLOYS FOR BICYCLE FRAMES; GENERAL CONSIDERATIONS

Fig. 12.17. (a) A pre-World-War I wooden bicycle frame with steel-reinforced joints. (Courtesy of *Cycling Weekly*, IPC Magazines, London.) (b) A 19th-century frame made of bamboo. (from A. Sharp loc. cit.)

The first consideration for a frame is that, under any foreseeable conditions, it should never be loaded above the yield strength of the material at any location on the frame. That is, the yield strength must be high enough that the behavior is always elastic. The next consideration is that the frame have enough elastic stiffness that it does not distort excessively during hard riding. The most critical regions in this regard are, first, the juncture of the front fork and the head tube and, second, the area of the bottom bracket, where the seat tube, down tube, and chain stays meet and where the forces on the pedal crank are supported.

Some elastic deflection in the plane of the frame is desirable for rider comfort. If the frame were absolutely rigid, every road bump would deliver a shock. The desirable amount of in-plane deflection depends on where the rider is in the spectrum between world-class racer and weekend recreational rider. Whatever the application, every rider values the minimization of weight. Thus, a premium is placed on low density materials, with high strength and adequate stiffness. This is the reason for the introduction of light alloys for frames. A comparison of the relevant properties of these light alloys with traditional carbon steel and Cr-Mo alloy steel is made in Table 12.1.

Table 12.1 Comparison of the Properties of Metallic Frame Materials.

Material	Young's Modulus, E (GPa)	Density, ρ (g/cm^3)	0.2% yield stress, σ_{yld} (MPa)	E/ρ	σ_{yld}/r
Carbon steel	200	7.8	240	25.6	30
Cr-Mo steel	200	7.8	485	25.6	62
6061-T6 Al	70	2.7	260	25.9	95
Ti-3Al-2.5V	110	4.5	700	24.4	156

Note that the aluminum alloy has about one-third the density of the steel but only about one-third the stiffness, so the stiffness-to-density ratio is about the same for each. The titanium alloy is intermediate in both properties, and, again, the ratio is about the same. However, if the materials are compared on the basis of yield strength, the titanium alloy is about 40% stronger than the Cr-Mo steel, and the strength-to-density ratio is more than twice as large. The aluminum alloy is considerably weaker than the titanium alloy and the Cr-Mo steel, but its strength-to-density ratio is still greater than steel. The carbon steel compares very badly with the other alloys in strength-to-density ratio.

For performance of bicycle tubing, however, tube geometry is as important as material properties. As shown in Section 12.2, for a tube of outer and inner diameters D and d, the maximum stress for a given bending moment, M, is

$$\sigma_{max} = \frac{M\,D}{2\,I}$$

where I, the moment of inertia, is

$$I = \frac{\pi}{64}(D^4 - d^4)$$

Therefore, a low yield strength in the material can be compensated for by an increase in tube diameter and/or thickness.

The use of alternatives to steel must be based on some set of design criteria that include yield strength, weight, and the stiffness of the frame as a whole. The stiffness of a tube can be characterized by the curvature, $\frac{1}{r}$ imposed by a given bending moment according to

$$\frac{1}{r} = \frac{M}{E \cdot I}$$

and the weight can be characterized by the mass per unit length of tube, given by

$$\text{Mass per unit length} = \text{density} \cdot \text{area} = \text{density} \cdot \frac{\pi}{4}(D^2 - d^2)$$

Therefore, flexing of tubes can be reduced by increasing their moment of inertia by increasing tube diameter and/or thickness, but this is done at the expense of lightness. With the use of a given thickness of a low-stiffness material like aluminum, the moment of inertia increases as D^4, but the weight increases only as D^2.

A number of specimen frames are compared in Table 12.2 with regard to tube geometries and three measures of performance: mass per unit length, curvature produced by a bending moment of 100 N-m (885 in-lb), and the maximum stress at this bending moment, expressed as a fraction of the yield stress of the tube material.

Table 12.2

Material		D (OD) (in)	d (ID) (in)	wall thickness (in)	cross-section $\pi/4\,(D^2-d^2)$ in²	cross-section cm²	moment of inertia $\pi/64\,(D^4-d^4)$ in⁴	moment of inertia cm⁴	weight/ unit length g/m	curvature for a given bending moment $1/r = M/EI$ mm/100N-m	max. stress for a given bending moment $\sigma_m = \frac{MD}{2I}$ MPa/100N-m	$\%\sigma_{YLD}$
Road Bikes												
Carbon steel		1.130	1.030	0.050	0.17	1.10	0.025	1.041	860	48	138	58
Cr-Mo steel		1.133	1.060	0.036	0.120	0.77	0.018	0.750	601	67	190	39
6061-T6 Al		1.134	1.020	0.056	0.019	1.23	0.027	1.124	332	127	128	50
Mountain Bikes												
6061-T6Al A	top tube	1.50	1.40	0.05	0.23	1.47	0.06	2.50	397	57	76	30
	down tube	2.00	1.87	0.06	0.40	2.55	0.185	7.70	688	19	32	12
	seat tube	1.25	1.05	0.10	0.36	2.33	0.06	2.50	629	57	63	25
6061-T6Al B	top tube	1.375	1.213	0.081	0.33	2.13	0.069	2.87	575	49	61	24
	down tube	1.575	1.413	0.081	0.38	2.45	0.106	4.41	662	32	45	18
	seat tube	1.375	1.213	0.081	0.33	2.13	0.069	2.87	575	49	61	24
Ti-3Al-2.5V	top tube	1.25	1.16	0.045	0.19	1.22	0.034	1.41	549	64	45	18
	down tube	1.25	1.16	0.045	0.19	1.22	0.034	1.41	549	64	45	18
	seat tube	1.25	1.16	0.045	0.19	1.22	0.034	1.41	549	64	45	18

12.6 Comparisons of Steel, Aluminum-Alloy, and Titanium-Alloy Frames

12.6.1 Road Bikes

The Cr-Mo steel can be used as a standard for comparison. It can be seen that the carbon steel tubing is about 40% heavier, and the σ_{max}/σ_y for a given bending moment would be 50% higher than for alloy steel. The carbon steel tubing is stiffer, however; this is a result of the greater wall thickness, which is necessary to compensate for its low yield strength. Thus, the alloy-steel tubing is lighter and more resistant to fatigue cracking (owing to the higher strength) but is less stiff than the carbon steel.

These steel tubes were from road bikes, also known as racing-type bikes. They are intended to be light in weight and to be ridden over smooth surfaces. A 6061-T6 aluminum alloy tube from the same type of bicycle is also listed in Table 12.2. It has the same outer diameter as the steel tubes, and its wall thickness is about the same as the carbon steel tube. However, owing to the low density of aluminum, its weight is only 55% of the alloy steel tube. Because its yield strength is higher than the carbon steel, the maximum stress for a given bending moment would fall in between that for the alloy steel and carbon steel. Aside from its lightness, a rider would also feel the greater flexibility of the aluminum frame, because tube curvature caused by any given bending moment would be almost twice that found in the alloy steel tubing.

12.6.2 Mountain Bikes

A mountain bike is meant for riding over rough terrain; therefore, it must withstand larger impact forces than a road bike. The design goal should be to minimize σ_{max}/σ_y for a given bending moment without sacrificing lightness or optimum stiffness. The solution is to use tubing with a large diameter but with about the same wall thickness used for road bikes. However, manufacturers have applied this solution in different ways, as indicated in Table 12.2. One of the two Al-alloy frames has much fatter tubing than the other, but in both, the down tube has the maximum diameter, with the top tube and seat tube having the same smaller diameter in one, but different diameters in the other. The two frames have more or less the same characteristics in terms of maximum stress and weight, but they may have noticeable differences in riding character as a result of different combinations of tube stiffness.

The Ti-alloy tubing is superior to the Al-alloy tubes in terms of maximum stress and is similar to them in weight and stiffness. It might also be noted that the diameter of these tubes is much closer to those found in a road bike than the diameters of the Al-alloys; this is possible as a result of the greater strength and elastic modulus of titanium compared to aluminum.

The values for yield strength given in Table 12.2 are those for tubing, *per se*. However, these values cannot be applied in the vicinity of joints made by thermal processes, i.e., brazing or welding, because of the softening effects of the heat. The remainder of this chapter will be focussed on the microstructure of each metallic material (which is the source of the yield strength) and on the effects of thermal joining processes on microstructure and strength.

The alternative joining process is adhesive bonding, which employs an epoxy adhesive and a component at each joint that serves to define the joint angles and to form the space containing the adhesive. (In the use of such components, brazed and bonded joints are similar.) Bonded joints require only low temperatures for curing, and these do not affect the strengths even of aluminum alloys. However, bonded joints are a relatively recent innovation in frame building, and experience with joint durability is still being evaluated. Bonded joints will be discussed in Chapter 14.

12.7 Steel Frames

Several factors led to the dominance of steel for building bicycle frames. The rapid development and spread of large-scale steel making and the growth of the machine-tool industry that followed enabled the production of steel tubes. The development of alloy steels enabled the achievement of high strength in tubes joined by brazing. Moreover, steel has a high Young's modulus in comparison to its potential metallic competitors and can be strengthened by several processes, including the fine ferrite/carbide microstructures obtainable from the decomposition of alloyed austenite. The challenge has been to achieve or maintain high strength at the joints of frames assembled by brazing or welding.

12.7.1 Effects of Brazing on Steel Tubing

The most economical material for a bicycle frame would be thin-walled, cold-drawn carbon-steel tubing, if the tubing could be joined together without heating. The strength achieved by strain hardening of the mostly ferritic steel would permit the tubing wall to be thin, achieving lightness, and the high modulus would allow great leeway in tube design to achieve the desired frame stiffness. This is actually achievable as a result of recent developments in adhesives. However, at present, brazing is still generally used for frames joined with lugs. Even with silver-based brazing alloys, this requires heating the tubing in the vicinity of 700°C, which would recrystallize, and thereby soften, a cold-drawn tube. If a Cu-Zn brazing alloy were used, the steel would be completely transformed to austenite, because, as the phase diagram in Fig. 8.21(b) indicates, the 60Cu-40Zn alloy has a liquidus above 900°C.

The conclusion is that carbon steel tubes cannot be permanently strengthened by cold work if they are to be used in a brazed frame. After furnace brazing, the microstructure of the tubing would be in the normalized (austenitized and air-cooled) condition. In order to achieve the required strength, the wall-thickness of the carbon steel tubes must be greater than that of strain-hardened tubing. This means that a carbon-steel frame must be heavier than one made from thin-walled tubing.

A lightweight, brazed steel frame can be made from thin-walled tubing if an appropriate alloy steel is used. A steel commonly chosen contains chromium and molybdenum as the major alloy elements, along with about 0.3 to 0.4% carbon. These two alloy elements are BCC transition metals, and one of their effects in iron is to stabilize the BCC α phase. The result is an increase in the eutectoid temperature and a decrease in the carbon content at which the eutectoid reaction occurs, as illustrated schematically in Fig. 12.18. Thus, for a given carbon content, a slowly cooled Cr-Mo steel should have more of the eutectoid microstructure than would a plain-carbon steel.

An even more important property of these alloy elements is that they form carbides that are significantly more stable than Fe_3C; this is especially true for molybdenum. This effect has already been noted with regard to the phenomenon of secondary hardening during tempering, as illustrated in Fig. 10.10. As pointed out in Section 10.4.3, this is the result of the formation of fine carbides at elevated temperatures (i.e., where the diffusion of the substitutional alloy elements can occur).

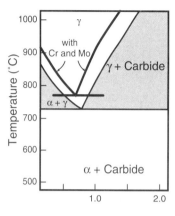

Fig. 12.18 Schematic illustration of the effect of the addition of ferrite-stabilizers, like chromium and molybdenum, on the eutectoid region of the iron-carbon diagram.

Another property of these elements, especially molybdenum, is that they have a strong affinity for carbon in solid solution in iron. The effect is, first, to retard the kinetics of the transformation of austenite to ferrite-plus-carbide (i.e., both elements increase hardenability) and, second, to alter the morphology of the eutectoid transformation product from lamellar pearlite to a much finer distribution of globular carbides in a ferrite matrix.

The result of these changes can be seen by comparing the microstructures of the seat stays from a lightweight, alloy-steel frame and a heavyweight low-carbon-steel frame, both of which had been assembled by furnace brazing, as shown in Fig. 12.19. The brazing alloy was Cu-Zn, which had to be heated above 900°C to be liquified; thus, the tubes would have been austenitized during brazing. When the fixture holding the frame was removed from the furnace and allowed to cool in air, the alloy steel, Fig. 12.19(b), began to transform by forming Widmanstätten ferrite and blocky ferrite on a very fine scale. Then the remaining austenite transformed eutectoidally to a ferrite-plus-carbide mixture, which is the analog of pearlite in this alloy steel. (Some of this two-phase constituent may be bainite, but the microstructure is too fine to be sure.) In any case, this fine two-phase constituent comprises 70 to 80% of the microstructure and is responsible for the high strength, compared to what would be found in a plain carbon steel with the same carbon content. (Figure 9.1(b) shows an air-cooled 1040 steel, which has 0.4% carbon, instead of the 0.3% in the 4130 alloy steel shown in Fig 12.19(b), but it still provides a useful comparison.)

In contrast, the low-carbon steel shown in Fig. 12.19(a) consists almost entirely of ferrite, with only a very small amount of pearlite. Thus, the carbon steel has the same ferritic matrix as the alloy steel, but without the dispersed carbides. The fact that it has only about half the strength of the alloy steel necessitates a greater wall thickness, as shown by the comparison of cross sections of the tubes in Fig. 12.19.

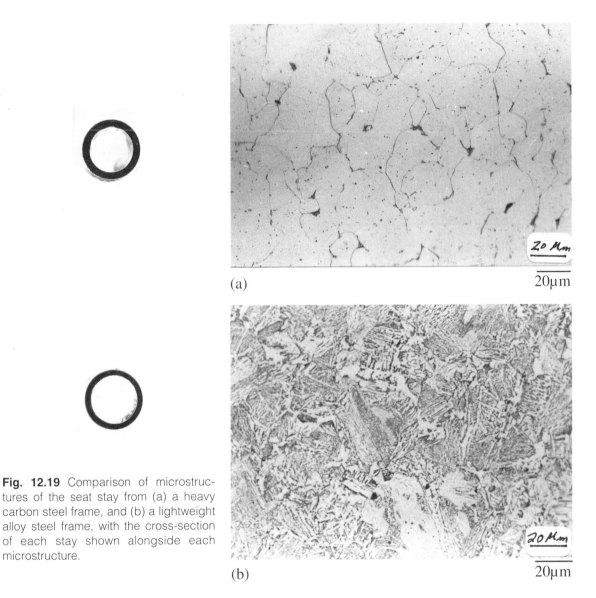

Fig. 12.19 Comparison of microstructures of the seat stay from (a) a heavy carbon steel frame, and (b) a lightweight alloy steel frame, with the cross-section of each stay shown alongside each microstructure.

Figure 12.20 shows load *vs.* elongation curves obtained from tensile specimens of tubes from the same two bicycles. Note that load, not stress, is plotted here, and the geometry of the tensile specimens cut from the frame was the same, except for the greater wall thickness of the carbon-steel tube. It is remarkable that, even though the alloy steel tube had a thinner wall than the carbon steel, it still carried about 50% more load before necking occurred. This means that the design thickness of the alloy steel tube was determined not by its yield strength, which is much more than adequate, but by the requirements for elastic stiffness of the frame.

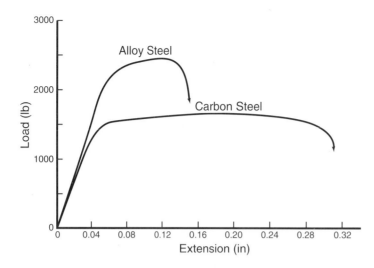

Fig. 12.20 Comparison of the load *vs.* extension curves of an alloy-steel tube and a carbon-steel tube from the same bicycle manufacturer.

12.7.2 Lugs; the Fabrication of Complex Shapes

The function of a lug is to hold the tubes at a particular angle and to provide a precisely dimensioned channel along which liquid brazing alloy can spread. In order to minimize weight, the lug should have a thin wall, and the joint area should be kept to a minimum, consistent with obtaining a sufficiently strong joint. Examples of two kinds of lugs are shown in Fig. 12.21. The arms of the lugs are designed to provide a large joint area where needed to resist the loading of the frame at that location. For example, in the top-tube-to-head-tube lugs shown in Fig. 12.21, the joint must resist the bending forces in the plane of the frame. Hence, there is more joint area at the top and bottom of the top tube and at the front and rear of the head tube than at the sides of these tubes.

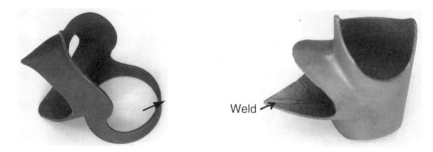

Fig. 12.21 Examples of top-tube-to-head-tube lugs: (a) investment cast, (b) fabricated from sheet.

One common method of producing a lug is the process of *investment casting*, also known as the lost-wax process. This process has been used to make artifacts dating to the time of the pharaohs. The first step is to make a replica of the desired article out of wax, which can be carved and molded easily at room temperature. Next, the wax replica is invested, or encased, in a ceramic mold by repeatedly coating it with a slurry, or paste, of a ceramic powder mixed with water, allowing each coat to dry before applying the next. In actual production, a kind of tree of wax replicas, connected by branches of wax, is constructed before the investment process. The tree is connected to the outside of the large ceramic mold by a larger wax branch. The mold is then heated in a furnace to densify it by sintering (cf. Section 10.6.3), i.e., to make it a solid ceramic with a minimum of porosity.

During the firing of the mold, the wax is vaporized, therefore "lost," leaving cavities into which molten metal may be poured. The mold may be heated before the casting is poured to allow the hot metal to flow throughout all the branches before solidifying. When the mold is cool, it is broken apart, and the cast pieces are cut from the tree. Usually only a minimum of finishing is needed, for example by grinding to remove excess metal, before the part is ready for use. The special art required is the knowledge of how to allow for the shrinkage of the mold during firing and of the metal during solidification to produce lugs of the desired dimensions (e.g., with the required clearances between the lugs and the tubes).

Any metal casting will contain some unavoidable amount of porosity from two sources. First, most materials contract upon solidification. When a liquid freezes, as most do, by the nucleation and growth of solid dendrites (cf. Section 8.2.4), the spaces between the dendritic arms contain the last liquid to freeze. Because of the contraction, there will not be enough liquid to fill these interdendritic spaces with solid; the result is so-called shrinkage porosity.

The other source of porosity is the evolution of gases, mainly oxygen and nitrogen, which are more soluble in liquid metals than in solid. The same phenomenon occurs when water freezes; thus, the porosity due to formation of gas bubbles can be seen in ice cubes. An example of the porosity found in an investment-cast lug is shown in Fig. 12.22. This low-carbon steel lug also contains a dispersion of fine particles; these are inadvertent impurity inclusions arising, most probably, from the entrainment of ceramic particles in the liquid metal as it flowed into the mold.

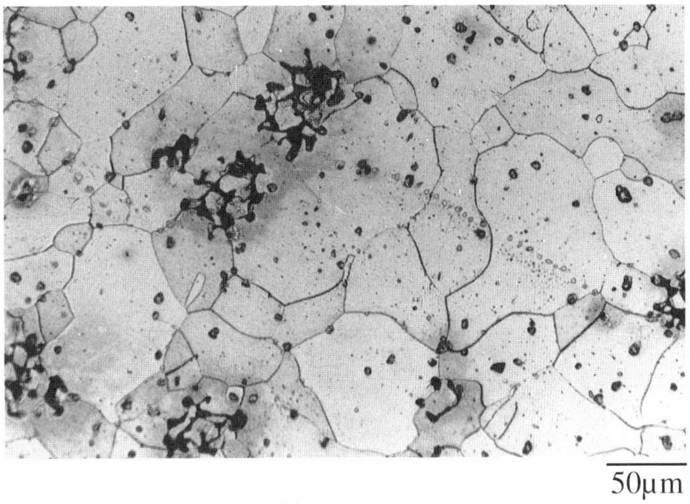

Fig. 12.22 Example of the porosity in an investment-cast low-carbon steel lug.

In bicycle lugs the casting porosity is generally not a problem, since the solidified brazing alloy is likely to have even larger cavities, due mainly to insufficient flow of the alloy to fill the space entirely, and in any case the brazing alloy is probably softer than the steel. However, in general, one must be aware that castings contain some degree of porosity and may also have coarse grain sizes if they solidify slowly. Thus, a casting is often not as strong as a wrought object made of the same material. A wrought object is something which is mechanically shaped, for example by rolling or forging, after the material was initially produced (usually by solidification). During high-temperature deformation processes, the porosity initially present in a cast ingot is removed by pressure-welding.

The second method of producing a bicycle lug is to shape it out of sheet and to use welding to close it into the final form. An example is shown in Fig. 12.21(b). Here, a low-carbon steel sheet produced by cold-rolling, i.e., a wrought product, was press-formed in a die to a shape which could then be curled and welded into the form of a lug. This is obviously a faster process than investment casting, and it produces a stronger and thinner lug, but involves the use of large expensive capital equipment and is only employed when a sufficiently large number of lugs of the same design are to be made.

12.7.3 Fabrication of Steel Frames by Welding

An alternative method of assembling lugless steel frames is welding by the tungsten inert gas (TIG) process. This was used originally for mountain bikes, which have non-standard tube sizes and angles, but it is now being applied more generally. The fundamental difference between welding and brazing is that the pieces to be joined by welding do not remain solid. They are fused along the joint, together with a filler metal.

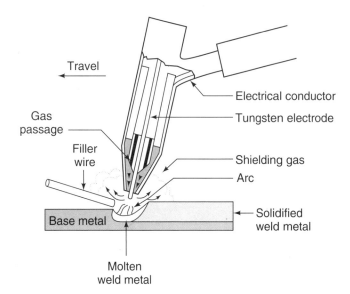

Fig. 12.23 Schematic illustration of a TIG welding gun in operation. (From S. Kalpakjian, *Manufacturing Engineering and Technology,* Addison-Wesley, 1989, p. 829.)

In the TIG process, which is simply a special variant of arc welding, an electric arc is struck between the tungsten electrode in the welding "gun" and the article to be welded. An arc is a high-temperature plasma of ionized gases, generated when two conductors with a large voltage difference are brought near each other. The voltage difference is created by an electrical generator attached to the tungsten electrode, shown in Fig. 12.23, while the "work," e.g., bicycle frame, is grounded. The gun is also connected by a hose to a container of argon or helium gas; this gas is blown over the joint to prevent oxidation of the weld pool. Filler metal is added to the joint by a separate welding wire, also shown schematically in Fig. 12.23. The process allows precisely located and controlled welding, and the heat-affected zone, the HAZ, can be kept quite small by using a small gun and carefully controlled heat input.

Weld joints at the head tube of an alloy steel mountain bike are shown in Fig. 12.24, along with the results of microhardness measurements in the vicinity of the joints. (Microhardness tests are carried out by using a small diamond-pyramid indenter attatched to a special microscope.) There are two important features of these joints. The first is that the weld metal and HAZ are both harder than the tubes themselves. The reason for this is evident from the microstructures shown in Fig. 12.24.

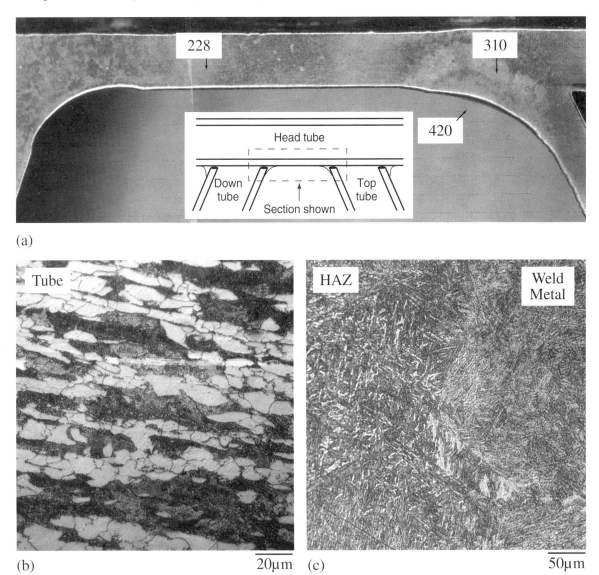

Fig. 12.24 (a) Weld joints between the head tube and the top and down tubes of an alloy steel mountain bike, showing microhardness values at various locations and the microstructures of (b) the original tube as well as (c) the juncture of the weld metal and heat-affected zone.

The tubes exhibit a banded microstructure, (Fig. 12.24) characteristic of a hot-worked alloy steel in which microsegregation of the alloy elements (Mn, Cr, and Mo) took place during solidification of the original ingot. The microstructure comprises alternating bands of ferrite and high-carbon (alloy-rich) bainite mixed with quasi-pearlite. (The eutectoid decomposition product in a Cr-Mo steel is not lamellar pearlite, but rather a mixture of ferrite and rod-like or globular carbides.)

In contrast, the weld metal is fully bainitic. The bainite was formed during rapid cooling from the austenite, which formed during solidification. The cooling was rapid because a small volume of molten metal was surrounded by the cold metal of the tubes, which conducted heat away quickly. The HAZ is also fully bainitic, but the bainite was formed from austenite of a much smaller grain size than in the weld metal, because it had not reached such a high temperature in the austenitic phase field. The boundary region between the HAZ and the weld metal is also shown in Fig. 12.24.

The second important feature is the "lack-of-penetration" type of defect in the inside of both weld joints; one is shown at higher magnification in Fig.12.25(a). This type of defect arises when the fusion zone has not penetrated completely through a joint. In this particular case, only a portion of the original gap between the two pieces of tubing was actually closed by fusion. When the joint was made, a weld bead was deposited alongside the joint, forming a bridge between the two pieces of tubing, as shown schematically in Fig. 12.25(b) and (c). This is a common type of defect in a butt joint, and it acts as a stress concentrator in the joint. It can lead to premature fatigue fracture, for example. In this case, the problem arises from poor welding technique; the amount of energy input to the joint was insufficient to fuse the joint all the way through.

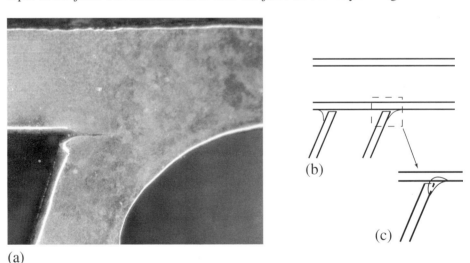

Fig. 12.25 (a) A lack-of-penetration defect in the interior of the joints shown in Fig. 12.24. (b) and (c) A schematic representation of how that joint was made.

With thin-walled tubing, it is not a trivial matter to get the energy input just right, and particular skill is required of the welder. In this situation, it is advisable to weld test pieces occasionally, so that they can be sectioned and checked to ensure that the welding technique is producing full-penetration welds. Alternatively, the joints can be subjected to non-destructive testing, using radiography, for example. This adds expense to the manufacturing process, but it would lessen the probability of a joint failure.

The insidious aspect of this type of defect is that it is on the inside of the tube and is, therefore, invisible to the user of the bike. The elimination of such defects is important in bicycle parts that are highly stressed in bending, such as around bottom brackets and in the joints of the head tube. The latter is particularly important in mountain bike frames, which, by virtue of their elongated-diamond design, are subject to large bending moments at the head tube, particularly when ridden over rough terrain.

12.7.4 Butted Tubing

To compensate for the softening during heating of the joints of hand-brazed or welded frames, the wall thickness of the end-sections of the tubes can be left larger in the tube-drawing process, as shown in Fig. 12.26. Such tubes are called butted tubes. To make them requires the use of a special mandrel, which is a solid cylindrical tool placed inside the tube during final drawing through the dies. This is obviously more expensive than making tubes with uniform wall thickness. It is often used to save weight on high-priced bikes joined by brazing.

Fig. 12.26 A schematic comparison of (a) conventional and (b) double-butted tubing.

Butted tubing is unnecessary for furnace-brazed bicycles, since the effects of heat are uniform over the entire frame. It is also unnecessary for frames to be joined by adhesive bonding, where only low-temperature heating is used to cure the adhesive. In this relatively new method of bicycle construction, a lightweight steel frame can be made using unbutted alloy-steel tubing assembled by bonding the tubes to lugs with an epoxy adhesive. Weight savings come from the elimination of the butted ends and the brazing alloy and sometimes from the substitution of aluminum for steel lugs. The nature of adhesives for this purpose will be covered in Chap. 14.

12.8 Joining of Light-Metal Frames

12.8.1 Al-Alloy Frames

The aluminum alloy of choice for frames is 6061 T6. As indicated in Chap. 11, the strengthening of the alloy is achieved by a solution treatment followed by water quenching and an aging treatment above room temperature. The tubing can be joined by either adhesive bonding or welding. In welding of tubing that was precipitation-hardened prior to welding, the heat-affected zones alongside the fusion zone of a weld would be over-aged, i.e., softened. Of course, the weld metal would also be soft, since it would be in the as-cast condition, but this weakness could be offset by a large volume of weld metal, so that the cross-section through the weld metal is larger than the thickness of the tube. The softened HAZ has no such simple remedy, and the frame designer must keep the extent of this region small by limiting energy input during the welding process.

One of the weld joints in mountain bike B (Table 12.2) was sectioned and polished metallographically so that the hardness at various points in the joint region could be measured. The joint is shown in Fig. 12.27(a), and the distribution of hardness values is shown in Fig. 12.27(b). The diamond-pyramid hardness (DPH) in the weld metal averages 73, whereas the hardness of the tubing far from the joint exceeds 150. The HAZ extends quite far from the weld metal, due to the large thermal conductivity of the aluminum alloy and the fact that overaging occurs rapidly in this material at a few hundred degrees centigrade. Thus, contrary to the case of the weld joint in the alloy steel, the joint in a welded Al-alloy frame is much softer than the tubing, and the soft HAZ is quite extensive. This must be considered by the designer when specifying the diameter and wall thickness of the tubing.

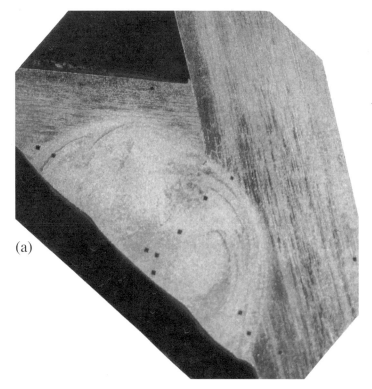

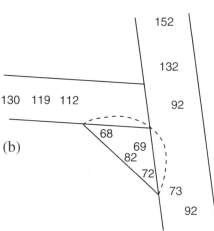

Fig. 12.27 (a) Cross-section of welded joint between the top tube and head tube of mountain bike B (cf. Table 12.2). (b) Diamond-pyramid hardness at various points in the joint region, determined by microhardness tests.

One way to eliminate the softened HAZ is to heat-treat the entire frame after welding is completed. Since this would involve rapid quenching from the solution-treatment temperature, it carries the strong possibility of distortion of the frame as a result of thermal stresses (cf. Section 10.4.4). The frame would have to be constrained in a rigid fixture during heat treatment, adding significantly to the manufacturing cost. The aging treatment applied after quenching would presumably relieve much of the residual stresses due to quenching.

It can be seen in Fig. 12.27, and more clearly in Fig. 12.28, that the joint in this mountain bike suffers from the same kind of lack-of-penetration defect that afflicts the steel joint in Fig. 12.24. This crack-like flaw in the unwelded section of the joint could lead to fatigue failure through the joint, particularly in such a soft region. This defect occurs on the inside of a tube, where it is invisible.

Fig. 12.28 The juncture of the lower side of the top tube and the head tube depicted in Fig. 12.27(a), showing the crack-like defect from the almost complete lack of penetration of the weld metal into this juncture. This is the result of an inadequate amount of energy delivered by the welding gun.

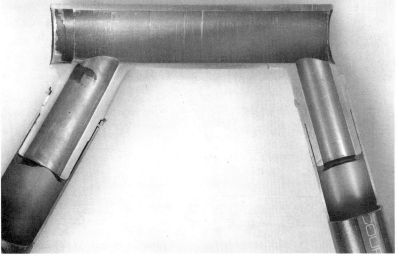

Fig. 12.29 Cross-section of joints between the top and down tubes and the head tube of the aluminum-alloy road bike listed in Table 12.2, showing the use of epoxy adhesive for the bonding of the joints.

The softening associated with welding of aluminum-alloy tubes can be avoided by using adhesive bonding. The Al-alloy road-bike tubing listed in Table 12.2 was joined together in that manner. The joints of the top tube and down tube to the head tube are shown in the sectioned frame in Fig. 12.29. This kind of head tube with tubular arms can be fabricated either by casting or by forging and machining. The heat-treated tubes are simply pressed over the arms, which have ridges to keep the tube-to-arm spacing uniform, and the epoxy adhesive (applied to the arms before assembly) is cured.

The cured epoxy is a glassy network polymer (see Chap. 14) and, as such, is a brittle material. That means that a bonded joint does not have the "give" that is found in a brazed joint; a brazing alloy can deform plastically if a joint is overstressed, but the epoxy cannot deform plastically without cracking. For this reason, it is imperative that the bonded joints be designed to carry all conceivable loads without plastic deformation.

12.8.2 Titanium Alloy Frames

Titanium alloys present a different set of challenges to a frame builder, primarily because titanium is a highly reactive metal. It forms a very stable oxide, carbide, and nitride, and it readily absorbs oxygen, carbon, and nitrogen in solid solution. It is not practical to heat-treat titanium in air, because the surface layers become hardened and embrittled by the uptake of these interstitial impurities. Thus, vacuum heat treatment is necessary for titanium and its alloys.

The attractiveness of titanium alloys for certain applications, in spite of the high processing costs, is that they are less dense than alloy steels, while having comparable strengths, and they can be highly resistant to chemical attack, as long as the surface is passivated by the TiO_2 film. In addition, they have melting temperatures higher than that of iron. As a result, titanium alloys have found applications in the aerospace and chemical industries.

Titanium undergoes an allotropic transformation from HCP to BCC at 883°C (analogous to the α to γ in iron at 910°C), as indicated in Fig. 12.30. It is common to formulate titanium alloys by adding at least two elements, one of which dissolves preferentially in, and stabilizes, the α (HCP) phase and another which dissolves in the β (BCC) phase. In this way one can get a two-phase alloy at room temperature with solid-solution strengthening in each phase. If the two-phase mixture can be produced on a fine scale, the effect is analogous to refinement of grain size, and this can produce substantial additional strengthening.

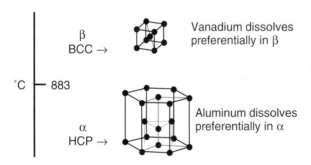

Fig. 12.30 Schematic representation of the allotropic transformation in titanium at 883°C, indicating the preferences of the alloying elements aluminum and vanadium for the respective phases.

The fact that aluminum is an α-stabilizer and vanadium is a β-stabilizer is reflected in the respective phase diagrams, as shown in Fig. 12.31. For either alloy element, it can be expected that, because of the β-to-α transformation, one can heat a titanium alloy into the all-β region and then produce a variety of microstructures by varying the cooling rate to room temperature. This will be illustrated for the case of the Ti-3%Al-2.5%V alloy used for bicycle tubing.

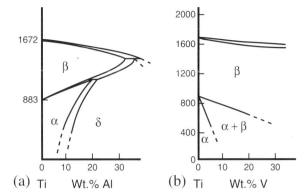

Fig. 12.31 (a) Titanium-rich side of the titanium-aluminum phase diagram and (b) the titanium-vanadium phase diagram.

The most common high-strength titanium alloy is Ti-6Al-4V, in which one can produce a fine microstructure having a high volume fraction of β in α, with large solid-solution strengthening in each phase. This alloy is not ideal for bicycle tubing, however, because it strain hardens so rapidly that tube drawing would require frequent anneals in a vacuum and would be prohibitively expensive. Hot working is also impractical, since it would have to be done in a protective environment.

Seamless tubing is produced for aerospace applications, however, from the Ti-3Al-2.5V alloy, which has only half the solid-solution strengthening and much less β at room temperature than the Ti-6Al-4V alloy. The availability of this tubing is the main reason for its selection for bicycle frames, but the properties of this alloy make it well-suited for frames, as evidenced by Tables 12.1 and 12.2. The high strength is made possible by the combination of: (a) solid-solution hardening of α by aluminum and of β by vanadium, (b) the fineness of the microstructure, achieved by heat treatment, and (c) the strain hardening from the tube-drawing process.

In order to interpret the microstructures found in the bicycle, it is necessary first to examine the relatively simple microstructure obtained by slowly cooling the alloy from the all-β region, shown in Fig. 12.32(a). The alloy was heated to 1000°C, and this produced a rather coarse-grained β. When the alloy was cooled below the β/α+β phase boundary, plates of α formed along certain crystallographic planes in the β grains. This is Widmanstätten α, and it is analogous to the Widmanstätten ferrite formed in carbon steels (cf. Section 9.3.3).

The images in Figs. 12.32 and 12.33 were obtained with a scanning electron microscope (cf. Fig. 14.26). In these figures the α phase appears dark, and the β phase, which stands out on the etched surface, appears white.

CHAPTER 12 THE BICYCLE FRAME

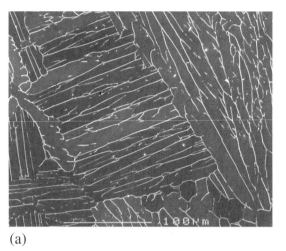
(a)

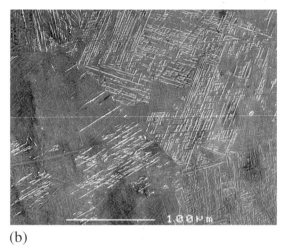
(b)

Fig. 12.32 (a) Microstructure of Ti-3Al-2.5V furnace-cooled from 1000°C, showing Widmanstätten α with residual β between the α plates. (b) Same alloy air-cooled from 1000°C; the α is now in very thin plates. Scanning electron micrographs using secondary electrons; courtesy John Marcon and Jeff Pfaendtner, Univ. of Penn.

As the α formed, the vanadium, which had been in solid solution in the β (along with the aluminum), was rejected into the remaining β, since the vanadium is more soluble in β than in α (Fig. 12.31b). Thus, the rate of formation of the α was controlled by the rate of outward diffusion of vanadium (a substitutional solute) in the same way that the formation of Widmanstätten ferrite is controlled by the diffusion of carbon into the surrounding austenite. The reason for the plate-like morphology of the α is that the rejected vanadium "piles-up" along the flat faces of the α and slows the rate of motion of that α/β boundary, whereas the rate of motion of the tip of the plate is much less affected by this piling-up.

The final condition is a microstructure consisting almost of entirely Widmanstätten α, with residual β in the spaces between the α plates. The fineness of this microstructure, even in the furnace-cooled condition, is the direct result of the slowness of diffusion of vanadium in the β phase.

An even finer microstructure is formed by air-cooling from the β region, as shown in Fig. 12.32(b). Here again the α has formed in very thin plates, which are the analog of bainite in steel. The strength of the air-cooled alloy is higher than that of the furnace-cooled because of the finer microstructure.

The microstructure of the cold-drawn tubing is shown in Fig. 12.33(a) and (b). It comprises flattened α grains elongated in the drawing direction. The microstructures of weld metal and the adjacent HAZ are shown in Figs. 12.33(c) and (d), respectively. Both are characteristic of an alloy cooled from the β region rather quickly, as seen from a comparison with Fig. 12.32(b). Although the weld metal and HAZ are not strain hardened, the microstructure is still quite fine; this produces a high strength in the joint. The effect of welding on the joint strength can be judged from the microhardness measurements shown in Table 12.3.

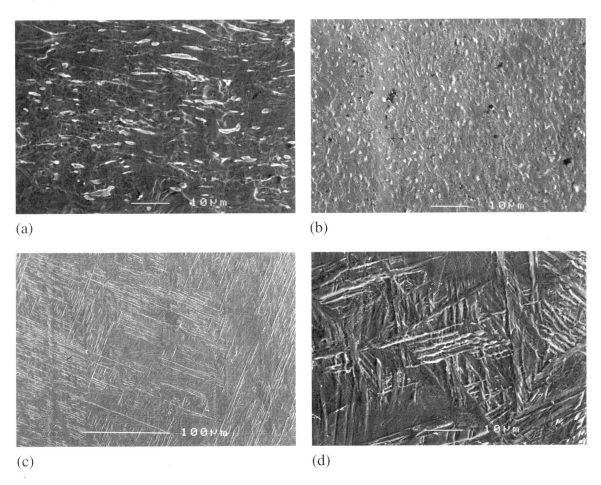

Fig. 12.33 Microstructures of tube and weld joint in Ti-3Al-2.5V bicycle frame. (a) Longitudinal section of as-drawn tube. (b) Transverse section of as-drawn tube. (c) Weld metal. (d) HAZ. SEMs by Marcon and Pfaendtner, Univ. of Penn.

Table 12.3 Microhardness Values at the Weld Joint in the Ti-Alloy Frame

Location	Average Hardness (DPH)
Base metal (tube)	250
HAZ	215
Weld metal	214

It is apparent that the HAZ and weld metal have about 85% of the hardness of the cold-drawn tube. Thus, strain hardening accounts for only a minor fraction of the strength of the tubing, and the loss of this hardening from the heat of welding is not a major concern.

It must be realized that the titanium-alloy frames are assembled by welding under a protective blanket of argon gas. The inside of the tubing also needs to be filled with argon to prevent uptake of the interstitial impurities oxygen, nitrogen, hydrogen and carbon from the air. If this is not done, the weld metal and HAZ can be embrittled. In effect, the interstitials pin dislocations to the extent that the resultant strengthening produces a large loss of ductility.

Because the surface of the titanium alloy is passivated by an extremely thin layer of titanium oxide, it is impervious to atmospheric corrosion. Therefore, the frame can be left unpainted, with its distinctive dull-grey metallic sheen. This appearance is one of the factors that make titanium-alloy bikes attractive to many people. However, owing to the high manufacturing costs, it is unlikely that welded all-titanium-alloy frames will ever be used for mass-market bicycle production.

12.8.3 Hybrid Frames

Except for welded all-steel frames, virtually all frames are constructed from combinations of materials. For example, because of the critical need for high strength and toughness in the front fork, many non-steel frames are made with steel front forks. (In principle, a titanium alloy could be used, but the material is not yet available commercially in the required shape.) One can also find frames made from Cr-Mo steel tubing adhesively bonded to aluminum-alloy joint fittings, such as head tubes. This trend is likely to continue. For example, frames with titanium-alloy tubing could be made much less expensively by use of adhesive bonding to aluminum-alloy or stainless steel fittings. The evaluation of such frames is still generally made in a subjective manner by a wide variety of riders. Progress in this field would probably be accelerated by the development of a systematic testing protocol to provide quantitative determinations of some critical features of frame behavior.

Summary

A bicycle frame can be modelled as a pin-jointed truss in order to get an approximation of the forces in the various members. This is done using the methods of engineering statics. It is seen that the most important stresses come from bending moments. These occur where the front section joins the head tube, at the bottom bracket, and in the chain stays. The moment of inertia of the components, as well as the elastic modulus, yield strength, and density, are the factors which must be integrated for proper design.

Most bicycle frames are made from metallic materials, as opposed to carbon-fiber-reinforced polymers (discussed in Chap. 14). The three common alloy types, steel, aluminum, and titanium, all have the same stiffness/density ratio, but vary widely with respect to strength/density ratio and response to brazing or welding. The traditional manner of assembly of tubes into frames is brazing, which is applicable only to steels, owing to the stability of the oxide films on the other two alloy types. Most brazing employs lugs, which are either investment-cast steel or fabricated from formed and welded sheet steel. Inexpensive (heavy) brazed frames are made from steel with a carbon content of about 0.1% or less. Lightweight steel frames are made from Cr-Mo steel with about 0.3% carbon. The microstructure of this alloy steel comprises mainly the eutectoid-decomposition product, which is a fairly fine dispersion of carbide particles in ferrite, even in furnace-brazed frames (which are air-cooled from the austenite region).

TIG welding is now a common joining method for alloy steel frames, particularly in mountain bikes, and this technique can work very well, because the weld metal and HAZ have a bainitic microstructure, due to the high hardenability of the Cr-Mo steel. However, the potential for welding defects stemming from lack of full penetration of the fusion zone through the joint must be recognized. The same potential exists for welded frames of other alloys, of course, but welded aluminum frames suffer from the presence of very soft weld metal and an HAZ softened by overaging. The HAZ is also large, due to the high thermal conductivity of aluminum.

The titanium alloy used for bicycle frames is Ti-3%Al-2.5%V; it was originally developed for aircraft hydraulic lines and is therefore available as tubing. Welding must be done in an inert environment, including the insides of the tubes, but the weld metal and HAZ have about 85% of the strength of the cold-worked tubing, due to a fine, two-phase microstructure in which each phase has substantial solid-solution strengthening.

In the future, it can be expected that an increasing proportion of frames will be assembled by means of adhesive bonding, in which an epoxy adhesive is used in conjunction with either internal or external lug-like fittings. This, of course, eliminates all problems associated with heating, such as annealing or overaging.

APPENDIX 12.1 - Bending *vs.* Axial Loading

It was pointed out earlier that the normal stresses (tensile or compressive) developed in bending are usually significantly higher than those developed in axial loading. This can be demonstrated quantitatively with the aid of Fig. A12.1.1, which shows a bar being loaded in uniaxial tension and as a cantilever beam.

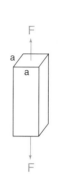
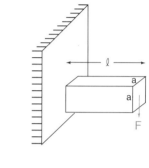

Fig. A12.1.1 A bar of length *l* and cross-section *a* x *a* loaded (a) in uniaxial tension and (b) as a cantilever beam.

The bar has a square cross-section, *a* on a side, with a cross-sectional area of a^2. The stress in the tensile specimen due to the axial force F is given by

$$\sigma = \frac{F}{a^2}$$

The maximum tensile stress in the same bar loaded in bending is given in Section 12.2 by

$$\sigma_{max} = \frac{M_{max}\, y_{max}}{I}$$

The maximum bending moment occurs where the cantilever is attached to its mounting and is given by $F \cdot l$. The value of y_{max} is $a/2$, and the moment of inertia is found from the value for a rectangular beam as follows:

$$I = \frac{b\,h^3}{12} = \frac{a^4}{12}$$

Thus, the maximum stress is given by

$$\sigma_{max} = F \cdot l \cdot \frac{a}{2} \cdot \frac{12}{a^4} = \frac{6Fl}{a^3} = \frac{6l}{a}\left(\frac{F}{a^2}\right)$$

Thus, the tensile stress is a factor of $6l/a$ higher than for the case of uniaxial loading. Since *l* is usually much larger than *a*, this factor can be very large.

Appendix 12.2 - Buckling of a Thin Strut under Axial Compression

A thin strut, like a spoke, cannot be loaded axially in compression to any substantial load without danger of buckling. It is common experience that any slender body loaded by compressive forces at its ends can easily suffer the elastic instability known as buckling. The reason for this can be seen from Fig. A12.2.1. As the axial force F is gradually increased from zero, the strut is initially straight and stable, as shown in (a). However, if the strut becomes even slightly curved, as shown in (b), the sections AB and BC become levers which exert a bending moment on the strut. The maximum moment occurs at point B and is equal to $y_0 F$. A further increase in F would cause more bending, leading to elastic collapse of the strut. The slenderness of the strut allows small strains to produce a large overall change in shape. One can determine the force at which a strut (having length l and moment of inertia I) becomes curved and, thus, on its way to instability.

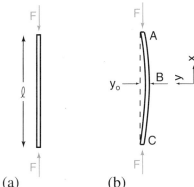

Fig. A12.2.1 A thin strut loaded in axial compression (a) while still stable and (b) just as buckling has commenced.

The relationship between the bending moment and the radius of curvature of a beam is given by

$$\frac{M}{EI} = \frac{1}{\rho}$$

In the present case, M at any point y along the strut is given by $M = F \cdot y$. Therefore,

$$\frac{1}{\rho} = \frac{F\,y}{EI}$$

In terms of the coordinate system shown in Fig. A12.2.1, the equation representing the curved strut, approximating it as a curve in the x,y plane, can be written as

$$\frac{d^2 y}{dx^2} = -\frac{1}{\rho}$$

Therefore, one can construct a differential equation with a known solution as follows:

$$\frac{d^2 y}{dx^2} + \alpha^2 y = 0$$

where α is defined as

$$\alpha^2 = \frac{F}{EI}$$

The general solution to this differential equation is

$$y = m \sin \alpha x + n \cos \alpha x$$

where m and n are arbitrary constants. This is the equation for the curved strut.

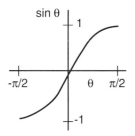
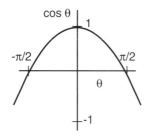

Fig. A12.2.2 Plots of sin θ and cos θ vs. θ.

Now if one examines the shapes of the sine and cosine functions, as shown in Fig. A12.2.2, it is obvious that the cosine function is consistent with the shape of the strut, while the sine function is not. Therefore, the constant m is set equal to zero and n = y_0, and the equation for the shape of the strut becomes

$$y = y_0 \cos \alpha x$$

It is also clear that y = 0 at x = l/2 or at x = -l/2. Thus,

$$y_0 \cos \alpha l/2 = 0$$

which means, if y_0 is to be finite, $\alpha l/2 = \pi/2$, or $\alpha l = \pi$. Therefore, from the definition of α,

$$\frac{F}{EI} = \frac{\pi^2}{l^2}$$

or, finally

$$F = \frac{\pi^2 EI}{l^2}$$

This gives the critical force to start buckling (i.e., to have a curved strut that satisfies the equation $y = y_0 \cos \alpha x$). Obviously, F decreases greatly as l increases and as the strut becomes thinner in the direction perpendicular to the axis of bending (i.e., as I decreases). This expression would give the critical force for the buckling of a bicycle spoke if the spoke were loaded as shown in Fig. A12.2.1, i.e., with both ends pin-jointed.

Exercises

Terms to Understand

Normal stress
Reaction force
Moment
Bending-moment diagram
Shear stress
Moment of inertia
Buckling
Cyclic loading
TIG welding
Furnace brazing

Cr-Mo steel
HAZ
Investment casting
Shrinkage porosity
Gas porosity
Butted tubing
Bonded joints
Ti-Al-V alloys
Mandrel
α-stabilizers *vs.* β-stabilizers

Problems

12.1 Using the results from Section 12.1.4, calculate the forces F_3 and F_4 acting at point B (cf. Fig. 12.4), which is shown isolated below. Check your answers with the values given in Fig. 12.5.

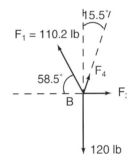

12.2 Using the results from Problem 12.1, calculate the forces F_5 and F_6 acting on point A, using the diagram below. Again, check your answers with Fig. 12.5.

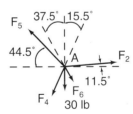

12.3 Recalculate the forces F_1 through F_6 with the new assumption that the rider applies 20% of his weight on the handlebars with a vertical line of action, while keeping 20% symmetrically on the pedals.

12.4 Recalculate these forces again with a second new assumption that the rider applies all of his weight symmetrically on the pedals.

12.5 Discuss the advisability of fabricating the front-fork tubes from a lightweight, high-strength, but brittle, material.

12.6 The cross-section of the rim of a particular design of bicycle wheel is shown here; give the rationale for this rim design from what you now know about bending of beams.

12.7 Prove that the beam shown in Fig. 12.9 has no shear stresses in the portion between the inner two loading points.

12.8 In an alloy that solidifies dendritically in a casting (or a weld or a braze), explain where in the microstructure the shrinkage and gas porosity will be found and why.

12.9 Suppose you are preparing an investment casting pattern out of wax for a tubular shape, such as one leg of a lug. The ceramic to be placed around the wax is known to undergo a volume shrinkage of 10% when fired. If the finished dimensions of the lug are to be 1.1in ID and 1.2in OD, what should the dimensions of the wax pattern be?

12.10 Assume that the eutectoid composition of a 4140 (Cr-Mo) steel is reduced from 0.77% to 0.65% by the alloy elements, how much proeutectoid ferrite and how much eutectoid constituent would be expected in an austenitized and slowly cooled sample?

12.11 If a titanium-alloy road bike were to be made with tubing of the same OD as the Cr-Mo steel bike (cf. Table 12.2), and if it were to have the same tube stiffness (i.e., the same EI product), how would the two types of tubing compare on the basis of weight per unit length?

12.12 Compare the three types of high-performance tubing: 6061 aluminum alloy, Ti-2.5Al-2V, and Cr-Mo steel, with regard to the strengthening mechanisms employed, describing how they work and the effect of welding on the tube strength.

12.13 Explain how it is possible to make a bicycle tube lighter by using an aluminum or titanium alloy while maintaining the same or greater stiffness as with steel, even though the ratio of the Young's modulus to density is virtually the same for all three alloys.

13

POLYMERIC MATERIALS; TIRES

13.1 Characteristics of Polymers

Polymers have been used in bicycles in the form of rubber tires for a century. More recent applications include adhesives and fiber-reinforced materials for frames, wheels, and other components. Polymers represent a class of materials entirely different from the metallic and ceramic materials considered so far in that they are long-chain molecules made of repeated units called *mers*. In metals and crystalline ceramics, individual atoms or ions can move about independently and take part in a variety of reactions. The properties of these aggregates of atoms or ions can be considered three-dimensionally, since one type of bonding predominates in each material, and it is essentially uniform in all directions.

Polymers, on the other hand, exhibit a much wider range of structures and properties. At one extreme, they can be similar in structure to a glassy ceramic, like an oxide glass, in which atoms are connected in a three-dimensional network of covalent bonds. At the other extreme, they can exist as long-chain, linear molecules with a strong, covalently bonded backbone. The chains are highly flexible and, at sufficiently high temperatures, can writhe about like snakes. In fact, one useful conceptual model of a linear polymer in the amorphous state is that of a collection of earthworms, as shown in Fig. 13.1. In an actual linear polymer there is relatively weak (i.e., secondary) bonding between chains, depending on the interchain spacing; this type of bonding is discussed in Appendix 13.1. If the temperature of the polymer becomes low enough, the wriggling of the chains practically ceases, and the secondary bonds then give the polymer the character of a three-dimensional solid.

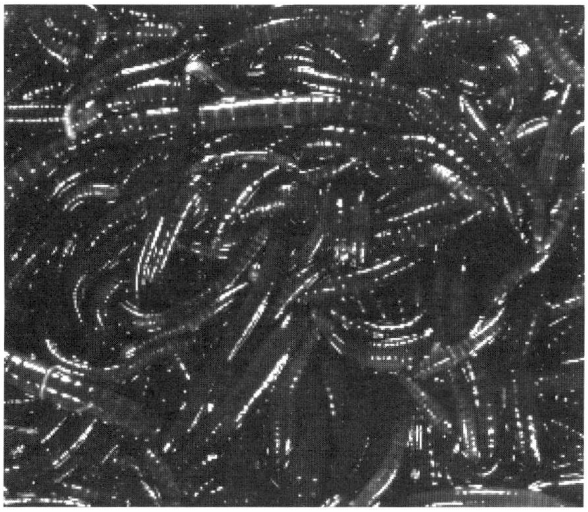

Fig. 13.1 A collection of tangled earthworms, serving as a model for a linear polymer at a temperature where viscous flow is possible. (Photograph of *Lumbricus rubellus* by Runk/Schoenberger; from Grant Heilman, reproduced with permission.)

Rubber used for tires is made from one of several possible linear polymers that are subjected to a special treatment, called *vulcanization,* which converts the viscous, amorphous mass into a rubbery material. To understand the structure and properties of a rubber, it is first necessary to understand the characteristics of linear polymers in general.

The most important characteristic of a linear polymer is the flexibility of the long-chain molecule. The simplest such molecule has a backbone of singly-bonded carbon atoms, as shown in Fig. 13.2(a). In addition to the two bonds per carbon which constitute the backbone, each carbon bonds to two other atoms, or groups of atoms, which attach themselves to the backbone. However, it is the character of the single carbon-carbon covalent bond that gives the backbone its flexibility. The only constraints on this bond are its fixed length and the fixed angle between neighboring carbon bonds. This allows a bond to rotate and take any position along the conic surface that preserves this angular relationship, as shown in Fig. 13.2(b). At sufficiently high temperatures the chains are free to become coiled and kinked to such an extent that they comprise an amorphous glob, like the "can of worms" depicted in Fig. 13.1.

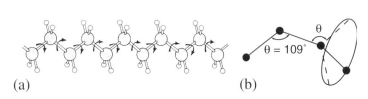

Fig. 13.2 (a) A linear polymer with the simplest possible backbone, a singly-bonded carbon chain. (b) The freedom of rotation in a single carbon-carbon bond. (Adapted from N. G. McCrum, C. P. Buckley, and C. B. Bucknall, *Principles of Polymer Engineering,* Oxford University Press, 1988, pp. 50 and 51; reprinted by permission of Oxford University Press.)

When the temperature is relatively high, the chains have a high kinetic energy and can wriggle about rather freely. Entropy tends to be maximized by the randomness of the chain configurations. The usual way to describe this is the path of a "random walk," which would be characterized by a large number of steps, all of the same length, but of completely random directions in space. That is, each step has an equal probability of being in any possible direction. Because of the easy motion of the chains, they can slide past each other readily, and the polymeric mass acts as a viscous fluid. The viscosity decreases, i.e., the ease of flow increases, as temperature increases.

Another important characteristic of a linear polymer is the length of the chains. The lengths, of course, are not all the same, but are distributed about some average. At any given temperature, viscosity increases as average chain length increases, because the chains are more likely to get tangled as they become longer. (The increase in viscosity with chain length is not linear; it becomes increasingly steep as length increases.) A classical example is the paraffin, or n-alkane (C_nH_{2n+2}), family of molecules in which each of the lateral carbon bonds is occupied by a hydrogen atom, as shown in Fig. 13.3. Methane, CH_4, is a gas at room temperature, but decane, $C_{10}H_{22}$, is an oily liquid (i.e., a paraffin oil). When the chain length is increased to $C_{36}H_{74}$, for example, the material has become a paraffin *wax*. When the average chain length is of the order 10^2 to 10^4 carbon bonds, the material is called *polyethylene*. In this condition, there are so many entanglements along each molecule that the material behaves like a solid at room temperature. That is, it is able to support a shear stress without significant viscous flow.

(a) a paraffin

(b) polyethylene

Fig. 13.3 The molecular structure of (a) a paraffin and (b) of polyethylene.

Although polyethylene acts like a solid at room temperature, it becomes a viscous liquid when heated; the increasing thermal energy allows increasing amounts of chain wriggling. If it is cooled back to room temperature, it again acts like a solid. This is called *thermoplastic* behavior. It is one of the characteristics that gives many polymers their great commercial value, because the temperatures at which large-scale flow occurs are generally much lower than the liquation temperatures of metallic alloys having equivalent strength. Thus, many articles of commerce are fabricated from thermoplastics by means of *injection molding*, in which the fluid polymer is forced into a die cavity and allowed to cool, producing a "plastic" part having the desired, often very intricate, shape.

A simple linear polymer, like polyethylene, does not usually assume a homogeneous, amorphous configuration upon cooling. Wherever possible, regions of neighboring chains tend to align themselves so that interchain spacing is minimized and the secondary bonds can exert maximum influence. These regions take on a three-dimensional long-range order and become, therefore, crystalline. The crystalline regions are mechanically strong in the direction parallel to the chain axes, because their strong backbones are aligned and because the secondary bonding inhibits sliding between chains. The crystalline regions of a partly crystalline polymer are more dense than the amorphous regions because of the tighter packing of the chains.

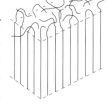

The character of a long-chain polymer changes if the chains become branched during the polymerization process. Branches, depicted schematically in Fig. 13.4 using polyethylene as an example, can vary in length, depending on the conditions of polymer synthesis. Since they interrupt the periodicity of aligned chains, branches retard the tendency toward crystallization. Thus, one would want to minimize branching when making high-density polyethylene milk containers, for example. Because branching also inhibits sliding between chains, it imparts strength to an amorphous polymer, and for this reason branching is sometimes designed into a polymerization process.

Fig. 13.4 Schematic representation of branching of a polyethylene molecule.

A further step up in complexity occurs when covalently bonded links are formed between neighboring chains by single atoms or short segments of chains. In the extreme, a three-dimensional network can be formed. This can be envisioned as a giant three-dimensional supermolecule, as illustrated schematically in Fig. 13.5. It does not soften or melt when heated, since it is held together tightly by covalent bonds. Because it does not flow and change shape when heated, a network polymer is called a *thermoset*. If it is heated excessively, it decomposes as the covalent bonds are broken chemically; e.g., by oxidation.

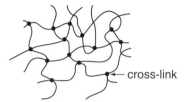

Fig. 13.5 Schematic representation of a network polymer, a thermoset.

An intermediate architecture is found in an *elastomer*, which is the technical name for a rubber-like material. Here, a polymer which is initially linear is treated to introduce cross-links by the formation of occasional bridges between chains. These consist of atoms or small molecules that form covalent bonds on adjacent parts of chains, as shown in Fig. 13.6. An elastomer is analogous to a network polymer with a relatively low density of cross-links. When an elastomer is in the relaxed state, the chain segments between the cross-links tend to be curled-up and kinked in a random arrangement. However, they can be straightened out when a stress is applied, so that overall extensions of several hundred percent are possible. Upon release of the stress, the elastomer springs back to its initial shape. A rubber band is the most common example.

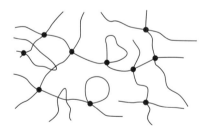

Fig. 13.6 Schematic representation of an elastomer.

The presence of covalent cross-links prevents melting when an elastomer is heated. If the temperature becomes too high, only decomposition occurs. Thus, an elastomer is intermediate in behavior between a thermoplastic and a thermoset. It might be noted that the covalent cross-links present a formidable obstacle to the recycling of rubber tires. So far no viable procedure has been worked out to break the cross-links selectively, leaving the main chains intact. If such a process were discovered, it would have major commercial significance.

One of the goals in this chapter and the next is to understand the structure, properties and behavior of fiber-reinforced composite materials. Tires are examples of such a composite, having an elastomeric (rubber) matrix. The distinguishing characteristic of elastomeric behavior is the enormous amount of recoverable strain, as illustrated in Fig. 13.7.

SECTION 13.2 POLYETHYLENE

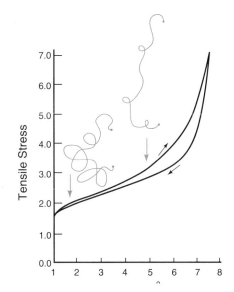

Fig. 13.7 Stress-strain behavior of a lightly cross-linked natural rubber at 50°C showing hysteresis during a full loading cycle. (Adapted from L. H. Sperling, *Introduction to Physical Polymer Science,* Wiley, New York, 1996, pp. 305 and 306. Reprinted by permission of John Wiley & Sons, Inc.)

The unloading part of a stress-strain curve of an elastomer does not follow the loading curve exactly. This effect is known as *hysteresis,* and the extent of it depends on the amount of crystallization that occurs in the rubber when it is stretched. (The crystallization occurs in regions where the stretched chains become aligned enough for the secondary bonds to promote the formation of crystallites; these crystallites "melt" again when the stretching force is relaxed.) To understand elastomeric behavior, it is best to begin by considering the simplest linear polymer, the thermoplastic polyethylene.

13.2 Polyethylene

13.2.1 Molecular Structure

As shown in Fig. 13.3, polyethylene, PE, is simply a carbon chain with two hydrogens attached to each carbon. The basic unit is the C_2H_4 mer, as illustrated in Fig. 13.8(b).

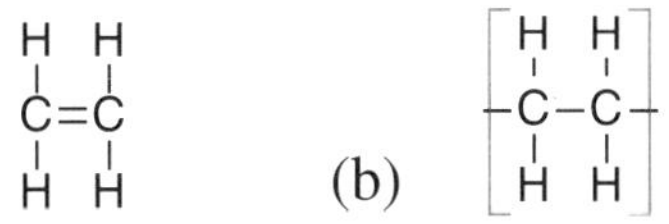

Fig. 13.8 Structure of (a) the ethylene monomer and (b) the polyethylene mer.

Polyethylene is made by polymerizing ethylene, C_2H_4, which has a double bond between the carbons (Fig. 13.8a). This is accomplished by reacting a large amount of ethylene with a small amount of an "initiator," which breaks the double bond in the ethylene to form a mer. An example of an initiator for PE is hydrogen peroxide, which can act as two H-O molecules, each of which has an unpaired electron, called a *free radical.* The free radical, or dangling bond, on the mer then converts another ethylene molecule into a mer, which then becomes attached to the first one, as illustrated in Fig. 13.9.

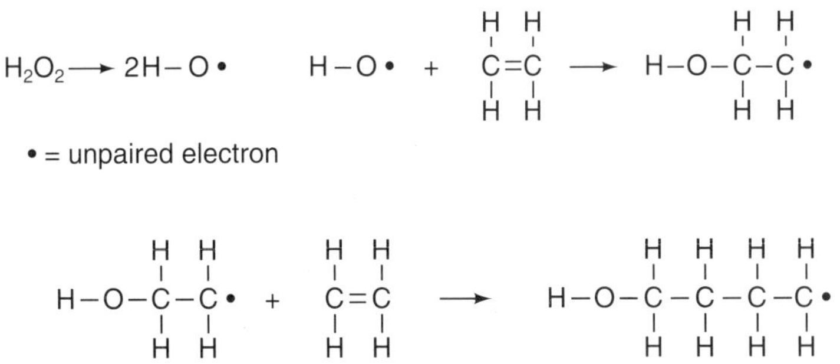

$H_2O_2 \rightarrow 2H-O\bullet$

$H-O\bullet + \begin{matrix} H & H \\ | & | \\ C=C \\ | & | \\ H & H \end{matrix} \rightarrow \begin{matrix} H & H \\ | & | \\ H-O-C-C\bullet \\ | & | \\ H & H \end{matrix}$

• = unpaired electron

$\begin{matrix} H & H \\ | & | \\ H-O-C-C\bullet \\ | & | \\ H & H \end{matrix} + \begin{matrix} H & H \\ | & | \\ C=C \\ | & | \\ H & H \end{matrix} \rightarrow \begin{matrix} H & H & H & H \\ | & | & | & | \\ H-O-C-C-C-C\bullet \\ | & | & | & | \\ H & H & H & H \end{matrix}$

Fig. 13.9 Action of hydrogen peroxide as the initiator in free-radical polymerization of PE.

This process, called *addition polymerization,* repeats itself until the growing chain runs into an initiator, as illustrated in Fig. 13.10, or to the end of another growing chain.

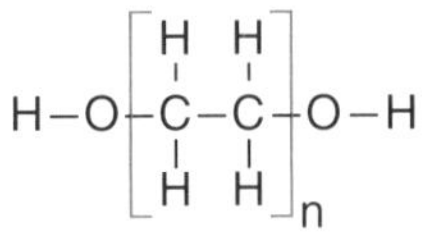

Fig. 13.10 An n-mer chain of PE terminated by an OH group.

$H-O{\left[\begin{matrix} H & H \\ | & | \\ C-C \\ | & | \\ H & H \end{matrix}\right]}_n O-H$

The termination of a chain is a statistical event. Therefore, addition polymerization results in a distribution of chain lengths. The physical and chemical behavior depend very much on both the average chain length (i.e., the molecular weight of the average chain) and the spread in the values of chain length. The distribution of chain lengths can be characterized, for example, in terms of the number-average molecular weight of the population of chains. This is defined by taking n_i as the number of molecules with a molecular weight M_i and writing the number-average molecular weight as

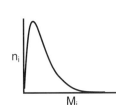

$$\overline{M}_N = \frac{\Sigma n_i M_i}{\Sigma n_i}$$

One also speaks of the *number-average degree of polymerization,* N, given by

$$N = \overline{M}_N/M_0$$

where M_0 is the molecular weight of one mer, which is 28 in the case of PE.

13.2.2 Crystallization

Most liquids tend to crystallize when cooled past the melting point. With polymers, the length, variability in length, and flexibility (and entanglement) of the molecules makes crystallization much more difficult than in the case of atomic solids, like metals, or solids comprising small molecules. In addition, the reliance on secondary (e.g., van der Waals) forces to bind the molecules together means that the melting point would be rather low, and at low temperatures the molecular mobility, needed for crystallization, is also low (i.e., the viscosity is high). Despite these facts, the simplicity of the PE molecule allows it to crystallize, at least partially, unless the liquid is quenched very rapidly. The unit cell is orthorhombic, as shown in Fig. 13.11, and the crystals are formed by the folding of the chains to create a sheave with a thickness equal to the fold length, as indicated in Fig. 13.12(a).

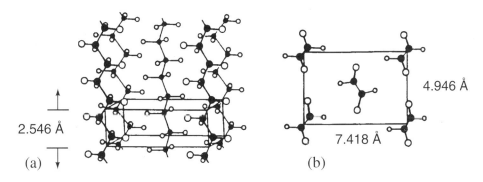

Fig. 13.11 Crystal structure of orthorhombic polyethylene; (a) unit cell, (b) projection of the unit cell along the chain direction. (From *Introduction to Polymers*, R. J. Young, Chapman and Hall, London, 1981, p. 159.)

When liquid PE is cooled fairly slowly, crystallization occurs by the growth of crystalline lamellae in three dimensions outward from a nucleation point to form "spherulites" (cf. Fig. 13.12a). If the cooling is slow enough, the spherulites grow until they contact one another, much in the manner of crystallization of a metal. If there is any material left between the spherulites, it remains amorphous. The spherulites themselves are not fully crystalline, because the lamellae are separated by the parts of molecules which were not incorporated in the folded-chain crystalline lamellae. Some of these act as tie-molecules between adjacent lamellae, because they have an end embedded in each. The gross morphology of the spherulites can be observed in thin sections shaved from a bulk specimen with a special knife, called a microtome, or by crystallization of a film of PE or a similar polymer on a microscope slide. For illumination in the optical microscope one uses transmitted polarized light, as was done to obtain Fig. 13.12(b).

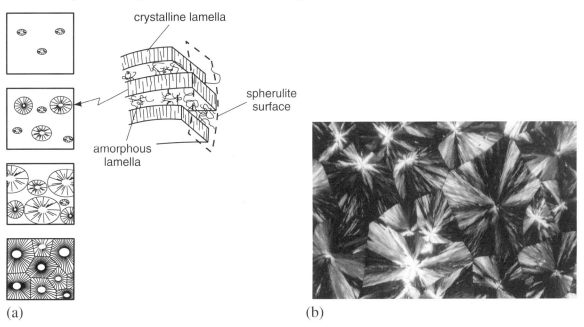

Fig. 13.12 (a) The stages of crystallization of a polymer in the form of spherulites and the morphology of the crystalline and amorphous lamellae. (Adapted from M. G. McCrum, et al., loc. cit.) (b) The spherulitic microstructure of crystalline polyethylene oxide, PEO, revealed by transmitted polarized light. (Courtesy of R. J. Composto, Univ. of Penn.)

The properties of polyethylene depend on the degree of crystallinity, as shown in Table 13.1. Thus, high-density PE (or HDPE) is two to three times stronger than normal low-density PE. That is, the increase in the degree of crystallinity is the analog of a strengthening mechanism in a metallic alloy. The strengthening comes from the fact that in the crystalline regions the molecular chains are aligned and packed tightly together. The covalently bonded backbone of the chains is extremely strong. That is, a very large force is required to stretch the length of a carbon-carbon bond or to change the bond angle from 109°. Therefore, the crystalline regions act like hard inclusions in the amorphous matrix, in which inter-chain sliding and chain straightening can occur. The overall strength of the material obviously would increase with the percent crystallinity.

Table 13.1 Comparison of Properties of Low-Density and High-Density PE.

	Low-Density PE	High-Density PE
Crystallinity, %	~ 65	~95
Density, g/cm³	0.92 - 0.93	0.95 - 0.96
Tensile Strength, MPa	6 - 18	20 - 38
Tensile Modulus, GPa	~ 0.2	~ 1.0

13.2.3 The Glass Transition

As with any liquid, the process of crystallization of a linear polymer can be followed by observing the volumetric change during cooling. The melting temperature T_M of a polymer can be determined from the discontinuous decrease in volume which accompanies crystallization, as shown in Fig.13.13(a).

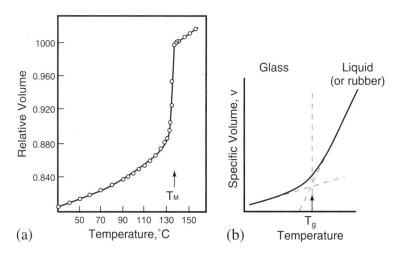

Fig. 13.13 a) Discontinuous decrease in volume which accompanies crystallization when polyethylene is cooled through the melting temperature. (From L. Mandelkern, *Chem Revs,* vol. 56, 1956, p. 911, reproduced with pemission of the American Chemical Society.) (b) Schematic representation of the variation of the specific volume upon cooling through the glass transition.

A linear polymer usually fails to crystallize completely when cooled below T_M, because the chains are tangled in the melt, and there is not enough mobility in the chains to allow them to reorganize completely into crystalline nuclei and for the nuclei to grow as complete crystals. The tendency to crystallize depends on the viscosity of the polymer at T_M, and this depends on the architecture of the chains. (The simpler the chain, the lower the viscosity at T_M.) In the supercooled amorphous regions that fail to crystallize, a different kind of transition occurs. This is the *glass transition,* and it is considered to occur at a temperature, T_G, defined by extrapolation of the two branches of the curve depicting thermal contraction above and below the transition range, as shown in Fig 13.13(b). Molecular motion involving the backbone of a chain (as opposed to oscillation of side groups) essentially ceases below the glass transition. Therefore, the coefficient of thermal contraction or expansion in the glassy state is smaller than in the supercooled-liquid state above T_G. Heating a glassy (amorphous) linear polymer above the glass transition effectively "melts" the secondary bonds between molecules.

In contrast with this behavior of an amorphous polymer, a crystalline region in a polymer undergoes the classical melting transition, just like a metal. This is because the chains in a polymer crystal are aligned close together, and the chain spacing is uniform, as opposed to the spectrum of chain spacings found in an amorphous polymer. Therefore, all of the secondary bonds give way at the same temperature, instead of over a range of temperatures. This temperature is significantly higher than that at which melting begins in the amorphous regions, where the chain spacing is, on average, greater. A partly crystalline polymer becomes rigid gradually as temperature is reduced, as depicted in the low-temperature branch of the curve in Fig. 13.13(a).

Example 13.1

Q. What structural features are necessary for a polymer to be a thermoplastic?

A. The polymer must be a linear polymer or a branched linear polymer and must have a flexible backbone. That is, the bonds along the backbone must be able to rotate so the chain can coil and uncoil.

13.3 Viscoelasticity

The glass transition in a thermoplastic polymer is somewhat analogous to a ductile-brittle transition in a metallic material. Below T_G plastic deformation is essentially nonexistent, and the material can be cracked in a brittle manner. A common example is a polystyrene drinking cup, which is glassy at room temperature, and for which T_G is around 100°C. Above T_G, the polymer can deform plastically; i.e., it is a "plastic." However, the fundamental difference between such a thermoplastic and a completely crystalline material, like a metal, is that the deformation of the thermoplastic is time dependent. (In metals this only occurs at temperatures where self-diffusion is rapid.) The time dependence arises from the fact that the amorphous regions are really part of a supercooled viscous liquid. The plastic deformation occurs by the translation of one part of a molecule past another, and this requires molecular mobility, which is thermally activated. That is, the rate depends exponentially on temperature.

A polymer above T_G behaves like a mixture of two kinds of materials. One is an elastic material in which reversible deformation occurs by bond bending. It follows Hooke's law (strain is proportional to stress) and can be modeled as a spring with a certain spring constant, as in Fig. 13.14(a). The other is a viscous material, following Newton's law of viscosity, which states that strain rate is proportional to stress; it deforms by disentanglement of the chains. A viscous material can be modeled by a dashpot, which is a piston-and-fluid arrangement of the type used to control the rate of closing of a door, as shown in Fig. 13.14(b). Here, the proportionality constant η is the *viscosity*.

(a) $\varepsilon = \dfrac{\sigma}{E}$ (b) $\dfrac{d\varepsilon}{dt} = \dfrac{\sigma}{\eta}$

Fig. 13.14 (a) The spring model of an elastic region in a polymer. (b) The dashpot model of a viscous region.

In principle, a viscoelastic polymer can be modeled as some combination of springs and dashpots in order to devise a mathematical description of its deformation behavior. Some simple versions of such models are depicted in Fig. 13.15. They are too simple to mimic the behavior of a real polymer, but they are useful for discussion of some aspects of time-dependent polymer deformation.

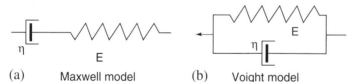

(a) Maxwell model (b) Voight model

Fig. 13.15 Some simple models for viscoelastic behavior.

To illustrate the use of such a model, consider the effect of a sudden force applied to a Maxwell solid (a spring and dashpot in series). The spring would extend immediately, but the dashpot would at first not respond at all. If the force were applied for only a very short time and then removed, the solid would seem to be completely elastic. This is the behavior of Silly Putty® when it is bounced on the floor like a rubber ball. On the other hand, if a ball of Silly Putty is allowed to rest on a surface for a long time, it will eventually flatten out under its own weight. Here, the force is small, and the elastic deformation is negligible, but, given enough time, the viscous flow can be appreciable, even with a small force. This, again, would correspond to the behavior of a Maxwell solid.

Two important aspects of polymer deformation are creep and *stress relaxation*, which are illustrated in Fig. 13.16. To illustrate creep one can choose a Voight model (a spring and dashpot in parallel) and hang on it a constant load. Initially there is no extension, because the dashpot does not respond quickly. However, extension does occur with passage of time, as shown in Fig. 13.16(a), until the spring is fully extended to the amount allowed by Hooke's law.

To illustrate stress relaxation, the Maxwell model is better. Here, the polymer is pulled to a certain extension, and the pulling grips are fixed in place. Initially, only the spring responds, and the stress has the value indicated by Hooke's law. However, as the dashpot responds with time, the extension is accommodated by viscous flow and the spring relaxes, as shown in Fig. 13.16(b). The time-dependent deformation of a real polymer is much more complicated and depends not only on the viscosity (which varies exponentially with temperature), but also on the details of the molecular architecture and the microstructure of the polymer.

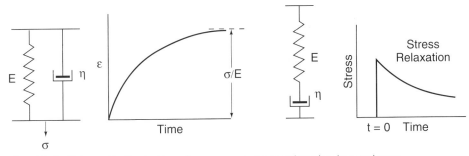

Fig. 13.16 Schematic illustration of creep and stress relaxation in a polymer.

Just as the elastic response to an applied stress in a metal or ceramic can be characterized by Young's modulus, the response of a polymer is characterized by the *viscoelastic modulus*. This is defined to take stress-relaxation into account; it is the stress corresponding to a given strain after a specified time of loading, e.g., 10 seconds. The viscoelastic modulus must be strongly dependent on temperature, since it involves viscous flow. This can be illustrated by the behavior of amorphous polystyrene. Polystyrene, the mer of which is shown in Fig. 13.17(a), is an example of another *vinyl* polymer, which refers to the family of mers in which at least three of the side groups of the carbon pair are hydrogen. Because the mer contains a benzene ring, polystyrene is also an example of an aromatic (i.e., smelly) hydrocarbon. (Polyethylene is also a vinyl; here, the fourth side group happens to be hydrogen.) Some other types of simple linear polymers, vinyl and otherwise, are described in Appendix 13.2. The dependence of the viscoelastic modulus of polystyrene is depicted in Fig. 13.17(b).

It can be seen that the behavior of the polystyrene varies enormously over a rather small temperature range. As already noted, the glass-transition temperature is around 100°C, so it is glassy at room temperature. Thus, a polystyrene drinking "glass," when squeezed too much, cracks in a brittle manner. However, in the vicinity of T_G the viscoelastic modulus falls by several orders of magnitude, and upon heating the behavior changes rapidly from glassy to "leathery" to rubbery before viscous flow begins.

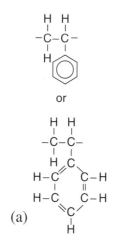

(a)

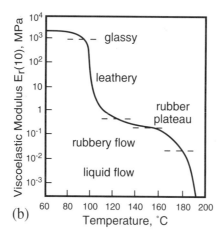

Fig. 13.17 (a) The polystyrene mer; (b) the dependence of the viscoelastic modulus (for 10-second relaxation) on temperature for amorphous polystyrene. (From A. V. Tobolsky, *Properties and Structure of Polymers,* Wiley, 1960, p. 73. Reprinted by permission of John Wiley & Sons, Inc.)

As with polyethylene, certain kinds of polystyrene can be produced with a high volume fraction in the crystalline state. In this case, the viscoelastic modulus falls by a much smaller amount at T_G, reflecting the small volume fraction of amorphous polymer, but it falls precipitously at T_M. This behavior is illustrated by the top curve in Fig. 13.18.

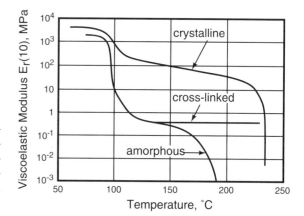

Fig. 13.18 Temperature dependence of the viscoelastic modulus of polystyrene in the amorphous, crystalline and cross-linked states. (From A. V. Tobolsky, loc. cit., p. 75.)

Whether or not polystyrene can crystallize depends on the *tacticity* of the chains, as described in Appendix 13.2. If the large (benzene-ring) side group is always on the same position for every mer in each chain, then the polystyrene is called *isotactic*. This kind of polystyrene can crystallize, although the large aromatic-ring acts as a retardant. However, if the large side group is located randomly in the mers in any chain, the polystyrene is *atactic,* and crystallization can never occur because long-range order is impossible in a polymer in which the chains themselves are disordered.

The regions labelled "leathery" and "rubbery" between the glassy and liquid states in Fig. 13.17 represent behavior that is controlled by entanglements in and between chains. The entanglements act as temporary blocks to the translation of one part of a chain past another. The duration of the blockages depends on the kinetic energy in the chains, therefore on the temperature. Thus, the transition from glass to liquid is retarded and spread over a range of temperatures. The extent of this range, particularly of the rubbery plateau, increases with the average length of the chains.

The effect of some cross-linking in amorphous polystyrene is shown in Fig. 13.18. Here, covalent bridges are formed between linear chains, and these prevent the chains from sliding past each other after they have uncoiled. Thus, the transition to viscous flow is prevented. Instead, the rubbery region is extended, and the polystyrene becomes an elastomer; i.e., it is capable of being stretched by large amounts without permanent deformation. This is the kind of behavior found in rubber.

13.4 Rubber

The rubber used in tires can be made by the addition polymerization of monomers called conjugated dienes. Important examples are *isoprene, chloroprene,* and *butadiene,* the structures of which are given in Fig. 13.19. Note that in each case the central bond is a single carbon-carbon bond and that the two carbons are free to rotate 360° about this bond axis. However, there can be no rotation about the axes of the double bonds.

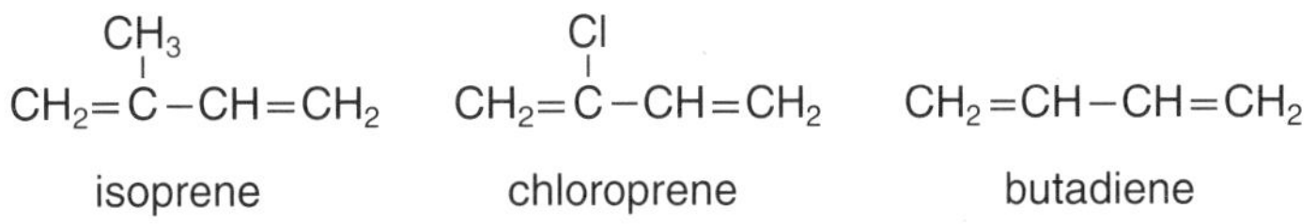

$$CH_2=\overset{\overset{\displaystyle CH_3}{|}}{C}-CH=CH_2 \qquad CH_2=\overset{\overset{\displaystyle Cl}{|}}{C}-CH=CH_2 \qquad CH_2=CH-CH=CH_2$$

$$\text{isoprene} \qquad\qquad \text{chloroprene} \qquad\qquad \text{butadiene}$$

Fig. 13.19 Structures of isoprene, chloroprene, and butadiene monomers.

When these monomers polymerize, the double bonds at each end become single bonds, and the single bond in the center becomes a double bond. The doubly bonded central carbon atoms are then no longer free to rotate about the bond axis. Thus, there are two possible configurations that each mer can assume, as shown in Fig. 13.20, and this configuration becomes locked in during polymerization. In each case, two *isomers* are shown. The two have the same chemical composition, but the structure of the mers differs in an important way. That is, the two central side groups can be either on the same side or on opposite sides of the central carbon atoms. These two cases are called *cis* and *trans,* respectively.

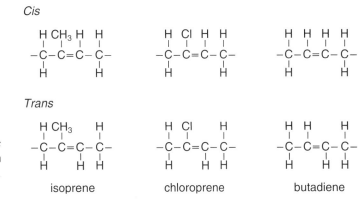

Fig. 13.20 The two isomeric forms, *cis* and *trans,* of the mers which can form from each of the monomers shown in Fig. 13.19.

Taking polyisoprene as an example, an alternative representation of the isomers which conveys the contrast more clearly is given in Fig. 13.21, along with the resulting configurations of the two polymer chains.

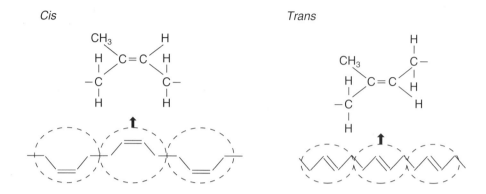

Fig. 13.21 The cis and trans mers of polyisoprene and the configuration of the polymer chains formed from each.

The substance known as natural rubber is the cis form of polyisoprene. It is obtained from the the *Hevea Brasiliensis* tree as a *latex,* meaning a suspension of polymer molecules in water, and it can also be made synthetically. The trans form of polyisoprene is called *gutta percha*. It is found in the latex of a different family of tropical trees, and its properties are quite different from the cis form. The more compact conformation of the chains of the trans form (Fig. 13.21) allows it to crystallize more readily, as indicated by the higher melting temperature:

	T_G (°C)	T_M (°C)
cis	-73	28
trans	-58	74

The cis form has a much lower percent crystallinity at room temperature, due to its less compact chain configuration. Because crystalline regions do not participate in elastomeric behavior (which involves uncoiling and recoiling of chains), the trans form with its greater crystalline content does not produce a good elastomer.

Raw natural rubber, i.e., cis-polyisoprene, is not elastomeric, because there is nothing to prevent the chains from flowing past one another interminably when the chains are uncoiled by stretching. That is, the raw rubber would behave like soft taffy when stretched. Elastomeric behavior can be imparted to the rubber by vulcanization, which means reacting the rubber with sulfur mixed in with the hot liquid cis-polyisoprene. The sulfur breaks double bonds in adjacent chains and forms a sulfur bridge cross-linking the two chains, as shown in Fig. 13.22.

The stiffness, or "hardness," of the rubber depends on the amount of sulfur mixed in; i.e., on the density of cross-links. Hence, the properties of the rubber can be varied over a wide range, as indicated by Fig. 13.23, which shows the effect of the degree of vulcanization on the viscoelastic modulus.

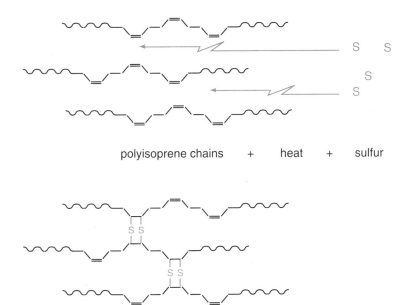

Fig. 13.22 Location of sulfur in a vulcanized rubber.

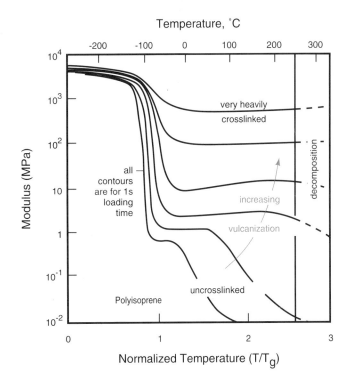

Fig. 13.23 Influence of increased cross-linking by sulfur (increasing vulcanization) on the temperature dependence of the viscoelastic modulus of polyisopene. (From M.F. Ashby and D.R.H. Jones, *Engineering Materials, Vol. 2,* Pergamon Press, Oxford, 1986, p. 227.)

Since oxygen is in the same periodic group as sulfur, it can also cross-link, and thus harden, polyisoprene. This is the cause of the hardening and surface cracking in aged rubber tires. The process is hastened by the presence of ozone, a more reactive form of oxygen, and also ultra-violet (UV) radiation (e.g., sunlight), which activates the breaking of the double bond. Tires are made UV-resistant by the addition of up to 30 wt% lampblack, which absorbs UV and thereby stabilizes the polymer.

The unique behavior of rubber makes it an excellent shock absorber. It deforms extensively and easily, but not permanently. Thus, it can absorb a great deal of energy and then release it in a controlled way. Its elastomeric behavior can be tailored to various requirements by adjusting the degree of cross-linking, i.e., the amount of sulfur mixed in with the rubber (cf. Fig. 13.23).

In addition to polyisoprene, synthetic rubbers can be made by having hydrogen or chlorine in the place of the methyl group, which gives polybutadiene or polychloroprene, respectively (cf. Fig. 13.19). Tires are often made of copolymers of polystyrene and polybutadiene, in which the two mers are distributed randomly along the chains. This is the analog of a solid solution in a metallic alloy.

An elastomeric rubber can be extended elastically by several hundred percent. This stretching occurs mainly by uncoiling of the chains, which remain interconnected by the cross-links. Since there is not much bending or stretching of covalent bonds (except at very high strains), there is not much change in the bond energy (i.e., enthalpy). Therefore, the increase in free energy when an elastomer is stretched an amount ΔX by force F comes from the decrease in entropy as the chains become more nearly aligned. Thus, the work done in stretching an elastomer can be expressed as:

$$F \Delta X = \Delta G = -T \Delta S \qquad (\Delta H \text{ is small})$$

Therefore, the restoring force at constant temperature is proportional to the change in entropy with extension, ΔX:

$$F_T \propto \frac{-\Delta S}{\Delta X}$$

As the temperature increases, the restoring force for a given ΔX, or $T\Delta S$, obviously increases, which is why the viscoelastic modulus increases with temperature in the region of rubber elasticity (cf. Figs. 13.18 and 13.23).

When natural rubber is stretched sufficiently, the chains become so aligned that a significant degree of crystallization occurs. This can be demonstrated by x-ray diffraction, as shown in Fig. 13.24, or by detecting the heat of crystallization. One can detect the release of this heat by wetting the lips and quickly stretching a rubber band held between the lips; that is, the rubber band is felt to warm up. If the rubber band is allowed to relax quickly, the melting of the crystalline regions is accompanied by an input of heat from the lips; that is, the rubber band is felt to become cool.

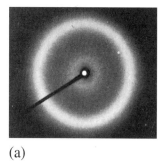 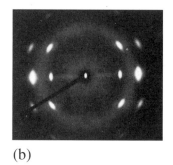

(a)   (b)

Fig. 13.24 X-ray (Laue) patterns of polyisoprene in (a) the relaxed (i.e., amorphous) condition and (b) the stretched (crystalline) condition (from S.D. Gehman, *Chem. Revs.*, vol. 26, 1940, p. 203.)

Exercise 13.2

Q. Can a thermoset be melted so that it can be reshaped?

A. No. Once the chemical reaction to form a thermoplastic component has taken place, the primary covalent bonds are very difficult to break. Thermosets show a weaker glass transition than thermoplastics; the modulus change at T_G decreases as the density of cross links increases.

Summary

Tires are fiber-reinforced composites in which the matrix is an elastomeric rubber based on a cross-linked linear polymer like polyisoprene. The subject of linear polymers is introduced by consideration of the simplest one, polyethylene. The flexibility of chains, by virtue of easy rotation around single C-C bonds, allows coiling and tangling of chains. This increases entropy at the expense of raising enthalpy. The latter can be reduced by crystallization, which is relatively easy in PE, due to the simplicity of its mer. The structure of highly crystalline PE comprises spherulites consisting of crystalline and amorphous components, the spherulites being arranged in a way analogous to the microstructure of a polycrystalline metal. The amorphous component undergoes a glass transition upon cooling, involving "melting" of van der Waals bonds over a small range of temperature, whereas the crystalline component undergoes a classical melting-crystallization transition at a specific temperature.

Polymers are viscoelastic above their glass-transition temperature, and their mechanical behavior must be defined in terms of the viscoelastic modulus, which is time dependent. The glass transition can be observed in a plot of the viscoelastic modulus (on a log scale) *vs.* the temperature. The location of this transition depends on the molecular architecture. Thus, the T_G for PE is about -130°C, whereas for polystyrene, with its large aromatic side group, it is about 100°C.

Polymer chains that contain double bonds can be cross-linked; a classic example is the vulcanization of rubber with sulfur. Cross-linking can convert a linear polymer to an elastomer, which exhibits a rising viscoelastic modulus with increasing temperature over an extended temperature range (above T_G). This behavior arises from the inhibition of the sliding of chains past one another; they can only uncoil and recoil as a tensile stress is applied and relaxed. Thus, the restoring force in a stretched elastomer is the entropy increase associated with recoiling.

The extreme versatility of polymeric materials arises from the great variety of polymer architecture and possibilities for polymer mixing, as well as factors such as branching, cross-linking, and different degrees of crystallization.

Appendix 13.1 Secondary Bonds

In addition to the strong, primary bonding types, i.e., metallic, ionic, and covalent, there exists a much weaker type of secondary bonding, called Van der Waals bonding, which is electrostatic in origin. Such bonding arises whenever the center of negative charge in an atom or molecule does not coincide with the center of positive charge. In such a case an electric dipole is formed, and bonding occurs between the negative end of one dipole and the positive end of the next. This bonding is not only relatively weak, but it acts over a much shorter range than does a primary bond. The energy of the Coulombic field around a dipole falls off as $1/r^6$, in contrast with that around a single charge, which falls off as $1/r$.

There are two classes of Van der Waals bonds. The first arises from fluctuating dipoles, and is best exemplified by the inert gases. The magnitude and direction of the dipole of a given atom fluctuates with time as the center of charge of the electrons fluctuates. The dipole of one atom can, at any instant, induce a dipole in an adjacent atom, allowing an electrostatic interaction to build up among neighbors, if atomic vibrations due to thermal energy are minimized. The strength of such bonding necessarily increases with the number of electrons, as illustrated by the melting and boiling points of inert gases in Table A13.1.1.

Table A13.1.1 Melting and Boiling Points of Some Atomic and Molecular Gases

Gas	Number of Electrons	M.P. (K)	B.P. (K)
He	2	1.6	4.2
Ne	10	24	27
Ar	18	84	87
H_2	2	14	20
O_2	16	54	90

Similar effects can occur in molecules, like H_2 and O_2 (cf. Table A13.1.1), but the description is more complex because of the nature of the electron clouds and the relative motion of the nuclei in each molecule.

The second class of Van der Waals bonding arises from permanent dipoles. An example, depicted schematically in Fig. A13.1.1, is the water molecule, which has a fairly large dipole moment and, of course, much higher melting and boiling points than the molecules in Table A13.1.1.

Van der Waals bonding accounts for the weak cohesion between polymer chains and atomic planes in crystals like graphite, in which all the primary bonds lie in the planes. Such bonding is the reason for the excellent lubricating property of graphite and other materials with a similar layered structure.

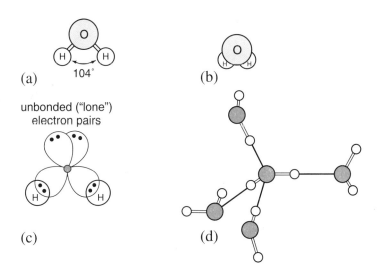

Fig. A13.1.1 Schematic representation of water molecule (a) in a ball-and-stick model, (b) in a space-filling model, and (c) showing the molecular orbitals of the oxygen, pointing to the corners of a tetrahedron, giving rise to a highly polar molecule. (d) Bonding between water molecules; the lone pairs of the oxygens of one molecule are attracted to the hydrogens of others.

Appendix 13.2 Architecture of Some Simple Polymers

Polyethylene is the simplest of the vinyl-type polymers. In the other vinyls, shown in Fig.A13.2.1, one of the hydrogens is replaced by another element or chemical group, such as chlorine in polyvinyl chloride (PVC), a methyl group in polypropylene, or a benzene ring in polystyrene. If all the hydrogens are replaced by fluorine, the result is polytetrafluoroethylene (PTFE). In polymethyl methacrylate (PMMA), a *vinylidene,* two of the hydrogens are replaced, one by a methyl group and the other by an acetate group. All of these linear polymers are thermoplastic, meaning that they soften upon heating.

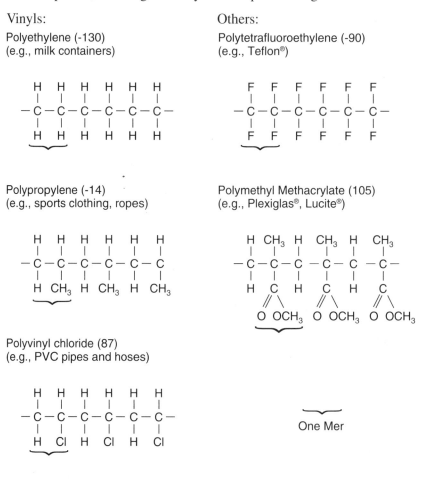

Fig. A13.2.1 Examples of some simple polymers, with T_G (in °C) shown.

As the mer becomes more complicated, crystallization becomes more difficult, and as the side groups become larger, molecular motion becomes more difficult and T_G rises.

Linear polymers with side groups may form in several configurations, the properties of which may be quite different; these are shown in Fig. A13.2.2. If the side groups are all on the same side of the chain, it is called *isotactic*. If they alternate from side to side in a regular way, it is *syndiotactic*, but if they are arranged randomly, it is *atactic*. The first two can crystallize, but the third cannot. Atactic polymers generally have poor mechanical properties, and until recently it has not been possible to produce consistently syndiotactic polymers. Therefore, most commercial polymers are isotactic.

The tacticity, as well as the extent of branching and the distribution of molecular weights, depends on the manufacturing process, which varies from one producer to another. Because of that, generic materials properties given in handbooks are not very precise, and the data supplied by each manufacturer must be used for design purposes.

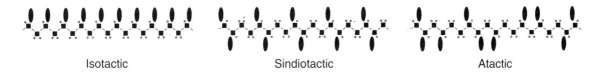

Isotactic   Sindiotactic   Atactic

Fig. A13.2.2 Schematic representation of three configurations of a vinyl-type polymer formed from a monomer of the type $CH_2 = CHR$. (From M.F. Ashby and D.H.R. Jones, loc. cit., p. 212.)

Exercises Chapter 13

Terms to Understand

Linear polymer	Random walk
Network polymer	Paraffin
Branched polymer	Polyethylene
Entanglement	Hysteresis
Thermoplastic	Unpaired electron (free radical)
Thermoset	Dashpot
Elastomer	Creep
Cross-linking	Isotactic
Monomer	Atactic
Mer	Syndiotactic
Addition polymerization	Chloroprene
Degree of polymerization	Butadiene
Spherulitic crystallization	Latex
High-density polyethylene	Gutta percha
Glass transition	Van der Waals bond
Viscoelasticity	Polystyrene
Stress relaxation	cis-Polyisoprene
Viscoelastic modulus	trans-Polyisoprene
Vinyl polymers	Vulcanization

Problems

13.1 Calculate the molecular weight of a single mer of polystyrene.

13.2 If a sample of polystyrene has a (number-average) degree of polymerization of 27,000, what is the number-average molecular weight?

13.3 Given that the length of a C-C single bond is 1.54 Å (= 1.54×10^{-8} cm), what is the maximum possible length of an average molecule in problem 13.2? (Remember what the carbon chain (molecular backbone) looks like.)

13.4 A copolymer is one made up of two different mers. An example is poly (butadiene-styrene). The different mers may form some pattern in each molecule, or they may appear at random along each molecular chain. (a) For the case of regular alternation of single (different) mers, sketch the repeating unit in a chain of this copolymer. (b) If the number-average molecular weight of this copolymer is 2.5×10^6 g/mole, what is the degree of polymerization?

13.5 Consider the creep curve in Fig. 13.16(a). After the system has been allowed to equilibrate under the imposed stress σ, i.e., to reach a strain of σ/E, let the stress suddenly be removed. Draw the entire strain vs. time curve from the initial application of stress to the final equilibration after the stress has been removed.

13.6 Perform the same exercise for the combination of springs and dashpots shown at left, and relate the curves to the effect of carbon redistribution in a stressed crystal of BCC iron (cf. Example 7.2.).

13.7 For polyisoprene with a degree of polymerization N, what is the minimum number of sulfur atoms needed to vulcanize the polyisoprene to saturation? Describe the expected mechanical behavior of this material.

13.8 Consider a strap of vulcanized rubber suitable for use in a tire that would exhibit an elongation (before fracture) of about 300%. A weight is hung from one end of the strap, the other end of which is attached to a horizontal beam. The extension caused by the weight is 30%. If the strap were warmed with a hot blower, would it get longer or shorter? Explain your reasoning.

13.9 Compare the behavior of the heated rubber strap in Problem 13.8 with the behavior one would find with a metal strap, and explain the difference.

14
COMPOSITE MATERIALS; TIRES, FRAMES, AND WHEELS

Composites are multi-phase materials, the phases being combined to produce a behavior superior to that of any one of them acting independently. Tires are a good example of a composite in which fibers are used for reinforcement. If a tire were produced without fibers, it would expand like a balloon upon being inflated; the fibers are necessary for the tire to sustain the internal pressure. That is, the stress in the wall of the tire is carried mainly by the fibers. However, a tire obviously could not be made from fibers alone. The rubber is necessary to seal in the air and to give the tire its shock-absorbing properties. Thus, both the fibers and rubber matrix play essential roles in tire performance. The fibers have a high strength and a relatively high elastic modulus, which enable them to carry the stress due to the internal pressure. The rubber matrix has a much lower strength and stiffness, which enable it to deform elastomerically and absorb road shocks in concert with the enclosed high-pressure air.

Composites have been encountered before, in the form of pearlitic steel, and everyone is familiar with natural composite materials in the form of wood, especially bamboo, a composite of cellulose fibers in a matrix of lignin, and bone, a composite of collagen fibers in a matrix of apetite crystals. Most fiber-reinforced composites used as structural materials are like tires or wood; they have strong, high-modulus fibers in a softer, ductile, low-modulus matrix. However, some are like bone, having a brittle matrix. A common example of the latter is reinforced concrete, in which the brittle concrete matrix is strengthened by strong, high-modulus steel rods. (It may be noted that this idea is far from new; pueblo-dwelling Native Americans have been using straw to reinforce mud bricks and plaster for many centuries, and gypsum-based plaster has been reinforced with horsehair for perhaps as long a time.)

Bicycles are making increased use of fiber-reinforced, polymer-matrix composites for frames and wheels. The object is to produce a high-strength lightweight structure, with stiffness tailored to fit the application. The light weight comes from the fact that only elements with low atomic weight are used, and the high strength comes from the strong covalent bonding along the backbone of the fibers. Bicycle manufacturers have also recently employed metal-matrix composites, in which, for example, an aluminum alloy is strengthened by a dispersion of ceramic fibers or particles, such as silicon carbide or aluminum oxide, which have high strength and stiffness.

14.1 Principles of Fiber Reinforcement

A rope is a material consisting completely of fibers, with no matrix. The main strands may be continuous, running end to end, but these strands are often made up of shorter, fine, discontinuous fibers. For tensile strength the strands rely not only on the strength of each fiber, but also on the friction between the fine fibers, which prevents their slipping past one another and coming apart. A rope has extremely anisotropic tensile behavior: It is strong in tension along its axis, but very weak in transverse tension. Only the friction arising from the braiding of the main strands keeps them from separating. In addition, a rope has no resistance to compression or bending (excluding steel-wire ropes). The reason is the very small moment of inertia of the constituent fibers, as can be understood from the simple beam theory of Chapter 12, which showed that the curvature of a cylindrical beam subjected to a bending moment, M, is given by

$$\frac{1}{\rho} = \frac{M}{EI}, \text{ where } I = \frac{\pi D^4}{64}$$

Because the fiber normally has a very small diameter, D, I is very small, and a small M can produce a large curvature, as indicated schematically in Fig. 14.1(a), even if the fiber has a large elastic modulus, E.

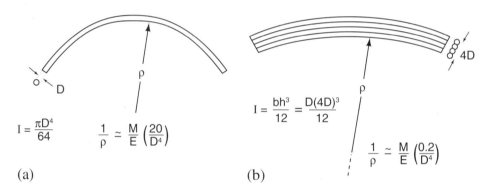

Fig. 14.1 (a) A single fiber with diameter D and elastic modulus E bent by a moment M. (b) A stack of four such fibers bonded together and loaded by the same bending moment.

How, then, is it possible, for example, to employ a bundle of parallel, fine glass fibers to make a pole for pole-vaulting? The answer is that one must glue the fibers together to increase I. The principle is illustrated by Fig. 14.1(b), in which a stack of four bonded fibers comprises a rectangular beam of cross-section Dx4D. (It is assumed that the composite is constrained to lie in a plane.) Due to the increase in I, the curvature for a given M and E is reduced by a factor of 100 by combining the four fibers. The essential factor, however, is that the glue must be strong enough to keep the fibers from sliding past one another. That is, they must all be forced to bend together. If the bond between the fibers does not have enough shear strength, then they could bend independently, and the composite effect would be lost. It is not difficult to achieve a high shear strength between fibers, because the surface area of a fiber is large. Thus, even for a large shearing force, the shear stress in the interface would be small.

SECTION 14.1.1 LONGITUDINALLY STRESSED FIBERS IN A CONTINUOUS-FIBER COMPOSITE

In actual practice, of course, the fibers are laid up in two lateral dimensions, not just one, and the bonding agent forms a matrix which completely encases the fibers, as shown schematically in Fig. 14.2(a). The matrix of a vaulting pole is a polymer, and it has a Young's modulus that is much smaller than the modulus of the glass fibers. Therefore, the modulus of the composite must be some kind of average between the moduli of the matrix and the fibers.

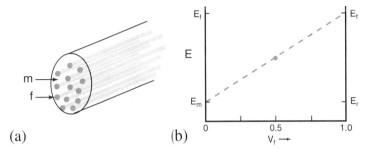

Fig. 14.2 (a) A composite of fibers (f) and a matrix (m) in which the volume fraction of fibers is 0.5. (b) The linear relationship between the elastic modulus of the composite and the volume fraction of fibers, depending on the moduli of the fibers and the matrix, respectively.

Intuitively, one would expect that, for such a composite stressed in simple tension along its length, the modulus of the composite would depend on the volume fraction of fiber *vs.* that of the matrix. If the pole were made entirely of the matrix polymer, the modulus, E_m, would be low. If it were entirely glass, the modulus, E_f, would be high, but the pole would be too brittle to use. If the volume fraction of fibers, V_f, were, say, 50%, one would guess that the modulus of the composite, E_c, would lie halfway between the two, as shown in Fig. 14.2(b). This guess would be exactly correct, as shown by the argument in the following section.

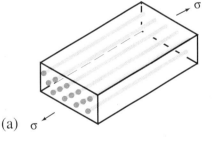
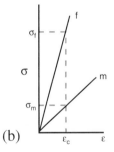

Fig. 14.3 (a) A parallel-fiber composite with continuous fibers stressed along the direction of the fibers. (b) The stress-strain curves of the matrix and of the fiber, tested in isolation.

14.1.1 Longitudinally Stressed Fibers in a Continuous-Fiber Composite

As long as the fibers cannot slide relative to the matrix, and, thus, to one another, the tensile strain in the fibers and the matrix must be the same in the longitudinally stessed composite shown in Fig. 14.3(a). Therefore, if only elastic behavior is considered, the stress in a fiber and in the matrix would depend on the strain in the composite (according to Hooke's law), and, due to its higher modulus, the stress in a fiber would necessarily be higher than in the matrix, as indicated schematically in Fig. 14.3(b).

In any composite stressed in a direction parallel to the fibers, the total force on the composite must be the sum of the forces on each component:

$$F_c = F_f + F_m$$

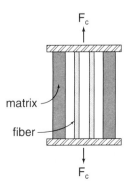

For a unit area of composite, taking the force to be stress times area, A, this can be written

$$\sigma_c = \sigma_f A_f + \sigma_m A_m$$

where A_f and A_m are the respective area fractions. Since for parallel phases, the area fraction is the same as the volume fraction, V, and since $V_f = 1 - V_m$,

$$\sigma_c = \sigma_f V_f + \sigma_m (1 - V_f)$$

or, from Hooke's law,

$$\varepsilon_c E_c = \varepsilon_f E_f V_f + \varepsilon_m E_m (1 - V_f)$$

Because all the strains are equal and can be cancelled out,

$$E_c = E_f V_f + E_m (1 - V_f)$$

which gives the linear "rule of mixtures" shown in Fig. 14.2(b).

Thus, the elastic properties of a composite can be tailored to fit the application. For example, the "spring" imparted by a vaulting pole can be adjusted to fit the weight of the athlete by varying the volume fraction of fibers, and/or by varying the modulus of the polymer matrix (i.e., by varying the degree of polymerization, the amount of cross-linking, etc.). The stiffness of the wall of a tire can be controlled in a similar manner.

It should be noted that in a fiberglass composite, the polymer matrix also plays the important role of protecting the surfaces of the glass fibers against damage. Glass, being a brittle material, is sensitive to the presence of very small surface flaws, which can act as Griffith cracks (cf. Appendix 10.2). Such cracks can be caused by surface contact between fibers or with another solid. Another function of a relatively tough matrix is to act as a crack-arrestor in the event that a fiber breaks. When the crack reaches the matrix, it can either blunt by forming a void, or it can be deflected into a direction parallel to the stress by traveling along the fiber/matrix interface, which is often relatively weak in tension. These possibilities are illustrated in Fig. 14.4.

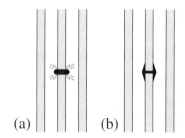

Fig. 14.4 Mechanisms of crack arrest by a tough matrix when a fiber breaks: (a) blunting, and (b) deflection into the fiber/matrix interface.

14.1.2 Transversely Stressed Fibers

As in the case of a rope, a composite can be weak in the direction perpendicular to the fibers. Thus, it is common experience for a piece of wood to be split easily along the direction of the "grain." In the case of transverse loading, Fig. 14.5, the matrix and the fibers are loaded in series, each element transmitting the full load to the next. Therefore, the forces must be equal in the matrix and the fibers. For continuous fibers, the fibers and the matrix can be considered geometrically similar. Hence, the force per unit area, or stress, rather than the strain, must be equal in each:

$$\sigma_c = \sigma_f = \sigma_m$$

In this case the strains are distributed linearly, according to the volume fractions of fiber and matrix, and the strain in the composite is given by

$$\varepsilon_c = \varepsilon_f V_f + \varepsilon_m (1 - V_f)$$

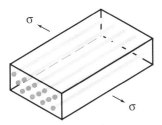

Fig. 14.5 A transversely loaded composite.

Using Hooke's law and setting the stress in the composite, fibers, and matrix equal to σ, this can be written as

$$\frac{\sigma}{E_c} = \frac{\sigma}{E_f} V_f + \frac{\sigma}{E_m}(1 - V_f)$$

which, upon dividing through by σ gives

$$\frac{1}{E_c} = \frac{V_f}{E_f} + \frac{1 - V_f}{E_m}$$

or, rearranged,

$$E_c = \frac{E_f E_m}{E_f (1 - V_f) + E_m V_f}$$

The expression for the modulus of the longitudinally stressed continuous fibers gives an upper bound for the modulus of any composite, and the expression for the transversely stressed continuous fibers is a lower bound for the modulus of any type of composite. For example, a composite with short (discontinuous) fibers, even if not aligned, would have a modulus between the two bounds. These upper and lower bounds for the Young's modulus of a composite are plotted as a function of the volume fraction of the reinforcing phase, whether fibers or particles, in Fig. 14.6.

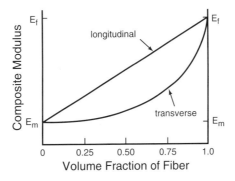

Fig. 14.6 Upper and lower bounds for the elastic modulus of composites as a function of volume fraction of the reinforcing phase.

The modulus is the most important mechanical property of a composite loaded in the elastic region (below the yield stress). The spacing between the upper and lower bounds gives some idea of how much the elastic modulus of a material strengthened by a set of parallel fibers can vary with the direction of loading. This anisotropy cannot always be tolerated. For example, when wide sheets of wood are required in a structure, one normally specifies plywood, in which thin layers of wood are stacked and glued so that each successive layer has its grain oriented at 90° to that of its neighbors. Here, one gets effective stiffening in two directions in the plane. (It is not isotropic in the plane, however; the modulus goes through minima at 45° to the fiber directions. To make such a material isotropic in the plane, the fibers would have to be aligned randomly; this would give a modulus somewhere in between the two bounds.)

14.1.3 Discontinuous Fibers

Sometimes, materials are strengthened by aligned but discontinuous fibers, as shown schematically in Fig. 14.7(a). Here, the reinforcement is less effective than in the continuous-fiber case, because the fibers are not uniformly loaded; the ends of the fibers carry less stress than the middle portions. This can be understood by examining what happens near the ends of a fiber. For this, consider an isolated fiber embedded in a matrix with a much lower elastic modulus.

As shown in Fig. 14.7(b), for a given applied stress the elastic strain in the matrix beyond the ends of the fiber is relatively large. However, the matrix strain along the central region of the fiber is much smaller, being equal to that in the fiber itself. (In the central region the situation is the same as in a composite with continuous fibers.) Over a certain distance from each end of the fiber there is a transition from the higher to the lower value of tensile strain in the matrix. As is evident from the figure, this variation in tensile strain creates a shear strain and stress in the matrix. The shear stress acts on cylindrical surfaces in the matrix that lie parallel to the axis of the fiber. The most important of these surfaces is the fiber/matrix interface. To get the maximum reinforcement from the fibers, this interface must be strong enough that debonding does not occur.

As indicated in Fig. 14.7(c), the shear stress in the fiber/matrix interface acts over a region at each end of the fiber and serves the same function as the grips on a tensile specimen in a tensile test. That is, it serves to load the fiber in tension by transfering load from the matrix to the fiber. This load is given by $\tau(x)\pi r x$, where r is the radius of the fiber and x is the length along the fiber measured from its end. The shear stress $\tau(x)$ falls from a maximum at the fiber end, where the strain mismatch between fiber and matrix is largest, to zero at the point along the fiber where this mismatch disappears.

The interfacial shear stress at the fiber ends must be balanced by the tensile stress, $\sigma(x)$, in the fiber. As the shear stress decreases to zero, the tensile stress builds up to a constant value in the central region, as shown in Fig. 14.7(c). The region of tensile stress buildup is called the *transfer length*. Over this length the fiber is carrying less of the load than would a continuous fiber, and this represents a loss of reinforcement efficiency. The effect is negligible for fibers with a long aspect ratio (i.e., length/diameter), but it becomes important as the aspect ratio decreases. Thus, significant reinforcement is lost in the case of short fibers.

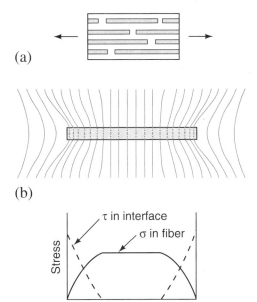

Fig. 14.7 (a) Schematic representation of a portion of a composite strengthened by discontinuous fibers. (b) Displacements in and near a discontinuous fiber. (c) Schematic plot of the tensile stress in the fiber and the shear stress in the fiber/matrix interface as a function of distance from the fiber ends.

It should be obvious that, if the fibers are replaced by elongated particles, the stress in the particle may never reach the maximum possible value. For this reason, particles are not very efficient in raising the elastic modulus. The composite modulus for short fibers and for particles is generally near the lower bound in Fig. 14.6.

14.1.4 Composite Behavior

Although the principles of fiber strengthening described above apply equally to tires and bicycle frames, these two structures are at opposite ends of a spectrum of mechanical behavior. In tires the matrix is highly viscoelastic. In frames, the matrix should have minimal viscoelasticity to avoid energy dissipation during the cyclic loading associated with pedaling. (Tires, of course, are required to dissipate large amounts of energy from road shocks.) Thus, while yielding and fracture are not a consideration with regard to tires, they are for frames. The matrix materials used for frames are glassy, brittle network polymers. Therefore, if the composite is loaded so much that fibers begin to break, then brittle fracture of the composite can occur.

Composites in which the matrix is capable of some plastic flow can have high toughness. This is because a crack which forms in a fiber would not propagate easily into a tough matrix, as indicated schematically in Fig. 14.4. Fracture of the composite would occur after a sufficient number of fibers have broken, but much energy would be absorbed by the separation of the fracture faces because of the frictional forces associated with fiber-pullout. This behavior is found in glass-fiber-reinforced composites like vaulting poles and fishing rods. The energy dissipated through viscoelastic deformation is of less concern in these items than in bicycle frames and tennis rackets.

14.2 Fibers For Reinforcement

14.2.1 Organically Based Fibers

Fibers have been used to reinforce tires for many decades. The first fibers used in tires were cotton, which is essentially cellulose, a natural polymer for which the formula is shown in Fig. 14.8. This is the fiber that strengthens plants, including woods. Later, the first fiber produced (partly) by man was introduced in the form of *rayon,* obtained by replacing one or more of the OH groups on cellulose fibers by the acetate group, also shown in Fig. 14.8. Depending on the number of OH groups replaced, a mono-, di-, or tri-acetate mer is produced. These mers can be found mixed in a rayon fiber. Rayon can exhibit stronger inter-fiber secondary bonding than cotton does, due to the larger polarity of the acetate group compared with the OH group. To make rayon fibers out of cellulose triacetate, water can be used as a plasticizer for the purpose of drawing fibers from the liquid. It would be removed by heat after the fiber is drawn.

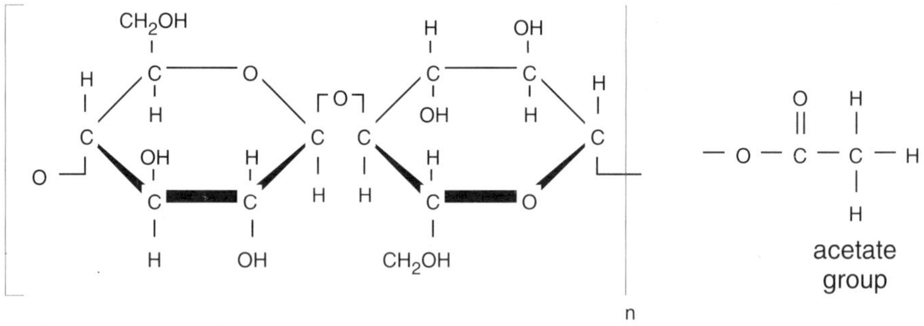

Fig. 14.8 The molecular structure of cellulose, showing the OH groups which act as reaction sites for acetate groups in the manufacture of cellulose acetate (rayon).

The first completely man-made fiber was *nylon,* invented by Wallace Carothers of the DuPont company and commercialized in 1939. Nylon is a polyamide and comprises, in effect, short paraffin-type chains connected by amide groups, which have the form

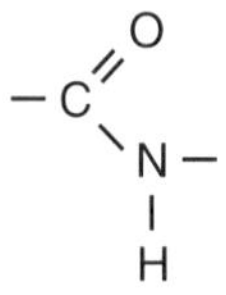

as shown in Fig. 14.9. It is made by a *condensation reaction* in which two different molecules react to give a long-chain polymer plus a by-product. In the production of nylon 6.6, one reactant, adipic acid, has an OH on each end, and the other, hexamethylene diamine, has a hydrogen at each end. When these react, the by-product is water. The mer of this polymer can be viewed as having two parts, each with six carbons; hence the designation 6.6.

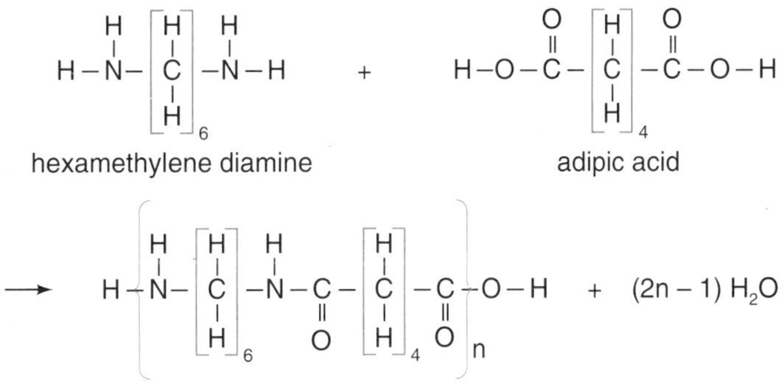

SECTION 14.2 FIBERS FOR REINFORCEMENT

$$H-N\genfrac{[}{]}{0pt}{}{\genfrac{}{}{0pt}{}{H}{|}}{\genfrac{}{}{0pt}{}{|}{H}}-\genfrac{[}{]}{0pt}{}{\genfrac{}{}{0pt}{}{H}{|C|}}{\genfrac{}{}{0pt}{}{|}{H}}\genfrac{}{}{0pt}{}{}{6}-N-H \quad + \quad H-O-C-\genfrac{[}{]}{0pt}{}{\genfrac{}{}{0pt}{}{H}{|C|}}{\genfrac{}{}{0pt}{}{|}{H}}\genfrac{}{}{0pt}{}{}{4}-C-O-H$$

hexamethylene diamine   adipic acid

$$\longrightarrow \quad H-N-[C]_6-N-C-[C]_4-C-O-H \quad + \quad (2n-1)\ H_2O$$

Fig. 14.9 The reaction which produces nylon 6.6.

To make nylon fibers, the polymer is heated to the liquid state and forced through a spinnerette, which is a plate with a myriad of small holes. The fibers are partly crystalline as they exit the holes in this extrusion-type process. After this they are extended (cold drawn) to increase the molecular alignment and the degree of crystallization, as shown schematically in Fig. 14.10. In this state further extension of the fiber is resisted almost entirely by the strong covalent bonds along the backbone of the molecular chain. The Young's modulus of nylon, for example, is of the order of 3 GPa (3000 MPa); this can be compared, for example, to that of an average tire rubber, which is less than 100 MPa.

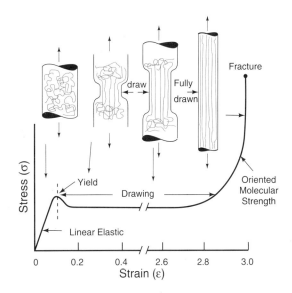

Fig. 14.10 Schematic illustration of cold-drawing of a linear polymer. (From M.F. Ashby and D.R.H. Jones, loc. cit., p. 229.)

The architecture of a nylon molecule, particularly with regard to the absence of a large side-group, is almost as simple as that of polyethylene. For that reason it crystallizes readily, and the van der Waals bonding between the closely spaced chains is strong. In addition, the molecule can form two hydrogen bonds per mer, as indicated in Fig. 14.11. The high strength of a nylon fiber arises from these two features, both of which make sliding between chains relatively difficult.

Nylon was originally developed for fabrics, and this is still its primary application, although it is also widely used for ropes and single-strand applications like fishing lines. It found widespread application as automobile tire cord for a number of years.

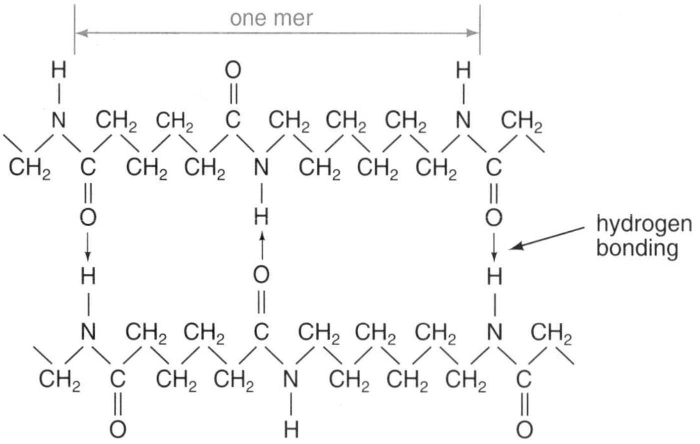

Fig. 14.11 Structure of crystalline nylon 6.6.

Another high strength organic fiber, developed originally for tire cord by DuPont, has the trade name *Kevlar®*. It is an aromatic polyamide with the chemical formula shown in Fig. 14.12(a). It is produced by extrusion, stretching, and drawing, so that the molecules line up in planar sheets, as shown in Fig. 14.12(b). The aromatic rings result in a fairly rigid chain, and again there is the opportunity for two hydrogen bonds per mer. In the fiber the sheets are stacked in radial lamellae, as shown in Fig. 14.12(c). The tensile modulus is about 130 GPa, or more than forty times that of nylon. This is a direct result of the greater stiffness of the six-carbon aromatic ring, compared with a simple carbon chain. Although not commercially successful as tire cord, Kevlar has found many applications in high-performance composite materials.

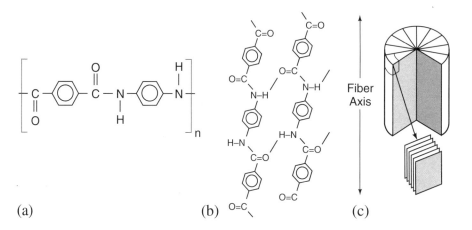

Fig. 14.12 (a) The mer of poly(p-phenylene terephthalamide) molecules. (b) The interchain hydrogen bonding which occurs when these molecules are aligned. (c) Schematic representation of the supramolecular structure of Kevlar® aromatic polyamide fibers depicting the radially arranged sheets. (After D. Tanner, J.A. Fitzgerald, and B. R. Phillips, *Verlag Chemie*, 1989.)

The stiffest fibers available for reinforcement of composites are made from graphitization of molecules of a polymer precursor. Their structure is based on layers of the hexagonal arrays of carbon atoms characteristic of graphite, as shown in Fig. 14.13(a). A graphite crystal would consist of such planes stacked in an ABAB... sequence and would have a calculated modulus of 910 GPa in an a direction. In a carbon fiber, the layers are stacked less regularly into small semi-crystalline units, which are, in turn, aligned with the fiber axis, but in a very complex structure, shown schematically in Fig. 14.13(b).

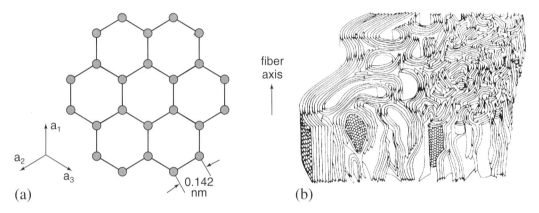

Fig. 14.13 (a) Arrangement of carbon atoms in a layer plane of graphite. (b) Schematic representation of a graphite fiber. (a) From D. Hull, *An Introduction to Composite Materials*, Cambridge University Press, 1981, pp. 9 and 10. Reprinted with permission of Cambridge University Press. (b) From S.C. Bennett, PhD Thesis, Univ. of Leeds, 1976.

Graphite fibers can be made, for example, from polyacrylonitrile, PAN, shown in Fig. 14.14(a), by stretching the polymer and heating it in oxygen to cause the nitrile groups (-C≡N) to interact to produce the chain of rings shown in Fig. 14.14(b). Further processing then produces the graphite structure.

Carbon fiber can be processed into two types, with tensile moduli of 390 GPa (Type I) or 250 GPa (Type II). Type I has the lower tensile strength: 2.2 vs. 2.7 GPa. The tensile strengths of the carbon fibers are clearly limited by the voids and other irregularities in the fiber structure. This could presumably be improved with further development.

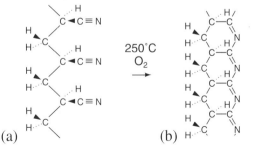

Fig. 14.14 (a) Structure of PAN. (b) Product of first stage of reaction in the formation of a carbon fiber. (From D. Hull, loc. cit. p. 12; reprinted with the permission of Cambridge University Press.)

At present, carbon fibers are too expensive to be used as tire cord, but they are the primary reinforcement in composite bicycle frames, as will be discussed below, as well as for sports equipment and military aircraft.

It is instructive to compare the stiffness of four kinds of organically based fibers, as in Table 14.1, and to consider the reasons for the differences from one fiber to the next. The simplest fiber is polyethylene, which has a backbone of aligned chains of singly bonded carbon atoms. Because the chains have only hydrogen atoms as sidegroups, each chain has a very small cross section. Therefore, they can be packed together at a very high density, and the large number of chains per unit area of cross section of the fiber results in a very high modulus.

Table 14.1 Comparison of the Elastic Moduli of Four Fibers

Fiber	Polyethylene	Nylon 6.6	Kevlar	Carbon
Elastic Modulus (GPa)	>3*	3	130	250 or 390
Structure	Carbon chains, small cross section	Carbon chains, lateral hydrogen bonds	Chains of carbon rings, lateral hydrogen bonds	Sheets of interlocked carbon rings

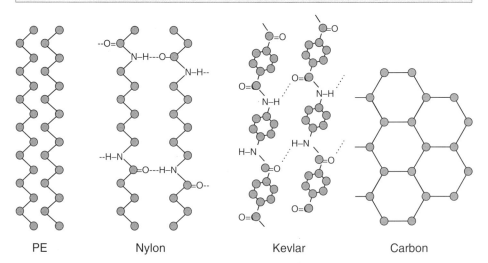

PE   Nylon   Kevlar   Carbon

The chain structure of nylon 6.6 is not too different from that of polyethylene, but the effective cross section of each chain is larger, due to the size of the amide groups. However, these groups provide the sites for hydrogen bonding between chains, and this retards the tendency for sliding between chains. This effect is responsible for the high modulus of nylon fibers.

The Kevlar fiber structure has chains of hexagonal rings of carbon, which are considerably stiffer than a simple linear carbon chain, and these rings are bonded laterally by hydrogen bonds. Therefore, it is not surprising that the modulus of Kevlar is much higher than that of nylon.

Finally, the carbon fibers are also composed of six-carbon rings, but these rings are interlocked in a two-dimensional array, which is, of course, much more effective than the hydrogen bonding of Kevlar. Therefore, carbon fibers have the highest modulus of all.

* The modulus of polyethylene can reach 170 GPa in the case of *SPECTRA*® fibers, which are made from PE chains with extremely high molecular weights and are drawn to a high degree of crystallinity.

14.2.2 Inorganically Based Fibers

Most present day automotive tires are reinforced with inorganic fibers, either glass or steel. Silica-based glass can be drawn into fibers from the molten state. The building block of such a glass is the SiO_4^{4-} tetrahedron (as noted in Section 10.6.3), with the silicon covalently bonded to four neighboring oxygens, each oxygen sharing two silicon atoms, as shown in Fig. 14.15(a). The three-dimensional network (Fig. 14.15b) formed by pure silica makes a strongly bonded glass that has a low coefficient of thermal expansion and is, therefore, highly resistant to thermal shock. However, it is also extremely viscous at high temperatures; hence, it is difficult to work into shapes and virtually impossible to draw into fibers. Silica glasses are made less viscous (and less resistant to thermal shock) by the addition of other oxides, which act to break up the silica network and are called network modifiers. An example of such an additive is Na_2O. A sodium ion can bond to only one oxygen (which is also covalently bonded to a silicon), and this does not form a bridge to a neighboring SiO_4^{4-} tetrahedron, as shown in Fig. 14.15(c).

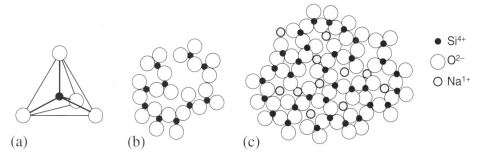

Fig. 14.15 (a) The SiO_4^{4-} tetrahedron. (b) Schematic representation of SiO_2 glass in which only three of the four tetrahedral oxygens are shown. (c) Network modification by the addition of Na_2O.

Viscosity decreases with the amount of network-modifying oxides added to the silica, making it possible to adjust the trade-off between workability and thermal shock resistance to fit the application. An indication of the range of viscosity possible in silica glasses is given in Fig. 14.16.

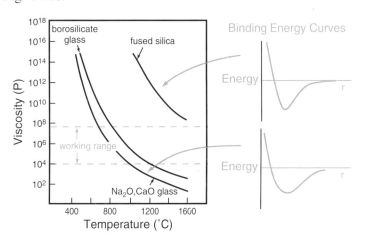

Fig. 14.16. Effect of network modification on viscosity. (Adapted from E.B. Shand, *Modern Materials*, Vol. 6, Academic Press, N. Y., 1968, p. 262.)

Obviously, workability is of paramount importance for making glass fibers; hence, they normally contain only about 55% silica. The fibers are made by heating the glass in a platinum container with holes in it through which the fibers are drawn. The glass is a Newtonian-viscous fluid, meaning that it obeys Newton's law of viscous flow in which the rate of shear strain, $d\gamma/dt$, is proportional to the shear stress, τ.

$$d\gamma/dt = \tau/\eta \qquad \eta = \text{viscosity}$$

This also means that the tensile strain rate in a fiber being drawn is proportional to the tensile stress.

$$\frac{d\varepsilon}{dt} \propto \sigma$$

or

$$\frac{dA}{Adt} \propto \frac{P}{A}$$

Therefore, the rate of area reduction is proportional to the load, P.

$$\frac{dA}{dt} \propto P$$

The result is that the rate of thinning of the fiber, dA/dt, must be constant along the length of the fiber, since P is constant along the length. That is, the fiber can be drawn as much as desired without the formation of a neck, contrary to the case of a metallic wire (for which Newton's law does not hold, even when the wire is hot).

Glass fibers have an elastic modulus of about 70 GPa, compared with about 3 GPa for nylon, for example, but they are quite fragile. A freshly-drawn fiber is free of surface flaws, and in this state its fracture strength approaches the ideal value of about E/10. However, even slight contact with another solid can induce surface cracks due to the high stresses generated by even light loads when the contact area is very small. A surface crack can extend when the tensile stress reaches the value given by the Griffith equation. (See Appendix 10.2.)

Freshly-drawn glass fibers are coated with size (an organic film) to protect their surfaces, but are further protected when embedded in a polymer. As noted earlier, this is one of the main functions of the matrix in a fiber-reinforced composite material: to protect the fibers from surface damage.

Wires made from eutectoid steel are also used in tire construction. For bicycles, they are used along the two beads that fit against the rim in a clincher-type tire, shown in Fig. 14.17(a), but they are also used for tread and casing reinforcement in automobile tires. The wire-drawing process is the same as used to make piano wire and is called *patenting*. It involves a succession of drawing dies separated by baths of molten lead. The wire is drawn through a die and then passed into a lead bath so that it is heated above the eutectoid temperature and transformed to austenite. It passes into a second lead bath held at a temperature below 560°C, in which the wire transforms to fine pearlite. It is then cold drawn. The process may be repeated several times, and results in a microstructure in which cementite lamellae a few nanometers thick and spaced as closely as 40 nm are aligned with the wire axis, as shown in Fig. 14.17(b). The tensile strength can be as high as 1.5 GPa, and the Young's modulus is over 200 GPa. It is one of the strongest metallic materials available commercially and is surprisingly ductile, having a typical reduction in area of about 20%.

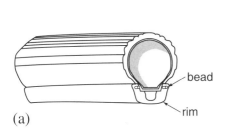

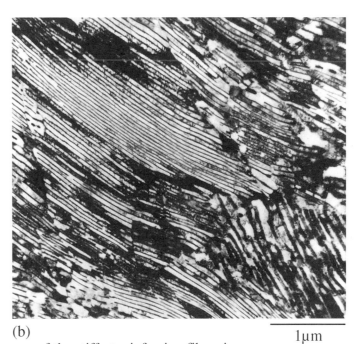

Fig. 14.17 (a) Steel wires are embedded in the rubber to hold the tire bead onto the wheel rim. (b) Section of cold-drawn wire of pearlitic steel (transmission electron micrograph, 20,000X). (G. Langford, *Metall. Trans. A,* vol. 8A, 1977, p. 861.)

A comparison of the stress-strain curves of the stiffest reinforcing fibers is made in Fig. 14.18. It is apparent that patented steel wire competes very well with carbon and Kevlar on the basis of elastic modulus. It is much more dense, however, and is not a viable choice where weight saving is important.

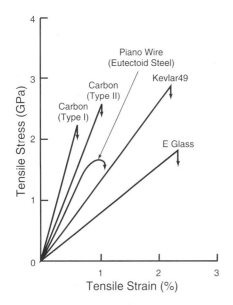

Composition of E Glass	
SiO_2	52.4
Al_2O_3, Fe_2O_3	14.4
CaO	17.2
MgO	4.6
Na_2O, K_2O	0.8
Ba_2O_3	10.6

Fig. 14.18 Comparison of the stress-strain curves of several high-strength fibers. (Adapted from D. Hull, loc. cit., p. 15.)

14.3 Network Polymers for Composites and Adhesives

14.3.1 Formation of Network Polymers

Network polymers are used in bicycle construction both as adhesives and as the matrix materials in fiber-strengthened composites. The most common are epoxy resins. These applications require a material with enough fluidity to fill the spaces between fibers, or between parts to be adhesively bonded, that will cure into a stiff solid. Ideally, the solid would exhibit essentially no viscous behavior. That is, creep and stress relaxation are undesirable in this application. This can be achieved by forming a three-dimensional network with a very large number of links between the basic structural units. In this way the units are prevented from sliding past each other, and the solid becomes essentially one giant molecule. It is then a glassy material. Elastic deformation can occur only by distortion of the molecular cage. It can deform plastically by shear (which involves breaking of covalent bonds) only when loaded to high stresses in compression. The yield stress is so high that, when loaded in tension, the material would crack before yielding. Thus, such a network polymer would have a low fracture toughness.

The first commercially successful network polymers were the phenolic resins called Bakelite, invented by Leo Baekeland in 1907. The basic network-forming reaction employs formaldehyde, H_2CO, to link together molecules of phenol, which are mainly benzene rings and which can form links at up to three sites around the ring, as shown in Fig. 14.19. (Note that the H atoms lie at three types of sites in the phenol molecule, considering the symmetry with respect to the OH group.) This illustrates a general requirement for the formation of a network: The primary molecular building block must be multifunctional; that is, it must have more than two reaction sites per molecule. (A molecule having only two reaction sites would produce a linear polymer, rather than a network.)

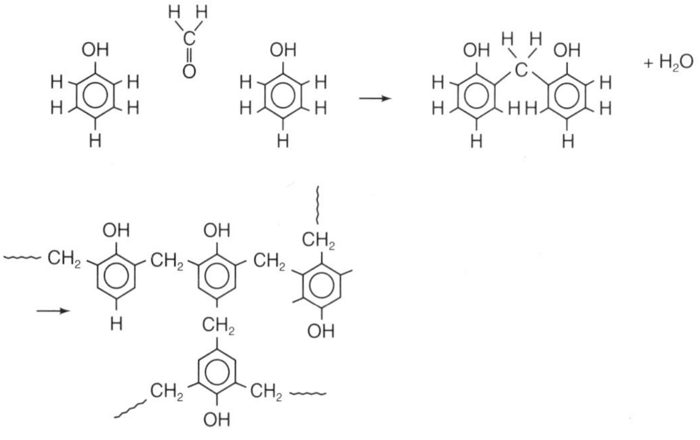

Fig. 14.19 Phenol-formaldehyde reaction. The phenol (C_6H_5OH) contributes hydrogen, and the formaldehyde (CH_2O) contributes oxygen to produce water as a by-product. The rings are joined by a CH_2 bridge. When this occurs in three dimensions, a network polymer is formed.

Epoxy resins are much more complicated chemically but also much more versatile. To produce an epoxy resin one starts by forming a prepolymer by reacting epichlorohydrin and bisphenol A in an aqueous alkaline environment, as shown in Fig. 14.20. The former contains an epoxy group, which has the configuration

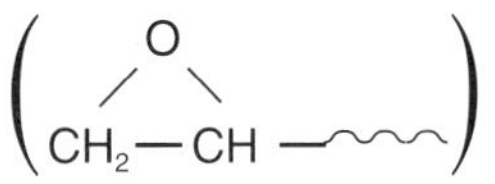

SECTION 14.3.1 FORMATION OF NETWORK POLYMERS

$$\left(\underset{CH_2-CH}{\overset{O}{\diagup \diagdown}} \sim\!\!\sim \right)$$

and this provides the key factor in the later network formation. In order to have this group at both ends of the chain of the prepolymer, excess epichlorohydrin is used. The sodium hydroxide is needed to neutralize the HCl, which forms as a by-product of this condensation reaction.

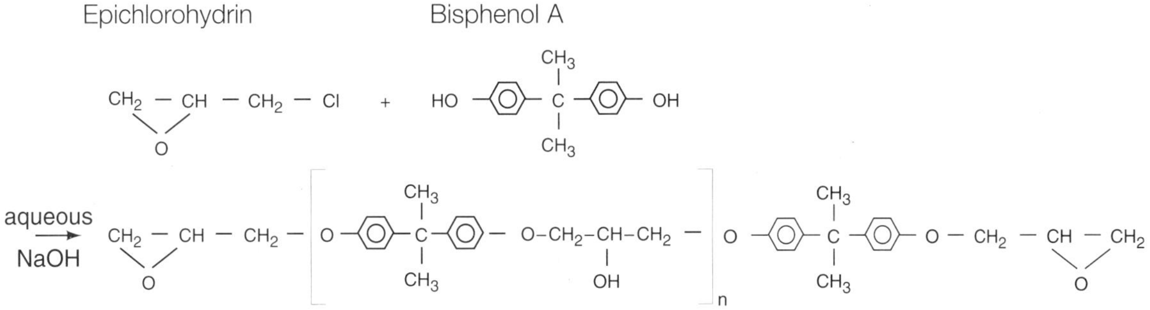

Fig. 14.20 Reaction of epichlorohydrin and bisphenol A to form the epoxy prepolymer.

This prepolymer can be produced as a solid (which may then be ground into a powder) or as a viscous fluid, depending on the degree of polymerization. The network is formed by reacting the prepolymer with a multifunctional curing agent that reacts with the terminal epoxy groups, as shown in Fig. 14.21.

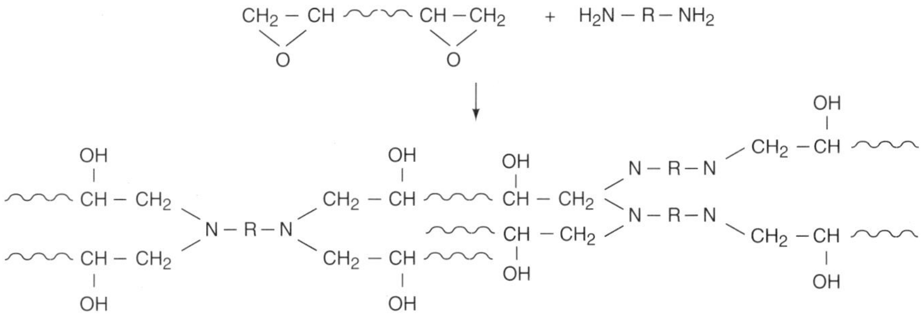

Fig. 14.21 Formation of an epoxy resin by reacting the prepolymer with a diamine.

Because an epoxy prepolymer molecule may have an extended length, and because the network of the epoxy resin is formed by tying together only the ends of these molecules, the resin may exhibit a glass transition, in contrast with the behavior of a network polymer like Bakelite. However, room temperature is normally below the T_G of epoxies, so they are classified as brittle materials.

14.3.2 Adhesive Bonding

With regard to adhesive bonding with an epoxy resin, it is essential that intimate molecular contact be made between the resin and the substrate so that bonding forces are established between the atoms or ions on the substrate and the molecules of the resin. These bonding forces are usually of a secondary nature. Typical bond energies are compared in Table 14.2. Calculations of bond strengths in the case of the weakest secondary bonds, i.e., dispersion forces (due to induced dipoles arising from internal electron movements, apart from any permanent dipole moments), indicate that they would be higher than the bond strengths observed experimentally. It appears, therefore, that actual bond strengths are controlled by factors such as voids or other kinds of defects in the joint, or the action of stress raisers, or the failure to achieve intimate contact, rather than by the type of the molecular level bonding achieved.

Table 14.2 Bond types and Typical Bond Energies. (From A. J. Kinloch *Adhesion and Adhesives,* Chapman and Hall, New York, 1987, p. 79.)

Bond Type		Bond Energy (kJ/mol)
Primary		
	Ionic	600 - 1000
	Covalent	60 - 700
	Metallic	110 - 350
Secondary - Hydrogen		
	Hydrogen bonds involving fluorine	Up to 40
	Hydrogen bonds not involving fluorine	10 - 25
Secondary - van der Waals		
	Permanent dipole-dipole interactions	4 - 20
	Dipole-induced dipole interactions	<2
	Dispersion (London) forces	0.08 - 40

For intimate contact the resin should wet the surface; that is, the contact angle should be zero (cf. Fig. 9.2). When it is not zero, pressure is required to force the resin to spread over the surfaces of the joint. The surface of the substrate may be roughened to provide increased area for bonding. (This also increases the fracture energy of the joint by increasing the amount of plastic or viscoelastic deformation accompanying the propagation of a crack along the joint.) The roughening may be done by a chemical process, like anodizing of aluminum, or by mechanical abrasion. A benefit of the latter is that it removes weak surface films like oil, grease and wax. However, surface roughening increases the contact angle, and thus it can complicate the achievement of intimate contact between the resin and substrate. In the case of an anodized surface, polymer molecules can actually diffuse into the surface pores by the process of *reptation* (i.e., snake-like penetration of the voids by the long-chain molecules), assuming wetting occurs or sufficient pressure is applied to the joint. This can give a very strong bond.

The design of an adhesive joint is important to its strength. It should avoid peeling-type forces, aiming rather for shear or compressive loading. In a proper joint the stresses are well distributed, and good resistance to fatigue failure can be achieved.

There are two types of epoxy resin in common use. In one, often referred to as a two-part adhesive, the curing agent is added to the prepolymer at the time of use, and the reaction occurs at ambient temperature. In the other the curing agent is mixed with the prepolymer by the manufacturer, and the mixture must be heated for the curing reaction to occur. In either case, the curing reaction is designed to occur with no by-product; that is, it is not a condensation reaction.

14.4 Carbon-Fiber Reinforced Composite Frames

14.4.1 Case Study of a Hybrid Composite Frame

A bicycle frame made from carbon-fiber reinforced plastic (CFRP) can have a Young's modulus comparable to metallic tubing, superior yield strength, and a much lower weight. This is still a relatively new engineering material, however, and the techniques for applying it to bicycles are still being developed. An example is a frame made of CFRP tubes, joined with components of cast aluminum alloy, as shown in Fig.14.22(a).

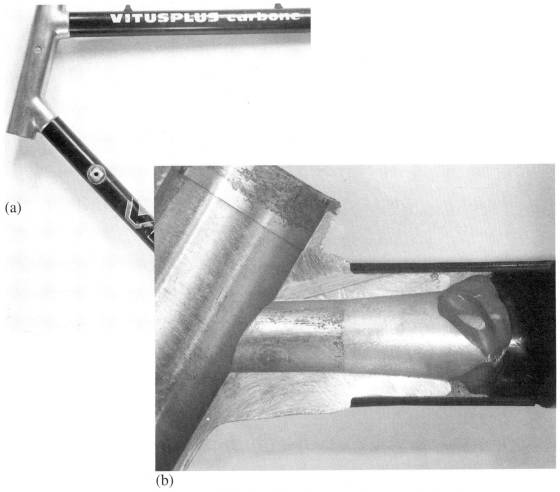

Fig. 14.22 (a) Part of a frame made from CFRP tubes joined to a cast-aluminum-alloy head tube. (b) The epoxy-adhesive joint between the CFRP tube and the head tube.

As shown in Fig. 14.22(b), the CFRP tube is attached to a protrusion on the aluminum-alloy head tube by means of an epoxy adhesive, as was done with the aluminum frame shown in Fig. 12.29. The potential problem with this design is that the abrupt change in cross-section of the frame at the joint produces a large localized change in stiffness of the frame. Such a discontinuity tends to act as a point of stress concentration; therefore, it could be a point where a fracture could occur.

The microstructures of the three materials at the joint are shown in Fig. 14.23. The casting is a hypoeutectic Al-Si alloy, shown at higher magnification in Fig. 14.24(a). The α-Al dendrites make up about one-fifth of the microstructure, indicating that the alloy composition is about 10.5% silicon, as can be deduced by application of the lever rule to the Al-Si phase diagram shown in Fig. 14.24(b). This type of alloy is commonly used for cast components of bicycles, such as pedal cranks, handlebar stems, etc.

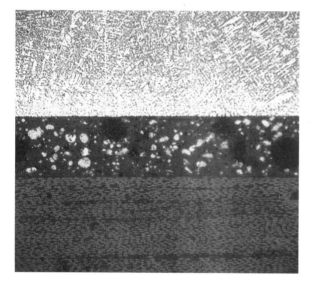

$\leftarrow$ Al alloy

$\leftarrow$ Epoxy

$\leftarrow$ CFRP

Fig. 14.23 Microstructures of the three materials across the joint shown in Fig. 14.22: the cast aluminum alloy, the epoxy adhesive, and the CFRP tube.

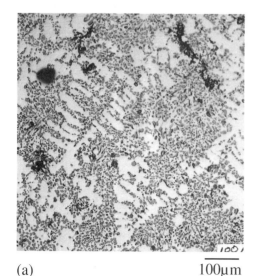

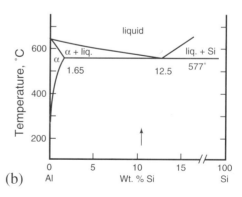

(a)   100μm

Fig. 14.24 (a) The aluminum alloy at a higher magnification and (b) the Al-Si phase diagram, the arrow showing the probable composition of the alloy.

The epoxy used for the adhesive contains particles which appear to be metallic. It was hypothesized that the particles are from an aluminum powder added to the epoxy to act as an inexpensive stiffener. This can be confirmed by an examination of the microstructure of the epoxy in a scanning-electron microscope (SEM) equipped with an energy-dispersive x-ray detector. The scanning-electron image of the epoxy and the x-ray map of this area, made with the characteristic x-rays emitted by aluminum, are shown in Fig. 14.25.

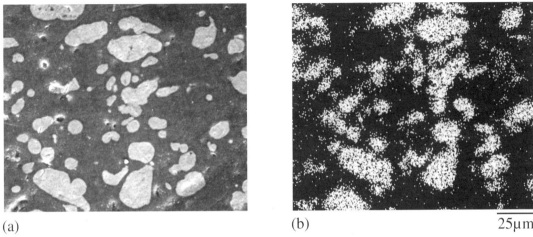

Fig. 14.25 (a) Scanning-electron image of the epoxy, showing the dispersed particles assumed to be aluminum. (b) Map made with x-rays having a wave-length characteristic of emission by aluminum, confirming that assumption. (Courtesy of Deborah Ricketts, Univ. of Penn.)

As shown in Fig. 14.26, a scanning electron microscope bears a great similarity to the transmission electron microscope shown in Fig. 5.15(a). The principal difference is the addition of deflection coils, similar to those used on a TV picture tube, which enable the electron beam to be scanned across the specimen. An image of the surface can be formed by collecting the back-scattered electrons in a detector and sending the signal from the detector to a cathode-ray tube (CRT), the beam of which is synchronized with the scanning electron beam. This is how the image in Fig. 14.25(a) was formed. The aluminum particles scatter more electrons than the surrounding epoxy, which is made of elements with lower atomic numbers than aluminum; hence, the signal from the particles is stronger, and they appear lighter than the epoxy matrix.

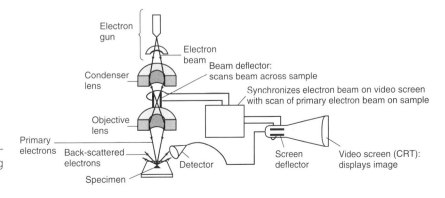

Fig. 14.26 Schematic representation of a scanning electron microscope.

The scanning electron beam knocks electrons out of their customary energy levels in the atoms near the surface of the specimen. This gives rise to the emission of x-rays as the excited electrons fall back to their former energy levels. Since each element emits its own characteristic x-ray spectrum, the x-rays emitted by the specimen can be used for chemical analysis of the surface regions. For this purpose the SEM must be equipped with an x-ray detector, the signal from which can be displayed on a CRT that is synchronized with the scanning electron beam.

Because it was suspected *a priori* that the particles were aluminum, the x-ray detector was tuned to an aluminum x-ray peak, and the signal was used to form the image on the CRT. The x-ray map thus formed, Fig. 14.25(b), shows a one-to-one correspondence between the particles in Fig. 14.25(a) and the regions emitting characteristic aluminum x-rays. Thus, the particles are aluminum-rich.

As shown in Table 3.1, the Young's modulus of aluminum is more than ten times that of an epoxy resin. Therefore, the addition of aluminum particles stiffens the epoxy to some extent. However, this is much less effective than stiffening by fibers; the modulus would be fairly close to the lower bound in Fig. 14.6. Nonetheless, the properties of the epoxy are improved somewhat by the addition of this "filler," which is actually less expensive than the epoxy. This works well as long as the epoxy bonds well to the particles. In the present case this is not a problem, since one of the two components being joined by the epoxy is, in fact, aluminum.

The epoxy contains a substantial amount of porosity, which is probably the result of air entrained when the epoxy was mixed. It may also have come from gases evolved during curing, if this was done at an elevated temperature. The epoxy adhesive appears to be bonded well to the epoxy matrix of the CFRP, as might be expected; no seam can be seen in the microscope. On the other side of the joint, the epoxy follows the contours of surface roughness of the aluminum casting, which gives a kind of mechanical locking along that interface.

The microstructure of the CFRP tubing is shown at low and high magnifications in Fig. 14.27. It is apparent that the tube was made in about 20 layers, with the carbon fibers in different orientations in the various layers. In the section cut transverse to the longitudinal direction, Fig. 14.27(a) and (b), a good deal of porosity can be seen between the layers where the bonding between them was incomplete. This is probably not a serious defect, since there is only a stress across these interfaces when the tube is twisted.

The layers were probably made from tape in which the carbon fibers were held in a partially cured epoxy. The layers of tape were probably laid up on a cylindrical form, called a mandrel, and the completed tube was then cured, probably at an elevated temperature. In some of the layers the fibers run along the longitudinal direction, since their cross-sections are circular in Fig. 14.27(b). The diameter of the fibers can be estimated to be about 6μm. The layers in which the fibers exhibit an elliptical cross-section are obviously ones in which the tape was wound in a helix around the mandrel. The longitudinal fibers are meant to resist bending of the tube, and the helically wound fibers are meant to resist twisting. Presumably, an equal number of right- and left-hand helices are included.

The layers in which the tape runs longitudinally can be seen more clearly in Fig. 14.27(c), where these layers comprise fiber sections that are very long ellipses. (They would be infinitely long if the fibers were straight and the plane

SECTION 14.4.1 CASE STUDY OF A HYBRID COMPOSITE FRAME

of the section were exactly parallel to the fiber axis.) It can be seen that in this tube, which was the down tube, the longitudinal fibers are in the first six layers, counting from the inside diameter (bottom of Fig. 14.27c), and in the twelfth and thirteenth layers. Presumably, this tube design was based on a stress analysis for this particular tube and the expected loads to be applied.

Note that this tube could not be analyzed using the beam theory outlined in Chapter 12 because it is not uniform. That is, the elastic modulus varies from layer to layer due to the different fiber orientations. This is one of the great advantages of using fiber-reinforced composites; the properties can be tailored to fit the requirements within the cross-section of the component itself.

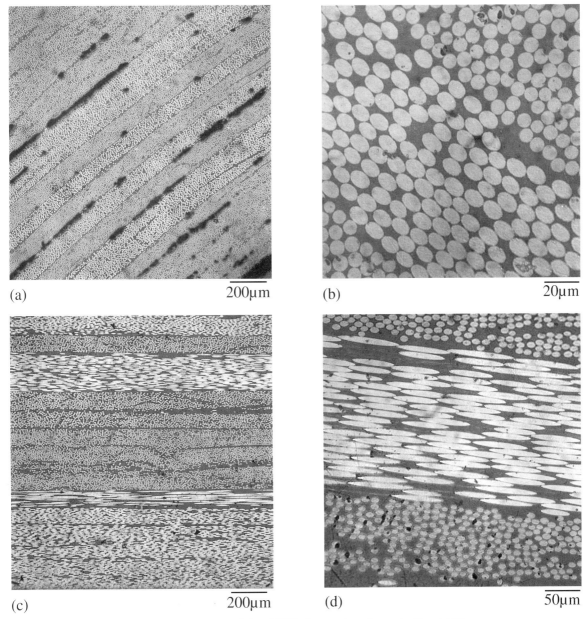

Fig. 14.27 (a) and (b) transverse section of the CFRP down tube shown in Fig. 14.22. (c) and (d) longitudinal section of the same tube.

The tensile behavior of the CFRP tube can be demonstrated by the stress-strain curve shown in Fig. 14.28(a). It is apparent that the tube is completely brittle; it failed in tension without any detectable plastic deformation. Thus, although the CFRP is stiff and light, it has almost no capacity to absorb energy when it fractures, as a metallic material would. A portion of the fracture surface of the CFRP as observed in the scanning electron microscope is shown in Fig. 14.28(b). The main feature is the protrusion of broken fibers above the fracture surface of the epoxy. This means that the fibers broke while embedded in the epoxy and were then pulled out of the matrix. This fiber-pullout phenomenon is a common feature of the fracture of composites, and provides a mechanism for some energy dissipation during fracture and makes a contribution to the fracture toughness of a composite material.

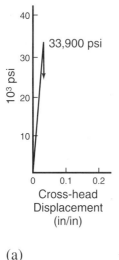

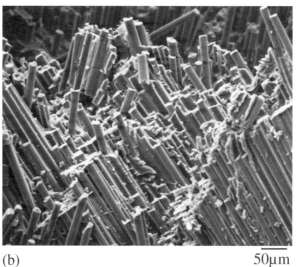

Fig. 14.28 (a) Stress-strain curve of CFRP tube shown in Fig. 14.22. (b) Scanning-electron micrograph of the fracture surface, showing the results of fiber-pullout.

(a)   (b)

Brittle behavior is an unavoidable feature of the CFRP. It is necessary that the polymer matrix be glassy; otherwise, the flexing of the tubing during use would dissipate energy. That is, the bicycle would have a "spongy" feel to a rider. Because the polymer matrix is not viscoelastic, it cannot act as a microcrack arrestor when the fibers start to break. The result is that the CFRP used for bicycle tubing is inherently brittle (as opposed to fishing rods or vaulting poles, which are intended to be flexible). Because of this behavior, it can be risky to make a component like the front fork of a bicycle out of CFRP. Front forks can receive large, sudden loads, and a brittle failure here would obviously threaten rider safety.

14.4.2 Monocoque Frames

A logical next step after CFRP tubes bonded to metal connecting pieces was to construct a one-piece frame using a continuously wound carbon-fiber tape impregnated with partially cured epoxy, with metal reinforcement at critical places. The first commercial-scale bicycle of this kind was the Kestrel, an example of which is shown in Fig. 14.29. Except for the elongated cross-section of the members, the frame has essentially the same shape as a conventional road bike.

Fig. 14.29 A monocoque CFRP frame by Kestrel.

A radically different frame design, permitted by use of the monocoque CFRP approach, was the Zipp frame shown in Fig. 14.30. The rear section has the shape of a boomerang, and the seat is cantilevered over the back wheel by a CFRP bar which acts somewhat as a shock absorber. Obviously, one must have complete confidence in the fracture resistance of this bar, because a failure here would be disasterous for the rider.

Fig. 14.30 The Zipp CFRP frame with cantilevered seat.

Perhaps the most famous monocoque bike is the one developed by Lotus of England for the 1992 Olympic Games in Barcelona, based on an original 1982 design by Mike Burrows. On this bike, shown in Fig. 14.31(a), a new world record was set in the 4000m pursuit race by gold medalist Chris Boardman of England. The frame has an air-foil cross section, and the wheels are held by axles cantilevered from the frame and monoblade front section, respectively. The monocoque was formed from carbon-fiber cloth impregnated with epoxy resin over a polyurethane foam core with Kevlar inserts. The steering column was titanium, and the chainset, pedals, and rear sprocket were made from titanium and steel. The stress analysis needed for the optimization of the geometry was done by the finite-element method. In this technique the frame and wheels are modeled by a mesh of elements of known strength and stiffness, each of which bears its part of the total load. When a load is applied to the frame or wheel in the computer model, the stress on any element can be calculated by the computer program, and this reveals the necessary carbon-fiber configuration at that location. This is a good example of how modern engineering design can be coupled with the custom tailoring of a material to give the required properties at every point in a structure.

Fig. 14.31 The Lotus monocoque pursuit racing bike.

14.4.3 Composite Wheels

In support of the adage that there is nothing new under the sun, one can find in the book by Archibald Sharp (1896) the picture that is reproduced here as Fig. 14.32. It shows both a disc wheel on the rear and an aerodynamically configured four-spoke wheel on the front of an early diamond-frame bicycle. Thus, what are taken today as recent innovations had actually been introduced within a few years of the development of the modern diamond frame. The disc wheel is intended to reduce the aerodynamic drag caused by spokes. It is suitable as a rear wheel, but would produce difficulties in steering if placed on the front, because the center of aerodynamic pressure would be forward of the steering axis. This would create unwanted torque, especially in strong cross winds. Therefore, a more open design is needed for the front wheel. It can be seen that a disc wheel and a three-spoke aerodynamic wheel were used in the Zipp and Lotus bikes (Figs. 14.30 and 14.31).

Fig. 14.32 A rendering of an early bicycle with a rear disc wheel and an open-spoke wheel in the front. (A. Sharp, 1896, loc. cit. p. 352.)

Another version of the open-spoked wheel was introduced by the DuPont Company in 1990 and is shown in Fig. 14.33(a). Its design utilized wind-tunnel experiments to minimize drag on the spokes and rim, arriving at the shapes shown in Fig. 14.33(b). The rim acts both as the leading edge and the trailing edge of an airfoil as the wheel rotates. The fact that the wheel has about 50% open area gives it good steering characteristics and stability in cross winds. The wind-tunnel experiments showed that this design had the lowest drag of any wheel tested and only about one-third the drag of a conventional 36-spoke wheel and one-half of that of a 24-spoke wheel.

The wheel is fabricated from CFRP fabric impregnated with epoxy resin. The number, orientation, and packing density of the carbon filaments required to give the wheel the desired stiffness and strength, especially when loaded between the spokes, was determined by computer modeling using the finite-element method. The type of mesh used is shown in Fig. 14.33(c).

CHAPTER 14 COMPOSITE MATERIALS; TIRES, FRAMES, AND WHEELS

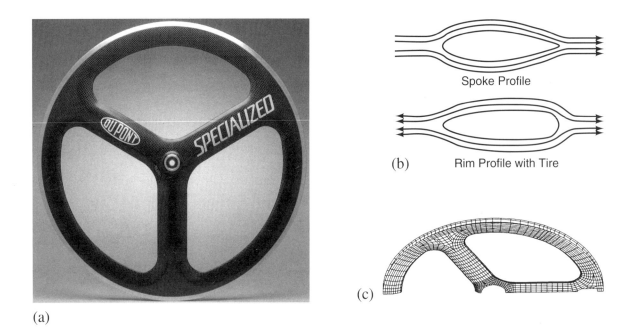

Fig. 14.33 (a) The Du Pont composite wheel. (b) The profiles of the spokes and rim as optimized to minimize drag. (c) Mesh used for finite-element model of the wheel. (M. W. Hopkins et al., 35th Intnl. SAMPE Symposium, April, 1990.)

Summary

In a fiber-reinforced (FR) composite material such as a rubber tire or bicycle frame, the function of the fibers is to carry the load. This is achieved by employing a fiber with an elastic modulus greater than that of the matrix. The main function of the matrix is to deform and, by so doing, to transfer load to the fibers.

The stiffness of a FR composite depends on the orientation of the fibers relative to the stress. Thus, when stress is applied along the fiber direction, the modulus of the composite follows a rule of mixtures in which the modulus increases linearly with the volume fraction of fibers. When stress is applied at right angles to the fibers, the modulus of the composite depends mainly on that of the matrix until the volume fraction of fibers is well in excess of 50%. When the fibers are not continuous, the load transfer occurs over a certain distance at each end of a fiber, and this means that the efficiency of stiffening of the composite by the fibers is significantly reduced from the continuous-fiber case.

The first fibers used for tires were completely natural, namely cotton. Later, a partially man-made fiber was invented; this was rayon, or cellulose acetate. The first completely man-made fiber was nylon, which is made by a condensation reaction, as opposed to the addition polymerization used to make PE. Nylon has a backbone of single covalent bonds, almost the same as PE, but it also has hydrogen bonds, which retard interchain sliding more effectively than the van der Waals bonds between PE chains. A more recent invention was Kevlar, which is a polyamide like nylon (and therefore has interchain hydrogen bonds), but which has aromatic rings along the chains that give the chains considerably more stiffness than the single covalent bonds of PE and nylon. The ultimate in stiffness is found in fibers made from graphitized polymers. Here, the aromatic rings extend in two dimensions, and the chain-like structure is almost completely converted to one of irregularly interleaved sheets.

A common reinforcing fiber is silica-based glass. This is a three-dimensional amorphous structure in which covalently bonded and linked SiO_4^{4-} tetrahedra are mixed with network modifiers in sufficient quantity to allow fiber drawing at reasonably low temperatures. The drawing is done simply by pulling on the fibers (above T_G), which works by virtue of the Newtonian-viscous nature of molten glass. Another inorganic fiber used in tires is patented eutectoid steel, or "piano wire." This can compete with graphite fibers in terms of stiffness and strength but not density.

The opposite end of the spectrum from a thermoplastic linear polymer is a thermosetting network polymer, the prototype of which was Bakelite, or phenol formaldehyde. Toward the latter end of this spectrum are the highly cross-linked polymers used for adhesives and as the matrix for fiber-reinforced components of very high stiffness, such as CFRP bicycle tubes. Epoxies are the most common example. They employ a prepolymerization by a condensation reaction and then a multifunctional curing agent that ties the ends of the prepolymer chains together into a three-dimensional network. The T_G and the extent of viscoelastic behavior depends on the lengths of the chains of the prepolymer.

Part of a bicycle frame fabricated from CFRP tubes adhesively bonded to aluminum-alloy fittings was dissected and examined metallographically, including the use of an SEM with x-ray analysis. A tensile test of a specimen cut from a tube illustrated the low toughness of this material. The use of CFRP to make monocoque frames and wheels was illustrated by means of recently developed products.

Exercises Chapter 14

Terms to Understand

Fiber-reinforced composites
Bamboo, cellulose
Bone, collagen
Longitudinal *vs.* transverse loading
Composite elastic modulus
Crack arrest in the matrix
Discontinuous fibers, end effects
Transfer length, aspect ratio
Fiber pullout
Rayon
Nylon
Cold-drawing of a linear polymer
Kevlar
Condensation reaction
Spinnerette
Polyacrilonitrile

Carbon fiber
Silica glass
Network modifiers
Newtonian viscosity
Drawing of glass fibers
Workability *vs.* thermal-shock resistance
Patenting of steel, piano wire
Bakelite
Epoxy adhesives
Adhesive bonding
CFRP tubes
Scanning-electron microscopy (SEM)
Energy-dispersive x-ray analysis
Curing agent
Prepolymer
Reptation

Problems

14.1 A fiber-reinforced composite consists of continuous, parallel strands of glass fiber (E = 70 GPa) embedded in a polyester resin (E = 4.0 GPa). The fibers occupy 65% of the total volume. Calculate the elastic modulus of the composite.

(a) in the direction parallel to the fibers, and

(b) in the direction perpendicular to the fibers.

14.2 A fiber composite is to be made from Kevlar polyamide fibers (E = 130 GPa) in an epoxy matrix (E = 5 GPa). (a) What volume fraction of Kevlar fibers is needed to make the longitudinal modulus of the composite equal to 45 GPa? (b) What volume fraction of fibers would be needed if one substituted for the Kevlar either Type I or Type II carbon fibers? (Use Table 3.1.)

14.3 Repeat the calculation of the curvature in a fiber bundle, shown in Fig. 14.1, for the case of a cylindrical bundle with 100 and 1000 fibers bonded together, instead of one fiber or a four-fiber bundle.

14.4 With regard to the choice of matrix for a fiber-composite, compare the expected crack-arresting capabilities of a glassy thermoset *vs.* a cross-linked linear polymer above its glass-transition temperature. What is the opposite side of the trade-off in this choice? (That is, what are the advantages and disadvantages of each type of matrix?)

14.5 Why, in a composite strengthened by discontinuous parallel fibers, should the fiber length be made as long as possible?

14.6 Use Fig. 14.16 to explain why Pyrex®, a borosilicate glass, is used to make kitchenware that can be transferred from the refrigerator to the stove without breaking, as opposed to pure silica, which would have the ultimate thermal-shock resistance, or soda-lime glass, which would have the ultimate workability. Include an explanation of what is meant by thermal shock and workability, and how and why these are changed according to the composition of the glass. Which kind of silica-based glass is used for making fibers, and why?

14.7 (a) Explain why a multifunctional polymer molecule is needed to make a network polymer. (b) There is an important difference between the reaction used to make Bakelite and that used for curing of an epoxy resin (not in making the prepolymer). What is it and why would it be important for some applications?

14.8 Explain the important factors in the functioning of an adhesive.

14.9 Using Fig. 14.27 as an example, describe the considerations that a bicycle designer would use in specifying how the layers of so-called pre-preg tape (carbon fibers in partially cured epoxy) are to be laid-up in the manufacture of a bicycle tube.

15

THE WALKMAN*; MAGNETIC RECORDING AND PLAYBACK

15.1 Historical Background

The first successful sound recording is credited to Thomas Edison. The Edison talking machine, or phonograph, was introduced in 1877 and was entirely mechanical; no electricity was used. Variations in air pressure (that is, sound waves) caused a thin diaphragm to vibrate, and these vibrations were transmitted through a mechanical coupling system to a needle. The vibrations of the needle were recorded on a rotating cylinder covered with a layer of tin foil; the needle scratched a wiggly helical track on the foil. The sound was played back by the reverse process: the needle, following a prerecorded track, transmitted the wiggles in the track to the diaphragm, which (more or less) reproduced the original sound waves. The diaphragm thus served as the microphone for recording and the speaker for playback.

Later developments allowed for the conversion of the diaphragm vibration to an electrical signal, the electronic amplification of this signals, and the use of polymer disks to record the signal. In the rapidly disappearing long-play stereo disk, both vertical and lateral motions of the needle are recorded, allowing for separate right and left stereo channels. This type of recording is an analog process, since the wiggles recorded in the groove are an analog of the original sound.

Magnetic recording was introduced during World War II, primarily to increase the length of uninterrupted recording of radio transmissions. It was quickly developed as a commercial product after the war, and has been widely used since about 1950. In addition to providing longer unbroken recording and playing times, magnetic recordings are much easier to edit and are less vulnerable to damage from wear and rough handling than disk recordings. They permit the ordinary user to make recordings, and sound quality can be excellent. The same basic technology is used for video recording, although the much higher density of information requires various modifications.

Ordinary audio and video tape recordings are still analog, although digital technology has begun to appear. Computer data storage on magnetic tapes, floppy disks, and hard disks is entirely digital. Digital recording is conceptually simpler, since only on-off (0 and 1) information must be saved and retrieved. However, enormous volumes of digital information need to be stored to reproduce sound, and even larger quantities for video. Audio and video recordings can tolerate some errors without noticeable degradation of output, but computer data recording cannot. We will restrict our attention to analog recording, since all the materials principles can be illustrated in this way.

* The trade name "Walkman" is the property of the Sony Corporation.

Technical folklore says that the Walkman originated in the desire of Akio Morita, founder and president of Sony, to have an audio tape player he could carry with him to listen to tapes whenever he wished. Sony engineers were able to make such a device because of the development in the late 1970s of miniature motors and headphones, which were made possible by new and better permanent magnet materials. Once the prototype Walkman was made, Morita wanted to make it a commercial product. There is said to have been serious opposition within the company, based on the assumption that the public would not buy a tape recorder that did not record but only played tapes. Morita prevailed (it helps to be founder and president), and his judgement was spectacularly justified by sales and profits.

15.2 General Operation of the Walkman

The original Walkman and its immediate successors were portable, battery-powered tape players with a built-in amplifier capable of driving headphones. Only later was recording capacity and AM/FM radio reception added. Figure 15.1 is a schematic illustration of the major parts of the device. The tape and its supply spools are, of course, contained in a standard audio cassette. The tape-transport mechanism, including the drive motor and the tape playback head, is shown in Fig. 15.2. (This is a playback-only model dating from the mid-1980s.) The electronic components are mounted on a separate board, and they will be treated in Chapter 16.

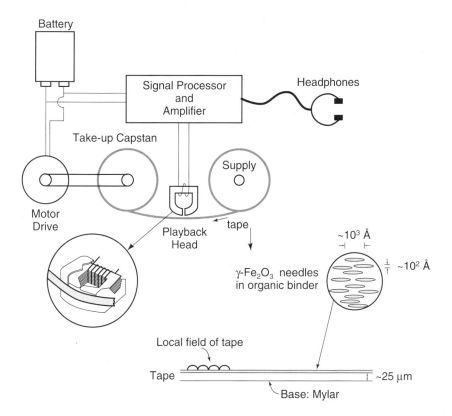

Fig. 15.1 The parts of a Walkman portable tape player (schematic).

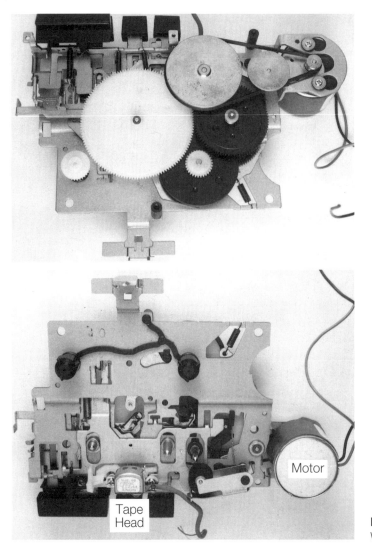

Fig. 15.2 Mechanical components of a Walkman

The function of the battery-powered dc motor is simply to pull the tape past the head at a constant speed. The headphones, which, like the motor, depend on permanent-magnet materials for their operation, convert the amplified electrical signal from the tape head into sound. The components unique to magnetic recording, the tape itself and the tape playback head, are shown in some detail in Fig. 15.1.

Ordinary audio tape is a polymer ribbon, usually mylar, about 25 μm thick, on which is carried a layer of microscopic needle-shaped particles of a magnetic iron oxide in an organic binder. The varying magnetization of groups of these particles carries the recorded information. The magnetized particles create just above the tape surface a locally varying magnetic field, which is sensed when the tape passes by the gap in the ring-shaped head. The head is made of a material which can be very easily magnetized and demagnetized, called a *"soft" magnetic material.* The head material is magnetized to varying levels by the field of the recorded tape, and these changes in magnetization generate a time-varying voltage in the copper windings on the head. This voltage is processed and amplified to drive the headphones. To understand magnetic recording, it is necessary to know something about magnetic materials and their response to magnetic fields.

15.3 Magnetic Materials

The first known magnetic material is believed to have been natural lodestone, or magnetite, a compound of FeO and Fe_2O_3 with the chemical formula Fe_3O_4. It was discovered in antiquity that pieces of this ore were attracted to iron objects and that, if supported so they were free to rotate (by placing them on a floating chip of wood), they always aligned themselves in the same direction. This was the origin (apparently in China) of the magnetic compass.

By 1600, iron or steel (the distinction did not then exist) magnets and compass needles were in regular use, and in 1644 René Descartes described how to use iron filings to map out the magnetic field lines around a permanent magnet, as in Fig. 15.3(a).

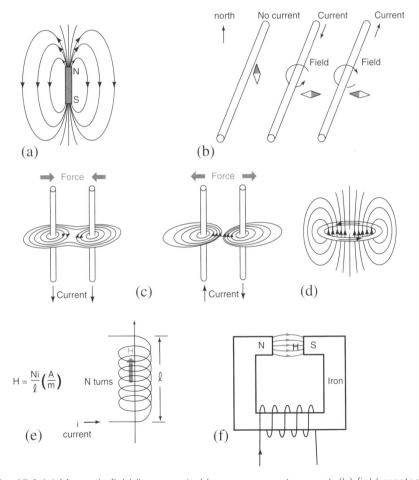

Fig. 15.3 (a) Magnetic field lines created by a permanent magnet; (b) field created by a current, showing deflection of a magnetic compass needle; (c) force between parallel currents; (d) field of a current loop; (e) field of a solenoid; (f) field of an electromagnet.

The connection between electricity and magnetism was not made until 1820, when Hans Christian Oersted found that a compass needle is deflected when brought near a current-carrying wire (Fig. 15.3b). This established that an electric current generates a circumferential magnetic field and led to an understanding of the force between current-carrying wires (Fig. 15.3c). If two parallel wires carry currents in the same direction, the fields in the region between the wires tend to cancel; this lowers the energy of the system, so the wires are attracted towards one another. If the currents are in opposite directions, the fields add, raising the energy of the system and causing the wires to experience a repulsive force.

A loop of wire carrying an electric current produces a field with the same geometry as a very short bar magnet; compare Figs. 15.3(a) and (d). A current loop or a bar magnet has a *magnetic moment,* expressed in Am^2. A series of adjacent loops, all carrying the same current, is a *solenoid* (Fig. 15.3e); it can be used to generate modest magnetic fields for measurements or engineering devices. A solenoid, or winding, wrapped around a soft magnetic material in a C or U shape (Fig. 15.3f) can produce a large field across the gap. This is an electromagnet. A tape recorder head can be regarded as a miniature electromagnet.

15.4 Magnetic Quantities

The *magnetic field* H at the center of a solenoid is given simply by the number of turns or coils, N, per unit length, *l*, times the current, i, in each turn:

$$H = \frac{Ni}{l}$$

In SI units, magnetic field is measured in amperes/m or in tesla; $1\ T = \mu_0(A/m)$, where μ_0 is the magnetic constant, $4\pi \cdot 10^{-7}$ T m/A (or henry/m). The magnetic state of a material is specified by its *magnetization,* M, which, depending on the applied field and the history of the sample, may take any value from zero to a saturation value, M_s. Magnetization may be thought of as magnetic moment/volume, $Am^2/m^3 = A/m$. The limiting value, M_s, depends on the composition of the material and also on temperature.

The quantity B, known as the *magnetic induction* or *magnetic flux density,* is proportional to the sum of the field and the magnetization:

$$B = \mu_0(H+M).$$

Flux density, B, is measured in weber/m^2. (One Wb/m^2 = one tesla.) In free space, and for practical purposes in air, M=0, so $B = \mu_0 H$. Thus, a magnetic field can be expressed either in units of H or B. The quantity B is useful, because it appears in Faraday's Law for the generation of voltage in a coil of N turns and area A, in which B changes with time:

$$E = -NA\ (dB/dt).$$

Here E is in volts, A in m^2, and dB/dt in tesla/sec. A little manipulation shows that a tesla is a Vsec/m^2, and, since $1\ T = 1\ Wb/m^2$, $1\ Wb = 1\ V \cdot sec$. Faraday's Law describes the generation of voltage in a generator, transformer, or tape recorder head.

15.5 Magnetic cgs Units

Much of the published literature on magnetic materials uses the older centimeter-gram-second (cgs) system of units, in which magnetic field is in oersteds and flux density is in gauss. The cgs or Gaussian system has some virtues, including the fact that it sets the magnetic constant equal to unity, and is still widely used in the U.S. and Japan. Unfortunately, the conversions are not completely straightforward. A table of conversions is given in Appendix 15.1.

15.6 Types of Magnetic Behavior

Most materials are very weakly magnetic. They have no magnetization in the absence of a field and only a very small magnetization when a field is applied. They are not noticeably attracted to a permanent magnet and are loosely described as nonmagnetic. Most solids, including almost all polymers, develop a small negative magnetization in a positive field; they are called *diamagnetic*. In some materials, mainly metals, there is a competing and opposite effect which may be large enough to give a positive magnetization in a positive field; such materials are called *paramagnetic*. Except at very high fields and very low temperatures, the magnetization is proportional to the field, and the magnetic behavior is specified by the proportionality constant, called the (volume) magnetic susceptibility: $\chi = M/H$. Diamagnetic (negative) and paramagnetic (positive) susceptibilties are usually almost independent of temperature and are insignificant in comparison to the effect of ferromagnetism, as shown schematically in Fig. 15.4.

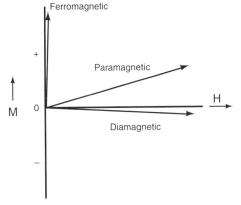

Fig. 15.4 Diamagnetic and paramagnetic susceptibilities, compared schematically to the much larger ferromagnetic susceptibility.

A relatively few elements, alloys, and compounds are strongly magnetic. These are classified as *ferromagnetic* or *ferrimagnetic* (the distinction will be explained later), and their properties make possible the technology of electric power generation and transmission, magnetic information storage, refrigerator door seals, and a large number of other applications. Only three elements are strongly magnetic at room temperature: iron, nickel, and cobalt.*

---

*One of the first uses of x-ray diffraction was to see whether these elements shared a common crystal structure that might explain their magnetic properties. It was found that iron is BCC, nickel is FCC, and cobalt is HCP; thus, the explanation lay elsewhere. We now know that amorphous (noncrystalline) solids can also be ferromagnetic.

Many of the rare-earth elements are ferromagnetic, but only below room temperature. Some alloys and compounds of manganese and chromium are strongly magnetic, as are a number of metal oxides and other compounds containing iron, nickel, and cobalt, such as Fe_3O_4 and other spinel ferrites, magnetic garnets, etc.

Understanding the origins of magnetic behavior would require venturing rather deeply into quantum mechanics. For present purposes, it is sufficient to note that in most strongly magnetic materials, magnetization arises from a property of the electron known as *spin*. If the electron is thought of as a sphere of electrical charge spinning about a fixed axis, it acts like a tiny conducting loop carrying a constant current, as in Fig. 15.5. Such a structure has a permanent magnetic moment which is also known as a *dipole moment*, because it behaves exactly like a pair of magnetic poles, North and South, separated by a fixed distance. The moment of a single electron spin is quantized and is called a *Bohr magneton* and is designated μ_B. It is often used as a unit to measure the net magnetic moment of an atom or a unit cell.

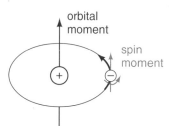

Fig. 15.5. Magnetic moment of an electron spin and of an electron orbit.

In most atoms, the net spin moment is zero, but in the ferromagnetic elements there can be a non-zero net moment. For obscure quantum-mechanical reasons, the net moment per atom need not be an integral number of Bohr magnetons. For example, pure iron has a net magnetization of 2.2 Bohr magnetons per atom.

There can also be a contribution to the magnetization per atom from the motion of the electron around the atomic nucleus. This is called the orbital moment (cf. Fig. 15.5) and is significant mainly in rare-earth atoms. An atom with a net magnetic moment, whether spin, orbital, or both, is loosely but commonly referred to as a magnetic atom. The magnetization of a solid, as noted above, is simply the net magnetic moment per unit volume: M= (m/V), where m is the net atomic moment.

The existence of magnetic atoms is not sufficient to produce a strongly magnetic material. If the atomic moments are oriented at random, thermal energy resists alignment by an external field, and the result is a more or less strongly paramagnetic material for which magnetization returns to zero when an applied field is removed. (This paramagnetism is strongly temperature dependent, in contrast to the paramagnetism discussed earlier.) In many cases, however, there exists a quantum-mechanical effect called the *exchange interaction*, which acts to align neighboring atomic moments parallel (positive exchange) or antiparallel (negative exchange). When positive exchange aligns all the atomic moments parallel, the material is said to be ferromagnetic (Fig. 15.6a). This is the magnetic structure of iron, nickel, and cobalt. If the exchange interaction is negative, and the atomic moments are all equal in magnitude, the net magnetization is zero, and the material is said to be *antiferromagnetic* (Fig. 15.6b). If the moments are not equal, a negative exchange interaction can lead to an antiparallel alignment of moments with a net magnetization not equal to zero (Fig. 15.6c or d). For example, magnetite, Fe_3O_4, may be regarded as $FeO \cdot Fe_2O_3$, containing Fe^{2+} and Fe^{3+} ions, each with its own atomic (or ionic) moment. The alignment of moments is antiparallel, but the net moment is nonzero. Such a structure is called *ferrimagnetic*.

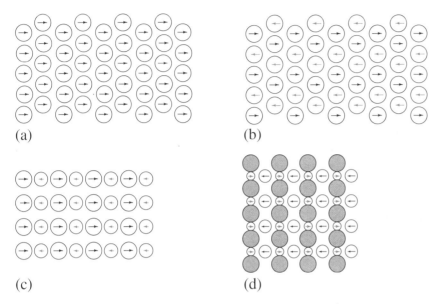

Fig. 15.6 (a) Ferromagnetic structure; (b) antiferromagnet; (c) and (d) ferrimagnets; shaded atoms have no magnetic moment, like oxygen, for example.

15.7 Magnetic Domains

The existence of positive exchange energy and the resultant ferromagnetic or ferrimagnetic structures would seem to require that strongly magnetic materials like iron should always be magnetized to saturation, with all the atomic moments parallel. That is, any magnetic material should be a permanent magnet. Clearly this is not the case. The reason lies in the existence of magnetic domains. Magnetic materials are normally subdivided into regions, smaller than the grain size, called domains. Each domain is magnetized to saturation, but the direction of magnetization can vary from domain to domain, so that the net magnetization can be zero.

Magnetic domains exist because the bar containing only a single domain (Fig. 15.7a) creates a large field in the surrounding space. There is an energy associated with this field, which can be substantially reduced by introducing domain walls (Fig. 15.7 b & c). However, the domain walls themselves have an energy cost, and an equilibrium number of domain walls exists when the reduction in field energy produced by adding one more domain wall just equals the additional energy of the added wall. In real samples with many crystals and many domains in each crystal, there is no single stable domain configuration. Very many possible domain arrangements have essentially equal energy and probability.

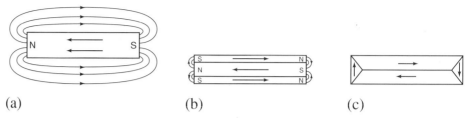

Fig. 15.7 (a) A uniformly magnetized rod creates an external field; (b) the external field is reduced by division into domains; (c) a domain structure producing no external field.

15.7.1 Magnetic Anisotropy

Magnetic materials may be anisotropic (different properties in different directions) for several reasons. Magnetic samples prefer to be magnetized parallel to a long direction as opposed to a short direction. This is known as *shape anisotropy*. Also, elastic stresses can produce anisotropic behavior, and the crystallographic arrangement of atoms in a solid can produce directional effects. The latter phenomenon goes by the cumbersome name of *magnetocrystalline anisotropy*, or sometimes just *crystal anisotropy*. It simply means that in crystalline materials the magnetization prefers to lie along one type of crystallographic direction rather than other types. In iron, the favored directions are the cube edges, <001>; in nickel, they are the body diagonals, <111>. Fig. 15.8 shows the curves of M *vs.* H for three crystallographic directions in iron. The favored directions are called the *easy directions* or *easy axes*. The crystal anisotropy normally determines the direction of magnetization of the individual domains; in Fig. 15.7, the edges of the single-crystal iron rectangle are <001> directions.

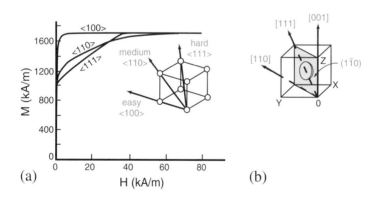

Fig. 15.8 (a) Magnetization curves for three principal directions in single-crystal iron. (b) shows how a single disk sample can contain three principal directions. (After K. Honda and S. Kaya, *Sci. Rpts., Tohoku Imp. Univ.*, Vol. 15, 1926, p. 721.)

15.7.2 Domain Walls

The walls separating domains are not infinitely thin. Their thickness is determined by a balance between two energies: the exchange energy acts to keep each moment parallel to its neighbors and is minimized when the domain wall is very thick; the anisotropy energy acts to keep all moments parallel to a particular crystallographic direction and is minimized when the wall is very thin. The calculation of wall thickness is not particularly difficult, but will not be given here. For common magnetic materials like iron and nickel, the wall thickness is several hundred atom diameters. Figure 15.9 is a sketch of the orientation of the atomic moments along one line of atoms extending perpendicular to a domain wall. The dashed lines indicate the location of the wall, and it is clear that the wall does not have sharp boundaries. Therefore, its "thickness" is not a precise number.

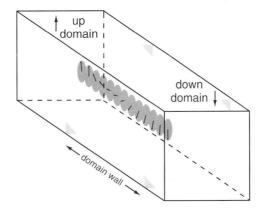

Fig. 15.9. Arrangement of atomic moments along one atomic row passing perpendicular to a domain wall.

Because the spins in a domain wall are not all parallel and are not aligned in the easy crystallographic directions, they are in a higher energy state than the spins in the bulk of a domain. Therefore, a domain wall has an energy per unit area that depends on the magnitude of the anisotropy and the exchange energies. Domain-wall energies are typically a few mJ/m^2 or a few erg/cm^2, although in special cases they can be hundreds of times larger.

15.7.3 Magnetization Processes

In pure, single-phase, stress-free materials, domain walls are relatively free to move. Thus, if a magnetic field is applied to such a material, the domains move and rearrange to produce a net magnetization, as sketched in Fig. 15.10. Eventually, the remaining domains will be magnetized along the local easy axis most nearly parallel to the applied field. If an even higher field is applied, the domain magnetization will rotate into the direction of the applied field.

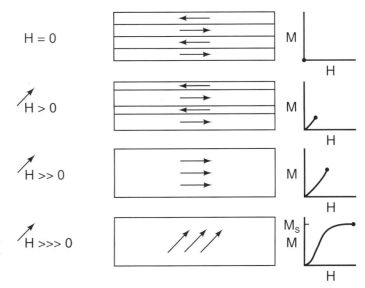

Fig. 15.10 Magnetization of a single crystal by domain wall motion, followed by magnetization rotation.

When the field is decreased, the rotation process is reversed, but the domain wall motion is not. A plot of magnetization *vs.* field then looks like Fig. 15.11, the familiar *hysteresis loop*. The magnetization remaining in the sample after a saturating field has been applied and then removed, M_r, is called the *remanence*. Often it is expressed as the *remanence ratio*, M_r/M_s. The negative field required to reduce the magnetization to zero is called the *coercive field* or *coercive force*, H_c. (Strictly, this is the intrinsic coercive field H_{ci}. The ordinary coercive field is the equivalent point on a plot of $B=\mu_0(H+M)$ *vs.* H. The two values are effectively the same except in the case of permanent-magnet materials with very high coercive fields, where they may be very different.)

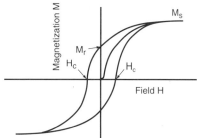

Fig. 15.11. Magnetic hysteresis loop, with reference points labelled. Hysteresis loops are usually plotted for dc magnetization, or very low frequency, but can be measured for any frequency.

The area enclosed by the hysteresis loop is proportional to the energy lost as heat when the magnetic material is magnetized first in the positive direction, and then in the negative direction, and then again in the positive direction. In ac devices such as motors and transformers, this energy loss can be substantial; it is part of the reason that motors and transformers run hot and may need provision for cooling.

15.8 The Curie Temperature

When a ferromagnet is heated, the exchange energy is increasingly offset by thermal energy. This causes the saturation magnetization to decrease with increasing temperature, slowly at first but then rapidly, until it reaches zero at a fairly well-defined temperature called the *Curie temperature,* as shown in Fig. 15.12. For most applications, a high Curie temperature is desirable. If the Curie temperature is too near the working temperature, the magnetic properties of the device will be strongly temperature dependent. Above the Curie temperature, a ferromagnetic material behaves as a paramagnet, with susceptibility decreasing rapidly with increasing temperature.

Fig. 15.12 Magnetization *vs.* temperature for iron, nickel, and cobalt, showing the Curie temperatures T_C.

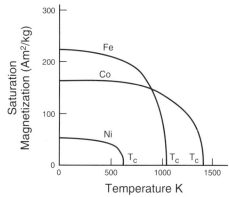

15.9 Magnetic Materials in the Walkman

15.9.1 The Tape Head

The recorded signal on the tape produces a small magnetic field above the surface of the tape, varying in magnitude and direction. When the tape passes by the gap in the playback head, this field magnetizes the material of the head, and the resulting change in magnetic flux density, B, generates a voltage in the winding on the head. This voltage signal, appropriately manipulated and amplified, drives the headphones and recreates the original recorded sound. The magnetic material of the head must be easily magnetized and demagnetized and must lose very little energy as heat. Therefore, it needs to have a steep and narrow hysteresis loop, as in Fig. 15.13.

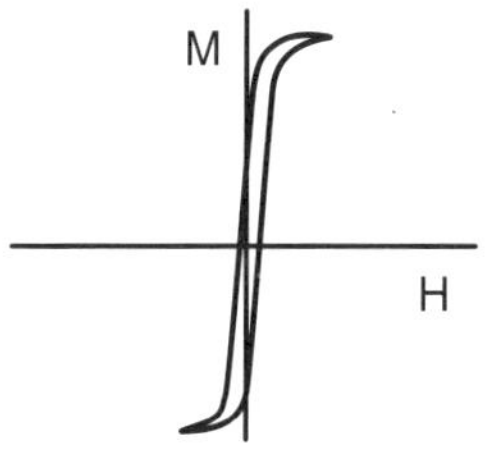

Fig. 15.13 Ideal hysteresis loop for a tape head.

The ease of magnetization is expressed by the magnetic *permeability*, defined as the slope of the line joining the origin and a point on the magnetization curve of B *vs.* H. The value of permeability varies with position on the magnetization curve, as shown in Fig. 15.14. The values usually quoted are the maximum permeability μ_m and the initial permeability μ_i. Permeability is usually given as a dimensionless relative permeability, μ/μ_0, which is equivalent to $B/\mu_0 H$.

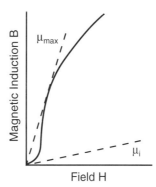

Fig. 15.14 Early part of a magnetization curve, showing the definitions of maximum permeability and initial permeability.

A material with high permeability and low coercive field is called a *soft magnetic material*. To obtain the required behavior, the domain walls must move easily and reversibly. This means generally high-purity materials, no second-phase particles, and minimum dislocation density. Face-centered cubic nickel-iron alloys called *permalloys* are often used. They have excellent properties but are generally too expensive for heavier equipment such as power generators or large motors. Ferrimagnetic oxides may also be used for tape heads. They have lower saturation magnetization, because the magnetic moments are partly cancelled and because they contain non-magnetic oxygen, as in Fig. 15.6(d). However, they retain their good magnetic properties to high frequencies (see the following section). The oxides also have better resistance than the soft permalloys to mechanical wear and abrasion by the moving tape.

Amorphous alloys or metallic glasses can also be used for tape heads. These are alloys of iron or iron and cobalt with about 15 to 20at% silicon and boron, produced by casting onto a rapidly rotating copper wheel, as in Fig. 15.15. The process is called *melt-spinning*. Under these conditions, solidification occurs so rapidly that the liquid structure is retained in the solid state for some special compositions. Amorphous alloys are metastable and will crystallize if heated enough, but they can be used at temperatures up to 150 to 200°C.

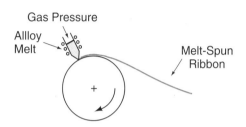

Fig. 15.15 Schematic representation of the melt-spinning process for producing an amorphous alloy (glassy metal).

15.9.2 Eddy Currents

When the flux density changes in a magnetic material, as when a domain wall moves, the application of Faraday's Law shows that internal voltages are produced. In a conducting material, these voltages drive local circulating currents known as *eddy currents* (from their mathematical similarity to eddies in water moving past an obstacle). Eddy currents flow in circular paths perpendicular to the direction of the changing flux density, and they generate heat that is the major source of energy loss in magnetic devices. Any device that operates at high frequency, like a tape head, may suffer from eddy-current losses.

Eddy-current losses can be reduced in two ways: by increasing the electrical resistivity of the material, e.g., by adding alloying elements, as in Ni-Fe permalloys and amorphous alloys, and by subdividing the material into electrically separate layers to break the eddy-current path. All metallic magnetic materials for use in ac devices such as motors, generators, and transformers are laminated for this purpose (see Fig. 15.1). The thickness of the laminations must be decreased as the operating frequency increases.

15.9.3 Ceramic Tape Heads

Some oxides are ferromagnetic; a schematic representation is shown in Fig. 15.16. Chromium dioxide CrO_2 is the best-known example. In this material, the chromium atoms are magnetic, and their moments are aligned parallel; the oxygen atoms carry no moment. Chromium dioxide particles are used in some audio and video tapes.

The most common magnetic oxides are iron-containing compounds known generically as *ferrites* (not to be confused with ferrite, the metallurgist's name for BCC iron). Most oxide ferrites have the *spinel* structure (Fig. 15.17), named for the mineral $MgAl_2O_4$.

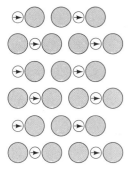

Fig. 15.16 Schematic representation of a ferromagnetic oxide. Shaded atoms are non-magnetic oxygen.

The magnetic ferrites have the general formula MFe_2O_4, where M can be any divalent transition metal atom. If M is Fe^{2+}, the resulting ferrite is magnetite, Fe_3O_4. The cubic unit cell contains eight formula units.* In the spinel structure the oxygen atoms, which are relatively large, are in an FCC arrangement. The smaller metal atoms occupy interstitial sites, which are of two kinds: octahedral (at the center of an octahedron of oxygens), and tetrahedral (at the center of a tetrahedron of oxygens, as shown in Fig. 15.17). There are 64 tetrahedral sites and 32 octahedral sites in each unit cell, but only 8 tetrahedral and 16 octahedral sites are normally occupied by metal atoms. In the normal spinel structure, divalent atoms occupy the tetrahedral sites and trivalent atoms occupy the octahedral sites. However, most of the magnetic ferrites are *inverse spinels,* with divalent atoms occupying half of the octahedral sites and trivalent atoms divided equally between the octahedral and tetrahedral sites. Intermediate cases are also possible. All of this is important, because the arrangement of the metal atoms, especially iron, controls the resultant magnetization of the ferrite.

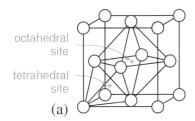
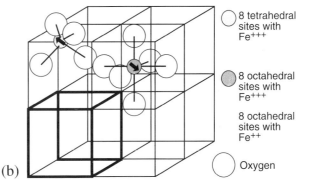

Fig. 15.17 Structure of ferrimagnetic ferrites. (a) Oxygen ions in an FCC array (b) The unit cell of Fe_3O_4 contains 8 of the cells shown in (a). Examples of iron-ion sites are shown.

The exchange interaction in ferrites is indirect, meaning that it operates not between neighboring metal atoms, but through an intervening oxygen atom. The exchange is predominantly negative, meaning that the moments of metal atoms separated by an oxygen atom tend to point in opposite directions. The strength of the exchange depends on the interatomic distances and also on the angle between the two metal-oxygen bonds. There are competing effects on the atomic moments, and working out the resultant magnetic structure is complicated.

---

* A formula unit comprises three iron and four oxygen atoms.

The net result, however, is simple. All moments of the tetrahedral sites are parallel, and all moments of the octahedral sites are parallel, but the two sets are antiparallel. The magnetic structure is, therefore, ferrimagnetic, and the resultant saturation magnetization is the difference between the total moment of the octahedral sites and the tetrahedral sites.

Taking the case of Fe_3O_4, an inverse spinel, we find the tetrahedral sites contain 8 Fe^{3+} ions, each with a moment of 5 μ_B, for a total moment of 40 μ_B, and the octahedral sites contain 8 Fe^{3+} (40 μ_B) and 8 Fe^{2+} ions, each with a moment of 4 μ_B (32 μ_B), giving a total octahedral site moment of 72 μ_B. Subtracting the tetrahedral site moment from the octahedral site moment gives a theoretical net moment of 32 μ_B per unit cell. The measured value (which depends to some extent on the exact method of preparation of the sample) is about 32.8 μ_B, an agreement with theory within 2.5%.

Since most of the atomic moments cancel, and since much of the volume of the unit cell is occupied by nonmagnetic oxygen atoms, the resulting saturation magnetization of ferrites is fairly small, less than one-third the value for metallic iron. However, the electrical resistivity is at least 10^8 times higher than that of iron, so the eddy-current losses are extremely small, and ferrites need not be laminated for high-frequency uses. The resistivity of ferrites, although always high, can vary greatly. An important mechanism of conductivity is *electron hopping;* an electron jumps from a divalent to a trivalent iron, exchanging the charges of the ions and moving an electron through the lattice.

One can make an infinite number of magnetic ferrites containing varying amounts of manganese, cobalt, nickel, copper, magnesium, lithium, etc., in addition to iron. Most of the ferrites used for engineering purposes fall into two groups: nickel-zinc ferrite, $NiZnO \cdot Fe_2O_3$, and manganese-zinc ferrite, which can be written $MnZnO \cdot Fe_2O_3$. (Manganese can substitute for trivalent iron as well as for divalent iron.) It seems odd to add nonmagnetic zinc atoms deliberately. The reason for it is that zinc strongly prefers octahedral sites and forces divalent iron from octahedral to tetrahedral sites, giving a net increase in saturation magnetization. Here, the addition of a nonmagnetic atom actually increases the magnetization.

15.10 Materials for Magnetic Recording Tape

Information is recorded on tape by the varying field from the gap in the recording head as the tape passes by the head. The time for a point on the tape to pass across the gap is $t = l_g/v$, where l_g is the length of the gap (measured parallel to the tape length) and v is the velocity of the tape. The maximum frequency that can be recorded is given approximately by $1/t$. For a maximum recorded frequency of 10 kHz (not very hi-fi) and a standard audio cassette tape speed of 1⅞ in/sec, the gap length works out to be 2×10^{-4} inch or 5 μm. Therefore, rather special manufacturing techniques are needed for mass production of tape heads. To record higher frequencies (or to pack more information onto a digital recording), fast tape speeds and small gaps are required.

The varying magnetization of the recorded tape creates regions of varying magnetic field above the surface of the tape. To retrieve the recorded information, the tape is pulled past a playback head, as shown in Fig. 15.18, and the field from the tape slightly magnetizes the magnetic material of the head. Changes in the magnetization of the head material generate voltage in the winding on the head, according to Faraday's Law, E=-NA dB/dt, as given previously. Note that the output of the head is the time derivative of the recorded signal, so the output voltage must be integrated to recover the signal. The electronic circuitry of the Walkman does this and also modifies the signal to correct for the frequency dependence of the playback voltage and amplifies the resulting signal so that it can drive the headphones. The power for all this comes from the battery, as does power for the miniature drive motor.

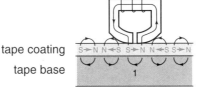
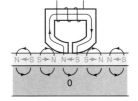

Fig. 15.18. A recorded signal produces a magnetic-flux change in the playback head.

tape coating
tape base

The most common magnetic material for recording tape is γ-Fe_2O_3, *maghemite*, used in the form of tiny elongated particles imbedded in a polymer binder. The particles are mixed in a liquid polymer that is coated onto the mylar base, and a strong magnetic field is applied parallel to the tape axis. The particles physically rotate in the field so that their long axes are parallel to the field and to the tape axis. When the binder hardens, the oriented magnetic particles are locked in place. Magnetic particles occupy about 40% of the volume of the magnetic layer on the tape. The binder also contains additives to reduce wear by abrasion as the tape passes over the playback head.

Maghemite is a metastable compound made by low-temperature oxidation of magnetite (Fe_3O_4). It retains the spinel structure, but the divalent iron sites are unoccupied. Maghemite transforms to α-Fe_2O_3 (hematite) when heated to about 400°C. (Magnetite itself is not used as a recording material, because it absorbs water from the air, changing its physical properties and density.) The recording particles, about 1000Å long and 200Å in diameter,* are made by a fairly elaborate series of chemical-process steps, often starting with sulfuric acid that has been used to clean steel sheet, and so contains a high concentration of Fe^{2+} ions. This liquid is known in a steel mill as *pickle liquor*.

Other materials can be used for recording tape. Chromium dioxide, CrO_2, is sometimes used, but it is expensive, since it must be formed under pressure. So-called *metal tape* consists of particles of iron or FeCo imbedded in a polymer binder. Evaporated metal tape, ordinarily used for video recording, comprises a thin continuous layer of a metal alloy, often FeP, coated onto the tape base by evaporation or some other method. Floppy disks for computer-data recording are made by the same process as tape, except the particles are not aligned. Hard disks are similar, except that the substrate is a rigid glass or aluminum alloy plate, and the magnetic coating is often a continuous metal film. The head of a hard disk drive does not actually touch the disk surface, but "flies" less than 1 μm above the disk surface supported on a dynamic layer of gas molecules. The disk surface must be extraordinarily flat and smooth and is usually permanently sealed into a dust-free case.

* One Angstrom unit, Å, equals 10^{-10} m.

The properties required in a recording tape are the following:

1. The remanent magnetization must be high, so that the recorded information produces a sufficient field above the tape. This in turn requires a material with a reasonably high saturation magnetization and remanence ratio.

2. The coercive field must be high enough so that the recorded information is not easily erased, but low enough so that the recording operation does not require an unreasonably high field. Higher recording densities require higher coercive fields, since small recorded regions tend to self-demagnetize. (That is, the field produced by the region acts to demagnetize it.) The coercive field of $\gamma\text{-Fe}_2\text{O}_3$ particles can be significantly increased by treating the particles with a hot aqueous solution containing cobalt ions. The surface layers of the particles are enriched in cobalt, which increases the coercive field, presumably because the crystal anisotropy is increased.

Each variety of tape requires somewhat different recording conditions and also different modifications of the playback signal. High-quality tape players have switch settings for different kinds of tape, and professional machines may allow for continuous "tuning" of the drive field and other parameters for optimum results.

Note that tape particles of maghemite are a few hundred Å in the smallest dimension. Typical domain wall thicknesses are about 1000 Å, so the particles are smaller than the thickness of a domain wall. This means that no domain walls can exist in a particle; it is a *single-domain particle* which is always magnetized to saturation. Just as the yield stress of a metal is high if no dislocations exist, the coercive field of a magnetic particle can be high if no domain walls exist. The coercive field of a single-domain particle depends on the anisotropy of the particle, since all the moments must rotate together and must pass through a hard anisotropy axis. The theoretical coercive field can be calculated quite simply, and in a few cases the calculation agrees with experiment. Usually, however, the measured coercive field is lower than the calculated maximum. The difference is attributed to a magnetization-reversal mechanism in which all the atomic moments are not strictly parallel.

15.11 Magnetic Materials for Motors and Headphones

Small high-efficiency motors, as well as loudspeakers and headphones, make use of permanent-magnet materials. These materials are used to create a permanent magnetic field in some (usually small) region of space. Examples are shown in the sketches of Fig. 15.19. A wire carrying a current in a direction perpendicular to a magnetic field experiences a mechanical force in a direction perpendicular to the wire and to the field. This force is used to make a motor spin or a loudspeaker cone vibrate to produce sound waves. Permanent magnets need high remanence to produce strong fields and high coercive fields to resist demagnetization. As in the case of recording materials, the strongest field acting to demagnetize the magnet is usually its own field. (The field produced by a magnet always acts to reduce the magnetization of the magnet; see Fig. 15.7b and c). Since the field of a permanent magnet does not change with time, the electrical resistivity is usually unimportant.

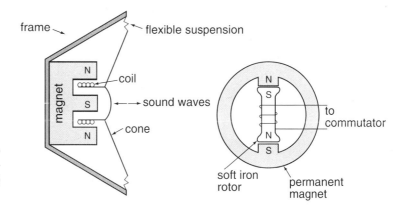

Fig. 15.19 Use of permanent magnets in (a) a loudspeaker (headphones are miniature loudspeakers) and (b) a simple motor.

Since a permanent magnet always works in its own demagnetizing field, the important part of its hysteresis loop is the second quadrant (B positive, H negative). Examples of this part of the M *vs.* H demagnetizing curves of some important permanent magnet materials are shown in Fig. 15.20.

At each point on the hysteresis loop, one can calculate a quantity called the *energy product,* BH. This product has a maximum value (somewhere between H=0 and B=0), called the maximum energy product, $(BH)_{max}$, in joule/m^3. This is the common figure-of-merit for permanent magnet materials. The volume of magnetic material required to produce a known field in a known space is inversely proportional to $(BH)_{max}$.

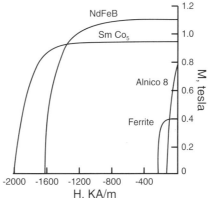

Fig. 15.20 Demagnetizing curves of some permanent magnet materials.

A permanent magnet material obviously needs high remanence and high coercive field. The remanence depends mostly on the saturation magnetization, which in turn depends on composition. The coercive field, however, is highly structure-sensitive. To increase the coercive field, there are two approaches. First, as noted above, the material can be subdivided into single-domain particles, which in theory have a coercive field determined entirely by the anisotropy of the material. In practice, this limit is rarely attained. A further difficulty with fine-particle magnets is that the space between particles, needed to insure that they behave as separate particles, necessarily dilutes the strength of the magnetization. A second approach is to produce a bulk (not fine particle) material, but treat it to make the domain walls immovable, or movable only in high fields. This is most often done by introducing a fine distribution of second-phase particles, comparable in size to the thickness of a domain wall, and having a magnetization very different from that of the bulk material. The particles act to pin the domain walls, and raise the coercive field.

The first man-made permanent magnets were steels, hardened by quenching. They were replaced, starting in the 1930s, by a family of complex precipitation alloys containing iron, aluminum, nickel, and cobalt, and often some zirconium or titanium. They are known generically as *Alnico,* from the chemical symbols Al, Ni, and Co. The best Alnico materials are solution treated and then precipitated in a strong magnetic field. This causes the strongly magnetic precipitating phase (rich in Fe and Co) to form as single-domain particles elongated in the direction of the field. The resulting magnets are anisotropic, with the best properties in the direction of the precipitating field. The Alnicos had a $(BH)_{max}$ 10 times higher than steel magnets; and they quickly replaced steel.

The most common permanent-magnet materials today are a class of iron oxide ferrites with a hexagonal crystal structure containing barium or strontium and known as *hard ferrites.* They have relatively low remanence, like all ferrites, but high coercive fields and are very inexpensive to make. Since they are ceramic materials, hard and brittle, they must be ground to size if exact dimensions are required. If lower magnetic properties can be tolerated, hard ferrites can be ground into a coarse powder and embedded in a matrix of rubber or a rigid polymer. These *bonded magnets* can be easily formed to almost any shape by conventional polymer injection-molding or extrusion equipment. This allows, for example, the production of magnetic door gaskets for refrigerators, replacing a complicated mechanical latch with a safe, simple, cheap, and foolproof sealing system.

The best available permanent magnets are based on intermetallic compounds of rare-earth metals and transition metals. Many of the rare-earth elements are ferromagnetic at low temperatures, with large atomic moments and high crystal anisotropy. Their compounds with iron and other transition metals are often ferromagnetic, with Curie temperatures well above room temperature and high anisotropy. The rare-earth magnets fall into three families: those based on $SmCo_5$, on Sm_2Co_{17} with added iron and zirconium, and on $Nd_2Fe_{14}B$. Of these, FeNdB is the newest, the least expensive, and the best, with a $(BH)_{max}$ five times that of the best Alnico. Its major weakness is a relatively low Curie temperature, just over 300°C, which makes the room-temperature properties rather temperature-sensitive. It is also hard and brittle and needs a protective coating to prevent degradation in humid atmospheres.

The rare-earth magnets are normally prepared by grinding the starting material into single-crystal particles a few micrometers in size, aligning the particles in a strong field, pressing the particles together under pressure, and then sintering at a high temperature. The particle size is larger than the single-domain size, but high coercive fields are obtained nevertheless. The rare earth elements form very stable oxides, so the particles need to be protected from exposure to air. The ease of oxidation also makes it difficult and expensive to produce very fine powders; the high surface-to-volume ratio leads to rapid oxidation.

In addition to their uses in miniature headphones, these high-quality permanent magnets are widely used in drive motors and stepper motors in hard-disk drives, printers, and other devices where small size and high efficiency are important.

Summary

The Walkman was made possible by the development of permanent magnet materials of high maximum energy product (high remanence and high coercive field) for miniature motors and headphones, as well as by the development of semiconductor electronics. In addition to permanent magnets, the Walkman uses a soft magnetic material for the playback head (either laminated NiFe alloy, laminated amorphous FeSiB, or bulk oxide ferrite), and a semihard magnetic material, usually ferrite particles, for the magnetic tape.

The magnetization curve and hysteresis loop of a magnetic material can be understood in terms of domain wall motion; the walls move so as to increase the volume of material magnetized in the direction of the applied field. At high fields, the magnetization rotates gradually away from the local easy direction toward the applied field. In soft magnetic materials, the aim is for the domain walls to move as easily as possible. This means using high-purity materials, minimizing elastic and plastic strains, and in some cases aligning the easy crystallographic directions. In hard magnetic materials, or permanent magnets, the aim is to prevent domain wall motion. This is done by introducing small second-phase particles to pin the domain walls, or by subdividing the material into regions too small to permit a domain wall to exist. Materials for use as recording media (tapes and disks) must have magnetic properties intermediate between soft and hard; they are sometimes called semihard materials.

Appendix 15.1 Some Units Employed for Magnetism

Quantity	Symbol	Gaussian	Conversion factor **	SI
Magnetic flux density, magnetic induction	B	gauss (G)	10^{-4}	tesla (T),
Magnetic flux	Φ	maxwell (Mx), G·cm²	10^{-8}	weber (Wb), volt second (V·s)
Magnetic field strength, magnetizing force	H	oersted (Oe)	$10^3/4\pi$	A/m (ampere-turn per meter)
Volume magnetization	M	emu/cm³ *	10^3	A/m
Volume susceptibility	X	dimensionless	4π	dimensionless
Permeability	μ	dimensionless	$4\pi \times 10^{-7}$	Wb/(A·m)

* emu means electromagnetic unit; strictly speaking, it is not a unit.
** Multiply a number in Gaussian units by the conversion factor to convert it to SI units.

Exercises Chapter 15

Terms to Understand

Magnetization
Flux density
Permeability
Dipole moment
Spin
Diamagnetic
Paramagnetic
Domain
Domain wall
Magnetic susceptibility
Hysteresis curve
Ferromagnet
Coercive Field
Ferrimagnet
Remanence

Recorded signal
Curie temperature
Playback head
Magnetic domain wall
Head gap
Single-domain particle
Antiferromagnet
Magnetic anisotropy
Alnico
Permalloy
Metallic glass
Laminations
Spinel
Inverse spinel

Problems

15.1 Does a tape recorder head made of ferrite need to be laminated (made from a stack of thin sheets)? Explain.

15.2 Canadian nickels (5 cent coins) are pure nickel, whereas U.S. nickels are about 70%Cu-30%Ni. The dimensions are identical. How do you think U. S. telephone companies guard against the use of Canadian nickels in pay telephones?

15.3 On the same set of M *vs.* H coordinates, draw a hysteresis loop for a good tape-head material and for a good permanent-magnet material. Explain your answers.

15.4 Given a piece of iron, describe possible treatments that would make it a soft or a hard magnetic material.

15.5 Describe the distinguishing features of the following: a paramagnetic metal, a ferromagnetic metal, a ferromagnetic oxide, and a ferrimagnetic oxide.

15.6 On the same set of B *vs.* H coordinates, draw (schematically) a hysteresis loop for iron at room temperature, 1020K, and 1060K. (The Curie temperature for iron is 1043K.)

15.7 Shown in Fig 15.8 are the magnetization curves for an iron single crystal in three low-index crystallographic directions, along with a hypothetical slice of a crystal cut parallel to a {110} plane. Given below is the domain structure in this slice at zero magnetization. (a) What phenomenon do the curves represent? (b) Show by further schematic sketches how the domain structure would change with increasing field intensity if the field is applied along the [001] direction. (c) Do the same for a field applied along the [01$\bar{1}$] direction.

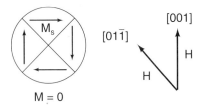

16
ELECTRONIC MATERIALS IN THE WALKMAN

16.1 Introduction

The Walkman in particular, and miniature electronic equipment in general, depend on semiconductor devices, mainly transistors. The first transistor was invented in 1947 at the Bell Telephone Laboratories, and the first commercial devices using transistors appeared in the early 1950s. Individual transistors are still made and used, but most transistors now are packed together in large numbers on small wafers known as integrated circuits. All electronic functions of the Walkman (integration of the voltage signal from the playback head, frequency compensation, and amplification) are accomplished in two small integrated-circuit packages on the circuit board shown in Fig. 16.1, along with indicator lights, the volume control, the headphone jack, and other components.

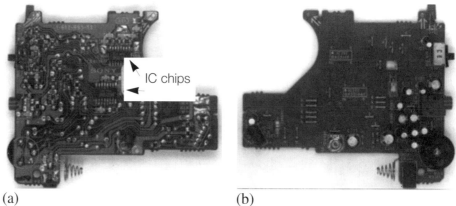

(a) (b)

Fig. 16.1 The circuit board of a Walkman, showing (a) the two packaged integrated-circuit chips and the wiring that connects the chips and (b) the associated resistors and capacitors on the reverse side of the board.

The operation of transistors depends on the electronic properties of semiconductors, which are a relatively small class of materials that are neither electrical conductors, like metals, nor electrical insulators, like most ceramics. Only two elements, germanium and silicon, are regularly used for transistors. However, a number of compounds have properties that are useful for other semiconductor devices; these include gallium arsenide (GaAs) and gallium phosphide (GaP). Such compounds comprise equal numbers of atoms from Group III and Group V of the periodic table, while the elemental semiconductors are found in Group IV. Compound semiconductors can also be made by combining one atom from Group II with one from VI (CdS, ZnTe). The reason for this will soon become clear.

The electrical properties of semiconductors are critically dependent on minute amounts of specific elements which are added deliberately and are generically known as dopants. That is, a semiconductor is said to be doped with B or P or some other appropriate element. Both the understanding of semiconductor behavior and the commercial production of semiconductor devices were made possible by advances in materials processing that allowed the production of materials with previously impossible levels of purity. Semiconductor-grade silicon is now routinely produced in large quantities with purity levels of 99.9999999 at%.

16.2 Electrical Conductivity

For reference, one should start with the basic equation of electrical behavior, Ohm's Law:

$$V = IR$$

where V is the electrical potential difference across a conductor (volts); I is the electric current (amperes), and R is the electrical resistance of the conductor (ohms).

The electrical resistance R depends on the dimensions of the conductor and on the material of which it is made, as given by

$$R = \rho \frac{L}{A}$$

where L (m) is the length of the conductor and A (m²) is its cross-sectional area. The quantity ρ is a property of the material, known as the *electrical resistivity;* its units (SI) are ohm-meters.

In dealing with semiconductors, it is more useful to work with the reciprocal of the resistivity, known as the *conductivity,* σ. The unit of conductivity is officially the siemens, but it is usually expressed as (ohm-m)$^{-1}$ or reciprocal ohm-m. Thus,

$$\sigma = \frac{1}{\rho}$$

The electric current in a metal, for example, is a flow of electrons and may be defined as the total electric charge passing a fixed point in unit time:

$$I = \frac{q}{t} = \frac{ne}{t}$$

where the total charge q is the number of electrons n times the charge of one electron e (1.6x10^{-19} coulomb).

Some of the electrons in a conducting or semiconducting material are in constant random motion, even in the absence of a potential gradient. When an electric field is applied, this random motion of the electrons becomes biased, so that there is a net drift in the direction away from the negative side of the applied field. The electrons can be said to have an average velocity, usually called a drift velocity, that is proportional to the applied field:

$$v = \mu E$$

where v is the drift velocity (m/sec); μ is the *mobility* of the electron [(m/sec)/(V/m)], or [m²/Vsec]; and E is the electric field gradient (V/m).

The conductivity can be defined in a fundamental way as

$$\sigma = n e \mu$$

where n is the number of free (or conducting) electrons per unit volume. As will be seen, conductivity in semiconductors can be controlled by manipulating the number of charge carriers n. In this case n can refer to either electrons or empty sites, called *holes*, into which electrons may move in a way analogous to self-diffusion in crystals (cf. Chapter 6).

In very pure metals, the resistivity is controlled almost entirely by thermal vibrations of the structure, referred to by physicists as *phonons*. A phonon is a localized lattice vibration (the energy of which is quantized). The number of phonons in a crystal lattice increases with temperature. Because a phonon represents a perturbation in the periodicity of the crystal lattice and, therefore, of the periodic electron density (i.e., of the "electric field") in the crystal, conduction electrons are scattered by phonons. Experiments have shown that the resistivity varies approximately linearly with temperature, starting from a small value at zero degrees Kelvin, as shown in Fig. 16.2. This means that the phonons act as individual scattering centers. The variation of electrical resistivity with temperature in a metal follows the relationship

$$\rho(T) = \rho_0 (1 + \alpha T)$$

where α is the temperature coefficient of resistivity.

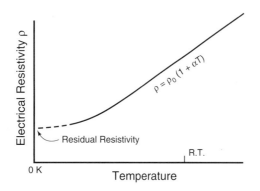

Fig. 16.2 Schematic representation of the variation of electrical resistivity with temperature in a good metallic conductor.

The addition of alloy or impurity solute elements to a pure metal raises the resistivity, usually in proportion to the amount of added element, indicating that the impurities scatter independently; an example is shown in Fig. 16.3.

When large amounts of alloying elements are added to a metal, not only does the resistivity become large, but the temperature dependence of resistivity becomes small. This is the desired behavior in materials used for heating elements in stoves, ovens, and furnaces, where a high electrical resistance in a conductor is desired. Since these elements typically operate at high temperatures, the alloy selected must also have good resistance to oxidation.

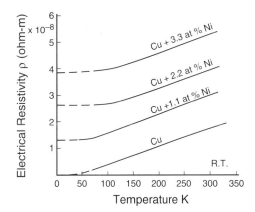

Fig. 16.3 The effect of additions of small amounts of nickel to pure copper on the resistivity vs. temperature behavior. Note that the slope of the curve remains unchanged, indicating that the nickel atoms and phonons all scatter independently. (From J. O. Linde, *Ann. Physik,* vol. 5, 1932, p. 219.)

16.3 Electron Energy Bands and Conductivity

To understand the behavior of semiconductors even in the most elementary way, it is necessary to use the band model of solids, which was introduced in Fig. 10.2. Recall first the conventional picture of an isolated atom, with electrons occupying a series of discrete energy levels which are occupied starting from the lowest level and working up as electrons are added to build an atom of a given atomic number. One of the basic principles of quantum mechanics is that only two electrons, which must have opposite spin, can occupy each energy level; this is known as the Pauli Exclusion Principle. When a large number of (identical) atoms are assembled into a solid, the number of valence, or non-core, electrons in the system is equal to the atomic valence times the number of atoms in the solid. To avoid violating the Exclusion Principle, the number of energy levels has to be large enough to accommodate all these electrons. The result is that the energy difference between adjacent levels becomes so small that the discrete levels essentially degenerate into a quasi continuum, known as an *energy band*.

One of the two possible configurations of a metal was represented in Fig. 10.2, in which the valence band is only partially filled; this is represented in a different way in Fig. 16.4(a). In this configuration, electrons in the uppermost occupied levels have empty levels very close by, into which they can be excited by the application of an electric field. This is what enables metals to be good electrical conductors.

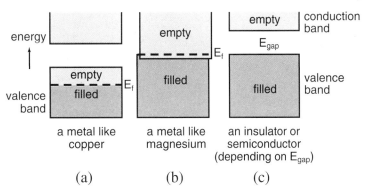

Fig. 16.4 Schematic representation of energy bands (a) and (b) in metals and (c) in insulators or semiconductors.

The energy level at the top of the filled states of the valence band in a metal is called the *Fermi level*. At zero degrees Kelvin the states are fully occupied up to the Fermi level, but at any higher temperature the thermal energy promotes some electrons into higher levels, and the energy-level occupancy near the Fermi level becomes diffuse.

The other possible configuration for a metal is a filled valence band that overlaps with the bottom of the next higher-level band, as shown schematically in Fig. 16.4(b). In this case, good electrical conductivity can still occur, due to the availability of unoccupied energy levels near the Fermi level.

Nonmetals are distinguished from metals by the characteristic that the valence band is completely filled and there is no easy access to empty energy levels above the Fermi level. This is represented by the band gap in Fig. 16.4(c), which separates the filled valence band from the empty band above it, called the *conduction band*. If this band gap has a width greater than 2eV, the material is considered to be an insulator.

The electron-volt, eV, is the energy imparted to an electron by a potential difference of one volt:

$$1 \text{ eV} = 1.6 \times 10^{-19} \text{ coulomb} \times 1 \text{ volt} = 1.6 \times 10^{-19} \text{ Joule}$$

If the band gap is less than 2eV, the material is considered to be a semiconductor, because, at temperatures of a few hundred degrees centigrade and below, there is a reasonable probability that electrons at the top of the valence band will be sufficiently excited by thermal energy to cross the gap into the conduction band.

Once in the conduction band, the electron can migrate in an electric field, as in a metal, because there are available unoccupied states into which it can be excited by an applied electric field. At very high temperatures or at very high fields, this can also happen in an insulator, in which case the insulator is said to have reached the "breakdown" voltage.

16.4 Intrinsic Semiconductors

In a pure, or intrinsic semiconductor, an electron is excited from the valence band to the conduction band by thermal energy, as noted above and as illustrated in Fig. 16.5. The electrical conductivity increases with increasing temperature, as increasing numbers of electrons gain sufficient thermal energy to cross the band gap. This is in contrast to metals, which show decreasing conductivity with increasing temperature. Referring to the definition of conductivity ($\sigma = ne\mu$), we note that in metals the number of conducting electrons is independent of temperature, but the mobility decreases with increasing temperature (due to scattering by phonons). Thus, the temperature dependence of the mobility controls the temperature dependence of the conductivity in metals. In semiconductors, the number of charge carriers increases strongly with increasing temperature, and this effect outweighs the much smaller (and opposite) temperature dependence of the mobility.

Because the number of electrons that jump the energy gap in an intrinsic semiconductor depends on the thermal energy, it depends exponentially on the temperature. The equilibrium number of electrons in the conduction band n_e at any temperature, T (in degrees Kelvin), is given by

$$n_e = n_0 \exp\left(-\frac{E_g}{2kT}\right)$$

where E_g is the magnitude of the energy gap, and n_0 is the number of electrons that cross the gap at infinite temperature (i.e., the maximum possible number). The factor of 2 in the denominator comes about because, in a semiconductor, the Fermi level is considered to be located not at the top of the valence band, but in the middle of the energy gap, halfway between the top of the valence band and the bottom of the conduction band.

It is important to realize that, when an electron jumps from the valence band into the conduction band, it leaves behind an unoccupied energy state, i.e., a hole. Since a hole is merely the absence of an electron, it can be regarded as having a positive charge; this is illustrated schematically in Fig. 16.5(b). Any other electron in a nearby energy state in the valence band can move into this hole, just as an atom in a crystal can hop into an adjacent vacancy. In a manner analogous to self diffusion in a crystal, upon the application of an electric field, the hole can move through the crystal by way of energy states in the valence band, just as a mobile electron can move through the crystal by way of energy states in the conduction band.

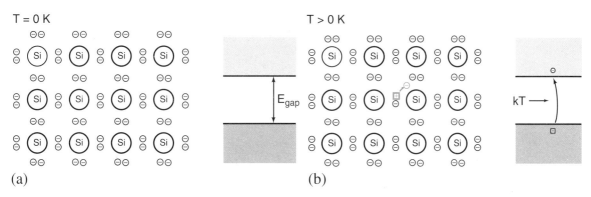

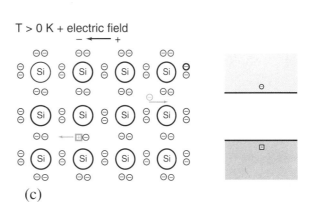

Fig. 16.5 Schematic representation of silicon: (a) At zero degrees Kelvin, where all valence electrons are localized in covalent bonds (i.e., the valence band is filled, and the conduction band is empty); (b) at a higher temperature, where thermal energy has created an electron-hole pair by the excitation of a valence electron into the conduction band; (c) at the higher temperature upon the application of an electric field, which causes the hole to drift in the negative direction and the now-delocalized electron to drift in the opposite direction. (Adapted from W.D. Callister, Jr., *Materials Science and Engineering,* John Wiley, New York, 1991, p. 620. Reprinted by permission of John Wiley & Sons, Inc.)

It is apparent that in an intrinsic semiconductor there are two kinds of charge carriers: electrons and holes. The number of (mobile) electrons must be exactly equal to the number of holes. In fact, the excited electrons are continually dropping back across the energy gap and recombining with holes, but new electrons and holes are continually being created by the thermal energy (kT), so the equilibrium concentration of electrons and holes remains the same (at constant temperature).

Because there are two kinds of charge carriers with different mobilities, the definition of the conductivity for an intrinsic semiconductor must be written as

$$\sigma = n_e |e| \mu_e + n_h |e| \mu_h$$

Because $n_e = n_h$, this can be simplified to

$$\sigma = n |e| (\mu_e + \mu_h)$$

The electron mobility, μ_e, is always significantly greater than the hole mobility, μ_h.

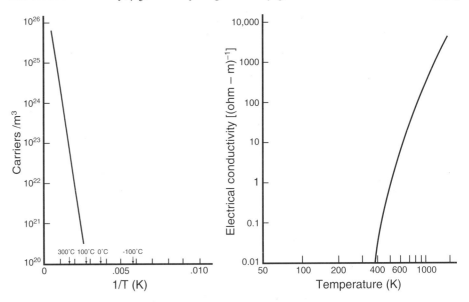

Fig. 16.6 Variation of (a) carrier density with inverse temperature and (b) electrical conductivity with temperature for intrinsic (i.e., pure) silicon. (Adapted from G. L. Pearson and J. Bardeen, *Phys. Rev.* vol. 75, 1949, p. 865.)

The temperature dependence of the conductivity of an intrinsic semiconductor, ignoring the relatively minor effect of temperature on μ, is essentially the same as the temperature dependence of n. This can be written as

$$\ln n = \ln n_0 - \frac{E_g}{2kT}$$

or

$$\log n = \log n_0 - \frac{E_g}{2(2.3)kT}$$

The temperature dependence of the conductivity for pure silicon is shown by the log-log plots of carrier density *vs.* 1/T in Fig. 16.6(a) and σ *vs.* T in Fig. 16.6(b). The slope of the former plot gives the value of E_g, and the intercept on the ordinate gives n_0.

16.5 Extrinsic Semiconductors

If a small concentration of foreign atoms (from some group of the Periodic Table other than Group IV) is added to a semiconductor, important changes in the electronic structure occur. Consider first the addition of an element from Group V, like phosphorus, to pure silicon. The added atoms have one more electron per atom than does silicon, so the band structure is perturbed in the immediate neighborhood of the phosphorus atom, and the extra electron occupies its own particular energy level, called a *localized state,* somewhere in the energy gap, as shown schematically in Fig. 16.7(a). The usefulness of phosphorus as a dopant in silicon lies in the fact that the localized state is near the top of the band gap (i.e., near the bottom of the conduction band) of silicon. The electron in the localized state needs only a small amount of activation energy to jump the intervening energy gap into the conduction band (Fig. 16.7(b)). That is, this electron is bound relatively weakly to the phosphorus atom, because it does not participate in the covalent bonding of the crystal. Once in the conduction band, it is free to move in an applied electric field, and, since one electron cannot be distinguished from another, it has the same mobility as the intrinsic electrons in the conduction band. The dopant is said to be a *donor* atom, because it donates an electron to the energy level near the conduction band. This level is called a *donor level,* or *donor state.*

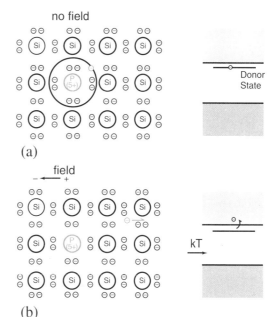

Fig. 16.7 Schematic representation of (a) a silicon crystal doped with phosphorus, showing the extra (donor) electron weakly bound to the phosphorus atom, and (b) the drift of this donor electron in an applied electric field. (Adapted from W.D. Callister, loc. cit., p. 622. Reprinted by permission of John Wiley & Sons, Inc.)

The electron hole left behind in the localized state is unable to move in the applied field, because too much energy is required to promote an electron from the valence band into that state (i.e., almost as much activation energy as for intrinsic conduction). Therefore, the number of mobile electrons is not equal to the number of holes. The level of doping can be made large enough that the number of conduction electrons coming from the donor states is much larger than the number of intrinsic conducting electrons. Under these conditions, the conductivity is controlled by the doping level and is almost entirely due to mobile electrons, rather than to holes. In other words, a larger, extrinsic, effect has been superimposed on the intrinsic behavior, and the doped silicon is called an *extrinsic semiconductor*. Phosphorus-doped silicon is called an *n-type* extrinsic semiconductor, because the charge carriers are primarily negative.

When the energy gap between the localized states and the bottom of the conduction band is small, thermal energy (kT) at room temperature can be sufficient to excite essentially all of the donor electrons into the conduction band. This is called the exhausted state, and it has the important consequence that the conductivity becomes independent of temperature over an extended range of temperature. Figure 16.8(a) shows schematically the temperature dependence of conductivity of an n-type extrinsic semiconductor compared with that of a pure (intrinsic) semiconductor.

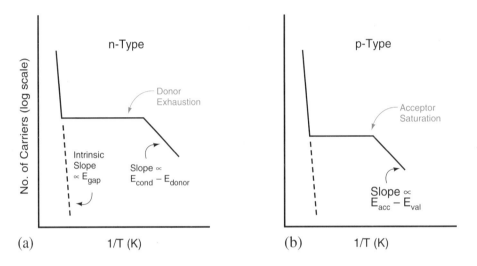

Fig. 16.8 Schematic representation of the temperature dependence of the number of carriers in (a) n-type silicon and (b) p-type silicon, compared with pure silicon.

Silicon may also be doped by the addition of a Group-III element, like boron for example. In this case, as depicted in Fig. 16.9, one of the covalent bonds linking a boron atom to the surrounding silicon atoms is missing an electron, because the boron atom has only three valence electrons to contribute to these bonds. This missing electron is nearly equivalent to a built-in hole in the valence band, and this hole is weakly bound to the boron atom. An electron from a neighboring Si-Si covalent bond could jump into the hole at the boron atom if its energy were raised a small amount from the top of the valence band. In other words, the boron atom has introduced a new energy level just above the valence band which can accept one of the valence electrons. This new state, illustrated schematically in Fig. 16.9(a), is called an *acceptor state*. If this process were to occur, a "normal" hole would then have been created in the valence band, and this hole could migrate in an electric field, as illustrated in Fig. 16.9(b), just as any hole would in pure (intrinsic) silicon.

When silicon is doped with a Group III element, it becomes a *p-type* extrinsic semiconductor, because the dominant charge carrier is the hole introduced with each Group III atom, and a hole is equivalent to a positive charge. Thermal activation can supply the energy needed for electrons to move into the acceptor levels, and when the temperature is reached at which all the acceptor levels are filled, the semiconductor is said to have reached the saturated state. Here, as in the exhausted state in n-type silicon, conductivity becomes essentially independent of temperature, as shown in Fig. 16.8(b).

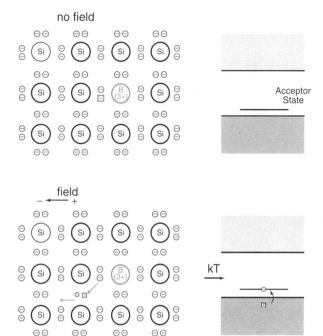

Fig. 16.9 Schematic illustration of (a) a boron (Group III) impurity in a silicon crystal showing the missing electron in one of the covalent B-Si bonds, which constitutes a special built-in hole into which a valence electron could jump if its energy were raised to the acceptor level, just above the valence band. (b) shows how the newly created "normal" hole (i.e., in a Si-Si bond) would then migrate in an applied electric field. (Adapted from W.D. Callister, loc. cit. p. 624. Reprinted by permission of John Wiley & Sons, Inc.)

The integrated circuit chips in the Walkman employ both n-type and p-type silicon, and it is important that the performance of the device not vary between summer and winter. That means the semiconductors should be in their exhausted and saturated states, respectively, in the temperature range of, say, -40 to +40°C. Therefore, the energy gaps between the donor levels and the conduction band, and between the acceptor levels and the valence band should be such that there is sufficient thermal energy at 233K to achieve exhaustion and saturation. It is also important that intrinsic behavior of the silicon not become significant until well above 313K. Otherwise, the extrinsic behavior would be swamped by the intrinsic electron-hole pair creation, and the performance of the device would become highly temperature dependent. Fortunately, this would require temperatures of several hundred degrees centigrade for normally doped silicon.

16.6 Purification and Crystal Growth of Silicon for Semiconductors

The room-temperature conductivity of silicon is extremely sensitive to the concentration of acceptor or donor impurities, as indicated by Fig. 16.10 for the cases of boron and phosphorus. Typical applications call for dopant concentrations in the range 10^{-4} to 10^{-2} at%, which corresponds to the range 1 to 100 parts per million (ppm). Therefore, it is necessary to start out with extremely pure silicon before doping. Otherwise, the electronic behavior would be uncontrollable, due to the effects of unwanted impurities.

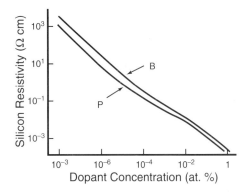

Fig. 16.10 The variation of room-temperature electrical resistivity of silicon with the concentration of phosphorus or boron. (From C.R.M. Grovner, *Materials for Semiconductor Devices*, Inst. of Metals, London, 1987, p. 17.)

It is also important not to have crystal defects that trap (i.e., immobilize) the charge carriers, because this would also make the conductivity uncontrollable. Such traps occur wherever there is a "dangling" silicon bond, which is an electron not participating in covalent bonding. Dangling bonds occur in the cores of dislocations and in grain boundaries, so it is necessary to produce fairly large pieces of monocrystalline silicon having a very low dislocation concentration.

The methods to produce large, pure silicon crystals with low dislocation concentrations were worked out by some of the early pioneers in materials science. William Pfann, a metallurgist at the Bell Telephone laboratories, invented zone refining, and large single crystals can be grown by a related method. These procedures will be described briefly here, because they are classic examples of major developments in the area of materials processing, which is often the route connecting the physically possible with the commercially available.

16.6.1 Zone Refining

This method of purification relies on the fact that, when a liquid freezes, the solid that forms often is very different from the remaining liquid with regard to the concentration of impurities. This is because a solute can either lower or raise the liquidus temperature, as depicted in Fig. 16.11. The melting temperature of silicon is so high that most solutes lower the liquidus, as in Fig. 16.11(a). In this case, the solid that forms at a certain temperature contains much less solute than the starting liquid, and the liquid which remains is enriched by the solute rejected from the solid, as discussed in Chapter 8. At a fixed temperature of solidification the solid forms with a fraction, k, of the concentration of the solute in the liquid, where k is called the *distribution coefficient* and is defined by

$$k = \frac{C_s}{C_l}$$

where C_s and C_l are the equilibrium solute concentrations in the solid and liquid, respectively, at the particular temperature. If the solidus and liquidus curves can be approximated as straight lines over a certain temperature range, then k is constant over this range. The smaller the value of k, the greater is the purification effect upon solidification. For example, potable water can be obtained by solidifying sea water, due to the very small value of k in the H_2O-NaCl system (cf. Fig. 8.5).

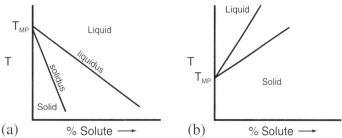

Fig. 16.11 A solute can either (a) lower or (b) raise the liquidus temperature.

The zone-refining process consists of passing a concentrated heat source along a solid bar such that a molten zone is made to pass from one end of the bar to the other, as in Fig. 16.12(a). The solid that forms behind the molten zone is depleted of impurities with k<1 (as in Fig. 16.11 (a)); $C_s = k C_l$ and, therefore, $C_s < C_l$). Because the molten zone is continually being enriched in rejected solute, C_s rises gradually along the bar, as shown schematically in Fig. 16.12(b). (It is assumed here that diffusive mixing in the liquid is rapid, so there is only a negligible buildup of solute at the advancing solid/liquid interface.) The end result is a sweeping of the solute along the bar into the end that solidifies last.

After one pass the concentration at the end of the bar that solidified first is kC_0, where C_0 is the initial concentration of the impurity. Every time this process is repeated, this end of the bar is reduced in solute concentration by a factor of k, so that after, say, ten passes, most of the solute has been moved from this end to the other, as shown in Fig. 16.12(b). The number of passes employed and the length of the purified end of the bar which can be used for semiconductor-grade silicon depends on the particular purity requirements.

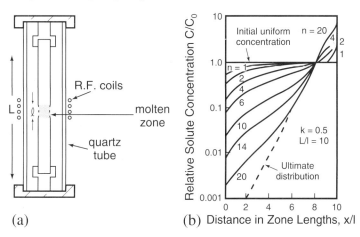

Fig. 16.12 (a) Apparatus for zone refining of silicon. (b) Distribution of solute with k<1 in a bar after one or more zone-refining passes. ((a) From A. Bar-Lev, *Semiconductors and Electronic Devices,* Prentice-Hall International, 1984, p. 3. (b) After W.G. Pfann, *Zone Melting,* John Wiley, N.Y., 1958, p. 290.)

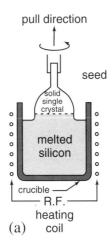 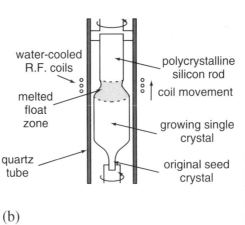

Fig. 16.13 Methods for single crystal growth. (a) is the Czochralski pulling technique, and (b) is seeded zone refining. (From A. Bar-Lev, loc. cit., p. 3. Reproduced with permission.)

16.6.2 Growth of Single Crystals

There are two common ways to grow a large single crystal of silicon. Each employs a starting "seed" crystal, which in one case, Fig. 16.13(a), is dipped into a molten bath of silicon and then slowly extracted so that the silicon solidifies onto the seed crystal. In this way the orientation of the large crystal can be controlled by the orientation of the seed. The other method utilizes the same apparatus as zone refining, as shown in Fig. 16.13(b), and the large single crystal is obtained by starting the melting from a seeded end. (In the latter method, purification and crystal growth are combined.) Wafers are sliced from the single-crystal "logs" produced by either method and are used as the starting substrate for integrated circuit chips.

16.7 Semiconducting Devices

Although the vast majority of devices based on semiconductors use extrinsic semiconductors, there are some important applications of intrinsic semiconductors. Two common uses are the thermistor, a kind of electronic thermometer, and the photoresistor, which can be used in photocells (e.g., "electric eyes" and light meters). Because the electrical conductivity of an intrinsic semiconductor is such a strong function of temperature (Fig. 16.6), a simple measurement of the electrical resistance can be used to measure temperature. This is the idea behind the thermistor. Because the resistance changes so enormously with temperature, there are practical difficulties in measuring a wide range of temperatures. For this reason, thermistors are most useful in biological work, where very small changes in temperature may be important, but wide temperature ranges are not encountered.

If the band gap of an intrinsic semiconductor is of an appropriate magnitude, the energy of a photon of visible light may be sufficient to excite electrons across the gap. This can increase the number of charge carriers well above the number produced by thermal excitation alone, and the effect on the electrical conductivity is large.

A semiconductor for which conductivity changes with the intensity of light falling on the material is called a photoresistor or a photoconductor. Such materials function as sensitive light meters and are used in many automatic-exposure cameras. A photoresistor needs a source of current in order that the resistance can be measured, and this requires that the camera be equipped with a small battery (cf. Section 16.11). There is another kind of semiconductor light meter, based on a solar cell (cf. Section 16.10.2), that actually generates power proportional to the intensity of the incident light and so does not require a battery.

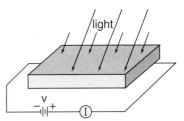

16.7.1 Junction Devices; The Rectifier

Most semiconductor devices rely on the properties of a boundary between an n-type and a p-type semiconductor. This is called a semiconducting *junction,* and the resulting device is a junction device. Consider first a single p-n junction, as illustrated in Fig. 16.14(a). The right-hand side is n-type, with electrons as the majority charge carriers; the left-hand side is p-type, with holes as the majority charge carriers. Consider what happens when a voltage is applied, as in Fig. 16.14(b), with positive potential at the left and negative at the right. The mobile electrons are attracted to the positive electrode on the left, and, since nothing prevents them from crossing the junction (it is nothing but a very slight change in composition), they do so, trying to reach the positive electrode. However, the high concentration of electrons in the conduction band of the n-type material is out of place in the p-type; i.e., it is greater than the equilibrium concentration. Once they are across the junction, they start to drop across the band gap and combine with holes in the p-type material. In a similar way, when the holes in the p-type material cross the junction, heading for the negative electrode, they are annihilated by electrons dropping from the conduction band on the n-type side. Thus, electrons and holes are continually recombining in a narrow region on each side of the junction. That is, the extrinsic carriers annihilate one another and disappear in the vicinity of the junction. However, as electrons are annihilated by recombination, more enter the device at the negative electrode, driven by the voltage source. Also, as holes are eliminated at the junction by recombination, more are generated at the positive electrode as electrons are removed from the device (i.e., as current flows). Thus, electrons enter the device at one electrode and leave at the other, and the device acts as an electrical conductor.

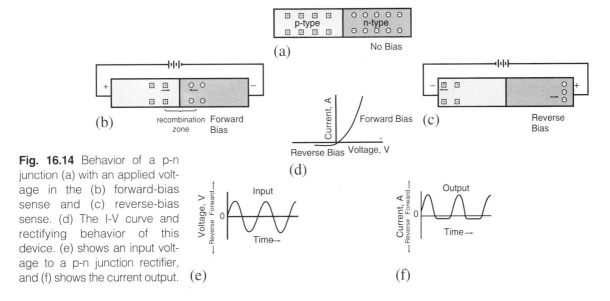

Fig. 16.14 Behavior of a p-n junction (a) with an applied voltage in the (b) forward-bias sense and (c) reverse-bias sense. (d) The I-V curve and rectifying behavior of this device. (e) shows an input voltage to a p-n junction rectifier, and (f) shows the current output.

If the applied voltage is reversed, as in Fig. 16.14(c), a completely different situation arises. The mobile electrons are attracted to the positive electrode and so move to the right; the positive holes are attracted to the negative electrode at the left. At the junction, there is created a region in which there are virtually no available charge carriers. Thus, after the initial rearrangement of the charge carriers, no current flows, and the device is almost an electrical insulator (except for the very small intrinsic conduction).

This device, therefore, is one that conducts electricity in one direction but not the other; it is called a *rectifier*. The current *vs.* voltage characteristics of a rectifier are shown in Fig.16.14(d). Rectifiers are used to run a Walkman from the 120 volt ac wall outlet. Since the Walkman is designed to operate with small dc voltages, such as are provided by dry cells, the ac voltage must be rectified to dc (as well as reduced from 120 to 6 or 9 volts). A set of four semiconductor rectifiers, connected in a bridge circuit and combined with a transformer, as shown in Fig. 16.15, accomplishes the rectification and voltage reduction needed to power a Walkman from house current.

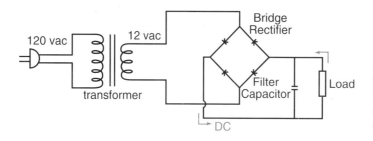

Fig. 16.15 A low-voltage dc power supply for a Walkman. The transformer converts the 120 volts to 12 volts, and the bridge rectifier converts the ac to dc. The filter capacitor smooths the pulsating voltage.

16.7.2 The Junction Transistor

A *transistor* is a device comprising two semiconducting junctions, arranged either as p-n-p or n-p-n, with an electrical connection made to each section, as depicted schematically in Fig. 16.16. It is a three-terminal device which can be used to amplify a small signal into a larger one. A series of transistors amplifies the small signal picked up from the recorded tape by the playback head of the Walkman and ultimately produces a signal strong enough to drive the small headphones.

To understand how the transistor in Fig. 16.16 operates, first observe that the (left-hand) junction between the p-type *emitter* and the n-type *base* is biased in the forward direction, and that the (right-hand) junction between the base and the p-type *collector* is reverse-biased. This means that, if any holes get from the emitter side to the collector side before being annihilated in the base, a current would flow through the transistor. Now, turn on the input (e.g., a signal from the tape head); this is a varying voltage which either adds to or subtracts from the forward-bias voltage on the emitter side. When the bias voltage is increased, the probability that a hole makes it through the (thin) base is increased, and more current flows through the transistor. Conversely, when the bias voltage is opposed during the negative part of an input cycle, less current flows through the transistor. The doping concentrations, the bias voltages, and the thickness of the base can be arranged so that a change in the small input voltage can produce a much larger change in the (much larger) output voltage, which is applied to the load (e.g., the headphones). This is how the transistor acts as an amplifier. The same effect can be produced by an n-p-n arrangement; the main difference is that electrons, rather than holes, flow from the emitter across the base into the collector.

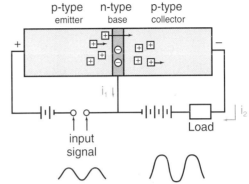

Fig. 16.16 Schematic representation of a p-n-p junction transistor. The number of holes crossing the thin n-doped region (the base) is controlled by the input signal voltage. The flow of holes into the collector is equivalent to the output current which flows through the (load) resistor. The small input signal, therefore, modulates the output signal.

16.7.3 The Field-Effect Transistor

Another type of transistor used as an amplifier is shown in Fig. 16.17. Two islands of p-type silicon are embedded in n-type silicon and are connected by a shallow p-type channel. Electrical connections are made directly to the islands, which are known as the *source* and the *drain*, respectively. A third electrode is placed above the channel, but is separated from it by a thin layer of SiO_2; this part of the device is called the *gate*.

The output of the field-effect transistor (FET) is governed by the current (of holes in Fig. 16.17) which flows from the source to the drain. This current is modulated by a small varying voltage (the input) between the metal contact at the gate and the n-type silicon substrate. When the input voltage is positive, the holes flowing along the channel are driven across the channel and into the substrate, where they are lost by recombination. Thus, the current flowing along the channel is reduced. This setup can be arranged so that a change in the gate voltage produces a much larger change in the output. The presence of the oxide layer at the gate keeps the current in the gate circuit very low. Thus, the FET is designed for applications in which the input current must be kept low, as is the case with the Walkman.

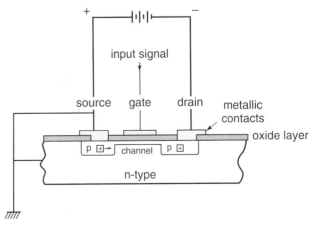

Fig. 16.17 Schematic illustration of a metal-oxide-silicon field-effect transistor (a MOSFET).

The gate depicted in the FET in Fig. 16.17 consists of layers of metal (usually aluminum), oxide, and silicon; therefore, the device is called a MOSFET. One reason why silicon is so widely used for FETs is that it is easy to grow a layer of SiO_2 on silicon in a controlled and reproducible manner.

16.8 Integrated Circuits

The invention of the semiconductor devices described in the previous sections ushered in the age of solid-state electronics, as distinguished from the previous age of vacuum-tube technology. This would not have been sufficient to produce the Walkman, which had to wait not only for advances in permanent-magnet materials, but also for the development of microelectronics, or integrated-circuit technology, in which a very large number of electronic components can be crowded together in a very small space and operated with very low power. The main driving force for this technology was the aerospace program, in which these features were absolutely essential.

Miniaturization was needed not only to conserve weight and space, but also to minimize the conduction paths between components in order to increase the operating speed and also to minimize the heat generation, or I^2R heating. The integrated circuit required a new kind of materials processing. First, new levels of cleanliness were essential, because the components were so small that dust particles could easily result in defects. Second, the close spacing required high precision in the location of the components. Third, high reproducibility and reliability were necessary, because a defect in any one of a large number of components could mean that the whole assembly would be rejected. These problems were gradually solved by the development of ultra-clean manufacturing facilities, automated systems, the frequent use of high-vacuum chambers, and the process of high-definition photolithography to locate components that were formed by deposition from a vapor phase. This new kind of manufacture was called thin-film technology, because the thickness of the deposited layers was often on the order of tens or hundreds of nanometers (10^{-9} meters).

To illustrate the kinds of materials processing involved in making integrated circuits, the MOSFET depicted in Fig. 16.17 will be used as an example. Keep in mind that this would be only one component of the hundreds or thousands that are formed simultaneously on the substrate of the integrated circuit. The substrate could be a thin wafer cut from a silicon crystal, or it could be a wafer of some inert material, such as a sapphire crystal (Al_2O_3). Each component would be connected by a conducting strip consisting of a thin film of metal, like aluminum.

It will be assumed that the starting condition is a flat, smooth, and clean layer of pure silicon. The manufacturing process must then proceed through the following stages:

1. Introduce phosphorus atoms into the region to make n-type silicon.

2. Introduce boron into the source and drain regions.

3. Introduce more boron to form the channel between the source and drain.

4. Apply the metal contacts to the source, drain, and gate regions.

There are other steps, like applying thin wires by soldering to metal contacts and encasing the final chip in a protective coating, which will not be discussed. The intention here is merely to illustrate the kinds of processes involved.

The kinds of steps required can be described as follows:

1. **Oxidation** - Form a thin layer of SiO_2 by heating in a vacuum system which has been back-filled with a low pressure of oxygen.

2. **Application of photoresist** - Apply a thin layer of a polymer that is light-sensitive in the sense that light causes cross-linking of the polymer, which makes it resistive to dissolution in a solvent. The polymer film can then be exposed to light passing through a mask. The mask is transparent except in regions where it is desired to dissolve the film selectively.

3. **Expose the photoresist** - Use ultraviolet light to cross-link the photoresist through the mask. The precision of the pattern on the mask determines the precision of the placement of the layers of the component.

4. **Dissolve the unexposed (masked) regions of the photoresist** - Apply a solvent to the polymer film to dissolve the regions that are not cross-linked.

5. **Dissolve the oxide layer in the uncovered regions** - Apply hydrofluoric acid to the surface to dissolve the SiO_2 that was uncovered when the masked regions of the photoresist were dissolved.

6. **Burn off the remaining photoresist** - Heat in oxygen to decompose the remaining polymer film into gaseous combustion products.

7. **Introduce dopant** - Use ion implantation (from a particle accelerator) or vapor deposition to introduce the dopant elements in the regions not covered by oxide film.

8. **Diffusion anneal** - Heat the chip for a specific time at a specific temperature to allow the dopant to diffuse into the desired depth below the surface.

9. **Deposit metal contacts** - Use a heated metal source to form a vapor that deposits in the desired regions, defined by a mask.

The stages of the process, starting with the clean silicon, are illustrated schematically in the following sequence:

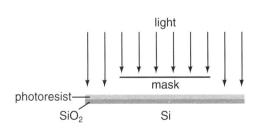

1. Oxidize
2. Apply photoresist
3. Expose photoresist through mask
4. Dissolve unexposed photoresist
5. Dissolve uncovered oxide
6. Burn off remaining photoresist

CHAPTER 16 ELECTRONIC MATERIALS IN THE WALKMAN

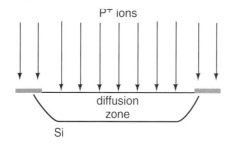

1. Ion-implant with phosphorus
2. Diffusion anneal
3. Dissolve oxide

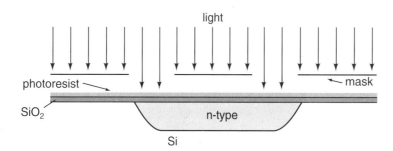

1. Oxidize
2. Apply photoresist
3. Expose photoresist through mask
4. Dissolve unexposed photoresist
5. Dissolve uncovered oxide
6. Burn off remaining photoresist

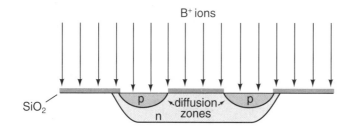

1. Ion implant with boron
2. Diffusion anneal
3. Dissolve oxide

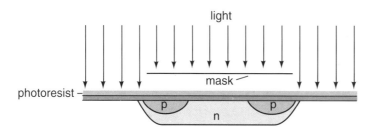

1. Oxidize
2. Apply photoresist
3. Expose photoresist through mask
4. Dissolve unexposed photoresist
5. Dissolve uncovered oxide
6. Burn off remaining photoresist

1. Ion implant with boron (small amount)
2. Diffusion anneal (short)
3. Dissolve oxide

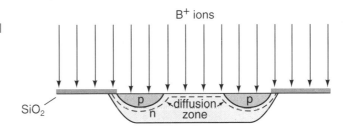

1. Oxidize
2. Apply photoresist
3. Expose photoresist through mask
4. Dissolve unexposed photoresist
5. Disolve uncovered oxide
6. Burn off remaining photoresist

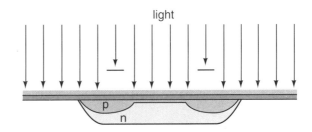

Deposit metal contacts

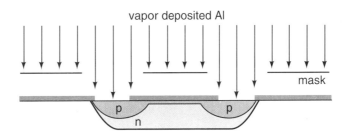

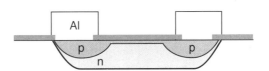

16.9 Other Semiconductor Materials

Although silicon is by far the most common material used for semiconductor devices, a variety of other materials are available for special purposes. Examples of these and their important properties are given in Table 16.1

Table 16.1 Examples of Elemental and Compound Semiconductors with Selected Room-Temperature Properties.

Material	Band Gap (eV)	Electrical Conductivity (ohm-m)$^{-1}$	Mobility of Electrons (m^2/V-sec)	Mobility of Holes (m^2/V-sec)
Elemental				
Si	1.11	4×10^{-4}	0.14	0.05
Ge	0.67	2.2	0.38	0.18
III-V Compounds				
GaP	2.25		0.05	0.002
GaAs	1.35	10^{-6}	0.85	0.45
InSb	0.17	2×10^{-4}	7.7	0.07
II-VI Compounds				
CdS	2.40		0.03	
ZnTe	2.26		0.03	0.01

It is apparent that germanium has significantly higher electron and hole mobility than silicon. This makes it preferable for high-speed applications. However, it does not form oxide layers as silicon does, and this makes the formation of insulating layers in integrated circuits more difficult. Because of its smaller band gap, it has a much higher intrinsic conductivity at room temperature than silicon.

The compound semiconductors exhibit a range of intrinsic conductivities and carrier mobilities. Examples of some applications will be given in the next section. The high carrier mobility of GaAs makes it attractive for high-speed applications, but it is considerably more difficult to work with than silicon.

16.10 Examples of Related Devices

In the context of sound reproduction, several additional devices can be described to give further illustrations of materials used for device applications. They are used in Walkman-type equipment, compact disc players, and in the all but extinct phonograph.

16.10.1 The Light-Emitting Diode

The LED is an optoelectronic device which is, in effect, a low-intensity solid-state analog of a light bulb. It draws very little current and is used as an indicator light on the Walkman to show when the power is on (as a useful reminder to preserve the battery). Its construction is illustrated in Fig. 16.18.

The LED is essentially a single p-n junction biased in the forward direction; energy is emitted in the form of visible light when electron-hole recombination occurs. Table 16.1 shows that GaP has some attractive properties for this application. The band gap is so large that the intrinsic conductivity is negligible, and the energy of the band gap corresponds to a photon of light in the visible spectrum. In addition, the mobility of electrons is 25 times the mobility of holes. Therefore, when the voltage is on, the main effect is the introduction of electrons into the p-type material. As they cross the junction, the electrons are annihilated by recombination with holes, and for each recombination a photon is emitted with an energy roughly equal to that of the band gap. Thus, the light is emitted at the junction. The current can be kept low by controlling the doping level in the GaP.

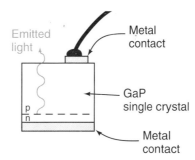

Fig. 16.18 Schematic illustration of the construction of a light-emitting diode.

When electron-hole recombination occurs, the energy released can be dissipated in the form of either photons or heat. The latter comprises quantized packets of kinetic energy, i.e., phonons, which are essentially localized regions of enhanced lattice vibration, as noted earlier. The efficiency of operation of a GaP LED depends on the presence of impurities which provide energy levels that act as temporary traps for electrons before they recombine with holes. While they are trapped, the electrons dissipate momentum in the form of phonons, and this makes the radiative (i.e., photon emitting) type of recombination more likely in GaP. Additions of CdO or ZnO are used to produce efficient emission of red light, and sulfur or nitrogen can be used to shift the wavelength of emitted light toward green. In the latter case, a significantly lower efficiency is compensated for by the fact that the human eye is more sensitive to green than to red.

16.10.2 The Solar Cell

The solar cell can be used to power a Walkman or to recharge its battery. It acts like an LED in reverse; instead of a voltage generating photons, photons generate a voltage. The operation of a solar cell is illustrated by Fig. 16.19. The principle relies on the fact that, in the immediate vicinity of a p-n junction, electron-hole recombination occurs, and this leaves a thin positively charged region on the n-side of the junction and a corresponding negatively charged region on the p-side, Fig. 16.19(a). This must be so, because each side would be electrically neutral in the absence of this recombination. Therefore, the loss of electrons from the n-side must leave an excess of positive charge, and the loss of holes from the p-side must, similarly, leave an excess of negative charge. Thus, a potential gradient is set up across the junction.

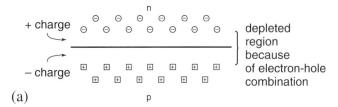

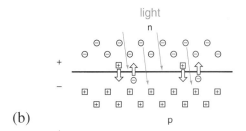

Fig. 16.19 Schematic illustration of a p-n junction showing (a) the effect of electron-hole recombination in the vicinity of the junction, leaving a charged double layer at the junction, and (b) the effect of light-induced electron-hole pair creation in the vicinity of the junction, the recombination-induced potential difference causing electrons to move into the n-type material and holes to move into the p-type.

Now, let a light with photon energy greater than the band gap fall on the region of the junction. Each photon creates an intrinsic electron-hole pair. The potential gradient at the junction drives the electrons away from the p-side (into the n-type material) and holes away from the n-side (into the p-type material) (Fig. 16.19(b)). This in turn sets up gradients of carrier concentration on each side, and causes a current to flow. This current is then collected by contacts on the two surfaces of the cell.

The layer of n-type material must be thin enough that the sunlight can penetrate to the junction. The electrical contacts on the n-side must be made of a grid of fine wires to minimize blockage of the sunlight.

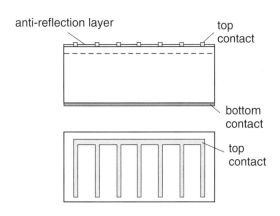

Silicon has been the material of choice for solar cells, because sunlight contains wavelengths corresponding to photon energies greater than the 1.1eV band gap, and it is available and relatively economical. The band gap of GaAs (1.35eV) is also small enough for it to function as a solar cell, and the carrier mobility is greater than in silicon. This would permit greater currents and correspondingly greater output for a given amount of sunlight. Silicon is, at present, less costly to produce, however.

16.10.3 The Laser

The recording medium that is replacing the tape cassette is the compact disc. This has a digital format in which microscopic pits on the polished surface of a thin metallized disc are used in conjunction with a light beam, the diameter of which is approximately the same as one of the pits. If the beam strikes the polished surface between pits, it is reflected into a light detector, and this can be "read" as a one. When the beam strikes a pit, it is scattered, and the detector produces no signal; this lack of signal is read as a zero.

The most common discs constitute "read-only memory," or ROM. That is, they only play back; one cannot record on them as with a tape. Their advantage is a very high recording density, very high signal-to-noise ratio, and freedom from noise like tape hiss and the noise resulting from surface defects on a phonograph record.

A portable compact-disc player uses a light beam produced by a semiconductor *laser,* which is an acronym for *l*ight *a*mplification *b*y *s*timulated *e*mission of *r*adiation. This laser utilizes the radiative recombination that occurs near the p-n junction in a forward-biased GaAs diode, along with special geometrical features and operating conditions that produce the lasing effect.

In a normal light-emitting diode the photons are emitted randomly, and there is no phase relationship among the electromagnetic waves that comprise each photon; that is, the emitted light is incoherent. Lasing, on the other hand, involves the emission of light that is coherent, or in phase, and monochromatic, meaning that the wavelengths of the emitted photons are the same to within a few angstroms. This phenomenon is based on the fact that photons can interact with electrons in semiconductors in either of two ways, as illustrated in Fig. 16.20. The photon can be absorbed by the promotion of an electron to an excited state (i.e., from the valence band to the conduction band, Fig. 16.20(a)). Alternatively, if the electron is already in the excited state, the photon can stimulate the electron to return to the lower energy state, and this event would be accompanied by the emission of another photon having the same wavelength as the first. In the latter process the original photon is amplified, and this is the basis of the lasing phenomenon.

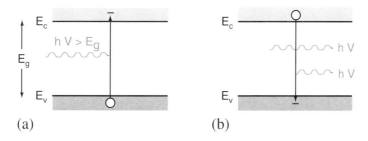

Fig. 16.20 Two possible electron-photon interactions: (a) Excitation of electron by the absorption of the energy of the photon; in (b) electrons are already in the excited state; i.e., a population inversion exists (hatched areas indicate energy levels occupied by electrons), and the photon stimulates the return of an excited electron to the lower-energy state, giving rise to the emission of a second photon. (From A. Bar-Lev, loc. cit., p. 173.)

Lasing requires that a so-called population inversion exist. This can be achieved near a p-n junction (Fig. 16.21(a)) by use of a forward bias and a high doping density. On the p side of the junction, a high density of electrons is injected into the conduction band, and on the n side of the junction, a high density of holes is injected into the valence band, as shown schematically in Fig. 16.21(b). As the injected electrons recombine spontaneously with holes, the emitted photons will stimulate the emission of more photons. If opposite ends of the crystal are highly reflective, some of these photons will be reflected back into the region of the junction and stimulate further photon emission. Continued back-and-forth reflections from the ends causes the light intensity to increase. If the photon creation by this stimulated emission exceeds the photon loss by radiation to the outside of the crystal and to absorption in regions of the crystal remote from the junction, then the light intensity can increase rapidly to the level where the rate of photon creation is just balanced by the rate of loss. A fraction of these photons is emitted from the end of the crystal in the form of intense, coherent, monochromatic light (Fig. 16.21(c)).

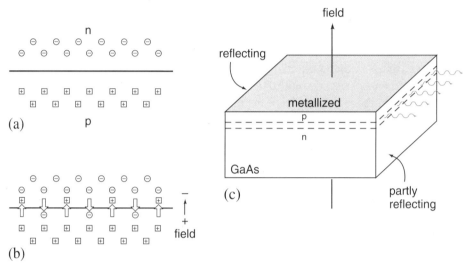

Fig. 16.21. (a) A p-n junction in the absence of a biasing voltage. (b) The same junction after a forward bias has been applied to create a population inversion; recombination is accompanied by photon emission, and these photons stimulate more photon emission. (c) The back-and-forth reflections of photons created by photon-stimulated emission results in a fraction of these photons being emitted externally as laser light. (From A. Bar-Lev, loc. cit., p. 174.)

16.10.4 The Piezoelectric Phono Pickup

The recording medium that preceded the magnetic tape was, of course, the phonograph record, which, as noted earlier, operated through the motions of a needle that tracked along the grooves on the surface of the record. One of the ways to convert this mechanical motion of the needle into an electrical signal is to use a *piezoelectric* crystal. Piezo is from the Greek and means pressure.

The piezoelectric effect can be illustrated by consideration of the unit cell of barium titanate, $BaTiO_3$, as shown in Fig. 16.22. Above 120°C, the crystal is cubic, but below that temperature it becomes tetragonal, because the four oxygen ions on the side faces move a bit below the center of the faces, and the central titanium ion moves a bit above the center of the cell. Thus, the centers of positive and negative charge no longer coincide, and the unit cell takes on the character of an electric dipole. Neighboring unit cells tend to align their dipoles cooperatively, and the crystal is referred to as a *ferroelectric,* by analogy with a ferromagnetic material.

SECTION 16.10.4 THE PIEZOELECTRIC PHONO PICKUP

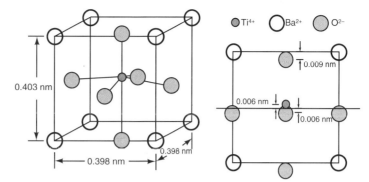

Fig. 16.22 The barium titanate unit cell in the ferroelectric state. (Adapted from L.H. Van Vlack, *Elements of Materials Science and Engineering,* 4th Ed., 1980, Addison-Wesley, p. 304.)

A piezoelectric crystal can be used to convert a mechanical force into an electrical signal (i.e., to act as a transducer) in the way illustrated in Fig. 16.23. A crystal with a metallic film on each of two opposing faces (normal to the axis of the dipole) will experience a buildup of electrons in the metal film on one face and a deficit of electrons in the other if the ends are electrically connected, as shown in Fig. 16.23(a). This is a response to the internal electric dipole of the crystal.

If a compressive force is applied to the crystal, as shown in Fig. 16.23(b), the dipole strength is decreased, because the centers of positive and negative charge are forced closer together. This causes electrons to flow in the circuit to reduce the charge on the ends; thus, the pressure creates a voltage in the external wire. A varying pressure would create a varying voltage. Thus, if the phonograph needle were connected to such a crystal of barium titanate, its movements would produce a signal that could be amplified and heard as music. This is the basis of the ceramic phonograph cartridge.

The alternative use of a piezoelectric crystal is to generate mechanical displacements by applying voltages to the crystal to force the length of the dipole, and thus the crystal, to increase or decrease. Figure. 16.23(c) shows a voltage which forces the crystal to elongate; the opposite voltage would force it to shorten. If an alternating voltage were applied, as in Fig. 16.23(d), the crystal would vibrate. This effect is used for many applications, such as the generation of sound waves (sonar, ultrasonic equipment, etc.).

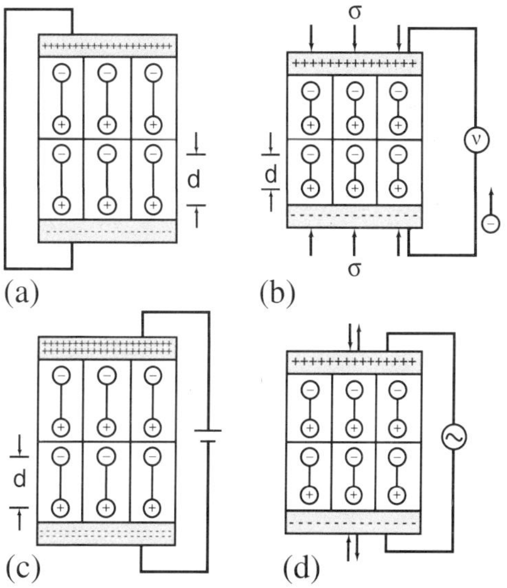

Fig. 16.23 Schematic representation of (a) a crystal of barium titanate with metallized ends connected by a conducting wire, (b) the voltage generated by pressure which forces the dipole moment to decrease, (c) deformation of the crystal by application of an external voltage, and (d) the generation of vibrations in the crystal by an alternating voltage. (Adapted from L.H. Van Vlack, loc. cit., p. 305.)

16.11 Batteries

There are two different types of batteries used in the various models of the Walkman, and these are widely used in other consumer products. The most common type is a so-called primary cell, the name given to a battery that is discharged once and then discarded. When such a battery has no free liquid electrolyte, it is called a "dry" cell and is relatively inexpensive and convenient to use. There are various batteries of this type, but the most common ones are the zinc-carbon (or Leclanche) cell and the alkaline-manganese-oxide battery.

The other type is called a secondary cell. Such a battery can be recharged after having been discharged, and this can be done repeatedly. One that can be used for a Walkman is the sealed nickel-cadmium battery, and it can be discharged and recharged hundreds of times. It is more expensive than an alkaline-manganese-oxide battery, but the lifetime cost is much lower.

These two battery types have different discharge characteristics, the Ni-Cd battery having a much flatter discharge curve, as shown in Fig. 16.24(a). The Ni-Cd discharge curve actually decreases gradually with the number of charging cycles, as shown in Fig. 16.24(b). The discharge curve depends on the current being drawn; an example is shown in Fig. 16.25.

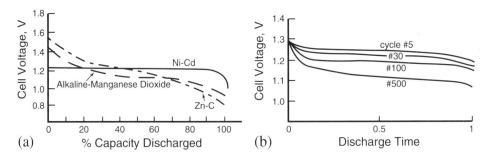

Fig. 16.24 (a) Comparison of the discharge profiles of a zinc-carbon and an alkaline-manganese oxide battery *vs.* a Ni-Cd battery. (b) Effect of repeated discharges for one hour at the C/5 rate on the discharge profile of a Ni-Cd battery. (From the *Handbook of Batteries and Fuel Cells,* David Linden, Ed., McGraw-Hill, NY, 1984, pp. 3-24, 18-13.)

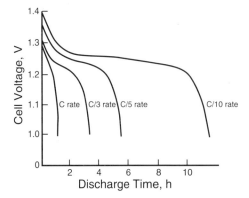

Fig. 16.25 Discharge curves for different discharge rates for an alkaline-manganese oxide battery. (From Linden, loc. cit. p. 18-6)

The discharge current is often designated by the C-rate, as follows:

$$I = \frac{C_n}{N}$$

where I is the discharge current in amperes, N is the number of hours of discharge, and C_n is the capacity rating of the cell, in ampere-hours, at the n-hour rate. (The capacity generally decreases as the discharge current increases.) For example, if a battery is rated at 2Ah for a 20-hour rate of discharge, then the C/10 discharge current is

$$I = \frac{C_{20}}{N} = \frac{2}{10} = 0.2A$$

Any battery is an electrochemical cell in which the flow of anions to an anode, and cations to a cathode, through an electrolyte, produces a discharge current when an external circuit exists, as shown schematically in Fig. 16.26(a). For a rechargeable battery, a dc power supply can be used in place of the load to reverse the current flow in the external circuit. This makes the positive terminal the anode and the negative terminal the cathode, and the internal flow of ions is reversed from the discharge case, Fig. 16.26(b).

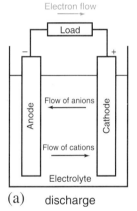
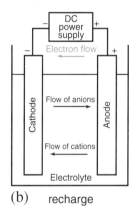

Fig. 16.26 Schematic representation of a battery under (a) discharge and (b) recharge conditions. (From Linden, loc. cit., p. 1-5.)

(a) discharge (b) recharge

16.11.1 The Zinc-Carbon Cell

This is the generic household dry cell; it has been available in a variety of sizes and shapes for many decades. The construction of a cylindrical version is shown in Fig. 16.27. The anode is a zinc can, formed by deep drawing from a sheet; it serves also to contain the other parts of the cell. The sheet is produced from 99.99%-pure zinc to which is added about 0.1% cadmium to increase strength and to make the zinc more resistant to corrosion in the electrolyte. Lead is also added in somewhat greater amounts to aid in the forming process.

The cathode is a cylindrical bobbin formed from powdered manganese dioxide mixed with carbon black to improve electrical conductivity. It is separated from the anode by a porous electrical insulator made of a gelled paste or a coated paper. The electrolyte is mixed in with the cathode bobbin and with the separator so that electrolytic conduction can occur between the anode and cathode. An example of a paste separator would be a mixture of ammonium chloride, zinc chloride, water, and starch or flour as the gelling agent.

The electrolyte may be either a mixture of ammonium chloride and zinc chloride in water, with a small amount of zinc-corrosion inhibitor, or this mixture with the ammonium chloride omitted. The zinc chloride is essential in order to minimize corrosion of the anode.

The cathode current is collected by a porous carbon rod inserted through the center of the bobbin. The porosity, which is adjusted by additions of oil or wax, is necessary for the venting of hydrogen during heavy discharges.

As the cell is discharged, the zinc is oxidized and the manganese oxide is reduced. The overall cell reaction is

$$Zn + 2MnO_2 \rightarrow ZnO + Mn_2O_3$$

The details of the reaction depend on a number of factors, including the rate, extent, and temperature of discharge, the electrolyte concentration, cell geometry, and the type of MnO_2 used for the cathode (ranging from natural MnO_2 ore with attendant impurities to high-purity MnO_2 produced electrolytically). The capacity of the battery depends on the quality of the MnO_2 used.

Some of the major advantages and disadvantages of the zinc-carbon cell are summarized in Table 16.2.

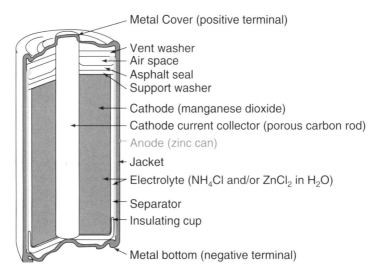

Fig. 16.27 Cross section of a cylindrical zinc-carbon cell. (From Linden, loc. cit., p. 5-5.)

Table 16.2 Advantages and Disadvantages of the Zinc-Carbon Cell (From Linden, loc. cit., p 5-2)

Advantages	Disadvantages	Comments
Low cell cost	Low energy density	For best capacity, the discharge should be intermittent
Low cost per watt-hour	Poor low-temperature service	
Large variety of sizes, shapes, voltages, and capacities	Poor leakage resistance under abusive conditions	
Wide distribution and availability	Low efficiency under high current drains	Capacity decreases as the discharge drain increases
Long tradition of reliability	Comparatively poor shelf life; voltage falls steadily with discharge	Improved shelf life if refrigerated

16.11.2 The Alkaline-Manganese Oxide Cell

This cell utilizes the same materials for the anode and cathode as the zinc-carbon cell, and the overall reaction during discharge is the same. However, the electrolyte is highly conductive potassium hydroxide, which results in a lower internal resistance in the cell. The cell construction is illustrated in Fig. 16.28.

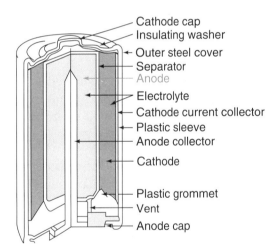

Fig. 16.28 Construction of a cylindrical alkaline manganese dioxide cell. (From Linden, loc. cit., p. 7-6.)

In this cell the anode is not used as a container, but rather the anode and cathode positions are reversed from the zinc-carbon cell, with the anode now in the form of a porous zinc powder mixed with 4 to 8% mercury, which covers the zinc particles to suppress hydrogen gassing. (Zinc in contact with an alkaline solution is unstable and reacts to form hydrogen.) The KOH electrolyte is suffused throughout the anode. Sometimes it is mixed with a gelling agent; in other cases it is mixed with a plastic binder. The large surface area of the zinc particles permits the anode to carry very high currents. The cathode employs electrolytic MnO_2, which has higher reactivity and capacity than natural ores.

The advantages and disadvantages of the alkaline manganese dioxide cell are summarized in Table 16.3.

Table 16.3 Advantages and Disadvantages of the Alkaline Manganese Dioxide Cell (From Linden, loc. cit., p. 7-1)

Advantages	Disadvantages
Good high-rate discharge capability	Higher initial unit cost than zinc-carbon cells
Higher energy output than zinc-carbon cell (depending on load)	If shorted or abused, cell temperatures could rise to high levels (e.g., 100°C)
Good shelf life	Sloping discharge curve, but less pronounced than with zinc-carbon cell
Good leakage resistance	Mercury content presents more severe waste-disposal problem with expended cells
"Rest periods" not necessary; efficient when used continuously	
Good low-temperature performance	
Good shock resistance	
Low gassing rate	

16.11.3 Sealed Nickel-Cadmium Batteries

These cells contain design features that eliminate buildup of gas pressure during charging and, therefore, can be permanently sealed. In the charged state, cadmium is the negative electrode (i.e., the anode), and nickel oxide is the positive. A KOH solution is the electrolyte, and the overall reaction is

$$Cd + 2NiOOH + 2H_2O \rightarrow Cd(OH)_2 + 2Ni(OH)_2$$

When the cell is charged, the positive electrode reaches full charge first and begins to emit oxygen, which then migrates to the negative electrode and oxidizes (discharges) the cadmium to produce cadmium hydroxide:

$$Cd + 1/2\ O_2 + H_2O \rightarrow Cd(OH)_2$$

The construction of a cylindrical cell is shown in Fig. 16.29. The positive electrode is highly porous sintered nickel impregnated with nickel hydroxide. The cadmium is deposited or plated onto a substrate to make the negative electrode, and the separator can be unwoven nylon or polypropylene, either of which is highly absorbent to the KOH electrolyte and is permeable to oxygen; there is no free electrolyte. The electrodes with the separator in between are wound like a jelly roll and placed in a nickel-plated can. The connection to the negative electrode is welded to the can, and the positive connection is welded to the top cover.

During operation, the materials of the electrodes change only their oxidation state, not their physical state, and there is little if any change in the electrolyte concentration. Both electrodes remain insoluble in the electrolyte throughout discharging and recharging. For this reason the discharge profile remains relatively flat over a wide range of discharge current, and the cells have a long life in both cyclic and standby use. The advantages and disadvantages are given in Table 16.4.

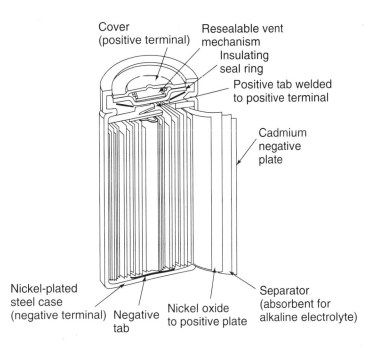

Fig. 16.29 Construction of a sealed cylindrical nickel-cadmium cell. (From Linden loc. cit., p. 18-4.)

Table 16.4 Advantages and Disadvantages of Sealed Nickel-Cadmium Batteries (From Linden, loc. cit. p. 18-1)

Advantages	Disadvantages
Sealed cells, no maintenance required	Higher cost than sealed lead-acid battery
Long life cycle	"Memory" effect
Good low temperature and high-rate performance capability	
Long shelf life in any state of charge	

One advantage of a nickel-cadmium battery compared with the primary cells described earlier which should receive special note is that the need to dispose of spent batteries containing cadmium, lead, and mercury is greatly reduced. Thus, the nickel-cadmium battery is much more environment-friendly.

Summary

The electronic components of the Walkman depend on the properties of semiconductors. In contrast with metals, which have accessible empty energy levels into which conduction electrons can be promoted by an applied voltage, semiconductors have an energy gap (usually <2eV) between the filled valence band and the next higher band, called the conduction band. If energy is supplied by means of temperature, light photons, or high voltage, electrons from the covalent bonds can cross the energy gap. This results in two kinds of carriers: electrons in the conduction band and holes in the valence band. The holes, having the effective charge of +1 electron, drift in an applied electric field by a process analogous to that of self diffusion in a crystal. This behavior is called intrinsic, because it occurs in an undoped semiconductor. The increase in conductivity with temperature is opposite to that in a metal. The phonon scattering of conduction electrons in a metal, which decreases the carrier (i.e., electron) mobility, occurs in semiconductors as well, but this effect is insignificant compared to that of the exponential increase in the number of carriers with increasing temperature in the semiconductor.

Doping of semiconductors means substituting impurities having higher valence (n-type) or lower valence (p-type) for silicon atoms. The n-type impurity (e.g., phosphorus) should have an energy state just below the bottom of the conduction band of silicon, so that only a relatively small energy is needed to promote the impurity electron into the conduction band of the silicon. This type of impurity is an electron donor, and the extra energy state is called a donor state. The conductivity of n-type silicon is higher than that of pure silicon, and it increases with temperature until all the donor electrons have been promoted into the conduction band. After that, the conductivity remains almost constant with increasing temperature (phonon scattering exerts a relatively small negative effect) until the temperature range is reached where intrinsic behavior becomes significant.

A p-type impurity (e.g., boron) provides a built-in hole, the energy level of which should be just above the top of the valence band of silicon, so that a silicon electron can easily be promoted into it. This, therefore, is called an acceptor state. The plot of conductivity *vs.* temperature is similar to that of n-type silicon, becoming flat when all the acceptor states are filled. Since doped, or extrinsic, silicon is normally used at temperatures where intrinsic behavior is negligible, only one kind of charge carrier is relevant: either electrons for n-type or holes for p-type. This, and the fact that these carriers move in opposite directions in an electric field, form the basis for the usefulness of semiconductors in devices.

Semiconductor-grade silicon is purified by zone refining and is processed into single crystals by directional solidification (to eliminate carrier-trapping grain boundaries and dislocations) before doping is carried out by vapor deposition or ion bombardment followed by a diffusion anneal. Integrated circuits, which contain many device elements together on one silicon crystal substrate, are fabricated by a series of steps involving photolithography to situate the elements, and the formation of SiO_2 is an important factor in this processing. This is the main factor which makes silicon the choice over germanium, for example. Examples of device elements discussed include a p-n junction rectifier, a junction transistor, and a MOSFET. Other related devices employing p-n junctions are described; these include the LED, the solar cell, and the semiconductor laser (used in the Discman®, for example).

The Walkman is powered by batteries, either the disposable alkaline-manganese oxide cell or the rechargeable Ni-Cd cell. The construction and operating principles of these are described.

One other useful device for sound reproduction is the piezoelectric crystal, used for example in some phonograph cartridges. Barium titanate is used to explain the piezoelectric effect found in crystals having separate centers of positive and negative charge density. In such a crystal an applied stress can cause a change in the electric dipole strength (which can produce an external voltage), or a voltage applied to the crystal can cause it to change shape.

Exercises Chapter 16

Terms to Understand

Electrical conductivity, resistivity
Phonon
Carrier mobility
Hole
Semiconductor
Dopant
Energy band
Band gap
Fermi level
Conduction band
Valence band
Intrinsic semiconductor
Extrinsic semiconductor
Donor level
Acceptor level
Recombination
n-type
p-type

Donor exhaustion
Acceptor saturation
Zone refining
Photoconductor
p-n junction, rectifier
Junction transistor
MOSFET
Integrated circuit
Photolithography
LED
Solar cell
Semiconductor laser
Piezoelectric crystal
Primary cell
Secondary cell
Zinc-carbon cell
Alkaline-manganese oxide cell
Nickel-cadmium battery

Problems

16.1 A piece of annealed aluminum wire has a measured electrical resistance of 0.01 ohm. The wire is stretched in tension to a plastic strain of 40%. (The required stress was less than the UTS.)

   (a) How will the electrical resistance of the wire be affected by this deformation, if at all? Explain your answer.
   (b) If the wire is again annealed to recrystallize it, how would the final electrical resistance compare with the original value? (Give the percent difference.)

16.2 When phosphorus is added as a dopant to pure silicon, do the following quantities increase, decrease, or remain the same? Explain your answers.

   (a) the hole concentration
   (b) the mobility of electrons
   (c) the conductivity
   (d) the temperature coefficient of conductivity

16.3 The silicon rectifier shown below will conduct electricity when the voltage is applied as shown by the +/- symbols.

(a) Which part of the rectifier is n-type and which is p-type?
(b) Where in the rectifier are holes created?
(c) Where in the rectifier does recombination occur?

16.4 What element(s) could be added to InSb to make an n-type semiconductor?

16.5 A semiconducting sample is found to have a conductivity of 3100 (ohm-m)$^{-1}$ at room temperature. This value remains nearly constant over the temperature range from -50 to +150°C.

(a) Is this an intrinsic or extrinsic semiconductor? Explain.
(b) If the temperature were raised to, say, 200°C, would you expect the conductivity to increase or decrease? Explain.
(c) If the temperature were reduced to, say, -100°C, would the conductivity be higher or lower than at room temperature? Explain.

Appendix 1
Conversion of Units

Length
1 Å = 10^{-10} m = 10^{-8} cm = 0.1 nm = 3.937 x 10^{-8} in
1 cm = 10^{-2} m = 0.3937 in
1 ft = 12 in = 0.3048 m
1 in = 0.0254 m = 2.54 cm = 25.4 mm
1 µm = 10^{-6} m
1 nm = 10^{-9} m

Volume
1 cm^3 = 0.0610 in^3
1 in^3 = 16.3x10^{-6} m^3
1 gal (US) = 3.78x10^{-3} m^3

Mass
1 g = 0.602x10^{24} amu = 2.20x10^{-3} lb_m
1 kg = 2.20 lb = 10^3 g
1 lb_m = 0.454 kg
1 oz (Avoirdupois) = 0.0625 lb_m = 28.37 g

Density
1 g/cm^3 = 62.4 lb_m/ft^3 = 10^3 kg/m^3 = 1 Mg/m^3 = 1 mg/mm^3
1 lb_m/ft^3 = 16.0 kg/m^3
1 $lb_m/1in^3$ = 27.68 g/cm^3 = 27.68x10^3 kg/m^3

Energy or Work
1 Btu = 1.06x10^3 J
1 gram cal = 4.18 J
1 eV = 0.160x10^{-18} J
1 ft·lb_f = 1.355 J
1 in·lb_f = 0.113 J
1 dyne·cm = 10^{-7} J
1J = 1N-m

Force
1 N = 0.224 lb_f = 98.0 g_f
1 lb_f = 4.44 N
1 dyne = 10^{-5} N = 2.24x10^{-6} lb_f

Stress or Pressure
1 Pa = 1 N/m^2 = 0.145x10^{-3} lb_f/in^2
1 lb/in^2 = 6.89 kPa

Appendix 2
Physical Constants

atomic mass unit (amu) = 1.66×10^{-24} g = 1.66×10^{-21} kg
Avogadro's number (N) = 6.022×10^{23} atoms or molecules/mole
Boltzmann's constant (k) = 13.8×10^{-24} J/K = 8.63×10^{-5} eV/K
Gas constant (R) = Boltzmann's constant x Avogadro's number = 8.31 J/K·mole
electronic charge (q) = 0.16×10^{-18} Coulombs
Bohr magneton (magnetic moment of one electron spin) = 9.27×10^{-24} A·m²
electron volt = 0.160×10^{-18} J
acceleration of gravity (g) = 9.80 m/sec²

Appendix 3
SI Prefixes

10^9	giga	G
10^6	mega	M
10^3	kilo	k
10^{-3}	milli	m
10^{-6}	micro	µ
10^{-9}	nano	n
10^{-12}	pico	p

Appendix 4

Physical Properties of Common Metals and Semiconductors (Room Temperature)

element	atomic number	atomic mass (amu)	atomic diameter (Å)	density (g/cm^3)	crystal structure
Ag	47	107.87	2.89	10.5	fcc
Al	13	26.98	2.86	2.70	fcc
Au	79	196.97	2.88	19.32	fcc
Be	4	9.01	2.28	1.85	hcp
Cd	48	112.40	2.96	8.65	hcp
Co	27	58.93	2.50	8.9	hcp
Cr	24	52.00	2.50	7.20	bcc
Cu	29	63.54	2.56	8.92	fcc
Fe (α)	26	55.85	2.48	7.88	bcc
Ge	32	72.59	2.44	5.35	diamond cubic
Li	3	6.94	3.04	0.534	bcc
Mg	12	24.31	3.22	1.74	hcp
Mn	25	54.94		7.2	complex
Mo	42	95.94	2.73	10.22	bcc
Ni	28	58.71	2.49	8.90	fcc
Pb	82	207.19	3.50	11.34	fcc
Pt	78	195.1	2.78	21.45	fcc
Sc	14	28.09	2.36	2.33	diamond cubic
Sn	50	118.69		7.31	diamond cubic
Ti(α)	22	47.90	2.93	4.51	hcp
V	23	50.94	2.63	6.11	bcc
W	74	183.85	2.73	19.4	bcc
Zn	30	65.37	2.78	7.14	hcp
Zr	40	91.22	3.24	6.51	hcp

Appendix 5

Elastic Constants of Common Metals

metal	Young's Modulus E MPa	Poisson's Ratio ν	Shear Modulus G MPa
W	345x10^3	0.28	134x10^3
Ni	210	0.31	75
Fe	205	0.28	82
Cu	110	0.35	41
Ti	107	0.36	45
Al	70	0.34	26
Mg	45	0.29	17
Pb	14	0.45	4.8

Appendix 6
Ionic Radii (Å)

ion	CN=4	CN=6	CN=8
Al^{2+}	0.46	0.51	
Co^{2+}		0.99	1.02
Cl^-		1.81	1.87
Cr^{3+}		0.63	
F^-	1.21	1.33	
Fe^{2+}	0.67	0.74	
Fe^{3+}	0.58	0.64	
K^+		1.33	
Li^+		0.68	
Mg^{2+}	0.60	0.66	0.68
Mn^{2+}	0.60	0.66	0.68
Na^+		0.97	
Ni^{2+}	0.63	0.69	
O^{2-}	1.28	1.40	1.44
OH^-		1.40	
P^{5+}		0.35	
S^{2-}	1.68	1.84	1.90
Si^{4+}	0.38	0.42	
Ti^{4+}		0.68	
Zn^{2+}	0.67	0.74	
Zr^{4+}		0.79	0.82

CN=Coordination number

Appendix 7
Specific Modulus Values

bulk materials	density, ρ g/cm^3	modulus, E MPa	E/ρ
Al	2.7	70x10^3	26x10^3
Fe (steel)	7.8	205	26
Mg	1.7	45	26
soda-lime glass	2.5	70	28
softwood	0.4	11	26
hardwood	0.6	17	27
PVC	1.3	4.5	22
reinforcing materials			
Al$_2$O$_3$	3.9	400x10^3	100x10^3
B	2.3	400	170
Be	1.9	300	160
BeO	3.0	400	130
C	2.3	700	300
SiC	3.2	500	160
SiN	3.2	400	120

Appendix 8

Properties of Some Engineering Materials

material	density g/cm^3	thermal conductivity (W/mm^2)/(K/mm)	thermal expansion coefficient K^{-1}	electrical resistivity ohm-m	elastic modulus MPa
Al alloys	2.7	0.16	22x10^{-6}	45x10^{-9}	70x10^3
brass (70-30)	8.5	0.12	20	62	110
grey cast iron	7.2	—	10	—	140
Cu	8.9	0.40	17	17	110
Fe	7.88	0.072	12	98	210
steel 1020	7.86	0.05	12	170	210
steel 1040	7.85	0.05	11	171	210
steel 1080	7.84	0.04	11	180	210
stainless steel	7.93	0.015	16	700	210
Mg	1.74	0.16	25	45	45
Al$_2$O$_3$	3.8	0.029	9	10^{12}	350
concrete	2.4	0.001	13	—	14
brick	2.3	0.006	9	—	—
glass					
window	2.5	0.0008	9	10^{12}	70
silica	2.2	0.0012	0.5	>10^{15}	70
Vycor	2.2	0.0012	0.6	10^{18}	70
graphite	1.9	—	5	10^{-5}	7
MgO	3.6	—	9	—	205
quartz	2.65	0.012	—	10^{12}	310
SiC	3.17	0.012	4.5	—	—

INDEX

8-N rule, 200
Abrasion by recording tape, 345
Abrasive wear, 184
Acceptor levels, 360
Acceptor state, 360
Activation energy, 88
 for diffusion, 196
 for substitutional diffusion, 116, 195
Active metal in corrosion, 15
Addition polymerization, 282
Adhesive bonding, 265, 316ff
Adhesive joints, design of, 317
Adhesive wear, 182, 184
Age-hardenable aluminum alloys, compositions, 228
Aging of rubber, 291
Airplane, relationship to the bicycle, 3
AISI/SAE system of steel designation, 171
Alkaline-manganese oxide cell, 378, 381ff
α-Si_3N_4, 206
α–stabilizer in titanium alloys, 266
Allotropic transformation
 in iron, 163
 in tin, 156
 in titanium, 266
Alloy carbides in tempered steel, 190
Alloy elements
 Cr and Mo in steel frames, 255ff
 in steel, hardenability, 188ff
Alloy steel
 in bicycle frames, 252ff
 AISI 4130, 190
 AISI 4340, 189
 AISI 52100, 187
 vs. carbon steel tubing, strength comparison, 258
Alnico, 348
Alumina, 200
Aluminum alloy 6061
 heat treatment, 229
 microstructure (TEM), 229
 phase diagram, 229
Aluminum-lithium alloy, 217
Amorphous alloys, 342
Amorphous polymer, 285
Analog recording, 330
Anelastic strain due to interstitial motion, 117
Anisotropy, 25
Annealing, 87ff
Annealing twins, 209
Anode
 in corrosion, 13,
 in batteries, 379

Anodic reaction, 13
Antiferromagnetic, 336
Arc welding, 260
Aromatic polyamide (Kevlar), 308
Arrhenius equation, 92
Aspect ratio of a fiber, 305
Asperities (in friction), 183, 199
Atactic, 288, 296
Atomic moments, 336
Audio tape, 332
Austenite
 decomposition of, 164ff
 in stainless steel, 18
 in carbon steel, 163
 stabilizer, 178, 188
Austenitic stainless steel
 composition, 18
 crystal structure, 49
Automobile, relationship to the bicycle, 1,2
Bainite, 168, 174, 262
Bakelite, 314
Bamboo bicycle frame, 250
Band gap, 356
Band model of solids, 182, 355
Banded microstructure in an alloy steel, 262
Barium titanate, 376
Base (in a junction transistor), 366
Batteries, 378ff
Beam, pure bending of, 240ff
Bending, 23
 vs. axial loading, 272
Bending moment, 23, 242
Bending moment diagram, 239, 241
β-SiAlON, 206
 microstucture (TEM), 207
Bias voltage 366
Bicycle frames
 comparisons of different materials, 253
 nomenclature of components, 233
 wooden, 251
Bicycle, history of development, 1ff
Bicycle wheel, 10ff, 324
Binding-energy curve and the elastic modulus, 30, 31
Body-centered-cubic crystal, BCC, 17
 slip systems in, 112
Body-centered-tetragonal, BCT
 unit cell, 129, 156
 in martensite 177, 179
Body diagonals in cubic crystals, 18, 53
Bohr magneton, 336
Boltzmann's constant, 92

vii

INDEX

Bonded joint
    in composite frame, 317ff
    in Al-alloy frame, 266
Bonded magnets, 348
Bonding in ceramics, 201
Boneshaker (early bicycle), 4
Bottom bracket, stresses due to pedaling force, 247
Boundary lubrication, 185
Bragg's law, 67, 82
Braiding of rope, 300
Branching in polymers, 279
Branching of a polyethylene molecule, 279
Bravais lattices, 128, 129
Braze welding, 154
Brazed bicycle frames, carbon steel *vs.* alloy steel, 255ff
Brazing, 131ff
    alloys, 153
Breakdown voltage, 356
Bridging oxygens, 202 311
Brittle fracture
    in carbon steel, 126
    in ceramics, 199
    in CFRP tubing, 322
Bubble growth, 97
Bubble raft, 48
Buckling of a thin strut under axial compression, 273
Burgers vector, 60
    of a screw dislocation, 60
    of an edge dislocation, 60
Butadiene, 289
Butted tubes, 263
C-curve, eutecoid steel, 167
C-rate of a battery, 379
Capillarity
    in brazing, 131, 133
    in sintering, 204
Carbide-forming elements in steels, 192
Carbon fibers
    elastic modulus, 309
    PAN precursor, 309
Carbon-carbon bond, flexibility, 278
Carbon steel *vs.* stainless steel spokes, 13ff
Carbon steel in frames, 252ff
Carburization, 193
Carriers in semiconductors, 358
Carrier
    mobility, 353, 358
    density, 358
    trapping, 362
Case depth, 194
Case hardening, 193ff

Cast irons, 162
Cathode
    in corrosion, 13
    in batteries, 379
Cathode:anode area ratio, 15, 16
Cathodic reaction, 13
Cavities in a tensile specimen, 38
Cellulose, 299, 306
    acetate, 306
Cementite
    in carbon steel, 103
    stucture of, 163
Ceramic
    bearings, 199ff
    tapeheads, 342
    phonograph cartridges, 377
Ceramics
    charge neutrality in, 200
    dislocation mobility in, 201
CFRP (carbon-fiber reinforced plastic), 317
    tape, 322
    tube, tensile behavior, 322
    tubes, 317ff
CGS units in magnetism (Gaussian system), 335
Chain stays, stresses due to pedaling force, 249
Charge carriers in semiconductors, 356
Chemical potential, 151, 158
Chloroprene, 289
Chromium dioxide, CrO_2, 342
Chromium plating, 15
Circuit board of a Walkman, 352
cis-Polyisoprene, 290
Clay, 203
Climb of dislocations, 95
Clincher-type tires, 312
Close-packed direction, 57
Close-packed plane, 49, 50
Coarsening
    of cementite, 105
    of precipitates, 219
Coefficient of friction, 183
Coercive field, 340
Coercive force, 340
Coherent light from a laser, 375
Coherent precipitates, 224
Cold welding, 183, 199
Cold-drawing of a linear polymer, 307
Cold-worked state, 87
Collagen, 299
Collector (in a junction transistor), 366
Compact disc, 374

Components in phase diagrams, 134ff
Composite
    frames, 317ff
    materials, 299ff, 305
    wheels, 324ff
Composites
    elastic modulus, upper and lower bounds, 303
    load distribution, 301
Compound semiconductors, 372
Compressive stress, 22
Condensation reaction, 306
Conduction band, 356
Conductivity, 353
    of silicon, effect of impurities, 362
    vs. temperature, n-type *vs.* p-type silicon, 360
Conjugated dienes, 289
Constant-temperature solidification, 141
Constituents of a microstructure, 142
Contact angle in brazing (wetting), 131
Continuous-fiber composites, 301ff
Continuous-transformation diagram, 168ff
    for a 1040 steel, 173
    for eutectoid steel, compared with IT diagram, 169
Contrast in optical microscopy, 47
Cooling curves for phase diagram determination, 140
Coordinate systems, Cartesian *vs.* cylindrical, 69
Copolymers, 292
Copper-zinc alloys for brazing, 153
Core
    of a dendrite, 150
    of a screw dislocation, 61
Corrosion current, 16
Corrosion of iron, 13
Corrosion protection of carbon steel
    sacrificial anode, 14
    chromium plating, 15
Covalently bonded ceramics, 199
Crack arrest in composites, 302
Creep
    of a viscoelastic material, 287
    of solder, 152
Critical radius of particle, 220
Critical stress for fracture, 211
Cross slip, 61, 77
Cross-links, 280
    in vulcanization of rubber, 291
Crystal anisotropy, 338
Crystal growth, 364
Crystal structure
    of silicon nitride, 205
    of ceramics, 200

Crystal symmetry, 128
Crystallization
    of polyethylene, 282
    of rubber, 281
Crystalline regions in polymers, 279
CT diagram, eutectoid steel, *vs.* IT diagram, 169
Cubic crystals, 51
Curie temperature, 340
Curing agent, 315
Curvature of a beam, 243ff
Cyclic loading of spokes, 11
Cylindrical tube as a beam, 241, 245
Dangling bonds
    in ceramics, 200
    in polymerization, 281
Dark-field TEM, 217
Dashpot, 286
Decarburization and fatigue, 124
Deformation of a beam, grid pattern, 242
Deformation twins, 210
Degree of polymerization, 282
Degrees of freedom in the phase rule, 151, 157
Demagnetizing curves, 347
Dendrites, 138, 147
    in Cu-Sn and succronitrile, 147
Dendritic solidification, explanation of, 147
Densification in sintering, 202ff
Diamagnetism, 335
Diamond-cubic crystals, unit cell and covalent bonds, 156, 200
Diamond frame of a bicycle, 233
Diffraction contrast, 66
Diffraction of xrays and electrons by crystals, 82
Diffusion anneal for IC fabrication, 369
Diffusion coefficient, diffusivity 194
Digital recording, 330
Dilatational stress terms, 70, 71, 83
Dilatometer, 170
Dimensional stability in heat-treated steels, 192
Dipole moment, 336
Direction indices, 53
Disc wheels, 324ff
Discharge current, 378, 379
Discharge profiles of batteries, comparison, 378
Discontinuous fibers, 304
Dislocations
    by-passing of particles, 218
    bowing, shear stress for, 76
    cell structure, 96
    climb, 95
    concept, 34

core
   stresses in, 70
   of a screw, 113
cutting of particles, 217
density, 73, 78
energy, 71
glide, 60
   resistance to, 65, 201
imaging of, 66
line, 59
loop, 64, 76
mobility
   in BCC crystals, 113, 126
   in ceramics, 201
multiplication, 76
   need for, 74
pinning, relation to fatigue limit, 116, 124
positive vs. negative, 63, 64
source, 77
types, 59ff
Disorder vs. order, 89
Displacement transducer, 28
Distribution coefficient, 362
Domain, magnetic, 337
Domain walls, 338ff
   energy, 339
   motion, 339, 347
   pinning, 348
Donor atom, 359
Donor level (state), 359
Dot product of vectors, 55
Double cross slip, 77
Double-butted tubing, 263
Drain (in a MOSFET), 367
Drift velocity, 353
Driving force
   for austenite decomposition, 165
   for coarsening, 105
   for grain growth, 97, 100
   for precipitation, 220
   of a reaction, 88
   for recovery, 95
   for recrystallization, 90
Dry cells, 378ff
Ductile-to-brittle transition, 113, 126
Ductility, 17, 39, 201
Easy directions (easy axes) of magnetization, 338
Eddy current losses, 342
Edge dislocation, 34, 59
Elastic modulus, 23
   physical basis, 30

   of composites, 301ff
   of fibers, 310
Elastic
   stiffness, 23
   strain, 23
Elastomer, 280
Elastomeric behavior, 281, 292
Electrical conductivity, 353
Electrical resistivity, 353
   effect of phonons, 354
   effect of solutes, 355
   use in monitoring recovery, 95
Electrochemical nature of corrosion, 14
Electromagnet, 333
Electron hole, 357
   spin, 336
Electron-hole pair
   creation in silicon, 357
   recombination, 358
Elemental and compound semiconductors, 372
Emitter (in a junction transistor), 366
Energy band, 182, 355
Energy-dispersive x-ray analysis, 319
Energy gap 357
Energy of a dislocation, 71ff
Energy product of a permanent magnet, 347
Engineering stress, (tensile or compressive), 25
   stress-strain curve, 37
   strain, 25
Enthalpy, 89
Entropy, 89
   in crystals, 109
   in polymers, 278
   in rubber, 292
Epoxy group, 315
Epoxy resins, 314
   formation reaction, 315
   in adhesive bonding, 266
Equilibrium state of a system, 87
   criterion for, 89
Equilibrium defects in crystals, 117
Equilibrium of surface tensions
   in brazing, 132
   in grain growth, 99
Error function, 195
Etching, 49
Eutectic composition, 134
   constituent, 143, 148
   divorcement, 149
   point, 134
   temperature, 134

Eutectoid
    reaction, in Fe-C, 163, 165
    steel wire, 312
Excess free volume in grain boundaries, 48, 49, 100
Exchange interaction, 336
Exhausted state, 360
Extensometer, 28
External field of a magnet, 337
Extra half-plane of an edge dislocation, 59, 95
Extrinsic semiconductors, 359ff
Extrusion of wheel rims, 227
Face diagonals in FCC crystals, 18, 54
Faraday's Law of magnetism, 334, 345
Fatigue, 13, 118ff
    carbon steel *vs.* stainless steel spokes, 124, 125
    crack growth, 118
    crack initiation, 118
    fracture surfaces, 119, 123
    life, scatter of, 121
    limit, 124
    resistance, 121
Face-centered-cubic crystal, FCC, 18
Fe-C phase diagram, 162
Fe_3O_4, 343
Fermi level, 356
Ferrimagnetism, 335, 337
Ferrite (BCC iron), 103, 162
Ferrite-stabilizing alloy elements in steels, 256
Ferrites, magnetic, 343ff
Ferroelectric, 376
Ferromagnetism, 335
Ferromagnetic oxide, 343
Fiber/matrix interface strength, 304
Fiber-pullout, 306, 322
Fiber-reinforced composites, 299ff
Fibers
    for composites, 306ff
    with carbon backbones, comparison of, 310
Fick's first law, 194
Fick's second law, 195
Field-effect transistor (FET), 367
Filler metal
    in welding, 260
    in epoxy resin, 320
Fillet brazing, 154, 155
Finite-element method, 324, 326
Firing of ceramics, 203
Flaws in brittle materials, 199, 211
Flow stress, 72
Fluctuating dipoles, 294
Flux in brazing and soldering, 133

Force on a dislocation, 75
Forces on links of pin-jointed frame, calculated, 236ff
Forsterite, 202
Fraction transformed
    in recrystallization, 93
    in austenite decomposition, 167
Fracture mechanics, 197ff
Fracture toughness, 198, 211
    of CFRP composites, 322
Frame geometry, road bike, 233ff
Frame, rear section, ideal geometry, 240
Frank-Read process, 77
Free energy, 89
Free-body, 23
    diagram, 239
Free-radical polymerization, 282
Friction, 182ff
Frictional force, 183
Front fork,
    calculation of stresses, 246ff
    moments of inertia, 246
Front section, resolved forces, 238ff
Fully annealed carbon steel, 106
Furnace brazing, 154
Furnace-cooled *vs.* air-cooled carbon-steel spokes, 174
GaAs 352, 372, 374, 375
GaP, 352, 373
Galvanic series, 15
Gas constant, 92
Gate (in a MOSFET), 367
Gauge section, 28
Gibbs free energy, 89
Gibbs phase rule, 151ff, 157ff
Glass fibers
    elastic modulus, 312
    drawing of, 312
Glass transition, 284
    temperature, 285
    of epoxy resins, 315
Glassy state, 285
Glide
    of a dislocation, 35
    of a mixed dislocation, 62
    of a screw dislocation, 61
    of an edge dislocation, 60
Grain boundary, 48
    energy, 48, 98, 99, 100
    mobility, 100
    motion in grain growth, 98
Grain growth, 97ff
    effects of impurity atoms and particles, 100, 101

kinetics of, 99
Graphite
    as a lubricant, 185
    fibers, 309
    structure and bonding of, 163
Graphitization of PAN, 309
Griffith crack, 211, 302
Griffith equation, 211
Guinier-Preston (GP) zones, 223ff
Gutta percha, 290
Habit relationship, 176
Hand brazing, 153
Hard ferrites, 348
Hard-sphere model, 18, 30, 50, 57, 58
Hardenability of steel, 188ff
Hardened steel, applications in the bicycle, 186ff
Hardening
    effect of grain size, 102
    of rubber (by vulcanization), 291
    of steel, 175
Hardness
    of martensite, 187
    decrease during tempering, 189
    tests, 108
Headphones, 331, 346
Heat of fusion, 141, 145
Heat treatment of steels, 176, 191ff
Heat-affected zone (HAZ), 261, 263, 270
Hematite, 345
Heterogeneous nucleation, 221
Hexagonal close packed crystal, HCP, 209, 210
Holes, 354, 357
Homogeneous nucleation, 221
Hooke's law
    in compression, 22
    in shear, 24
    in tension, 31
Hot working, 102
Hot-isostatic-pressing (HIPing), 206
Hot-pressing, 206
Hydrodynamic lubrication, 185
Hydrogen bond
    in water, 136, 295
    in nylon, 308, 310
    in Kevlar, 308, 310
Hydrostatic compression, 68
Hypereutectic, 140
    lead-tin alloys, 150
Hypoeutectic, 140
    lead-tin alloys, 146
Hysteresis

in an elastomer, 281
in magnetism, 340
I-beam, 241
Initial permeability, 341
Initiator in polymerization, 281
Injection molding, 279
Insulator, 356
Integrated circuits 352, 361
    fabrication, 368ff
Interfaces
    coherent, semi-coherent, incoherent, 222, 224, 225
Interfacial energy
    in precipitation hardening, 219, 222
    in sintering, 204
Intermetallic compounds, 136
Internal energy, 89
Interstices, 17
Interstitial solute, 17
    diffusion, 116
    in a BCC lattice, 114
Intrinsic semiconductior, 356
    conductivity *vs.* temperature, 358
Inverse spinels, 343
Investment casting, 258
Ionically bonded ceramics, 199
Ion implantation, 369
Iron-carbon phase diagram, 162ff
Isomers, 289
Isoprene, 289
Isotactic, 288, 296
Isothermal transformation diagram, 168
    for 1040 steel, 171
    for eutectoid steel, 167
Isothermal transformation in a eutectoid steel, 164ff
Isotropy, 25
Jominy bar, 168
Jominy end-quench test, 188
Jominy hardness curves 1040 *vs.* 4340 steels, 189
Junction devices, 365ff
Junction Transistor, 366
Kestral monocoque CFRP frame, 322
Kevlar, 308
    fibers, structure of, 308
    elastic modulus, 310
Kinetic energy of a particle, 92
Kinetics
    of grain growth, 99
    of recovery, 95
    of recrystallization, 91
Lack-of-penetration defect, 262, 265

Lamellar morphology
   of a eutectic, 148
   of a eutectoid, 164, 170
Laminations in tape heads and motors, 342
Large plastic strains, 26
Laser, 374ff
Latent heat of fusion, 141
Latex, 290
Lath *vs.* plate martensite, 189
Lattice parameter, 51
Lattice vacancies, 87, 94
   equilibrium concentration of, 109
Lattice-friction stress, 231
Lead-tin alloy system, 140ff
Leclanche cell, 378
Leathery behavior in a polymer, 287
Light-emitting diode, LED, 372ff
Light meter, 364
Lignin, 299
Linear polymer, 277
Lever rule, 138ff
Line tension of a dislocation, 76
Linear elastic fracture mechanics, 198, 211
Link, in a framework, 236
Liquid-phase sintering, 203
   of Si_3N_4, 207
Liquidus lines, 139
Load cell, 28, 42
Lodestone 333
Localized state in the band gap of a semiconductor, 359
Longitudinal strain in bent beam, 242
Longitudinal stress in bent beam, 241
Longitudinally stressed fibers, 301ff
Lost wax process, 258
Lotus monocoque pursuit bike, 324
Lubricants, 185
Lubrication, 185
Lugged joints for brazing, 153
Lugs, investment cast *vs.* fabricated, 258
Maghemite, 345
Magnesia, 200
Magnetic
   anisotropy, 338
   compass, 333
   constant, 334
   domains, 337ff
   ferrites, 343
   field lines, 333
   flux density, 334
   induction, 334
   moment, 334
   particles, 331, 345, 347
   permeability, 341
   recording, 330ff
   susceptibility, 335
   units, 335, 349
Magnetite, Fe_3O_4, 333, 336
Magnetization curve, 338ff
Magnetization processes, 339ff
Magnetization, 334
Magnetocrystalline anisotropy, 338
Mandrel
   for tube drawing, 263,
   for CFRP tubing, 320
Martensite, 165, 168
   defects in, 176, 187
Martensitic transformation, 175ff, 178
Martensite-start temperature, 178
Martensite-finish temperature, 178
Maximum
   energy product in a permanent magnet, 347
   permeability, 341
   stress in a beam, 244
Maxwell solid, 286
Maxwell-Boltzmann distribution, 91
Melt-spinning, 342
Metal tape, 345
Metal-matrix composites, 299
Metallic bonding, 182
Metallic glasses, 342
Metallographic
   examination, 45ff
   microscope, 47
   specimen, 45
Metastable precipitates, 223ff
Methane, 278
Mica, 203
Microcracks in brittle fracture, 126
Microhardness tests, 261
Microstructure
   of a brazed joint, 154
   of a partly martensitic steel, 177
   of Al-Li (TEM), 217
   of AL-Si casting, 318
   of a eutectic Pb-Sn alloy, 141
   of a hypereutectic Pb-Sn alloy, 151
   of a hypoeutectic Pb-Sn alloy (10% Pb), 149
   of a hypoeuthectic Pb-Sn alloy (30% Pb), 148
   of carbon steel spoke
      fully annealed, 106
      recrystallized, 104
   of cold-drawn carbon-steel spoke, 103, 175

of quenched 1.4%C steel, 177
of tempered 4130 steel (TEM), 190
of the CFRP tubing, 321
of the pin from a bicycle chain, 192
Microstructures
   in hypothetical simple eutectic system, 137
   of titanium alloy Ti-3Al-2.5V, 268
   of weld joint, alloy-steel frame, 261
   of weld joint in titanium-alloy frame, 269
   carbon-steel spoke, furnace-cooled *vs.* air-cooled, 161
Miller indices, 51
Mixed dislocation, 62
Mobility
   of dislocations as related to bonding, 65, 66, 201
   of electrons and holes, 353, 356, 372
Molecular weight of a polymer, 282
Molybdenum disulfide, 185
Moment arm, 23
Moment of inertia of beam, 243ff
Moments of inertia, various cross sections, 245
Monocoque CFRP frames, 322
Monomer, 202
Morita, Akio, 331
MOSFET, 368
Motor of Walkman, 331, 346
Mountain bikes, 254ff
Muntz metal, 153
n-alkane 278, 281
n-type extrinsic semiconductor, 360
n-type silicon, 360
Necking, 38
Necking criterion, 42
Network modification in silica, 311
Network modifiers, 311
Network polymers, 280, 314
Neutral axis of beam in pure bending, 241
Newton's law
   in mechanics, 235
   of viscous flow, 286, 311
Nickel-cadmium batteries, 378, 382ff
Non-equilibrium solidification, 149
Non-equilibrium solidus, 150
Normal spinel, 343
Nucleation and growth, 8
Number-average molecular weight, 282
Number-average degree of polymerization, 282
Nylon, 306
Nylon 6.6, formation reaction, 307
Nylon, elastic modulus, 310
Octahedral site
   of a BCC lattice, 114

of an FCC lattice, 343
Ohm's law, 353
Optical metallography, 45
Optical microscopy, 45
Orbital moment, 336
Order *vs.* disorder, 89
Ordered precipitates, 230
Ordinary (early bicycle), 6
Organically based fibers, 306
Orowan mechanism, 218
Overaging, 219, 227
Oxide films
   as lubricants, 185
   in passivation, 16
Ozone, effect on rubber, 291
p-n junction, 365, 373, 375
p-type silicon, 360
p-type extrinsic semiconductor, 360
Parabolic
   grain-growth law, 100
   strain hardening, 78
Paraffin, 278, 281
Paramagnetism, 335
Passivation
   in corrosion, 16
   of stainless steel, 16
Patenting, 312
Pauli Exclusion Principle, 182, 355
Pearlite in carbon steel, 106, 162
Pedaling forces, effects of, 247ff
Peierls stress, 231
Percent elongation, 39
Percent reduction in area, 39
Perfectly braced frame, 235
Perfectly plastic material, 183
Permalloys, 341
Permanent dipoles, 294
Permanent magnet, 346
Permanent magnet materials, 347
Permeability, 341
Phase diagram
   aluminum-copper, 216
   aluminum-silver, 223
   copper-silver, 153
   copper-zinc, 153
   hypothetical simple-eutectic system, 135ff
   iron-carbon, 162
   lead-tin, 140
Phase diagrams
   NaCl and $CaCl_2$, 134
   terminal solid solutions, 135

Ti-Al and Ti-V, 267
    use of composition axis, 134
Phase rule, 151ff
    derived, 157ff
Phase, defined, 17
Phenol-formaldehyde reaction, 314
Phonograph, 330
Phonons, 354
Photocell, 364
Photoconductor, 364
Photolithography, 368
Photomicrograph, 47
Photoresist, 369
Photoresistor, 364
Piano wire, 312
Pickle liquor, 345
Piezoelectric crystal, 376
Piezoelectric transducer, operation, 377
Pin joints, 235
Pin-jointed frame, 235
Plain carbon steel, 171
Plane strain, 198
Plane stress, 198
Plastic deformation, 32, 35
    process summarized, 81
Plastic-blunting mechanism of fatigue-crack growth, 119
Plastic
    constraint, 39
    instability in tension, 38, 42
    strain, equation for, 74
    work in fracture, 213
Plasticizer, 306
Playback head, 341
Plywood, 304
Point defects, 87, 94, 109
Poisson's ratio, 30
Polyamide, 306
Polycrystalline aggregate, 47, 99
Polycrystalline material, 47ff
Polyethylene, 278, 281ff
    crystal structure, 283
    crystalline spherulites, 283
    elastic modulus, 310
    low density *vs.* high-density, 284
    molecular structure, 281
    properties, 284
Polymer
    architecture, 296
    stabilizers, 291
Polymer-matrix composites, 299
Polymers, general characteristics, 277ff

Polymethylmethacrylate, 296
Polymorphs of Si_3N_4, 205
Poly(p-phenylene terephthalamide (Kevlar), 308
Polystyrene, 287, 296
Population inversion at a p-n junction, 374, 375
Porosity
    in castings, 259
    in ceramics, 204
Precipitation hardening, 216ff
    in tempered steel, 190
    heat treatment, 218
    mechanism, 217ff
    requirements for, 218ff
    stages of, 226
Precipitation of a second phase, 139, 148
Precipitation-hardened aluminum alloys, 228
Prepolymer for epoxy resin, 314
Primary cell, 378
Primary phases, 146
Pro-eutectoid ferrite, 171ff
Pure shear, 69
Purification of silicon, 362ff
Quench cracking, 192
Radius of curvature of beam, 243
Random solid solution, 115
Random walk, polymer chain configuration, 278
Rare-earth magnets, 348
Rate of substitutional diffusion, 116
Rayon, 306
Reaction bonding, 205
Reaction forces on a bicycle, 236
Recalescence, 175
Recombination of electrons and holes, 373
Recording tape, 344
    magnetic properties, 346
Recovery, 94ff
Recrystallization, 90ff
    temperature, 93
Rectifier, 365ff
Recycling of rubber tires, 280
Relative amounts of phases, 137
Remanence ratio, 340
Remanence, 340
Remanent magnetization, 340
Reptation, 316
Resistance strain gauge, 42
Resolution of force components, 234
Resolved stress components, 36
Restoring force
    in a spring, 22
    in rubber, 292

Retained austenite, 177, 178, 192, 193
Road bikes, 254
Rocksalt, 200
ROM, 374
Rope, 300
Rover (early bicycle), 6
Rubber, 278
Rubber elasticity, role of entropy, 292
Rubbery behavior of a polymer, 287
Rupture by void formation and coalescence, 38, 211
Rust pit, 14
S-N curve, 120ff
S-shaped curve, 8, 92, 166
Sacrificial anode, 14
Salt-water phase diagrams, 134
Sapphire, 368
Saturated state, 360
Saturation magnetization, 340
Scalar product of vectors, 55
Scanning electron microscope, 319
Scatter in the fatigue life, 121
Scattering of electrons by phonons, 354
Schmid's law, 36
Screw dislocation, 60ff
Screw dislocation
    core of, 113
    stress field of, 68
Secondary bonding, 294
Secondary cell, 378
Secondary hardening, 190
Second moment of area, 243
Segregation of solutes
    to dislocations, 115ff
    to grain boundaries, 100
Seizing (cold welding), 183
Self diffusion, 95, 115
Semicoherent precipitates, 224
Semiconducting devices, 364ff
    junction, 365
Semiconductors, 352
Semiconductor laser, 375
Semiconductor light meter, 364
Semiconductor-grade silicon, 353, 363
Shape anisotropy, 338
Shear modulus, 24
    relation to Young's modulus, 30
Shear strain, 24
Shear stress, 24
Sheet-like silicates, 203
Shrinkage porosity in castings, 259
SiN_4^{8-} tetrahedron, 205

Silica-based glass, 311
Silicate structures, 202
Silicon carbide, 200
Silicon nitride, 205ff
Silly Putty, 286
Simple cubic lattice, 51, 129
Single-domain particle, 346, 347
Sintering, 202ff
SiO_4^{4-} tetrahedron, 202
Size, for coating fibers, 312
Sliding friction, 183
Slip, 32
    and close-packed directions, 57
    and close-packed planes, 58
    dislocation motion, 34
    in FCC crystals, 57ff
    lines in deformed brass, 33
    step, 33
    system, 59
    vector, 60
Small-scale deformation, 25
Small-scale-yielding approximation, 198
$SmCo_5$ and Sm_2Co_{17} magnets, 348
Soap-bubble model for grain growth, 97
Soft magnetic material, 332, 341
Soft solder, 140
Softening during recrystallization, 90ff
Softening during grain growth, 102
Solar cell, 373ff
Soldering, 131ff, 141
Solenoid, 334
Solid lubricant, 185
Solid/liquid interface stability, 146
Solid solubility, 135
    factors that control, 144
    in Pb-Sn alloys, 143
Solid solution, 17
    in phase diagrams, 135ff
Solid-solution hardening, 113
Solid-state sintering, 204
Solid-state *vs.* liquid-state transformations, 162
Solidus lines, 139
Solubility limit, 136
    of sugar in water, 136
Solute atom/dislocation interactions, 71, 80, 113
Solute elements, effect on resistivity, 354
Solution treatment, 218
Solvus line, 136
    experimental determination of, 145
Source (in a MOSFET), 367
Space lattice, 128

Spheroidization of pearlite, 166
Spherulites in polyethylene, 283
Spinel ferrites, 336
Spinel struture, 343
Spinnerette, 307
Spokes
    corrosion of, 13ff
    stress-strain curves of, 28
    tensioning of, 11
Spring constant, 22
Stability of a particle, 220
Stabilizers in polymers, 291
Stacking fault, 209
Stacking-fault energy, 209
Stainless steel
    passivation, 16
    vs. carbon steel for spokes, 13ff
Stages of annealing, 90
Static friction, 183
Stiffness
    of bicycle frames, 252ff
    of rubber, 291
Stored elastic energy, physical basis, 31
Stored energy of cold work, 90
Strain, 23
Strain energy per unit length of a dislocation, 72
Strain gauge, electrical resistance, 28, 42
Strain hardening, 34, 38, 41, 56ff, 78ff
    in wire drawing, 41
    mechanism of, 78
Stress at a point in a stress field, definition, 67, 68
    components, 67, 68
Stress concentration
    factor, 198
    in fatigue, 118, 120, 122
    in a plate, 197ff
Stress field of a dislocation, 67
    of a mixed dislocation, 71
    of a screw dislocation, 68ff
    of an edge dislocation, 70, 83, 84
Stress relaxation in a viscoelastic material, 287
Stress, bending *vs.* axial loading, 272
    types, 22
Stress-concentration factor, 197, 198
Stress-intensity factor, 198, 211
Stress-strain curves, 37ff
    annealed *vs.* cold-drawn spoke, 41
    carbon steel *vs.* stainless-steel spoke, 29, 57
    parabolic hardening, 78
    of fibers, 313
Strip-chart recorder, 37

Substitution transformation, 8, 92
Substitutional solute, 17
    diffusion, 115ff
    interaction with dislocations, 113
    solid solution, 17
Substrate of precipitate nucleation, 221
Surface energy
    in fracture, 212
    in sintering, 204
    in wetting, 131ff
Surface hardening, 193ff
Surface tension
    in brazing, 132
    of a bubble, 97
Syndiotactic, 296
Synthetic rubbers, 292
Tacticity, 288, 296
Tangent-spoked wheel, 12
Tape head, 341
Transmission electron microscopy, TEM, 66
Temper designations for aluminum alloys, 228
Temperature coefficient of resistivity, 354
Tempering of a hardened steel, 178, 189ff
Tensile
    strain, 23
    stress, 22
    test, 27ff
Tensioning of spokes, 11
Terminal solid solutions, 143
Tetragonal distortion
    in BCC crystals, 114
    in martensite, 177
Tetragonal unit cell, 129
Tetrahedral site in FCC crystal, 343
Thermal
    activation, 88
    analysis, 143
    energy, 91
    shock, 311
    stresses
        in quenched steels, 191
        in quenched Al-alloy bicycle frames, 264
Thermistor, 364
Thermoplastic behavior of polymers, 279, 285
Thermoset, 280
Thin-film technology, 368
Three-phase equilibrium, 152
Tie line, 137ff
TIG welding, 260
Tin, crystal structures, 156
Tire cord, 307ff

xvii

INDEX

Tires, 299
Titanium-alloy frames, 252ff, 266ff
Toughness, 17, 211
trans-polyisoprene, 290
Transfer length, 305
Transistors, 352, 366, 367
Transverse compression in a wire-drawing die, 40
Transverse tension in a necked tensile specimen, 39
Transversely stressed fibers, 302
True strain, 26
True stress, 26
Tube geometry in bicycle-frame design, 252
Twinning in crystals, 209ff
Two-phase equilibrium, 138ff
Undercooling
    eutectic, 148
    eutectoid, 167
Unit cell, 51, 129
    of polyethylene, 283
Ultimate tensile stress, UTS, 37
Vacancy, 87, 94
    concentration, 109
    migration, 95
Van der Waals bonding, 294
Vinyl-type polymers, 287, 296
Viscoelasticity, 285ff
Viscoelastic modulus, 287, 292
Viscosity
    of a lubricant, 186
    of polymers, 278
    of silica glasses, 311, 312
Viscous flow, 312
Voight model, 286
Vulcanization, 278, 290
Walkman, 331ff
    magnetic components of, 332
    operation, 331
Wear, 184
Welding of bicycle frames, 260ff
Welding defects, 262, 265
Wetting in brazing, 131ff
Wheel rims, 227
Wheels
    bicycle *vs.* wagon, 10
    open spoke (aerodynamic), 325, 326
    tangential *vs.* radial spoking, 12
Widmanstätten α in TiAlV, 267
Widmanstätten ferrite, 174
Wire drawing, stresses in the die, 40
Wöhler curve, 120ff
Wood-frame bicycle, 250

Workability of glass for fibers, 311
X-ray
    analysis, 145
    diffraction
        Bragg's law, 82
        from rubber, 292
    map, 319
    scattering, 82
X-Y recorder, 28
Yield stress, 26
    characterization of, 28
Young's modulus (see also Elastic modulus), 23
    of various materials, table, 27
Zinc on carbon steel, 15
Zinc-carbon cell, 378, 379ff
Zipp monocoque CFRP bicycle frame, 323
Zone refining, 362ff

FIRSTBORN

The Life of Luis Fred Kennedy 1908–1982

as told by his son
Fred W. Kennedy

# FriesenPress

One Printers Way
Altona, MB R0G 0B0
Canada

www.friesenpress.com

Copyright © 2022 by Fred W. Kennedy
First Edition — 2022

Cover design by Sarah Kennedy

All rights reserved.

No part of this publication may be reproduced in any form, or by any means, electronic or mechanical, including photocopying, recording, or any information browsing, storage, or retrieval system, without permission in writing from FriesenPress.

As a biographer, I have re-told the story of my dad's life by adhering to the best of my ability to actual events and experiences as I remember them. The essential purpose of the narrative has been to portray the essence of the man in the truest way possible. If anyone or anything has been misrepresented, this has not been my intent.

ISBN
978-1-03-914291-6 (Hardcover)
978-1-03-914290-9 (Paperback)
978-1-03-914292-3 (eBook)

1. BIOGRAPHY & AUTOBIOGRAPHY, BUSINESS.

Distributed to the trade by The Ingram Book Company

For my brother

Francis (Paco) Xavier Kennedy
1940–2014

In Praise of
FIRSTBORN

This is a must-read book for anyone interested in the history of Jamaica over the last 100 years and the role played by GraceKennedy Ltd. during this period, especially in helping Jamaica survive the turbulent 70s and early 80s. There is no one more qualified to write this history than son of Luis Fred Kennedy, author Fred W. Kennedy, who takes us on an intriguing journey as GraceKennedy morphed from a very small company into a large dominant conglomerate having a staff complement of over 2,000.

Errol Donovan Anderson
Former Director/Group Corporate Secretary,
GraceKennedy Ltd.

In *Firstborn: The Life of Luis Fred Kennedy*, Fred Kennedy has successfully painted an authentic portrait of a complex man, our father. The historical, political, and social context offer a riveting backdrop to the timeline of Luis Fred Kennedy's growth and development to become the catalyst for the expansion of a company that would become one of the Caribbean's largest corporate entities.

Dr. Mary Cameron
Professor Emeritus of Education, |The University of Findlay.

FIRSTBORN is meticulously researched and masterfully curated. As though guiding the reader on a multimedia museum tour, Fred Kennedy brings to life some of Jamaica's rich history in his account of the lives and legacies of generations of his family. I wish I had had access to historical accounts as vivid and engaging when I was a student.

Dr. Mary Anne Chambers OOnt
Retired business executive and former cabinet minister; corporate director, university chancellor and author.

An intriguingly readable book that meticulously weaves together the life of Luis Fred Kennedy, family histories, personal emotions, GraceKennedy's century long journey, and Jamaica's transition from colony to independent nation. I joined GraceKennedy after Luis Fred Kennedy's retirement, and so never met him personally, but felt his enduring presence through our company's oral history. *Firstborn* has now given me a vivid three-dimensional view of the man and his times. I am so grateful.

Douglas Orane
Retired Chairman and CEO, GraceKennedy Limited.

Based on a rich array of sources, Luis Fred Kennedy's family and business story, as lucidly narrated by his son, is a significant contribution to Jamaica's social and economic history, post slavery. Coming from two generations of Jamaicans in commerce, Luis Kennedy, with visionary leadership and a dynamic management team, built a commercial entity that is the pride of modern Jamaica. A rewarding read for scholars and the general public.

Dr. Swithin Wilmot
Caribbean historian.

Was Luis Fred Kennedy a great man?
Absolutely!
He was a giant of a man.[1]

Was he perfect?
No!

He had faults like everyone else,
but his good qualities far outweighed them.

Did I love him?
You bet!

Lydia Loinaz Kennedy
January 7, 1997

---

1 Lydia M. Kennedy, "A Giant of a Man," 1997. My mom gave credit to Aaron Matalon, who used the phrase to describe Luis Fred. See Chapter 18, note #37.

Lives of great men all remind us
We can make our lives sublime,
And, departing, leave behind us
Footprints on the sands of time.

From "A Psalm of Life"
by Henry Wadsworth Longfellow

The greatness of a man is not in how much wealth
he acquires,
but in his integrity and his ability to affect those around
him positively.

Bob Marley

Foreword

This biography tells the story of my father, Luis Fred Kennedy. The motivation came from a conviction of the importance of memorializing his life, to share his history and legacy. He contributed significantly to Jamaica's national development, and in this pursuit demonstrated the character of a man who became a staunch nationalist and fierce advocate of private enterprise.

Once I decided to undertake the project, a day did not pass without recalling vignettes of my dad's life. Everything started to jog my memory: places we had been together, shared hobbies of gardening and boating, vacations and trips to New York and Montreal, casual, daily conversations on the verandah at home, and the more serious business-like and philosophical discourses that he relished.

As a firstborn child, Luis Fred's position within the family granted him privileges that shaped his character. It affected his personality and his relationships with his siblings and the entire Kennedy clan. Physically, he appeared older than his years, and intellectually, he was superior to his peers. As he matured, he grew to be ambitious and courageous, undaunted by challenges. At age twenty-one, after the untimely death of his father, he became head of his family and co-manager and part owner of the family business. The expectations were high and the opportunities to succeed limitless.

In telling his story, I re-discovered him to be a man who was spiritual, intellectual, and passionate, a "giant of a man," as my mom called him. He was raised in colonial Jamaica, a society riddled with prejudices, which he spent a lifetime striving to overcome. He was complex, driven by opposing forces of nature: of courage and fear, of love and hate, of forgiveness and anger. He was motivated by a fierce loyalty to family and strong religious beliefs, which tempered his otherwise volatile nature.

As I composed chapters, I became conscious of "narrative distance."[1] I intentionally switched between a more objective vantage point (describing the man, Mr. Kennedy, as others would have seen him) and a more subjective one (describing him as "Dad" from the point of view of his son). In addition to expository prose, I included conversations

---

1 Jack Hart in *Telling True Stories* (2007), 103, uses the phrase "narrative distance" to describe the stance of the writer as the story's narrator —deciding how close to stand to the action.

and stories that, although not verbatim, are re-creations of actual dialogues and events.[2] Employing differing narrative styles enabled me to show, in addition to others' perceptions of Luis Fred, my own interactions and relationship with him as my father.

Volumes of family archives, pictures, letters, and mementos were rich sources of information. I am indebted to my sister, Mary, for sharing collections of personal letters that our parents wrote to one another during and after their courtship years in the 1930s and '40s. Other letters included those written by my grandfather, Fred William Kennedy, to his son in the 1920s, and in more recent times, those that I received from my dad in the 1960s. I was fortunate to have records of Tillie Kennedy's stories. My mom wrote a detailed tribute to Dad in 1997 for the seventy-fifth anniversary of Grace, Kennedy, and at the age of ninety-one, two months before her passing, recorded her memoirs in an interview with my wife, Georgianne Thompson Kennedy.

My brother Paco had a father-son relationship more complex than mine because he knew Dad in both spheres of his life, business and personal. He loved and admired him like no one else. I regret that I never asked Paco to record his memoirs before he passed in 2014. He had a photographic memory of both the family and of Grace, Kennedy's history. I have dedicated the biography to him.

Primary sources included newspapers online (in particular, *The Daily Gleaner*), archival material of Grace, Kennedy & Co. Ltd., and genealogical information—birth, marriage, death, and immigration records. Available on the Internet were journal articles, digitized government records, and publications of rare book selections, formerly accessible only in hard copy. DNA technology helped. As a specialist in the field, my wife, Georgie, uncovered clues to my paternal family history and contributed information about other families germane to the story—the Graces, Moss-Solomons, and Alexanders.

Douglas Hall's *Grace, Kennedy & Company Ltd: A Story of a Jamaican Enterprise* was the quintessential secondary source. Published in 1992, it records the history of the company, combining research with anecdotal accounts of the lives and events of seventy years of the history of Grace, Kennedy. Frank Cundall's histories of Jamaica at the turn of the nineteenth century, Richard Hart's and Frank Hill's firsthand accounts of the 1930s, books by Michael Manley, biographies of Norman Manley, Alexander Bustamante, and Edward Seaga, and Patrick Bryan's extensive treatises on race, class, and culture in Jamaica are among the many other sources.

Luis Fred Kennedy's legacy will forever be tied to the name of Grace, Kennedy.[3] Detailed accounts are given of its history and the role he played in its development. He was a director of the company for fifty-two years, chairman of the board for

---

2   These recreations, both of my mom's and dad's stories and sayings, appear in italics.

3   Grace, Kennedy & Co. Ltd. officially changed its name to GraceKennedy Ltd. with the Registrar of Companies in 2005. Original name will be used in this biography, which covers the period in the company's history, 1922–1982.

twenty-nine, and governing director for twenty-six of those years. He was an entrepreneur by nature, a risk-taker, innovative and creative, motivated by a genuine desire to build wealth for the common good.

Grace, Kennedy changed from a private company to one publicly listed on the Jamaica Stock Exchange in 1986. It has grown over a one-hundred-year period to be one of the largest in the Caribbean—a multinational corporation with shareholder equity valued at over J$65 billion. It is a beacon of private enterprise, founded on a core set of values and committed to the welfare of Jamaicans at home and abroad. GraceKennedy has withstood the test of time—one hundred years of integrity and prosperity. This is the legacy for which all those who have been and are associated with Grace can be proud.

For those who knew Luis Fred Kennedy, I trust that this biography will shed new light on the kind of person he was, and for those who did not know him, that they will be inspired to read about a man whose leadership helped shape the political and economic landscape of Jamaica. He ranks among the best that this nation of ours has produced. He empowered others to succeed. He celebrated success, and with a magnanimous heart mentored others to realize their own potential. Neither money nor class nor privilege made him who he was; he had little regard for these. He was driven by a deep love of family and by values for which he never made compromises.

I am blessed for having had such a father.

Happy reading. Think of the biography as a work of the imagination, in the words of nineteenth century English poet Edmund Gosse, "a faithful portrait of the soul in its adventures through life."[4]

Fred W. Kennedy.

---

[4] Excerpt from an article by Sir Edmund William Gosse (1849–1928), the Lloyd Roberts Lecture (1923), published in *Encyclopedia Britannica*. His autobiography is entitled *Father and Son* (1907).

Contents

INTRODUCTION (1976)	8
PART ONE First and Second Generations Post-Emancipation (1839–1908)	17
PART TWO Coming of Age (1909–1930)	47
PART THREE Intrepidness of Youth (1930–1946)	125
PART FOUR Prime of Life (1947–1961)	197
PART FIVE Prosperity (1962–1971)	251
PART SIX Transitions (1972–1976)	315
PART SEVEN Exile (1977–1982)	371
POSTSCRIPT (1982—2022)	415
APPENDICES	443

INTRODUCTION

(1976)

At home, 45 Lady Musgrave Road, Kingston (1976)

Troubled Times

June 1976

"Manley has declared a f *** ing State of Emergency."[5]

I seldom heard my dad swear, least of all at home.

He sat in his special wicker rocking chair on the front verandah. No one else in the family used it; there was no spoken or written rule, it's just that everyone knew it was reserved for him, and for him alone. He had a unique way of sitting, with one leg draped over the arm of the chair, and the other leg tucked underneath. Next to him on a table was a transistor, battery-powered radio, tuned into Jamaica Broadcasting Corporation (JBC) 5:30 PM news broadcast. Other constant companions were an Agatha Christie novel and the Time magazine, which he read faithfully every week but with difficulty because of a loss of vision in his right eye from a mild stroke he had suffered.

Our two-storey home at 45 Lady Musgrave Road in Kingston, Jamaica, faced north with an unobstructed view of the Blue Mountains. Dr. and Mrs. Mosley, from whom my parents had bought the house in 1949, had strategically placed the verandahs, upstairs and down, to capture the beauty of the landscape.

The mountains took on a different aspect when the rains cleared in the afternoon. The water washed them clean, and when the storm clouds passed and the sun came out, they turned velvet blue. Dad said they had a grandeur akin to the Alps, tall and majestic, with a climate and geography altogether different from the rest of Jamaica. A day did not pass without him remarking, "What a magnificent sight; one never tires of it."

He turned down the radio.

"Sorry to disturb you, Dad."

"That's no trouble. I will catch the 6:00 news on RJR." He listened to news broadcasts at every opportunity he had, morning and evening. "We live in troubled times, son. I'm being faced with one hell of a decision."

"What's that?"

---

5   Governor General, Sir Florizel Glasspole, on advice of Cabinet, declared a National State of Public Emergency, effective noon, June 19, 1976. Security forces were given instructions by the government to use their powers to "lock up and keep locked up all persons whose activities are likely to endanger the public safety." *The Sunday Gleaner*, June 20, 1976.

"The future of our family and the company. I'm positive Manley will win a second term. It could even be a landslide victory, for he has the masses behind him. He's stuck between a rock and a hard place."

"How's that?"

"He has made promises that he can't go back on. The more he talks, the deeper the hole he digs for himself. He's driving away the middle class that he needs so desperately."

"His social reforms have helped vast numbers of people, though, Dad."

"We can't afford everything he's done ... buying bauxite companies, sugar factories, land to give to farmers."

"Jamaicanization is a good thing, don't you think?"

"Yes, I've always said so, but private capital is a funny animal. It gets scared easily and starts to run like hell."

"What about Grace, Kennedy?"

"Carlton[6] is the right man for the job but the challenges are enormous–government monopolies on imports, shortage of foreign exchange, scarcity of basic food supplies. People can't even get flour and rice to buy, for Christ's sake. We're in a lot of trouble. You can't create wealth without employment, son. And then there is the violence. Look, for example, at your situation–you haven't been able to report to classes."[7]

"Yes, we have been holding fifth-form classes at Kingston and St. Andrew Parish Church. Trench Town has become a no man's land."

"I could never for once have imagined the atrocities that have been committed. We have seen horrific incidents[8] over the last couple months: attacks on police and gangs setting fire to buildings."

"He had no choice but to declare a State of Emergency."

"Well, that's because he may be losing control of the country; the people he supports have become too powerful. He knows it too. It feels as if we are living in a bloody police state–helicopters flying overhead, curfews. It's as if Jamaica is under siege."

---

6   Carlton Alexander had been appointed group managing director of Grace, Kennedy & Co. Ltd. three years earlier in August 1973. Dad remained as chairman of the board.

7   In 1975, I had taken up a position to teach at Trench Town Comprehensive High School. Classes had been suspended and the school closed for several weeks because of increased violence in the area. In July 1976, when I returned to the school to deliver final report cards, my car was surrounded, and I was robbed by young men entering the school yard.

8   On May 19, 1976, the residential property at 182–184 Orange Street in Kingston was the scene of a devastating fire in which eight adults and three children died. Gangs blocked police and firemen from entering the premises to put out the fire. It was reported that one man grabbed a baby from a woman fleeing the scene and tossed the child back into the fire. The governor general appointed a Commission of Enquiry to report on the circumstances of the fire. The conclusions were that the cause of the fire was arson, an operation politically instigated and carried out by gangs of youth.

"Do you think there's any truth to the rumour that there are plots to overthrow the government? He's pleading with people to help him find out who is destabilizing the country. The threats could be real, Dad."

"I expect so, but he has only himself to blame. The real threat in my mind, son, is the erosion of civil rights–what we are seeing with the new powers of arrest and indefinite detentions. The Gun Court Act[9] that was enacted into law a couple of years ago has got to be unconstitutional, I don't care what the Privy Council says. And now the State of Emergency gives the government even more powers."

"Desperate measures for desperate times."

"It's aimed to quell violence, but my hunch is it will make matters worse. There will be nothing to stop the government from mass incarcerations." He stared off into the distance. The silence between us was not awkward. He would often break the silence by humming the tune of his college anthem, "Holy Cross, O Holy Cross, Thy Spirits Loyal, True and Strong," lyrics written to the tune of the Christmas carol, "O Christmas tree, O Christmas tree / How lovely are thy branches." He was tone-deaf, but for some reason he could always stay on key for the hymn of the "Holy Cross."

"I have finally told your mother. I didn't want to worry her more than was necessary–doctors are saying I should consider medical treatment abroad."

He still looked robust. He had lost weight, which improved his physical appearance, the muscle tone in his arms and legs seeming youthful.

"What are you thinking?"

"I want us to explore options."

"What do you mean?"

"If rumours are true, we're in for trouble. Our family is being threatened, your mother has told me of verbal attacks against her, and the company's future is uncertain. There's a lot at stake."

"We can't just walk away."

"I'm not walking away, son. The writing is on the wall; Jamaica is moving towards a totalitarian state."

"What are you thinking can be done?"

"We should think of ways to move money and to establish residence abroad. Not for my sake, for I won't be around much longer. It's for the company, for your mother, and for you all, to ride this out until there is a change of government."

I couldn't help thinking that, given his bullish nature, he would find a way to succeed, to put safeguards in place.

"Perhaps there is a way for family members who are resident abroad to get money out for us. We'll see."

---

9   The Gun Court was established by Parliament in 1974 to combat rising gun violence. It was empowered to try suspects *in-camera* without a jury. On June 25, 1976, it was ruled constitutional by the Privy Council.

He noticed me looking down at the floor. The marble tiles took on a purplish colour as they caught the evening light. The sun slipped below the horizon in an instant, and the evening turned dark. Twilight was short in Jamaica, even in the summer months. We sat in the semi-dark.

Draco came running up from the garden on to the verandah. The Doberman pinscher headed straight for Dad–resting its snout on his lap, waiting to be petted. (Our two previous dogs, Bruno, a stray Jamaican mongrel, and Flippo, so called because it had only three legs, both had a special affection for Dad.)

"Imagine it has come to this, needing a watch dog. Our home is like Fort Knox: lights, iron bars." He shifted his legs in the rocker and ordered Draco to lie down. "Your Aunt Louise and Uncle Simon are thinking of packing up; they feel the country is being infiltrated with commie spies. Maybe they're right. Trouble is you never know what is fact from what is fiction. Manley says he's not a communist, but then we've seen headlines in the papers like, "Days of Capitalism are Over."[10]

"It's difficult to know for sure, Dad."

"You remember when Manley returned home after a visit to Cuba? You remember what he told us? 'No more millionaires–there are five flights a day to Miami.'[11] If that isn't an invitation to leave Jamaica, my name isn't Fred Kennedy."

"I don't think he's a communist, Dad."

"He could bloody well have fooled me."

I got up to go. "I should see if Mom needs any help with supper."

"A question for you, son."

"What's that?"

"How much is a haircut these days?"

"Not sure, it's been a while." I chuckled. "I thought you just got one, though."

"Yes, I did," he said with a twinkle in his eye.

"Oh, I understand."

"Why don't you let your ol' man treat you?"

"Not to worry, Dad, I'm fine."

"No, I'm serious, son. You're starting to look like one of those radicals at the university."

We both laughed.

August 17, 1976

Dad had a medical emergency, a blood clot in his right leg, which confined him to bed. My Uncle Che[12] and I assisted paramedics to carry him downstairs from his bedroom to the ambulance that transported him to the Norman Manley International

---

10   *The Daily Gleaner,* November 18, 1974.

11   Ibid., July 16, 1975, page 1 headline, "No one Can Become a Millionaire Here—PM"

12   Uncle Che, my dad's brother, José.

Airport. Doctors advised that he be flown to Miami for treatment of a thrombosis caused by peripheral arterial disease. Mom accompanied him, and the rest of family joined her a few days later.

Carlton Alexander made the announcement at a meeting of the board of directors on August 26. He "reported with sadness that Mr. L. F. Kennedy was ill and that he was under observation at the Cedars of Lebanon Hospital in Miami."[13]

The chairman of Grace, Kennedy & Co. Ltd. had fallen critically ill.

---

13   Grace, Kennedy & Co. Ltd., minutes of the Board of Directors Meeting, August 26, 1976.

PART ONE

First and Second Generations Post-Emancipation (1839–1908)

(Photo, taken by author, of coconut palms at Discovery Bay.)

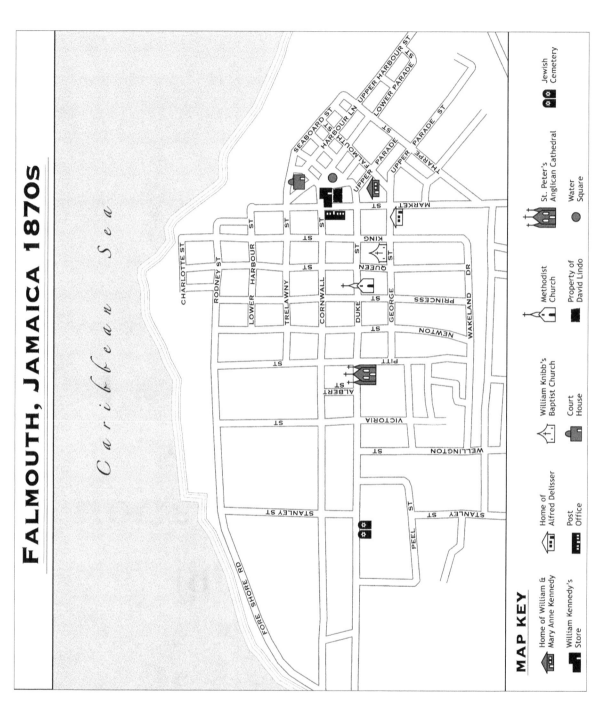

Map of Falmouth, Jamaica, designed by Mapping Specialists Ltd.
© Fred W. Kennedy

1

My Great-Grandfather (19th Century)

The story of the Kennedy family begins with Luis Fred Kennedy's paternal grandfather, William. He was born a free man in 1839, one year after full Emancipation, in Stewart Town, located in the hills of the Parish of Trelawny, Jamaica. Information gathered from residents of Falmouth establishes a family connection to Biddeford and Hopewell Estates in the Parish. In the 1830s, the twin properties contained a sugar plantation with a population of 197 slaves, located close to Jackson Town, Trelawny, where Alexander Kennedy Esq. had been an overseer. Residents of Falmouth[1] who knew of the Kennedy family in Trelawny have suggested that Alexander was an ancestor of William Kennedy. To date, no genealogical records have verified the connection.

What is certain is that William Kennedy was of mixed heritage, African and Irish,[2] a progeny who became a member of the first generation, post-Emancipation, of free

---

1 My sister, Celia Kennedy, lived and worked in the Parish of Trelawny during the 1970s. She was interested in family history and came to know elders in the community, who told her about the William Kennedy family.

2 My Y chromosome DNA testing reveals that my patrilineal ancestry is Irish. Y chromosome DNA testing is used to explore direct male-line ancestry; the Y chromosome, like the patrilineal surname, passes down virtually unchanged from father to son.

"brown"[3] men born in Jamaica. His father's ancestral line can possibly be traced back to Long Andrew Kennedy in Tipperary, Ireland.[4] His mother was likely of African ancestry—a free Black, free coloured,[5] or person formerly enslaved.

Dr. Swithin Wilmot, who specializes in post-Emancipation Jamaican history, commented in an interview:

> This would have meant that your great-grandfather was born into what would have been known then as a free coloured class. The free coloureds prior to 1830 were recognized as free men, free people but denied basic civil rights. There were limits on which property they could inherit, they couldn't give evidence in court against white people, they couldn't get jobs in government, and they couldn't vote.[6]

William Kennedy grew up in the post-Emancipation period, a time that was challenging for the ex-slaves. Planters were recalcitrant, paying out low wages and charging high rents for accommodations and use of provision grounds.

> The planters do not offer continuous labour and are in the habit of dismissing their labourers on the slightest interruption ... Were the wages more adequate to the growing wants of the Creole Negro, and work more regular and constant, there can be no doubt that large numbers of the young people who possess no land might be attracted to the estates.[7]

---

3   "Brown" and "coloured" are used in the historical context of post-Emancipation Jamaica to designate people of mixed race. "Coloured" was used in the Jamaican census of 1844 and "Brown" in the census of 1861 along with "White" and "Black" as categories of race.

4   Long Andrew Kennedy (1736–1824) was the ancestor of many Irish Kennedys. He was reputed to have been six feet, ten inches tall—hence, his first name, Long. It is yet unknown which, if any, of his descendants migrated to Jamaica. Christopher Kennedy, who lives in Australia, contacted me and my cousin Tony Kennedy, to confirm this connection of a common Irish Kennedy ancestry from Tipperary.

5   Racist ideologies inherent in terminology (mulatto, sambo, quadroon, and mestee) during slavery to describe persons of varying shades of colour carried over into the post-Emancipation period with the usage of labels such as "brown" and "coloured." Anita Kalunta-Crumpton (2019) and others argue "for the elimination of the term people of color and related labels from popular usage."

6   Interview with Swithin Wilmot, Discovery Bay, May 12, 2019.

7   Edward Bean Underhill, *The West Indies: Their social and Religious Condition* (1862), page 375. Underhill, joint Secretary of the Baptist Missionary Society (BMS), made a missionary tour of Jamaica in 1859–1860.

As a result, labourers with newfound liberty left the plantations in droves and set themselves up as freeholders. By the 1840s, close to 200,000, more than half of the population, were engaged in agriculture.[8]

Villages and towns sprouted up throughout the island and demand for goods and services likewise increased. "Old market towns expanded, and new market towns were created, in response to the increase in trade which involved the free people as consumers for dry goods."[9] "There had been both a geographical and functional spilling-over from estate population."[10] Skilled workers, carpenters, blacksmiths, masons, painters, and plumbers, previously employed on the plantations, sought work in the towns. "Coloureds," Jews, and later Chinese were the ones who predominantly became retail and wholesale merchants.

Merchants were well established within their communities. In 1861, the Jamaica Census listed the number of merchants in Jamaica to be 433 and the number of storekeepers, 544.[11] The average annual income for a shopkeeper was estimated at £300 and that of a merchant, £700. The incomes were among the highest, comparable to attorneys, bankers, and public servants. In contrast, a freeholder who owned and farmed an acre of land could not expect more than an annual net yield of £20.

Little is known of William Kennedy's schooling and upbringing in the years after Emancipation, but whether by way of financial support from wealthy families in his childhood or through association by marriage, William Kennedy was able to secure the means to establish himself as a merchant.[12]

He married Mary Ann Eberall, December 18, 1867, in the Wesleyan Methodist Church in Falmouth, Trelawny. She too was of mixed heritage, eldest child of Francis Lunan Eberall, occupation, printer, and Johannah, originally from Spanish Town, St. Catherine, where Mary Ann had been born in the early 1830s.

William was baptized as an adult at the Methodist Church in Falmouth, at age twenty-one, in 1860,[13] and Mary Ann at St. Peter's Anglican Cathedral, May 1851, when her family moved to Falmouth from Spanish Town.

---

8 In the 1830s after Emancipation, "the total black and coloured population numbered 380,000 souls and free coloureds, 60,000." By the middle of the century many of the 15,000 whites had left the island. The 1844 census showed a total population of 361,657, a decline caused by disease (cholera and smallpox), high rates of child mortality, and lack of medical resources. William G. Sewell, *The British West Indies* (1861), 245–246.

9 Swithin Wilmot (2019).

10 Douglas Hall, *Free Jamaica 1838–1865* (1959), 212.

11 Ibid., 266.

12 Original land deeds and bills of sale have been passed down from my great-grandfather. It is unclear what the association was between him and Catherine and Peter Dublin, the names which appear on the deeds. The Dublins purchased land from Mary Reeves, resident of Falmouth, from as early as 1852.

13 Baptismal Certificate, Methodist Church: Kennedy, William, adult, res Falmouth, bap 9/30/1860 by Samuel Smyth, p. 50 #1733. http://www.jamaicanfamilysearch.com/Members/MethodistFal07.htm.

My dad was proud of the fact that we were related to the Lunan family, prominent landowners, and printers by profession, residents of the Parish of St. Catherine. He showed a keen interest in family history and wanted to share with us what his father had shown him at the Anglican Cathedral of St. Jago de la Vega in Spanish Town. I remember it was a bright sunny day, March 1978. Inside, on the south wall of the nave of the church, was a monument in the name of one of his ancestors, John Lunan. Dad pointed to the inscription engraved on a tall marble tablet, discoloured by age. *"My dad brought me here when I was a boy,"* he said. *"John Lunan is an ancestor on his mother's side."*

> Sacred to the memory of
> the Honorable John Lunan
> who departed this life
> On the 23rd December 1839
> in the 69th year of his age.
> He was for 44 years a resident
> of the parish of St. Catherine
> and at different periods filled the
> situations of member of Assembly,
> Custos Rotulorum and
> Assistant Judge of the Grand Court.
> Few men have lived more honoured,
> respected and beloved, not only for
> his superior talents but also for his
> amiable and charitable disposition …

Even though Dad's maternal great-grandfather was named Francis Lunan Eberall, and one of his uncles, Lunan Kennedy, the Eberalls are nowhere to be found in the John Lunan family trees. More genealogical clues need to be uncovered. It is evident that John Lunan was a direct ancestor but his wife, Elizabeth Lunan, was not.

William and Mary Ann Kennedy raised seven children (born over a span of fourteen years), each baptized as infants in the Wesleyan Methodist Church located on Duke Street in Falmouth. The eldest was Henrietta Eliza (born 1870) followed by Lunan Dexter (1873), Fred William Kennedy (my grandfather, 1874), Fillan Mable Josephine (1877), Theodore (1880), Rose Eliza (1882), and the youngest, Lena Stewart (1884).

Their home was situated on Falmouth Street close to Market Street—a two-storey wooden structure of Jamaican Georgian architecture, with an upstairs verandah decorated in fretwork. In the 1960s I visited the home with my dad; it was then dilapidated and termite-ridden and has since been demolished.

Founded in 1769, Falmouth had flourished as a market and seaport town in a period when sugar was king in Jamaica, then one of the largest sugar producers in the world.[14] It attracted commerce and trades of all types, becoming home to planters, import merchants, wharfingers, builders, and retailers.

In William Kennedy's time, Falmouth was a town bustling with activity. On a tour around the island in the 1860s, Edward Bean Underhill, Baptist described the town:

> Falmouth does not exhibit the decayed appearance of Kingston. During the season for exporting sugar, its harbour is sometimes filled with ships. The houses are well built, and the stores display an abundant stock of merchandise. More than one storekeeper informed me that the retail trade with the black population is a flourishing one.[15]

The grand courthouse, constructed in 1817, housed plays, musical concerts, and lectures. The magnificent building attracted hundreds of visitors from neighbouring parishes to attend elaborate balls. It was the finest in the island, with thick cut-stone walls, extravagant furnishings of Jamaican mahogany and cedar, ornate, gilt-edged decor, and glittering chandeliers.[16] Other forms of entertainment in the town included cricket matches at the Grass Piece, horse races at Cave Island, rifle-shooting contests, crab hunting at Salt Marsh, and bachelor nights at Miss Campbell's on Duke Street.[17]

In 1861, Jamaica Census listed the population at 3,127, the fourth most populous town after Kingston (27,359), Spanish Town (5,362), and Montego Bay (4553).[18]

Genealogical records show that William Kennedy was working as a clerk in 1870 and as an accountant in 1873 in the town of Falmouth. By 1876, he was a partner in a dry goods business. An advertisement appeared in the *Falmouth Post,* November 17, 1876:

CO-PARTNERSHIP
J.R. YOUNG & CO.

Begs to notify that from the 13th November, 1876, Mr. William Kennedy of this Town is admitted a partner in his business.

---

14   The town took its name from Falmouth in England, the birthplace of Sir William Trelawny, Governor of Jamaica (1767–1772), who also founded and named the parish.

15   Underhill (1862), 370.

16   Carey Robinson, *The Rise and Fall of Falmouth* (2007), 16. This book offers vivid descriptions of life in nineteenth-century Falmouth..

17   Ibid., 65–66.

18   Statistics and details of post-Emancipation period sourced from *Free Jamaica* by Douglas Hall.

> Mr. Kennedy is well known to this community and will attend specially to the DRY GOODS DEPARTMENT.
>
> In reference to the above, MR. KENNEDY'S best efforts shall be put forth to please, so that by courtesy to our customers attention to their orders, and by keeping up a supply of Articles required by the community, we may secure a full share of the patronage of the Gentry, and of the public generally, both in this and the adjacent Parish.[19]

By the 1880s, he acquired sufficient capital to expand his business. "Mr. Kennedy opened an extensive merchandise business, dry goods and provisions, in Stewart Town, Brown's Town and Jackson Town, with a considerable produce branch in the days when there was money in it."[20]

The adjacent towns to Falmouth in the hills of Trelawny were prosperous:

> In the 1880s. Stewart Town was still a reasonably sized market town with several thriving businesses. There were three churches—Anglican, Methodist and Baptist—and the town was in two parts, the older part called Bottom.[21] Kennedy's house and store were in Bottom ... In 1889, Kennedy also leased premises at Jackson Town—a house with a store underneath and a run of land of fruit and coffee trees behind ...[22]

The main branch of his businesses was in Falmouth, where he established his residence. Water Square in Falmouth was at the vibrant core of the town, where prominent merchants operated their businesses. David Lindo owned a drug store, and Alfred Leopoldo Delgado, a hardware and farm goods store.[23] William Kennedy's store in Falmouth was located opposite to the telegraph and post offices, on the corner of Cornwall and Market Streets. His outlets carried farm produce as well as imported dry goods: salted beef, flour, coal, dried fish, pickled herring, rice, and soap.

With the expansion of his business, he spent his time in the towns where his stores were located. Roads in Jamaica had not seen much improvement since the days of

---

19   From Falmouth Post,1860, quoted in http://www.jamaicanfamilysearch.com/Members/f/falpst12.htm.

20   Obituary of William Kennedy, February 1915, from a Falmouth newspaper, exact source unknown. My dad gave a copy of the printed obituary to my wife, Georgianne, in the 1970s.

21   The Anglican church is in the area referred to as "Bottom." The Methodist chapel is located on top of a hill that overlooks this area of town.

22   Peta Gay Jensen, *The Last Colonials* (2005), page 14. See Note Chapter 5. William Kennedy owned both stores and residences in several towns in Trelawny.

23   Robinson (2007), 109.

slavery—narrow tracks and bridle paths, muddy and impassable when rains fell in the hills of St. Ann and Trelawny. Commuting by horse and buggy proved difficult.

As a Methodist, William Kennedy was known to have had a special devotion to the Good Book and was renowned in his community for works of charity.[24] The core beliefs of his religion were based on the absolute authority of the Old and New Testament scriptures. Throughout his life, "he rendered splendid service to the Church of his choice, and filled important offices such as Sunday school teacher, Superintendent and Circuit Steward." [25]

William Kennedy's merchant business thrived even though not all was well with the Jamaican economy. Swithin Wilmot commented:

> FK: Towards the end of the 19th century, Jamaica experienced an economic depression, and people became poorer as a result. Can you comment on this?
>
> SW: The simple truth is Emancipation did not destroy the plantation economy entirely. When sugar estates and the sugar economy went into crisis in the 1870s and 1880s because of beet sugar production with which we could not compete, the sugar economy went into crisis again and that had implications for wages and market conditions in Jamaica.[26]

The full effects of Britain equalizing duties for sugar and coffee imports into the UK, with the Free Trade Act of 1846, signaled the gradual decline of the sugar industry throughout the nineteenth century. Jamaica, once "the brightest jewel in the British crown," was no longer the largest producer of sugar in the world. Industry and manufacturing were slow in developing to replace the sugar plantations. Hardships also led to political upheaval and a reactionary move by the British to institute Crown Colony government following the Morant Bay Rebellion of 1865. The ruling classes feared that Black and "brown" people would form a majority by election to the Legislative House of Assembly.

---

24   By the middle of the nineteenth century, Methodism was well established in Jamaica with chapels built in almost every parish; in urban areas such as Falmouth, free "coloureds" were mainly associated with the sect. It was one of the oldest Protestant churches in Jamaica, a mission begun by Rev. Thomas Coke in 1789. Methodists initially met fierce resistance from the planter class because they were sympathetic to the cause of abolitionism.

25   Obituary of William Kennedy, a Falmouth newspaper (name unknown), February 1915.

26   Interview with Swithin Wilmot (2019).

The Old York Castle School, Alderton, St. Ann, 1876–1906.

(Courtesy of York Castle High School)

https://www.yorkcastlehighschool.com/about/
Original image from an old postage stamp.

Joseph Henry Levy (1843–1927)

Fred William Kennedy's first employer in Brown's Town, St. Ann, 1890s.

Who's Who Jamaica 1916, page 160

(Photo, courtesy of © The Gleaner Company (Media) Ltd.)

2

MY GRANDFATHER

York Castle

(1880s)

In 1885, Fred William Kennedy, William and Mary Ann's third child, qualified for entry to York Castle High School, the prestigious Methodist boys' boarding school located in Alderton, St. Ann, fifty miles from his hometown of Falmouth.

Secondary schools like York Castle were funded mainly by the churches for sons of the first generation of middle-income parents who could afford tuition. The very wealthy continued to send their children to England for secondary and tertiary studies, just as members of the slave owner class had done before 1838.

Pre-Emancipation, colonial governments resisted the provision of formal education for enslaved persons. Towards the end of the eighteenth century, foreign sectarian ministers established schools on some of the plantations with permission of owners, but these were largely intended for religious instruction. It was a common belief among slaveholders that literacy would prove a threat to the system of slavery, weaken their control as masters, and increase the risk of rebellion. After 1838, British colonies were faced with masses of freed persons with little or no formal education.

In the years immediately following Emancipation, missionaries established a system of elementary schooling to be taken over by the Jamaican government in the 1860s when funds were made available by the local Legislative Assembly to build and staff elementary schools. In Jamaica, 232 schools were opened between 1868 and 1877.[1] However, the quality of education was appalling; only fifty percent of elementary age children were registered, and of those, only sixty percent were in regular attendance. The majority of the first generation of Jamaican children after 1838 left grade six with sub-standard levels of literacy and numeracy and were ill-equipped to enter the work force.

Fred William was eleven the year he enrolled at York Castle as a boarder. Established as both a secondary school and theological college by the Wesleyan Methodists in 1876, the institution boasted some of the brightest young minds of the period. It rivalled The Jamaica School (Jamaica College), founded 1789 in Kingston, and the Potsdam School[2] (Munro College), founded 1856 in St. Elizabeth. York Castrians, as the boys used to call themselves, represented the largest number of Jamaica scholars graduating during the period 1881 to 1895.

The original school was situated on a 500-acre property at the end of a buggy trail at an elevation of 3,000 feet above sea level, in the garden parish of St. Ann. It was seven miles from Finger Post, known today as the Town of Claremont. Three large wooden structures housed a student dormitory, masters' quarters, classrooms, dining rooms, and dwelling for the governor's family. The old house had been the original great house of the coffee plantation, formerly owned by the Curtis family.

Changed little from the days of horse and buggy, the road from Claremont to the Old York Castle is not much different now than it would have been in the 1880s. In May 2019, I went in search of the old school in Alderton, St. Ann, and discovered a set of ruins hidden amidst overgrown grasses on a steep hillside. Approaching the site along a high mountain ridge is a narrow road, crowded with cows, donkeys and goats that block the passage of vehicles. The scenery is breathtaking from the summit where the ruins of the school buildings stand, commanding the highest point for miles around, with a 360-degree panoramic view of the verdant hills and valleys of the Garden Parish.

Ansel Hart,[3] an alumnus who spent seven years at the school in the 1890s, makes mention of the Kennedy brothers in his memoirs. Fred William's younger brother, Theodore, two years older than Ansel Hart, also attended York Castle. The "older Kennedy," whom he mentioned in his memoirs, is a reference to my grandfather:

> The older Kennedy who was there before my time, as the younger was to be, was in the banana business as agent of the United Fruit

---

1   Ruby King, Education in the British Caribbean: The Legacy of the Nineteenth Century (1998), 6.

2   The name Potsdam was changed to Munro to disassociate the school from the rise of Nazism in Europe.

3   Ansel Henry Lester Hart (1878–1973). Distinguished attorney, resident of Montego Bay. Son of Samuel Hart and Constance Nunes, he had four children: Samuel, Helen, Herbert, and Richard.

Company. He later became a large Kingston merchant in association with Grace. He had great commercial ability.[4]

Hart described Theodore as a rabble-rouser who once spouted racist remarks publicly during a school debate on the topic of "Public Education," an act for which he was duly punished by the governor of the college. He was said to have excelled in mathematics but paid little attention to other subjects.

Fred William attended YC at its peak of academic performance. He was schooled in Latin, mathematics, science, English, French, history, and religion—subjects taught by Oxford-trained masters and ministers. My dad mentioned that he performed exceptionally well at school, excelling in both the arts and sciences; he had a brilliant mind for figures but was also talented with his hands. He was an artist, poet, and carpenter.

The rigorous curriculum included a program of sports. Soccer was played every Friday evening, and cricket matches were held on Saturdays. In addition, they had free time to roam the surrounding properties and to play without supervision. Hart recalled:

> A narrow track led circuitously round a mountain to the extensive playing fields flanked by pimento trees and hidden from view by the fold of the hills. Thence ... led tracks to woodlands, caves and sinkholes, the delight of exploring youngsters. The venturesome might squeeze their way into the underground "Barker's Cave," wondering if they might possibly get stuck there, some might collect the exquisite land shells from the cliffs, or the famous "blue emperor" from the decayed bastard-cedar. Some secured ironwood to be smoothed and polished into walking sticks, some obtained lancewood for bows and arrows, others played military games over an extensive terrain, or scrounged for fruit ... in the adjoining orange groves at Rhoden Hall and Barrett's Grass Piece.[5]

Student life at York Castle was one of "freedom with discipline,"[6] a hallmark of the old Y.C. The boys often got into mischief, but corporal punishment was used only for the more serious offences. The Methodist philosophy was to build character, to instill moral virtues of honesty and generosity. The boys were taught by the choicest schoolmasters and achieved excellence in academics comparable to the top grammar schools in England.

---

4   Ansel Hart, "Memories and Reflections: More on Old York Castle," (Volume 6, No. 5, 1968). This is a detailed, vivid account of student life in Alderton during the 1890s. Part of a larger publication, *Monthly Comments Jamaica Volume 6*. Thanks to Robbie Vernon for this source material.

5   Ibid.

6   Ibid.

School enrolment never exceeded 150 students, the majority of whom boarded on campus. Students came from parishes across the island: from St. Ann, the Covers, Arscotts, Levys, and Clarkes; from St. Mary, the Goffes; from Trelawny, the Kennedys and Milliners; from St. James, the Vernons;[7] and from Hanover, young Johnston.[8] Fred William would encounter the Goffe brothers in later years when he worked in the banana industry.[9]

The student body consisted of boys from different racial and cultural backgrounds—white, Black, and Jamaicans of mixed race; Latin Americans (Dominicans, Cubans, Nicaraguans, Colombians); and Haitians. They formed a community in which race and discrimination were not an issue.[10] Ansel Hart commented: "There was no colour (pigmentary) sense or sensitiveness whatever at York Castle. McFarlane, a black boy, the son of a policeman from Sandy Bay, was a very desirable companion, and David Norman (father of Dr. Norman)—Sally—one of the most charming and popular boys, of outstanding integrity."[11]

Tuition was not cheap.[12] By the 1890s, annual fees per student amounted to £50, a sizeable portion of the income of even the wealthiest merchants and professionals of that period. The Kennedy household felt the financial strain. Fred William was forced to leave school in his fifth year at YC to find a full-time job to help support the family. (It is not certain whether he sat his end-of-school examinations.) In years following, the family was able to afford tuition for Theodore to attend York Castle.

Fred William's desire was to have attended university, but to qualify for a tertiary scholarship, he would have needed to remain in school, like many of his classmates, for another year or two. The Jamaica Scholarship, instituted in 1881 by Governor Sir Anthony Musgrave, required that the candidate "will have completed his seventeenth, but not completed his nineteenth year, on the fifteenth day of December of the year in which he is a Candidate."[13]

---

7   Alexander Apfel Vernon (1875–1925) was a contemporary and classmate of my grandfather at York Castle. He studied medicine in the United States and opened a practice in Montego Bay. He is the grandfather of two friends and former St. George's College classmates, Robbie Vernon and Swithin Wilmot, who are first cousins.

8   Lewis (nicknamed Daniel) Ewart Johnston (1883–1934), younger brother of Charles Edward Johnston (see Chapters 5 and 11). Charles Edward (born 1873) was closer in age to my grandfather. He attended Ruseau's High School (established 1777) in Lucea, Hanover.

9   See Chapter 3.

10   Ansel Hart (1968).

11   Ibid.

12   Schools held a limited number of vacancies for "foundationers," students eligible to receive a "foundation," or reduced fee. Recipients of grants would normally reside within parish limits, but those from out-of-parish could also qualify.

13   "Jamaica Scholarships" in *Handbook of Jamaica* 1886–87, by A.C. Sinclair, 315.

My Grandfather (1880s)

My grandfather entered the world of work at age fifteen. His parents wanted him to seek employment in the civil service, but he chose instead to join a private business owned by the Levy family in Brown's Town. Joseph Henry Levy,[14] originally from the parish of Manchester, operated a wholesale store, J. H. Levy and Son, located on Wesley Crescent close to the post office and Methodist church. The Levys also exported coffee obtained from a factory in Aenon Town and owned a rum bar near to the police station.[15] (Brown's Town is famous for its bars and churches.) Joseph Levy was prominent in the community; he was a justice of the peace, chairman of the local parochial board, and member of the Legislative Council. His son Charles Levy (born 1878) was my grandfather's contemporary and attended YC and later studied medicine.

---

14   Joseph Henry Levy, born 1843, Manchester, educated at Ridgemont Elementary School, became a prominent citizen of Parish of St. Ann. Biographical information gathered from *Who's Who in Jamaica 1916*.

15   Details, courtesy of Norma Walters, JP and educator, whose family members are long-time residents of Brown's Town, St. Ann.

Lorenzo Baker

© Library of Congress
https://www.loc.gov/item/2002712356/
No known restrictions on publication.

Loading bananas, Port Antonio, 1900.

Source: New York Public Library
https://digitalcollections.nypl.org
In the public domain.

3

OCTOPUS
(1890s)

In 1895, at the age of twenty-one, Fred William left Brown's Town for Port Antonio to work in the banana industry. As a young adult, he learned the business, and with hard work and keen business acumen, proved himself a well-respected employee of a firm that was to become the United Fruit Company.

Fred William was a member of an emerging class of educated "browns" who were positioned to climb further up the social and economic ladder than the generation before them. With dwindling numbers of Europeans and colonials, the second generation acquired certain privileges and came to regard themselves as the rightful inheritors of the Jamaican soil.

It was hard to conceive of a large multinational corporation growing out of the cultivation of bananas, but so it was with the birth of a mega enterprise at the turn of the twentieth century.

Three young men gave the industry a start: Lorenzo Dow Baker, Minor C. Keith, and Andrew W. Preston. The small yellow fruit, previously unknown to most North Americans, found its way to the United States from Jamaica:

> Baker was the captain of an 85-ton schooner, *Telegraph*, out of Wellfleet, Massachusetts, and one day in June of 1870 he pulled into the harbour at Port Maria, Jamaica for a load of bamboo and a sip of the local rum. While sipping, he was approached by a local impresario who had several stems of green bananas to sell. Baker considered the proposition for the length of a second cup of planter's punch, then agreed to

pay the man twenty-five cents a bunch. Eleven days later, Baker sold those same bananas in New York for prices ranging between ten and fifteen times what he paid for them.[1]

Thus began an industry that would become an empire with more economic power than their host countries in the Caribbean basin.

In the early years, Baker experienced horrendous losses because of slow travel by sail between Jamaica and New York and Boston. This resulted in dumping tons of rotten fruit. But by the 1880s, the *Lorenzo Dow Baker* vessel was built, heralding a new age of transport by steam engines, enabling the carriage of 10,000 stems of bananas within a ten-day period from Jamaica to Boston. With increased volume, he built his own marketing enterprise and, in 1885, formed the Boston Fruit Company with Andrew Preston and eight others. Captain Baker filled orders as fast as he could load his boats.[2]

Fred William Kennedy was initially employed by the Boston Fruit Company to work in Port Antonio, where ship captain Lorenzo Dow Baker served as president of the company and manager in Jamaica. By 1895,[3] when he joined the company, "the corporation owned nearly 40,000 acres of land, including 35 plantations and deep-water frontage in the harbours of Port Antonio and Port Morant. They operated their steamship lines between Jamaica and the United States, the ports of Boston, Philadelphia, and Baltimore."[4] The company's headquarters was based in Boston.

Fred William worked with the accounting department, and before long was appointed auditor, entrusted with the responsibility of reviewing accounts and ensuring the legality of financial records. He also served in an advisory capacity to recommend risk aversion and cost-saving measures and oversaw the movement of money going in and out of the organization.

Port Antonio was the ideal location for the centre of Baker's operations. The lands of the parish of Portland were fertile for the growing of bananas, containing rich, black alluvial soil formed from the deposits of a vast numbers of rivers. In addition, the average rainfall, which exceeded 130 inches per year, created perfect growing conditions.

By the late 1880s, the banana industry had transitioned from being a small holder crop to a planation industry. In 1879, Baker owned one large banana estate in the parish; by the early 1890s, "more than one hundred banana plantations existed."[5] His company acquired properties close to shipping ports, including Buff Bay, Orange Bay,

---

1 Thomas P. McCann, *An American Company: The Tragedy of United Fruit* (1976), 15.
2 Ibid., 15.
3 This is an approximate date.
4 https://en.wikipedia.org/wiki/Boston_Fruit_Company.
5 Dr. Jenny Jemmott, *The Parish History Project: A History of Portland* (2018), 130.
Published as part of The Jamaica Parish Histories Project (2015), planned and executed by Dept. of History and Archaeology, UWI, in collaboration with The Jamaica National Building Society Foundation.

Hope Bay, St. Margaret's Bay, Blue Hole, and Manchioneal.[6] The extension of the railroad from Kingston to Port Antonio and the opening of the new railway station in Port Antonio helped expand the business with transport of fruit from various parts of the island including Bog Walk.

The company had strict controls. It used measures to reduce waste and to maintain standards of quality in what became a highly competitive market. By the 1890s, Boston Fruit was raising the standard "bunch count" to eight and nine hands, which included varieties such as the Gros Michel, introduced to Jamaica in the 1830s.[7] Agents graded bunches by the number of hands, the fullness of individual fruits, and the outward appearance of the peels.[8] Top priorities for Baker were safeguarding the loyalty of local growers and remaining competitive for buyers in foreign markets.

Tensions were high between the US and local entrepreneurs who were making inroads into the industry. Conflicts arose between Baker, who wanted full domination of the banana industry in Jamaica, and local agents and shippers, who formed their own banana export businesses. They were "economic nationalists," Jamaicans attempting to take back control of the banana trade, "building trade links with other people, buyers abroad."[9] The Goffes were one such group of pioneers who established a banana-purchasing firm, Messrs. Goffe Bros., in Port Maria in 1897.[10] This marked the beginning of a banana war to be waged for generations.

Baker did everything he could to safeguard control and supremacy. It is no surprise that when A. C. Goffe and his brothers partnered with Italian American merchant Antonio Lanasa to form the Lanasa and Goffe Steamship and Importing Company, Captain Baker's response was forceful and caustic:

> We have got to take charge of every banana that reaches the wharf. This is for our own safeguard, just to form a nucleus and shut out the element of danger in a crabbed and jibbing market. There is no danger of our spreading our tentacles like an octopus and grabbing all the land in sight.[11]

---

6   Ibid.

7   Gros Michel was introduced to Jamaica in 1835 from Martinique. It became the dominant variety for export up until the 1950s, when Panama disease spread and devastated crops in the Caribbean. It was favoured because of its dense bunches and thick peel, resilient to bruising when shipped.

8   John Soluri (2002), quoted in https://en.wikipedia.org/wiki/Boston_Fruit_Company.

9   Swithin Wilmot, interview with author (2019).

10   *When Banana was King: A Jamaican Banana King in John Crow America* (2007) by Leslie Gordon Goffe is a fascinating, well-researched narrative of a family banana business in Jamaica.

11   Baker's comments appeared in *The Daily Telegraph*, March 8, 1906 (British newspaper founded 1855), quoted by Leslie Gordon Goffe (2007), 6–7.

Goffe posed a threat to the foreign control of the banana trade in Jamaica. His company coordinated the growing and purchasing of bananas and chartered steamships for transport of millions of bunches of bananas to the United States.[12] But competition was fierce.

Broken relationships with independent growers in Jamaica meant "an unstable link in the supply chain."[13] One of Baker's objectives was to negotiate and ease tensions with the banana growers, both small and large. By the 1890s the novelty of the tropical fruit in North America had worn off.[14] Consumers were demanding a high grade of fully ripened fruit. This meant, on the supply side, smooth operations, guaranteed volumes, and correct timing of harvest. Logistics were of paramount importance when dealing with a highly perishable fruit.

From March to November a steamer arrived every ten days, and during the winter months, every two weeks, carrying on average of 115 thousand bunches of bananas. Due to volume and advanced technology, Boston Fruit was able to undersell its competitors with prices ranging from fifty cents to $1.50 depending on grade.

In March 1899, Andrew Preston and Lorenzo Dow Baker of Boston Fruit Company formed an agreement with Minor Cooper Keith to create the United Fruit Company. Keith had consolidated huge interests in the banana trade in Central America by building railroads and amassing properties in Costa Rica and Colombia.[15] The amalgamation of these enterprises ensured increased hegemony in the region with control over production, shipment, and marketing of bananas.

Fred William was employed by a North American company that adopted hard-nosed, competitive business practices. How well did he fit in with the culture of the typical *banana cowboy* of his time?[16] It's hard to say. The idea of a genteel, polite, and cultured Jamaican man working for UFCO seemed out of sync with the image and persona of the banana agent "from foreign."

United States banana agents who were assigned to work in the Caribbean viewed themselves as part of a white macho elite, hardy enough to endure the rigours of a tropical climate and to deal with the rough conditions of the management of local labour. Fred William's position as a member of staff was complicated further by the fact that he was not a white North American, but a "brown" man, native of the British West Indies. Notwithstanding these cultural and racial differences, Captain Lorenzo

---

12   Leslie Gordon Goffe, *When Banana was King* (2007), 2.

13   James W. Martin, *Banana Cowboys* (2018), 22.

14   Ibid., 23.

15   Descendants of Jamaicans who helped construct the railroad in the 1880s live today in the Puerto Limón region of Costa Rica.

16   Although behaviour of office personnel would have been different from those out on the field, the American way of doing business pervaded the whole organization.

Baker and others were impressed by Fred William's work in Port Antonio. They recommended him to be assigned to the Dominican Republic in 1901.

Captain Lorenzo Baker had become the largest producer of bananas in the world. With the advent of large-scale production, the export of bananas increased from 329,000 stems in 1879 to 8,000,000 in 1900.[17] Plantation bananas accounted for seventy-four percent of land under banana cultivation. By 1890 bananas outstripped sugar as Jamaica's major export, and by 1900 contributed 25.6% compared to sugar's 10.8% of exports.[18] Port Antonio became a bustling town, centre of operations and home to most of Captain Baker's employees.

After six years of employment, Fred William would now become a pioneer of the banana frontier to face US corporate culture in a Caribbean country altogether different from his own.

---

17   Patrick Bryan (2002) *The Jamaican People 1880–1902: Race, Class and Social Control*, 7.
18   Ibid.

Fred William Kennedy, bachelor, early 1900s.

The Arzeno Family (1888)

José Antonio Arzeno Rodríguez (1848–1902) and Carmen María Abreu Imbert (1851–1928) with their children:

L–R: María Aminta (1871–1953), Máximo José, Rafael Arzeno (1881–1962), **Luisa Matilde (1888–1962)**, Luís Sebastián (1873–1922), and **Carmen Celia (1876–1963).**

Not in photo: Two youngest children born after photograph was taken: Ana Sofía (1889–1961) and Gloria Mercedes (1891–1971), and the eldest child of José Arzeno, Adolfo Arturo Arthur (1868–1928).

(Genealogical records indicate that José Antonio Arzeno may have been adopted.)

4

THE BANANA FRONTIER (1901–1908)

In 1901, Fred William Kennedy was a resident of Sosúa, Dominican Republic. At the age of twenty-seven, the young, adventurous Jamaican accepted the position of bookkeeper with the United Fruit Company, which had recently begun operations with the purchase of large tracts of unpopulated lands surrounding the town of Sosúa.

At the beginning of the twentieth century, the Dominican Republic was sparsely populated, with only three communities that could be classified as cities: Santo Domingo, Santiago, and Puerto Plata. With fewer than 300,000 inhabitants, it had the lowest population density of the Caribbean islands.[1] Immense opportunities existed for economic development of large, uncultivated and forested areas, especially in the northeast areas of the island, a region called, *el Cibao*. In 1900 the United Fruit Company acquired thousands of acres on which they cultivated 1.5 million banana plants for export of fruit to the United States.[2] The operation averaged 230,000 bunches per year, which required seventeen ships of The Great White Fleet for transport to the United States. The small outpost of Sosúa, where the United Fruit Company opened its business, is located twenty-five kilometres from Puerto Plata, the capital of the Cibao region of Dominican Republic.

---

1 Bernardo Vega (ed.) *Dominican Cultures: The Making of a Caribbean Society* (2007), 170.
2 www.sosuamuseum.org.

The Banana Frontier (1901–1908)

The early twentieth century was a period of aggressive US expansionism in the region. During the Cuban War of Independence of 1898, the US had declared war against Spain in a so-called effort to liberate Cuba. Even though their stated intention was not to annex the country, the United States held Cuba as a protectorate until 1902. (Future US intervention in the internal affairs of the country lasted until the 1950s.) Neighbouring Dominican Republic witnessed a similar pattern of North American military intervention and control of economic interests, creating a volatile political climate involving the overthrow of successive governments. Because of the commercial importance of the Cibao region, Puerto Plata was a hotbed for rebel activity and US naval invasions.

The historical period is referred to as the "Crisis Phase" in the political history of the Dominican Republic. In 1902 an open rebellion was waged to overthrow President Juan Jimenez, and a provisional government was formed. This was followed by a military coup, which in turn overthrew the government of Felipe Vásquez in 1903. The following five years were rife with successive rebellions and changes of government, with intermittent regimes supported by North Americans who landed marines in Puerto Plata in 1904.

Fred William, as banana agent of the United Fruit Company, was a member of the American elite.[3] He lived in what came to be known throughout the Caribbean as a company town, a miniature US city inhabited by managers, agronomists, and engineers, with all the amenities of a North American community. Homes[4] were fully equipped with imported furniture, and commissaries supplied food, clothing, and footwear.[5] Company employees enjoyed the use of electricity and running water, which were luxuries of the time.

He joined a group of select first-class employees, responsible for a much larger workforce of Black West Indians and Hispanic *mestizos*.[6] Locals did the *machete* work, as it was called. Fred William, a "brown" man from Jamaica, was in an unusual position as a club-member of American whites in a Spanish-speaking country. His associates, mostly from the northeastern states, were college-educated and young. They perceived and treated Mr. Kennedy as different, not because of his personality or aptitudes, but of his ethnicity, colour, and place of birth.

---

3   Fred William served as a buying and selling agent whose business travels started soon after his arrival. He was listed in 1901 as a passenger, banana agent of the United Fruit Company, and resident of Sosúa, Dominican Republic, on board the *Ethelwold* Philadelphia Steamer, a UK vessel chartered to the United Fruit Company.

4   United Fruit constructed twenty homes for managerial and supervisory staff in Sosúa.

5   Atalia Shragai, "Do Bananas Have a Culture? United Fruit Company Colonies in Central America 1900–1960" (2011), 68. The company had as part of its operations a large merchandising and supply department responsible for distribution of food and supplies throughout its corporate empire. The company was so extensive that *Unitfruitcos* living and working in the Caribbean developed their own unique "banana culture."

6   James W. Martin, *Banana Cowboys* (2018), 5.

He was immersed in a frontier-flavoured macho culture. The term banana cowboy was coined in the early frontier days of United Fruit to characterize the white Anglo-Saxon pioneer who saw his assignment to the tropical regions as one of adventure and exploitation. He was "the tough, efficient herder of fruit." The caricature was of cowboy on horseback, wearing a Stetson and carrying a pistol.[7] In the culture of the banana world, the description referred not only to supervisors of natives in the field but to professionals who held office and technical positions with the company.

Whether or not Fred William encountered racial discrimination, overt or otherwise, is hard to say. Reports of the period do indicate that a racial divide existed in the workplace:

> The racial division of labour occasionally blurred when circumstances required. Necessity occasionally forced managers to promote blacks to supervisory positions, though the colour bar hung perilously over such promotions. Superintendent Mullins explained one such case of a vacancy, "At the present time there is a very competent nigger over the men," he wrote, "and I do not want to make any more experiments."[8]

Mr. Kennedy was valued by his superiors. He had work experience with the company in Jamaica and exhibited few of the vulnerabilities of US immigrants. He was not susceptible to tropical diseases as many foreigners were and was not high risk for excessive consumption of alcohol and other indulgences that hampered productivity. Adapting to the tropics for the North American was fraught with problems: nostalgia, rum-drinking, and general malaise. Letters home revealed cultural stereotypes of the Caribbean: "This is the laziest-feeling place on earth. It is not the fashion to hurry in this part of the world."[9]

The young Jamaican bachelor more than survived his tenure in the Dominican Republic. During his eight years, the banana industry boomed. Its annual shipment almost tripled to 640,000 bunches of fruit exported to the US in 1907. As an agent, he travelled extensively, on occasion returning to Port Antonio and representing the company on trips to Boston. In his leisure time, he looked for entertainment outside the confines of Sosúa. Spanish America was exotic and romantic. Like many foreigners, he was enamoured with the culture and became involved with local communities in Puerto Plata.

In the early twentieth century, Puerto Plata was the heartbeat of the Dominican Republic, the cultural and cosmopolitan centre of the country. Foreigners working in the outposts frequented the town on weekends to enjoy the entertainment the town

---

7 Ibid., 11.
8 Ibid., 41.
9 Ibid., 146.

had to offer, mixing with locals at dances, clubs, and civic events. A creolized culture developed among expatriates. US citizens grew to love the music, language, and food, and became so immersed in the culture that some distanced themselves from the more patriotic behaviours of Uncle Sam.[10] Mixing was made easier by UFCO employment requirements of a working knowledge of Spanish. Fred William became proficient in the language.

By 1900, Puerto Plata was considered "the most beautiful and the cleanest city of the Republic."[11] The cobblestone streets had adequate drainage; the wharves were modern, constructed of masonry and steel; and the homes were fashioned in Victorian architectural style and painted in vibrant colours. It was a town appealing and picturesque in every way.[12] *El Parque Central Independencia*, decorated with ornate fountains and flowers, was surrounded by the Town Hall, the Church of San Felipe, private homes, the Recreational and Cultural Centre,[13] and the *Club de Comercio*. Puerto Plata also boasted *el Teatro Central* (with a capacity of 600), where performing artists of the period debuted; the Central Train Station, where passengers and freight trains ran from Santiago; and the Fire Station, one of its most iconic landmarks, built in 1895.

Fred William frequented the theatre, joined the Masonic Lodge, and fraternized with members of business clubs and cultural societies. He met influential community members, among whom were members of the Arzeno family. Carmen María Abreu Imbert, widowed at age fifty, lived with her ten children in a Victorian-style home opposite to the cathedral in the main square. Fred William became smitten with one of her younger daughters, Luisa Matilde, a teenager at the time of their courtship.[14] She was fourteen years younger and thought to be "the most beautiful of the Arzeno family" (LMK).

Luisa Matilde's father, José Arzeno, died in 1902 at the age of fifty-five. He had been a prominent citizen of Puerto Plata, Venezuelan Consul for Dominican Republic, and owner and commission agent of several import/export businesses. He was the son of Sebastián Arzeno, an Italian Jew, who had migrated to the Dominican Republic in the nineteenth century from Genoa. Sebastián Arzeno and his wife, María Carmen

---

10   Shargai (2011), 71.

11   Zeller, Neici M., "Puerto Plata en el Siglo XIX" (1977), 42—*la más linda y la más limpia ciudad de la República*—

12   I visited Puerto Plata in 2019. Its homes, cathedral, public buildings, and fort in the centre of the city have been fully restored to their former glory as part of a UNESCO World Heritage Project.

13   *La Fe de Porvenir* was founded in Puerto Plata in May 1889 by prominent residents who were all related—Felipe Augusto Arzeno, Adolfo Arturo Arthur, and Charles Loinaz (my maternal grandfather). It was a club where men would assemble to socialize and talk politics, a place where plots were formed to overthrow dictators. Today it is a centre for planning and promotion of cultural events.

14   Luisa Matilde Arzeno was born February 3, 1888, Puerto Plata, Dominican Republic.

Maldonado, had eight children and numerous grandchildren. Most were educated and entered the fields of commerce, medicine, politics, the military, the arts, and engineering.

Fred William found himself not fully suited to marry into the Arzeno family. He was *mestizo* [15] and a non-Catholic. In the Dominican Republic, *mestizaje,* the mixing of races, is a phenomenon more complex than historical "shades of colour" in Jamaican society.[16] A strong caste system of social stratification existed along racial lines, with whites at the apex of the pyramid and Haitians and persons "of colour" at the base. *La primera clase*—creole whites—were descendants of the original Spanish colonists and European immigrants and held the more privileged jobs and positions in society. The Arzeno family saw themselves as being members of this group; they had intermarried with other families of equal social standing in their community: Brugal, Gómez, de León, Abreu, Imbert, Martínez, Rodríguez, Ortiz, Loinaz, and others.

Fred William's Methodism was the greatest obstacle to marriage. In the early 1900s, Roman Catholicism dominated every aspect of living. Over ninety percent of the population were practising Catholics. In the new republic, Church and state were entwined as one in the constitution, and even though Protestant churches had been established from as early as the nineteenth century,[17] the Catholic Church did not recognize marriage to a Protestant as a sacramental union.

Despite these hindrances, the Arzeno family consented to the marriage. Courtship rules were very strict. "A couple could never be alone together; they were always supervised under the watchful eye of a chaperon."[18] The custom was for the young suitor to bring a bouquet of white flowers on each visit and to sit and chat in the parlour in the presence of other family members. Couples were seldom allowed to go out in public unaccompanied by a chaperon, least of all at night.

Fred William agreed to convert to Catholicism and pledged that the children would be baptized in the Church and raised as Catholics. But even though the marriage received the blessing of the Catholic Church, it could not be performed in a church sanctuary. It needed to be a civil ceremony, held in the *Ayuntamiento* (Municipal Townhall) by the *alcalde* or notary public.

---

15 Skin colour in the colonial society of the Dominican Republic "took precedence over all the other factors of stratification within the social structure, and thus a relationship between ethnicity and occupation or social position was established." Bernardo Vega (ed.) (2007), 158.

16 In the *Mestizaje* lexicon of Hispanic America, there are historically at least eighty-two known terms to describe categories of persons of mixed race. Manuel A. García Arévalo, *"Orígenes del Mesitjaze y de la Multización en Santo Domingo"* (1995), 24. Some of these racialized representations of colour persist today.

17 The Methodist Church of England established a mission in Puerto Plata in 1822.

18 Lydia M. Kennedy, interview with Georgianne Thompson Kennedy, August 10, 2002. At Georgie's request, Mom recorded her memoirs at age ninety-one, two months before her passing.

The Banana Frontier (1901–1908)

*Luisa Arzeno Abreu y Fred Kennedy se casaron por la ley,
y el cura católico sólo los bendijo, por ser él protestante*[19]

Luisa Arzeno Abreu and Fred Kennedy had a civil wedding,
and the Catholic priest merely blessed them, for he (Fred Kennedy)
was Protestant.

Fred William Kennedy, aged thirty-three, and Luisa Matilde Arzeno, aged nineteen, were pronounced man and wife on January 22, 1908.

Among the guests at the wedding was Fred William's sister, Rose Eliza Kennedy. In 1904 she had migrated from Jamaica to the Dominican Republic, where she met Dr. William Morris[20] from St. John's, Newfoundland. Dr. Morris was the resident medical doctor attending to the employees of United Fruit Company in Sosúa. They married in 1906 and had an only child, whom they named Katherine Morris.[21] Rose died of diphtheria two years later at the age of twenty-six, the same year that her brother got married.

Within weeks of having been married, Luisa Matilde was pregnant. In anticipation of their firstborn, her husband built a mahogany chest to hold the infant's clothes and blankets.[22] He was a talented cabinet maker, a hobby he pursued through his adult life. On November 2, they welcomed to the world Luis Fred Kennedy, named for his mother, Luisa, and his father, Fred.

---

19   genealogy.comgenealogy.com *"Descendientes de Sebastián Arzeno"* https://www.genealogy.com/ftm/a/r/t/Luis-H-Arthur/GENE6-0005.html.

20   Dr. William Morris was born in 1863, nineteen years older than his wife Rose Kennedy. His brother was Edward Patrick Morris, Prime Minister of Newfoundland (1909–1917). William Morris remarried in the Dominican Republic and had other children from his second wife.

21   Katherine Roy (nee Morris) was my dad's first cousin. She was born on December 4, 1906 and died at the age of eighty-six in October 1993. Katherine was educated at home and later attended Northwestern University in Chicago and the University of Berlin. She settled in Montreal where she became an author, journalist, and publicist. She had a daughter, Simonne, whose descendants reside in Canada (Toronto and Montreal) and Sydney, Australia. (Information is courtesy of Georgianne Kennedy's genealogical research.)

22   The baby trunk has been passed down through the generations and has served many purposes, including a family hope chest and children's toy chest. My grandfather also made two other pieces of furniture, a bookcase and den table, from Jamaican mahogany.

PART TWO

Coming of Age (1909–1930)

Methodist Chapel, founded by William Kennedy in 1907, Stewart Town, Trelawny. (Photo by author)

Charles Edward Johnston

The largest and most extensive business enterprise of its kind in Port Antonio was C. E. Johnston & Co. Wholesale and Retail General Merchants and Commission Agents, established in 1891. (Photo, courtesy of his grandson, Charles H. Johnston, Chairman and Managing Director of Jamaica Fruit and Shipping Co. Ltd. and Chairman of Jamaica Producers Group Ltd.)

Luis Fred Kennedy, Montego Bay, circa 1912.

5

THE WOODEN LEG

(1909-1914)

In 1909 Fred William Kennedy was re-assigned to work in Port Antonio as company auditor with the United Fruit Company headquarters managed by Captain S.D. List.[1] On May 3, he travelled to New York, en route to Jamaica from the Dominican Republic with his wife, Luisa Matilde, aged twenty, their newborn, Luis Fred, and nanny Eliza Taylor.[2]

Within months of arrival, their second child was born on February 21, 1910. Carmen María (Mim) was named after her maternal grandmother, Carmen María Arzeno. Only fifteen months apart, she and her brother, Luis Fred, grew up as companions, a sibling friendship that lasted a lifetime.

Port Antonio was now a tourist resort town. Prior to his passing in 1908,[3] Captain Lorenzo Dow Baker had begun using his banana steamships of the Great White Fleet to bring visitors to Jamaica.[4] Guesthouses were built in the town to

---

[1] Sibrandt Duhn List, born Denmark 1867, manager of UFCO operations in Jamaica. His grandson Michael List recalls, "He was captain of a three-mast ship that covered South America and the Caribbean carrying cargo to England. It caught fire in the mid 1890s and he was lucky to be rescued at sea." S. D. List joined Charles E. Johnston in 1919 to establish and manage the newly formed Jamaica Fruit and Shipping Jamaica Ltd.

[2] Eliza Taylor was born in Turks Islands in 1877; her name was entered as "servant" on ship manifests.

[3] In 1908 the Cape Cod shipmaster returned to Massachusetts, where he died. He was laid to rest in the Town of Wellfleet, June 21 of that year.

[4] Space was made available on Baker's banana steamships for American tourists (twelve per passage) at a price of fifty dollars round trip. Jemmot (2018), 142.

accommodate tourists, many of whom employees of the company also hosted in their homes. In 1905 Baker re-built the Titchfield Hotel (situated on top of the Peninsula Hill in Port Antonio), which contained 600 feet of piazza and 400 guest rooms. The hotel became internationally acclaimed, famous for its bath houses and luxurious conveniences, an exotic getaway to the Caribbean for North Americans and Europeans. United Fruit developed travel packages that included carriage rides from the wharves when passengers arrived by ship from Boston, visits to Blue Lagoon, and rafting on the Rio Grande. Port Antonio rightfully earned its name as the birthplace of Jamaican tourism.

Due to the growth of the banana business, the population grew with a large influx of UFCO employees from the United States.[5] A concrete, two-storey office building of the United Fruit Company was under construction and near completion at the cost of $40,000 to become the most modern and well-equipped office building in Port Antonio.[6] The town saw a boost in commercial activity, shops, taverns, drug stores, and banking. School enrolment at the major secondary school, Titchfield High, had an enrolment of 500 students.[7] The two major banks doing business in the town were the Bank of Nova Scotia and the Colonial Bank. And the largest and most extensive business enterprise of its kind in town was C.E. Johnston & Co. Wholesale and Retail General Merchants and Commission Agents, established in 1891.[8]

In 1910 Fred William was promoted to manager of the Montego Bay Agency of the United Fruit Company. The post was a significant development, a testament to his managerial abilities and the reputation he held with the company.[9]

---

5   Estimates are as high as 400 expatriates out of a total town population exceeding 2000 in 1910.

6   More than half of the buildings, including UFCO offices, were leased from the government as part of the Tichfield Trust, which owned 350 acres of land in the town. Revenues went towards supporting the Tichfield Free School, founded in 1785 (US Department of Commerce and Labour, *Daily and Consular Trade Reports.* 1912).

7   The headmaster of the school was Major William Henry Plant. Years later, his son, Owen Cathcart Plant, married into the Kennedy family.

8   C. E. Johnston (born 1871) owned and operated several stores as a wholesale provision merchant in Port Antonio. Charles Johnston grew up in Lucea, Hanover, and left school early to work in the banana industry. His father, Patrick Johnston, was a merchant from that town. From as early as 1887, Charlie Johnston was involved in establishing the Jamaica Co-operative Fruit and Trading Company and was also part of the team to form the Atlantic Fruit Company in 1904.

9   A photograph of Luis Fred Kennedy's initials (LFK) is first used in this chapter as a paragraph divider.

The family purchased a home on the south side of Gloucester Avenue in Montego Bay about a quarter mile west of Doctor's Cave.[10] (My dad showed me his boyhood home in the 1950s, but by then, it was in disrepair with the upstairs verandah leaning to one side and the louvred windows boarded up. It has since been demolished.) Luisa Matilde named it *Quisqueya,* which in Dominican Spanish means "Mother of all lands," a reference to the Cibao region of the Dominican Republic.[11] The house was a majestic two-storey structure of Georgian architecture with its hip roof, fretwork, and louvred windows designed to catch the cool sea breezes. The street was lined with residences, each with a commanding view of the bay and the Bogue Islands.

Montego Bay had a population of approximately 5000 residents in 1910. Like Falmouth, it had suffered economically from the decline of the sugar industry in the latter half of the nineteenth century. However, commerce remained viable. The fertile lands of the Great River Valley produced sugar, bananas, and ground provisions for export. The town's seaport was the second largest to Kingston with its boating facilities and a fledgling tourist business. Montego Bay was famous as a "health resort with exceptionally attractive bathing grounds" at the celebrated Doctor's Cave Seawater Bath with "excellent hotel accommodations."[12]

The Kennedys were among the privileged who visited the private beach club on weekends.[13] Luis Fred recalled crawling through a narrow opening of the cave to get to the water on the other side, where he learned to swim at six years of age. His father threw him from the wharf into the ocean—to sink or swim.

Luis Fred did not enjoy elementary school. Report cards showed that he underachieved in most subjects except mathematics. From an early age he showed a rebellious streak, impatient with pursuits that did not interest him. He felt confined by rules and became upset if anyone told him how to behave. This put him in a difficult situation, not only at school but at home, where both parents were disciplinarians—a mother with traditional religious values, and a father with a quick temper.

He loathed the formal wear of the period: dress shirts with neck ties, jackets, leggings, laced-up shoes, and stockings that he found too hot and tight. Neither did he relish the idea of getting clean. He dreaded bath time. His mother insisted that after school each day he be scrubbed down with a coir brush made of the coarse bristles of coconut husks. Eliza struggled to keep the boy in check.

---

10 The Kennedy home in Montego Bay was located close to present-day Pelican Grill on the Hip Strip or Bottom Road, Montego Bay.

11 The Taíno name for the Island of Hispaniola is *Haiti* or *Ayiti*, meaning, Land of High Mountains.

12 http://www.jamaicanfamilysearch.com/Members/1910d07.htm

13 Doctor's Cave was originally owned by Dr. Alexander James McCatty, a physician, who donated the property to the town in 1906.

The Wooden Leg (1909-1914)

Luis Fred was a free spirit, choosing to spend his leisure time outdoors— barefoot in the yard. I remember him saying, *"As a child, I loved the smell and feel of the rain. I would run out in the yard stark naked, especially during a downpour. I'd stand under the gutters on the side of the house and splash through the puddles. If my mother caught me, she'd cry out, ' ¡Ay, dios mío, qué travieso, maldito eres!'* [14] *When my father came home from work, I'd receive a good thrashing with his leather belt."*

The family travelled regularly by horse-drawn carriage to Falmouth to spend time with William and Mary Ann Kennedy. My dad often told the story: *"I remember very clearly one visit, in particular. I was only six years of age. It was a big home near the market. As the front door opened, I beheld a one-legged man leaning on crutches. After we entered, he slammed the door shut, and it's then that I heard a loud banging noise. I turned around and, to my amazement, saw a wooden leg dangling from a hook on the back of the door."*

One of William Kennedy's apprentices, John Stockhausen (1861–1939), and his wife, Miriam, lived in Stewart Town when the amputation had taken place. Their granddaughter, Peta Gay Jensen, in *The Last Colonials, The Story of Two European Families in Jamaica,*[15] recounts in detail the story told to her of the terrors of that night:

> My grandmother Miriam (Brown) Stockhausen often told her own children stories…When they were young girls, Old Mass[16] Ken's (William Kennedy's) leg had to be amputated. A surgeon was sent for from Brown's Town, since Stewart Town had no resident doctor. Margaret, Miriam's mother, was asked to assist in the operation; she had become well known in the neighbourhood for her expertise in setting broken bones …
>
> The operation was to take place on Old Mass Ken's dining-table … My aunt Linda, in her nineties, could still remember her mother's description of their terror that night, when lying in bed with the sheets pulled

---

14   "Oh my God, what a naughty, bad little boy you are!" Spanish was spoken in the home.

15   Peta Gay Jensen, *The Last Colonials* (2005), traces and retells the history of two of her ancestral lines in Jamaica, the Stockhausen and Clerk families. One of these ancestors, her grandfather John Samuel Stockhausen, born September 1861, lived in Stewart Town and was a business partner with William Kennedy in the 1880s. I am indebted to Raf Diaz for this historical source.

16 Creole spelling of the word is Maas. "Formerly master; now, Mr., but rather more informal, and even intimate; and toward older people, respectful." Cassidy and Le Page (1967), 283.

up over their heads, they could hear in the street below a repeated tok-tok-tok as Old Mass Ken's amputated leg walked by.[17]

From that time onwards, William Kennedy received the unfortunate name of "Cork-legged Maas Ken" in the Stockhausen family.

His apprentice, John Stockhausen,[18] bought many of the old man's businesses when William Kennedy's health deteriorated. The young entrepreneur became a successful and wealthy merchant, acquiring the post office and other properties; among them was the site of the Methodist Chapel founded by William Kennedy in 1907. "The beautiful little Wesleyan Chapel at Stewart Town is a memento of his (William Kennedy's) untiring energy, unselfishness and liberality. His residence at that place was the home of the Ministers visiting that station."[19]

Peta Gay Jensen retells Aunt Linda Stockhausen's memories of visiting William Kennedy when he resided in Falmouth.

> Another treat for the younger children was being taken by their father [John Stockhausen] to Falmouth. On these buying trips, the children would be left at Old Mass Ken's house while their father handled his business in the town. Mass Ken had not completely retired. He made and bottled guava jelly at home. There were rooms full of shelves of guava jelly, in glass jars with tin lids. The children's mouths watered, but although Mass Ken was a kind man, he never offered a spoonful or a bottle to take home. I was amused to see an article about Jamaica's stand at the 1992 Expo at Seville in Spain that guava jelly was the first item mentioned in connection with his descendants' exhibit.[20]

Before the Great War began, Fred William and Luisa Matilde had two more children. On January 23, 1913, they welcomed the birth of their second son, José Antonio, who was named after his maternal grandfather, José Antonio Arzeno (Papa Che). A year later, on April 24, 1914, a fourth child was born, a second daughter, Louisa

---

17 Peta Gay Jensen, *The Last Colonials: The Story of Two European Families in Jamaica* (2005), 19.

18 Ibid., 23. John Samuel Stockhausen's (1861–1939) nephew was Dr. Joseph Stockhausen (b. 1899), who was the Kennedy family's doctor when we were children growing up in Kingston in the 1950s.

19 William Kennedy's obituary published February 1915. My wife and I visited this chapel in the 1980s and discovered the corner stone on the church's foundation: FWK 1907. Founded William Kennedy, 1907.

20 Peta Gay Jensen (2005), 38.

Matilde, who was named after her mother; her first name was spelled differently, and she was most often called Louise, not Louisa.

My grandmother, Luisa Matilde Kennedy, circa 1916

Luis Fred Kennedy, Montego Bay, circa 1916.

6

BRAVE HEARTS (1914–1917)

Britain entered the war in defence of Belgium's sovereignty when Germany marched its armies through allied territory to invade France on August 3, 1914. The British colonies of the West Indies pledged their support to King George V, who called for "men of every class, creed and colour"[1] to join in the fight against Germany.

Jamaica made large contributions to the war by sending men, money, foodstuffs, and other supplies.[2] Soon after Britain's declaration of war, a provisional War Contingent Committee was established, and with approval of Governor Manning[3] and elected members of the Legislative Council, approvals were enacted for an aggressive recruitment campaign and allotment of large sums of money in support of the war effort. The

---

1   https://www.iwm.org.uk/history/how-the-west-indies-helped-the-war-effort-in-the-first-world-war.

2   "A sum of £10,000 was voted for defence purposes and a gift of 1,300 tons or £50,000 worth of sugar was shipped to England in 1915. Jamaica also supplied England with cash to purchase airplanes and motor ambulances." "The Great War and a Small Island -Jamaica and World War I" by Dalea Bean PhD.

3   Brigadier-General Sir William Henry Manning (1863–1932), former British Indian Army officer and Governor of Jamaica, 1913–1918. Jamaica's successful war efforts were credited in large part to his resolve in promoting patriotism and to his calm, measured demeanour.

governor issued an invitation to the women of Jamaica to form organizations across the country to raise money and supplies for the troops. The response was overwhelming, with women groups raising over £80,000 in cash and supplying handmade woolen clothing, cigarettes, and bedding.[4]

Money poured in from all quarters of the island to support the Jamaica War Relief Fund. *The Daily Gleaner* edition of Thursday, July 8, 1915, carried the following headline:

FUND BEING RAISED TO SEND OFF A JAMAICA CONTINGENT TO MOTHER COUNTRY

Another Week's Subscription from Employees of the U. F. Co.
SPLENDID DONATION

Manager of the United Fruit Company, Captain S. D. List, was quoted as saying that the total collected to date from all agencies, estates, and stevedores, island-wide, was £476 10/2.[5] Some weekly deductions from pay were as small as 3d., but every penny counted. No coercion was used; donations were voluntary. Listed in the summary of the subscriptions of the week ending June 24 was the contribution of the Montego Bay Agency, headed by Fred W. Kennedy, in the amount of £1 16s. 4d. His personal ongoing contributions were 5/- a week or £1 per month.

As per the request of council, a Jamaica Reserve Regiment was established in every parish, where recruitment meetings were held with hundreds of volunteers in attendance to hear the patriotic speeches of Governor Manning, Brigadier General Blackden, and members of the Legislative Council. They called on "the Colony's young manhood" to come forward and serve the motherland in time of crisis. Large banners were hoisted and hung high for all to see: "England expects every man to do his duty, Sons of Jamaica, your King and Country need you, enlist at once." [6]

The battle cry was for "Men, more men, and yet more men." They were exhorted to remember that "it was not freedom to sit with folded arms while their liberty was being taken from them. Each man should ask himself what he was doing so that the Empire may be preserved from the iron heel of the ruthless, crushing power of Germany."[7] Hatred of the enemy was palpable and real. People were encouraged to report any suspicious behaviours of espionage, and persons of possible German origin were interrogated, interned, and deported.

On November 8, 1915, the first of nine contingents, consisting of 722 men and twelve officers, set sail for Europe in active service of the British Empire to fight in

---

4   Dalea Bean (1969).
5   Equivalent today of approximately £50,000.
6   *The Daily Gleaner*, Wednesday, October 13, 1915, 13.
7   Ibid.

Brave Hearts (1914–1917)

the Great War. The week preceding the departure of the brave sons of Empire was marked by colourful parades through the streets of Kingston, with crowds cheering on the soldiers as they marched to their port of embarkation. They formed part of the British West Indies Regiment, a recruitment by the British government in May 1915 of soldiers from Jamaica, Barbados, British Guyana, Trinidad and Tobago, and other territories of the British West Indies.

Although large numbers of Jamaicans were impoverished and disenfranchised, they held a curious affection for the British royalty—a latent belief that through continued loyalty, there would be protection and liberty for all, and that the monarch could be trusted as a "source of higher authority."[8] With England, Jamaica stood or fell.[9]

Over a four-year period, Jamaica sent over 10,000 men to fight in the Great War. Many more volunteered for service, with an estimated 14,000 rejected as medically unfit.[10] Recruits came from all walks of life: bakers, butchers, coachmen, shoemakers, cultivators, carpenters, engineers, masons, labourers, and shopkeepers. The majority (seventy-five percent) were between ages nineteen and twenty-five.

By 1914, bananas continued to be big business. UFCO owned and operated estates, shipping lines, and wharves, and over seventeen agencies island wide. The UFCO Montego Bay agency was a large-scale operation. Office staff included R. P. Gallwey, R. Parkinson, H. P. Hendricks, F. N. DaCosta,[11] M. Nairn, W. G. Hilton, A. A. Alexander, C. Watson, E. Stennett, T. Gale, J. Thompson, J. Thomas, R. Bowen, E. Cross, J. Crighton, C. Carrol, A. Leslie, and K. Gough. They were buying agents and shippers not only for bananas but for other produce, including citrus and sugar. Mr. Kennedy was also responsible for handling the importation and clearance of goods using United Fruit Company ships.

As lucrative as the banana industry was, it was vulnerable to tropical storms. Hurricanes of 1915 and 1916[12] devastated crops and infrastructure along the north coast. In August 1915, wharves and homes in Oracabessa, Ocho Rios, St. Ann's Bay, Falmouth, and Montego Bay received extensive damage. Eighty to ninety percent of the banana crop was lost. A second hurricane hit less than a month later, September 24–25. The centre of the storm, located south of Jamaica, affected parishes of Clarendon, Westmoreland, and St. James. The hurricane the following year, August 1916, was more intense, the eye of the

---

8   Richard Smith, *Jamaican Volunteers in the First World War* (2004), 39

9   University of Florida Digital Collections, *Jamaica and Great War,* Chapter III, "In Aid of England," 18.

10   Stephen A. Hill (ed.) *Who's Who in Jamaica, 1919–1920.* Kingston, Jamaica: Gleaner Co. Ltd., 1920.

11   R. P. Gallwey and F. N. DaCosta would both join Grace, Kennedy & Co. Ltd.

12   https://nlj.gov.jm, "History of Hurricanes and Floods in Jamaica."

storm passing directly over the island. Fifty-seven persons were killed, thousands injured and made homeless, and bananas, along with cocoa and coconuts, were destroyed.

The ruinous effects on banana production lasted throughout the war years. The annual average output for 1910–14 was approximately 14.5 million stems, which represented close to fifty percent of the total export trade of Jamaica. By 1916, production had shrunk to 3.5 million; in years leading up to 1919, bananas consisted of 20.5% of Jamaica's exports.

Setbacks were exacerbated by government policies enacted during the war. In a protectionist effort to assist with the economic solvency of British territories, the governor of Jamaica issued regulations prohibiting export of fruit to any country other than Britain or its colonies.[13] The measures were temporary, but they inflicted a serious blow to Jamaica's trade with the United States.

UFCO did not suffer losses to the extent of local growers. Its sheer size with access to markets in Central and South America, and to Britain by way of their acquisition (fifty percent) of the British firm, Elders and Fyffes (1903), helped protect it.

In 1917, the United States government established the War Trade Board, whose function was to license exports and imports, keeping trade out of enemy hands, and to conserve and protect commodities and shipping for US and Allied use.[14] The board's Caribbean committee set trade policies for the region, but it proved difficult for local growers to have a voice—they were marginalized by large US fruit companies like United Fruit, whose executives controlled representation.[15] Charles E. Johnston, one of Jamaica's fruit exporters affected by the restrictions, commented: "By pressure at Washington, a war measure was stretched to prevent us running a line of steamers to America, and thus bringing about healthy competition."[16] Very few licenses to carry fruit to the US were granted to Jamaican producers and shippers.

Fred William suffered the loss of three members of his immediate family, all from natural causes. His father, William Kennedy, died February 23, 1915, at his home in Falmouth. Causes of death were exhaustion, blood poisoning from inflammation, and other ailments. Present at his death was his friend, William FitzRitson,[17] resident of Falmouth Street and

---

13   Leslie Gordon Goffe, *When Banana was King* (2007), 139.

14   US National Archives https://www.archives.gov/research/guide-fed-records/groups/182.html. President Woodrow Wilson went before a joint session of Congress to request a declaration of war against Germany on April 2, 1917.

15   Goffe (2007), 142.

16   C.E. Johnston, as quoted by Goffe (2007), 142.

17   At the start of the war effort, FitzRitson recommended to the governor "the formation of a local defence force to free regular troops stationed in Jamaica for service overseas. FitzRiston stressed support for the Empire and a desire to defend Jamaica." Richard Smith, *Jamaican Volunteers in the First World War* (2004), 36.

town clerk at the Trelawny Parochial Board. In his obituary,[18] William Kennedy was described as "one of the oldest and most respected inhabitants of Falmouth"—a God-fearing man with a successful career in business. He was well-read and sought comfort from books, especially in his last days of enforced confinement. His favourite was *"The Book,* with the riches none was more familiar than he." Persons of every walk of life came to pay their last respects. A gathering was first held in his home on Falmouth Street, followed by service at the Wesleyan Church and internment in the Falmouth Cemetery.

His widow, Mary Ann Eberall Kennedy, died the following year, also at home in Falmouth, on May 17, 1916, with her youngest child, Lena Stewart, by her side. Paralysis was recorded as the cause of Mrs. Kennedy's death. She, like her husband, had been an invalid for some time, confined to bed for almost two years. Their daughter Lena, although not well herself, had cared for both parents during their illnesses. Surviving Mary Ann Eberall were two daughters,[19] Fillan and Lena; three sons, Lunan Dexter, Fred William, and Theodore;[20] and five grandchildren, Frederick Albert Kennedy, son of Lunan Dexter, and Fred William's four children. She had three siblings: Rosabella Eberall, who died as a child of nine in 1851; a younger brother, Gilbert Eberall (b. 1847); and another sister, Sarah Louisa Eberall (b.1850).[21]

The following year, Mary Ann's daughter, Lena Stewart, died in Port Antonio, January 21, 1917, at age thirty-one. She was visiting her sister Fillan, resident of Port Antonio, at the time of death. On the same evening of her passing, the body was conveyed to her hometown of Falmouth in a motor vehicle belonging to Charles Edward Johnston. Wesleyan minister Rev. King, who conducted the service the following morning, mentioned that "her early death had been hastened by her intense devotion and attention to her father and mother during their illness until their deaths."[22]

In 1917 Fred William's only sibling residing in Jamaica was his sister, Fillan Mable Josephine Kennedy, who worked in Port Antonio. His older brother, Dexter Lunan, had migrated to Canada at the beginning of the century (1900) and married Edna Roblin, resident of Hastings, Ontario, Canada, in 1903. He subsequently moved to Minneapolis, Minnesota, where their son, Frederick Albert Kennedy, was born in 1910. His wife died May 23 that year, after which he returned to

---

18   Obituary of William Kennedy, entitled, "Death of Mr. W. Kennedy," published in a Falmouth newspaper, February 1915.

19   A birth record exists for her eldest child, Henrietta Eliza, born August 10, 1870, but no further mention appears in genealogical sources.

20   No genealogical information has been recovered for Theodore Kennedy except for his place and year of birth, Falmouth, Jamaica, 1880. My dad knew of his Uncle Teddy, who allegedly moved to Australia and raised a family there.

21   Dates of death and names of descendants (if any) are still unknown.

22   *The Daily Gleaner,* January 25, 1917.

Jamaica on vacation.[23] He re-married a few years later to an American, Lucy Ellen McLaughlin of Minnesota, and settled in Vancouver, Canada.

In his early forties, Fred William Kennedy was at a crossroads in his career. He would first apply for promotion with United Fruit, and perhaps seek ownership of a stake in the company. If he was not granted either of these, he felt it would be time to find his fortunes elsewhere. My dad recalled, *"He was never told explicitly but always suspected that the reason he did not receive the promotion was that certain positions were reserved for foreigners only."*

Michael Sheffield Grace, nephew of the New York magnate William Russell Grace,[24] was starting up business on the island.

On December 15, 1916, *The Daily Gleaner* carried an article announcing that the business of Messrs. Wessels and Nephew,[25] which had operated in Jamaica for a period of twenty years, was changing hands. The newly named successors were Grace Bros. Ltd. of London, England, and W. R. Grace & Co. New York. A local company would be formed named Grace Ltd., a fully owned subsidiary of the Grace group of companies to be managed by Michael Sheffield Grace. The company would "take over the business of Wessels & Nephew in its entirety and extend it on behalf of the Grace interests."[26] There would be no changes to the personnel, and Mr. Walter Wessel would retain his position with the firm.

Grace Ltd. Jamaica BWI was incorporated March 1917, a fully owned subsidiary of the New York conglomerate. Grace asked Fred William to join him.

Michael Grace was the son of Morgan Stanislaus Grace, younger brother of William Russell Grace. His father, Morgan, was born in Ireland but settled in New Zealand after being posted there in service to the British Army. Unlike his brother William Russell, he did not build a commercial empire. He had a distinguished career as surgeon general

---

23   Dexter Lunan Kennedy's profession was listed as piano merchant on the ship manifest for the SS *Admiral Dewey,* Philadelphia to Port Antonio. A special physical marker given was an amputated left arm. There is no mention of his child, Frederick Albert.

24   W. R. Grace (1832–1904), born in Ireland of parents James Grace and Eleanor May Russell (née Ellen), founded, along with his father, W. R. Grace & Company in Peru in 1854. In 1860 he established a merchant steamship line to serve the Americas, and in 1865, relocated to New York to expand trade with Europe and Latin America. In his capacity as mayor of New York in 1885, he accepted the Statue of Liberty from the people of France. The company opened offices in London, England in 1890.

25   At the commencement of the war, the company changed ownership. Locally it was formally known as Wessels Bros. & Von Gontard in partnership with a New York firm called Wessels Kulenkampff. When the partnership dissolved, the new company came to be known as Wessels and Nephew. It is unclear whether a German connection was the reason for the dissolution of partnership and/or change of names.

26   *The Daily Gleaner,* December 15, 1916.

and member of the New Zealand Parliament. He married a New Zealander, Agnes Mary Johnston, and together they had nine children, of whom Michael Sheffield was the youngest. Michael's oldest brother was John Johnston Grace, born 1870[27] (one of the founders of Grace, Kennedy in 1922). When their father died, Michael went to Hawaii to be with his brother, Dr. John Grace, who had a medical practice there. The brothers invested in several businesses, including a soda water company, a bank, a garage business, and sugar cane.[28]

Before forming Grace Ltd. BWI, Michael Sheffield Grace had resided in England, where he had volunteered for the Metropolitan Special Constabulary. (He was disqualified from joining the British Armed Forces due to a hearing impairment.) Accompanying him on his return to Jamaica in 1915 was his new bride, Dorothy Ann Morgan-Brown. They settled in Bath, St. Thomas, where he trained military recruits. He became captain of B Company, Jamaica Reserve Regiment, and was a justice of the peace for the parish of St. Thomas.

A glowing tribute to Fred William Kennedy appeared in *The Daily Gleaner* edition of July 30, 1917, announcing his leaving United Fruit Company to become "General Manager of Grace Ltd. in Jamaica." The article praised him as "one of the best thought employees of the big corporation," proof of which was his promotion as manager to the Montego Bay Office, considered to be one of the most important in the island. "The energy and enterprise of Mr. Kennedy" accounted for the all-round improvement of the agency over the previous six years while he had been manager. "It will be a pleasure," the report said, to watch "this capable young Jamaican steer the Grace ship of business along the right channel and into success."[29]

---

27 Biographical information gathered from University of Florida Digital Services, *The Daily Gleaner*, and genealogical research by Georgianne Thompson Kennedy on ancestry.ca

28 Douglas Hall, *Grace, Kennedy & Co. Ltd.: A Story of Jamaican Enterprise* (1992), 5.

29 *The Daily Gleaner,* July 30, 1917, 11.

7

A QUESTION
(1917)

"Have you ever had a strong feeling that you've been in an identical place or situation before but cannot fully recall or explain it?

In 1917 my father wrote and illustrated a poem about this phenomenon called a déjà vu" (**LFK**).

A Question   (1917)

A Question (1917)

Foreword.

We have all, at some time or other, on seeing some beautiful landscape for the first time, experienced the feeling that the scene is a familiar one, — that we have visited the spot before.

The writer endeavors to describe this feeling, and expresses the hope that he voices the experience of a goodly number.

Fred W. Kennedy,
Montego Bay,
Jamaica, B.W.I.,
April 3rd, 1917.

A Question (1917)

'Tis day! — I journey on a road.
Dame Nature seems to deck her children
    In their most gorgeous garb.
I reach the crest of a commanding hill,
And pause to gaze, enchanted, on the scene.
When suddenly a weird, mysterious feeling
    Steals upon my Soul!
I seem to see what I have seen before,—
The trees, the flowers, the bright, the glorious sun,—
All seem but a picture thrown upon the
        canvas of my mind,
A reproduction of some past!
So real! So like the present,
That I,— I feel unreal! ....
    'Tis gone! This feeling's gone!
    I sink to earth once more!
But still that question rises in my mind,
Still fraught with import,— still unanswered!—
Have I,— the I within me,— had a past?

A Question (1917)

'Tis night! – I sit engaged in pleasing
    converse with my friends,
When – some chance word dropped,
Or some slight action on the part of one
Seem to vibrate some chord of memory.
Again that weird, mysterious feeling
    Steals upon my Soul!
I seem to see and hear
What I have seen and heard before,
In some dim, distant past;
    And the whole scene, –
        Friends, conversation, all, –
All seem but a reproduction of some past,
So real, – so like the present,
That I, – I feel unreal! ....
'Tis gone! This feeling's gone!
    I sink to earth once more!
But still that question rises in my mind,
Still fraught with import, – still unanswered, –
Have I, – the I within me, – had a past?

Kennedys' Home at 25 South Camp Road, Kingston, 1917–1938.
Luis Fred's mother is seen standing on front veranda.

1915 Overland Model 83B Open Tourer Car, similar to the one owned by
Fred William Kennedy and family when they first moved to Kingston.
Photograph © "Bruno from Belgium," owner of vehicle.

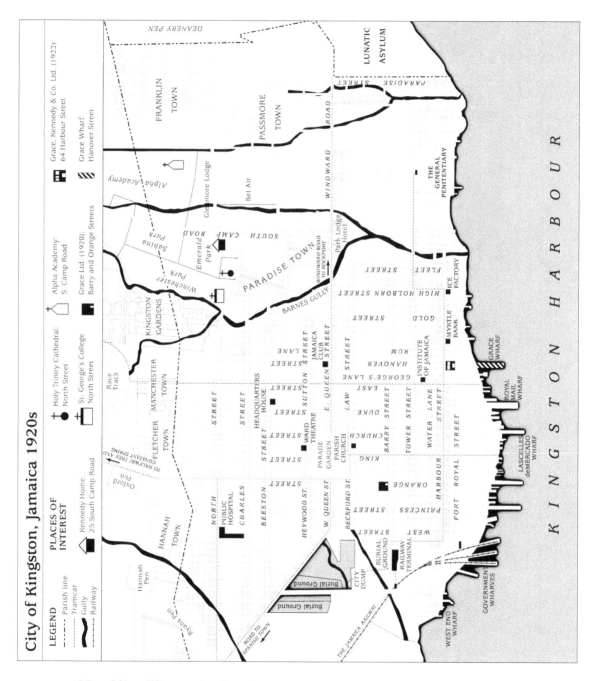

Map of City of Kingston, Jamaica 1920s created by Mapping Specialists Ltd., Fitchburg, WI. © Fred W. Kennedy. Adapted from composite of antique maps of Kingston.

8

KINGSTON (1917–1918)

Luis Fred's mother's dream of a move to the big city of Kingston came true. And his father's aspirations to become a manager of a company independent of United Fruit were realized. The world of commerce was at his doorstep:

> No commercial concern in Jamaica has greater potentialities nor more extensive connections and resources than that of Messrs. Grace Ltd.—a subsidiary of the vast and renowned Grace organization that influences markets in all parts of the world. Vessels on every ocean carry Grace cargoes.[1]

The Grace group of companies had over two hundred commercial centres in North America, South America, the Caribbean, India, China, Japan, Europe, and West Africa. Operations included myriad forms of business—banking, shipping, import/export of goods/services, management and ownership of cotton mills, nitrate and mineral works, and sugar cane factories.[2]

Local advertisements in 1917 showed that Grace Ltd. Jamaica BWI held its main offices at Olivier Place in Kingston at Barry and Orange Streets, with branches in Brown's Town and St. Ann's Bay. They were dealers in sugar and exporters of tropical produce, as well as importers and distributors of foodstuffs and manufactured products.[3]

---

1 Macmillan, Allister. *The Red Book of the West Indies* (1920), 68.
2 Ibid.
3 University of Florida Digital Collections https://ufdc.ufl.edu/UF00081175/00001/28j.

Along with Lascelles deMercado, they were among the largest general merchants doing business in Jamaica.[4]

Kingston was originally laid out on a formal grid with King Street, the main artery north to south, intersecting with Queen Street. Parade Square was at the intersection, with Kingston Parish Church on the south and Ward Theatre on the north, both rebuilt after the 1907 earthquake.[5] In 1917 the population of Kingston was about 60,000, ten times or more the size of Montego Bay.[6] Migration from other parishes had started from the turn of the century with people looking for jobs due to a shortage of lands available for cultivation.[7]

The population had begun to spread to the Liguanea Plains, but most of the residential areas were still in and around the downtown core. The old tramway of 1876 had been replaced by more advanced electrical cars running by rail on the principal streets of the city as far as Constant Spring and Papine Corner. Markets were plentifully supplied with fruit, vegetables, and fish, and haberdashery and dry goods stores carried a wide variety of goods. The city's hotels included The Jamaica Club, The South Camp Road Hotel, The Queen's Hotel, the more expensive Myrtle Bank, and over thirty lodging houses.[8] Myrtle Bank was the largest in Kingston with 205 guest rooms. Destroyed in the 1907 earthquake, it was renovated and bought by UFCO in 1918.

The family purchased a home at 25 South Camp Road,[9] close to Holy Trinity Cathedral and the Sabina Park cricket grounds. Swithin Wilmot commented:

---

4   Other large merchant firms included Fred L. Myers & Son (1879); Messrs. C. V. Percy & Stanley Lindo, Linda Bros. & Co.; J. Wray and Nephew Rum Distillers (1825); Isaacs and Brandon Ltd. (1909); Adolph Levy and Bro. Commission Merchants (1893); Bryden and Evelyn Manufacturers' Agents (1910); Cecil deCordova and Co. (1900); N. C. Henriques Ltd.; and E. A. Issa and Bros. (1894). Most of these businesses were located in downtown Kingston on Port Royal and Harbour Streets. (*The Red Book of the West Indies.* 1922)

5   Measuring 6.5 on the Richter Scale, an earthquake struck Jamaica at 3:00 PM on January 14, 1907, with its epicentre in the City of Kingston. One thousand people lost their lives. It spawned a fire that gutted most of the commercial district of the city, reducing many buildings to rubble. Water pipes were broken, tram lines twisted, and electrical posts bent and uprooted. Kingston became a ghost town.

6   In the 1881 Census, the population of Kingston was 36,846; in 1891, 46,542; and in 1914, 57,379. The population of Jamaica in 1917 was 850,000. As population spread north into the suburbs of St. Andrew, the two parishes were amalgamated in 1923.

7   G. E. Cumper, "Population Movements in Jamaica, 1830–1950," (September 1956), 272.

8   In the early 1900s, a week's stay cost 80/- (Ansel Hart, 1968).

9   In the 1950s, the property at 25 South Camp Road became the People's National Party Headquarters. The original name of the property was Nether Edge. Solomon Clifford Lindo (contemporary of Fred William) and his family lived at 29 South Camp Road, a property named Cricklewood. The Lindos of S. C. Lindo Ltd. were wholesale and wine merchants.

> The location is now part of Holy Trinity Secondary, and there is a concrete wall with an iron gate around where the entrance to the two-storey building was.
>
> The building was located on the western side of South Camp Road, just below and opposite to where Glenmore Road comes off South Camp Rd. It was around 50 yards to the north of North Street, around halfway between North St and Emerald Rd, by Sabina Park.
>
> Catholics lived in the vicinity (almost like Jews to their synagogue). My mom said that John Figueroa's family lived on Emerald Road.
>
> Elias Issa, father of Abe lived higher up South Camp, opposite to Alpha Boys entrance, where Stewart's Auto Sales now is. I would see him sitting in a pagoda when I was walking home from school in late 50s, early 60s.[10]

The house was of the late Georgian period, with Caribbean colonial treatments: deep verandahs, fretwork transom and railings, and decorative gables. It was a spacious, two-storey structure with a masonry ground floor, corner quoins, and a wooden upper level.[11] It was set on a large property with carefully manicured gardens, paths, and driveways. Luisa Matilde had the sign from the Montego Bay home mounted at the front gate. "It was made of brass; it was about twelve inches long and six inches high. The name *Quisqueya* was printed in black capital letters."[12]

It was an ideal location for a Catholic family with four young children. Luis Fred and his sister Mim attended Alpha Primary (run by the Sisters of Mercy), which was only a few hundred yards up the road. Holy Trinity Cathedral[13] was nearby, and adjacent to it, the newly built Jesuit high school, St. George's College (1914), which Luis Fred and his brothers would attend.

Fred William was teased by friends that he was more devout than if he had been raised Catholic. The exact date of his conversion is uncertain. In *The Daily Gleaner* edition of January 17, 1915, Fred W. Kennedy's name appeared as one of the brothers

---

10   Swithin Wilmot, correspondence, May 2020.

11   Description courtesy of Jamaican architect Kevin Bryan, https://kevinbryanarchitect.com.

12   Fred W. (Bill) Valliant, Chapter One, "The House" in *Jigga Foot Boy at Quisqueya–A Proud Jamaican Heritage* (2020). In his memoirs, my cousin Bill (eldest child of Carmen Valliant, née Kennedy) recalls that the sign was also used at the family's third home built in the late 1930s.

13   Built in 1911 to replace the original cathedral (1811) destroyed by the earthquake of 1907. Of Byzantine architecture with a magnificent copper dome, stained glass windows, and over 3,000 square feet of murals and mosaics, the cathedral is an impressive Kingston landmark.

in attendance at a meeting of Masons at the Masonic Temple, Church Street, Montego Bay—a highly irregular occurrence had he been a member of the Catholic Church, which considered Freemasonry irreconcilable with Catholic doctrine.[14]

The home was close to the hub of downtown, Harbour and Duke Streets, and the Palace Theatre, with the "Latest and Best Pictures from all the First Run Theatres of the World," was within walking distance. For outings about Kingston and for trips to the country, the family used their blue five-seater Overland, an English-manufactured motor vehicle of which Fred William was proud. It had a forty-eight horsepower, three-litre, four-cylinder engine. Pleased with its performance and the way in which he was able to overhaul it and make it look new again,[15] he kept the car for many years.

The proverbial saying, "Children should be seen and not heard," characterized Luis Fred's and his siblings' upbringing. Their mother did not attend personally to their daily physical needs, bathing or dressing them, or cooking their meals. Instead, she employed nannies, a cook, a butler, cleaners, and a gardener. Her role was to supervise the household, to ensure everything was in order and the children kept clean and well fed.

When the butler rang the dinner bell, the family assembled in the dining room, children bathed and in best attire: shoes and socks, jackets for the boys, and dresses for the girls. The meal started with a prayer, followed by finger-washing in small bowls. A waiter wearing white gloves proceeded then to serve a four-course meal.

Spanish was spoken in the home and customs and religious practices of the Dominican Republic were observed. The children grew up knowing Latin American foods: *aceite de oliva* (olive oil), *sancocho* (beef stew), *habichuelas guisadas con arroces* (stewed beans with meat and rice), *tostones* (double fried green plantain), *dulces* (desserts), especially *quesillo* (flan), which was their mother's favourite. On January 6, *El día de los Reyes* (Day of the Kings), they exchanged gifts and roasted a suckling pig on a pit over an outdoor fire. (This tradition was passed down to the next generation.)

At the eleventh hour on the eleventh day of the eleventh month of 1918, the Great War ended when Germany signed an armistice agreement with the Allies. The loss of human life was catastrophic. Nine million combatants worldwide had died with

---

14   Fred William Kennedy attended a meeting of Masons for the induction of Dr. Alexander A. Vernon, former York Castle student, as Master of the Friendly Lodge. Both were members of Friendly Lodge 383 in Montego Bay after 1910. See note in Chapter 2 re Alexander Vernon.

15   Fred W. Kennedy, letter dated October 19, 1924, to his son Luis Fred. He advertised the car for sale in *The Daily Gleaner* in 1923 but decided instead to overhaul it, after which he claimed it ran like new. His contemporaries joked about his not wanting to purchase a new vehicle. His friend Masterton in 1924 referred to it as a "has been." Fred William purchased a new car in 1925.

21,000,000 wounded. In addition, close to 5,000,000 civilians perished from disease, starvation, and other causes. Jamaica suffered 1,000 casualties.

Post-war, the world was ravaged by an influenza epidemic, with a death toll of approximately 50,000,000. It spread rapidly through the British West Indies, with Jamaica, British Guyana, and Belize suffering more casualties than the Eastern Caribbean. The total number of deaths in the region was reported to be 100,000; the virus spread rapidly in densely populated areas and in communities with the poorest living conditions.

Demands for self-government and a better standard of living grew more intense throughout a disgruntled British Empire. Jamaica was no exception. Labour unrest rattled the country in the last two years of the war and in the years to follow. Immediately following the war, a worldwide recession made it more difficult for workers to find employment. Ice factory workers and cigar makers were among those groups who went on strike in 1917. "In April 1918, fire brigade employees took strike action, followed by the longshoremen on the Kingston wharves in June."[16]

With growing discontent came the need for Blacks to gain economic power and racial solidarity. The United Negro Improvement Association founded by Jamaican Marcus Garvey called for a global "confraternity" of all members of the Negro race.[17] He summoned all Blacks to have pride in themselves, to unite, to assist in the creation of independent nations and of a "great racial empire."[18] The movement took root in New York, where he expanded the membership of the UNIA to include American Blacks returning from the war. His ideas and those of others like Claude McKay,[19] also a native of Jamaica, planted the seeds of Black nationalism and Pan Africanism, which influenced a change in the class structure, politics, and commerce of Jamaican society.

The mass-movement shook the foundations of British imperialism. His Majesty King George V's subjects were clamouring at the grass-roots level for better livelihoods and a greater participation in politics.

---

16   Arnold Bertram, *N.W. Manley and the Making of Modern Jamaica* (2016), 80.

17   Marcus Mosiah Garvey (1887–1940), Jamaica's first National Hero.

18   Marcus Garvey, *Message to the People* (1937), "Lesson Three: Aims and Objectives of the U.N.I.A."

19   Claude McKay was born in the parish of Clarendon 1889, his first work, *Songs of Jamaica*, published in England in 1912. He joined the UNIA in New York in the 1920s and became a prolific and influential writer famous for the part he played in the Harlem Renaissance.

A rare occasion of Luis Fred dressed in costume, circa 1919, Kingston, Jamaica.

The O'Hare Building, constructed 1914.
St. George's College, North Street, Kingston.
(Photo in the public domain, courtesy of St. George's College)

9

STGC

Ad Majorem Dei Gloriam[1]

(1919)

Fred William's sister Fillan had stayed with the family in Kingston for a while before migrating to the United States to pursue a career as a stenographer. She was forty-one. Described as "dark skin, 5 ft. 5 inches tall, resident of Port Antonio, Jamaica," Fillan Mable Josephine Kennedy arrived in New York on board the SS *Turrialba* from Kingston on July 12, 1918. She settled in California, became a US citizen, and did not return to live in Jamaica until her retirement in 1935. She was Fred William's last sibling to leave the island.

Fred William's and Luisa's fifth child, Gloria Mercedes, arrived June 2, 1919. (There had been no war babies.) Gloria was named after her mother's youngest sibling, Gloria Mercedes, who lived in the Dominican Republic. Each of the five Kennedy children had been named after relatives in the Dominican Republic.

Luis Fred's enrolment at St. George's College marked the beginning of a lifetime association with the Jesuits, whom he admired for their scholarship and commitment to social justice.

The Jesuit mission began in Jamaica when two priests arrived on the island from Europe in 1837; they were followed by a full complement of twenty-one Spanish Jesuits

---

1   Jesuit motto: For the greater Glory of God.

thirteen years later. They had been ousted from New Granada (present-day Colombia) by a newly elected president with a social agenda against institutionalized religion.

Doors opened at 26 North Street on September 2, 1850. Two years later, the South American men left the island and turned over the mission to their British counterparts. By the late 1800s, their numbers had grown to twenty, but in 1894 their assignment was terminated because of demands for missionaries in Africa. They were replaced by North American Jesuits,[2] who inherited a cathedral on Duke Street, a Jesuit residence, parishes throughout the island, and St. George's College, originally located at the corner of North and Orange Streets. They formed a new community of priests, who established themselves at Winchester Park: a Jesuit Residence, completed in 1910 and the O'Hare Building[3] of St. George's College in 1914.

A secondary-school education was a luxury in 1920s British West Indies. No more than one percent of the population enjoyed the privilege. Fewer than eight percent of students achieved satisfactory standards upon graduation and even a lower percentage qualified for (or could afford) a tertiary education.[4] Only a handful of government tertiary scholarships were offered island-wide;[5] school fees were prohibitive,[6] and in addition, parents were required to fund uniforms, books, meals, and travel. Secondary-school education was elitist and out of range for most Jamaicans to afford.

St. George's College was established as a private school that catered primarily to the Catholic population in Kingston. In the 1920s there were 30–40,000 Catholics,[7] approximately five percent of the population. Half of these resided in Kingston, with 6,500 attending Catholic elementary and secondary schools. St. George's struggled financially. It supported some students who could not afford fees, and it had not yet applied or qualified for grant-in-aid allowances from the government.[8]

At age ten, Luis Fred was young and impressionable but intellectually and physically advanced for his age. He was not a diligent student but learned quickly and with a

---

2   Jesuits from the New York Province of the Society of Jesus.

3   The O'Hare Building has become the landmark of St. George's College. The architect was Braham Taylor Judah (1871–1954), father of three St. George's College alumni, Douglas (attorney), and Charles and Sydney Judah (Jesuit priests).

4   Duncan James Jeffrey, "Education, Economy and Class in Jamaica, 1700–1944" (1980), 236–237.

5   Six scholarships were granted annually to the top performing three boys and three girls under the age of sixteen in the Cambridge Junior Local Examination, and one scholarship was awarded island wide each year to the top performing student (sixteen to nineteen years of age) in the Cambridge Senior Local Examination.

6   Fees for private schools like St. George's were in the range of £3–5. Between the wars, few working-class Jamaicans earned more than £52 per year. Altink (2019), 70.

7   Kathryn Wirtenberger, "The Jesuits in Jamaica" (1942).

8   St. George's College became a grant-in-aid school in 1936 when it received a government subvention for payment of teacher salaries.

flair. He enjoyed mathematics, had a facility for Spanish and Latin, and was a master of both oral and written English. He excelled in elocution, a required subject, and with this skill won many school debates. He was an impulsive learner, spending little time doing homework or labouring over details. His parents thought that if he had only applied himself more, he could have achieved higher marks and earned scholarships. He engaged in rough play and at home ran about barefoot and shirtless, or as his father termed it, "semi-nude."[9] Much to the dismay of his parents, neither did he match up to their standards of grooming and decorum.

The school curriculum was based on a model of classical education. Subjects taught in the senior forms at St. George's with their number of hours were as follows: Latin (5); English (4); Mathematics (5); Christian doctrine (2); History (2); Elocution (1); Modern Languages (2); and alternative subjects such as Greek, shorthand, bookkeeping, and elementary physics (1 ½).[10] St. George's also included sports in its curriculum: track, football, cricket, and tennis, which was his favourite. He became a proficient player, continuing the sport into his adult years.

Schoolmates Abe Issa and Douglas Judah would become business associates and lifelong friends.[11] Jesuits too were friends of the family: Fr. Emmet Cronin S.J.,[12] Fr. Francis Delany S.J., Fr. Ferdinand Wheeler S.J., Fr. Francis Kelly S.J., and Fr. Leo Butler S.J., who was known for his converting the Jamaican Chinese community to Catholicism. Fr. Delaney, a harsh disciplinarian, was headmaster in Luis Fred's first year and was succeeded by Fr. Wheeler, 1920–1923.

In the summer of 1921, the school held its annual Prize Day on campus. The occasion featured the student production of the Shakespearean play, *The Merchant of Venice*, complete with full orchestra. Upwards of a thousand spectators gathered on the lawns opposite to the entrance of the O'Hare Building, where a stage had been erected for the performance.[13]

Among the special guests were His Lordship, Bishop O'Hare, the minister of education, the acting US consul, Mr. Fred W. Kennedy, Mr. Bernard O'Toole, and Mr. Braham Judah. Fred William had officially removed himself from Methodist circles and become deeply entrenched in the Roman Catholic community as one of its stalwart members.[14]

---

9   Fred W. Kennedy, letter to Luis Fred Kennedy, Tuesday, August 26, 1924, referenced in Chapter 12.

10   Rev. Carl F. Clarke, "A Concise History of St. George's College," (2004). 16-17.

11   Douglas Judah. See biographical note, Chapter 19.

12   See Chapter 12.

13   *The Daily Gleaner*, July 8, 1921, 13.

14   This began a tradition of Catholic philanthropy that was passed down through the generations.

Many of Luis Fred's friends were listed in the program's *dramatis personae* of *The Merchant of Venice*: Donald DeLeon[15] played the duke of Venice; Roderick Francis,[16] prince of Morocco; Abe Issa, Tybalt, friend of Shylock; Joe Issa, Leonardo, servant to Antonio; and Douglas Judah, Portia, a rich heiress. Luis Fred's name was conspicuously missing from the list of actors. (He disliked dressing in costume and despised the idea of boys playing female roles.) The performance was termed "a brilliant success." "The acting of the boys reflected the highest credit on the College." Douglas Judah received special mention for his portrayal of the rich heiress.

In the School Annual Report given that evening, Acting Headmaster Fr. Wheeler S.J. mentioned that 183 names were on roll with an average attendance of 154.[17] In the Cambridge Local Examinations, ten students were successful, one gaining second class honours.[18] Scholarship examinations were held in December, and twenty-eight boys competed for the three scholarships offered by the faculty for needy students, "irrespective of religious denomination." The school football team won the Manning Cup that year and tied with Munro College for the Olivier Shield.

While Luis Fred was at St. George's College, his younger siblings attended Convent of Mercy schools. In 1919, his sister Mim was in her last year of elementary school and qualified the following year for entrance to high school. His brother José was starting the select preparatory school for boys, and his sister Louise was finishing kindergarten. Alpha had attained a reputation for high standards of excellence. The academy was a large complex on South Camp Road including a high school for girls, an orphanage, an industrial school, elementary schools for both boys and girls, and a boarding school "in which a high-class English education" was imparted.[19]

---

15   DeLeons, DaCostas, DeSousas, Henriques and many others who attended St. George's College were descendants of the Sephardic Portuguese and Spanish Jewish congregation in Jamaica.

16   Roderick Francis Sr. pursued a career as maritime pilot and became known affectionately as "Pilot." He was involved in the Pan-African movement in Jamaica in the 1930s and in the formation of the Inter-Trade Union Advisory Council. He described himself as "someone involved with every activity to do with the common man." He died in 1999. Bryan and Watson (2003), 41.

17   With an additional 70 students registered in 1923, the college had a total enrolment of 258, making it one of the largest secondary schools in the island.

18   Prominent boys' schools in the island at the time included Wolmer's Free School, Munro and Dickenson's Free School, Manning's Free School, The Jamaica High School (Jamaica College), Jamaica Baptist College (Calabar), Montego Bay Secondary School. Henrice Altink, *Public Secrets: Race and Colour in Colonial and Independent Jamaica* (2019), 69.

19   Frank Cundall (1902), 315.

GRACE FAMILY: Four Generation Descendant Chart

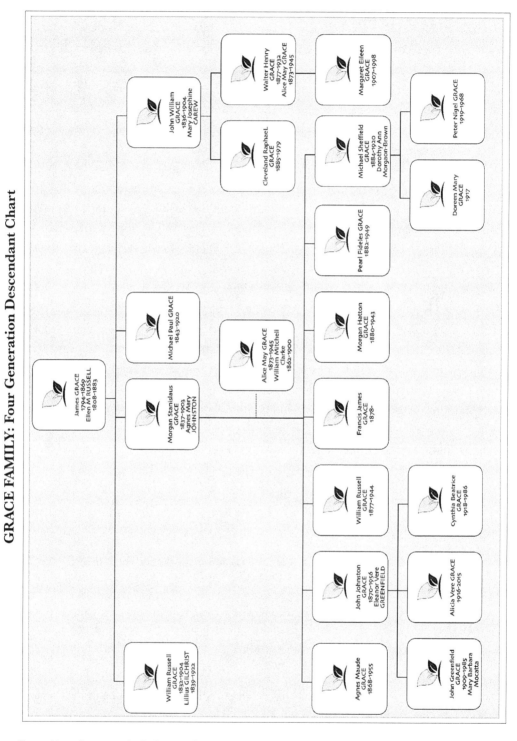

Created based on genealogical research conducted by Georgianne Thompson Kennedy and by use of © Family Tree Maker https://www.mackiev.com/ftm/

Dr. Morgan Stanislaus Grace (1837–1903), father of John, Morgan, and Michael.
(Photo in public domain, Wikimedia Commons)

Dr. John Johnston Grace (1870–1956)
(Photo, courtesy of the Moyls family private collection, posted by Angela Moyls on ancestry.ca)

Morgan Hatton Grace (1880–1943), circa 1920.
(Posted by June Barnes and Douglas Kellner on geni.com
Original source: Passport photo, ancestry.co.uk)

Michael Sheffield Grace (1884–1920)
The Daily Gleaner, January 24, 1920, 6.
(Photo, courtesy of © *The Gleaner Company* (Media) Ltd.)

10

THE GRACE BROS.
(1920s)

After Fred William assumed management of Grace Ltd., Michael Grace presented himself as a candidate for election to a vacant seat for St. Thomas in the Legislative Council in January 1918. "He is a man in full sympathy with the people of the island, and their interests. A better man could not be chosen."[1] Despite his reputation, Mr. Grace was defeated by contender J. H. Phillips[2] of Morant Bay.

Michael Grace was also involved in the car business. He invested in Jamaica Motor Car and Supplies Ltd., founded in 1918, the largest garage and motor car supply business in the West Indies. Their premises included two large blocks on Hanover and Harbour Streets, opposite to the Myrtle Bank Hotel. The large workshops on Hanover Street were equipped with the most up-to-date equipment for car and truck repairs, and the storerooms on Harbour Street contained "an enormous stock of accessories and spare parts of every description."[3] The company was agents for Cadillac, Buick, and Oakland cars, Federal trucks, and Cleveland tractors.

---

1   A. E. Holhs of Port Morant of Port Morant in a telegram to *The Daily Gleaner*, stated his withdrawal from his candidature and asked his friends to support "Captain Grace." *The Daily Gleaner*, Saturday, January 12, 1918, 3.

2   Mr. Phillips had a progressive record in promoting legislation calling for compensation of officers and servicemen who had returned from the Great War.

3   "Motor Car and Supplies Ltd., Harbour and Hanover Streets" in *The Red Book of the West Indies* (1922), 69.

The Grace Bros. (1920s)

Towards the end of 1919, Michael Grace stepped back from active management for health reasons. He was reported absent on many occasions when the company held functions, at which Deputy Manager Fred William Kennedy would represent him.

Mr. Grace was suffering from Bright's disease, a kidney ailment known today as chronic nephritis. It proved fatal. Michael Sheffield Grace, aged thirty-five, died at 6:00 AM on January 22, 1920, at his residence at Fullar Ewan on Hope Road, St. Andrew. *The Daily Gleaner* carried the announcement:

> THE LATE MR. M. S. GRACE
> ## THE DEMISE OF MR. M. S. GRACE
> The Death of the Head of
> The Well-Known firm of
> Grace Ltd.[4]

Surviving him were his young widow, Dorothy Ann Morgan-Brown, aged twenty-six, and two children, Doreen Mary Grace, three years, and Peter Nigel Grace, ten months.

Grace Ltd. shut its doors to business, and flags of various shipping companies were flown at half-mast. On the day of his death, a procession was held from his family residence to the Holy Trinity Cathedral. His Lordship Bishop Collins S.J. and Rev. Fr. Delaney, Acting Jesuit Superior, conducted the service before a congregation that packed the church. Fred W. Kennedy was one of six pall bearers, along with Thomas Kemp[5], J. C. Farquharson, J. H. Cargill[6], E. C. White, and R. E. Hollinsed.

Grace Ltd. continued to expand its operations under the capable management of Fred William Kennedy. The company's successes did not escape the attention of W. R. Grace & Co. of New York. (Michael Grace's first cousin, Joseph P. Grace Sr., was president of the company in 1920.)[7] Even though the US was feeling the effects of a post-war recession, the board of directors retained its interests in the Jamaican enterprise, due largely to the subsidiary's favourable financial prospects and track record. Michael's eldest brother, Dr. John Johnston Grace, who was residing in London, England, at the time, expressed an interest in filling the position of managing director of Grace Ltd. He ventured to Jamaica with his family in early 1920.

Dr. John Grace took up residence at Strawberry Hill, St. Andrew, in the cool foothills of the Blue Mountains with his wife, Eleanor Vere Greenfield, and three children,

---

4   *The Daily Gleaner*, Saturday, January 24, 1920, 6.

5   Thomas Kemp held a senior position as company secretary of Grace Ltd.

6   Captain John Henry Cargill had distinguished himself during the Great War for services in connection with Red Cross and recruitment of servicemen. He was an accomplished sportsman (tennis), prominent solicitor, and partner in the law firm, Cargill, Cargill and Dunn, whose offices were at 4 Duke Street, Kingston.

7   Joseph P. Grace was appointed president of W. R. Grace in 1909, five years after the death of his father, founder William R. Grace

John Greenfield Grace (eleven years), Alicia Vere (four years), and Cynthia Beatrice (two years).[8]

John Grace was a medical doctor, having obtained his certification, Fellowship of the Royal College of Surgeons (FRCS), in 1896. Two years later, he travelled to the United States and opened a medical practice in Hawaii. He married Eleanor Vere Greenfield, a resident of Hawaii, where their first child, John, was born in 1909. They resided in Hilo, an island with dramatic waterfalls, fertile rainforests, and gardens. Dr. Grace experimented with business ventures while residing in Hawaii, but his primary profession was physician. Their two daughters had been born in England before the family migrated to Jamaica in 1920.

On March 27, 1920, he was guest of honour at a welcome party hosted by Grace Ltd. at the South Camp Road Hotel. Special guests included Fred W. Kennedy and staff; representatives of W. R. Grace of New York; W. J. Wessels, former owner of Grace Ltd; J. H. Cargill and H. H. Dunn; Walter Fletcher and C. Ingram, managers of the Montego Bay and St. Ann's Bay offices, respectively.

Fred William spoke on behalf of the company. He said how relieved the staff felt at knowing that a member of the Grace family would fill Michael Grace's position.

Dr. Grace remarked that he was overwhelmed with the cordiality of the reception and had never witnessed people more capable of niceties than Jamaicans. He mentioned that he had employed his brother Michael in Hawaii and that they had remained in contact through frequent correspondence over the years. He praised his brother for the ways in which he had grown Grace Ltd., and admitted, "It will be some time before I can hope to get the position which he held—the position of esteem and affection which he occupied in your hearts."[9] He hoped that like the cool, gentle breeze, "the Doctor," as Jamaicans called it, he would always be "stimulating and invigorating, or seldom or never be boisterous." Speakers who followed wished the company continued success and that every employee be imbued with the "Grace spirit," which accounted for the successes of its world-wide undertakings.

El Carrrillo, a luxury steamship of the United Fruit Company White Fleet, sailed out of Kingston Harbour in June 1920, bound for New York City. On board were Fred William Kennedy, his wife, five children, and nanny Eliza Taylor.

The *Carrillo* was one of the largest of the fleet, 5,000 tons, a freight carrier with deluxe accommodations for one hundred passengers. United Fruit was proud of the modernity of its liners—150-foot promenade decks, cabins with ensuite bathrooms, formal, spacious dining rooms, a smoking room for gentlemen, and lounges decorated

---

8   See Grace Family Tree at the beginning of chapter.
9   *The Daily Gleaner,* April 2, 1920, 8.

in Edwardian style with the finest fabrics. They were vessels "with a total absence of vibration, jar or noise of machinery."[10] Gone were the days of dirty coal baskets and flying dust; a new era of shipping had dawned with the advent of oil-fired boilers.

The United States Bureau of Immigration required aliens to complete information regarding gender, marital status, familial relations, occupation, nationality, and a section called "Races or Peoples." Most Jamaicans entered British for nationality, and West Indian for race/people. Some US shipping lines carried their own reservation brochures, which included the instructions, "If colored, so state."[11] Travellers would not be eligible for first-class cabins if they self-identified as "colored"; these were reserved for "whites" only.

The family went to spend time with Luisa Matilde's older sister, Carmen Celia, in Mount Vernon, New York.[12] She was mourning the death of her husband, Charles Henry Loinaz, who had died three months earlier.[13] The Loinaz family had fled Puerto Plata during World War I due to political upheavals, the Civil War of 1914, and the United States occupation of the Dominican Republic in 1916. Charles Henry had been British consul in the Dominican Republic and associated with business enterprises, including sugar and chocolate. He left an inheritance sufficient for his widow to support the family, but this would suffer grave losses in the years of the Great Depression. He was survived by his widow, Carmen Celia (age forty-three), and their seven children:: Georgina (age twenty-one), Charles Eugenio (nineteen), José (eighteen), Carmen Celia (sixteen), Mercedes Amanda (fifteen), Lydia Mathilde (nine), and Altagracia María (two).

Before returning to Jamaica, the Kennedys visited Puerto Plata for the children to meet their grandmother, aunts, uncles, and cousins.

On July 1, 1920, *The Daily Gleaner* reported that the government was purchasing the block of buildings and premises from Grace Ltd. on Orange and Barry Streets to give extra space for the post office. The price agreed upon was £13,000. Grace Ltd. was expanding its operations:

> The extensive premises between Olivier Place and Orange Street, purchased by the new company, proved inadequate for the rapid development of their trade, and accordingly a larger and more suitable site was secured, extending from Harbour Street to the sea—a distance of about 800 yards, where big warehouses, offices, and other buildings

---

10   Catherine Cocks, *Tropical Whites: The Rise of the Tourist South in the Americas* (2013), 66.
11   Ibid., 63.
12   Carmen Celia Loinaz née Arzeno Abreu (July 21, 1876–March 11, 1963)
13   Charles Henry Loinaz, died March 3, 1920.

have been erected, to which the firm has recently removed. The new premises have a wharf running 550 out into the sea, where the Grace steamers and other vessels can discharge and receive their cargoes. The firm has the largest warehouse in Jamaica.[14]

The new wharf acquired by Messrs. Grace Ltd. had belonged to Soutar & Company,[15] a large merchant firm in Kingston.

Dr. Grace was not averse to taking risks to diversify the company's interests:

> Since their advent in Jamaica, Messrs. Grace Ltd. have developed their activities enormously, and these are still being extended. In 1921, they acquired the large works of The Jamaica Shoe and Leather Company in Windward Road, Rockfort District. The establishment is the chief of its kind in the island and is equipped with an up-to-date machinery and appliances pertaining to the tanning industry, the magnificent series of tanning pits and every detail of the factory being well fitted to sustain a very large production.[16]

World markets were uncertain due to the recession of 1921. It was deep and sharp but quick, lasting eighteen months, from January 1920 to July 1921. *The Wall Street Journal* reported "drastic prices recessions in foodstuffs," [17] with prices of cereals, meats, sugar, and coffee deflated below pre-war levels. Reductions also affected textiles, hides, leather, rubber, and metals, some of which were traded commodities of W. R. Grace & Co. Deflation had taken billions of dollars from prices of goods. Yet not all was doom and gloom. Confidence was in the air, a belief that a country like the United States with a more than abundant food supply could work itself out of the crisis. There was "no stress, but merely a slackening due to the readjustment from war to peace conditions … The future should be secure because there is an abundance of food."[18] The United States of America could feed itself. And this was its saving grace.

The W. R. Grace group was resilient in time of crisis. In December 1920, it announced substantial new investments with the purchase and expansion of shipyard construction at Hog Island in New York in partnership with the American International Corporation (AIC) and the Pacific Mall Steamship Company. W. R.

---

14   Allister MacMillan (ed.), *The Red Book of the West Indies*, "Grace Limited, General Exporters and Importers etc." (1922), 68.

15   Grace Ltd. purchased the wharf from D. Soutar (1837–1923), who was proprietor of Temple Hall Estate and Kingston merchant. He was father of William Donald Soutar (1888–1972) and grandfather of Simon Soutar (1912–1991), who married Luis Fred's sister, Louisa Matilde Kennedy in 1939. He was originally from the Isle of Man.

16   Macmillan (1922), 69.

17   *Wall Street Journal,* December 3, 1920, 8.

18   Ibid.

Grace also continued its trade with South America during this period with importation of gold and silver from Peru and nitrates from Chile. The company was hailed for the prominent role it played in the development of United States foreign trade.[19]

In response to the recession, W. R. Grace & Co. made a move to consolidate its business operations, giving room for some of its subsidiaries abroad to incorporate themselves at the local level. Grace Ltd. BWI was one such company. Grace was a multiple corporate structure, the parent company of about sixty corporations worldwide engaged in a wide variety of undertakings.[20] Most of the enterprises were fully owned and operated by the parent company, as was the case with Grace Ltd. BWI.[21] When W. R. Grace & Co. decided to pull out of the merchandising side of the business, John Grace and Fred Kennedy negotiated an offer to purchase. They saw the potential for retaining and expanding the import/export trade, manufacturing, and wholesale distribution of goods.

Dr. Grace approached three members of the Grace family to invest capital in the company: his first cousin, Cleveland Raphael Grace,[22] who was residing in Jamaica; his brother Morgan Hatton Grace;[23] and Walter H. Grace, his brother-in-law.

In early February 1922, *The Daily Gleaner* received a circular from Messrs. Grace Ltd. announcing their intention to withdraw from the merchandise business in the island, and the sale of their stock in trade and book debts to Messrs. Grace, Kennedy & Co. Ltd., who would take over all the accounts of the old firm.[24]

---

19   Ibid., December 25, 1920, 9.
20   Marquis James, *The Story of W. R. Grace* (1993), 317.
21   The registered name Grace Ltd. BWI (British West Indies) appears in archives. The name distinguished it from other international Grace Ltd. subsidiaries
22   Cleveland Raphael Grace, born in Valparaiso, Chile, 1885, was the son of John J. Grace's uncle, John William Grace (1836–1904).
23   Morgan Hatton Grace, born in Wellington, New Zealand, April 9, 1880, brother of John Johnston Grace and Michael Sheffield Grace.
24   *The Daily Gleaner*, February 3, 1922, 3.

Fred William Kennedy

(Photo, courtesy of ©
GraceKennedy Ltd.)

Dr. John Johnston Grace

(Photo, courtesy of ©
GraceKennedy Ltd)

Grace, Kennedy & Co. Ltd., 64 Harbour Street, Kingston, Jamaica.
(Photo by author)

11

Happy St. Valentine's Day
GRAKENCO (1922)

On Valentine's Day, February 14, 1922, the anniversary of the birth of Fred William Kennedy's mother, Mary Ann Eberall Kennedy, a new firm was incorporated: Grace, Kennedy & Co. Ltd. At age forty-seven, Fred William Kennedy, a Jamaican-born son of Falmouth, Trelawny, joined the merchant class as a stakeholder, part owner of an enterprise that would bring prosperity to his fellow citizens for generations to come. The new corporation was an investment by two families, forming a nucleus for the growth of multiple enterprises, local and international.

The Memorandum of Association gave the company the mandate "to carry on a general mercantile and commercial business and any other business (whether manufacturing or otherwise) which may seem to the company capable of being conveniently carried on in connection with the business … or calculated … to enhance the value of and render profitable any of the company's property or rights."[1]

The original signatories of the Memorandum of Association included E. V. Grace (wife of John Grace), Luisa M. Kennedy (wife of Fred W. Kennedy), John J. Grace,

---

1 "Memorandum of Association of Grace, Kennedy & Co. Ltd.," *The Companies Act* (1965).

Fred W. Kennedy, J. H. Cargill, H. H. Dunn, and S. R. Cargill. The authorized capital amounted to £25,000 of 250 shares valued at £100 each.

Dr. John Johnston Grace was named the first governing director, who "for the time being shall hold that office for life, and whilst he is Governing Director, the government and control of the company shall be vested in him."[2] The governing director could at any time resign office or relinquish his position at the written request of the directors, who represented fifty-five percent or more of the issued shares of the company.

Dr. John Johnston Grace published a letter dated February 14, 1922, with an announcement of the new firm:

> NOTICE is hereby given that we the undersigned have this day sold to GRACE KENNEDY & COMPANY LIMITED, whose registered office is at No. 64 Harbour Street in Kingston, the Merchandising Department of our business which we lately carried on in Kingston and at Montego Bay in the parish of Saint James, (Jamaica).
>
> All debts owing to the Merchandise Department of our said business sold aforesaid, are payable to Grace Kennedy & Company Limited who will carry on their business at No. 64 Harbour St., in Kingston, and at Montego Bay aforesaid.
>
> Dated Feb. 14, 1922
> GRACE LIMITED
> By J. J. GRACE
> Managing Director.

The first directors' meeting of Grace, Kennedy & Co. Ltd. was held the next day at the offices of Cargill, Cargill & Dunn, 4 Duke Street.[3] Present were John Johnston Grace, Governing Director, Fred William Kennedy, and G. F. Wallace. The first order of business was the appointment by John J. Grace of Fred William Kennedy as a director of the new company. The Colonial Bank was named as bankers, G. F. Wallace was appointed secretary, James S. Moss-Solomon, accountant, (friend and former co-worker of Fred William Kennedy), and Cargill, Cargill & Dunn, solicitors.

The lawyers were instructed to prepare an agreement between the company and Messrs. John Johnston Grace and Fred William Kennedy, who were to be appointed managers. Walter Fletcher[4] was to continue in his role as manager of the Montego Bay

---

2   Douglas Hall, *Grace, Kennedy & Co. Ltd: Story of a Jamaican Enterprise* (1992), 9.

3   Grace, Kennedy & Co. Ltd., minutes of meeting of board of directors.

4   Fletcher, Walter, born, Blackburn, Lancashire, England in 1880. He was son of the late J. R. Fletcher, Solicitor. He came to Jamaica in 1902 to take up banana and cane planting as proprietor of Virgin Valley Estate and manager of Latium Estate. He was J.P. for Parish of St. James.

office. A bank account was to be opened in the name of Grace, Kennedy & Co. Ltd. for which cheques could be signed by John J. Grace alone or by joint signature of Fred William Kennedy and James S. Moss-Solomon. Shares were allotted and the secretary instructed to prepare certificates:

John Johnston Grace	46
Fred William Kennedy	19
Eleanor Vere Grace (wife of John J. Grace)	1
Luisa Matilde Kennedy (wife of F. W. Kennedy)	1
Sidney Raynes Cargill	1
John Henry Cargill	1
Harold Herbert Dunn	1
Cleveland R. Grace	60
Walter H. Grace	40
Morgan Hatton Grace	30
Walter Fletcher	20

Subsequent meetings of the company were held that year at the registered office, 64 Harbour Street, on April 13, July 5, August 8 and 31, 1922. Dr. Grace increased his shareholding by an additional ten shares with a deposit of £1000 with the Colonial Bank. This completed the total number of 230 shares issued, which appeared on the balance sheet of January 31, 1923. R. P. Gallwey[5] replaced G. F. Wallace as secretary, and at the recommendation of Fred William Kennedy, Cleveland Raphael Grace was appointed a director of the company.

Fred William Kennedy held shares[6] in a company that was eighty-five percent controlled by members of the Grace family who, except for Governing Director Dr. John Grace, were for the most part removed from daily operations. As a newly appointed director and manager of the company, Fred William worked closely with Dr. Grace and leaned on the support of those like his friend James S. Moss-Solomon.

James Moss-Solomon was born in 1894 in Dry Harbour, St. Ann, the fourth of twelve children of Joseph Henry Moss-Solomon (1860–1945) and Theresa Adelaide Seivwright (1861–1934).[7] His father, Joseph, had been born in August

---

5   R.P. Gallwey had worked with Fred William Kennedy as assistant manager at the Montego Bay Agency of the United Fruit Company, 1910–1917.

6   In Memorandum of Association and Articles of Association of the Company, signed February 9, 1922, Fred William Kennedy, whose name along with others was subscribed as "desirous of being formed into a company," agreed to take twenty shares in the capital of the company. Dr. Grace's initial subscription totaled ten shares.

7   See Family Tree, Chapter 34. Official records show various spellings for Seivwright.

1860, months after his grandfather, Henry Moss-Solomon, proprietor of the Hague and Spring Estates in Trelawny, had died at the age of thirty-seven.

When James was a boy, the family moved to Brown's Town, St. Ann, where he attended elementary school. His father was a shopkeeper and shoemaker in the town, but by the time the last child arrived in 1905, because of illness, he was not able to financially support the family. James Moss-Solomon left school at the end of Grade 6 to seek employment.

From 1908–1914, he was employed as a clerk at J. H. Levy & Son, the wholesale merchandise store in Browns Town, where Fred William had also begun work as a young man in 1890.[8] From 1914–1916, he was Secretary of the Brown's Town Benefit Building Society, and then for two years joined the United Fruit Company's Montego Bay Agency where Fred William was general manager. In 1918, he secured a position with Grace Ltd., which had a branch in Montego Bay, where he worked as an accountant for four years before joining Grace, Kennedy. During his early working years, James contributed to family finances and paid for his youngest sister, Enid May (1905–1992), to attend St. Hilda's Diocesan High School in Brown's Town.[9]

James Moss-Solomon joined Grace, Kennedy with accounting and managerial experience together with the ambition characteristic of a self-starter. He engendered the trust of the Grace family, who incorporated him into the ownership and management of the company.

The Graces did not pull out of Jamaica altogether;[10] they continued to own majority shares in Grace, Kennedy and the company, W. R. Grace, owned the Grace Wharf, office buildings, and other fixed capital. Grace, Kennedy maintained ties to Grace's vast international network, its agencies, suppliers, and contracts worldwide, and tapped into the enormous resources of the House of Grace. W. R. Grace & Co. was known throughout South America as "the good employer."[11] Its name was untarnished, unlike other US multinationals that had been influenced by President Theodore Roosevelt's interventionist strategies in the region earlier in the century.

---

8   Biographical information is derived from interviews with members of the Moss-Solomon family and from *Who's Who Jamaica 1963*. See also Chapter 34, Uncle Jim.

9   St. Hilda's Diocesan High School, founded by the Anglican Church in 1906.

10   A notice appeared in *The Daily Gleaner*, September 5, 1927, page 2, under the heading "GRACE LIMITED" stating that their registered office was at No. 4 Duke St., Kingston, Solicitors for the company, CARGILL, CARGILL & DUNN.

11   Marquis James, *Merchant Adventurer* (1993), 296.

The Grace family was known to be clannish,[12] evidenced by John J. Grace's desire to build on his brother's legacy in Jamaica and by the support he was able to champion from his relatives to invest in Grace, Kennedy.

Cleveland Raphael Grace's influence in the company was significant. In 1922 he was the largest shareholder of Grace, Kennedy with sixty shares. Appointed a director of the company six months after its inception, he chaired board meetings as acting governing director in his cousin's absence. In August of 1926, he recommended that five shares be sold to James Moss-Solomon, and he put forward proposals to expand the business with the issuance of debentures as a means of raising capital.

Cleveland Grace was born in Valparaiso, Chile, 1885, the youngest child of John William Grace,[13] who became immensely wealthy by way of Grace's interests in the nitrate of soda and wool industries in Chile. He studied law in England and served as a lieutenant with the Royal Air Force in World War I. As a young adult, he travelled the world: Australia, Hawaii, Canada, and New Zealand. He listed himself in ship manifests as a resident of the United States, England, Austria, and Jamaica. In 1921 he was registered as a passenger from Kingston, Jamaica on board the SS *Turrialba* bound for Chile, his permanent residence listed as Strawberry Hill, Jamaica, his contact, cousin John Johnston Grace. He did not remain in Jamaica for long. He departed the island on November 14, 1923, on a visit to New Zealand and took up permanent residence in the UK. He remained single until his death in 1979.

Walter Henry Grace was Cleveland Grace's brother and John Johnston's first cousin and brother-in-law (married to John Grace's sister, Alice May Grace). His too was a life of adventures. Born in California and educated at Columbia University (president of his graduating class) in New York, he travelled extensively during his life—USA, England, France, Chile, Jamaica, and India. On none of the ship manifests is there a designation of profession, and there is no evidence of him having resided in Jamaica. His name appears on passenger lists of boats sailing from Santiago, Chile to Kingston, Jamaica during the 1920s.

Morgan Hatton Grace, John Grace's younger brother, was also a silent partner. He did not live in Jamaica but had substantial shareholdings in the company in 1922. He was married to Ruth Agnes Eden (1882–1959) from Long Island, New York, where Morgan made his new home after the death of his father in New

---

12   W. R. Grace was quoted as saying, "The problem was complicated by a desire to find them (CEOs) within the family ... I wish my boys were grown up and could be of some use." His son, Joseph P. Grace, succeeded him after W. R. Grace's death. Marquis James, *Merchant Adventurer* (1993), 287.

13   John William Grace (1836–1904) was the brother of William Russell Grace and Morgan Stanislaus Grace, father of John J. Grace.

Zealand in 1903. They had four sons. His profession was listed as merchant in a 1940 US Federal Census.

When starting out, Grace, Kennedy had more than ample office space at 64 Harbour Street, which they leased from Grace Ltd. of W. R. Grace of New York. In March 1923, the directors proposed that Grace, Kennedy invite Jamaica Fruit and Shipping Co. Ltd, which was in need of expansion,[14] to share office space and to put up half the cost of the lease. Dr. John Grace also recommended that C. E. Johnston be approached for a joint venture in the construction of a new pier. The agreement was ratified by August of that year, and the wharf completed by March 1924.

The connection between the Johnston and Kennedy families goes back to the days when Fred William was with United Fruit. Charles Edward Johnston, who had invested in Atlantic Fruit earlier in the century, branched out on his own to found Jamaica Fruit and Shipping Company Ltd. in 1919 with Captain Sibrandt Duhn List; they were pioneers in the struggle to establish "a Jamaican company for Jamaicans," to wrestle the monopoly away from the United Fruit Company and Atlantic Fruit. Their vision—to give small farmers and Jamaicans more control over the banana industry. C. E. Johnston rose in the public eye as a sharp-witted businessman, a Jamaican patriot with a drive to protect the interests of small landholders.[15] Like F. W. Kennedy, C. E. Johnston was a second-generation native-born Jamaican businessman, both vanguards of a new commercial and industrial independence for Jamaica.

Grace, Kennedy was profitable from the start, posting a profit in its first year of £800 with an eight percent interest paid to shareholders. Audited accounts in the early years of operation showed greater returns than expected, with retained earnings added to the company's reserve fund. Early setbacks included the closure of the Montego Bay business, which was sold on March 31, 1925, with notice given to debenture holders that loans would be repaid in full, with payment of interest.

Luis Fred completed fifth form at St. George's in December 1923. He did sufficiently well in Junior Cambridge examinations to apply for university in the United States, but for one of the more prestigious tertiary institutions in the United Kingdom, he said he would have needed to remain enrolled in school for a couple more years to sit more advanced level examinations (Senior Cambridge). He was accepted to Holy Cross College in Massachusetts to start in September 1924. To wait out the nine months before the start of the academic year in the US, he accepted a position in January 1924 to teach first-form Spanish at St. George's

---

14 Jamaica Fruit and Shipping Co. Ltd registered offices were at 75 Port Royal Street.

15 By 1921, Johnston's company was shipping twenty-one bunches of every hundred that left Jamaica for the United States. (Goffe, 251).

College. Given his fluency in Spanish, the teaching assignment came easily, but he disliked preparing lessons or marking papers. *"I lacked the patience to be a teacher,"* he often said.

His mother would miss him—she was seven months pregnant with her sixth child and wanted him home to celebrate the birth. She thought him too young (he was only fifteen) to be leaving for university and worried about him adapting to the cold climate. Besides, Christmas and Easter breaks were too short to allow him time to return home by ship. Commercial air travel did not start in Jamaica until the 1930s.[16] His father encouraged him to write weekly; he wanted "to know every detail of his new experiences."

---

16   Commercial aviation began with Pan American Clipper service of flying sea planes between Kingston and Miami in the 1930s.

Dr. John Grace with daughters Alicia (6) and Cynthia (4)
(Christmas greetings to the Kennedys, 1922. Photo, from family collection)

Visiting Town of Mandeville, Jamaica, 1925 while Luis Fred was at Holy Cross
L–R: Gloria, Carmen, Louise, their mom holding newborn Francis Xavier, and José sitting on front fender.

(Photo, courtesy of Lisa Valliant)

Grateful to Bozi Mohacek, Chairman, Surrey Vintage Vehicle Society, for following identification: "Looks like circa 1925 Willys Overland 65 Touring. It has front brakes, 6 rear and 12 front wheel studs, louvre panel with low mid handle on bonnet with wavy top surface, rear opening doors, drum headlights."

Kennedy family, Summer of 1926 at their home, *Quisqueya*, 24 South Camp Road, Kingston, Jamaica.
Parents, seated L–R:
Fred William Kennedy (aged 52), Luisa Matilde Kennedy (38)

Children, standing L–R:
José Antonio (13), Louisa Matilde (12), Luis Fred (17), Carmen María (16)

Children, seated L-R:
Francis Xavier (2) Gloria Mercedes (7)

(Photo from family collection)

Bob O'Connell, Red Coyle, Chuck Meaney, and Luis Fred at Holy Cross College, 1920s
(Photo from family collection)

"Too much partying"
With best friends Jimmy Quinn and Bill O'Connor, Holy Cross College.
The Holy Cross Purple Patcher (1928), 365.
(Photo, courtesy of © College of the Holy Cross)

12

Renaissance Man

College of the Holy Cross (1924–1928)

When Luis Fred Kennedy boarded the S. S. *Santa Marta*[1] in Kingston Harbour, August 23, 1924, bound for New York City, he could not have predicted what awaited him. The *Santa Marta* got caught in the path of a hurricane that travelled from the Caribbean to pass north of the Bahamas, headed for Cape Hatteras on the coast of North Carolina. The storm strengthened to a peak intensity of 125 mph[2] winds, creating fifty-foot waves that battered the ship. Another vessel, the *White Star*, which encountered the hurricane off the Nantucket Shoals, arrived in New York on August 27, with seventy-five injured. Luis Fred survived but was frazzled and shaken.

A concerned father wrote to him:

Quisqueya
25 South Camp Rod.
Kingston

Tuesday 26th August 1924

---

1 *National Archives and Records Administration.* Manifest of Alien Passengers for the United States Immigration Officer at Port of Arrival.
2 https://en.wikipedia.org/wiki/1924_Atlantic_hurricane_season. The year 1924 marked one of the worst hurricane seasons recorded. Of the thirteen storms that summer, six of them occurred simultaneously.

Renaissance Man   College of the Holy Cross (1924–1928)

> My Darling Boy: -
>
> How much we all miss you, you will never know.
> Every minute something crops up to remind us all of you,
> but as you know it is for your good. We are all trying to be
> brave and not fret too much.
> We are all anxious to hear of your safe arrival and
> hope that your vessel has escaped the rough weather that
> the recent cables have announced …[3]

As a minor two months shy of his sixteenth birthday, Luis Fred was discharged into the care of family friend Fr. Emmet Cronin S.J.,[4] who met him at port and made arrangements for him to stay with his parents in Manhattan. They helped Luis Fred shop for clothes and necessities for the first week of September, after which Fr. Cronin accompanied him to Worcester, Massachusetts to enroll him at the College of the Holy Cross.[5] The priest was happy to report to Luis Fred's father that all expenses had been covered with a remainder of $90, which he handed over to the treasurer at the college on account for his son.

Luis Fred received a second letter from his father:

> Remember that we are depending on you "To Make Good," not only in your studies but in your life. Always bear in mind what I told you about your actions. Keep this as your guide: "Never commit an act that you would be ashamed for your mother to know." If you follow this rule, you will be a clean man, a good man and a Real Man …
>
> Take good care of yourself and remember your good mother's admonitions in regard to keeping your chest well covered and your feet warm and dry. The US is not like Jamaica. The New England climate is

---

3   Fred William Kennedy, letter to Luis Fred Kennedy, August 26, 1924.

4   Rev. Emmet Cronin S.J., New York Province of the Society of Jesus, had served in Jamaica. He was one of the priests who presided at Michael Grace's funeral in 1920. Luis Fred's father had sent funds ahead to Fr. Cronin.

5   Founded in 1843 by the Rt. Reverend Benedict Joseph Fenwick S.J., second Bishop of Boston, The College of the Holy Cross is the oldest Catholic college in New England and one of the oldest in the United States. "The seat of the college is on one of the highest of the eminences surrounding the city of Worcester. Towards the north, this 'Hill of Pleasant Springs' commands an extensive and most delightful view of Worcester, at the time of the founding of the college a town of hardly 10,000 inhabitants, now a bustling city of more than 190,000." *College of the Holy Cross CrossWorks. 1925–1926 Catalog* https://crossworks.holycross.edu/course_catalog/37/.

very variable and treacherous so you must be careful. You cannot go around there as you used to here––No barefoot or semi-nude stunts.[6]

In his first month at Holy Cross, Luis Fred was called to the office of the president:

I was summoned out of class to visit the president's office. When I first arrived, I was introduced to Fr. Dinand,[7] who was anxious to meet me, for he knew of my family in Jamaica. My first thought was that I was in trouble, for it was highly unusual for a student to be called to meet with the president of the College. The other thought I had was that maybe something was wrong at home.

He was friendly enough, but the room was austere with low lighting and dark wooden furniture. He said that he had received a cable that day from Jamaica but was perplexed by the wording. All it said was, "BOY."

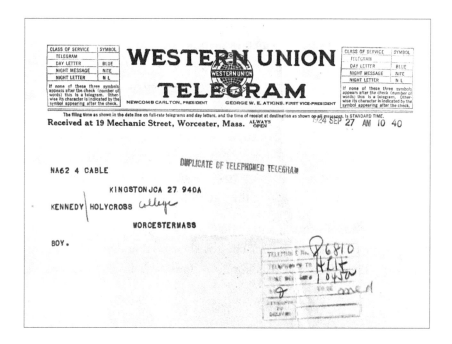

I told him that was wonderful news.

---

6   Fred William Kennedy (August 31, 1924).

7   The Right Reverend Joseph N. Dinand S.J. D.D. (1869–1943). Ordained a priest in 1903, his first apostolic mission was serving as headmaster of St. George's College (1906–1908). He was president of Holy Cross College for two terms, 1911–1918 and 1924–1927, and Vicar Apostolic (titular bishop) of Jamaica from 1927–1929. The Dinand Library, completed in 1927, at Holy Cross College was named after him.

He still did not fully understand. "Does this mean it is a baby boy? And whose baby boy is this, young man?"

When I told him he was my baby brother, he burst into laughter.

"Well, that is a relief! I was about to prepare your exit papers from the College."

He thought maybe I was the baby's father.

He politely dismissed me, saying that he would arrange for me to send my parents a cable of congratulations.[8]

His parents named the newborn Francis Xavier Kennedy. The day after Fred William sent the telegram, he penned a letter to his son.

28 Sept. 1924

My Darling Boy,

Well, you have another brother! We cabled you the glad news yesterday morning and, in the afternoon, got your cable sending us congratulations. This must have bored a hole in your funds, so I am sending $5.00 to replace. It was very kind and thoughtful of you to send the cable. The new boy is very much like his big brother. He looks just as you did when you were born. Needless to tell you we are all glad it is over, and everybody is pleased that it's a boy—3 Pairs—Enough don't you think …[9]

Luis Fred came to the US in a time of exciting and revolutionary change. He entered a world of glitz and glamour—everything became a wonder for him. It was the age of invention and new technology: the motor vehicle,[10] radio and mass communication,[11] the telephone, and aviation, all of which changed society in a rapid, unprecedented way. The Roaring Twenties were in full swing. It was a time of opulence, of clothing fashion, of art deco; it was the jazz age and the start of

---

8   Conversation recreated from an anecdote that my dad told many times. Although the incident was nerve-racking at the time, in retrospect, he found it humorous.

9   Fred William Kennedy (September 28, 1924).

10 The Model T had become fashionable and affordable for many, the cost in 1924, $260.

11 The first commercially licensed radio broadcast was heard in Pittsburgh, Pennsylvania, November 1920, the first North Carolina radio station started broadcasting in Charlotte, 1922. "1920s: A Decade of Change" by Barrett A. Silverstein (2004).

dance crazes like the Charleston. Young men had returned from the war in search of a better life, and women were seeking a new identity; they had gained the right to vote and were joining the workforce in large numbers.

Despite these modern trends, the more traditional values of Puritanism prevailed. A war of cultures erupted within the US, the more conservative elements of society condemning the permissiveness of a new wave of modernism. With ratification of the Eighteenth Amendment of the Constitution in 1919, Prohibition banned the production and sale of alcoholic beverages in the United States between 1920 and 1933. Massachusetts was one of the first states to enact the legislation.

This was not the only source of tensions during the 1920s. The Red Scare gave rise to discriminatory anti-immigration laws that favoured people from Great Britain and Northern Europe. Within the United States, the Great Migration of African Americans from the South into New York City and other Northern cities also threatened those who wanted to preserve the notion of a "White America."

The 1920s were the years of a resurgence of the KKK and the institution of Jim Crow laws in the South, which legalized segregation in churches, hospitals, schools, theatres, and transport, affecting almost every aspect of social life. Even though the laws did not exist in the northern states, Blacks still faced discrimination. Hundreds of thousands were forced to take low-paying jobs and live in crowded, sub-standard housing.

On ship manifest forms, US Immigration identified Luis Fred Kennedy as a West Indian of dark complexion. Out of a class of 223 young, mostly Catholic Irish from New York, New Jersey, Pennsylvania, and the New England states, Luis Fred was the only person of mixed race. Other Kennedys in his year were direct descendants of Irish immigrants, sharing no apparent commonality of race or culture with the young man from the British West Indies. At first, he was considered an oddity. Friends told him he spoke with an accent and had "quaint mannerisms." Lou was known for "his rich tongue and his smile, warm with the sun of his native Jamaica."[12] His friends at college called him Lou, a nickname used exclusively by his classmates at Holy Cross.

Dad never spoke of overt racism but he did tell stories of how challenging it was to adapt culturally and to adjust to being with classmates who were three years his senior. He was able to overcome these obstacles in ways that brought him academic success and lasting friendships.

Students started their day at 6:25 AM with celebration of the Liturgy, followed by a regimented schedule of classes and supervised recreational and sports

---

12   *The Purple Patcher* (1928), 136.

activities. Lights out were at 10:00 PM. He carried a full load of courses: scholastic philosophy, ethics, Latin, Greek, physics, chemistry, biology, astronomy, economics, English literature, history, religion, and psychology. He excelled in academics:

> At once, Lou took a leading place in his classes, and we all acknowledged the ability of the boy from the British West Indies. Horace, Plato, and Homer were like many incidentals to Lou judging from the way with which he mastered them. In like manner, he conquered Chemistry, the much-dreaded Physics, and the fear-bringing Orals.[13]

Orals were conducted in Latin and Greek.

He wrote home to ask his father if it would be all right to start smoking cigars, and whether he would be allowed to accept invitations to stay over at his friends' homes for holidays. With respect to cigars, his dad suggested that he might prefer trying the pipe, which he would find more delectable and becoming of a gentleman. Regarding his friends' families, his father would not be in a position to judge but said that he should seek the judicious advice of Fr. Dinand, and whatever the priest decided would be fine with him. He cautioned his son to choose friends wisely, to make sure they were of good character. His letters provided fatherly advice:

> Now my boy whenever you want anything do not be afraid to tell your old Dad about it ... but I want all my kiddies to have as good a time as can be reasonably expected. If, however I may have to refuse you any of your requests, I want you to feel, if I do, that it is not because I do not wish to, but that perhaps at the time it was not convenient ...

> Mother and Baby Francis are doing fine. If it is allowed by the Church, Mother and I would like you and Carmen to be Godparents for your little brother and we will let you know about this later. Keep cheerful my lad. Work hard! Play hard! Pray hard!

> God bless & keep you my Darling Boy is the prayer of your affectionate

> Dad.[14]

His father would not miss a birthday:

---

13 *The Purple Patcher* (1928), 136.
14 Fred William Kennedy (Sunday, October 19, 1924).

> **Memorandum.**
>
> **Grace, Kennedy & Co., Ltd.** Kingston 25th Oct. 1924
> Kingston & Montego Bay.
> JAMAICA.   To Luis F. Kennedy
>
> My Dear Boy,
> Just a hasty line to wish you a very happy Birthday when it comes — I have to rush to catch the Mail so you must excuse the scrawl.
> You can select to better advantage over there any little thing you want to enclose you will find $5.00 to use as you like
> Love from
> Your old Dad

It did not take long for Lou to become fully immersed in the social life of the college. He joined the marching band and the Sanctuary Society; he was secretary of the Historical Academy and member of the Sodality and Scientific Society; he served in the Civil Society and Nexus Club and was an editor of the school's first student newspaper, *The Tomahawk*. Students were allowed to form social clubs based on their home state or city, so as not to exclude him, one of his friends, Chuck Meaney, asked him to join the New Jersey Club as an honorary member.

Bill O'Connor, Jimmy Quinn and Chuck Meaney became his life-long friends.

William Francis O'Connor, from Rockaway Park, Long Island, had a cheery disposition with a winning smile and was voted the cutest looking guy of his graduating class. Bill was a leader among his classmates and excelled with top marks in almost every subject, which served him well in his chosen career of medicine.

James Francis Quinn from Brooklyn, New York, was a super athlete, unassuming, modest, and popular too among his classmates. Jimmy was a top performer of the track team and competed in the 1928 Summer Olympics in Amsterdam to win a gold medal in 4 x 100m relay race.

Charles Francis Meaney Jr. was from Woodcliff, New Jersey. Chuck was known for his "big-hearted, friendly and warm-hearted" ways, and was one of the most popular men on campus. He was the conscientious type, studying even when he did not have to, but was also a champ on the basketball court. Classmates loved Chuck for his "sincerity and good-heartedness."[15]

Luis Fred spent weekends, Christmas, and Easter vacations with his friends' families. He took an immediate fancy to his friends' sisters, whom he invited on double dates. On occasion, he would visit the Loinaz family in Mount Vernon and take his cousins, Carmen Celia (Chela) and Lydia, on dates. The young man from the British West Indies was popular with the ladies; family stories were told of several lining up to wave goodbye at the United Fruit Company docks in New York City at the end of each school year.

Although routines were rigid with not much free time to get into mischief, the college men found ways to break the rules. Lou learned to play poker and became the cardsharp with whom everyone wanted to partner. A meal out at the local speakeasy in the Hotel Warren, downtown Worcester, would be as serious as their friendly wagers would get. He repeated many times the story of their adventures off campus:

> *Those were the Prohibition days; all the saloons across America were shut down. We'd get leave from the college some Saturday evenings and head into town. The favourite spot was the Warren Hotel, which had as its advertisement, "Modern Improvement, Electric Lights and Steam Heat." The Hotel Vernon was also a hotspot. You went down a long wooden staircase to get access to the hidden entrance of the tavern. I am sure the bootleggers who supplied the liquor must have watered down the whiskey and gin, but we didn't know the difference.*
>
> *When funds were short, we looked up obituary listings for Irish wakes in the local papers. Irish funerals were a big social event at which they served lots of alcohol and good food. We made sure to dress in our finest attire and practised what best to say to comfort the mourners. We never got caught.*[16]

Wednesday, June 20, 1928

WORCESTER, June 20-Holy Cross College sent out a class of 223 graduates and awarded four honorary degrees today in its 79th annual commencement, which was attended by Gov. Alvan T. Fuller and members of his staff, hundreds of

---

15 *The Purple Patcher* (1928), 194, 205, 172. James F. Quinn died in Cranston, Rhode Island at the age of 97.

16 Recreated from anecdote that my dad related many times.

alumni from all over the country and relatives and friends of the young men who bade farewell to their alma mater.[17]

In the audience were Fred William and Luisa Matilde Kennedy, proud parents of their eldest born, Luis Fred.[18] This was a significant event—the first family member to receive a university degree. One of three orators was friend James Francis Quinn, who spoke on the topic of "The Liberty of the Citizen." Diplomas were presented to the graduates by Governor Fuller, who completed the program with an address:

> Established in 1843 on a foundation that gives positive recognition to the moral faculties; that provides definite training in character building; that gives high place to the importance of religious obligations, Holy Cross College has won an enduring place among the higher institutions of learning of our Commonwealth.[19]

Holy Cross meant for Lou the foundation of his future successes in life. He thought scholasticism to be not only a philosophy but a method of teaching that trained the mind.

He received offers to join business firms in Boston and New York, but even though he found these enticing, he decided to return to his native land to "conquer the business world of Jamaica."

A citation was published in the 1928 Yearbook:

LUIS FRED KENNEDY
Kingston, Jamaica B. W. I

Band 1; Sanctuary Society 1, 2, 4; Historical Academy, Secretary 2; BJF 1 2 3; B.V.M. Sodality 1,2,4; Scientific Society 3; Civil Service 4; Jersey Club, Honorary Member; Tomahawk 3, 4; League of S. H. 1 2 3 4; Nexus Club 4.

Across 1500 miles of land and water "Lou" heard the call of Alma Mater and began looking for timetables and geographies dealing with Worcester. After living through gales, typhoons, shipwrecks, and derailments he arrived at Mount St. James. At once "Lou" took a leading place in his classes and we all acknowledged the ability of the boy from the British West Indies. Horace, Plato, and Homer were like so many incidentals

---

17 *The Boston Globe*, Wednesday, June 20, 1928, 24.

18 They arrived from Jamaica in New York on June 7 aboard the United Fruit carrier the SS *Metapan* to be with their son on this special occasion.

19 Excerpt from Governor Fuller's speech from *The Boston Globe*, June 20, 1928, 24.

to "Lou" judging from the way with which he mastered them. In like manner he conquered Chemistry, the much-dreaded Physics, and the fear-bringing Orals.

Most of us will remember "Lou" for his quaint mannerisms, his rich tongue and his smile, warm with the sun of his native Jamaica. He did not leave the spirit of his slow happy home linger long in his blood, but with a gust of ambition he set sail for success on the turbulent seas of twenty-eight. His youth, for he is the youngest of the class, did not hinder the progress of his plans. Rather it helped to prove the adage, "Youth will be served."

Sedate, upright, progressive are the notes of Luis. With the disposition of his sunny home and the determination of the English to succeed, we leave him to conquer the business world of Jamaica.[20]

---

20   Citation from *The Holy Cross Purple Patcher* (1928), 136. Courtesy of © College of the Holy Cross Archives.

FIRSTBORN The Life of Luis Fred Kennedy 1908-1982

All the wiser for having received a Renaissance education.
Youngest member (aged 19) of his Graduation Class of The College of the Holy Cross, 1928.
(Photo from family collection)

AT THE WHARF.—Messrs. F. W. Kennedy, C. E. Johnston, Theodore Williams and R. B. Barker awaiting the arrival of the s.s. "Jamaica Producer" on Saturday morning.

The Daily Gleaner, Monday, April 15, 1929, page 21.

In April 1929, Sir Arthur Farquharson, C. E. Johnston, and Captain S. D. List formed the Jamaica Banana Producers Association with a membership of 6,000 growers to compete with the foreign owned fruit multi-national companies for the shipping and marketing of Jamaican bananas to the UK https://jpjamaica.com/timeline/

Fred William Kennedy, seen on left. He was not physically imposing—a person of slight build, just 5' 8" tall, and of quiet demeanor. Yet he had a commanding presence; he was a tower of strength and influence within his family, the business community, and Jamaican society.

His friend Robert Beacroft Barker (who appears in photo, far right) was younger. Born in 1890, he migrated from England to Jamaica in 1915 and served in WW I. He was appointed to the Legislature in 1940s, served as custos of Kingston 1952, and chairman of Institute of Jamaica. Knighted Sir Robert Barker, 1950. (*Who's Who Jamaica* 1957)

(Photo, courtesy of ©*The Gleaner Co. (Media) Ltd.*)

13

REQUIEM

(1930)

In Grace, Kennedy's first years of operation, significant changes occurred in the transfer and sale of shares.

In 1927 John J. Grace sold company shares, which he valued at £115 per share, to S. R. Cargill (3), J. H. Cargill (2) and to Edith Louise Zehnder (20 shares).[1]

The following year at an Extraordinary General Meeting, Harold H. Dunn moved, seconded by James S. Moss-Solomon, a proposal that the "Capital of the Company be increased by the creation of 60 new shares of £100 each to which shall be attached special rights and privileges."[2] At a subsequent meeting of directors, attended by Dr. John Grace, Fred Kennedy, and secretary R. P Gallwey, Fred Kennedy moved, and Dr. Grace seconded, that sixty (60) Cumulative Preference Shares [3] be issued to Dr. John J. Grace (51) and Mrs. E.V. Grace (9) valued at £100 per share.

---

[1] Edith Louise Zehnder (1893–1969), friend of the Grace family. She was married to Hugh Ransome Zehnder, barrister by profession who practised law in Singapore. They eventually settled in Jamaica, residents of Malvern, St. Elizabeth, where she died, September 25, 1969.

[2] Grace, Kennedy & Co. Ltd., minutes of Extraordinary General Meeting, April 19, 1928.

[3] Ibid. Agreements drawn up on the issuance of Preference Shares entitled holders to receive as a first charge on the profits of the company a cumulative preferential dividend of 8% per annum; the shares protected holders against depreciation of share value; they gave entitlements and protections in the event of the dissolution of the company. Preference shares would be issued to such persons and on such terms and conditions as the directors might think fit.

Requiem (1930)

One month later, Fred William acquired an additional 106 ordinary shares purchased at a cost of £100 per share from Morgan Hatton Grace, Cleveland Russell Grace, and John J. Grace.[4] With the sale of shares by members of the Grace family, Fred William Kennedy became the largest shareholder of Grace, Kennedy, but preference shares were held exclusively by the Graces.

Grace, Kennedy experienced a couple of bumper years with higher-than-expected profits. The directors recommended an additional fifteen percent dividend be declared in both 1928 and 1929.[5]

Luis Fred started at Grace, Kennedy as a ledger keeper, a job that he found boring at the start until he began to assume managerial duties. For his twenty-first birthday, November 2, 1929, he became a shareholder, receiving from Dr. Grace "$5000 of shares in the business, all free and unencumbered."[6] In his personal life, he savoured the freedom of bachelorhood, unhampered by family responsibilities. His father insisted that he change his signature from Fred Kennedy to Luis Fred Kennedy; bar bills kept landing on his desk at Grace, Kennedy.

Luis Fred was concerned about his father, who was taking time off work because of complaints of indigestion, fatigue, and poor eyesight. Doctors failed to diagnose the true cause of his ailments; one of their proposed remedies was to have all his teeth removed.

The unexpected happened the night of March 4.

> *I shall never forget the night. I always stayed up late reading. I was an avid reader, anything I could get my hands on. It was minutes to midnight. My dad knocked and entered the bedroom, which I shared with my brother Joey. I found it unusual, for he seldom got up late during the night. He didn't look well– his face was pale. He walked by my bed and headed straight to the adjoining bathroom. It wasn't long, maybe five minutes, and he re-entered the bedroom, held his chest, stumbled, and collapsed lifeless on my lap. Just like that–it was sudden, his heart took him. I called out to Joey to go get Mother. The events of that night replayed in my head all my life. Mamá never fully recovered from her loss.[7]*

His death sent shock waves through the business community.

---

4   Grace, Kennedy & Co. Ltd., minutes of meeting of board of directors, May 1 and May 10, 1928.

5   Grace, Kennedy & Co. Ltd., minutes of annual general meeting, February 21, 1928, and of directors' meeting, March 5, 1929.

6   Luis Fred Kennedy, letter to Lydia Loinaz, August 26, 1937. He was happy to report then that over an eight-year period, the shares had doubled in value.

7   Recreated, in my dad's words. He retold the occurrence of his father's death many times.

On March 6, 1930, *The Daily Gleaner* carried an announcement of his death on the front page:

MR. F. KENNEDY SUCCUMBS TO HEART ATTACK

Member of Kingston Commercial House Passes Away At Midnight Tuesday

FUNERAL YESTERDAY

HE ROSE Quickly in Business And Was Held in High Esteem Throughout the Land

Mr. Kennedy was a prominent figure in the commercial life of Kingston and the whole island, and the news of his death came as a shock to a large circle of friends and still greater number of acquaintances ...

He was a Jamaican who made his mark in his own country through robust energy and keen business aptitude. He was a member of the Jamaica Chamber of Commerce and Merchants Exchange and later became a member of the Council of that body. He was greatly respected by all for his sound opinions and foresight.

He knew, and was known in every part of Jamaica, and he had travelled a great deal in connection with his business in neighbouring countries and in Canada and America. He possessed the confidence and esteem of the local business community, having obtained a position in the Colony's commercial life of which any man would be proud.

A staunch supporter of the Roman Catholic Church in Jamaica he was always a keen worker for its advancement and every movement connected with it. Whenever there was a function held in connection with the Church Mr. Kennedy was indefatigable in his endeavours to make it a success.[8]

Funeral arrangements were made for service and interment Thursday afternoon. The cortège moved from *Quisqueya* to Holy Trinity Cathedral, the funeral service officiated by Jesuit Superior, Rev. Father Francis Kelly S.J. Luis Fred Kennedy, James

---

8   *The Daily Gleaner*, Thursday, March 6, 1930, 1, 3.

Requiem (1930)

Moss-Solomon, Captain Sibrandt D. List, Sidney R. Cargill, Dr. John Grace, and Vernon E. Manton[9] were pall bearers. Interment followed at Calvary Cemetery.

The Daily Gleaner published reports of a meeting of the Chamber of Commerce, March 12, 1930. Mr. A. D. Jacobs,[10] who opened the session, said that Mr. Kennedy would be remembered as a "live wire member who when he took up anything carried it out to a finality." He moved a resolution "That the Council of the Jamaica Chamber of Commerce and Merchants' Exchange learn with sincere regret of the death of Fred W. Kennedy on March 4, 1930[11] and desire to put on record their appreciation of his unbounded interest in the work of the Chamber of which Council he was member for nearly ten years."

> Mr. R. V. Butt[12] in seconding the motion said that the loss was not only to the Chamber of Commerce but to the business community of the island as a whole. Mr. Kennedy was one of the bright and active businessmen of the colony and they had certainly sustained a great loss.
>
> The members present stood for a minute in silence and the resolution passed unanimously.[13]

A copy of the resolution was sent to his "sorrowing widow and bereaved family."

The death of his father left Luis Fred with family responsibilities:

> He now faced the monumental task of settling his father's estate. His father died intestate, and he also faced the job of taking care of his mother and younger brothers and sisters. He was ever grateful to Dr. Grace and "Daddy" Dunn for their assistance through the painful and long process of settling the estate. Luis Fred Kennedy had just turned twenty-one years old when his father Fred W. Kennedy died.[14]

Mim was aged twenty, José, seventeen, and the three youngest were still school age: Louise, fifteen, Gloria, ten, and Francis, five. His sister Mim had returned from Canada,

---

9   Victor Evelyn Manton, born Mandeville 1875, son of S. H. Manton, Solicitor, Manton & Hart, 117 Duke St., Kingston.

10   Possibly proprietor of A. D. Jacobs & Co, established 1890, General Merchants, Shipping, Insurance.

11   Date of death recorded was March 5, 1930.

12   Mr. Reginald V. Butt (British) was the manager of Barclays Bank and representative of the Bank of Montreal, Canada in Jamaica.

13   *The Daily Gleaner,* March 12, 1930, 3.

14   Lydia M. Kennedy, "A Giant of a Man" (1997).

where she attended "finishing school"—studying art[15] and music at the Convent of the Sacred Heart, Sault-au-Recollet, Quebec.[16] She assumed an adult role along with her brother Luis Fred, attending official business functions and receptions, for example, at King's House. Shortly after his younger brother, José (aged eighteen), had completed studies at St. George's College, Luis Fred introduced him to work at Grace, Kennedy where he filled in as corporate secretary at directors' meetings in James Moss-Solomon's absence. In 1932 José started as an intern at Manton and Hart, Solicitors.

---

15   In *The Daily Gleaner,* Tuesday, October 28, 1930, Carmen Valliant, daughter of the late Fred William Kennedy, was commended for the execution and displays of her most exquisite artwork. Alpha Academy was hosting a function to honour the appointment of His Lordship Bishop Thomas Addis Emmet S. J., newly appointed bishop of the Roman Catholic Church to Jamaica. "Little Miss Gloria Kennedy" (aged 11) was also featured for her excellent delivery of the welcome address to His Lordship. As an adult, Gloria excelled in the study and teaching of English language and literature.

16   Name of school, courtesy of her son, Fred William Valliant. The boarding school for girls was run by The Society of the Sacred Heart, a religious order of nuns founded in France in 1800 by Madeleine-Sophie Barat. Mim (17) enrolled in 1927. She was accompanied by her mother and Luis Fred, travelling to New York in September on board the United Fruit vessel, the SS *Zacapa*. Her brother was starting his junior year at college.

PART THREE

Intrepidness of Youth (1930–1946)

I detest having to be careful (**LFK**).[1]

---

1 Luis Fred Kennedy, letter to Lydia Loinaz, June 22, 1937.

Grace, Kennedy & Co., Ltd.

64 Harbour Street, Kingston, Jamaica, B.W.I.

CABLE ADDRESS:
"GRAKENCO"
ALL CODES USED.

GENERAL IMPORTERS,
COMMISSION MERCHANTS,
WHARF OWNERS
AND
STEAMSHIP AGENTS.

TELEPHONE NO. 350.
P.O. BOX NO. 86.

FIRE INSURANCE AGENTS AND SPECIALISTS IN THE SUPPLYING OF ARTIFICIAL FERTILIZERS TO PLANTERS.

WE ARE AGENTS FOR THE FOLLOWING FIRMS AND CAN MEET YOUR REQUIREMENTS FOR ALL KINDS OF MERCHANDISE.

FIRMS:	ARTICLES:	FIRMS:	ARTICLES:
Arkadelphia Milling Co.	FLOUR AND MEAL.	Pan-American Export Co.	SUNDRIES.
Allfeld & Egloff	HOPS.	Pare's Confectionery Works	CONFECTIONERY.
Aluminum (IV) Ltd.	ALUMINUM WARE, PAINTS, MOULDINGS, ETC.	Paradisses Freres	SILKS.
		Petters Limited	LIGHTING PLANTS, MARINE ENGINES.
Aluminum S. S. Line	NEW ORLEANS, AND MOBILE, TO KINGSTON	Smith & Gill	TONIC WINE.
Beck's Beer Brewing Co.	BEER.	James Richardson & Son	OATS.
Buckersfield Ltd.	OATS.	Sinclair Cuba Oil Co.	LUBRICANTS, K/OIL.
British Extracting Co.	SOYA BEAN OIL.	Three Minutes Cereal Co.	NATIONAL OATS.
Bird & Son	RUGS, FLOOR COVERING, ETC.	Trading Co. "Cuveco"	SUNDRIES.
		R. C. Tait Ltd.	POTATOES.
California Packing Corp.	CANNED FRUIT AND VEGETABLES.	Trengrouse & Nathan Limited	BUTTER.
		Milam Grain & Milling Co.	CORN.
Cimenteries et Briqueteries Reunies	CEMENT.	Willards Chocolates Ltd.	CONFECTIONERY.
Doremans & Company	BEER.	Washburn-Crosby Co.	FLOUR AND MEAL.
Dominion Potash Supply Co.	FERTILIZERS.	North British & Mercantile Ins. Co.	FIRE INSURANCE.
Garnock, Bibby & Co.	ROPE.	Vadsco Sales Corp. (American Druggists Syndicate)	DRUGS & MEDICINES.
Galt Flour Mills	FLOUR.		
G. H. Hammond Co.	PACKING HOUSE PRODUCTS.	Grimault & Co.	DRUGS & MEDICINES.
Herring-Hall-Marvin Safe Co.	SAFES.	Rigaud & Co.	PERFUMERY.
Job Bros., & Co., Ltd.	CODFISH.	W. T. Owbridge & Co.	OWBRIDGE'S LUNG TONIC.
John Lucas & Co., Ltd.	PAINTS, VARNISHES, &C.	Construction Supply Co. of America	SANITARY FITTINGS &C. BLDG. MATERIALS.
McDonald & Muir	WHISKY.		
Nitrate Agencies	FERTILIZERS.	Dame Larsen & Parkin	BROKERS & COMMISSION MERCHANTS.
Nordal & Co.	BLACK PEPPER.		
Norseland Canning Co.	SARDINES.	Williams & Foulds	BROKERS & COMMISSION MERCHANTS.
Nordskog & Co.	PAPER.		

Herbert G. De Lisser (1930–31), *Planters' Punch: A Jamaica Magazine*, 63.
Digital Library of the Caribbean https://ufdc.ufl.edu/AA00004645/00003
(Photo, courtesy of © National Library of Jamaica)

James S. Moss-Solomon (1928)
(Photo, courtesy of the Moss-Solomon family)

Luis Fred Kennedy (1928)
The Holy Cross Purple Patcher, 136
(Photo, courtesy of © College of the Holy Cross Archives)

Grace, Kennedy & Co. Ltd. under new management:

Assistant Managers, Luis Fred Kennedy and James S. Moss-Solomon.

Minutes of meeting of board of directors of Grace, Kennedy & Co. Ltd., May 9, 1930. (Courtesy of © GraceKennedy Ltd.)

14

ASSISTANT MANAGER (1930s)

Within five days of his father's death, Luis Fred was invited to attend the Annual General Meeting, at which he served as acting secretary. Governing Director Dr. J. J. Grace referred to the "death of Mr. F. W. Kennedy, late Manager of the Company, expressing the great loss sustained by the business in general and himself personally."[1] He also announced the passing of Mr. R. P. Gallwey, former secretary, whom he praised for his contributions. In addition to Dr. Grace, others in attendance were H. H. Dunn, J. S. Moss-Solomon, and J. M. Nethersole,[2] personal representative of the deceased.

On April 15, Luis Fred attended his first directors' meeting, at which James S. Moss-Solomon was appointed secretary of the company. At a meeting the following month, Mr. H. H. Dunn moved "that the agreement of Mr. J. Moss-Solomon and Mr. Luis F. Kennedy as Assistant Managers be approved and sealed." Motion was seconded by Dr. Grace and carried. Both managers were given the scope to make decisions independent of Dr. Grace. They controlled day-to-day operations of the business with authority to sign individually on cheques and accounts without counter signatures, and to close deals on behalf of the company.

---

1   Grace, Kennedy & Co. Ltd., minutes of annual general meeting, March 10, 1930.

2   "J. M. Nethersole had had a unique career in the island's public service for one who was not white. He held positions of Administrator General and Trustee in Bankruptcy—public positions as high as any Jamaican attained in the modified Crown Colony era, with the exception of Sir Henry Brown who reached the High Court Bench." Bank of Jamaica, "Noel Newton Nethersole: A Short Study by James Carnegie," November 1975, 1.

Ongoing requests were made for transfer of shares. On December 9, 1930, H. H. Dunn recommended the transfer of shares formerly held by R. P. Gallwey to James S. Moss-Solomon and Gerald Arthur L. Mair, a new shareholder, who was a representative of Dominion Life Insurance Company of Jamaica. Mr. Dunn took a lead role in the company by assuming the chair of directors' meetings in the absence of Dr. Grace. Already in his fifties by the mid-1930s, he was senior to both Luis Fred and James Moss-Solomon.[3]

As governing director, Dr. Grace provided guidance for securing new contracts, for sale of major assets, for approving expenditures and acquisitions, and for monitoring the company's performance. He secured his position as head of the company by way of managerial contract agreements and as a holder of preference shares, terms agreed upon by the board of directors.[4] He also exercised his authority to determine the suitability of candidates for purchase of shares in the company.

The 1930s presented new challenges in the wake of the Great Depression. Many Jamaicans who had sought employment abroad returned to the island,[5] and the sugar industry, mainstay of the Jamaican economy, showed further decline due to a slump in prices. The economy suffered from what has been termed "extreme primary product export concentration;"[6] the two largest exports, bananas and sugar, together comprised 69.5% of total exports. When world demand for these commodities slumped, export earnings plummeted and in turn diminished the country's import capacity.[7]

Despite the effects of an economic downturn, the company made plans for new capital outlays. As early as September 1930, notice was given to exercise the option given under the lease of the Grace Wharf for the purchase of the wharf premises from W. R. Grace & Co. By March 30 of the following year, an agreement in the amount of £25,000 (50%) was signed with Jamaica Fruit and Shipping Company as joint tenants in common. It was also announced at the directors' meeting that both parties agreed to

---

3   Harold Herbert Dunn was born May 9, 1882, in Vere, Clarendon, Jamaica. He was married to Constance May Laidman.

4   Articles 67 and 69 of the Articles of Association were repealed and substituted with clauses defining conditions for the position of governing director. The resignation and/or appointment of the governing director was conditional on agreement of at least 55% of the issued ordinary shares in addition to 55% of the issued preference shares in the company's capital. (Emergency General Meeting of Shareholders of Grace, Kennedy & Co. Ltd., April 25, 1930)

5   It is estimated that almost 200,000 Jamaicans had migrated to other Caribbean and Latin American countries between the time of Emancipation and the start of the First World War in hopes of a better standard of living. Frank Hill (1976).

6   Richard L. Bernal, "The Great Depression, Colonial Policy and Industrialization in Jamaica" (1988), 36, 37, and 50.

7   "Export prices plummeted by 44% between 1929 and 1932, and the 1938 price index was still 28 per cent below the 1929 level. The price of sugar dropped by 31%, bananas by 24.5%, coffee by 28.4% and cocoa by 28.4%" (Bernal, 1988, 39).

buy the premises adjoining, known as Lindo's Lumber Wharf, to which they had made capital improvements.[8] The memorandum stated that both newly acquired properties, including concrete warehouses and the erection of a new pier[9] (expenses for which they had also shared equally with Jamaica Fruit), would be registered as GRACE WHARF.[10]

Messrs. Grace, Kennedy & Co. Ltd advertised themselves as General Importers, Commission Merchants, Wharf Owners, and Steamship Agents.[11] They imported a wide variety of products: flour and cornmeal, liquor, canned fruit and vegetables, fertilizers, codfish, butter, drugs and medicines, building materials, and tires. They were insurance agents representing North British and Mercantile Insurance Company, shipping agents for Aluminum SS Line, and partners with Lascelles deMercado as selling agents of the Jamaica Biscuit Co. Ltd. During the 1930s, hundreds of advertisements appeared in *The Daily Gleaner*, showcasing the wide variety of products and services offered by Grace, Kennedy & Co. Ltd.

The challenge for Grace, Kennedy was to re-establish relations with its trading partners and to devise new competitive advantages locally to retain their agencies:

> The first few years after Fred W. Kennedy died were a struggle with adjustments, changes, and competition. Competing firms tried to take Grace, Kennedy's agencies away, saying that this inexperienced young boy was going to ruin things at Grace, Kennedy and that it would soon go under.[12]

Luis Fred travelled extensively. His first trip in September 1930 was to Montreal, Canada. Through the efforts of C. E. Johnston and others who exported fruit from Jamaica, trade with Canada had expanded by 1930.[13] The Canadian-West Indian Trade Agreement of 1920 gave preferential treatment to products of the British West Indies, and in return for these concessions, the West Indies gave Canada reductions in tariffs ranging from twenty to fifty percent.[14] This created opportunities for Grace, Kennedy to do business with the Royal Bank of Canada and the Canadian Imperial

---

8   Grace, Kennedy & Co. Ltd., minutes of meeting of board of directors, March 30, 1931. Present were Dr. J. J. Grace, L. F. Kennedy, and J. Moss-Solomon.

9   This had been completed in 1924 when Fred William was manager.

10   The property assets of the company, including the wharves, concrete warehouses, office building at 64 Harbour Street, new pier, new extensions to the premises for Captain List and Mr. Johnston, were valued at £53,000 (minutes of directors' meeting, March 30, 1931).

11   Advertisement in *Planters' Punch* (1930–31), 63.

12   Lydia M. Kennedy, "A Giant of a Man" (1997).

13   In 1910 Jamaica's exports to Canada represented 7.8%. (Jamaica's two other principal trading partners were Great Britain and United States.) By 1930 this increased to 20%, and by 1938 to 26.6%. Richard L. Bernal, "The Great Depression, Colonial Police and Industrialization in Jamaica" (1988), 51.

14   Eisner, Gisela, *Jamaica 1830–1930: A Study of Economic Growth*, 281.

Bank of Commerce to expand its maritime business by becoming agents for shipping lines and opening import markets for flour, pharmaceuticals, and salt cod.

Luis Fred's travels included June 1931 to meet with executives of W. R. Grace & Co. of New York; April 1932 and April 1936 with SS Aluminum Line in New York; November 1932 with the National American Bank in New Orleans; May 1933 and November 1935 with Washburn-Crosby Flour Mills in Minneapolis;[15] and July 1937 to Louisville, Kentucky with Axton Fisher Tobacco Company.[16] His objectives were to introduce himself as a new representative of Grace, Kennedy, to make good on contracts that his father had negotiated, and to drum up new business for the company. His father had taught him the secret to success in business: create cordial relations with partners, customers, and competitors. He made routine reports at directors' meetings and received approval for signing off on new contracts.

Dr. Grace urged the assistant managers to be cautious about expanding too fast. Performance for 1931 and 1932 did not match expectations.[17] At the AGM of March 17, 1931, it was reported that the previous year had been one of "considerable difficulty." For the first time in the company's ten-year history, at a meeting of May 11, 1932, Dr. Grace recommended that remuneration to directors be reduced by one half.

But early hardships did not last long. With the robust efforts of the two assistant managers, the company began to show improved returns. Capital investment in the wharves was paying off, and Grace, Kennedy increased the rate of gross profit on sales of merchandise. Profits increased in 1933 and 1934; Harold Dunn recommended bonuses, charged to the profit and loss accounts, to be paid out to the joint managers to compensate them for their hard work.

The year 1933 brought the hiring of a young man named Selwyn Carlton Alexander,[18] who joined Grace, Kennedy at the age of seventeen to fill the position of book clerk at

---

15   General Mills was created in 1928 when Washburn-Crosby merged with three other mills with its main Gold Medal Factory in Minneapolis. Grace, Kennedy became distributors of Gold Medal Flour and Betty Crocker brand name products.

16   Information of travel dates from manifests of alien passengers for the United States Immigration officer at Port of Arrival, *National Archives and Records Administration*. Names of professions for L. F. Kennedy on ship manifests included Exporter, Merchant, Manager, Executive, Steamship Agent. He was described as West Indian, bi-lingual, speaking both English and Spanish, having a dark or ruddy complexion, 5ft 9.5 inches tall, black hair, brown eyes, and no distinguishing marks. By 1935 records show that he began to take Pan American Clipper air service (Sikorsky S-42 Flying Boat) to Miami instead of United Fruit Company vessels to New York.

17   The business experienced setbacks with early enterprises like the Pigeon Island experiment for production of salt by way of solar evaporation of sea water begun under Fred William's management; even with the purchase of a schooner, *Admiral Beatty,* and subsequent contracts with Turks Island (November 24, 1931), the enterprise was not profitable.

18   See Chapter 37.

a weekly salary of £2. Like Fred William Kennedy and James Moss-Solomon, he left school early to become the breadwinner of his family.

Anna Jarvis[19] commented:

> Uncle Jim (James Moss-Solomon) spoke to Mr. Fred who agreed that Carlton be interviewed and if he seemed a suitable recruit, he would work for twelve weeks to test his readiness for employment at Grace, Kennedy (without pay). Carlton was sent to work as a supervisor on the wharf. After the probationary period, Mr. Fred decided Carlton was made of the right stuff and offered him full time paid employment. The rest is history! [20]

Luis Fred saw himself in Carlton—rugged, hardworking, and exacting. Carlton had street smarts and was a man who developed into a salesman par excellence, one for whom customer relations became his lifelong pursuit.

Luis Fred maintained a vibrant social life (advice that he had taken from his father). He socialized at the Myrtle Bank Hotel; he was a member of the Royal Jamaica Society of Agriculture and Commerce and Merchants Exchange,[21] the St. Andrew Club, the Jamaica Club, where he played tennis, and Sabina Park Cricket Club, newly built and established in 1930. He enjoyed drinking and poker nights with his friends; his partner was a good friend named Jules Wolff.

On trips to New York, he visited his college friends and courted their sisters. Yet of all the women he dated in the 1930s, only one captured his heart—his cousin, Lydia Mathilde Loinaz, who lived with her family in Mount Vernon. He would make numerous trips to New York in the 1930s to win her heart and affection.

At home he was the breadwinner, providing his mother with his monthly earnings to support the family.[22] Education remained a top priority for both him and

---

19 Dr. Anna Jarvis née Figueroa is Carlton Alexander's niece. Anna's mother, Dorothy Grace Murray, was Carlton's sister. The Moss-Solomons are also related to Carlton Alexander. Carlton's mother, Rosina Grace Murray's youngest sibling, Minna Louise Murray, was married to James Moss-Solomon's younger brother, Copeland. See Family Trees, Chapters 34 and 37.

20 Anna Jarvis, "Luis Fred Kennedy -Memories." Note: Sources say Carlton started as a book clerk but he was likely quickly transferred to the wharves to work as supervisor.

21 Established in 1885, its purpose was to offer facilities and membership to those wanting to discuss proposals and schemes for the development of industry for the colony. *Handbook of Jamaica* (1902), 458.

22 In 1935 the family home had a new occupant. Aunt Fillan returned from California after seventeen years to live at 25 South Camp Road.

his mother; they made funds available for all his siblings to pursue studies at the tertiary level.

Mim worked briefly at Grace, Kennedy as her brother's secretary and later at the Canadian Imperial Bank of Commerce in Kingston.[23] She was a professional tennis player, in later years winning the Melbourne Invitational Tournament.[24] She frequently travelled abroad, accompanying her mother on trips to New York and England.[25]

José spent five years articled at Messrs. Manton and Hart and was called to the bar in 1937. A news story with his picture appeared in the *Daily Gleaner* of July 31, 1937, announcing that Mr. José Kennedy was added to the roll of local solicitors. Mr. Justice Cannon welcomed Mr. Kennedy and wished him success in the practice of his profession as he started work as a solicitor at the firm of Messrs. Manton and Hart. José was twenty-four years of age when articled.

In a letter dated August 1, 1937, Luis Fred wrote to Lydia Loinaz in New York:

> On Friday Joe was formally admitted before the Chief Justice to practise as a lawyer. It was a very short but impressive little ceremony. He had to swear allegiance to the King, and all sorts of things. Needless to say, all the family was present. He received many congratulations and got his picture in the local papers as per clipping enclosed. Last night he and Luisita went out to a party to celebrate the event, and when I was leaving for 7:00 o'clock Mass, they were just coming home! Mamá had something to say about that![26]

His sister Louise (whom he called Luisita) possessed a sharp wit and infectious sense of humour. Her eldest daughter, Mary Lou, details this part of her life in 1930s:

> The family went into a period of deep mourning that lasted three years. Louise would spend those years completing her secondary studies and teaching herself Latin—her sights were set on becoming a Barrister at Law, and Latin was one of the required subjects. In 1932, at eighteen years old, she sat the Jamaica Scholarship Exam and received the highest score. Because of her young age, officials decided that since she could have another chance to sit the exam, the scholarship would be awarded to the person with the second highest score—who at twenty-one, would not have another chance.

---

23 September 17, 1937, on board Pan American clipper to Miami, Florida. Carmen Valliant's profession is listed as stenographer.

24 Notes from Fred William (Bill) Valliant, Mim's eldest child.

25 Carmen Kennedy's name appears on ship manifests for 1934 to Mississippi, passenger lists of Pan American Airways to Miami in 1937, and to England in 1935 and 1937.

26 Luis Fred Kennedy, letter to Lydia Loinaz, August 1, 1937.

Louise would not sit the exam again. For the next few years, she taught at Alpha Academy, her *alma mater*, and in 1935, at the age of twenty-one, was accepted to study law in England, at Kings College, London. Her years at King's College were engaging and fulfilling. She made new and long-lasting friends and loved the study of law with its rituals and traditions, such as attending dinners at the Inns of Court. Sadly, for reasons of health, she was unable to complete her degree.[27]

Gloria graduated from Alpha Academy in 1937 and was admitted to the College of New Rochelle, a private institution founded in 1904 by the Ursuline Order of nuns. It was the first Catholic women's college in New York State, established at a time when women did not have easy access to tertiary level education. The youngest, Francis Xavier, was a student at St. George's College in the mid-1930s.

At age twenty-eight, Luis Fred would soon leave his bachelorhood behind. During his courtship period with Lydia Loinaz, friends and family teased him about his sudden interest in physical fitness. He signed up for a series of rigorous lessons at a local boxing club in Kingston—he was a heavyweight with a bulky physique weighing in at 185 lbs.

Luis Fred, at age twenty-six, on board the S. S. *Bremen*. He was travelling from Southampton to New York on one of his numerous business trips. Gentleman on far left is Erik Bastion. (Photo, 1935, from family collection. Biographical details of Erik Bastion, unknown.)

---

27 Excerpts from "Biographical Details -Louisa Matilde Kennedy Soutar," contributed by her eldest daughter, Mary Lou Soutar-Hynes, who lives in Toronto, Canada.

Carmen Celia Arzeno, 20
Mid-1890s, Puerto Plata,
Dominican Republic.

Charles Henry Loinaz, 28
Mid-1890s, Puerto Plata,
Dominican Republic.

Lydia Loinaz with brother, José,
Mount Vernon, circa 1917
"I was very close to my brother and devastated by his untimely death" (LMK).
(Photos from family collection)

La Glorieta, Parque Central de Puerto Plata, built in 1872.
"One of my first memories was dancing in the park in Santo Domingo where families came together" (LMK).
(Photo by author, Puerto Plata, DR, 2019)

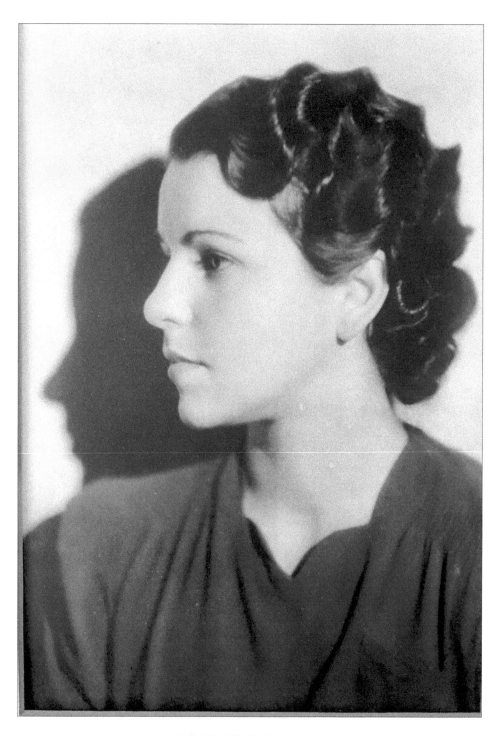

Lydia Mathilde Loinaz Arzeno,
New York, circa 1935, aged 25.
She declined offers to be photographed as a model for a fashion magazine.
(Photo from family collection)

15

TILLIE

Lydia Mathilde Loinaz was born in Puerto Plata, Dominican Republic, December 20, 1910, sixth child of Carmen Celia Arzeno and Charles Henry Loinaz. Her ancestors were Basque, migrating in the early 1700s from San Sebastián, Spain, to Cuba, where her grandfather, Diego Loinaz Arteaga (1841–1909), was born.[1]

In the 1860s, Diego Loinaz migrated from Cuba to Nassau Bahamas, where he married Georgianna Henry, who was of English descent. The eldest was Charles Henry Loinaz (my mom's father), born a British subject in New Providence, November 17, 1865. In the 1870s, after the birth of five more children, the family settled in Puerto Plata, where they joined others of the Loinaz clan.

The Loinaz were among the 400 or more Cubans who fled their country during and after the Cuban War of Independence (1895-1898). Many made their new home in the Dominican Republic, where they sought opportunities to buy and operate horse ranches, sugar and chocolate factories.[2] The period was called the Grand Reconstruction in the Dominican Republic at a time when immigrants poured in not only from Cuba, but also from England, Germany, and Holland.

Diego Loinaz had a large family of eleven children.[3] His siblings, cousins, and children acquired large properties and operated import and export agencies in Puerto Plata. The eldest born, Charles Henry Loinaz, founded a commercial enterprise,

---

[1] "We are descendants of a brother of a saint, Loinaz del Castillo, and I went to visit his birthplace. It's in the Basque country, Beasaín, a small town. They weren't expecting us, they were so surprised" (Lydia M. Kennedy, 2002). He was born Martín de Loinaz y Amunabarro, July 16, 1566. He became a Franciscan friar, and on mission to Japan was crucified along with twenty-five others, February 5, 1597.

[2] See Chapter 4.

[3] Diego Loinaz's first wife, Georgianna Henry, died in 1886. He remarried the following year and had three more children from Wilhelmina Nugent.

C. H. Loinaz Compañía Comercial, in 1887, general agents and commissions, with a profitable import/export trade with the United States.[4] He married Carmen Celia Arzeno Abreu in 1898,[5] and together they had seven children, of whom Lydia Loinaz (my mom) is the sixth child.[6] Her father served as the consul for Great Britain in Puerto Plata and obtained British passports for his wife and children. Through his business and family connections, Charles Loinaz knew Fred William Kennedy, who married his wife's younger sister, Luisa Matilde, in Puerto Plata in 1908.

Lydia Loinaz was born in a period when the country was coming under the control of the United States. The Marines invaded on May 5, 1916, with fierce resistance from local militia who could not hold their ground against the superior American artillery.[7]

She recalled both happy and scary times when she was a girl of five years of age:

> One of my first memories was dancing in the park in Santo Domingo where families came together.[8] Older girls dressed up to be admired by the boys. Dowagers, dressed in black, from head to toe, kept a strict eye from the edges, chaperoning. Girls had no freedom in those days and had to be watched. Every town had these parks.
>
> Everyone went to church and the cemetery. Women were almost constantly burying children and husbands. They married too young. My own grandmother lost at least one child to typhoid fever. She lived down the street. I would visit her for as long as I wanted. It was while I was there one time the house caught fire, a huge conflagration, from a gas lamp in another house, which spread quickly to all the houses. My uncle Tío Rafael,[9] very strong, came to get me, and on his way, he fell on the street and broke his leg.
>
> My grandmother, Carmen María was a lovely little woman. Her husband, José Arzeno, had his own export company, J. Arzeno and

---

4   The offices were located on La Calle Comercio No. 3. Germán Camarena, *Historia de la ciudad de Puerto Plata,* (2003), 177–178.

5   Prior to getting married, Charles Henry Loinaz had two children, Federico Antonio García and Enrique Loinaz Cuza. In 1900, he had another child, Clementina Filomena García, daughter of Antonia García Peralta.

6   Names of seven children: Ines Georgiana (1899–1994), Charles Eugenio (1990–1975), José (1902–1928), Carmen Celia (1903–1986), Mercedes Amanda (1905–1970), Lydia Mathilde (1910–2002), and Alta Gracia María (1918–2006).

7   See Chapter 4 for details of this crisis period in the history of the Dominican Republic.

8   Puerto Plata was one of the first places in the Dominican Republic in the nineteenth century to have a central park for recreation. Musicians and actors performed concerts in a circular pavilion in the centre of the piazza, *la Glorieta,* constructed in Victorian style by the famous English architect, Roderick Arthur. Camarena (2003), 275–276. See also, Chapter 4.

9   Her mother's brother, Rafael Arzeno Abreu (1881–1962).

Company, for local produce, cacao, which is the cocoa bean. We would take refuge in his warehouse during the dangerous times when the rival gangs fought. I remember hiding there with the smell of cacao in the bags.[10]

The Loinaz family fled the Dominican Republic, July 5, 1917, on board the SS *Iroquois* destined for Ellis Island, New York. Blanca Santana, listed as "servant" on the ship manifest, accompanied the family. On the passage from the Turks Islands, Lydia remembered the frightful nights when the ship's lights were extinguished for fear of being spotted by the Germans during World War I. She recalled hugging her teddy bear, which she hid under her pillow, to hide the bright light of its eyes, powered by two small batteries.

The family moved into 134 Chester Street in Mount Vernon, a city suburb of Manhattan, immediately north of the Bronx. Founded in 1850, Mount Vernon was one of the first communities established in the United States "explicitly planned and designed to enable middle- and working-class people to have an affordable home of their own."[11] The settlement mushroomed because of its proximity to New York City (a commuting time of thirty minutes by boat, train, or stagecoach);[12] it "attracted shopkeepers, artisans, intellectuals, artists, scientists, and bankers."[13] With a population of 42,000 in 1917 when the Loinaz family arrived, Mount Vernon was a vibrant, cosmopolitan city, with all the most modern services and conveniences: telephone, gas, electrical, water, and sewage.[14]

Lydia reminisced about her childhood:

> I lived on a lovely block, we played games with the neighbourhood kids, it was a really nice neighbourhood. I was too young for my older sisters, and older than my baby sister, so I had to go off and make friends of my own age. Luckily, I had a lot of friends my own age. My best friend, her name was Dorothy Hall, she and I were very close, we were like sisters. They were two houses down and she would come every day, or I would go down there. We even went to the same school. Even though he was older, my other good friend was my brother José. Dorothy was a great comfort when José was sick, she would come

---

10   Lydia M. Kennedy, interview with Georgianne Thompson Kennedy, August 10, 2002.

11   Larry H. Spruill, *Images of America: Mount Vernon Revisited* (2014), 6.

12   Ibid., 8. The village became a city in 1892.

13   Ibid., 13.

14   Ibid.

over and visit. I was a bit of a tomboy, I loved to play outside, playing marbles. Baseball was big, we used to go to Yankee Stadium.[15]

Lydia's dad, Charles Henry Loinaz, was absent during her childhood for much of the time, travelling back and forth on business to the Dominican Republic. (I don't recall her talking much about him.) He died at age fifty-four when she was only nine years old.

She praised the public school system in New York.[16] Schools not only offered academic subjects for entry to college but also practical courses such as shorthand, stenography, and technical subjects that prepared students for the workforce. The curriculum had changed in response to the growing numbers of women seeking employment as telephone operators, store clerks, bank tellers, and factory workers.

Lydia excelled in mathematics, English, and especially languages; she received the highest mark, ninety-eight percent, in Latin in her graduation class. Her schoolmates called her Tillie, short for Mathilde. She preferred that name because she did not like the anglicized pronunciation of either of her first names, Lydia or Mathilde. They sounded harsh to her. The Spanish way of saying them was romantic, with a smoother, more lyrical cadence. She enjoyed sports and music, joined the netball team, and was a shining star in the school choir. After graduating, she enrolled in a post high school course to train as a stenographer.

She and her brother José were best friends. Even though he was eight years older, he spent time with her, reading and playing games. They both had a passion for music; he taught her how to dance the tango, the merengue, and the waltz.

At age twenty-six, José succumbed to tuberculosis and died on August 7, 1928, the year that Tillie graduated from high school.[17] "He was young, handsome, and in his prime. I missed him terribly."[18]

Tillie's ambition was to attend university and train as an opera singer with the Manhattan Metropolitan Opera, but her family was strapped for money. In September 1929, her brother Charlie invested the family fortunes in the stock market and lost everything, including their home in Mount Vernon, in the Great Crash a month later. Tillie needed to forgo college plans and start looking for a job:

> I was the only one of my family working during the Depression. I had a job down at the Dominican Consulate, but when the government

---

15   Lydia M. Kennedy (2002).

16   The Women's Club in Mount Vernon had been the driving force behind the establishment of Mount Vernon High School in 1897 and the public library in 1904—one of the finest in the State. Larry H. Spruill (2014), 42.

17   Her grandmother, Carmen Abreu Imbert, also died in 1928 on June 6 in Puerto Rico. She was very fond of Mamá Carmita.

18   Lydia M. Kennedy (2002).

changed, I lost my job. I got lucky though, there was a girl from the Brazilian Embassy I knew, she had heard of this job with an insurance company, and she gave me the name of the person to see. They needed a stenographer, a Spanish stenographer. I got an interview and landed the job. What am I going to do? I don't know how to do shorthand in Spanish. I got a little book, and studied it up and stuff, and luckily my boss was impressed.[19] He thought he could speak Spanish, but he couldn't, he was terrible. He dictated his letters very slowly because he couldn't speak Spanish that well. They were full of mistakes, so I would take these down, and then correct his grammar. And that's how I got ahead.

They had a lot of business with South America, so we were busy all the time. I thought I wouldn't last a week, I was faking it, but it worked out for me. The vice-president was a very nice man, Chapman. We shared the same office and he helped me a lot because I had to learn insurance terms. It was all foreign, they didn't mean anything to me. So, he taught me a lot about insurance. He liked me and he promoted my staying there, he backed me up.

And the big boss, the head office was in Shanghai. It was called The American Asiatic Insurance and the US office was just a branch. They eventually changed the name to American International. He came over once from Shanghai to visit. I remember he was a great big, tall man, he was Dutch, Vander Starr[20] his name was. I remember he had terrible arthritis, and he went swimming in the Labrador Current and he was nearly crippled as a result. He was the one who promoted me to Secretary of the Company. So, it was me, the president, and the vice-president. I got a big raise.[21]

It was exceptional, almost non-existent during the 1930s, for a woman to hold an executive position at the corporate level. Only twenty-five percent of all women in the

---

19   George Moszkowski, a Polish refugee, was transferred from Shanghai in 1929 to head up the New York office.

20   Cornelius Vander Starr (1892–1968) founded the American Asiatic Underwriters (AAU) in 1919 in Shanghai, China, opening offices in New York in 1926 under the name of American International Underwriters Corporation (AIU). Those were the founding years of the American International Group (AIG), which has become a multinational finance and insurance corporation with operations in more than 80 countries.

21   Lydia M. Kennedy (2002).

US were gainfully employed in 1930,[22] and they worked for half the average annual salary of men. "Out of every ten women workers in 1940, three were in clerical or sales work, two were in factories, two in domestic service, one was a professional—a teacher or a nurse—and one was a service worker."[23]

Tillie was prudent with her earnings. She handed over a portion of her monthly earnings to her mother to run the household, kept back some for her own spending, and saved the balance. The Depression taught her a basic lesson about life and money, which she held sacred throughout her life: waste not, want not.

Her mother worried that she was getting too old to marry. Tillie had postponed the idea of marriage with a view to advancing her career and supporting her family. By 1937, when the courtship with Luis Fred began, she was already twenty-six, past the age when women in her culture would normally wed. She faced major dilemmas, the greatest of which was to choose between marriage and a career.

---

22 "Women in the 1930s and 1940s, SRJC: https://canvas.santarosa.edu/courses/24761/pages/women-in-the-1930s-and-1940s.

23 Ibid.

Lydia Mathilde Loinaz Arzeno, October 2, 1937, New York City.
(Photo from family collection)

Candid photo of newlyweds, New York City, 1937.
(Photo [cropped from original], courtesy of Lisa Valliant)

16

WHITE LINENS

(1937)

Qué amante siempre te adoraré [1]

Luis Fred visited New York in the early months of 1937, but despite his exhortations, Tillie did not initially agree to his hand in marriage.

> My dearest Tillie,
>
> I don't think that you know how much of a disappointment it was to have sailed without seeing you again. I thought that I would have heard from you, even by letter, before sailing, and when that didn't happen, I consoled myself with the thought that I would surely hear from you in Havana ... I am already making plans for a trip to New York sometime between the middle and end of March ... I suppose there isn't much hope, but should you change your mind about marrying me, before I get there, send me a cable, and I'll take the first plane that is available.[2]

---

[1] "Loving you, I will always adore you." From song, *"Quiéreme Mucho"* by Gonzalo Roig (1890–1970), composed in 1916 and first performed, 1922. During their courtship and continuing through their married life, they would take turns singing the verses. Luis Fred was tone deaf and Tillie had an operatic voice.

[2] Luis Fred Kennedy, letter to Lydia Loinaz, March 5, 1937.

When he returned to Jamaica, too impatient to wait for her replies (which would take weeks or months to arrive), he wrote her copious letters. He tried phoning through an international operator, for he would have been satisfied with even the briefest exchange of words at hearing her voice, but long-distance telephone signals in the 1930s were unreliable.

When she did write, her letter did not contain the news he was longing for. He was quick to reply:

> Thanks for telling me everything about how you feel. I shall not hide the fact that it was a bit of a blow, for I had been conceited enough to think that I had been making some progress. Anyway, I still love you and want you to marry me. I don't think that there is anything right now that I want more. In addition, I am a rather optimistic person, and although it is a bit hard to be that way today, I can say that I haven't given up hope that the possibility which you have mentioned of your changing your mind, will come about.
>
> One thing stands out very clearly, and that is, that on my last night in New York, you said that you were trying not to love me, because of all those infernal (excuse me) difficulties.[3]

He placed a framed picture of her on his desk at work. "It stares back at me all day long. I wish it would talk. The first few days it was there I thought I had made a mistake putting it on my desk, for I just couldn't do any real concentrating."[4]

By the middle of April, she agreed for him to visit her in New York. He penned a letter, April 19, 1937, filled with poetic professions of love, akin to those of a Shakespearean sonnet: "These infernal days have suddenly appeared to have as many hours as they used to have minutes."[5] He drowned his sorrows in work and late-night poker games with one consolation only—that his boss had taken leave.

Two months later, Tillie agreed to his proposal of marriage. She recalled their engagement of the summer of 1937 in New York:

> On his business trips, he would stay on, and when our relationship got more serious, he would lengthen the stays. I had never been to so

---

3 Luis Fred Kennedy (March 13, 1937). In this letter, he mentioned that one of his projects at Grace, Kennedy was management of a cigarette factory. He included in the envelope one of the cigarettes for her pleasure: "I am sending you one of the cigarettes from the first pack that we made. I don't like them, but I hope the people here will. They are after the English style -trashy- but that is what is smoked here." The factory used a blend of Virginian and Jamaican tobacco.

4 Luis Fred Kennedy (March 23, 1937).

5 Luis Fred Kennedy (April 19, 1937).

many good restaurants in my life. He proposed to me on the 9th of June, and we were to get married the following year, but we moved up the date. They had a sidewalk cafe at the St. Moritz.[6] They had tables outside and when it got warmish, they put them out…and we were sitting there out on the sidewalk, and he proposed to me. He said, "I love you. Would you like to marry me?" [7]

She said yes, but on one condition—that she return each year from Jamaica to visit her mother. It was a difficult decision for her to make. In 1937 Cornelius Vander Starr, president of American International Underwriters (AIU), had offered her a new position:

> I was going to head up a Life Insurance Company in the Philippines. I had to decide whether to get married or go there. There was also a doctor who was in love with me at the time. He was a medical doctor and his brother, who was an engineer, helped to design the New York subway. His brother's name was Miguel. We were a foursome, he dated my sister Chela, and the four of us would go out. It was a surprise to him that I was getting married to someone else. I chose Luis Fred instead. I left it till the last minute to tell him. He was heartbroken. He said that he was going to ask to marry me, but he hadn't asked. They were Puerto Ricans.[8]

The original plan was to get married in 1938, for she wanted to see Jamaica before making a final commitment. Luis Fred did not want to wait, but his personal finances were not in the best shape. "Yesterday (June 21, 1937), my bank manager friend called to say that I had reached my credit limit at his bank, and if I wanted more money to come and see him. I said I didn't want any more. I wanted to pay off my debts, which was alright with him."[9]

Luis Fred was determined to find a way. "In fact, my dear, I don't know that it will be possible for me to leave New York again without you. You may laugh, but I am dead serious when I say that it is going to be extremely difficult for me to wait until next year to marry you."[10]

A few weeks later, he wrote her to say that although he was not proud of his net worth, he was not "insolvent." His situation was complicated by his father's estate,

---

6   The Hotel St. Moritz was a luxury hotel located at 50 Central Park South, on the east side of Sixth Avenue, in New York City.

7   Lydia M. Kennedy, interview with Georgianne Thompson Kennedy, August 10, 2002.

8   Lydia M. Kennedy (2002).

9   Luis Fred Kennedy (June 22, 1937).

10 Ibid., June 26, 1937.

which had still not been settled. Before getting married, he needed to negotiate a steady income for both his mother and siblings and would do "his darnedest to straighten out everything in time." He met with his lawyers and did just that. He suggested moving up the wedding to October, and she agreed. He needed only to clear the date with his boss, Dr. John Grace.[11]

Tillie's greatest worry was that she would leave her mother destitute. She did not feel comfortable raising the issue with her fiancé and did not want to ask him for money to support her family. She did not know how to tell her boss, and at the start kept secret the news of her engagement. To make matters worse, in July she came down with a case of whooping cough. Doctors suggested that she was over-worked and recommended a month's period of recuperation at a lakeside resort.[12] By mid-August, she wrote to tell Luis Fred that she was fully recovered and reaffirmed her love for him and her wish to proceed with wedding plans.

She finally confided in him to say that the thought of leaving her mother was tearing her apart and making her ill. She had not known how to broach the subject of financial support, so kept all her thoughts and feelings locked inside. Opening her heart allowed them to commiserate and draw closer:

> Please do not say, darling, that you ought not to have written me that you had been worrying. I want you to promise me that, starting right now, you will write me or tell me whenever anything bothers you, and also that you will tell me what it is. I should hate to think, my love, that you were unhappy about something, which if given the opportunity, might well be something that I could easily arrange to have fixed. Promise?[13]

After several weeks, Luis Fred heard from Dr. Grace, who gave him leave "for as long as he wanted," but his boss did not agree to either a raise in pay or an advance in bonuses,[14]

---

11 When Dr. Grace heard of his engagement, he wrote Luis Fred to say, "I am very glad indeed that you are engaged to be married. If the girl is right, there could be nothing better for you. You will not approve of my "If", but of course all marriages do not go right. I hope that the lady is Catholic. A common religion makes things so much easier … and again my congratulations to you, my dear Fred" (John J. Grace, letter to Luis Fred Kennedy, July 3, 1937).

12 She booked to stay three weeks at Basin Harbour Lodge on Lake Champlain, Vermont. Her sisters, Carmen Celia (Chela) and Alta Gracia (Gracie), visited and kept her company for a few days, at the suggestion of Luis Fred.

13 Luis Fred Kennedy (August 15, 1937).

14 In 1937, Luis Fred's earnings were $10,000 per year. His guaranteed salary was $5,000 with an equivalent amount in bonus payments. He had written to his boss for an increase in salary to $7,500 with a reduced percentage bonus to be paid in January. In his letters during the 1930s and 1940s, he quotes his earnings in dollars, not pounds sterling.

White Linens (1937)

which were normally paid in January. Luis Fred told Tillie that the "old boy didn't approve of young men earning too much money."[15] Had he spoken to his boss face to face he thought he would have had better luck, but Dr. Grace's stay in England was longer than expected.

Luis Fred found sufficient funds to support Tillie's family and to pay for the wedding and honeymoon by convincing his bank manager to extend his credit limit. "The honeymoon will be expensive … but we shall be married only once, and I want to celebrate that occasion in a fitting way."

He was concerned about conditions in Jamaica. "We are all working overtime here and business is booming, but I am afraid that a reaction and a serious one will very soon set in. When 'reactions' happen in Jamaica they are serious, and everything stops—suddenly and many times disastrously, so we have to be awfully careful, and that is the part of my job I hate. I detest having to be careful!"[16]

He confessed to Tillie that he had forced himself out of bed every morning to accompany his mother to Mass at the cathedral to pray that the wedding day would one day come true. He thought himself a "lucky devil" and that he did not deserve the breaks he received, "least of all, you, my darling, the best bit of luck of all." When he started losing at poker games (which was highly unusual), he became convinced that the adage, *Lucky at cards, unlucky in love*, might be true after all.

Once plans were underway, he could hardly wait.[17] "You know I had a crazy idea of catching Thursday's plane, meeting you in New York, and coming back together … I am dancing inside. I could not eat any dinner tonight. Mother and the family made fun of me, but I just couldn't hold myself down."[18]

They were married on October 2, 1937, at the Roman Catholic Church of the Incarnation[19] in upper Manhattan. The bride was radiant in her 1930s wedding gown—satin with long sleeves, high neckline, tiara headdress, and tulle floor-length veil. The groom was dressed in formal daytime wear: a morning coat, a cutaway with striped trousers, and a vest—all at the behest of his fiancée. (He had not been amenable to the idea. After a week of deliberations, he wrote to her, "Yes, darling, I

---

15   Luis Fred Kennedy (June 22, 1937).

16   Ibid.

17   "We had to get special papal dispensation because we were first cousins. We also had doctors check our medical history to determine if there were any weaknesses in common ancestry" (Lydia Kennedy, 2002). Contrary to what they thought, the process was not complicated; they received permission within weeks.

18   Luis Fred Kennedy (September 8, 1937).

19   Church of the Incarnation is located at 1290 St. Nicholas Avenue in Washington Heights, close to where her mother's family lived. The magnificent, large neo-Gothic building with marble interior and vaulting stained glass windows opened its doors in 1928. It has come to be known in New York City as St. Patrick's Cathedral of Washington Heights.

will wear a morning coat at our wedding. I'd even wear a veil, a kilt, and no shoes, if I had to!")[20]

The maid of honour was the bride's eldest sister, Carmen Celia (Chela, aged thirty-four), who was dressed in midnight blue.[21] She assisted with preparations for the wedding, including purchase of dresses from Macy's in New York, the décor of white flowers that draped the altar and pews, and reception planning. In the absence of Tillie's father,[22] her brother, Charles Eugenio,[23] escorted her down the aisle. Other members of the immediate family present were her sister Georgina Thomen, her husband and two children,[24] and two sisters, Mercedes Amanda (Mecha, aged thirty-two), and Altagracia María (Gracie, aged nineteen).[25] Her uncle, the Rev. Diego Loinaz, and his family, who lived in Brooklyn, were also present, as were the bride's friends and co-workers from American International Underwriters in Manhattan. "I invited all the staff from AIU. They filled the Church, but I couldn't have them to dinner. I was sorry about that. They couldn't have held, we just had a few."[26]

Representing the groom were friends from Holy Cross and family from Jamaica—his mother, brother José, and sisters Carmen and Gloria. (Gloria had arrived earlier in the month to enroll at College of New Rochelle in New York.) His best man was Holy Cross College classmate Bill O'Connor.

Tillie insisted on another condition of marriage—that her wedding vows not contain the word "obey." She wanted them to be identical to his: "to love and to cherish, till death do us part."

Just months earlier, the Manhattan Section of the May 26, 1937, edition of *New York Daily News* flashed the following headlines across its front page:

---

20   In a letter dated July 21, 1937, he wrote, "I even told Mamá that you wanted me to wear a morning coat if you wear a veil. She thinks you are quite right, unfortunately; but the girls refuse to believe that I shall wear it. The mere thought of me in one those outfits makes them laugh. Carmen said that if I would take castor oil for you, it should be an easy matter dressing up for you."

21   Chela was close to both Tillie and Luis Fred. She had always wanted to get married, but despite having had many admirers, she shied away from relationships. She had contracted tuberculosis in her earlier years and feared that she remained contagious.

22   See Chapter 15. Charles Henry Loinaz died in 1920.

23   Charles Eugenio Loinaz, aged 37, was married to Frances Thiesmeier, and they had a son, Charles H. Loinaz.

24   Her eldest sister, Ines Georgina, aged 38 in 1937, was married to Emanuel Jacob Thomen, and they had two sons, Charles and Louis Thomen.

25   In 1947, Altagracia would marry Luis Fred's brother, José.

26   Lydia Kennedy (2002).

White Linens (1937)

TWO FLIGHTS BLAZE TRAIL OF BERMUDA LINE

The cover story announced the test run of the four-engine sea boats, the Pan American Bermuda Clipper[27] and the British Imperial's Cavalier, which would compete for the fastest flight time on the 773-mile hop between Port Washington, New York and Hamilton, Bermuda. The Pan Am plane, which flew out of New York, made it in four hours and fifty-four minutes, the Cavalier flying in the opposite direction from Bermuda to the United States, in six hours and fifty-six minutes. Trade winds, it was reported, had given the Clipper the advantage.

US newspapers were filled with stories of the novelty of air-travel and the opening of the flight route to Bermuda as a honeymoon destination:

> You know you're in a wonderland as soon as the keel of the great plane ruffles the smooth water of Hamilton Harbour; you know you're going to be in a wonderland from the moment you can first see the color of the ocean bottom sparkling up through the depths of the blue Atlantic.[28]

The Clipper was luxurious, with passenger windows circular like the portholes of a ship and interior furnishings like the trappings of a lavish passenger liner. Brochures advertised the trips as cruises with pictures of flight decks that resembled the bridge of a ship.[29] On board, sumptuous, full-course meals were served on tables with place settings of silverware and fine linens. Accommodations were comfortable, with armchairs arranged in spaces the size of large living rooms.

On October 6, 1937, newlyweds Luis Fred and Lydia Kennedy were on board a scheduled flight of the Bermuda Clipper bound for Hamilton, Bermuda to begin their exotic island honeymoon.

> He wanted to go somewhere where no one knew him, Jamaica was full of his mother's old friends and curious people. He wanted to be alone; Jamaica was much more provincial in those days.

> No motorcars were allowed in Bermuda in those days, we used bicycles to go everywhere, and we hired one of those buggies, a horse and buggy. The driver let me sit up top with him, you know, and gave me the reins but then the horse lifted its tail, and the breezes were blowing,

---

27   Sikorsky S-42 was a commercial flying boat designed and built by Sikorsky Aircraft (Pan American World Airways, 1931). The S-42 was often described as the "airliner that changed aviation history."

28   *The Brooklyn Daily Eagle* (Brooklyn, New York), Sunday, October 3, 1937, 24.

29   https://airwaysmag.com/best-of-airways/sikorsky-s-42-airliner-changed-aviation-history/.

> you know, in my direction (laughter). I came right down after that to join Luis Fred.
>
> Bermuda was lovely, it's full of coves, and we'd ride around, and we would have our own private little coves all to ourselves, swimming, and stuff (laughter). You're free to imagine (more laughter).[30]

They had thought to stay longer than their eight days, but Luis Fred experienced a mishap:

> Six white linen suits he had specially made, he had them washed before we left but hadn't tried them on. When he went to put them on the first day, oh my, (chuckles) they were halfway up his leg. They had shrunk in the wash. He was so embarrassed. Luckily, he had taken an extra dark suit with him, so he had one suit to wear. He didn't have that kind of money to go out and buy a whole new set. We were broke by then, we had spent too much money on the wedding, he had to borrow money from me, as a matter of fact. You know how that must have humiliated him; he was very proud. But none of this prevented us from having a good time.[31]

They arrived in Kingston Harbour, November 4, 1937, on board the United Fruit vessel SS *Ulua*. To Tillie's surprise, a photographer from a local newspaper was on hand to greet them at the docks. Two days later, their picture and story headlined, "Honeymooners Arrive from United States," appeared on page eighteen of *The Daily Gleaner*. Their first stop and place of residence was the Manor House Hotel in Constant Spring.[32]

Tillie knew that her decision to marry would change everything. How radical an adjustment she did not fully recognize at the time. A New Yorker, who had been a career professional in the financial district of Lower Manhattan, would miss the cosmopolitanism of the city with its lure and glitter and the hustle and bustle of millions. Island life would be different—married life with its norms and expectations and a British colony seething with civil unrest.

---

30   Lydia M. Kennedy (2002).

31   Ibid.

32   Located on Constant Spring Road owned and operated by Captain Rutty and his wife. It became a tourist attraction in 1930s and 40s. Gemma Romain, *Race, Sexuality and Identity in Britain and Jamaica: The Biography of Patrick Nelson, 1916-1963* (2017), 37.

Labour unrest in Jamaica, 1938, over a pay dispute. Photograph shows a suspected strike-breaker being pursued and heckled by others.

(Photo, courtesy of © Illustrated London News Ltd./Mary Evans)

17

WHARFINGER

(1938)

When Luis Fred Kennedy returned with his bride to Jamaica from their honeymoon in Bermuda, the country was on the brink of a civil uprising. As co-manager of Grace, Kennedy Co. Ltd., which represented overseas manufacturers and employed large numbers of workers on the wharves, he was thrust into the middle of the fray that engulfed the island. He strove to strike a balance, defending the rights of labour and protecting the best interests of the company.

The political climate was volatile. Blacks were held in low-paying jobs, whereas most top positions in the churches, government, and industry were filled by those sent out from the United Kingdom. Local whites and those who "passed for white" were admitted to these posts but were never fully accepted into colonial social circles.

Most people remained in abject poverty with marginal wages and little opportunity for advancement. In 1935, the average labourer earned 1s. 6d., some as little as 9d. per day.[1] And tens of thousands were unemployed; not enough jobs existed on the wharves, on sugar estates, or in other industries.

In addition to being unemployed and underpaid, people were disenfranchised. In 1935, out of a population of 1.25 million, only five-and-a-half percent, or 68,637 people, were registered voters.[2] Political and economic power and prestige rested in the hands of a few. A mass movement erupted for the improvement of working and

---

1   Colin A. Palmer. *Freedom's Children: The 1938 Labour Rebellion and the Birth of Modern Jamaica.* (2014), 14. Labour leaders claimed that a living wage needed to support a family of four was more than four pounds sterling a month, double the amount of what a labourer earned.

2   Ibid., 10.

living conditions that resulted in strikes, civil disturbance, and violence. Jamaica was not alone; upheavals occurred in British Honduras (present day Belize) in 1934, followed by St. Kitts and St. Vincent (1935), St. Lucia (1936), Barbados and Trinidad and Tobago (1937), and British Guyana (1938-39).[3]

The period saw the rise of populist leaders, the most influential of whom was William Alexander Bustamante.[4] After his return from world travels in 1934, he joined the labour movement by aligning himself with the Jamaica Workers and Tradesmen's Union in 1936. But this did not last long. Conflicts arose between himself and the president of the organization, Allan Coombs.

In 1937, Bustamante appeared in a new independent role as "mediator between capital and labour"[5] when he represented striking workers at Serge Island Sugar Estate. They had not received pay increases for seven years—earning what Bustamante called "starvation wages."[6] Although prone to hyperbole, he was not far from the truth when he claimed in one of his letters that "about 50,000 children are roaming the country parts, not being able to go to school, chiefly because of lack of food and clothing."[7]

In 1938, labour protests began at Frome Estate in Westmoreland.

The British company Tate and Lyle acquired seventeen sugar estates, and in March were building a central factory to consolidate operations. Promises of new developments splashed across the front page of *The Daily Gleaner*, March 26, 1938: "Mr. Chamberlain Pleases Europe and the Empire/Tate and Lyle Probably Spend £500,000 in Westmoreland This Year." The hope was for the creation of thousands of jobs for artisans and labourers. Plans were in place to build cottages, a church, a hospital, and rail to link the estates.

When the company could not deliver on all its promises, trouble started.

On Friday, April 29, one thousand workers went on strike. Cane fields were set ablaze in protest, and police were brought in the following day to quell the violence. *The Daily Gleaner* reported that men and women, armed with sticks and stones, were demanding

---

3   Ibid.

4   Born William Alexander Clarke, February 24, 1884. "I was born in Hanover. At a very tender age, Spain became my home. I served in the Spanish Army as a cavalry officer in Morocco, North Africa. Subsequently, I became an inspector in the Havana Police Force. Recently I worked as a dietician in one of New York's largest hospitals." *The Best of Bustamante* (Ranston and Jones) 1977, 31. He said he acquired the name Bustamante to honour a sea captain who adopted him in his early years when he found safe passage to Spain. Historians have not been able to verify all details of his life story.

5   Richard Hart, *Rise and Organise: The Birth of The Workers and National Movements in Jamaica* (1989), 30.

6   Ibid., Chapter 7, 31–34.

7   Bustamante's letter quoted by George Padmore https://www.marxists.org/archive/padmore/1938/unrest-jamaica.htm.

a dollar a day, not less.[8] Strikers supplied themselves with anything they could get their hands on—pieces of wood, iron pipes, and axles. Four people were killed, twenty-five persons wounded, nine in hospital, and eighty-nine arrested. Headlines of the day read:

POLICE FORCED TO SHOOT DOWN RIOTERS IN WESTMORELAND
Dollar a Day Demand Ends in Death[9]

Gleaner reporters sent telegrams saying they were in the mob when "Hell broke loose at Frome. Strikers Mad with Lust to Wreck."[10]

Panic spread fast through the island. In Kingston, a large protest meeting was held at North Parade, and in the days following, hundreds demonstrated in the streets. Bustamante made public appearances before the crowds to hear their grouses and represent their interests.

Trouble started on the waterfront on May 19 when workers refused to unload cargo at the United Fruit Company Wharf. The company declined to meet their demands for an increase in wages. The following day, the Grace Wharf stopped operations. Bustamante was steadfast in his defence of the workers; he would make no concessions. Luis Fred remembered that day, which he described as *mob rule*. He desperately wanted to find a solution; he thought a deputation needed to be appointed so that order could be restored and proper labour negotiations conducted through a process of collective bargaining.

On Sunday, May 22, Bustamante met on Port Royal Street with a crowd of striking workers from Grace Wharf. Employers told the labour leader that they would be willing to negotiate terms only if dock workers returned to work. Bustamante would have nothing of it. He would not tolerate a situation in which shippers employed strike-breakers (clerks and others) to work the ships.

Strikers from the United Fruit Co.'s No. 2 Pier said that they were prepared to "suffer hunger and deprivation rather than resume work at under one shilling per hour."[11] After an impassioned harangue from their chief (as his followers called him), workers unanimously voted for a general strike. The following headline appeared in *The Daily Gleaner* of Monday, May 23:

---

8   *The Daily Gleaner,* Tuesday, May 3, 1938, 1.
9   Ibid.
10 Ibid.
11 *The Daily Gleaner*, Monday, May 23, 1938, 10.

SUNDAY'S DAWN BRINGS STRIKE TO NO. 2 PIER

Labour Leaders Harangue Cheering Crowd Urging Unity and Courage

BREAKER BEATEN

Dock Hand Vote Solidly for Immediate Strike When Asked to Decide.[12]

Bustamante addressed the crowd:

> I am deeply sorry for the people—the masses of this country … If I were not, I would not be here this morning. I am asking you in the name of God and in the name of decency, to be peaceful. But it must be understood that this strike must not be broken. A man will not stand by and see strike-breakers take his work. You cannot expect a man to stand by and allow others to destroy him. You must expect that when that happens, disorder is about to be created … Why are those shipping companies in a position always to tell us that they will work a boat whether we want it or not? They cannot load ships. They could not load a canoe together![13]

Bustamante held his ground. "Until they give in, I shall tie up the wharves. I am going to tie up every store in Kingston. The time when these managers could tell you they will not do anything until you return to work is past and gone."[14]

Discontent spread through the capital and to different parts of the island: Spanish Town, Old Harbour, Montego Bay, and sections of the parishes of St. Ann and St. Mary. Confrontation with police and soldiers erupted in Kingston on "Empire Day," Tuesday, May 24. *The Daily Gleaner* headlines of May 25 rang out an alarm of pure mayhem: "Monday Night That Will Long Be Remembered."

"Growing to proportions never before witnessed in the city, the demonstrators moved through commercial Kingston ordering all shop premises to shut down."[15] Tramcar operators, bus operators, city cleaners, and shirt manufacturers all joined in

---

12   Ibid.
13   Ibid.
14   Excerpt from speech by Bustamante, quoted by Richard Hart (1989), 48.
15   *The Daily Gleaner,* Wednesday, May 25, 1938, 7.

sympathy with the dock workers. Protestors blockaded streets, and fires were set to Chinese retail shops. Everything was in turmoil. The military was called out when strikers threatened to shut down the pumping station, which would have caused sewage to overflow into the streets of Kingston.

Ships that docked at the wharves had to leave, with cargoes remaining in the holds;[16] some diverted to the north coast, others to ports of origin in the United States, Canada, and the UK.

Demonstrations culminated on Tuesday with the arrest of Alexander Bustamante, William Grant,[17] and seventy strikers on Sutton Street in Kingston for "disorderly conduct." They were charged for inciting crowds to assemble unlawfully and for obstructing the police in execution of their duties. Bustamante was reported to have said that he would go to prison but that "worse would follow." Bail was refused on the grounds "that on their return to freedom, they might restart their unsocial activities."[18]

Norman Manley[19] intervened. He not only pledged his service to represent labour, but also influenced the authorities to free his cousin, Alexander Bustamante, from jail.

Manley helped ease the labour strife. "*A voice of sanity he was,*" my dad often said, amidst the chaos that enveloped the city. (Dad respected Manley as "*the most brilliant criminal lawyer Jamaica had ever seen.*") The Daily Gleaner quoted Norman Manley:

> The workers themselves will have to recognize…the putting of garbage tins in the streets, pulling up of trees, the smashing of windows, and being shot at and killed is not going to give them better wages or better hours or fairer conditions of employment. Equally well, the employing classes must realize that the country is moving forward, and that they must give every assistance to labour organizations and to movements which are designed to strengthen the position of labour and that the people must be encouraged to come forward under proper leadership.[20]

---

16   Ibid.

17   William Wellington Wellwood Grant OD (1894–1977) had returned to Jamaica from New York as a member of the United Negro Improvement Association (UNIA) headed by Marcus Garvey. He held gatherings every Sunday night at Victoria Park. A charismatic leader and orator, he was instrumental in encouraging Bustamante to join the labour movement.

18   *The Daily Gleaner,* "No Bail for Bustamante and Seventy Others Held at Sutton Street," Wednesday, March 25, 1938, 10.

19   Norman Washington Manley MM, QC (July, 4, 1893–September 2, 1969), leader of the People's National Party (which was formed in 1938), future premier and national hero of Jamaica.

20   *The Daily Gleaner,* Wednesday, May 25, 1938, 7.

Wharfinger (1938)

Manley negotiated with the shippers, Grace, Kennedy, Jamaica Banana Producers Association, and others for wage increases. "I have been in conference with wharf owners and shipping agents and representatives of the government all day yesterday (Wednesday, May 25), and I have put forward the case of the shippers and wharf labourers."[21] He met the following morning with 2,000 workers at No. 1 Pier, but they refused to accept the shippers' offer of an increase, demanded a minimum of 1/- per hour and 2/- per hour for overtime, and said that they would not return to work until their chief was released from prison.

Strikes became more widespread across the island. The wharves remained closed, and businesses shut their doors; everything was at a stalemate. As a remedy, Manley sought the support of labourers to form a trade union, whose president would be Alexander Bustamante, but they would hear nothing of it until their chief and St. William Grant were released.

On Saturday, May 28, on the strength of affidavits provided by Norman Manley and the defendants' counsel, Bustamante and Grant were released on bail with the surety that neither would "commit a breach of the peace in any form whatever."[22]

Once freed, Bustamante persuaded the workers to agree to increases negotiated by Manley with the bonus of double time at the new rate. *Gleaner* headlines on Monday, May 30 announced temporary relief to labour unrest.

DOCK STRIKE SETTLED SATURDAY AFTERNOON
ORDERLINESS: BACK TO WORK:
LABOUR PLEDGE

The longshoremen now had a "saviour" in the person of Alexander Bustamante. St. William Wellington Grant addressed thousands at Victoria Park after his release from prison, "Long Live Bustamante, God save Bustamante ... Bustamante is now our God ... for leading Jamaicans into perpetual freedom."[23] The governor had known it was the courts' only choice, and Busta knew it also. "If they ever send me back to prison, Kingston will be a mass of ruins."[24]

---

21   Ibid., Thursday, May 26, 1938, 1.
22   Ibid., Monday, May 30, 1938, 7.
23   Ibid., 1.
24   Jack Ranston and Ken Jones, *The Best of Bustamante: Selected Quotations 1935–74* (1977), 34.

The second half of 1938 and early months of 1939 saw the formation of trade unions, albeit disorganized, under the leadership of Bustamante.[25] But matters were still unsettled on the wharves. Wages did not increase to a level satisfactory for longshoremen to make a decent living, disputes arose concerning allegiance of workers to different unions, and the government was slow to settle grievances taken to arbitration. Bustamante did not have exclusive bargaining rights with the employers, and this caused dissension not only among the labourers but also between the shippers themselves.

In a fit of anger,[26] Bustamante, then leader of a group of industrial unions, called another general strike island-wide of all dock workers to begin Monday, February 13, 1939. Everything was again thrown into chaos. Governor Sir Arthur Richards[27] responded quickly, declaring a State of Emergency that banned marches and public meetings, and put "a law-and-order machinery into motion."

Stories and pictures appeared in *The Daily Gleaner* of February 18, 1939, of workers at Grace Wharf unloading lard from *The Jamaica Planter* under the protection of armed police guard. Luis Fred Kennedy refused to comply with Bustamante's rash decision to shut down the ports and negotiated with the government to break the strike.

Governor Richards' intention was to arrest Bustamante for calling a national strike of dockworkers. Norman Manley again intervened. The governor agreed with Manley's recommendation that a Trade Union Advisory Council be formed to bring together rival unions under one umbrella. It was a tenuous arrangement; unions remained fractured, and the longshoremen became pawns in a game played by competing employers.

Luis Fred Kennedy and other business leaders organized employers to form a central bargaining unit. The trade union law was defined broadly enough to permit the inclusion of employers' associations.[28] Along with Charles Edward Johnston, Sibrandt Duhn List, and Thomas Bradshaw, he pioneered the formation of the Shipping Association of Jamaica, January 27, 1939.[29] "The Association established very clear goals, including a commitment to improve conditions of employment, establish uniformity of rates paid to labour on wharves and ships, as well as to ensure that labourers be given a fair day's work for a fair day's pay."[30]

---

25   The Bustamante Industrial Trade Union opened its offices at 30 Duke Street in June 1938.

26   *The Daily Gleaner,* February 15, 1939, reported that the decision was sudden, based on a dispute with the United Fruit Company over Bustamante's wishes to have George Reid, member of a rival union led by A. G. S. Coombs, dismissed from employment.

27   In June 1938, Sir Arthur Frederick Richards was appointed Governor of Jamaica to replace Sir Edward Denham. He became notorious for his autocratic style of governance. He and Luis Fred Kennedy came into conflict in later years. See also Chapters 18 and 20.

28   Phelps (1960), 419

29   Captain Sibrandt Duhn List became the first chairman of the Shipping Association of Jamaica. He was aged seventy-two.

30   "History of the SAJ", https://www.jamports.com/the-history-of-the-saj.

Wharfinger   (1938)

The Shipping Association was a new and unique institution, functioning as chief negotiator for wharfing interests in Jamaica. It functioned as an employer of labour on the waterfront to represent shipping agents and stevedores and later became a prime architect in the development and expansion of the ports.[31]

The days of subsistence wages and the sporadic hiring of casual labour, often at the whim of the wharfinger, were coming to an end. Violent demonstrations of the late 1930s had forced management to transform their employment practices and policies. New institutions were formed, albeit fledgling and scrambled.

But changes did not bring an immediate settlement to labour disputes. Unions were not satisfied with initial wage increases and bonuses put forward by the Shipping Association;[32] they did not constitute what the unions considered a living wage.

In spite of these early setbacks, fundamental changes occurred in the representation of workers' rights. Transparency and public disclosure helped the process of collective bargaining, and a forum now existed for deputations from both sides, employers and labourers, to meet at the table. The establishment of the Shipping Association of Jamaica also enabled the government to intervene as arbitrator if necessary.

Luis Fred Kennedy's interventions in 1938 laid the foundation for a lasting change in the structure of labour negotiations on the waterfront. He saw the need to bargain in good faith to defend the rights of both management and labour and he helped in the formation of an institution that would respect these. He believed in an economic system in which property and business were owned privately, but he often said that if employers without a social conscience were left unchecked, they would exploit labour for profit. He was wary of what he called, *"unbridled capitalism."*

After short stays at hotels in Kingston, the newlyweds moved in to stay with family at 25 South Camp Road. They felt crowded residing with Luis Fred's mother, Aunt Fillan, and siblings.

> Luis Fred had to wear a tie every evening for supper, there was too much etiquette. I was often told how to behave and was taught silly things like

---

31 "It has developed as a body that has brought and kept stability to labour relations in the port of Kingston. It continues to live up to the vision of its founders and has also emerged as a beacon of excellence for several industry and national associations, which have followed its development." *The Jamaica Gleaner,* "The Shipping Association of Jamaica is 80," January 29, 2019.

32 Formation of a union of ex-servicemen and conscription of a pool of dockworkers, who were non-members of the BITU, ensured a steady supply of labour for employers in the event of disputes. The arrangement stabilized conditions on the wharves but for obvious reasons did not please Bustamante. Letters discussing the disputes were published in *The Daily Gleaner* by Richard Hart, secretary of the Jamaica General Trades Union Council and Captain S. D. List, chairman of the association.

how to prepare tea the proper way, "Always pass the milk around after the tea is poured so people can help themselves." He didn't fully know what it was like for me. It was his Aunt Fillan, who called him aside one day, "Why don't you get your wife your own place?" That made him realize. I hadn't said anything, I guess she just observed things while she lived there with us. So, we moved out, and for the first while, we rented different houses, the first was on Upper Montrose Road. We were far more independent and could start a family of our own.[33]

In 1938, Luis Fred's mother and family moved from South Camp Road to a new, more spacious home at 8 Seaview Avenue.[34]

Luis Fred took time off work, touring the island to introduce Tillie to people and places. She was enchanted by it all. "Jamaica was beautiful. We even visited some of Grace, Kennedy's customers in the towns we travelled through. I was amused by how the shopkeepers greeted me, 'Glad to meet you, Mistress.' Not being familiar with the language, I thought perhaps they had met other 'mistresses' on previous occasions."[35]

One of Tillie's first introductions was to Dr. John Grace at his home in Strawberry Hill. She recalled their visit:

> Dr. Grace invited us up to Strawberry Hill where he lived, for tea. Strawberry Hill was a rambling wooden structure on a very small flat piece of land but with a beautiful garden. The tea was traditionally English, tiny cucumber sandwiches, scones, jelly, and the best tea I've ever had. Conversation was polite and proper. One disconcerting note was Dr. Grace's hearing aid—a huge horn that he pointed at you, and I felt very foolish shouting small talk into his contraption. He was essentially a very kind man despite his stern bearing and he loved Fred and took him under his wing. At this time Dr. Grace was partially retired, went to office two to three time a week and made "a nuisance of himself" he liked to say.[36]

Tillie tried her best to remain in contact with her mom. "It was hard, I missed her. I tried to stay in touch, but telephone calls were garbled, they weren't worth it. We would correspond by letter. I would write every week."[37] She would take any opportunity to travel to New York to see her. "We made two in 1938 (I would try to

---

33 Lydia M. Kennedy, interview with Georgianne Thompson Kennedy (2002).
34 See Chapter 19.
35 Lydia M. Kennedy (2002).
36 Ibid.
37 Ibid.

accompany Luis Fred on his business trips), one in February by ship to New York and the other in September by Pan Am Clipper to Miami, but when the War broke out, they wouldn't allow us to travel as regularly, you needed special permission."[38]

---

38 Ibid.

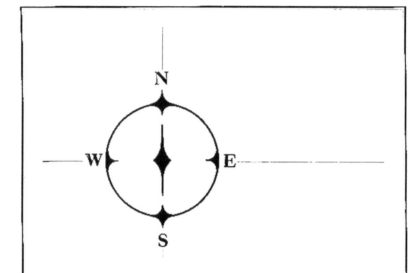

Planters' Punch 1938-1939, 94.
Digital Library of the Caribbean
https://ufdc.ufl.edu/AA00004645/00019/images
Courtesy of National Library of Jamaica)

18

WARTIME MERCHANT (1939)

On August 19, 1939, Tillie was on board Pan American Airways to Miami, en route to New York to receive medical attention for kidney ailments. While in hospital, she wrote to say that she had not been short of visitors, but the one whom she really wanted to see was not there. "Sweetheart, I miss you terribly. When I think of you so far away, I get an ache not caused by any double kidney business; I've got "*Quiéreme Mucho*" running through my head all the time today."[1]

With the advent of war, Luis Fred's letters expressed concerns about her safety:

> Business has been terrible in the past week. The war scare makes it impossible to do any business as people abroad are afraid to quote prices for fear they may go up, and locally they are afraid to buy less prices slump (August 30, 1939).[2]

> As you must know by now, Germany has attacked Poland, and it is now only a matter of time before England and France go to Poland's aid. No one has been able to do any work all day. We have been just sitting about talking about the whole disastrous business and speculating on the outcome. A waste of time, but I at any rate, just couldn't do

---

1 Lydia M. Kennedy, letter to Luis Fred Kennedy, August 25, 1939.
2 Luis F. Kennedy, letter to Lydia M. Kennedy. The pound sterling was slipping rapidly against the US. He started to buy up currency at the rate of $4.35 to pound sterling in anticipation of further declines. The day following his purchase, sterling closed at $3.90.

any work today. My nerves were all jittery especially because you are off the island.

> You know, my darling, I definitely think you should not come back by boat. Although you would be on an American ship, still many an American vessel was lost in the last war before America joined up. I think that you should fly back (September 1, 1939).[3]

Luis Fred also worried about his friend, Jules Wolff:

> Poor Jules is in a pretty tough spot. He has closed his office and is now only waiting for the government to walk in and take over the Co.'s money etc., which it will do if England fights. He personally is okay as he is a naturalized British subject, but he will be out of job, the poor guy! Today we are keeping each other company! —he without a job and I without a wife (temporarily of course, but still terribly).[4]

The Manchester Guardian, Monday, September 4, 1939, carried the alarming headline: WE ARE AT WAR/FIGHTING EVIL.[5] British Prime Minister Neville Chamberlain's broadcast from the Cabinet Room at 10 Downing Street reported that Germany had refused the ultimatum given to withdraw troops from Poland. The United Kingdom's declaration of war meant that its Crown Colonies, its protectorates, the Indian Empire, and the Dominions of Britain, Australia, Canada, South Africa, and New Zealand would join the Allied war effort.

Jamaica took swift action to protect its own borders and to incarcerate those who were considered enemy aliens. Twenty-two Germans[6] were rounded up and incarcerated at Up Park Camp. Headlines appeared in the *Daily Gleaner,* September 6, 1939: "ALIENS INTERNED/The Government today ordered all enemy aliens in Jamaica interned and their businesses turned over to the official receiver."[7]

---

3   Ibid., September 1, 1939.

4   Ibid. Jules Wolff's name appears often in Luis Fred's personal correspondence. Luis Fred had asked Jules to be his best man at his wedding in 1937, but because of illness, Jules was unable to travel to New York.

5   *The Manchester Guardian,* 10. Monday September 4, 1939.

6   Joanna Newman, *Nearly the New World: The British West Indies and the Flight from Nazism, 1933–1945* (2019), 201–202. Arnold Von der Porton gives a riveting account of daily life in prison at Up Park Camp in *The Nine Lives of Arnold* (2003).

7   *The Daily Gleaner,* Wednesday, September 6, 1939, 24. Besides holding German prisoners of war, countries of the Caribbean opened its borders to accommodate Jewish refugees from Europe. Second only to Cuba, which received between 8000 and 12000, Jamaica accommodated between 2,000 and 4800 Holocaust refugees. Excellent source: Suzanne Francis-Brown, "Jamaica: Fixed-Term Haven and Holding Tank during World War II". In Eccles (2017), LOC 5982 of 895

"I am afraid, my love, that I am going to have to insist on your returning by plane from Miami instead of boat. There are reports of German submarines in the Caribbean."[8] Luis Fred departed Jamaica on September 14 to fetch her. They returned together on Pan American Airways ten days later, Tillie fully recovered.[9]

Although the British colonies of the Caribbean were not directly engaged in combat, the German U-boat campaign[10] caused havoc to merchant shipping and threatened the United States' strategic interests, its supplies of oil, bauxite, and sugar, and affected its global trade routes through the Panama Canal.[11]

Restriction on the movement of ships[12] had dire effects on merchant firms like Grace, Kennedy. Import licenses needed to be obtained, and companies "were allotted quotas based on their share of the market over the eighteen months prior to the outbreak of war."[13] Along with a shortage of basic foods, cornmeal, flour, rice, milk, oil and pickled meats, the average citizen experienced an increase in the cost of living.[14] The poorest became the most vulnerable.[15]

Jamaica and other countries of the region were placed under the Defence of the Realm Act, which was aimed at regulating the prices of all commodities to prevent profiteering. The governor, Sir Arthur Frederick Richards,[16] also imposed censorship

---

8   Luis Fred Kennedy, September 5, 1939.

9   She did not need an operation. A team of doctors led by urologist Dr. Louis R. Kaufman determined that there was no kidney damage, and that the pain she was experiencing could be alleviated through drugs, exercise, and diet. Doctors discovered in later years that she had a duplex kidney.

10 Lovell Francis, "A Dance with Death: Labour Problems and the Sugar Crisis of World War II in Trinidad." In Eccles (2017), LOC 444 of 8958.

11 German U-boats sank 114 ships in the Caribbean in the first half of 1942, accounting for the loss of over 350,000 tons of cargo.

12 The government cancelled its subsidy to the British steamer the *North Star*, which was commissioned to bring tourists to Jamaica during the winter seasons. The Tourist Trade Development Board also stopped its advertisements in the United States.

13 Douglas Hall (1992), 31. Grace, Kennedy's quota of counter flour was approximately 25% of the market, and for sailfish, it was given a 7% allocation. Because of its prior involvement in the salt trade, it was able to supply the country with some of its needs from its Pigeon Island solar salt operations.

14 "By August 1942 the cost-of-living index, based on prices in 1939 equated to 100, had risen to 159." Richard Hart (1999), 183. Hart provides a detailed list of increases in prices of clothing, basic foods, miscellaneous items, imported and domestic, 181-183.

15 Ibid., 78.

16 Luis Fred had a particular distaste for Governor Richards. Patrick Bryan (2000) quotes the governor as saying, "Jamaicans are constitutionally unteachable, the home of poisonous misinterpretations, ignorant prejudice and untaught conceit." Jamaicans had "an incredible ignorance and capacity for being misled." He was threatened in particular by the trade union movement and by what he considered the radical elements within the PNP. He was governor of Jamaica from August 19, 1938 to July 1943.

of the press, mail, and telegraph services.[17] Orders went out as early as the second day of September 1939, "ordering that the wholesale and retail prices of foodstuffs and certain other commodities shall not be increased beyond those prevailing on August 31, 1939—F.E.V. Smith, Competent Authority and Food Controller." [18]

Exporters were also hit hard. The Competent Authority required licenses for the exportation of foodstuffs and other locally produced goods. Centralized control of commerce was evident in almost every sector of the economy:

> NOTICE IS HERBY GIVEN that we have by Orders under our hands prohibited the exportation of any article except bananas and the personal effects of *bona fide* passengers on any ship, vessel, or aircraft, unless a licence has been applied for and granted by a Competent Authority.[19]

Despite special allowances, banana exports dropped. Many of "the ships belonging to the Jamaica Banana Producers Steamship Company were commandeered by the British Admiralty to carry food (other than bananas) from North America to Britain."[20] European and US ports were inaccessible, and the fruit became a scarce luxury item; Britain, US and other nations made the transport of essential goods top priority—staple foods and war supplies.[21] Although a banana subsidy brought relief to the industry towards the end of 1940, the industry also suffered from the onset of the Panama leaf-spot disease.

As early as December 1939, the governor issued authorization with orders by the Competent Authority to establish a Coconut Industry Clearing House through which "all soaps, coconut oil, compounds, lard, margarine and other coconut products" would have to be sold.[22] Factories were licensed by the government, and no coconut products were to be exported from the island except through the explicit permission of the Clearing House. They would be sold through the usual agents but at fixed prices set by the Competent Authority.

The centralizing of governance drew the ire of Luis Fred Kennedy, who minced no words in his public criticism of the governor:

---

17   Rebecca Tortello (2007), 220.

18   *The Daily Gleaner,* Saturday, September 23, 1939, 19. In some instances, wholesale prices, for example of flour, had to be reversed. The cost per bag had risen from 22/- on August 31 to 27/- on September 2, 1939. (*The Daily Gleaner,* Wednesday, September 6, 1939, 8)

19   Ibid.

20   Richard Hart, *Towards Decolonisation* (1999), 178. All but one of these ships were sunk by German U-boats.

21   Annual export of stems of bananas exported from Jamaica declined from 18,772,000 in 1939 to 289,000 in 1943. Quoted by Hall (1992), page 34 from D. W. Rodriguez (1955), *Bananas: An Outline of the Economic History of Production and Trader with Special Reference to Jamaica.*

22   *The Daily Gleaner.* December 30, 1939, 19.

Wartime Merchant (1939)

THE EDITOR: Sir, your recent editorial "Coconut Deal" and the Coconut Marketing Board's statement published in your issue of the third inst. miss completely or deliberately avoid the main point at issue - the complete control over the coconut industry vested in the Governor personally.

The establishment of a Coconut Control Authority to direct the industry is optional to the Governor alone. If he does establish it, he alone has the power, in his absolute discretion, to define the authority's powers and duties…

I cannot think of a better description of the condition created by the present statutes than Nationalization, or, if the Marketing Board prefers, the complete control by the State of the private enterprises of the growing, processing and distribution of coconuts and their by-products.[23]

Grace, Kennedy had a vested interest in the coconut industry, as it was the sole distributor under the auspices of the Jamaica Coconut Producers' Association Ltd. for Guinea Gold Superior Laundry Soap.

Britain, already stretched to its limits in waging war against the Germans in Europe, was unable to provide the resources necessary to protect its colonies in the Caribbean. The German forces proved formidable. The Hitler war machine had knocked out Poland in twenty-seven days, Norway in twenty-three days, Holland in four days, Belgium in eighteen days, and France in seventeen days. The Blitz, the German bombing campaign against the United Kingdom, began in August 1940. The UK was helpless in racking up adequate defences elsewhere in the world.

In 1940, Britain and the United States negotiated the Anglo-American Destroyers for Bases Agreement, by which land was leased to the United States in the region for ninety-nine years, rent free, to establish naval and air bases in exchange for fifty old destroyers. The agreement authorized the United States to construct military and naval bases in the British colonies of Antigua, the Bahamas, Bermuda, British Guyana, Jamaica, Newfoundland, St. Lucia, and Trinidad. The sovereign nations of Cuba, Haiti and Dominican Republic, the

---

23   Ibid., Monday, December 20, 1939, 4. The Jamaica Coconut Producers Association was formed in the 1930s when the coconut industry was operated by independent growers. Legislation was passed to protect the local industry from open, external competition. Government came under increasing criticism with a law passed in 1937, which enacted compulsory quotas, giving the Jamaica Coconut Producers Association the lion's share of 46% of the total output of manufactured oil. During the war years, government control of the industry became even more stringent.

US territory of Puerto Rico, and the British colonies (Jamaica and island nations of the Eastern Caribbean) played a role in defence of the Caribbean basin and participated in a convoy system for the safe passage of merchant vessels between the islands and the mainland. US presence in the Caribbean kept links open in the supply chain.

Tillie recalled:

> The War years were lean ones indeed. More and more ships were being sunk and imports came almost to a standstill. The RAF established a base on the Palisadoes road and quite a few German submarines were sunk off the Jamaican coast. This base was enclosed by barb wire and in order to get to the airport you had to have a priority pass which was almost impossible to obtain. The US had its base at Vernam Field. Everything was controlled—food, gas, foreign exchange, travel. The story goes that an officious young British lieutenant went around to schooner owners telling them that they must paint their hulls a sea-green colour as camouflage. One owner pointed out, "Yes Sar, but what about dat white sail up there?" The order was soon rescinded. Another story going around was that Spanish Town was safe because if the Germans flew over it, they would leave it alone because they would think that it had already been bombed.[24]

With sea routes opening, Luis Fred seized the opportunity to expand Grace, Kennedy's shipping. He travelled to New York, leaving Kingston for Miami on board Pan American World Airways on November 22, 1939. He met with executives of Ocean Dominion Steamship Corporation. His mission was to secure Grace, Kennedy as Dominion's agent, to ensure continued and expanded transport of freight from the US to Jamaica and the Caribbean. Dominion was withdrawing its Aluminium Line passenger and freight ships, which plied from New Orleans, Mobile, and Tampa to Kingston, Aruba, Curacao, Barbados, British and Dutch Guianas, and Trinidad,[25] and it was consolidating with two other companies to form the Alcoa Steamship Company.

A year later, on December 12, 1940, the 5,500-ton American steamer, *Alcoa Shipper*, arrived in Kingston Harbour. This was the inauguration of its fortnightly service from Mobile and New Orleans to Jamaica and the rest of the Caribbean. The local agents were Messrs. Grace, Kennedy & Co. Ltd. The vessel, much larger than the previously used

---

24 Lydia M. Kennedy, interview with Georgianne Thompson Kennedy, 2002.
25 *The Daily Gleaner.* November 27, 1940, 16.

Vangen Aluminium passenger and freight ships, brought well needed supplies of general cargo and lumber from the Gulf States.[26]

To de-regulate the markets, Luis Fred fought the government on import controls of purchase of goods from the United Kingdom, which he claimed was anxious to expand its export market. He called for local officials to negotiate a larger quota from the UK (as Trinidad was able to do) and accused the Competent Authority in Jamaica of restricting trade.

British Trade Commissioner Mr. Gick denied the accusations. The trade commissioner remarked that merchants' statements in the press were both "scurrilous and inaccurate."[27] His comment caused a sharp reprisal from Luis Fred Kennedy, one of the Jamaica Chamber of Commerce's younger members, who called for the resignation of the trade commissioner:

> Mr. Kennedy: If the Trade Commissioner is sent here to deal with the merchants and if the merchants disapprove of him, I don't see why we don't have the right to demand his recall …
>
> I would ask him to leave my office, and I suppose other merchants could not deal with a man if they held him in that regard. I feel the first step we should take is to ask Mr. Gick if he did make that statement. If he did not make that statement, he should take the necessary steps to deny having made it. If he says he did make the statement, I don't think even an apology would satisfy me. I should say the Imperial government should be requested to remove the man. He cannot absolutely be again allowed to remain here if he made that statement.[28]

Mr. Kennedy concluded that before drafting a resolution for the commissioner's immediate resignation, the chamber should give Mr. Gick an opportunity to clarify his statement. The chamber complimented Mr. Kennedy for his outspokenness, leadership, and "moral courage" shown before council.[29]

---

26   Ibid., December 13, 1940, 3.

27   Ibid., Saturday, January 25, 1941, 6.

28   Ibid., Monday January 27, 1941.

29   Mr. Kennedy became increasingly frustrated with the chamber. Statements were published, *The Daily Gleaner* January 30, 1941, page 7, of rumours of "a threat of the radical element in the Chamber of Commerce to oust all the older merchants from office." The movement had been launched by Messrs. Kennedy and Breakspeare. Outgoing president, Mr. Kieffer, met with the gentlemen prior to the chamber's annual general meeting to explain that even though the chamber would not be in a position to accept the terms of their memorandum, that their membership and contributions were highly valued. The memorandum called for changes in the structure and governance of the council, giving it more independence from the president and body of general membership.

One day after the release of the chamber's deliberations, the commissioner requested an audience of a select committee, of which Luis Fred Kennedy was a member. *The Daily Gleaner* carried headlines the following day:

```
Mr. Gick, On Reflection,
Finds "Scurrilous" Was
"Rather Ill-Chosen" Word
Epithet That Angered City Merchants
Retracted Before Council Of
Chamber of Commerce. 30
```

Mr. Gick retracted his statement and hastened to assure the merchants that he had not intended to slander their good name.[31]

Luis Fred was appointed an advisor to the governor on matters of price controls and was assigned on a mission to Santiago, Cuba, where he remained several weeks in 1943 to arrange for transport of goods in transit from the US to Jamaica. His fluency in Spanish gave him a distinct advantage for conducting business in Cuba. Tillie commented:

> When things became worse and worse, the government sent Fred on a mission to Cuba to try to get food for us. He was able to arrange, after much delay and talk, to have essential supplies sent to Cuba from the United States, railed across the island to Santiago which is only ninety miles away from Jamaica and sent here by schooner, and that way we did get some food at least. Grace, Kennedy acquired a luxury yacht named *Dauntless*, which had been previously owned by an American millionaire. It had been fitted with chandeliers, golden doorknobs and marble counters. It was stripped and converted into a schooner to carry cargo and participated in the movement of basic food items from Cuba.[32]

By 1943 economic conditions in Jamaica started to improve. After entering the war with the bombing of Pearl Harbour (December 1941), the United States was more aggressive in protecting the convoy system for the safe passage of merchant shipping in the Caribbean.[33]

---

30   Ibid., Wednesday, January 29, 1941, 5.
31   Ibid.
32   Lydia M. Kennedy, "A Giant of a Man" (1997).
33   *Gleaner* headlines of June 21, 1943, announced, FIERCE U-BOAT OFFENSIVE SMASHED IN ATLANTIC/ Convoys Safe after 5 Day-Night Battle.

The Jamaican government still had in place a stringent economic policy to curtail both exports and imports, but complementary to this were price subsidies amounting to £200,000 and a vigorous campaign to encourage local agriculture to lessen the country's dependence on imports.

At the opening of the 1943 spring session of the Legislative Council, Governor Richards provided a comprehensive review of "Conditions Obtaining in the Island."[34] He reported an opening surplus of £321,297 on April 1, 1943. He emphasized a general shift in economic policy that meant future prosperity for the colony.

> Greater importance is now being placed on economy in the use of shipping which has become a vital matter in consequence of the developments of the war. The direction of trade points to Canada and the Americas with which trade has somewhat improved. Imports in general are restricted to commodities which are essential to the life of the community.[35]

Grace, Kennedy and other shipping merchants welcomed the news.

However, recovery was slow, and by no means immediate. Between 1937 and 1943, the number of steamships entering the Port of Kingston decreased from 1,284 to 267.[36] Trade quickly picked up when the war ended, and by the 1950s would resume to pre-war levels. In anticipation of an upswing in the volume of shipping, Luis Fred led a consortium of shippers to form Kingston Wharves Ltd. in 1945.

During the war years, Luis Fred Kennedy steered Grace, Kennedy through turbulent times—rebelling against excessive controls, negotiating fair quotas, and keeping supply chains open between North America and Jamaica. Aaron Matalon described him as a "giant of a man."[37] His stature as a Kingston merchant was indisputable.

---

34 Text of governor's speech printed in *The Daily Gleaner,* Thursday, March 11, 1943, 1.

35 *The Daily Gleaner,* Thursday, March 11, 1943, 8.

36 https://www.gracekennedy.com/where-we-have-been/history/the-world-war-years-and-immediately-after-1939-to-1950/.

37 At the book launch of Douglas Hall's *A Story of Jamaican Enterprise,* February 1993, Aaron Matalon told the story of a time when he was secretary of the Dry Goods Liaison Committee of the Trade Control Board during the 1940s. He defended the position of Grace, Kennedy, which had taken the risk to future large shipments of pickled mackerel—to profit from a devaluation of sterling. After the meeting, Luis Fred turned to Aaron Matalon and said, "That was excellent, Matalon. You won the day." Matalon reflected on that moment to say, "It is not easy for me to describe what this compliment from this big man, a leader in the food trade, meant to the youngster I was at the time. It gave me a major lift in self-esteem, that this big man would take the trouble to offer this encouragement or compliment, was to me quite unusual, and I have never forgotten it. I remember him as a giant in my time, and a leader of commerce in Jamaica." (Aaron Matalon was twelve years younger than Luis Fred Kennedy.) See Aaron Matalon's biography: http://jamaica-gleaner.mobi/20090406/news/news3.php for biographical information.

"I would meet Fred for lunch sometimes during the week at the Myrtle Bank Hotel" (LMK).
(Photo in public domain, picryl.com)

L—R: Three eldest grandchildren of Luisa Matilde and Fred W. Kennedy with their mothers: Paco with Tillie Kennedy, Mary Lou with Louisa Soutar, and Bill with Carmen Valliant. (Photo, 1941, courtesy of Lisa Valliant)

Luisa Kennedy's home, *Quisqueya,* 8 Seaview Avenue, 1940s.
(Photo from family collection)

Hope, a twin-engine motor launch, which Luis Fred and Tillie co-owned with the Judahs during the 1940s. George Munro, cashier at Grace, Kennedy, piloted the boat. Tillie seen here standing in the bow. (Photos from family collection)

Francis (Paco) Xavier) Kennedy, seen here
with Aunt Mim (L) and Mom (R)
in background.
Paco was born September 10, 1940.
(Photo from family collection)

Celia Jean seen here riding Nellie at Amberly.
Celia was born September 19, 1942.
(Photo, courtesy of Mary)

Mom's first car, 1938 Standard 8HP Saloon.
(Photo from family collection)
Vehicle identified by Surrey Vintage Vehicle
Society, kindness of Bozi Mohacek,
Society Chairman.
.http://www.svvs.org/help167.shtml

Mary Teresa, Palisadoes.
Mary was born October 20, 1945.
(Photo, courtesy of Mary)

19

FAMILY MAN

(1940s)

At first, Dad had a difficult time letting go of his bachelor habits. When he was first married, he hosted poker games at home that involved loud chatter and heavy drinking into the wee hours of the morning. His new bride was not pleased. Her sisters-in-law would often rescue her by inviting her out, but other times she would be confined to the house. Before long, she delivered an ultimatum: "Either you stop having your bachelor parties, or I return to New York." The shenanigans came to an abrupt halt.

His self-image improved as a married man. He grew more confident of his looks and agreed to start having his picture taken. (Mom always thought him handsome.) When Mom was away on trips to New York, she would ask him to mail her pictures of himself, which at first he did, but reluctantly. She wrote in August 1939:

> I haven't even thanked you for your wonderful letter with the snapshots and the check. The snaps are lovely, darling. You look so well, and so many pictures of you! And I have such a hard time getting you to pose for them ...
>
> I had forgotten to tell you that Tía Llolla[1] found you very handsome. Thinks your looks have greatly improved since last she saw you. Mother said, "*Se le pegó de Tillie.*" Ahem![2]

---

1 Her mother's youngest sister, Gloria Mercedes Arzeno Abreu (1889—1981).

2 Lydia M. Kennedy, letter to Luis Fred Kennedy, August 1939. *Se le pegó*, a Spanish idiom, is used in reference to Luis Fred. When Tillie's mother, Carmen Celia, learned of positive changes in his behaviour and appearance, she jokingly said that Tillie's good influence must have "rubbed off" on him.

Married life agreed with him! And he was the first to admit it. "Take good care of yourself, my darling. I can't get anyone else like you, and your poor hubby just can't get along without you."[3]

Not for want of trying, my parents were unable to conceive a child in their first two years of marriage. Mom thought she might be too old to start having children or that her medical conditions were preventing her from becoming pregnant. As devout Catholics, they resorted to prayer and hoped for a miracle. They recited a Novena of Grace to the Jesuit saint, Francis Xavier, for their wish to be granted. The local parish priest supplied them with a devotional medal, which he suggested she pin to her nightie. Miraculously, their prayers were answered. On September 10, 1940, a beautiful baby boy was born. They named him Francis Xavier Kennedy but called him Paco, the Spanish nickname for Francis, to distinguish him from his namesake, his uncle Francis Xavier. After having two more children, Celia Jean, born September 19, 1942, and Mary Teresa, October 20, 1945, Dad recommended cancelling the novenas and returning the medal to the priest.[4]

Dad's three sisters also started to have children. Louise (aged twenty-five) was married to Simon Soutar[5] in September 1939, Carmen (aged thirty) to John William Valliant[6] in July 1940, and Gloria at the age of twenty-one to Owen Cathcart Plant[7] in June 1941. "Carmen, Louise and I were pregnant all at the same time,[8] and right through the 1940s and 50s, it seemed there wasn't a year that at least one or more of the Kennedy families weren't having children" (LMK).[9]

His brother José was financially independent as a practicing solicitor. The youngest, Francis Xavier, graduated from St. George's College in 1942, after which he followed in his big brother's footsteps to attend Holy Cross College in Worcester, Massachusetts. He and José would both get married in 1947.

---

3   Luis Fred Kennedy, letter to Lydia Kennedy, August 1939.

4   Lydia M. Kennedy, interview with Georgianne Thompson Kennedy (2002).

5   Simon Soutar (1912–1991), son of William Donald Soutar and Hilda Blanche Vendryes. Simon joined Grace, Kennedy, where he worked as a manager and director of the company. Fred William Kennedy had known William Donald Soutar and his father, Simon D. Soutar (1837-1923), from whom Grace Ltd. acquired its wharf in 1920. (See Chapter 11.)

6   John William Valliant (1911–1988), Canadian born, Fort William, Ontario, son of William Valliant and Bella May Condie. He worked as manager of the Bank of Nova Scotia, Kingston, Jamaica.

7   Owen Cathcart Plant (1907–1989), son of William Henry Plant and Ethel Grace Cathcart. Fred William Kennedy had known Major William Henry Plant, headmaster of Titchfield High School, Port Antonio (1885–1927). See Chapter 5.

8   Luisa M. Kennedy's first three grandchildren were Francis Xavier Kennedy, born September 10, 1940; Mary Louise Soutar, October 28, 1940; and Fred William Valliant, April 28, 1941.

9   Lydia M. Kennedy (2002).

Family Man   (1940s)

Their mother remained the matriarch of the Kennedy family, but her responsibilities had lessened and the financial burden of supporting her children eased considerably. Ten years after the passing of her husband, the estate was finally settled and her former home at 25 South Camp Road was advertised for sale in *The Daily Gleaner,* May 1940.[10] Her major source of income was dividends from shares held by the estate and transferred to her name.[11]

Dad would stop by at his mother's home at 8 Seaview Avenue every afternoon after work. If he left downtown early, he would collect Mom and the children to accompany him. The house was on a three-acre property with rose gardens, fruit trees, expansive lawns, and a tennis court. The home, designed by a young Spanish architect, Vicens Oliver, was grand in every way: a porte-cochere with steep, imposing arches graced the entrance; bedrooms and hallway upstairs opened out to wrought iron balconies; thick, white exterior walls accented a terracotta-tiled roof; and the interior had floors of exquisite marble tiles imported from Italy.[12] The home was a specimen of Mediterranean revival architecture that earned Luisa Matilde Arzeno the reputation among her Dominican relatives as *la duquesa de la familia* (the duchess of the family).

Dad was enchanted by the Blue Mountains—an affection that became part of his family life for years to come. In the 1940s, he and Mom purchased a property in Irish Town, a small cottage called Amberly, in co-ownership with Douglas and Mary Judah, their lifelong friends.[13]

My sister Mary recalls:

> I remember Amberly, but only snippets. The pool was round, and in the shallow end had a square platform and then steps going down. I remember loving to jump off that square step into the water. Also, there were lots of big trees, the yard was shady and cool. The grass was that cool, flat, very green, broad bladed grass (crab grass?) and I loved

---

10   *The Daily Gleaner,* Friday, May 10, 1940, 19. A parcel of land known as *Belle Isle,* including # 25 (property of Fred William Kennedy), # 27 and # 29 South Camp Road (property of S. C. Lindo) was advertised for sale.

11   In 1937, income from shares amounted to approximately $3,000 per annum.

12   Batholomew Vicens Oliver was born in Mallorca and migrated to Jamaica in 1936. He married Millicent Mary Desnoes and had one son, Michael. He designed Tower Isle Hotel, Red Stripe Brewery, Sts. Peter and Paul Church, and Caymanas Park Racetrack. He died at age fifty-one in 1962. *Jamaican Family Search Genealogy.*

13   Douglas Judah, solicitor and company director, partner, Judah and Randall Solicitors (founded 1919); Member of the Legislative Council (1942–1958), chairman of numerous companies in Jamaica. Born, Kingston, July 4, 1906, son of Braham Taylor Judah and Mary Mennell. Married Mary Horn 1935. *Who's Who Jamaica 1963.* They had three sons, Bruce, Peter, and David, whom I contacted, for purposes of this biography, after a lapse of fifty years.

> the feel of it under my bare feet. Paco and Celia would run off and play without me as I couldn't keep up with them. And yes, there was Nellie, the donkey, and I do remember riding her but was scared![14]

The mountain retreat was a playground for Paco, Celia, Mary, and their cousins. Mary Lou, Aunt Louise's eldest daughter, who spent summers at the hillside cottage with the Kennedys, has vivid memories, seventy-five years later, of roaming the hills and taking hikes in the bush.[15] It was a place too for the adults to socialize and, in true Jamaican style, engage in verandah chat about every imaginable topic, social, political, and economic.

Dad was equally (perhaps even more) enamoured with the sea. The Kennedys and Judahs co-owned a boat called *Hope*, a twin-engine outboard motor launch they purchased from Dr. Mosley.[16] George Munro, who was cashier at Grace, Kennedy in late 1940s, piloted the boat for the two families and their friends.[17] They would venture out on weekends from Kingston Harbour on fishing expeditions to the outer banks.

During the war, Dad abandoned his car and resorted to cycling because of gas rations. He bicycled daily from home to office at 64 Harbour Street, and while Mom was at St. Joseph's Hospital (Deanery Road in Kingston) for the births of the children, he would take a ride down Mountain View Avenue. But he would always try to find a ride home. He would say, *"If you don't think Kingston is on a steep slope, try riding a bicycle up Mountain View Avenue from St. Joseph's Hospital or up South Camp Road from Breezy Castle."*

Mom had difficulties adjusting to life in Jamaica in the 1940s.

> Social life in Jamaica was very different then from now, very formal and full of etiquette and protocol. Dress was formal for most occasions and cards were dropped at homes, these counting as a visit, as well as

---

14   Mary Cameron, personal correspondence (2020).

15   See Mary Lou Soutar's tribute in Appendices.

16   Dr. John Grimson Mosley and his wife, Rachel Hoag Dempsey, lived at 45 Lady Musgrave Road during the 1940s. He was the son of Alicia Cecilia Sheridan and Charles Albert Mosley, who was district medical officer for Port Antonio for over forty years. John Mosley was born in Port Antonio, October 25, 1886 and died January 11, 1949. His obituary is printed in *The Daily Gleaner,* January 13, 1949, 3.

17   George Munro is a family friend of the Moss-Solomons. He started as a bookkeeper at Grace, Kennedy in the 1940s, and was soon to be promoted to head cashier. After working at the company for ten years, he apprenticed as a sea pilot to work for the United Fruit Company. He began piloting banana boats on the north coast during the 1960s. In January 2021, I was able to contact George Munro through Joan Belcher, James S. Moss-Solomon's daughter. He lives in the coastal Parish of Portland where he enjoys a beautiful life at the age of 93.

at Kings House if one wanted to be invited to any function. Women wore hat and gloves to go anywhere in the daytime and men wore full suits to the office.[18]

In sharp contrast to the modernism of North American society, Jamaica was the bastion of British colonialism, not far removed from the imperial rule of Queen Victoria. Mom knew that Dad was not a keen admirer either of the hypocrisies inherent in the way of life. "Fred hated all the pomp and ceremony of those days and refused to have cocktail parties at home that were the vogue."[19] He had echoed these sentiments in letters written during their courtship: "I belong to no 'society' as you put it. I suppose that we did move pretty high up in it when Father was alive, but after his death, Mamá stopped going out, and I, who never liked that, took advantage of the situation."[20]

At Dad's urging, she took her driver's test, and in September 1938, he bought her a two-door Standard 8HP Saloon, the smallest of the line launched by the Standard Motor Co. Ltd. Trips to Cross Roads included visiting the butcher's shop and the local open market. "I remember going to a Chinese wholesaler and seeing them work an abacus quicker than any calculator. He would give credit, no questions asked. A man's word was his bond, and many a deal was sealed with a handshake. Home-life was simple then."[21]

She stored fresh meat, butter, cheese, and other perishables by packing them in an icebox at home. Even though Kelvinators had been introduced to the local market in the late 1930s, electric refrigerators did not become a common household appliance until much later.[22] (As a child in the 1950s, I recall the iceman making weekly deliveries of twenty-five and fifty-pound blocks of ice that were covered in sawdust and sprinkled with water to prevent them from melting in the scorching heat.)

Hired help took care of gardening, cooking, cleaning, and laundry, and nannies provided infant care. Mom wanted to try out cooking, to experiment with recipes from *Good Housekeeping's Book of Menus, Recipes and Household Discoveries,* which her mother had given her as a wedding present, but she soon discovered being a housewife in the kitchen was *infra dig.* Family and friends informed her that this was

---

18   Lydia M. Kennedy, "A Giant of a Man," (1997).
19   Ibid.
20   Luis Fred Kennedy (1937).
21   Lydia M. Kennedy (1997).
22   Kelvinators were marketed and serviced by The Jamaica Public Service Ltd. in the 1930s. Initially they were not reliable, with owners complaining of seal leaks of sulphur dioxide. My parents did not purchase an electric refrigerator for the home until the early 1950s. Old-timers in Jamaica often use the word "icebox" to refer to refrigerators.

not her role. Besides, the kitchen was not part of the main house but in a separate building with a wood stove, chimney, and charcoal-stained walls.

Jamaican English proved difficult:

> I couldn't understand the helpers, they couldn't understand me. I looked up all these recipes, mother gave me … the cook couldn't understand my accent … all these fancy things I wanted to cook … we got beef balls all week long. She always said to me, "Yes, ma'am, yes ma'am." I finally told Fred that he would have to translate for me, so every morning he'd talk to her and explain what we wanted. My ear just wasn't attuned to it. I have more exposure now. I was a businessperson; it was a culture shock. I wrote to mother in Spanish every week.[23]

She joined the Catholic's Women's League and hosted canasta parties with the church ladies in the afternoons. She took up sewing and found pleasure in growing a garden. Sporting activities included playing tennis at St. Andrew's Club and socializing with young sailing aficionados at the Royal Jamaica Yacht Club. The Soutars and the Plants often won regattas in the Kingston Harbour.

"I would meet Fred for lunch sometimes during the week at the Myrtle Bank Hotel, and on weekends, we would attend live productions at the Ward Theatre[24] and movies at the brand-new Carib Theatre, opened in Kingston the year after I arrived in Jamaica."[25] Sound motion pictures were a novelty. The first feature-length movies carrying synchronized sound appeared in theatres at the end of the 1920s. By the 1930s, "talkies" marked The Golden Age of Hollywood.[26]

Mom and Dad indulged also in livelier times at the Glass Bucket Nightclub in Halfway Tree.[27] On Saturday nights, they partied into the wee hours of the morning, stopping for Sunday morning Mass at church on their way home.

If Mom did not accompany Dad on business trips, she would stay with his mother and Aunt Fillan at 8 Seaview Avenue. According to Dad, his aunt had

---

23 Lydia M. Kennedy (2002).

24 In 1944, Dad's sister, Gloria Kennedy Plant, took part in the island's Drama Festival with an acting role in *The Barretts of Wimpole Street*. She was among an "outstanding line-up of actors and actresses, that reads like a WHO'S WHO of the local theatre." *The Daily Gleaner,* April 12, 1944, 5.

25 Lydia M. Kennedy (2002).

26 Famous actors mentioned in their personal correspondence included Fred Astaire and Ginger Rogers, Spencer Tracy, Cary Grant, and others.

27 The Glass Bucket Club, owned by Bob Webster, opened in Half Way Tree, Kingston on December 22, 1934. On opening night, 700 patrons packed the club to see The Rhythm Raiders, a new dance orchestra under the direction of Dan Williams. http://old.skabook.com/foundationska/tag/glass-bucket-club/.

Family Man   (1940s)

grown old and lonely, with seemingly little purpose of her own. She died on April 6, 1944, at the age of sixty-seven, the cause of death reported by the US Foreign Service to be breast cancer.[28]

By mid-1944, Mom was again not feeling well. Dr. John Grace assisted by referring her to a Dr. Chapman whom he knew in New York City. On Paco's fourth birthday, September 10, 1944, she wrote:

> Thought of our little Paqui all day. I hope Mary Lou was able to come and you had a little party for him with all those things he loves to eat. Poor sweetheart—did you give him a big kiss for me? What did you finally get him for a present?
>
> And Celia—I hope she hasn't the fever again, my darling little girl.
>
> Gosh, I'm getting blue, and if I get started on the subject of you, Pops, I'll be a hopeless mess. Do write me as often as you can, if only a few lines, as I'll get a thrill from even seeing the envelope addressed in your handwriting.[29]

She had a minor operation, but the doctors agreed to suspend more treatment. She wrote in September 1944: "He also finds an enlargement of the uterus but says it's useless doing anything about it till we are sure there are not to be any more children ... And who can be sure we will have no more children?"[30] Mary Teresa was born the following year, October 20.

She sent a cable for their seventh anniversary on October 2, followed by a lengthy love letter replete with terms of endearment: "My darling, seven years of the most wonderful happiness any woman has a right to in this world, and I won't be there to tell you how much I love you."

Mom returned to Jamaica on November 10. She had wanted an earlier return but found it difficult to get train reservations from New York to Miami, as Americans were returning to their home states for a presidential election. Democratic President Franklin D. Roosevelt won an unprecedented fourth term.

Jamaica also held elections that month. The first General Parliamentary Elections under universal adult suffrage took place on November 20, 1944. And Luis Fred Kennedy played a role in that significant political movement in Jamaican history.

---

28   *Report of the Death of an American Citizen, American Foreign Service,* Form 192. Fillan Mabel Josephine Kennedy. She was reported to have been in the care of her sister-in-law, Mrs. Carmen Valliant at #8 Seaview Avenue. Her remains are interred in Providence Church Graveyard, Matilda's Corner, Hope Road, Parish of St. Andrew, Jamaica.

29   Lydia M. Kennedy, letter to Luis Fred Kennedy, September 10, 1944.

30   Ibid.

YOUR PARTY

EVERY country gets the Government it deserves. With the New Constitution YOU will be the Government.

The New Constitution, just by itself, cannot make a better Jamaica. No people, or country, can be handed a better life. If you want a better Jamaica you have to work for it. The aim of the Jamaica Democratic Party is a better Jamaica . . . a Jamaica better for Labour, better for Agriculture, better for EVERYBODY; for the Jamaica Democratic Party is EVERYBODY'S PARTY.

EVERYBODY has a say in the Jamaica Democratic Party. It is run on Democratic principles, and the policy of the Party is dictated not by one person, not by one group of persons, but by the will of the majority of the members.

The Jamaica Democratic Party has pledged itself to work for the greatest good for the greatest number. It is the party of the people, for the people!

The Jamaica Democratic Party Is EVERYBODY'S Party

..This is the eighth of a series to be published in the interest of Jamaica by the JAMAICA DEMOCRATIC PARTY. Membership to the J.D.P. is open to all who agree with democratic principles, and who want a BETTER JAMAICA. For further information, write to The Jamaica Democratic Party, 41 Port Royal St. Kgn.

The Daily Gleaner, November 27, 1943, 8.
(Courtesy of © The Gleaner Company (Media) Ltd.)

20

ADULT SUFFRAGE

(1944)

Party affiliation was seldom talked about in the home, only politics in general. Dad defended political positions based on what he considered was good for business and for Jamaica. He leaned on Mom for her sage advice: *"Make sure you don't show favouritism to one party over the other; if you support one and not the other, it may be seen as bribery."* In the 1940s, he liked neither the PNP nor the JLP.[1] So he joined a political movement that best represented the interests of private business—the Jamaica Democratic Party.

The People's National Party

Although Luis Fred identified with Norman Manley's intellect and respected his nationalism, he was wary of the socialist bent of the People's National Party. Manley was criticized for inviting Sr Richard Stafford Cripps,[2] a British Labour politician, to the inauguration of the People's National Party in 1938. In the minds of many, this invitation branded him a "revolutionary and communist."[3] He defended his actions by claiming that his prime objective in forming a political party was "to secure self-government and nationhood."[4]

---

1   Although Bustamante first indicated plans to form a political party in July 1942, the Jamaica Labour Party was not officially launched until July 8, 1943, after the formation of the Jamaica Democratic Party.

2   Sir Richard Stafford Cripps (1889–1952) became a leading spokesman for the left-wing and was said to be a supporter of communism and Fabian socialism.

3   Rex Nettleford (ed.), *Manley and the New Jamaica*, 12.

4   Ibid., 92.

Norman Manley was a private man who did not mix his personal life with politics or business. Originally, he had no interest in politics, but the 1938 riots gave him a change of heart; a movement for constitutional change was necessary to bring about the economic and social changes Jamaica needed.[5] He saw himself as the person to make this happen.

The People's National Party was launched at the Ward Theatre on September 18, 1938. Mr. Manley's oratory captivated the crowd. He reiterated his position as a friend of the labour movement and no enemy of Alexander Bustamante. He spoke of a new Jamaica, one whose only path to prosperity was self-government:

> ONE STRAIGHT CHOICE
>
> As I see it today there is one straight choice before Jamaica. Either make up our minds to go back to Crown Colony government and have nothing to do with our own government at all, either be shepherded people, benevolently shepherded in the interests of everybody, with as its highest ideal the contentment of the country; or have your voice, and face the hard road of political organization, facing the hard road of discipline, developing your own capacities, your own powers and leadership and your own people to the stage where they are capable of administering their own affairs [loud applause].[6]

The formation of the PNP was the catalyst for full self-government in Jamaica. A few months after the founding of the party, the Colonial Office sent a Royal Commission[7] to "investigate the social and economic conditions in the West Indies."[8] The commission, chaired by Lord Moyne, broadened its mandate to hear "evidence on the Jamaican Constitution so far as may be necessary to elucidate social and economic problems."[9]

Manley and his party played a major role in drafting Jamaica's Constitution. "The mandate was clear and simple: universal adult suffrage, higher elective component in the Legislative Council and governmental recognition of trade unions."[10] At its first con-

---

5 Arnold Bertram (2016), 163.

6 *The Daily Gleaner,* September 19, 1938, 7.

7 The Moyne Commission arrived in Jamaica for its first visit, November 1, 1938. The full report was published in 1945. Recommendations included health and education reform along with sugar subsidies and allocations of funds through the Development and Welfare Act. Unfortunately, £1 million allotted to the West Indies over a twenty-year period did little to alleviate hardships.

8 Bertram (2016), 179.

9 Ibid., 180.

10 Paul C. Bradley. "Mass Parties in Jamaica: Structure and Organization" (1966), 378.

vention, held in April 1939, the party passed a resolution that it should make demands of the British government for full self-government and universal adult suffrage.[11]

A long battle ensued with a recalcitrant governor who wished to retain control over governance. In 1942 Richards launched an offensive against those whom he considered radicals (who became known as the Four Hs)[12] within the People's National Party. He used his wartime emergency powers to shut down the Jamaica Government Railway Employees Union and the Postal and Telegraph Workers Union—stripping them of their rights to operate as trade unions. Richards's letter to the secretary of state revealed his paranoia:

> There is in Jamaica a considerable and growing subversive element now hardly troubling to conceal its real aim, the overthrow of all ordered and honest government and its replacement by some form of local administration independent financially and otherwise of the United Kingdom.[13]

One of Governor Richards' objectives in arresting the PNP "radicals" and of delegitimizing trade unions was to frustrate the movement towards self-government.[14] His tactics did not work. The secretary of state, although sympathetic to the situation Richards found himself in, ordered him to withdraw regulations banning union activities and forced the hand of the governor to suspend internment orders. By 1943 the detentions became a public embarrassment for the British government.[15] The way was open for the People's National Party to proceed with negotiations for a new constitution, to provide adult suffrage for a fully elected House of Representatives.

The Jamaica Labour Party

Luis Fred also distanced himself from the Labour Party when it was formed in 1943, even though it came out on the side of "capitalists." He had not fully made peace with its leader, who kept shutting down the waterfront.

Despite his frustrations, Luis Fred was the first to admit that despite Bustamante's "erratic management style and unfamiliarity with the etiquette of the negotiating table,"[16] the labour leader had done more than anyone for the cause of trade unionism and social justice among the working class of Jamaica.

---

11   Alex Zeidenfelt, "Political and Constitutional Developments in Jamaica" (1952), 515.

12   The Four Hs. On November 3, 1942, Governor Richards ordered the detention of Ken Hill, Frank Hill, Richard Hart, and Arthur Henry, who were arrested and interned in Up Park Camp.

13   Quoted by Hart (1999), 198.

14   Ibid., 207.

15   Ibid., 215.

16   Colin A. Palmer, *Freedom's Children* (2014), 359.

Bustamante became the target of the governor. Sir Arthur Richards, known for his vehement opposition to the "nationalist stirrings of the Jamaican people,"[17] took advantage of the power vested in him by the Jamaica Defence Regulations. He again imprisoned Bustamante in 1940, sentencing him for "inciting bloodshed, racial war and revolution." Bustamante was quoted as saying:

> It will be bloodshed ... we will let those employers respect us. We will take away their land and give them to the workers. We shall be ruthless ... Bradshaw, Johnston, Mercier, and Captain List we shall fight with vengeance ... we want our own government, and it must be self-government too. The niggers of this country shall rise ... this will be war. We want revolution ... before the whites destroy us, we will destroy them.[18]

Governor Richards refused bail. He reported, "Bustamante is not quite normal. He was very calm and reasonable before us, but it is obvious that he's quite a different person on the platform addressing a public meeting."[19] He interned Bustamante for seventeen months until his release on February 8, 1942.

Norman Manley's intervention in negotiations restored temporary order to working conditions on the waterfront and in factories across the island, the total membership of the BITU more than doubling in the time that Bustamante was in prison.[20] When he was released, Bustamante broke ranks with Manley and the Trade Union Advisory Council, with aspirations to establish the BITU as a "general workers' union wherein he sought to organize all categories of workers within a single union under his control."[21] Relations worsened between the cousins. Bustamante accused Manley of wanting to destroy him and take charge of the union for political gain, and Manley alleged that Busta's attack on the PNP was the *quid pro quo* paid to Governor Richards for his release from prison.[22]

In 1943 Bustamante formed his own political party, the Jamaica Labour Party. At a meeting at the Ward Theatre on July 8, 1943, the JLP was launched with a resolution adopted to endorse its policies and principles. Bustamante pledged that the JLP would "keep within a moderate conservative policy in order not to destroy

---

17   Ibid., 253.

18   (PRO: CO 137/840 -Richards to secretary of state December 13, 1940) Quoted by Richard Hart (1999), 82.

19   Colonial office file Co 968/68/7 file 14463 Report of advisory committee January 1942. Quoted by Hart (1999), 154.

20   George E. Eaton, *Alexander Bustamante and Modern Jamaica* (1975), 79. BITU membership increased from 8,133 to 20,612 between February 1941 and 1942.

21   Hart (1999), 223.

22   Eaton (1975), 80.

the wealth of the capitalists to any extreme that will eventually hurt their economical inferiors."[23] A resolution passed at the July inauguration stated:

> This meeting records its full and implicit confidence in the Leader-designate of the Party who for a considerable number of years in all vicissitudes of life, in sunshine and in storm has been a rock to lean upon and a fortress in the working people of this country, whose untiring zeal, sacrificial devotion and courage in the cause of Labour is a tradition and model worthy of emulation.[24]

Bustamante was both leader of the party and president of the union. He had successfully transformed the social philosophy of his trade union movement into political action.[25]

Busta's leadership among the masses was unrivalled. His base extended into every corner of Jamaica, where he had established branches of the BITU. The appeal of the man himself was integral to the success of his labour movement. "There are men and women working in the fields, on the farms, in the factories and every sphere of human activity willing to follow Bustamante's dictates to the last man."[26]

His split with the PNP suited the governor. In a curious sense, Richards might have viewed the Labour Party as his ally. Bustamante initially expressed little interest in self-government, which he denounced as "brown-man" government and a return to slavery.[27]

Mr. Manley's party platform, on the other hand, had a nationalist agenda and aimed to consolidate politics into a socialist movement, perceived on both counts as threats to the *status quo* of colonial rule. Bustamante knew how to play the socialist card to his advantage. He understood the psyche of the small-land holder: "Land symbolized his status of independence and removal from slavery."[28] For the small farmer, a socialist government would not work; it would be asking him to divide and share what little he had. The PNP became the common enemy of Bustamante, the governor, and for that matter, the Jamaica Democratic Party.

---

23 Quoted by Arnold Bertram, *N.W. Manley and the Making of Modern Jamaica* (2016), 196. Article appeared in *The Daily Gleaner*, Friday, July 9, 1943, 3.
24 *The Daily Gleaner*, Friday, July 9, 1943, 3.
25 Eaton (1975), 85.
26 Ibid.
27 Ibid., 89.
28 Ibid.

The Jamaica Democratic Party

Peter Moss-Solomon[29] recalls eavesdropping on conversations between his father, James Moss-Solomon, Luis Fred Kennedy, and Abe Issa[30] about the formation of a new political party:

> Yes, I recall the times my father and your father would go out, and a group of private sector people used to meet for lunch at Myrtle Bank. They represented the merchant class. They would also meet at the Yacht Club, go out on the boat[31] your dad owned with Douglas Judah, and stop in at our house or at your dad's house for drinks and shoot the breeze. They wanted to see if there was enough strength to challenge Busta and Manley.[32]

A plea for supporters to join a new political party appeared in a letter by Thomas Hicks Sharp[33] to *The Daily Gleaner*, Tuesday, March 3, 1942. Sharp felt that the country was in disarray because of the war, and even though measures were in place to alleviate hardships, they were short-term solutions, not based on sound economic policy. He floated the idea of a new political party as a trial balloon to see if those who stood "on middle ground" would show interest. It received widespread response from Kingston merchants.

On Wednesday, March 17, 1943, The Jamaica Democratic Party held a public meeting at the Ward Theatre at which delegates representing the parishes of Kingston and St. Andrew, St. James, St. Ann, Manchester, and St. Catherine were present. JDP Executive thanked delegates of the People's National Party and the Federation of Citizens Association who collaborated with the legislature and the governor to approve a draft of the party's new constitution. A group of fifty People's National Party supporters attempted to break up the meeting with jeers of "Capitalist and Capitalism."[34]

---

29   Peter is the firstborn of three children of James S. Moss-Solomon Sr. and Ismilda Norton.

30   Born 1905, Issa was educated at St. George's College and Holy Cross College. He was son of Elias Abraham Issa and Mary Brimo, and a business executive, industrialist, sportsman, and philanthropist, chairman, E. A. Issa & Bros. Ltd., Issa Retail Ltd., Jamaica Electrical Supplies Ltd., Issa Hardware Co. Ltd., Motor Sales & Service Co., and other Issa companies. The family acquired Myrtle Bank Hotel in 1943 and completed Tower Isle Hotel in 1949. Issa married Lorraine Shaouy, 1937 and had two sons and four daughters (*Who's Who Jamaica 1963*, 207).

31   The yacht, *Hope*.

32   Peter Moss-Solomon, interview with author (Feb. 20, 2019).

33   Thomas Hicks Sharp, a solicitor, was secretary of Trout Hall Ltd., a former member of the Parochial Board of Manchester, and founder of Jamaica Building Society.

34   *The Daily Gleaner*, Thursday, March 18, 1943, 9.

It was later learned that neither Mr. Manley nor his party leaders had prior knowledge of the agitators.

A committee was named to finalize a draft of the constitution in January of the following year. Principal organizers met at their headquarters at 41 Port Royal Street with plans to bring recommendations before the Legislative Council. The main executive consisted of the Hon. Douglas Judah, Messrs. J. Stanley Lyon, T. H. Sharp, Douglas Fletcher,[35] Abe Issa, and W. Robert Fletcher.[36] They established an auxiliary committee, comprised of Messrs. L. F. Kennedy, F. A. Foster, H. R. Sharp, S. A. Phillips, and F. V. Nunes, to advise and assist the executive in "planning and executing intensive organization programmes."[37] The gentlemen with whom Luis Fred worked were prominent business directors, company chairmen, solicitors, and, in the case of Howard Roberts Sharp, a citrus planter and chairman of the All-Island Cane Farmers' Association Ltd.

The party espoused the merits of free enterprise and a rejection of socialism, "state ownership and control of all the means of production." One of its objectives was full employment to be achieved without the sacrifice of essential freedoms. Another tenet of their constitution was compulsory elementary education, which the party viewed as the foundation for the future prosperity of the country. It called for the opening of teacher-training institutions to include a centre for teachers of infant schools.

Mr. Walter Robert Fletcher, General Secretary of the Party, was a heavyweight in the election campaign. He drew crowds "representing a cross-section of the ethnological and economic life of the island."[38] On September 22, "a crowd of over 6000 people blocked the entire parade square in Montego Bay, and listened, cheered themselves hoarse and even encored speeches." As Mr. Fletcher walked up to the platform, he was greeted with a loud applause and the singing of "For he's a jolly good fellow."[39]

Business associate and friend of Luis Fred, Gerald Mair,[40] was the party's spokesperson on the economy. He claimed, "The Island was heading for bankruptcy unless

---

35   The Hon. Douglas Valmore Fletcher (1917–2001) was Member of the Legislative Council, solicitor, legislator and company director, partner of Myers, Fletcher and Gordon.

36   Walter Robert Fletcher (1912–1989), secretary and director, Fletcher & Co. Ltd., was General Secretary of The Jamaica Democratic Party. He was the son of Walter Fletcher, JP, who was Managing Director of Fletcher & Co. Ltd and a founding member of Grace, Kennedy & Co. Ltd.

37   *The Daily Gleaner,* Wednesday, January 25, 1944, 6.

38   Ibid., Monday, October 30, 1944, 7.

39   Ibid., Friday, September 29, 1944, 8.

40   Gerald Arthur Lothian Mair, certified accountant, chartered secretary and economist, senior partner, Gerald Mair, Goldson & Co., former branch manager, Dominion Life, director of many Boards including Jamaica Chamber of Commerce.

some way was found of producing more goods and services for ourselves and for exportation. How can we take care of unemployment unless we build up agriculture and industries?"[41]

An *avant-garde* position taken by the Jamaica Democratic Party was the publication of a *Charter for Women*. The party recognized that "our whole future must in fact depend on the proper recognition of their contribution to society, on the frank admission of society's past neglect of women's interests and disregard of women's legitimate status."[42] The charter publicly advocated for the rights of women in seven areas:

> Equality of Opportunity
> Equal Pay for Equal Work
> No Discriminatory Laws
> Widowhood Grant and Child Maintenance Benefit
> Safeguarding Women's Special Interests
> Representation of Women on National Boards
> Permanent Women's Commission to press implementation of the Charter.[43]

The JDP manifesto incapsulated Luis Fred's philosophy and politics: his desire to build a better Jamaica through promoting democracy, equality of opportunity, and safeguarding private enterprise, tenets of a philosophy that defined both his personal and business life.

> We believe that the island's resources can best be developed by the enterprise and initiative of the individual as opposed to state ownership, which involves monopoly and bureaucracy—arch enemies of progress.
>
> We believe in freedom of enterprise, subject to regulation by government for the common good, and are therefore opposed to all monopolistic and bureaucratic activities.
>
> We believe in the freedom of the citizen to plan and manage his own affairs, to seek employment in any field of endeavour, and to enter any business or profession which he chooses.[44] (Principles 13, 14 and 15 of the Jamaica Democratic Party)

---

41 *The Daily Gleaner*, Monday, October 30, 1944, 7.
42 *The Daily Gleaner*, Tuesday, November 28, 1944, 5
43 Ibid.
44 Ibid., Wednesday November 22, 1944, 7.

Adult Suffrage (1944)

Governor Sir John Huggins was appointed to begin duties, November 1943. On November 20, 1944, a new constitution came into effect. The island had been governed under a constitution (1884) that "limited the participation of the elected members of the legislature … and vested sufficient power in the Governor to enable him to exercise the duties of his office in an autocratic manner."[45] The new constitution replaced the old Legislative Council with a bi-cameral form of government with an elected House of Representatives and a nominated Legislative Council. It ensured the right to "equitable and humane treatment" guaranteed to all regardless of race, colour, place of origin, class, gender, religion, or political allegiance. Jamaica was granted full adult suffrage. The aspirations of three generations of freed people since Emancipation were now being realized.

Norman Manley did not win everything he had fought for, but he encouraged others to support the new constitution[46] with a provisional acceptance of a five-year trial period. Jamaica would have the opportunity to negotiate for complete self-rule with the institution of a ministerial system of government.

Sir John Huggins called for the first general parliamentary elections under universal adult suffrage to be held on December 14, 1944, for thirty-two seats in the House of Representatives.[47] Norman Manley's People's National Party came in a poor second to Alexander Bustamante's Jamaica Labour Party, which won twenty-two seats and 41.67% of the popular vote. Out of a total of nineteen PNP candidates, only five were elected[48] with 23.5% of total votes counted. The Jamaica Democratic Party did not win representation of any of nine seats contested, ending up with a total of 14,123 votes cast, or 4.1% of eligible voters. The party lacked the broader base of support necessary to contest seats.

The mandate was clear. Jamaicans voted for the party they believed would improve the social and economic conditions of the masses. Bustamante promised change for 140,000 persons who had no work. (Jamaica had an unemployment rate of twenty-five percent.) And for those who were employed, seventy-seven percent earned less than £1 per week.[49] Their vote in Jamaica's first general election was "an instrument of the

---

45   Zeidenfelt (1952), 525.

46   "We have got a constitution as you are witnessing that gives us very considerable powers, but it has some major defects, as the People's National Party foresaw." Norman Manley in *Manley & The New Jamaica: Selected Speeches and Writings* (1971), 122.

47   Jamaica was "only the third state in the British Empire to conduct elections on the basis of universal adult suffrage, preceded only by New Zealand and the United Kingdom … and the only colony, the only state with a black majority." http://digjamaica.com/m/blog/a-moment-in-history-jamaica-is-granted-universal-adult-suffrage/.

48   The great misfortune for the PNP was Norman Manley's loss of his seat in St. Andrew Eastern. The vote was split three ways: JLP with 5,253 votes, Manley with 4,858 and Gerald Mair of the JDP proving to be the spoiler with 3,135. Bertram (2016), 202.

49   1942 Census. Quoted by Bertram (2016), 199.

power of the people,"⁵⁰ the means to secure their rightful place in the economic and political life of the British colony.

The party's platform was the union platform. In the words of Norman Manley, "The Labour vote ... hardened and a sudden miracle of class-conscious solidarity was wrought. 'Vote Labour' became a magic slogan. It ran like a fire through the island and cast a spell on all the poor, labourer and small farmer alike."⁵¹ People had voted not only for the message, but for the messenger, the man who "derived his legitimacy from his personal charisma."⁵² His populist appeal was undeniable.

Alexander Bustamante became Jamaica's first chief minister. He was a force to be reckoned with, never before seen in Jamaican politics. His actions of shutting down the wharves and factories in the 1930s had taken the island by storm. He had shaken the status quo. And 1944 was his vindication; the masses whom he represented gave him his just reward by way of the ballot box.

Although each of the three political parties had different platforms, they shared a common objective: to build a better Jamaica with a representational government elected on the basis of universal adult suffrage.

The Jamaica Democratic Party was disbanded shortly after the election. The year 1944 ended Luis Fred Kennedy's affiliation with a political party, but his drive to create opportunities for a more prosperous Jamaica strengthened to become his life-long mission.

---

50   Robert Buddan, "Universal Adult Suffrage in Jamaica and the Caribbean since 1944," (2004), Abstract, 1.
51   Quoted by Bertram (2016), 203.
52   Eaton (1975), 89.

PART FOUR

Prime of Life (1947–1961)

Luisa Matilde Kennedy seated in the middle with her six children and their families. Seated L–R: Gloria Plant, Louise Soutar, and Carmen Valliant with their husbands standing: Owen Plant, Simon Soutar, and Jack Valliant. Standing from L–R: Luis Fred Kennedy, José Kennedy, and Francis Kennedy with their wives seated: Lydia Kennedy, Altagracia Kennedy, and Cecile Kennedy. Grandchildren are grouped with their respective families. (Photo from family collection. Photo was taken before the birth of Michèle Kennedy, Elizabeth Kennedy, Kevin Kennedy, Patrick Kennedy, and Mary Lou Kennedy.)

"In my Prime of Life" (LFK). 1950s.

(Photo from family collection, original by Wally Allen)

21

GOVERNING DIRECTOR (1947)

After 1945, a mood of jubilation in Jamaica contrasted with that of the sombre war years. The following article appeared on the front page of *The Daily Gleaner,* Tuesday, October 2, 1945:

New Wharf Co.
Takes Over
Appointments Made

With a view to improving conditions generally on the waterfront, the quarter million-pound company known as Kingston Wharves Limited commenced operations yesterday.

The new company has taken possession of four city wharves: Grace Wharf, George and Branday Wharf, Princess Street Wharf and Henderson's East Street Wharf ...

Governing Director   (1947)

> Purpose of new company is to afford importers and exporters better services and it is hoped that these new facilities will be taken full advantage of.[1]

The initial authorized capital was £249,000, of which £25,000 were preference shares. Directors of the board included Chairman D. J. Judah; Managing Directors, Messrs. L. F. Kennedy and C. L. George; and Messrs. C. E. Johnston, P. A. George, T. P. Evelyn, and J. S. Moss-Solomon.

In December 1945, *The Daily Gleaner* reported that since October all wharves buzzed with activity, with "large tonnage of imported goods and island produce for export passing over them, and thousands of waterfront workers, stevedores and common labourers, resuming their places in ships' holds and on docks." One central organization under the auspices of Kingston Wharves Ltd. was in place to meet the challenges. "Labour also played its part in bringing about a rapid revival … they have been spurred on by the much-advanced wage rates now in force … and ships have been quickly dispatched, a feature of high efficiency in port organization."[2]

Mom travelled to New York in 1946, again for medical reasons, this time for a lengthier stay of five months. Accompanying her on the flight to Miami and New York, July 29, 1946, were her two children, Paco and Celia, two sisters, Altagracia (Gracie) and Mercedes (Mecha), and brother-in-law Francis, who months earlier had graduated from Holy Cross College.[3] Her sisters had arrived in Jamaica some weeks earlier on their first trip to the island. With adventure tours around the island in the summer of '46, a romance blossomed between Altagracia Loinaz and Luis Fred's brother, José Kennedy.[4] Mom recalled those times:

> I loved José you know. We had a good time, travelled the island. We were a foursome (Luis Fred and Tillie, Gracie and Joey). After a couple

---

1   *The Daily Gleaner,* Tuesday, October 2, 1945, 1.

2   Ibid., December 19, 1945, age 9. War in Europe ended, May 8, 1945, and in August, US atomic bombings forced Japan's surrender. Six million Jews had been murdered in the Holocaust with over 50,000,000 casualties, civilian and military, worldwide. From Caribbean, 236 volunteers died in service with British units.

3   By 1946, air travel became faster and more convenient. Pan American World Airways now offered a direct service to Miami on board the four-engine Boeing Stratoclipper. Passengers could confirm bookings in Kingston on a connecting flight from Miami to New York (*The Daily Gleaner,* July 18, 1946, 1.) The last Pan Am 314 sea boat to be retired was the *California Clipper* NC18602 in 1946. It had accumulated more than a million flight miles.

4   José and Gracie would have met each other on previous occasions: my parent's wedding in 1937 and subsequent business trips that José made for Grace, Kennedy to New York after 1942.

weeks she (Gracie) wanted to stay on, so contacted her boss in New York to tell him that she had contracted malaria. When it was time for her to go back, they announced that they were going to get married.[5]

José had left private law practice in 1942 to be employed with Grace, Kennedy, where he held a senior position in shipping. "He was a brilliant lawyer, had set himself up on Duke Street and received much of Grace, Kennedy's legal work."[6] Shortly after his engagement to be married, he travelled to New York in August 1946 on a business trip. Luis Fred wrote:

> His main job is to make personal contact with and friends of the Alcoa officials. He will be entertaining them and be entertained by them. By this time, he should be nearly through. He is to get on a sufficiently friendly basis so as to be able to ask one of them to get an airplane agency for us, that I have been working on.[7]

In Jamaica, his responsibilities increased with the boom in the shipping business. "Joe is really working hard. Has had several ships in the last week or two and has had little time for his own affairs. He wants to be planning for his wedding. Geoff Dodd is back and has started here again. He and Joe are together in the S.S. Department and Insurance."[8]

Dad's letters expressed a longing for his dear wife: "What a thrill to hear your darling voice once again—even on the telephone. If it was wonderful nine years ago, it was a million times more wonderful this morning to talk to you. Oh dearest, how I hated to hang up. It was like saying goodbye to you all over again."[9] He was at loose ends without her: "Don't worry about me dearest. I am really alright, although a bit tired. But I should soon take some time off. Trouble is I can't enjoy any time off without you alone … Work is about the only thing I have that makes it possible to exist without you here."[10]

In September, Mom registered Paco (aged six) and Celia (aged four) in a neighbourhood school in Woodside, Long Island. Mary, not yet a year old, remained in Jamaica under the care of a nanny at 8 Seaview Avenue. "You should see our darling—how she has grown and how sweet she is. She looks a lot like you. As I told you on the phone, she is crawling now all over the place, and stands occasionally. If you don't get better quickly, she will be walking before you are back."[11]

---

5   Lydia M. Kennedy, interview with Georgianne Thompson Kennedy, 2002.
6   Ibid.
7   Luis Fred Kennedy, letter to Lydia Kennedy, September 1946.
8   Ibid., October 2, 1946.
9   Ibid., September 1946.
10 Ibid.
11 Ibid., October 2, 1946.

Dad suffered a mishap. He wrote to explain his misfortune:

> The accident was a stupid affair. I had gone with Simon to the West End—the Cornmeal factory—to inspect an installation there in connection with a large shipment of corn we had sold the government, and which was about to arrive. I was all through and, stepping off a ledge not more than four feet high, my heel caught on the edge or somewhere and I fell, apparently on my left ankle. I knew immediately that something had broken as my foot was pointing one way and my leg another … By noon I was here at Saint Joseph's where I was first put in the Maternity Ward (I shall never live that down).
>
> I'm in plaster up to my knee and the pain was caused by the ankle swelling and causing the leg and toes to swell too … I expect them to put on a brace and walking iron to enable me to hobble about without any weight going on the ankle.
>
> I don't know how many bones are broken in my ankle. My impression is, however, that both the knobs on each side are busted off … The medical name for the injury is a double Potts fracture and dislocation. It has certainly put me in the soup for a while.[12]

With his leg in a cast, he hobbled about for months not wanting to be confined to bed. Among those who poked fun was Alexander Bustamante. Mom recalled the incident:

> Again, there was trouble on the waterfront. Bustamante was trying to re-unionize the workers. There were strikes, riots, discontent. Fred and Charlie DaCosta tried to oppose the move, and Busta retaliated with all the charisma and skill he had. This was the time he addressed a large crowd gathered outside Grace, Kennedy on Harbour Street and shouted in his best nasal twang, "Bruck-foot Kennedy (Fred had a broken ankle) and twis' neck DaCosta don't have a chance!"[13]

Very much on the mend after several doctors' appointments, Mom returned from New York with Paco and Celia before Christmas of 1946, and her sister Gracie arrived in January. Mom recalled: "And they (Gracie and Joey) were married from our house. We had a reception for them. They went to Amberly for their honeymoon. They lived with us for about a year. They were still living with us

---

12 Luis Fred Kennedy, November 16, 1946. The accident occurred on November 6.
13 Lydia M. Kennedy, "A Giant of a Man: A Tribute to Luis Fred Kennedy," (1997).

when Philip was born. She almost lost the baby, had to be confined to bed during her pregnancy."[14]

Grace, Kennedy's aggressive post-war development program saw the start of the new Alcoa Line run from USA to Jamaica in January 1947. The Alcoa Steamship Company called them the "super-liners in miniature."[15] They carried ninety-six passengers in luxurious, air-conditioned rooms, and had a freight capacity of 8,500 tons of deadweight cargo.

In June 1947, *The Daily Gleaner* published a story of the arrival of the new steamship from Florida:

New Northland Starts New Cruise Service
Enthusiastic Reception Given on Arrival

An enthusiastic reception was given to the SS *New Northland* of the Seaway Steamship Line, her passengers and crew, as she arrived in Kingston on Saturday to start a new fortnightly tourist cruise service between Jacksonville, Florida, and Jamaica.

Grace Wharf, where the vessel docked on arrival shortly after 8:00 a.m. was decorated with flags down its whole length while from a flagstaff at the head of the pier the Union Jack flew in the morning air saluting the visiting cruise ship.[16]

Tourism boomed after the war. The value of the trade increased from £11 million in 1938 to £31 million in 1948,[17] which was good news for shipping agents and wholesale merchants.

Dr. Grace's assets in the company had appreciated significantly since 1922. It was now time to sell. In 1947, at age seventy-seven, Dr. John Johnston Grace divested his stock holdings in Grace, Kennedy & Co. Ltd. He transferred his family's preference shares in equal amounts to Luis Fred Kennedy and James Seivwright Moss-Solomon, who became heirs

---

14   Lydia Kennedy (2002). Altagracia María Loinaz (aged 28) and José Antonio Kennedy (34) were married, February 8, 1947, in St. Andrew, Jamaica. Philip Thomas Kennedy was born December 23, 1947.

15   Announcements of the arrival of the new Alcoa Line ships appeared in the *Daily Gleaner,* September 23, 1946 (page 10) and in *The Boston Globe,* October 6, 1946. Names of the ships were SS *Alcoa Clipper, Alcoa Cavalier,* and *Alcoa Corsair.* They were a postwar development in shipping that recognized the futility of competing with aviation (on basis of speed), offering instead luxury passenger accommodations and large freight capacity.

16   *The Daily Gleaner,* Monday, June 9, 1947, 12.

17   Colonial Office, *Colonial Annual Reports Jamaica 1948,* 27.

Governing Director   (1947)

to the legacy of the House of Grace that had helped shape the culture of Grace, Kennedy for twenty-five years.

By entrusting both Kennedy and Moss-Solomon with governorship of the company, he left both gentlemen in an untenable position. Peter Moss-Solomon commented:

> When Dr. Grace decided to retire, the company was controlled by preference shares, and Dr. Grace had all the preference shares. When Dr. Grace was retiring, he called them together (Luis Fred Kennedy and James Moss-Solomon) and gave them each an equal amount of preference shares. Well, they both felt that ownership of an equal amount of preference shares would be a disaster for the company. If there was ever any disagreement the company would be at a standstill. So, by agreement my father transferred one preference share and your father gave him an ordinary share in return and that meant that Kennedy now was the chief, your father, the Governing Director.[18]

The mutual respect the two men held for one another led to a smooth transition and an amicable working relationship that would last another thirty years. Although different, their business styles and personalities complemented each other: Kennedy was aggressive and irascible, Moss-Solomon, reserved and even-tempered. Their bond was a shared vision, an unwavering commitment to promote what was good for Grace, Kennedy and for Jamaica.

In September 1947, John Grace travelled with his wife, Eleanor, to New York City for medical treatment. His hearing was impaired, and he had lost sight in his right eye. The couple later migrated to California where their youngest child, Mrs. Cynthia Beatrice Muir, resided with her husband. (Cynthia had trained in England to be an electrologist.) Their son, John Greenfield Grace, also lived in California after having migrated from Jamaica to the United Kingdom, where he trained and qualified as an architect at Christ's Church, Cambridge. He had married in 1938 but divorced before moving to the United States in 1940.[19]

John and Eleanor's middle child, Alicia Vere Grace, managed to secure a passage from Halifax to Bristol during the war in 1942. "Jamaica was a little bit of an outpost. So, we would dream of big cities, like Paris, dream of wonderful places where the action was going on. But *en dehors de cela*, when war broke out, there was no question—my place was there [Britain]."[20] She married Charles De Brosses in 1951 to become

---

18   Peter Moss-Solomon, interview with the author, February 20, 2019.

19   John Greenfield Grace married Mary Barbara Mocatta, October 1938, Westminster, United Kingdom.

20   https://www.alliancefrancaise.london/Alicia-Grace-Alliance-Francaise-de-Londres.php. Her story is a fascinating tale of a cross-Atlantic passage from Halifax to Bristol on board the SS *Tuscan Star,* which was torpedoed by German submarines.

Countess Alicia De Brosses. They had two children, born in Paris: Charles (1952) and Anne De Brosses (1953).

Dr. Grace lost three siblings during the 1940s: Morgan Hatton Grace (1880–1943), William Russell Grace (1877–1944), and Alice May Grace (1873–1945). Morgan Hatton Grace, an original shareholder of Grace, Kennedy, had returned to the United States, where he became president and director of the Phosphate Export Association of New York City and headed an export firm bearing his own name. He was survived by his wife, Ruth Eden Grace, and five sons.[21]

Dr. Grace's other brother, William Russell Grace,[22] who was not directly involved with Grace, Kennedy, except for business ties between New Zealand and Jamaica, had retired as a cattle and sheep farmer who managed his father's 6,700-acre property in North Island, New Zealand.

Dr. Grace's sister, Alice May, died a widow on June 26, 1945, having spent most of her life in England. Before her marriage to her cousin, Walter Henry Grace, she had been previously married to William Michell Clarke, who died at the young age of thirty-eight (1862–1900). She had no children from her first husband, but she and Walter Grace had a daughter, Margaret Eileen (1907–1998). Alice May survived both husbands—Walter Henry died in 1932 in Cannes, France.

With the expansion of W. R. Grace & Co., the Graces built their commercial empire by integrating three core areas of business: export and import trade, shipping, and industry.[23] In its first hundred years (1854–1954), it remained a company managed and governed by descendants of William Russell Grace, sons and nephews, and friends and associates of the Grace family.[24] Although the founder's explicit wish was for family succession, he realized that his business "was outgrowing the capacities of the Grace clan to handle" its rapid growth, so he "brought young men into his organization with the idea of advancing them as far and as fast as their talents would justify."[25] As younger members of the family became interested in the business, succession within the Grace family was held as a priority.

W. R. Grace & Co. (founded 1854) maintained its reputation:

---

21 Morgan H. Grace's obituary appeared in *The Brooklyn Daily Eagle*, July 3, 1943, 5.

22 Named after his uncle William Russell Grace (1832-1904).

23 Marquis James, *Merchant Adventurer: The Story of W. R. Grace* (1995), 317. The biography of the Grace family was completed in 1948 but not published until 1995 by W. R. Grace & Co. The manuscript was discovered in the firm's archives thirty years after its completion.

24 "When W. R. Grace died in 1904, the presidency of the corporation went to founder's nephew, Irish-born Edward Eyre." J. P. Grace, son of the founder, succeeded his cousin three years later, assuming the chairmanship in 1929. His best friend and Columbia classmate, Stewart Iglehart, became president. In 1945, both Grace and Iglehart retired, and William Grace Holloway, the founder's first grandchild became chairman. Marquis James (1995), 319–320.

25 Ibid., 292.

Governing Director (1947)

> Let any veteran reflect how differently any business is conducted today (1948) from what it was when he started out. W. R. Grace & Co. has changed. Its airplanes cover the distance between the United States and Peru in twelve hours whereas in 1904 its sailing ships, a good many of which were still in service, took a hundred days. Yet basically the business has altered very little. This is not because Mr. Grace's successors have fallen behind the times. It is because the founder was ahead of his time; and because his house had been built upon principles that are enduring.[26]

Abstention from politics in foreign countries and a forward-thinking approach to profit sharing helped to build its character. Under the presidency of J. P. Grace, son of W. R. Grace, "the management base was broadened, and the founder's principle of compensation based on participation in profits was expanded … Opportunity for ownership of common stock was also given to the staff through sale at attractive prices on easy payment terms."[27]

Governing Director Luis Fred Kennedy was free to be his own boss. Leaning on the traditions inherited from the House of Grace, he shaped the future of Grace, Kennedy with his own style of directive and visionary leadership.

One of the first measures he put in place was a medical scheme, free of cost to all employees of Grace, Kennedy. In 1947 he introduced a medical plan for all staff of newly formed Kingston Wharves Ltd., of which Grace, Kennedy held majority shares.[28] It was a progressive, innovative business practice, even for US companies that were starting to offer employer-sponsored health insurance in the years following World War II.

Dad's youngest brother, Francis Xavier, joined Grace, Kennedy, and like José married in 1947. He met his fiancée, Marie Margaret Cecile Lemay, in Worcester, Massachusetts, where he attended college.[29] They were married in St. Andrew, Jamaica on November 21. Their first child, Lareine Agnes Kennedy, was born the following year.

On November 2, 1947, Dad's birthday, Mom suffered the loss of a child: "I don't think I ever got over it. I went into a deep depression. Luli was worried about me. Fortunately, the priest was at the hospital, and he had time to baptize her. She only lived for a couple hours. They named her Mary. I never saw her alive."[30] She was their fourth child.

---

26   Ibid., 317.

27   Ibid., 320.

28   Grace, Kennedy board meeting minutes are not available for 1947. It is assumed that the free medical scheme introduced for employees of Kingston Wharves Ltd. in 1947 included all employees of Grace, Kennedy.

29   Marie Margaret Cecile was known by everyone as Cecile. She was born December 1, 1924, to August Lemay and Marie Della Rondeau, the youngest of six children.

30   Lydia M. Kennedy (2002)

22

GUTSY VICE-PRESIDENT (1948)

In 1945 Great Britain had held general elections, which brought a surprise landslide victory for Clement Attlee of the Labour Party over Conservative incumbent Winston Churchill. This heralded a new era of socialism in the United Kingdom, beginning with the nationalization of industries like coal mining and the creation of a welfare state.

Luis Fred Kennedy was concerned about the negative impact the change of government was having on trade and commerce in Jamaica. He objected to price controls and import quotas imposed by government and fought battles in the open arena with both the chief minister and the governor. He viewed an imperialist trade policy as the root cause of the problem and believed that the solution was for traders and retailers to unite and speak with one voice. He was fearless in his attacks.

By 1947, the government of Jamaica faced shortages of foreign exchange, caused largely by the British government's decision in November 1946 to stop monetary contributions towards price stabilization. The Jamaican government continued bulk buying of commodities to maintain lower prices for the consumer. And it subsidized essential foods, flour, pickled meats, codfish, cornmeal, and milk, but because of local budget constraints, it was not able to afford subsidies on other imports. As a result, prices of both imports and exports rose, and with an adverse balance of trade, licenses were suspended and restrictions placed on imports. Measures hurt food merchants.

Gutsy Vice-President (1948)

At a meeting of food merchants at the chamber on October 10, 1947, Mr. Kennedy proposed the appointment of a food committee, to consist of two distributors, two wholesalers and two retailers for representation to the food controller.[1] He presented a comprehensive list of resolutions and suggested that the committee obtain details of London's instructions for the curtailment of imports, information that was necessary for formulating a new food trade policy. He recommended to government "substitution of locally grown foods for imported food and a scheme for the marketing and distribution of local foods."[2]

In January 1948,[3] Mr. Kennedy informed the Chamber of Commerce that published reports of food trade controls were inaccurate. The Competent Authority and the Imports, Exports, and Prices Board were not, in fact, fully responsible; general policies and decisions were instead being made by the Executive Council, which was usurping their powers. He requested an interview with Executive Council so that concerns could be brought forward and proposals made for reorganizing the Trade Controllers Department. He was disgusted with the lack of direction and policy enacted by government.

He next attacked the policies of the mother country. In November, *The Daily Gleaner* published his scathing criticism of British monetary and trade practices. Jamaica was hurting from a crippling cost of living, which he believed was enabled by punitive policies of the imperial government. He made special mention of the "Dollar Pool" to which Jamaica was forced to contribute its earnings. "It is controlled solely and entirely by the British government, and our local authorities take direct instructions from London as to what dollars we may have from the Pool."[4]

He found unconscionable the arrangement by which Jamaica was forced "to purchase essential goods from Britain at high prices and sell essential goods to her at cheap prices." Very little regard, he argued, was given in trade policy to reduce unemployment or to attend to the general welfare of Jamaicans. In a letter to the editor of *The Gleaner,* he wrote, "We therefore have in Jamaica a colony which is self-governed in all matters except the important ones; we have a government which takes orders direct from England on all matters affecting both its vital trades and the development of the island."[5]

He believed that imperial trade policy was stifling private enterprise with mismanagement of public funds. The solution lay in more local control and power invested in

---

1   After turning down nomination to council from as early as 1941, in years following, Luis Fred assumed responsibilities as head of committees and was appointed as second VP in 1948.

2   *The Daily Gleaner,* Saturday, October 11, 1947, 12.

3   *The Daily Gleaner,* January 21, 1948, 8.

4   *Ibid.,* November 9, 1948, 6. Headline was entitled, "Is London Forcing Socialism on Us?" By Luis F. Kennedy in a letter to the editor.

5   Ibid.

the legislature and private business. Jamaica should decide what was important for its own development, and Britain should permit merchants to purchase commodities on the open market at the most competitive prices. The stranglehold was bad for business and for Jamaica.

Governor Sir John Huggins did not take kindly to the criticisms. In a speech at Frome Agricultural Show, November 22, 1948, His Excellency made reference to Mr. Luis Fred Kennedy, who had charged the government with "gross mismanagement and inefficiency in the handling of bulk purchases of imported foodstuffs."[6] He said that Mr. Kennedy had furthermore accused the government of intentionally inflating the prices of flour and rice by levying an unauthorized tax in order to eliminate deficit payments. The governor was angered by Mr. Kennedy's misleading statements, which he viewed as "propaganda," "designed to force the hand of government to pay higher margins of profit."[7]

Mr. Kennedy defended merchants without apology and criticized the governor for smearing their good name:

> There is, for example, His Excellency's charge against this Chamber's Committee and myself of selfish motives. I will not dwell on the fact that we have never made any secret of our desire to be permitted by government to make a just profit or upon further fact that, our proposals to government to return our business to us have always stipulated that price control and profit control should continue in protection of the public and as a safeguard against abuse by unscrupulous traders.[8]

Luis Fred declared that the governor had resorted to *ad hominem* arguments, for which he had a particular distaste and found unbecoming of someone entrusted with the office of governor.

Kennedy claimed that government had recovered "enormous losses at the expense of the public in arbitrary fashion and without authority of the Legislature."[9] Unacceptable was the expectation that merchants should "subsidize the price of imported foods to the extent of trading at a loss in order to provide money to cover up Government's losses through mismanagement."[10] He would not accuse the governor of "impure motives, bad faith or scheming against the trade," but actions to regulate prices, he insisted, had forced merchants into a situation of claiming losses. Mr. Kennedy declared that the government had

---

6 *The Daily Gleaner,* November 24, 1948, 11.

7 Ibid., November 23, 1948, front page headline: "Govt. Not Profiteering Sir John Declares/Merchants' Margin Motive Blamed."

8 "Mr. Luis Kennedy Answers Governor" in *The Daily Gleaner,* Saturday, November 27, 1948, 13.

9 Ibid., Saturday, November 27, 1948, 13.

10 Ibid.

reneged on its promise to protect the consumer; it could not afford the high level of subsidies and seemed desperate to recoup losses. Merchants were the targets. The government, he maintained, deflected criticism of their own profiteering by blaming traders.

Mr. Kennedy recommended that a delegation be sent to the House of Representatives and to the United Kingdom with the grievance that the government of Jamaica was mismanaging funds for no other reason than that "the financial position of the country was being tampered with … Great Britain, whether the government be Socialist or Tory, must recognize that a colony as this, still without complete self-government, should have its affairs run properly and with due safeguard."[11]

The chamber was duly impressed by Mr. Kennedy's delivery. Chairman Charles DaCosta thanked him for his address, which "was dignified, restrained and moderate."

Luis Fred Kennedy's leadership reached beyond indictments of government malpractice. He shared his vision for the economic prosperity of Jamaica and became an advocate for a unified voice among merchants. In a report to council, June 2, 1949, he called for the country to fix its international financial position. Jamaica was borrowing too much and was in danger of being mortgaged to other countries. The answer to the problem was to "import less and export more."[12] He urged the chamber to start an export drive to boost the growth of local industry and ban importation of luxury items. Procurement of essentials, at the most competitive prices, should be Jamaica's priority.

He appealed to the social conscience of employers, urging them to institute training programs and pension schemes to guarantee the economic security of workers. He advocated for the private sector to be responsible for serving its employees and customers. And he led the debate to discuss an austerity plan for Jamaica at a special meeting of the Chamber of Commerce on June 16, 1949. Mr. Kennedy exhorted his colleagues to take collective responsibility as members of the business community for the economic development of Jamaica:

> Surely it must be the responsibility of the private enterprise to take positive action in these matters. We know how inadequate and generally unsuccessful is individual action in preserving to ourselves the rights which we claim under private enterprise; for the same reasons must action by each of us individually be inadequate and unsuccessful in accomplishing the discharge of the common obligations of private enterprise which are the counterparts of our rights.

---

11   Ibid.
12   Ibid., Thursday, June 2, 1949.

I therefore feel that acknowledging as we must a great measure of responsibility a way must be found for combined effort to meet it.

An alternative is to leave ourselves open to the charge that we unite for the protection of our rights but are content to leave to individual and unorganized effort the discharge of our obligations and responsibilities.[13]

An article entitled "With Proper Labour Unions Private Enterprise Will Provide Square Deal for All" appeared in *The Daily Gleaner* of October 21, 1949. The writer commented:

However unpopular with the government some of the remarks of Mr. L. F. Kennedy may have been, Mr. Kennedy has been the spearhead in the Chamber of Commerce of the school of thought that private enterprise must set its house in order and improve its services to its employees and to the public.[14]

Luis Fred Kennedy adopted for Grace, Kennedy many of the proposals he put forward to the chamber. Ironically, his measures revealed a socialist philosophy, not unlike Norman Manley's "Plan for Progress," an economic development vision for Jamaica, which formed the main message of the PNP platform for the 1949 general elections.

In December 1949, Alexander Bustamante was re-elected chief minister, but this time the People's National Party won most votes (representing an increase of almost twenty percent), falling four seats short of the Jamaica Labour Party's seventeen. An urban/rural split in voter support was evident. Manley gained support of the urban middle classes, and Bustamante remained fully entrenched in the rural areas, with voter support linked to union (BITU) membership.[15]

Although Manley had come under attack within his party for leftist leanings, he gained increased popularity in the post-war period. With the influence of intellectuals like Noel Nethersole[16] and others who put forward a strong economic plan for Jamaica, many began to see Norman Manley as "Father of the Nation."[17] He made a formidable Leader of the Opposition during Bustamante's second term.

---

13   Chamber of Commerce meeting reported in *The Daily Gleaner,* Friday, June 17, 1949, pages 1 and 12, "Austerity Plan Suggested for Island Stability." Luis Fred Kennedy's speech was also quoted in *S. Carlton Alexander: May 1933–May 1983* (April 1983).

14   *The Daily Gleaner* Friday, October 21, 1949, 8.

15   George E. Eaton, *Alexander Bustamante and Modern Jamaica* (1975), 134.

16   The Hon. Noel Newton Nethersole, solicitor, legislator, and sportsman, was elected member of the House of Representatives for Central St. Andrew, 1949; he was first vice-president of the People's National Party to become finance minister,1955. (Information from *Who's Who Jamaica 1957,* 361.)

17   Arnold Betram, *NW Manley and the Making of Modern Jamaica* (2016), 221.

Home, 45 Lady Musgrave Road
(Photo by author, 1960s)

Fred William, born July 28, 1950.　　Elizabeth Louise, born November 1, 1953.
(Photo from family collection)　　　　(Photo from family collection)

Sitting on the back fender of Dad's DeSoto: Mary, Liz, and me with Mr. Forbes, family chauffeur, circa 1955.

Celia and Mary in school uniform, circa 1957.

Our family in New York, 1955. Paco was in his second year at Georgetown Prep.
(Photos from family collection)

23

HIS CASTLE

(1950s)

In the summer of 1949, the family moved into 45 Lady Musgrave Road, which had been the residence of family doctor, John Mosley, his wife, Rachel, and their two sons.[1] Mom recalled the move:

> Every time we passed by it, and that was almost daily, Fred would say, "I love that house, I'm going to own that house one day." He admired its look from the outside but had never been on the inside. Old Mosley had died earlier that year, and his wife decided to rent the house. We moved in, and then shortly after, she changed her mind and sold it to us. We borrowed money to purchase it and took out a mortgage, which Fred didn't like. He didn't like owing money.[2]

The home was on an acre of land, which sloped at an elevation of four to six feet from front to back, none of it landscaped. The previous owners had ignored upkeep of both the interior and exterior of the home. Dad's first job was to terrace the land. He hired contractors to create three levels, each area bordered with low brick walls.

A grass tennis court was constructed on the front lawn. (Dad thought tennis should be played on a natural surface as it was done at Wimbledon, and he wanted also to avoid concreted or asphalted surfaces.) On the area surrounding the house he built a fishpond made of old-time Jamaica red brick, raised four feet above ground to prevent

---

1 See Note, Chapter 19.
2 Lydia Kennedy, interview with Georgianne Thompson Kennedy, August 10, 2002.

stray animals from disturbing the fish. On the third level at the back of the house, a badminton court and pool were constructed, the property line on one side defined by a six-foot grey stone wall and red brick barbecue. The back yard became a Sunday playground for our cousins and their families, who were treated to a barbecue, swim, and badminton matches. Uncle Vin George[3] also joined the family barbecues every Sunday.

Mom and Dad fancied the architectural style of the home but soon discovered that the layout left much to be desired. Over the years, they made structural changes to the interior by adding rooms and re-configuring the layout of the ground floor to make living space more practical. He used to say, *"If old man Mosley's duppy ever visited, he would be sure to get lost."*

A local architect described the home:

> It's a mix of Spanish, Georgian and local colonial styles. The clay, red-tiled roof, eave corbels and thick walls are typically Spanish. The formal entry, symmetrical arrangement, and window treatments are Georgian. The verandah, awnings above windows, and the roof slope are colonial adaptations for our local weather. It was a contemporary design of the period (1930s). The Spanish style added exotic and weather touches, the Georgian gave it a serious and grand feel, and the local colonial added comfort and sensibleness.[4]

Its twelve-inch-thick reinforced concrete walls gave the impression of a fortress, the interior cool as if it were air conditioned.

One year after the family moved in, I was born on July 28, 1950. I was named after my paternal grandfather, Fred William Kennedy. Three years later, on November 1, 1953, we welcomed the youngest, Elizabeth Louise, named after Her Majesty Queen Elizabeth II, crowned in June of that year, and her maternal grandmother, Luisa Kennedy. On the day I was born, Dad planted an Otaheite apple tree, and on the day my sister Liz was born, a dwarf coconut, his favourite plant.

"You can tell the young ones when they are on the tree because they are smooth on the outside, bright green with no discoloration." To drink the water of the jelly coconut, a glass would not do; instead, Dad tilted his head back and drank the water straight from the shell. To eat the jelly, a regular spoon would not do either; he scooped out the flesh with a spoon cut from the husk. He liked the jelly soft, delicate enough to dissolve in the mouth. Coconuts, not grapes, were the fruit of the gods.

Both fruit trees bore profusely for decades.

---

3   Vincent Louis George (Uncle Vin), family friend, managing director, George and Branday Ltd., director, Kingston Wharves Ltd. Born Kingston, 1888, son of the late Arthur George J.P. and Rose Emily Branday, educated, St. George's College, married to Ethel Catherine. *Who's Who Jamaica 1957,* 190.

4   Description, courtesy of Jamaican architect, Kevin Bryan.

His Castle   (1950s)

The three oldest children were away from home for long periods. Paco was absent the most because he was abroad, returning only for Christmas and summer holidays. He started at St. George's College in 1951, but Dad decided that he would receive a better education at Georgetown Preparatory, an all-boys Jesuit boarding school in North Bethesda, Maryland. Mom accompanied him to New York in 1954 to enroll him in freshman year at Georgetown Prep. He was fourteen.

Shortly after, Celia and Mary went to boarding school, Celia at age twelve in January 1955, and Mary at age ten in January 1956. They attended the Servite Convent of the Assumption for girls in Brown's Town, Jamaica.

Dad's idea of sending his children to boarding school was well-intentioned. Catholicism was a prime factor in the choice of schools, institutions run by priests and by nuns who had a reputation for academic excellence. The schools provided compulsory religious observances, a regimented set of rules, and a community intended to be a home away from home. From a parent's viewpoint, this made sense, but for a ten- or twelve-year-old, the experience of leaving home was filled with angst and resentment.

Church was a central part of living. Attendance at Mass was compulsory, every Sunday, 7:30 AM, at Sts. Peter and Paul Roman Catholic Church. Besides Sundays, Holy Days of Obligation were observed; the principal ones were New Year's (January 1), Epiphany (January 6), Ash Wednesday, Good Friday, Easter Sunday, All Saints Day (November 1), and Christmas. We were brought up in the days of pre-Vatican II (1962), when priests offered Mass with their backs turned to the congregation, who responded in turn with prayers spoken in Ecclesiastical Latin.

Church formed a community of family and friends, who attended bazaars, fundraisers, dances, and garden parties, and were members of clubs like Altar Boys, youth groups, and Catholic Women's League. At home, it meant strict Lenten observance and abstinence, prayers before and after meals and at bedtime. On Good Friday, we could not run about, play cards, listen to music, or speak loudly. Most observances were ritualistic; not much was spoken in terms of faith or spirituality. It was not until many years later that Dad would discuss with us his skepticism of Church doctrine.

Life was orderly: three-square meals a day with Dad home most days for lunch. His favourite was salt fish and ackee, or minced meat and rice with avocado and raw scotch bonnet pepper (seeds and all). He would return to the office for a few hours and be back before dark. Mom commented: "His home was his castle. Anyone who called at night about business he would tell them to get in touch with him the next day at the office. He was a great family man and usually came home at 4.00 pm when everyone had to be bathed and dressed and we would then go out for a drive somewhere or visiting."[5]

---

5   Lydia Kennedy, "A Giant of a Man: A Tribute to Luis Fred Kennedy," (1997).

There were no fancy gadgets, televisions, computers, or other devices. Evenings were spent reading and playing cards or parlour games like Ghost.[6]

Dad would often read to us in the evenings on the upstairs verandah. Dickens was his favourite author. (The living room contained the complete Dickens collection of thirty volumes.) Holding the leather-bound book high up so all his children could follow along, he would read the opening lines of *A Tale of Two Cities*:

> It was the best of times, it was the worst of times,
> it was the age of wisdom, it was the age of foolishness.[7]

He took care not to open the book too wide lest he damage the binding. The volume had gilt-edged lettering on the spine, laced with decorative crowns, and inside, detailed black and white illustrations for each chapter.

He would stop occasionally to paraphrase, explaining in succinct ways the complex ideas. Even if he kept on reading, it did not matter, for we were mesmerized by his lilting voice. Dad had a keen ear for the cadence of Dickensian prose.

As a young man, he had read all thirty volumes. He was fascinated by the realism of the author's portrayal of the underbelly of British society and of the oppressive conditions of nineteenth century industrialization.

Movie night was a regular occurrence. Dad owned a Keystone Model A-7 16mm projector, which he had purchased in New York in the 1940s. He would set up a screen and projector on the front verandah where we viewed the movies, mostly of family events: first steps, First Communion, Christmas at Grandma's, pool parties, and barbecues. (He kept taking movies up until the early 1960s, for almost a span of twenty years.) The biggest thrill was seeing short, black and white adventure films. Dad showed them repeatedly, but we never complained. My sister Mary has vivid recollections of these:

> They were old cowboy movies. I particularly remember in one scene the guy is dying in the wilderness, no horse, dehydrated and can hardly walk. A big buzzard is circling overhead, and he looks up, shakes his fist and says, "You ain't got me yet," and promptly collapses. There was also the film of the Queen's coronation.

Although incidents of break-ins were rare, for protection, Dad owned a gun, which he stored under his mattress. The windows, free of burglar bars, remained wide open at night; we had no watch dogs and no watchmen. All that came in later decades.

One night the unexpected happened. Mom used to tell the story:

---

6   Ghost was a family favourite, a word game in which players try to add a letter to a growing word fragment. The hilarious part came when one became a ghost, whose job it was to try to get the other players to talk. If one talked to a ghost, he/she would in turn become a ghost. It was great fun and worked well to foster family togetherness.

7   Charles Dicken, *A Tale of Two Cities* (1906), 1.

His Castle (1950s)

As you know, I am a light sleeper. One night, I was awakened by the creaking sound of the bedroom door opening. I remember a large shadowy figure crossed in front of our bed. I could make out the shape of a huge, muscular man. I was terrified. Your dad was in deep sleep, flat out on his back. The next thing I know I was screaming, and I then covered my face with the sheets.

Your dad shouted, "Get the gun! Get the gun!" I could not move. He leapt out of bed and lunged at the intruder to wrestle him to the ground but couldn't grab hold. The man's torso was slippery with grease and he was wearing no shirt. "The gun, the gun," your dad repeated. The prowler slipped away and bolted through the door to the upstairs verandah. He jumped off the balcony into the flowerbed below and disappeared into the night.[8]

A few weeks later, Mom surprised Paco brandishing the loaded gun above his head. We never saw the .45 calibre revolver again. For years after, Dad kept the bullets stored in little boxes on top of his chest of drawers. I remember, they had brass shell casings.

I looked forward to Christmas when my older siblings came home from boarding school. The season was joyous, our gift-giving neither extravagant nor ostentatious. It was Jonkonnu time. The costumed bands came by, playing drums and rattles and performing frantic dances. We were mesmerized and spooked by the masqueraders: Horsehead, who wore a mule's skull; the Devil with his pitchfork and horns; and Pitch-Patchy, the acrobatic trouper with layers of brightly coloured cloth dangling from his tall, athletic figure. Caroling groups from the orphanage also spread Christmas cheer, parading through the streets, and collecting alms.

Yuletide would not be complete without everyone gathered at my grandmother's home on Seaview Avenue. Happy to be free of the adult world, we children played hide and seek, discovering every nook and cranny of our grandmother's property. Sumptuous meals were served and presents exchanged, and towards evening, Dad organized fireworks with flares and Roman candles, bottle rockets shooting high above the lawns, and sparklers and firecrackers for the children to handle.

He was a big tease and loved to play pranks. One Christmas he had a few of the children pour a bottle of "disappearing ink" on his mother's white carpet, one of her treasures, which she kept in pristine condition in the large living room. "¡*Ay, Dios mío!*" she exclaimed as she looked at her ink-stained carpet. He did not let on right away but waited a while before revealing the trick. She was not amused.

Dad's moods were changeable. Sometimes he was a strict disciplinarian, lessons and habits learned from his father: "spare the rod, spoil the child." He had a way of acting

---

8   Recreation of the anecdote as I recall Mom telling it to us.

tough both at home and at work, but on other occasions, he was a sentimental, doting, fun-loving dad. The older children bore the brunt of his more authoritative behaviour.

Pressures built up in the home during the mid-fifties when Dad resorted to heavy drinking. It is difficult to know what precipitated his addictive behaviour. When the situation deteriorated, he made a solemn promise, for the sake of his marriage and family, that he would go on the wagon. Of contrite heart, he pleaded for Mom's forgiveness. He remained a teetotaler for over a decade, and when he did resume drinking, did not abuse it in the way he had done previously. He was aware of the damage alcohol could have on families and was determined that it would not destroy his. He found the will and grit to exercise restraint and pledged to provide the structure and love necessary to preserve the fabric of his family life.

During the summers, we often vacationed at Tower Isle Hotel, the first tourist resort of its kind on the north coast. It was opened in 1947 by Dad's friend, schoolmate, and business associate, Abe Issa. In 1949 the House of Issa was promoting Tower Isle to airlines and travel agents in the US and the UK as "the most beautiful hotel in the British West Indies." The campaign popularized Jamaica as a tourist destination.

During July and August, both locals and foreigners patronized the hotel. For us children, it was a week of swimming, listening to calypso bands, and taking part in crab races, a time of freedom to move about without adult supervision. It felt different, foreign, a life apart from Jamaica.

We would also frequent Bonnie View Hotel in Port Antonio.[9] One summer during the mid-fifties, Dad invited Bustamante along for the vacation. (By then, they had overcome old hostilities.) "Busta was the sweetest of men personally. He used to come to the house and play with the children and joke around."[10] He had a fondness for the Spanish ladies and loved to show off his use of Castilian in the most flamboyant fashion. That summer at Bonnie View, my grandmother was vacationing with us. Relishing the chance to practise his Spanish, Busta struck up long conversations with the *señora* from Dominican Republic. Mom recalled the encounter: "Busta was a wonderful old-time gentleman and very gallant with the ladies. I remember once when we took Fred's mother to Port Antonio and we were admiring the view from the balcony at Bonnie View Hotel, Busta came up, knelt in front of her and proposed. She protested very prettily but I think she was privately pleased and flattered (all in fun, Lady B)."[11]

---

9   Bonnie View Hotel was built in the 1940s, set in the foothills of the mountains, with a spectacular view of the twin harbours and Navy Island of Port Antonio. During the 1950s, it was a hotspot for entertainment and a favourite tourist destination for American and British visitors to the island.

10   Lydia Kennedy (1997).

11   Ibid.

24

THE HON. LUIS FRED KENNEDY MLC

(1953)

Luis Fred Kennedy joined the legislative branch of government with a temporary appointment of six months by Governor Sir Hugh Foot to the Upper House in 1953. He brought a wealth of experience to his position as member of the Legislative Council.

His previous involvement in industrial relations proved valuable to his role as a member of the legislature.[1] In July 1952, new ground was broken when the first Joint Industrial Council held its inaugural meeting.[2] The Hon. Hugh Shearer,[3] who represented the BITU, and Luis Fred Kennedy, the Shipping Association of Jamaica (SAJ), were elected co-chairmen of the council. Mr. Kennedy assured trade unions that "the employers were determined that there should be no question that would not receive their most reasonable consideration and, in all respects, the greatest help they could give to make the Council work."[4] Mr. Shearer remarked

---

1   In 1951, the Sugar Manufacturers Association nominated Luis Fred Kennedy to the government-appointed arbitration board to serve as assessor in labour disputes in the sugar industry.

2   *The Daily Gleaner,* Tuesday, July 22, 1952, 1.

3   Hugh Lawson Shearer (1923–2004), trade unionist and legislator; member of the House of Representatives [JLP] for Western Kingston, 1955. The Hon. Hugh Shearer would become Jamaica's third prime minister (1967–1972).

4   *The Daily Gleaner,* Tuesday, July 22, 1952, 11.

that the "unions would bend over backwards as far as possible to achieve the best results" and that he would depend on the full cooperation of both sides to make the council work. Mr. Kennedy and Mr. Shearer proved true to their words, and co-chairmanship of the council marked the start of a long-time friendship between Luis Fred Kennedy and Hugh Shearer.

The formation of the council was hailed as the beginning of a new era in industrial relations and received praise and endorsement from the governor, Sir Hugh Foot, for its innovative approach to industrial relations. Mr. Ken Hill expressed the hope that this would be the start of similar councils formed by other industrial sectors. Its formation heightened the expectation that trade unions would do more than negotiate pay increases—that they would improve working conditions in general.

In August, Co-Chairmen Kennedy and Shearer signed the draft form of the Constitution of the Port of Kingston Joint Industrial Council. Its first order of business was to settle work distribution. The United Port Workers Union was under contract with the Shipping Association for employment of fifty percent of the work on the waterfront, and another union, not represented on the council, employed ten percent. This left forty percent of the workers in need of representation by trade unions.[5] Full registration of all port workers along with agreements in wages and benefits was realized in the 1960s with The Port of Kingston Accord.[6]

The Upper House was composed of the colonial secretary, attorney-general and treasurer, and two official and ten unofficial members appointed by the governor. The Legislative Council was a clearing house for bills and functioned as an intermediary between the House of Representatives and the Executive Council. Its function was to debate legislation, but its power to amend or reject bills passed by the house was restricted. A simple majority of all present in both houses was sufficient to pass a bill for submission to the governor for final approval. In the case of insoluble differences between the two houses, the governor reserved the right of veto power.[7] Any bill passed twice by the House of Representatives and rejected an equal number of times by the Legislative Council could still become law if it received the approval of the governor.[8]

Hon. Douglas Judah was a member of the chamber; other nominated members included Sir Robert Barker, Houghton Burke, Allan Campbell, Kaleb Helwig, Robert

---

5   Ibid., Saturday, August 16, 1952, 1.

6   See Chapter 33.

7   "If a Bill were passed by the House of Representatives and rejected by the Legislative Council in two successive annual sessions, it would then be presented to the governor for his signature," Sires (1955), 164.

8   Alex Zeidenfelt *Political and Constitutional Developments in Jamaica* (1952), 526.

Kirkwood, Brian Nation, and Philip Sherlock. The Hon. Colonel Aldington Curphey[9] was president. Bills covered a variety of subjects, including income tax and custom tariffs, marketing regulations for banana growers, a Sugar Control Bill, the World Bank Farm Subsidy Scheme, local corn production, and traffic laws.

In August 1953, Luis Fred Kennedy delivered a lengthy rebuttal of a bill that proposed new marketing and shipping arrangements for the All-Island Banana Growers Association. The Honourable Member prefaced his remarks by commending the government for the role it had played in bringing back the banana industry to a state of prosperity. He objected on the grounds that the bill granted a monopoly to two shipping companies and excluded access to markets that competitors might wish to explore. Two other marketing organizations, he claimed, would be prepared to offer proposals more beneficial to the grower than those provided by the bill. He also criticized the power given to a minister to make regulations that could result in putting the whole industry under the control of the minister without the need for going to Executive Council. He feared that the bill set a precedent for making by ministerial regulation what was in effect a statutory obligation.[10] The attorney general commended him for his vociferous attacks on the bill in a debate of "very high order."[11] The House went into committee, which passed clauses without amendment. Council resumed, the third reading was taken, and the bill passed.

In September, the Legislative Council debated a bill to amend the Income Tax Law, which proposed an increase in corporate taxation. Financial Secretary Hon. Robert Newton introduced a bill designed to make "a contribution to raising revenue for the development of the island,"[12] a commitment by the government of Jamaica to the £34 million development programme of the World Bank. The Hon. L. F. Kennedy recommended a corporate tax as the source for additional funds, given the fact that thirty-five percent of ordinary revenue was derived from customs duties. The tax was to be levied on dividends, not on corporations, so as not to burden companies with an already steep corporate tax of 37.5%. The proposed levy was six pence (6d.) in the pound or 2.5%. He argued that the bill was in effect imposing a company surtax, a measure that was scrupulously avoided in the UK, US, and Canada.

The Honourable Member was "unalterably opposed to the measure." Mr. Kennedy felt that it should be deferred and recommended that a more equitable means of direct taxation replace what he termed a most obnoxious and dangerous proposal. As an

---

9   Hon. Colonel Aldington George Curphey, medical practitioner and custos rotulorum for St. Ann, president of Legislative Council 1952, member since 1945, joint president of Commonwealth Parliamentary Association. Born 1880, attended York Castle High School, Wolmer's and Queen's University in Canada.

10   *The Daily Gleaner,* Monday, August 24, 1953, 6.

11   Ibid., 13.

12   Ibid., Tuesday, September 8, 1953, 6.

aside, he criticized the tax system in general. He felt that people earning £2000 or £5000 a year were not making their proportionate contributions, and that people in lower income brackets were burdened by indirect taxation—forced to buy goods at excessively high prices. He thought government was at fault for limiting imports of cheaper commodities. Despite a lengthy debate, and admission by the financial secretary and other officials of the basic inequities of the bill, it was passed by simple majority and enacted into law.

Kingston Wharves

Luis Fred contended with congestion on the finger wharves that were unable to handle the volume of ships or accommodate larger vessels, for example, from New Zealand. In October 1953 meetings were held with government officials, who maintained that removing silt from the harbour was an expensive and technical operation and that engineers advised dredging could weaken the structure of existing piers.[13]

The Daily Gleaner in editorials published, January 3 and 4, 1954, accused the Minister of Communications and Works of "bureaucratic apathy" in dealing with shippers, and of ignoring complaints by New Zealand shipping lines of unacceptable harbour conditions. In a November 1954 meeting of the chamber, Luis Fred Kennedy spoke, on behalf of Kingston Wharves Ltd. and the Shipping Association, of the deplorable conditions. Traffic in the harbour had outgrown port facilities, and long delays in offloading, inadequate storage of goods, and pilferage were causing loss of revenue for shippers.[14]

One solution, initiated by the JLP government, was to open a new port area in Kingston's west-end at a cost of £8 million, but private business was reluctant to risk large outlays of capital without the assurance that government would not nationalize the waterfront or set up wharf facilities in direct competition with privately-owned wharves.

Government changed in 1955 with the election of Chief Minister Norman Manley,[15] who began a new set of consultations with wharf representatives. He scrapped the idea of a new port and invited shippers to invest instead in expanding existing facilities.[16]

In a meeting of the House of Representatives in November 1956, Mr. Manley outlined a plan to construct a foreshore road by completing a forty-foot fill along the entire waterfront. The scheme entailed the extension of several finger piers to provide deep water berths.[17] Each wharf owner would be responsible for his company's own area. It

---

13   Meetings included Chairman of the Shipping Association, Luis Fred Kennedy, the harbour master, representatives from the Ministry for Trade and Industry and the Citrus Growers Association, the director of Public Works, the acting general manager of the railway, and Mr. Hart of the Banana Producers Association. Luis Fred Kennedy was chairman of the Shipping Association, 1950–1953.

14   *The Daily Gleaner,* November 22, 1954, 15.

15   See Chapter 26.

16   *The Daily Gleaner,* Thursday, March 29, 1956, 1.

17   *Ibid.,* Thursday, November 15, 1956, 11.

was a less expensive project, aimed at relieving vehicular traffic congestion along the waterfront and providing better berthing and storage facilities. Mr. Manley regretted that Kingston was not ready or able to invest in the creation of a more modern port with advanced technology. The estimated cost of £8-10 million to fund the scheme was not feasible given the commitments he had made to the improvement of education and infrastructure.

Luis Fred realized that Manley's plan could only be a temporary solution given the growth of the shipping industry.

KENNEDY, Luis Frederick, B.A. (Holy Cross), Business Executive; Mng. Dir. Grace, Kennedy & Co., Ltd., Kgn. Wharves Ltd.; Dir. Ja. Rums Ltd., Grace Shipping Co., Ltd., Ja. Biscuit Co., Ltd., Sheffield & Co., Ltd.; Chairman Rice Industry Bd.; mem. Spirit Control Bd.; b. Puerto Plata, Dominican Republic, Nov. 2, 1908, s. late Fred William Kennedy, Merchant, & Luisa Matilda Arzeno, his wife; Ed. St. George's College (Ja.), Holy Cross College (U.S.A.); formerly Vice-Pres. Ja. Chamber of Commerce; acted as M.L.C., Ja., 1953; Chairman Shipping Assn. of Jamaica 1950-53; Religion: Roman Catholic; m. Oct. 2, 1937, Lydia Matilda Loinaz of N.Y.; 2 s., 3 d.; Clubs: Jamaica, Liguanea, Royal Jamaica Yacht, Kingston Cricket; Address: 64 Harbour St., Kingston, & 45 Lady Musgrave Rd., Half Way Tree P.O., St. Andrew.

Illustrated Biographical Record of Luis Fred Kennedy from *Who's Who Jamaica 1957*, page 267

(Photo, courtesy of © The Gleaner Company (Media) Ltd.)

"We would welcome you to Grace."

With Peter Moss-Solomon, Montreal, 1958.

(Photo from family collection)

25

BUILDING THE FAMILY BUSINESS (1950s)

I think of the company as a family (**LFK**).

The 1950s saw the phenomenal growth of Grace, Kennedy & Co. Ltd. Five years after assuming the position of governing director, Luis Fred reported significant increases in the net income of the company. In one year only, 1951–1952, revenue after all expenses grew from £26,783-0-0 to £76,146-19-1.[1] He would often say, "*I make no apologies for making a profit.*"

He aimed to increase the company's profitability and capacity for future growth. Strategies he employed were to increase market share through diversification and acquisition of new companies and to build an asset base by capitalizing part of General

---

[1] Details provided re Grace, Kennedy's balance of payments and operations are gleaned from the minutes of board of directors' meetings for the period, 1952–1961.

Reserve for issuance of bonus shares.[2] He believed that employee shared ownership was paramount to productivity.

Each year directors recommended payment of dividends in the amounts of eight percent on preference shares and five percent on ordinary shares. Distribution of profits were made, appropriating amounts for the payment of additional salaries or Christmas bonuses to staff, amounting to a total of £543, 732 over the period, 1952–1961, and of a ten percent commission to managers, capped at £3000 per annum.

Through capitalization of the General Reserve in 1952 and subsequent years, he recommended the issuance of bonus shares, each valued at £100, to ordinary and employee shareholders.[3] In 1952 the board reported that "part of the amount standing to the credit of the General Reserve was capitalized and distributed among the Shareholders in the proportion of one share for one share held on December 1951."[4]

Over a ten-year period 1951–1960, the company issued 3,900 ordinary shares valued at £390,000 and 480 employee shares valued at £48,000. Through capitalization of retained earnings, the capital of the company over this same period was increased from £26,783-0-1 to £800,000.

In 1952, effective September 1, "the Company instituted a non-contributory Group Life Insurance scheme for the benefit of employees on a cover of approximately one year's basic salary with an approved graduated scale."[5]

Luis Fred sought guidance from a board of directors. In the early fifties, Harold H. Dunn (Daddy Dunn), a contemporary of his father's, was still serving on the board. General Manager, James Moss-Solomon, was second in command, filling in for Mr. Kennedy when he was absent from meetings. A relative of the Moss-Solomons, F. N. DaCosta[6] was a long-time employee and director and served for a while as company secretary, appointed by Dr. Grace in 1936. He retired in 1957. Luis Fred's brother, José (Manager, Shipping Division), and brother-in-law, Simon

---

    2   In the minutes of the board of directors' meetings the term "bonus shares" is often used. Bonus shares were transferred out of capitalization of the General Reserve and issued in proportion to each shareholder's stake in the company. "Bonus issues do not dilute shareholders' equity, because they are issued to existing shareholders in a constant ratio that keeps the relative equity of each shareholder the same as before the issue." https://www.investopedia.com/terms/b/bonusissue.asp

    3   Employee shares were sold at par on a re-payment program, and on condition that, if the employee should leave the company for whatever reason, the company would buy back the shares. When employee shares were issued, employees would benefit in the same way as those who held ordinary shares. Employee shares were offered at the discretion of the governing director and the board.

    4   Grace, Kennedy & Co. Ltd., minutes of the board of directors, May 13, 1952.

    5   Ibid., April 23, 1953.

    6   Uncle Frankie, a relative of the Moss-Solomons.

Soutar (Manager, Agriculture, and Assistant Manager to Mr. Moss-Solomon),[7] were also on the board. Luis Fred appointed his youngest brother, Francis, director in 1961.

Carlton Alexander acquired shareholdings during the period; he was appointed a director of the board and promoted to assistant manager to James Moss-Solomon. Geoffrey Dodd[8] was also a trusted long-serving employee, who had joined Grace, Kennedy in 1934. He worked in shipping alongside José Kennedy and was appointed director in 1953. Gladstone Kamicka served as secretary until 1960, when Mr. H. M. Aarons was appointed.

Transfers of shares required the explicit permission of the board. During the period, Luis Fred's mother, Luisa Kennedy, transferred part of her shareholding to her six children.[9] In May 1958,[10] James Moss-Solomon and his wife, Ismilda, transferred shares to their son Peter and to Peter's first cousin, D. A. Moss-Solomon.[11] Joan E. Moss-Solomon, Peter's sister, received a portion of these with a transfer made two years later.[12]

Expanding Overseas

In February 1952, the chairman reported that it had been decided to form a Canadian subsidiary of the company called Grace, Kennedy & Co. (Canada) Ltd., fully subscribed by Grace, Kennedy and incorporated for the sum of Canadian $50,000. He appointed Mr. Jules Wolfe[13] as general manager, but the company did not get off to a good start. Mr. Wolfe passed away shortly after taking charge, and Mr. Skelton was promoted to fill the position.

---

7   José Kennedy joined the company in 1942 after leaving his law practice. Simon Soutar joined in 1941.

8   Geoffrey Evelyn Dodd, born May 17, 1912, was son of Edward Austin Dodd, civil engineer, and his wife, Hilda, née Sharp. He was educated at Wolmer's (1922–1928) and worked with Canadian Imperial Bank of Commerce (1928–1934) before joining Grace, Kennedy.

9   Grace, Kennedy & Co. Ltd., November 1 and 3, 1954.

10 Ibid., May 6, 1958.

11 Don Aidan Moss-Solomon (1923–1998), nephew to James Moss-Solomon, son of his older brother, Alfred. He was married to Marguerite Marie Robey with four children.

12 Grace, Kennedy & Co. Ltd., October 21, 1960.

13 Dad kept in touch with Jules Wolfe after he migrated. An earlier spelling of his name was Germanic, Wolff. He later used an anglicized, spelling of his surname, Wolfe.

In 1955, the company advertised for the job of assistant manager. A young man by the name of Boerries Terfloth,[14] who had migrated to Canada from Germany in 1953, applied for the position.

> I was very pleased to find this job as assistant manager for the Grace, Kennedy company in Montreal and was interviewed by Mr. Skelton's friend, who was a professor at the Edmonton University. I obtained the job, accepted out of 180 applicants, at a low salary, $180/month. I was then personally interviewed by Mr. Gladstone Kamicka who was the accounting head. This in turn led to Mr. L. F. Kennedy coming personally to speak to me.[15]

In April 1956, the chairman reported that the Canadian subsidiary, since its incorporation, "had suffered substantial losses in proportion to its shareholding in the amount of Canadian $20,000."[16] Mr. Kennedy offered the position of general manager to Mr. Terfloth, who accepted on condition of a sixty percent increase in salary to $450/month. "He consequently made me manager of the company, which had only two office rooms and a part time secretary. It followed that he invited me and my wife to Jamaica to see the operation there, and he suggested I take a Canadian shipping company ship from St. Lucia to Antigua to get to know the islands. I also went to Belize and spent a few days in Haiti."[17] "Mr. Kennedy let Mr. Skelton go because he really didn't know much about export, and from then I got the business going."[18]

"Your father listened to me. He was a good trader; he followed up what he was thinking about, and he put me in charge. I was happy I could wipe out the loss position and make the company profitable."[19]

One year later, the chairman reported that Mr. Terfloth had done "an admirable job to put the company's business on a footing whereby the profits were enough to cover the expenditure of running the business."[20] This was the beginning of Terfloth's

---

14   Boerries Terfloth, born, February 20, 1929, in Riga, Latvia, died on May 22, 2022 in Hudson, Quebec. He was a director of B. Terfloth & Co. USA (Inc). The history of B. Terfloth & Co., leader in food supply management, dates to 1774 in Greven, Westfalia, Germany, where Boerries's father was involved in the food and lumber trade and shipbuilding. See https://www.terfloth.com/en/history. I had the opportunity to speak with Boerries in 2020. He shared memories of Luis Fred Kennedy and his association with Grace, Kennedy.

15   Boerries Terfloth, excerpt from personal correspondence with the author, June 28, 2019.

16   Grace, Kennedy & Co. Ltd. (April 26, 1956).

17   Boerries Terfloth, excerpt from personal correspondence with the author, June 28, 2019.

18   Boerries Terfloth, interview with author, January 23, 2020.

19   Ibid.

20   Grace, Kennedy & Co. Ltd., April 16, 1957.

forty-year association with Grace, Kennedy & Co. Ltd. in trade and in international marketing of the Grace brand, which was launched in the early 1960s.

With the assistance of Boerries, Luis Fred established two other overseas subsidiaries: in 1958, Grace, Kennedy & Co. (Europa) in Rotterdam, Holland, and in 1962, Grace, Kennedy & Co. (UK) in London.

Promoting Young Management

Two shining stars hired during the 1950s were Bruce Rickards and Paul Scott. Bruce joined Grace in 1951 at the age of twenty-one as an accounts clerk. He gained experience as a collector and as a salesman before being appointed supervisor of the Order Department in 1956. Three years later, he was promoted to products manager for Cold Storage and Hotel Merchandise, and for the next three decades, he filled other top managerial positions.[21]

Laurence Paul Scott joined Grace, Kennedy in 1958 at age twenty-two.[22] After having first met with José Kennedy and Simon Soutar of the Shipping Division, and then with Carlton Alexander, he was interviewed by Luis Fred:

> I was ushered into Mr. Kennedy's office by someone whom I later came to know as Dolly Chevolleau,[23] his secretary. She was very friendly and kind to this somewhat intimidated candidate for employment. When I entered the office, there was a gentleman with thick glasses speaking on the telephone—very serious and gruff looking—the very opposite impression to the friendly secretary. He continued on the telephone for some time while I waited, in the meantime each of us sizing up the other as he continued to speak on the telephone.
>
> My interview began and Mr. Kennedy gave an overview of his concept of expanding the company and in particular its shipping interests, the planning for a new line (Canada-Jamaica Line) to be owned by Jamaican importers to challenge Saguenay Shipping from Eastern

---

21 Bruce Errol Anthony Rickards was born in Spanish Town, April 12, 1930. After graduating from Kingston College in 1947, he worked as a clerk of the Agricultural Loan Society's Board until 1951. He married Marjorie Elizabeth Fennell in 1956 and had one son and one daughter. He was the recipient of many awards, including the Order of Distinction (commander). He worked for Grace, Kennedy for forty-two years. He died February 23, 2002. See also note, Chapter 31.

22 Laurence Paul Scott was born June 19, 1936, son of Laurence Matthew Scott, managing director of Cecil B. Facey Ltd., and Clare O'Toole. He was educated at Munro College (1946–1948) and Priory School (1948–1951), after which he worked for Bank of Nova Scotia and Pan American World Airways. He married Marian Joan Orrett, May 3, 1958. Source: *Who's Who Jamaica* 1963.

23 Mary Dorothy Chevolleau. See photo, Chapter 28.

Canada into Jamaica and the Caribbean, and the expansion of Kingston Wharves Ltd., which GKCO managed and in which they also had a major shareholding.

> I was impressed by Mr. Luis Fred and greatly respected his vision for the company and the future. Fortunately, he apparently thought I was worth taking a chance on and I was hired as Grace, Kennedy's very first "Freight Solicitor," although exactly what this was would emerge only over time.
>
> When I joined GKCO in early 1958, the Shipping Department consisted of Geoff Dodd, José Kennedy, Ronny Rickards, Rosalee West and myself in a single office on the ground floor near the front door of 64 Harbour Street. So began life at what was to be my "business university" where I learned from the examples of Messieurs Moss-Solomon, Kennedy, and Alexander, but mostly Professor Luis Fred Kennedy.[24]

Luis Fred tested Paul's skills at labour negotiations by appointing him secretary of the Port of Kingston Joint Industrial Council.[25] Paul subsequently held the post of vice-chairman of the Shipping Association (1961) and was promoted to general manager of Grace, Kennedy & Co. Shipping Ltd.[26] in May 1963 at the age of twenty-seven.

The shipping agency and wharfing business were lucrative. In 1955, representatives of the Alcoa Steamship Co. Ld. hosted a special reception at the Myrtle Bank Hotel to celebrate twenty-five years of Grace, Kennedy serving as their agents. Both Luis Fred and his mother were presented with aluminum trays and gifts for their "faithful service."[27] Mr. Kennedy was praised for being a "wonderful person" with whom to do business.

Grace, Kennedy was the agent for several shipping companies: the French Line, with its fortnightly sailings to ports in Europe; the K Line with monthly schedules from Hong Kong, Canada, the US West Coast and Panama Canal to Kingston; the Grace Line with its passenger and freight service every Friday to New York from Kingston; for the Begona, a Spanish-owned steamship company from Kingston to Southampton; and the company was a joint agent with Jamaica Fruit & Shipping for the newly formed Canada-Jamaica Line (a first and only direct service from Eastern Canada to Kingston).

---

24   Laurence Paul Scott. "Memories of Grace." In *Memories of What Made Grace* (1997).

25   See Chapter 24 for information on the KJIC, which Luis Fred founded and co-chaired with the Hon. Hugh Shearer.

26   Grace, Kennedy & Co. Shipping Ltd. was formed in 1963 out of a consolidation of shipping activities with Royal Mail Lines. The company assumed the general agency for lines formally held by Grace, Kennedy and Royal Mail for all ports in Jamaica. Source: Laurence Paul Scott (1997).

27   *The Daily Gleaner,* August 3, 1955, 7.

Building the Family Business (1950s)

The growth of the shipping side of the business went hand-in-hand with Grace, Kennedy's core Merchandise Division, managed by Carlton Alexander. During the 1950s, Grace, Kennedy did not yet have its own Grace brand. Sixty percent of merchandise was importation of bulk items and forty percent products such as New Zealand Anchor butter, Anchor pure cream milk powder, Hellaby's corned beef, and General Mills Gold Medal flour.[28]

Attention was paid to grooming family members. The younger generation, Peter Moss-Solomon and Francis (Paco) Xavier Kennedy, firstborn sons of James Moss-Solomon and Luis Fred Kennedy respectively, started at Grace, Kennedy during the late 1950s. Their fathers took steps to include them as shareholders and to train them in all aspects of management.[29]

Peter commented in an interview about a conversation he had with Mr. Kennedy in 1958 in Montreal, where he was studying medicine at McGill University:

> This is where your dad comes in. I came home for Christmas and went back up at the end of January. Your Dad and Carlton usually came up to Montreal, and they used to meet with Terfloth. That year, he phoned me when he got up there and he discovered that I wasn't really doing very well, and he had me to dinner at his hotel. I told him my thoughts and he said to me, "Peter, if you ever decide not to proceed, we would welcome you at Grace," and that was the first time I had any thought about joining Grace. So I came down the end of February and decided to take a summer job at Grace on the first day of April 1958. I went there on a summer job, and I stayed for forty years.[30]

Having graduated from Georgetown Prep, Paco started at Holy Cross College in September 1958. But at the end of his first academic year, chose to return home to join the business. In a tribute to Luis Fred Kennedy, Anna Jarvis[31] commented about Paco's first day at work:

> Did Paco ever tell you that on his first day of work at Grace, Kennedy he got into his father's car to drive with him to the office on Harbour Street? Mr. Kennedy told Paco he was not earning enough to drive to work, he should take the bus or ride his bicycle! Mr. Fred never

---

28   Bruce Rickards, Grace Foods International Conference, Montreal. Canada, April 2, 1970. In *Memories of What Made Grace* (1997), xix.

29   Paco received a transfer of thirty shares from his father, September 1961.

30   Peter N. Moss-Solomon, interview with author, February 20, 2019.

31   Anna Jarvis (née Figueroa) is a close family friend. Dr. Dorothy Anna Jarvis is retired staff physician in the Division of Pediatric Emergency Medicine at SickKids Hospital in Toronto, and professor emerita at the Department of Pediatric University of Toronto. She is married to Dr. Del Jarvis. See other biographical note, Chapter 14.

showed any favouritism to family members at work. This was a trait that Carlton Alexander adopted and embraced fully. He never allowed his family members any slack either.[32]

Peter at age twenty-one and Paco at age nineteen embarked on a challenging career of a lifetime. They worked together as managers and became good friends, like their fathers.

Dr. John J. Grace died in San Rafael California in his eighty-sixth year.[33] The obituary in *The Daily Gleaner* of February 29, 1956, praised him as a man who had been held in high esteem in Jamaica. He outlived his eight siblings, having lost his last remaining sister, Agnes Maude Grace, aged eighty-four, in 1955. He was survived by his wife, Eleanor Vere Greenfield, and three children, John, Alicia, and Cynthia. His passing concluded a chapter in the life of Luis Fred, who bade farewell to one of his father's few remaining contemporaries and business associates, octogenarians, born in the 1870s.

The Kennedy family also suffered a loss that year. At noon hour on August 12, Cecille Lemay, wife of Dad's youngest brother, Francis, died at the age of thirty-one. She had been admitted to hospital for a routine operation after which she developed a high fever caused by a bacterial infection, later diagnosed as tetanus. It was never determined how she may have contracted the fatal disease, but the family suspected contaminated surgical instruments may have been the culprit. When the news broke, Mom gathered us to kneel and say a prayer before the statue of Our Lady, which stood on an altar in our living room. Cecille's immediate family was devastated. She left four young children, Lareine (eight), Stephen (six), Anne Marie (five), and Kevin (eighteen months) in the care of their father, Francis, who was also only thirty-one at the time.

The following year, Jamaica lost one of its stalwart business leaders. On April 3, 1957, *The Daily Gleaner* carried the headline:

Pioneer in banana co-op
movement in Jamaica

Mr. Charles Johnston dies
In his 86th year[34]

---

32 Anna Jarvis, "Mr. Luis Fred Kennedy: Memories," (2019).
33 February 11, 1956.
34 *The Daily Gleaner,* April 3, 1957, 1.

Building the Family Business (1950s)

At the time of his death,[35] Maas Charlie was the managing director of the Jamaica Banana Producers Association, a director of the Jamaica Fruit and Shipping Company Ltd., of Adolph Levy and Brothers Ltd., and of Kingston Wharves. He was survived by his wife, Madge, and two children, Lucille and Ernest, and grandchildren.

Governor Sir Hugh Foot hailed him as the giant of the banana industry in Jamaica. *The Daily Gleaner* reported that "Maas Charlie, who was a self-educated man of great capabilities, was also innately modest. Thus, although he gave very generously to any cause in which he believed, it was on a single condition: that he remained anonymous."[36] "Not only the banana industry but the whole country had lost a leader by his passing."[37]

Men like John Johnston Grace and Charles Edward Johnston had seen dramatic changes in Jamaica over a span of three generations, which witnessed the ravages of two world wars and the successes of rebuilding in their aftermath. They lived through what is commonly known as the Second Industrial Revolution (1870–1950) and "The Great Leap Forward"[38]—a time of rapid technological and economic change with advancements in steel, internal combustion engines, the railroad, electricity, rubber, and plastics. They saw inventions of the light bulb, the telephone, radio, cameras, airplanes, motion pictures, and television. Grace and Johnston were business moguls who had the ingenuity to succeed during a period of global transformation.

---

35   April 2, 1957, C. E. Johnston died at his home on Osbourne Road in St. Ann after an illness of some months.

36   *The Daily Gleaner,* Wednesday, April 3, 1957, 13.

37   Ibid.

38   Chris Vickers and Nicolas L. Ziebarth, "Lessons for Today from Past Periods of Rapid Technological Change." https://www.un.org/esa/desa/papers/2019/wp158_2019.pdf (2018).

SENATOR DOUGLAS JUDAH

Hon. Douglas Judah (MLC), one of two Jamaicans appointed by Governor General Rt. Hon. Lord Hailes, April 1958, as member of the nineteen-seat West Indies Senate.

He was named as Senator by Hon. Alexander Bustamante, Leader of the Opposition.

The Daily Gleaner, Saturday, April 12, 1958, 1.

(Photo, courtesy of © The Gleaner Company (Media) Ltd.)

Hon. Douglas Judah greets the newly appointed Governor General of Independent Jamaica, Sir Clifford Clarence Campbell.

A few months after Independence, December 1, 1962, Clifford Campbell was sworn in as Governor General, succeeding Sir Kenneth Blackburne.

(Photo, courtesy of Douglas Judah's sons, Bruce, Peter, and David)

26

MARCH TO FEDERATION

"One from ten leaves nought."[1]

(1955–1961)

Significant political changes had occurred in the previous decade to pave the way for a Federated British West Indies. With constitutional reforms of self-government that included elected legislators and diminished powers of the governor, the West Indies evolved to a stage within reach of independence from British rule. Norman Manley and other heads of state, Sir Grantley Adams of Barbados and Dr. Eric Williams of Trinidad, were strong proponents of a federated nation.

Jamaica went to the polls in 1955.

On election eve, thousands of PNP supporters across the island were seen roaming the streets of Kingston and rural towns with brooms in hand, sweeping, singing, and shouting, "Sweep them out!" Norman Manley's son, Michael Manley,

---

1 Dr. Eric Williams

was new to politics, and part of the campaign; he played a key role in winning electoral votes by broadening the base of membership of the National Workers Union.[2]

Norman Manley's victory on January 12, 1955, was decisive. The party won by eighteen to JLP's fourteen seats and gained the popular vote, 50.50% to JLP's 39%. Norman Manley became Chief Minister of Jamaica:

> I cannot find words to express thanks to the voters of Jamaica for the remarkable victory they have given the People's National Party this day. Tight enough, but acceptable and accepted.
>
> To the country as a whole I renew our pledge to work for self-government at the earliest possible date, and to seek to enhance the national spirit in every way. To the Caribbean territories I proclaim a party devoted to the interests of the area as a whole and to the march to federation and dominion status (Norman Manley).[3]

Bustamante was spirited in the face of defeat. "What I have done for the working class of this country in the last 15 years, no other man or other group of persons will be able to equal it, perhaps within 100 years."[4] *The London Times* reported that Bustamante had not lost due to any fault of his own. He had proven himself to be a shrewd, fair-minded leader, who had cooperated loyally with successive governors. The man who replaced him was a "more urban and middle-class Jamaican, less flamboyant and perhaps more contemporary."[5]

Historically, a federation had received the support of the British government. As early as 1909, discussions were held in London about forming a union. In the

---

2   Michael Manley became a member of the Executive Committee of the PNP and National Workers Union in 1952, and island supervisor and first vice-president of the union in 1955. He had recently returned from England where he studied at the London School of Economics (1946–1950) and later worked as a journalist with BBC (1950–51). He was born, December 10, 1924, and was educated at Jamaica College (1935–1943). After high school, he left for Canada, where he served as a pilot in the Royal Canadian Air Force (1943–45). Before returning to Jamaica, he married Jacqueline Kamellard, with whom he had his firstborn, Rachel Manley, (*Who's Who 1957*, 307; Bertram, 2016, 241).

3   Proclaiming victory after January 1955 general election, Norman Manley is quoted by Rex Nettleford (ed.), *Manley & the New Jamaica: Selected Speeches and Writings 1938–1968* (1971), 194.

4   *The Daily Gleaner*, Thursday, January 13, 1955, 1.

5   *The London Times*, January 14, 1955, as quoted by *The Daily Gleaner*, Saturday, January 15, 1955, 6.

1940s, the Moyne Commission[6] was instrumental in urging West Indian territories to begin negotiating. By 1944, a West Indies Labour Conference meeting was held in British Guyana, where member delegates from Barbados, Grenada, Trinidad, and British Guyana unanimously endorsed a resolution calling for federation. The following year, secretary of state for the colonies, Oliver Stanley, asked governors of the territories "to place the issue of federation before the colonial legislatures for debate."[7] The British government would cooperate in framing proposals if there was support. In 1947 at a conference in Montego Bay, politicians, labour leaders, and other groups representing twelve territories, appointed a standing committee to draft a constitution.[8] A series of drawn-out debates followed. In 1953 delegates met in London to reconcile differences, and three years later, formalized an agreement.

Private business played a role. Norman Manley's policies to encourage private investment in the economy won the support of Luis Fred Kennedy and others of the mercantile community. Luis Fred became an advocate of the PNP agenda through his representation on government boards (Spirits Control Board and chair of the Rice Industry Board), the Chamber of Commerce, and the Jamaica Imperial Association.[9] Jamaican entrepreneurs saw opportunities for investment in shipping, manufacturing, tourism, and construction industries[10] within a new union of states.

Douglas Judah chaired delegations for the signing of the official document in London on February 23, 1956, at Lancaster House. It represented a legal agreement for the creation of a new federated state, endorsed by West Indian

---

6 See Chapter 20. The Moyne Commission reported that social reforms could not be fully implemented in the West Indies "unless they were accompanied by the largest measure of constitutional development" and "greater participation of the people in the business of government." (Jesse Harris Proctor Jr., December 1961, 284–285, quoting *West India Royal Commission Report*.)

7 Jesse Harris Proctor Jr., *British West Indian Society and Government in Transition 1920–1960* (December 1962), 295.

8 "The British government called the conference of the leading politicians in the respective islands, and of other civil groups like the chambers of commerce, trade unions etc. Chief Minister Bustamante and Norman Manley, representatives of the Caribbean Commission, were both present. The Caribbean Labour Congress timed their meeting for a week before the MoBay conference so that they would be well represented" (Swithin Wilmot, editorial notes).

9 Jamaica Imperial Association was founded in 1917 by Sir Arthur Farquharson. It was an independent institution (not financed by the government) that represented the economic interests of Jamaica and the region. In its early years, it supported the growth of the sugar and rum industry by influencing the government in the 1920s to enact the Sugar Industry Aid Law to protect sugar planters. Fred William Kennedy had participated in JIA negotiations during the 1920s. (See *The Daily Gleaner*, May 19, 1923, 3.)

10 The 1950s saw the rise of the Matalon and Hanna dynasties (Bertram, 2016, 256–257).

territories.[11] Weeks after, the Jamaica Chamber of Commerce invited him to be guest speaker at its annual meeting held in March 1956. Mr. Judah exhorted the chamber to embrace change, to take that "leap of faith," as he called it. Jamaica had a chance to be independent of Britain, politically and economically, if it joined a federal union. The time was right, he said. Progress in science, the telegraph, cable, and the aircraft had shortened distance and time between islands, previously thought to have been isolated from one other. Mr. Judah reaffirmed his belief that a common language, colonial history, law, customs, and economies meant that cooperation was possible and desirable. But it would take hard work. The concept of federation by Order of Council were mere words on a page, futile without the good will and cooperation of businessmen, politicians, and the general populace. His speech was met with an unexpectedly loud applause and a standing ovation.[12]

Following the annual meeting of the chamber, Chief Minister Hon. Norman Manley was the guest of honour at a luncheon attended by over 200 prominent business leaders, including Carlton Alexander representing Grace, Kennedy. President Dudley Levy[13] gave the full support of the chamber: "To the Chief Minister, let me say that we are proud that you guide our destiny."[14] Mr. Manley praised the chamber's endorsement of federation and said too that he wished to dispel rumours that he and his government opposed a customs union for the region. He was confident that an agreement could be reached with respect to trade, commerce, and customs, to be included in the constitution.

Douglas Judah, with the support of Luis Fred Kennedy and other members of the Jamaica Imperial Association, promoted federation. They were confident that the House of Representations would receive proposals from the association with open arms. Mr. Kennedy recommended in 1957 that the JIA deal with affairs pertaining to the West Indies as a whole (for example, trade, taxes, tariffs, investments) and that plans be put in place to establish a branch in Port of Spain, Trinidad.[15]

The vigorous endorsement by the People's National Party, support of Trinidad and Tobago, and concessions made by the Leeward and Windward Islands brought plans to fruition. The West Indies Federation was formed on January 3, 1958, with the swearing

---

11   British territories to join the federation in 1958 included Jamaica, Cayman Islands, Turks and Caicos Islands, Barbados, Antigua and Barbuda, St. Christopher-Nevis-Anguilla, Montserrat, Dominica, St. Lucia, St. Vincent and the Grenadines, Grenada, and Trinidad and Tobago. Opting out were the Bahamas, Bermuda, Belize, the British Virgin Islands, and Guyana.

12   *The Daily Gleaner*, Friday, March 16, 1956, 16.

13   Dudley Emanuel Levy, J.P., President of Chamber of Commerce, Chairman, Levy Bros. Ltd., member of Jamaica Imperial Association, b. St. Andrew, September 9, 1900. (*Who's Who Jamaica 1957*).

14   *The Daily Gleaner*, Friday, March 16, 1956, 16.

15   *The Daily Gleaner*, Thursday, October 17, 1957, 6.

in of the governor general and, in April, of Prime Minister Sir Grantley Adams, who was vice-president of the West Indies Federal Labour Party (WIFLP)[16] and the first Premier of Barbados (1953). In April 1958, Mr. Douglas Judah, named by Alexander Bustamante, Leader of the Opposition, and appointed by the governor general, was one of two Jamaicans senators to complete the first West Indies Parliament.[17]

Member states of the Eastern Caribbean were disappointed that neither Mr. Manley nor Dr. Eric Williams had contested seats in the federal elections. The Federation was off to a difficult start: the PNP of Jamaica, one of WIFLP's strongest affiliates, won only 5 of 17 seats, and the People's National Movement of Trinidad and Tobago also failed to secure a majority of seats.

Manley believed his first duty was to Jamaica: "I came into politics to work and I have come to the conclusion, not without some bitter moments, but I have come to the conclusion to stay in Jamaica."[18] If federation were to move to the next stage, to be recognized as a nation within the Commonwealth, he claimed, "Jamaica and Trinidad must be developed to the fullest possible extent and they must be supporting Federation one hundred per cent."[19] The newly formed federation could not survive without Jamaica and Trinidad.

With the advent of a federated state came political and economic uncertainties that changed the mood of the mercantile community.[20] Its support waned. Jamaica's membership in GATT (General Agreement on Tariffs and Trade) was a foregone conclusion—Jamaica would be expected to sign agreements on internal free trade and a common external tariff.[21] The merchants were not happy.

---

16   West Indies Federal Labour Party (WIFLP) was founded by Norman Manley, who became its first president in 1956. In the 1958 elections, it narrowly defeated the West Indies Democratic Labour Party (WIDLP), 22-20 seats. The WIDLP was headed by Alexander Bustamante. Information gathered from: https://www.encyclopedia.com/history/encyclopedias-almanacs-transcripts-and-maps/west-indies-federal-labour-party

17   *The Daily Gleaner*, Saturday, April 12, 1958, 1. The other Jamaican, named by the Council of Ministers, was Mr. Allan George Richard Byfield, schoolmaster. He was named by the Council of Ministers headed by Norman Manley, leader of the Federal Labour Party.

18   Norman Manley, in Rex Nettleford (ed.), *Manley and the New Jamaica: Selected Speeches and Writings 1938–1968* (1971), 169.

19   Ibid.

20   Problems plagued the newly formed West Indies Federation: restricted movement of people between territories; inadequate funding (small federal budget); disagreements re levying of taxes; seat allocations in House of Representatives; excessive control by the British Crown (veto powers of the governor general).

21   GATT: an international multilateral trade agreement (1947), which evolved over a twenty-eight-year period to become World Trade Organization (WTO) in 1995. Luis Fred was opposed to GATT. In 1952 he drafted a memorandum to be submitted to the Conference of Commonwealth Prime Ministers on the economic implications of GATT on a federated union (*The Daily Gleaner,* Saturday, September 6, 1952, 1). And in 1953 he publicly expressed his concerns at a meeting of the Jamaica Imperial Association (*The Daily Gleaner,* Tuesday, February 8, 1953).

Luis Fred was troubled by the potential impact of a new Customs Union on local industry, which had historically enjoyed preferential tariff treatment from Britain and Canada.[22] Secondary industries had built up behind "high tariff walls,"[23] which both government and private business guarded, protections that GATT would put at risk. Jamaica's economy was working well. Diversification through mining, manufacturing, and tourism was a buffer against adverse world economic conditions. Size mattered. Jamaica, as the most industrialized country within Federation, afforded higher tariffs to protect local industry than the smaller islands would or could levy. With minority representation in parliament, Jamaica's chances of reconciling these differences Luis Fred thought to be improbable.

The Chamber of Commerce changed its tune. It criticized the government for its silence on "the material costs" to Jamaica, of "disproportionate representation" in the federal parliament, and for its failure to "afford Jamaican industries fuller protection."[24] The chamber welcomed the support of the Jamaica Manufacturers Association, who warned that "the future was menaced with the possible collapse of Jamaican industries."[25] The business community was alarmed. "We have gone into Federation with our eyes closed."

On January 13, 1958, the chamber recommended a telegram be dispatched to Mr. Manley urging him to make a statement about Jamaica and GATT. A committee of merchants (one of whom was Mr. Carlton Alexander) was formed to examine the issues.

Norman Manley was cautiously optimistic that an agreement could be reached with respect to trade:

> All I say in regard to that is this, that we in Jamaica will play our part and express our views when the time comes, because we know what we think our policy should be. We know that it is life and death to the West Indies as a whole and life and death to Jamaica if we do not industrialize.[26]

Even though an active proponent of federation, Mr. Manley was not blinded by "honours or glory."[27] He was reasoned in his approach: "We are not prepared to go

---

22 Under the Canada-West Indies Agreement, "margins of preference were granted to Canada and other Commonwealth countries by the West Indies in return for which the West Indies enjoyed valuable concessions in the Commonwealth markets including both Canada and the UK." *The Daily Gleaner,* Tuesday, December 30, 1958, 8.

23 Charles H. Archibald, "The Failure of the West Indies Federation" (June 1962), 237.

24 *The Daily Gleaner*, Tuesday, January 14, 1958, 11, article entitled, COC ALARMED OVER GATT.

25 Ibid.

26 Norman Manley, excerpt from Annual Party Conference Speech, September 28, 1958, in Rex Nettleford (ed.), *Manley and the New Jamaica,* 248.

27 Norman Manley, January 18, 1958, in Rex Nettleford (ed.), *Manley and the New Jamaica,* 169.

with anybody in attempting to build federation so fast as to destroy Jamaica. Sooner than that we will come out."[28]

General elections in Jamaica were held on July 28, 1959.

Bustamante's campaign slogan, "Manley Failed Jamaica," was not successful in swinging votes. Busta claimed that Manley had failed Jamaica because he signed a "Bad Federal Constitution" that provided no protection for Jamaica. The JLP would keep Jamaica free of "socialism, republicanism, radicalism and every other type of dangerous 'ism."[29]

Despite opposition from the labour party and the private sector, Manley won the popular vote (fifty-five percent) to become Premier of Jamaica.[30] The PNP won twenty-nine of forty-five seats in Parliament.

Manley remained conflicted—Nationalism or Federalism? When Bustamante reiterated the Jamaica Labour Party's opposition to federation and declared the JLP's withdrawal from the federal by-election in St. Thomas, May 1960,[31] Manley's response on public radio one week later was an impassioned plea for Jamaica to become part of a West Indian nation. He said smallness would not stand in a world that had become too big and too competitive, that strength and the hope of strength lay in unity and cooperation.[32]

Yet he conceded that Jamaica's destiny must not be decided by politicians but by the people:

> So, I have decided to put the matter to the final test in the only way such matters can be put to final test. I have decided to go to the people and ask them to vote on one single question all by itself—the question is: do we stay in the federation, or do we get out? Yes, we stay. No, we go.[33]

---

28   Ibid., 173.

29   Advertisement by the Jamaica Labour Party, Election Day, *The Daily Gleaner,* July 28, 1959, 8.

30   During his term as chief minister, Norman Manley promoted his agenda for internal self-government, which he saw as a prelude to the country's readiness for federation. In November 1957, Chief Minister Manley became the chairman of the Council of Ministers, a body which replaced the Executive Council. The governor's authority was reduced to a position of a ceremonial executive who could only invoke powers to address matters of national emergency. In 1959 the council became an official cabinet and the chief minister, premier of Jamaica.

31   Bustamante made the announcement, May 30, 1960, that the JLP would not be contesting the seat in the West Indies by-election in St. Thomas. (JLP Edwin Allen had been chosen to fill the vacancy created by Robert Lightbourne, who had resigned from Parliament.)

32   Norman Manley, May 31, 1960, in Rex Nettleford (ed.), *Manley and the New Jamaica,* 176.

33   Ibid.

A referendum was held, September 19, 1961: YES: forty-six percent; NO: fifty-four percent. The people's vote sealed the fate of federation.

Jamaica's secession from Federation signaled its collapse. The nation was not only the largest island geographically, but it also encompassed more than half the total population[34] of the West Indies and contributed more than forty percent of the federal revenue.[35] Trinidad and Tobago, the wealthiest territory of the group, soon followed suit. Dr. Eric Williams—who had come to power in 1956 with the formation of the People's National Movement (PNM)—succinctly expressed the reaction of his government to the withdrawal of Jamaica, "One from ten leaves nought."

True to Manley's predictions, the union did not survive without Jamaica. May 31, 1962, the day appointed for the Independence of the West Indies Federation, through an "irony of fate"[36] marked instead its final dissolution. It never achieved dominion or republican status within the Commonwealth.

The vote to secede from Federation in 1961 catapulted Jamaica towards political independence and a surge in the popularity of Alexander Bustamante and the Jamaica Labour Party. Manley did not regret holding the referendum. "It was the right thing to do whatever the outcome, and for the simple reason that it is the only way in which to start on the journey that we are about to start on now."[37] And that road was Independence. "One's main purpose must be to see that we get out on that road in good heart, in confidence, in unity, if leaders will allow it, so that we make the most of a start that we would never have a chance to make again."[38]

Post-federation, Luis Fred entered a phase of life more measured and tempered. He retreated from politics and from his involvement with the chamber of commerce, and encouraged Carlton Alexander to fill these roles. He turned his full attention instead to Grace, Kennedy and to family. He valued his privacy, his home, and hobbies, one of which was growing food and tending animals on a farm in the hills of St. Andrew.

---

34   In 1961, Jamaica's population was 1.6 M, Trinidad and Tobago, 800,000, and the grouping of smaller islands comprised another 700,000.

35   Charles H. Archibald, *The Failure of the West Indies Federation* (June 1962), 234.

36   Paul Firmino Lusaka, *The Dissolution of the West Indies Federation* (November 1963), 147.

37   Norman Manley, October 12, 1961, "Reflections on the Former Federation" in Rex Nettleford (ed.), *Manley and the New Jamaica*, 180.

38   Ibid.

27

BIRDS AND THE BEES

(January 1961)

Dad didn't give it a special name. It was known simply in the family as, "The Hill"–twenty acres of lush sloping hillside, a quarter mile below Irish Town, St. Andrew, with a breathtaking, panoramic view of Kingston. The family acquired the property in the early fifties after selling Amberly, located on a rise a few hundred feet above it. He cherished his retreat in the mountains.

We had a ritual. Every Saturday afternoon, we travelled together up the narrow, winding roads of the hills of St. Andrew. From Papine, we followed the Hope River, crossed the single lane bridge, and veered west at the Cooperage intersection. On nearing the hamlet of Maryland a few miles up the road, the mountain air became cooler, and the earthy smells of the countryside, pungent with the smell of guinea grass. The trip took another twenty minutes, or longer if it was raining. Heavy downpours caused landslides of large boulders and loosened rocks to block the New Castle Road.[1]

Dad was not a careful driver. He would careen around hairpin curves, grabbing the steering wheel with one hand while puffing a cigar held in the other. He didn't inhale the smoke but rolled it around in his mouth, and sometimes he'd blow smoke rings, which would drift about and linger for a while inside the car.[2] *If he were stuck behind a truck, he'd blast the horn and floor the accelerator of the V-8-cylinder machine to veer into the right lane. The DeSoto's left-hand drive made the situation even more precarious.*

---

1   Road named after Jamaica Defence Force (JDF) Newcastle Base Camp, another twenty minute drive above Irish Town.

2   As a child, I enjoyed the aroma of cigars, but I never smoked them. I was sixteen when my dad tried to teach me the technique of lighting and puffing a cigar. I ended up having a violent coughing fit.

His brand loyalty was to Chrysler. In the 1960s, he test-drove others, but none compared: the Rover he found bulky, with too small a trunk; the Buick Skylark was not his style–too sporty; and the Mercedes was expensive–a status symbol, a "show-off," he called it. He kept his Chryslers until they made whirring and banging noises caused from either a failed differential or damaged suspension.

I recall one sunny afternoon in January 1961. He was in a pensive mood, humming his usual tune, "O Holy Cross." It was best to wait for him to start the conversation.

My sister Mary had secretly told me Dad was on a special mission and that I was the right age for it, ten years.

"But why the Hill?" I had asked Mary

"To teach you about the birds and the bees."

Dad broke the silence. "You look worried, son. What's the matter?"

"Nothing at all."

"You're too serious sometimes for a boy your age." He chuckled. "It's a beautiful Saturday, no rain, I hope." He threw the cigar butt out the window. "How have your first few days at St. George's been?"

"I love it."

"I'm sorry that the school in England didn't work out. Boarding school would have been good for you."[3]

"I didn't want to leave home, Dad."

"I'm sure you will do well at St. George's. Have you given any thought to what you might like to be when you grow up?"

"I'm not sure. Did you always want to be a businessman?"

"My father wanted me to be a lawyer. He said I would win cases."

"What kind of lawyer?"

"Criminal lawyer. But I told him I didn't think I could defend someone I knew was guilty of a crime. He would then tell me about the presumption of innocence."

"What's that?"

"A person is presumed innocent until proven guilty." He paused for a moment. "My dad was a wonderful man."

"You have spoken a lot about him."

"You know how your ol' man is all thumbs. Well, he was different–he could draw, make furniture, write poetry. I wish you had known him."

"Me too."

"Well, here we are, coming up to the entrance." He made a sharp left turn into the property.

---

3   I had been pulled out of school in September to receive private tutoring in Latin and French, requirements for entry to a Jesuit boarding school in England. With the Bay of Pigs invasion looming, my parents cancelled arrangements to send me abroad. I was enrolled instead at St. George's College in January 1961. Campion College opened that year; they had given me the choice of which school to attend.

Birds and the Bees (January 1961)

The morning mist had burned off, the early afternoon clouds forming in the upper regions of the Blue Mountains. The hills in the northeast rose in full view, tall and majestic, and towards Kingston, the Long Mountain ridge framed a commanding view of the harbour, Portmore, and the plains of St. Catherine to the west.

Fruit trees lined the marl-covered driveway from the gate to the flat at the top of the hill–not Jamaican fruit as one would expect, but foreign, North American apples, pears, and plums. He imported seeds from Miami and experimented with grafting new varieties. They bore in abundance. The climate was temperate enough for growing species native to cooler zones.

From the top of the hill, the land fell off in steep slopes, marked by dirt footpaths leading to different parts of the farm. The earth was rocky and pebbly, the colour of naseberry, yet fertile enough for growing native conifers like juniper, cedar, and mountain yacca.

Dad wore an old pair of baggy trousers. Now that he had gained weight, they fell below his belly to rest on his hips, and regardless of how much he tried to pull them up or tighten his belt, they would keep falling. They made his legs look short. He wore an equally over-sized short-sleeve shirt (which wouldn't stay tucked in) and a pair of high-fitting rubber boots, the type farmers wear to wade through cow pastures and pig pens. He cherished the smell of the earth–paying no mind to getting his hands and clothes dirty.

A cow pen for dairy cattle was at the very bottom of the hill on the east side of the property. The cows were fed in their stalls, for the hillsides did not have much flat pastureland. They were a special dairy breed called Jamaica Hope, originating from Hope Farm in Jamaica. The livestock was bred to tolerate hot weather and to produce milk in poor pasture conditions. He kept chickens and turkeys in nearby coops and goats in pens. I was amazed at how smart the goats were, knowing how to find their way back in the evening after being allowed free range during the day. Theft was a problem, though. They didn't always return in full numbers.

Dad's special interest was in raising pigs. The pens were located on a stretch of level ground higher up from the stalls of the other animals. I never understood why he thought pigs to be clean; every time I saw them, they were rolling around in mud to keep themselves cool. He had names for each of them. Mom objected to his practice of naming them after people he knew, but he meant no harm or insult–he loved pigs and thought he was paying people a compliment. She agreed on one condition–that none of them be called Tillie.

His favourite was the British Hampshire breed. The animal was mostly black, except for a white belt that wrapped around the front shoulders and extended down to the forefeet. Among the herd was a young sow, about six to seven months old. "She ready now fe breed, Maas Ken," the pen keeper explained. She was restless, twitching her tail, and her ears were standing up straight. The boar that she was visiting in his pen was older.

The plan didn't work. Dad blamed the boar who, for whatever reason, was not interested. "Well!" He laughed and said in his typical matter-of-fact way, "That didn't quite go as expected."

It became a family joke that I never got my lesson about how babies were made.

We stayed late to watch the sunset. The prevailing winds died down, creating a lull before the gentler mountain breeze started up. "God, it's beautiful; the colours are different every time. Your mother and I are planning to build a retirement home, so we need to decide, the mountains or the sea. I think she prefers the sea."

Before leaving the farm, I helped the caretaker finish milking the cows. He had filled a couple ten-gallon aluminum containers with fresh milk–hair, bacteria, and all. When we got home, Dad needed to shower and change straight away. Mom's orders. I helped pour the milk into bottles, setting aside a few for our family. The others Mom delivered to my aunts and uncles the following day. Non-pasteurized and non-homogenized. Dad said it was more nutritional and delicious than store-bought milk, but Mom wasn't convinced. She was wary of the bacteria.

We had lengthy debates about which was the family favourite, the mountains or the sea. Mom and Dad searched for properties on the north coast, Port Antonio and Ocho Rios, finally settling for one in Discovery Bay, an acre of honeycomb rock and bush with its own private cove, white sands, and crystal-clear waters.

On opposite page, photo taken by author of hills of St. Andrew, Jamaica.

PART FIVE

Prosperity (1962–1971)

A special bond. Proud mother, Luisa M. Kennedy, presenting a commemorative silver tray to her firstborn on the occasion of the fortieth anniversary of Grace, Kennedy & Co. Ltd., February 1962. He had served the company for thirty-four years.

(Photo, courtesy of © GraceKennedy Ltd.)

Luisa M. Kennedy presenting award to Mr. Don Moss-Solomon, nephew of James S. Moss-Solomon.

Mr. James S. Moss-Solomon, Managing Director, receiving award for forty years of service.

Assistant General Manager Mr. S. Carlton Alexander, receiving award for twenty-nine years of service. Ms. Shirley Jackson and Mr. Harold Aarons, assisting.

Mr. Gladstone Kamicka, Head of Accounting and former secretary of the company, receiving award. (Photos, courtesy of © GraceKennedy Ltd.)

Mary Dorothy (Dolly) Chevolleau, Luis Fred Kennedy's secretary, presenting bouquet to Luisa M. Kennedy, in appreciation.

Luisa M. Kennedy presenting to her son, José Kennedy, Director, for twenty years of service.

(Photos, courtesy of © GraceKennedy Ltd.)

Manager of Merchandising Division, Francis X. Kennedy, Luis Fred's youngest brother, with Managers, Reginald Owen Byles and Ferdinand (Ferdie) Figueroa, Cecil deCordova & Co. Ltd. (1962).

28

Matriarch

(1962)

Luis Fred's mother, Luisa Matilde Kennedy, was matriarch of the family and major shareholder of Grace, Kennedy. Over the years, the bond between mother and son had strengthened—a filial relationship that formed the core of Dad's ambition to honour the name of his late father and to provide a livelihood for his family.

His mother was shy by nature, shunned public appearances, and avoided shareholders' meetings. Yet she maintained close relations with Grace, Kennedy and met often with her son to give advice on business and family affairs. Although she was now showing signs of physical weakness, the founder's wife exuded an executive presence not only with her family but with employees of Grace, Kennedy.

Photographs of the company's fortieth celebrations in February 1962 attest to her stature.[1] She was centre stage, presenting silver trays and awards to members of staff to commemorate their long service and dedication to the company: James Moss-Solomon (forty years of service); to her three sons, Luis Fred (thirty-four), José (twenty), and Francis (sixteen); Carlton Alexander (twenty-nine); her son-in-law, Simon Soutar (twenty-one); Geoff Dodd (twenty-eight); Don Moss-Solomon and Gladstone Kamicka (twenty); and others. For Luis Fred, the occasion was an opportunity to express his gratitude and to reward employees for their productivity and loyalty.

---

1 Photographs are courtesy of Peter Moss-Solomon and © GraceKennedy Ltd.

The gala was the last time that the Kennedys, headed by Luisa M. Kennedy and her six children and spouses, gathered as a family unit of company shareholders.

Less than three months after the celebrations, my grandmother was admitted to hospital after experiencing frequent falls caused from weakness in her legs. While vacationing on the north coast, Dad received a phone call on Wednesday, May 2, to hear that his mother had died at St. Joseph's Hospital. Post-mortem diagnosis was multiple myeloma, malignant plasma cells in the bone marrow that had spread through her body.

A notice appeared in *The Daily* Gleaner, May 4:

> Owing to the death of
> Mrs. Luis M. Kennedy
> WIDOW OF FRED W. KENNEDY,
> ONE OF THE FOUNDERS OF
> GRACE, KENNEDY & CO. LTD.
> The Offices and Warehouses
> of
> Grace, Kennedy & Co. Ltd.
> Grace Shipping Co.
> Grace Agricultural Co.
> Jamaica Rums Ltd.
> Sheffield & Co. Ltd.
> Cecil deCordova & Co. Ltd.
> Medical Supplies Co.
>
> ## WILL BE CLOSED TODAY
> (Friday, May 4)

Funeral services, officiated by Most Rev. John J. McEleney S.J., Catholic Bishop of Kingston, were held at the family's parish church, Sts. Peter and Paul, and interment was followed at Calvary Cemetery.

My grandmother had lived to see the dreams of her late husband come true: the prosperity of his merchant business and a family of six children, educated, married with children, and self-sufficient. She had been widowed at the age of forty-two after only twenty-two years of marriage, and for the remainder of her life, she dressed in black and purple, as was the custom in Spanish America. For her entire life, she remained a devout Catholic, attending daily Mass whenever possible.

Her youngest son, Francis Kennedy, had remarried in 1958[2] after the death of his first wife, Cecille, two years earlier. He and his new bride Anne Marie Angerhausen

---

2 Francis X. Kennedy and Anne Marie Angerhausen were married Wednesday, April 16, 1958, at Sts. Peter and Paul Church. Announcement appeared in *The Daily Gleaner*, Saturday, April 19, 1958, page 18, in article entitled, "Rum Company Manager Takes German Bride." Anne was originally from Trier, Germany and was teaching at Alpha Prep when they met.

had one child, Mary Lou Kennedy, who was the youngest of the Kennedy clan. The oldest grandchildren were already attending university: Bill Valliant graduated from Holy Cross College in 1962, and his brother, Jim, was enrolled at McGill in Montreal. Gloria's two eldest, Margaret Plant and her brother William, were both at university. Mary Lou Soutar, Louise's eldest daughter, had entered the Sisters of Mercy in 1957, and Celia, my eldest sister, had left for England in 1960 to join the Order of the Servants of Mary (Servites).

My grandmother's death changed the dynamics of both family and business. Her majority shareholding was distributed equally among her six children, who would chart their own lives independent of the company over the next couple of decades. No signed agreement existed among family members to prevent them from selling their shares, only a tacit understanding that investments in the company would remain intact. Sales and transfers were often restricted, requiring the explicit permission of the board of directors and of the governing director. In the first forty years of the company's history, my grandmother had given no thought to selling her shares, and my dad, in turn, carefully guarded her majority holdings.

Not long afterwards, Francis Kennedy migrated with his family to England, and Carmen Valliant to Canada. These were the beginnings of a shift away from a close-knit control of family holdings. Dad's siblings eventually sought to liquidate capital to meet their own personal and financial needs. He did not block their wishes; instead, he made accommodations, acknowledging the change as an inevitable step in the company's evolution.

Jamaica Military Band leading the march of the National Float Parade, King Street, August 11, 1962. (Photo by author with a Brownie Kodak camera, a gift for my twelfth birthday)

Mitsy Constantine (centre), Miss Jamaica 1964, flanked by Erica Cooke (left), second place, and Joy Gaussen (right), at the beauty contest held on August 1. She had been crowned Miss Kingston, Jamaica's Independence and became a "goodwill ambassador of the 1962 celebrations." Joan Crawford won the Miss World Contest in 1963.

Photo, courtesy of © Gleaner Company (Media) Ltd.)

29

OUT OF MANY ONE PEOPLE

(1962)

Premier Manley called the general election in February, even though one was not due until 1964. After his defeat in the referendum vote, he wished to determine which of the two major parties would form the government of an independent Jamaica. He miscalculated. On April 10, 1962, the JLP won twenty-six of forty-five seats with 50.4% of the popular vote. Bustamante was sworn in as Premier of Jamaica by taking the oath of office, April 24, 1962. With his personal secretary, Miss Gladys Longbridge, by his side, he replied to Governor Blackburne's congratulatory remarks:

> I pledge to be faithful to my Ministry and to the people and to do everything possible to reduce the sorrow of small people throughout the country, inclusive of the small farmer, and with the help of God and the prayers of all, I am confident of success.[1]

Three months later, on Sunday, August 5, 1962, minutes before midnight, the Union Jack was lowered in the darkness before a crowd of 20,000 at the newly built National Stadium. Seconds before the start of the new day, Jamaica's national flag was hoisted—"Down the Union Jack, up the Black, Green and Gold"[2]—and fireworks

---

1 *The Daily Gleaner,* Wednesday, April 25, 1962, 1.
2 Ibid., Wednesday, August 8, 1962, 1.

exploded to light up the sky with thousands of sparks. The Jamaica Military Band opened with a flourish, and the voices of thousands filled the stadium with the sounds of the new Jamaican national anthem. Princess Margaret, her husband, Lord Snowdon, Premier (soon to be Prime Minister) Alexander Bustamante, Governor General Sir Kenneth Blackburne and Lady Blackburne, Leader of the Opposition Hon. Norman Manley and Mrs. Edna Manley, parliamentarians, religious leaders, US Vice-President Lyndon Johnson, and representatives from Australia, Pakistan, and African nations were among the dignitaries present.

Attended by foreign delegates and hundreds of Jamaicans, the Independence State Ball was held at the Sheraton Hotel in New Kingston on Monday night. A picture appeared in the *Daily Gleaner* of Alexander Bustamante, tall and gallant in his black coattail suit, dancing with a charmed Princess Margaret, who was attired in formal wear of a silk skirt and lace bodice.[3]

The following morning, August 7, Her Royal Highness Princess Margaret, representing Her Majesty Queen Elizabeth II, opened the first Parliament of Independent Jamaica. On behalf of her sister, she welcomed Jamaica to the Commonwealth family of nations, and assured parliamentarians that the newly independent nation "will have a vital contribution to make to the cause of fuller cooperation, understanding and tolerance far beyond the immediate area of the world in which it is situated."[4] "My government in the United Kingdom," she said, "has laid down its responsibilities and has ceased to have any authority in and over Jamaica, after more than 300 years."[5]

Prime Minister Hon. Alexander Bustamante approached the dais to receive the constitutional instruments of Independent Jamaica from her Royal Highness, to whom he expressed his deepest honour and thanks. Leader of the Opposition Norman Manley paid homage to "the men who in the past and through all our history strove to keep alight the torch of freedom."[6] He hoped that one day Jamaicans would make their motto "Out of Many One People" come to speak the truth about themselves. His speech was met with loud applause.

"Throughout Jamaica the people of the villages and hamlets have been given an opportunity never before given to celebrate a national event at the national level."[7] Thousands turned out for flag-raising ceremonies, beauty-queen pageants, and St. John's Ambulance Brigade's marches. Church services were held across fourteen parishes with tens of thousands in attendance. Two hundred Maroons from Scott's Hall danced in the main square of Port Maria, and in the evening, people held candlelight

---

3 Ibid., August 9, 1962, 1
4 Ibid., August 8, 1962, 1.
5 https://www.caribbean-beat.com/issue-116/jamaica-freedomcome#axzz79kZfONER.
6 *The Daily Gleaner,* August 8,1962, 33.
7 Ibid., Friday, August 10, 1962, 14.

vigils in parks and town squares throughout the parish. Celebrations lasted into the night with folk and quadrille dancers dressed in colourful costume.[8]

On August 11, the Corporate Area was alive with the pageantry of the National Float Parade. Transport was snarled, roads closed, and buses overflowed with passengers. Throngs filled the streets to watch the floats travel from Cross Roads along North Race Course, East, Harbour and King Streets.

My family grabbed a spot on the rooftop of Nathan's, the two-floor clothing store on King's Street, where we had a bird eye's view of the parade.[9] Members of the Jamaica Military Band dressed in ceremonial scarlet tunics were among the first to receive cheers, followed by the mayor's float carrying Mitsy Constantine,[10] Miss City of Kingston.[11] Along King Street marched effigies of Governor General Sir Kenneth Blackburne, Prime Minister Alexander Bustamante, Louise Bennett, and Ranny Williams. We saw floats representing the Chinese and East Indian communities, the Association of the Deaf, the YMCA, Jamaica Federation of Women, Jamaica Manufacturers Association, Jamaica Industrial Development Corporation, and many others. The parade was a spectacle not before seen in Jamaica.

We had our own flag, our own coat of arms, our own Bank of Jamaica. In August, we cheered on athletes of the Ninth Central American and Caribbean Games at the National Stadium. In September, an exuberant Alexander Bustamante watched as the black, green, and gold flag was hoisted at the United Nations in New York. In December, we welcomed our native-born son, Sir Clifford Campbell,[12] to serve as His Excellency the Governor General of Jamaica, and in that month, we celebrated the inauguration of the National Dance Theatre Company.

The transition to Independence was peaceful—no rebellion, no bloodshed. A working-class aristocracy was emerging, evidenced by upward mobility in Jamaican

---

8   *The Daily Gleaner* published detailed descriptions of Independence celebrations by parish during the week of August 6, 1962.

9   Nathan and Co. Ltd, "Metropolitan House" was opened in 1908 as a large retail and wholesale clothing store, rebuilt after the 1907 earthquake. The history of the business dates to 1882, founded by Mr. A. M. Nathan, born in Jamaica, 1851. (*The Red Book of the West Indies*, 86).

10   Marie Elizabeth Constantine, nicknamed Mitsy, was aged eighteen in 1962. She attended the Queen's School and later migrated to England for a short period. She was crowned Miss Jamaica in 1964 and married the Hon. Edward Seaga, minister of Development and Welfare, in 1965. They had three children.

11   A contest was opened in July for unmarried girls born in Kingston and St. Andrew between the ages of eighteen and twenty-five. Crowning took place at the Mayor's Charity Ball at the Sheraton Hotel, Thursday, August 9.

12   Jamaican educator and politician, Sir Clifford Campbell (1892–1991), served as the first Jamaican born Governor General of an independent Jamaica from 1962 to 1973.

society. The new sector consisted of "craftsmen, technicians and production workers throughout industry and commerce but also included unskilled manual and service workers, for example, in bauxite and oil engineering."[13] Growth was significant "in the context both of other developing economies and of the subsequent performance of the Jamaican economy. GDP would increase at an annual average rate of 5.1 percent between 1960 and 1972, while inflation was kept to an average of 4.5 percent."[14]

Despite these signs of prosperity, the country faced its challenges. In 1960, eleven percent of those eligible to work were unemployed (49,000 out of a workforce of 454,100). And the number of young people who would be looking for work over the next decade was increasing rapidly; in 1962, forty-one percent of a population of 1.6 million were under the age of fifteen. Creating better access to education and providing employment became top priorities of the Five-Year Independence Plan proposed by the Labour government.

Jamaica's alliance with the United States, Europe, and the Western democracies after Independence helped secure traditional markets and provide confidence for private business to expand trade links. Bustamante took an "acquiescent" approach to foreign policy to the extent that he guarded "inherited ties of economic and political dependence."[15] His tactic was to petition "the British, US, Canadian, and other Western governments for commodity price-supports and markets."[16]

Jamaica was vulnerable as a small state. It risked losing preferential tariffs with Great Britain; it was threatened too by the rise to power of Fidel Castro, who nationalized US owned industries in Cuba and opened diplomatic relations with the USSR. To secure an economic and political future, Bustamante ensured that Jamaica remained within the Western alliance of nations[17] by forging closer relations with the United States. On August 7, 1962, the day after Independence, the prime minister extended an open invitation to the US to increase private investment and establish a military base in Jamaica.

North America's culture was all-pervasive in the region. Improved sea and air transport meant closer links, bringing tourists and businesspeople in droves to the island.[18] With affluence came new technology, mass production of goods, and the proliferation of advertising media. The change was good for local business, which widened trade routes to the United States and Canada, but North American acculturation would unalterably change the consumer habits of Jamaicans.

---

13   Terry Lacey, *Violence and politics in Jamaica, 1960–1970* (1977), 30.

14   Damien King, 'The Evolution of Structural Adjustment and Stabilization Policy in Jamaica,' (March 2001), 6.

15   Wendell Bell, "Jamaica Enters World Politics," (1977), 686.

16   Ibid., 687.

17   R. B. Manderson-Jones (1990), cited by Patrick Bryan (2004), 96.

18   Martin Ira Glassner, "The Foreign Relations of Jamaica and Trinidad and Tobago, 1960–1965." (October 1970), 125.

Despite the threats, Jamaica stood proud as a sovereign nation. "Me head still feel giddy," was the way Louse Bennett expressed the mood in her poem, "Jamaica Elevate."[19] Our own governor general was "Jus like one o' we own fambily/De very same complexion."[20]

---

19 Louise Bennett, "Jamaica Elevate," *Jamaica Labrish* (1972), 174–176.
20 Ibid., 176.

Aerial view of Puerto Plata, beachside retirement home,
Discovery Bay, St. Ann. (Photo from family collection)

Elizabeth (who was hanging on for dear life)
and Fred in new dugout from Falmouth,
1964. Dad in foreground.
(Photo from family collection)

Acrylic painting, original in colour.
Banky, the fisherman, Discovery Bay
© Sarah Kennedy 2005

30

PUERTO PLATA

(1963)

The Adirondack chairs in the gazebo cast long shadows in the fading light. The water reflected the soft orange of the sky, its colour spreading solid and thick across both land and sea. The gazebo, or the summerhouse, as Dad called it, was rustic, its uprights made from twisted logs of the red-mangrove, and its roof from fronds of the bull thatch palm.

"What a beautiful evening," my dad said, relaxed and pensive, looking out across the bay. "Have you noticed how the waves break over the reef?"

"The sea is rough today." I sat in the chair next to him, just the two of us.

My parents bought the seaside acre of land in Discovery Bay from the Tinglings of Port Antonio in April 1962 for the sum of £5000. They named it after their birthplace in Dominican Republic, Puerto Plata (Silver Harbour)–significant too because it was a gift to my mom for their twenty-fifth wedding anniversary. He wished to give it to her entirely, so included only her name, Lydia M. Kennedy, on the land deed. Dad often said his one regret was that his mother had not seen the property; she had died just weeks after the purchase was made.

"It's curious how the waves come into the bay. After they break over the reef from the open sea, they make a complete ninety-degree turn to ride the coral ridge until they hit the shore. Look."

The waves changed direction, as he said, the white caps picking up the colours of the evening light that suddenly turned purple.

"Yes, it's strange."

"I don't think I've ever seen water so crystal clear."

"The fishermen say it's the freshwater springs."

"Yes, that give it a mineral quality. Maybe a bit too late for a swim; I've discovered jelly fish come out in the late evening, but it's still my favourite time to go in."

He walked to the edge of the gazebo and plunged in, his movements not clumsy or awkward as one would expect because of his heavy weight; the dive was graceful, his body arched in perfect form, and his outstretched arms breaking the water with a ripple that was barely audible.

"God, it's beautiful," he shouted. "Makes one believe there's a God after all." He laughed then swam deep below the surface of the water.

Although a staunch Roman Catholic, he was a skeptic of Church doctrine. The story of Adam and Eve, for example, defied logic; creationism failed the test of science and of reason and it went contrary to the feelings of the heart.

Puerto Plata was a place of solace, a retreat from work and the hustle and bustle of Kingston life.

It was not a magazine-style house, just a simple structure with aluminum roofing and plenty of louvres to catch the sea breeze. Dad followed the construction every step of the way, ensuring that a weekend did not go by without visiting the site. He mistrusted Mr. Meghie's[1] knowledge of engineering; he thought the builder was making a dreadful mistake to drive steel directly into the volcanic rock instead of excavating to build a foundation. But the contractor proved to be right; the house has withstood the test of time, unscathed by earthquakes and Category 5 hurricanes.

Dad had given up farming in the hills of St. Andrew and turned his attention to creating a garden at Discovery Bay, despite the arid conditions and rough terrain.

The original plan was not to disturb the wild bush that covered the acre of land, but nightly onslaughts of mosquitoes changed his mind. He carefully supervised hired help to root out the shrubbery and to ensure that they did not harm the mature, indigenous trees: the flame of the forest, brasiletto and wild poui. He held the conviction that, *"Give a Jamaican gardener a machete, and he will chop down everything in sight."*

To build the garden, he chose indigenous plants, those tolerant of sea salt and dry, windy conditions. We would take treks through the bush located on the opposite side of the parochial road to hunt for suitable vegetation. We cut slips and stuffed our pockets with dried pods and seeds of trees and shrubs: the purple allamanda, wild orchids, flame of the forest, quick stick, dogwood, naseberry, and the tree with white blossoms, for which he did not have a name.

Seeds were quick to germinate in the nursery he built in the back yard—we could have had a forest with those that sprouted. Once the saplings were mature enough to plant, we found fissures by driving a machete into the crevices of the honeycomb rock. The fastest to grow was the tree with the white flowers.

Every weekend when we came from Kingston, he would take out his tape measure. *"I swear if you stare long enough at this tree, you can actually see it grow; it gets taller by the*

---

[1] Mr. Meghie was a local building contractor from Brown's Town.

minute." Sure enough, it must have grown a foot taller or more each time he measured it—no exaggeration. Before long, it became a mature tree, medium height, twenty feet tall, with a smooth, dark-grey bark, arching boughs, and a wide, umbrella-shaped crown. In the spring, it bloomed thick bouquets of white flowers, just as we had seen in the wild.

What bothered him most was that he did not know what to call it. He scoured his library[2] for information on its botanical name but he came up empty. He thought then to inquire from his friend Sullivan,[3] a local fisherman, whom he considered wise in every regard.

"Do you happen to know the name of this tree, Banky?" Most fishermen in the bay had a nickname.

"Yes, Maas Fred." The fisherman paused, looked high up into the tree, and with great deliberation exclaimed, "Yes, Sar Ken, dem grow all 'bout de place, de right and proper name fe it is 'Flowers Tree.'"

Flowers Tree became its official name and the subject of many amusing conversations. Sullivan had satisfied Dad's curiosity; he needed to inquire no further. In recent years my wife discovered its botanical name to be *moringa oleifera,* native to Northwest India. It is commonly known in Jamaica for its curative properties; moringa tea made from the leaves is said to cure stomach disorders and serve as a potent aphrodisiac.

The garden flourished despite the harsh conditions of Dry Harbour. Dad grew a vegetable garden in raised beds, grafted plants to create multiple varieties of hibiscus and crotons, and lined the driveway with fruit trees: limes, oranges, cherries, grapefruit, guavas, and naseberries.

He encouraged me to learn to fish. For my fourteenth birthday, we travelled together to Falmouth to purchase a dugout canoe, handmade by fishermen from the Jamaican silk cotton tree. It was eight feet long, much heavier than I would have expected, and painted green, red, and white. Before leaving Falmouth, we drove by Market Square so he could show me the home of his grandfather, William Kennedy.

Banky taught me to haul pots, cast nets, and do deep sea line fishing. We brought home lobster, queen crab, jack, tuna, parrot, soursop, doctor, soldier, grunt, butter, and goat fish. Dad's dearest was deep-water silk, the red snapper whose eyes popped out when it surfaced from the ocean depths. Its flesh was sweet and silken.

Banky's six-year-old son, Kaya, often joined us on fishing expeditions. When we returned with a boat full of catch, Dad would greet us at the pier with gifts in hand—a

---

2 His library of gardening books included *Botany for the Caribbean* (1963) by Robertson and Gooding; *Caribbean Gardening, With Special Reference to Jamaica* (1964) by Aimee Webster; *Flower Garden Primer* (1936) by Julia Cummins; *The Farmer's Guide* (1954) by the Jamaica Agricultural Society; and *Tropical Planting and Gardening* (1956) by H.F. MacMillan.

3 Arnold Sullivan. I only recently discovered the fisherman's birth name from his grandson, Sean Ascott. His nickname was Banky.

cigar and a shot of Appleton overproof white rum for Banky, and for Kaya, a couple ounces of red rum. "Dat is right and proper, Sar Ken, de red one is fi de bwoi, it come in like wata to me." Dad would light his own cigar to keep Banky company, and the two men would sit and chat on the wharf for hours, talking about Sullivan's eldest daughter, who was a nurse in England, about the ways fishermen forecast the weather by reading the stars and the wind, and about the good ol' time days in Jamaica. Banky reminded Dad of Hemingway's character in *The Old Man and the Sea*.

Dad also had a fascination for the stars. At night, we would sit out on the terrace by the sea to watch them shoot through the sky, and we'd make secret wishes in the split seconds that they flashed. (He kept a book about stars in the bookcase that his father had built.) He taught me to recognize stars and planets by their shapes and positions in the sky. We studied Orion's Belt with its threesome of evenly spaced stars, and the Big Dipper with seven in the constellation, whose shape was like a Great Bear. He told stories about Zeus and his wife, Hera, who was jealous of his affairs with the nymphs, and about how he turned them into constellations to protect them from her wrath. *"The planets are different,"* he would say. *"They look larger, and they do not blink; their lights are steady."*

Mom and Dad made ties with the community. They arranged for a Catholic church to be built in Discovery Bay. While it was being constructed, Fr. William Dwyer S.J., parish priest in Brown's Town, celebrated Sunday Mass at Puerto Plata for residents of Discovery Bay. For refreshments after service, our family served cupcakes, sweet drinks, and coffee for the congregation. Our home became the temporary site of the local Catholic church for about a year until The Immaculate Heart of Mary building was completed.

Dad also formed a Homeowners Association and drafted a constitution for the Beach Club, to which Fortlands Road residents and those on the hill in Discovery Bay had access. Mom was not pleased when he nominated her secretary of the club at its inaugural meeting.

Five years later, when I was studying abroad in Massachusetts, Dad wrote to keep me informed of the garden at Puerto Plata:

> We go to Puerto Plata this afternoon for the weekend. It is 2 weeks since we have been, and we are looking forward to the two days there. The garden there was in the best condition that I've ever seen it, and I brought back to Kingston a lot of plants that Douglas[4] caught for #45.

---

4 Fitzgerald Douglas was the gardener and caretaker at Puerto Plata. He was a young man of nineteen when hired, eager to learn what Dad taught him about gardening, pruning, and mulching plants.

They are now in the ground and the garden is recovering nicely now that we can use the water for the plants. The roses at the dining terrace have been particularly beautiful.[5]

He took pride too in upgrades to the hardscape:

When you come out, you will see many changes at work at Puerto Plata. I have paved with concrete most of the circular roadway in front of the house and this has been a great improvement. I am sorry now that I did not do the whole thing. We are also putting up two lampposts along the driveway. The summer house is being rebuilt as I told you on the phone and I have built a new walkway to the change-rooms along the Halsalls' property line and will install a lamppost there too. It all looks very lovely. Douglas is still with us and continues to do a good job in the garden.[6]

Their intention was to use Puerto Plata as a retirement home, but the politics of the 1970s and the burdens of ill-health changed their plans. Dad often quoted Robbie Burnes: *"The best laid schemes o' mice an' men gang aft agley."*

---

5   Luis Fred Kennedy, letter to son Fred W. Kennedy, April 25, 1969.
6   Ibid., December 13, 1969.

The 1960s brought to the local and global markets GRACE branded foods, FOR TABLES AROUND THE WORLD. Canned Vienna Sausage was the first.
The Daily Gleaner, Thursday, January 28, 1960, 6.
(Photo, courtesy of © The Gleaner Company (Media) Ltd.)

31

THE GOOD FOOD PEOPLE

For Tables around the World

(1960s)

In time for the fortieth celebrations came the changing face of the company in the local market—its own branded products. The name GRACE was being marketed for "Tables around the World," its trademark appearing on labels with the oval design and distinctive red background. Grace-canned Vienna sausages were the first. Boerries Terfloth in *Memories of What Made Grace* recalled the occasion on which the branding for the local market was first conceived:

> The GRACE Brand (1960) was introduced to Jamaica, not on the basis of a formal marketing plan, but rather from a good sense of the marketplace and its possibilities. This was not unusual for those times. Mr. Carlton Alexander and I dreamed up the pivotal strategy during a midnight whistle-stop at an inauspicious Montreal Airport Inn. He was en route from Kingston to Newfoundland in search of Jamaica's

staple of salted cod. By early morning, we had formulated the basis for the future development of the GRACE brand in Jamaica: GRACE Viennas in tins—the same product that is produced in Jamaica today.[1]

The Daily Gleaner carried advertisements not only for Grace Vienna sausages but for "vitaminized" apple juice, tomato juice, carrot juice, bacon, whole-kernel corn, cream-style golden corn, peas, and maraschino cherries. Grace was no longer only agent and distributor for other brand-named products and services but now owned and marketed its own.

At a Grace Foods International Conference in Montreal, 1970, Purchasing Manager Bruce Rickards commented about the shifts that had taken place in merchandising:

> The picture has now significantly changed. Bulk has declined as a proportion of the business, not because we are selling less but because we have developed more Grace brand items and more cold storage items. Cold storage is a very large and growing business in Jamaica. When I had the pleasure of heading that department between 1959 and 1962 it was very small, and we were doing about five to six carloads of "backs and necks" a month. Today, we are doing between 15 and 20 carloads each month. That's just one item.[2]

With the shift in focus, branding became an important feature of marketing for Grace, Kennedy. Arnold (Junior) Foote[3] was the genius behind the early advertising campaigns. Hired in 1955, he was soon promoted to sales manager and advertising manager to work with Carlton Alexander and Luis Fred Kennedy. In an interview, Junior shared memories of the start of advertising at Grace:

> Mr. Luis Fred Kennedy said that Foote would always be an advertising man, and I took his advice. I cannot say enough about that great gentleman. He was a good businessman who always put the company first. He always treated everyone with love and respect. We discussed the development of the Grace Kitchens, which became a huge success. One of the most enjoyable days I had working with your father was

---

1   B. H. Terfloth in *Memories of What Made Grace*. (1997), xiv.

2   Bruce Rickards, "Grace Foods International Conference" (April 2, 1970).

3   Arnold Foote Jr., born 1934, son of Arnold Foote, solicitor and notary public, took Luis Fred's advice to become an advertising man and formed his own company, Advertising & Marketing (Ja) Ltd. During his career, he made significant contributions to Jamaica in the areas of consular services, tourism, and international trade missions. The Hon. Arnold Foote JP received the Order of Jamaica and the Order of Distinction in the Rank of Commander in recognition of his outstanding service to Jamaica. Arnold Foote died at his St. Andrew home, December 2, 2021 at age eighty-seven.

the day when I presented the commercial "Try it nuh," the slogan that was used at the launch of the Grace brand. We had the greatest marketing person running the company, Mr. Carlton Alexander. Your father saw the talent and brought him forward. He never cramped his style, and he put him in charge. I have and will always have a great respect and admiration for your father. The time I spent with him I will cherish forever.[4]

At a conference in Montego Bay, Junior Foote proposed that the company adopt the slogan, GRACE, THE GOOD FOOD PEOPLE. Although hesitant at first, Luis Fred Kennedy soon embraced the suggestion; he had become fully impressed by Junior Foote's brilliant salesmanship. Sixty years later, Grace is still known as the Good Food People.

The advent of television and the emergence of supermarkets in Jamaica during the 1960s provided ideal platforms for advertisement and marketing. *The Daily Gleaner* of July 5, 1963, carried a picture of Mr. Paul Scott of Grace, Kennedy & Co. (Shipping) Ltd. receiving "The Largest Shipment of Television Sets to Jamaica."[5] Two thousand Westinghouse televisions had arrived in Kingston from New York on Saturday, June 29, by Grace Line SS *Santa Paula,* just in time for the Independence Anniversary celebrations. The Jamaica Broadcasting Corporation (JBC)[6] began the nation's first television broadcast on Sunday, August 4, 1963, at 6:00 PM, two days before the first anniversary of Jamaica's Independence.

I recall the family gathered around the rented black and white Westinghouse television to watch the ad that became a sensation. (Televisions or any other product that Dad thought had a high depreciation rate—including company cars—he would lease or rent, not purchase.)[7] The debut TV Grace commercial was aired with the "Try it Nuh" campaign. Three TV personalities created an immediate connection between viewers and the Grace brand: an Englishman exclaiming in an exaggerated British accent, "Goodness gracious, Grace is good"; an American in a typical Yankee twang,

---

4   Interview with Junior Foote, December 6, 2020.

5   *The Daily Gleaner,* Friday, July 5, 1963, 11.

6   The Jamaica Broadcasting Corporation was introduced in June 1959 as a public broadcasting firm, founded by Norman Manley. It was the first media outlet in Jamaica to be a state-owned statutory corporation.

7   John Issa recalls, "My first direct interaction with your dad was in 1961 when I was working at Motor Sales and went to see him at the Harbour Street offices because Grace wanted to lease some Morris Oxfords. I remember asking why he didn't just buy them as it would be cheaper. He said he could earn more using the funds trading than investing in cars. I was always aware of his reputation as a strategic investor." John Issa served as a director of Grace, Kennedy from April 1982–Dec 2011. As chairman of SuperClubs, he has often been called the "Father of the All-Inclusive Revolution"; he created the first all-inclusive hotel in Negril in 1976, Negril Beach Village, later to become Hedonism II.

"For cooking with taste, you can't beat Grace"; and a Jamaican in his native patois, "What a great way Grace taste." A new era of advertisement and marketing had begun. Grace was to become a household name in Jamaica.

The advent of supermarkets changed food distribution and the retail trade. The traditional system of internal marketing, dating back to a slave plantation economy, consisted of rural and city markets tied to small farming. Post-emancipation, retail grocery shops developed in communities across the island, and by the middle of the twentieth century were almost exclusively owned and operated by Chinese Jamaicans.[8] In the late 1950s and early 60s, with the increased purchasing power of an emerging middle class, entrepreneurs began investing in supermarkets, which afforded retailers the opportunity to buy and sell in larger volumes. For the customer, supermarkets meant convenience, in some cases lower prices, more choice, and the luxury of self-serve shopping.

A plethora of supermarkets sprung up around Kingston: John R Wong opened in April 1959, Brooklyn in December, Lane Supermarket one year later, Melbourne Supermarket, Welcome, A&M on Pechon Street, and others. *The Daily Gleaner* carried full-page ads for grand openings not only in Kingston but also in St. Ann's Bay, Montego Bay, Mandeville, and other towns across the island.

Carlton Alexander saw the opportunity to launch Grace-branded products by flooding the retail markets of both the Chinese retail shops and newly established supermarkets. He created the market, promoted Grace products, and got them lifted off the shelves. He grew the merchandise business by extending unsecured credit, without interest, to the corner shops. In today's context this would be a high-risk business strategy, but the personal relationship Carlton developed with customers paid dividends. His son, Philip Alexander, commented in an interview about his father:

> This was his reputation; he went above and beyond what others would do to support customers. The customer came first. He set up many a retail Chinese business in Jamaica because he was willing to give them credit. They were able to start up a business because he would give them extended credit, sometimes six-month credit. This gave them a chance to establish and finance their business, and it suited Grace because the company built customer loyalty and guaranteed sales.[9]

---

8   The majority of Chinese Jamaicans in the 1960s traced their origins to indentured Chinese labourers brought to Jamaica starting in 1854. They are known as Hakka Chinese, who speak a unique dialect from the provinces of Shanxi, Honan, and Shandong in Mainland China. In the 1950s, there were approximately 14,000 Hakka Chinese residing in Jamaica.

9   Philip Alexander, retired executive and director, Grace, Kennedy. Interview with author, February 15, 2021. See Chapter 37.

Grace, Kennedy heavily advertised both its GRACE brand food products FOR TABLES AROUND THE WORLD and traditional products like Anchor Butter and Gold Medal Flour. By the mid-1960s, GRACE, with its red label and crown[10] on top, carried a wide choice of products, including Grace ketchup, instant coffee, canned vegetables, juices, and canned meats. In 1969, Grace, Kennedy's sales exceeded £14 million, of which Grace-branded products accounted for £4.5 M.[11]

Acquisitions and Mergers

Luis Fred was aggressive in acquiring ownership of rival companies to create market access for products and to diversify Grace, Kennedy's operations. The company held a distinct advantage: its profitability gave it access to internal capital and debt financing to support its newly acquired subsidiaries.

At a board meeting of April 21, 1960, the chairman informed the directors that he had purchased majority shareholding in Cecil deCordova & Co. Ltd. in a private agreement with G. J. deCordova.[12] Mr. Kennedy, in whose name the shares were to be issued, would give Grace, Kennedy "a transfer in blank, transferring the shares to the Company."[13] Without a succession plan in place, Mr. Gabriel deCordova's intention was to retire from business and sell his family company to a trusted enterprise that could take over its operations. Luis Fred appointed his brother, Francis X. Kennedy, and Mr. Ferdie Figueroa,[14] managers of the newly acquired company. Luis Fred (Chair) and Mr. Moss-Solomon were both on the board of directors.

Dating back to its inception in 1898, Cecil deCordova was one of the prominent firms of commission merchants in Kingston.[15] It was one of Grace, Kennedy's competitors, and their range of business comparable in food distribution, but it had become difficult for them to compete with Grace, Kennedy's expanding enterprise.[16] The acquisition was successful. In later years, Grace, Kennedy was able "to integrate their sales and

---

10   The Crown insignia was first used with packaging of lobster shipped from Canada to Europe in the 1950s by Boerries Terfloth of Grace, Kennedy & Co. (Canada) Ltd. Early in 1967, Luis Fred Kennedy arrived at a tentative agreement with W. R. Grace for the continuing use of the trademark by the overseas subsidiaries. Discussions went on for many months until finally, in December 1969, the Grace trademark was registered in the United States.

11   Douglas Hall *Grace, Kennedy & Company Ltd.: A Story of Jamaican Enterprise* (1992), 49. Sales had moved from £1M in 1950 to £14 in 1969.

12   Gabriel Joshua deCordova, J.P., Chairman of Cecil deCordova & Co., born January 1902, son of Cecil deCordova, founder of the Company in partnership with Mr. A. H. Selwyn.

13   Grace, Kennedy & Co. Ltd., minutes of the board of directors, April 12, 1960. An announcement of the sale appeared in *The Daily Gleaner,* Thursday, April 28, 1960, 2.

14   Ferdie Figueroa, brother of Professor John Joseph Figueroa. John Figueroa married Carlton Alexander's sister, Dorothy Grace Murray Alexander, in August 1944.

15   *The Red Book of the West Indies* (1922), page 84, carries a full description of the history of Cecil DeCordova.

16   Terfloth & Kennedy (U.K.) Ltd. *Memories of What Made Grace.,* 1997, lv.

accounting staff into the Merchandising Division,'[17] which strengthened operations of the parent company. The merger also diversified the range of merchandising to include non-food items such as pharmaceuticals, paper, and greeting cards.

In 1964, Grace, Kennedy purchased the entire shareholding of George & Branday Company Ltd.,[18] a large import and export business, traders, shipping, and insurance agents. Mr. Vincent George was a director and former partner of the business, and his son, Geoffrey George, managing director. George and Branday Ltd. was heavily indebted. To keep it afloat, in 1964 Grace, Kennedy secured loans in the amount of £14,800 and in 1965 this was increased to £18,684.[19] With the astute management of Ed Muschett, hired in 1964, the insurance business, which was assigned to George and Branday, soon became a profitable division of the group.

Grace, Kennedy also acquired a hardware business, Messrs. Sheffield & Co., which was in a similar situation to George & Branday—in need of the group to secure loans from Barclay's Bank. Luis Fred became increasingly frustrated with its operations. At the board meeting of October 25, 1967, he proposed "the cessation of the hardware business … that the debts be transferred to the Merchandise Division of Grace, Kennedy for collection and the hardware stock be sold out."[20] He would reconsider his position in view of finding the right general manager. In November, Mr. Robert McConnell was appointed manager, and within six months, the company was showing improvements in sales and in the collection of debts. By October 1968, Sheffield & Co. was showing a net profit of £17,000.

Governance

As Grace, Kennedy expanded, operations became more sophisticated. Managers and directors relied more on legal counsel to deal with changing government regulations and complexities of business acquisitions, both local and foreign. José Kennedy, as a qualified solicitor, was an asset to the board in this regard. Luis Fred also sought technical advice from English and US consultants to modernize accounting systems, introduce the use of computer databases, establish a Quality Control Department, and advise on plant operations of factories.

In 1965 the board consisted of L. F. Kennedy, Governing Director, with five other internal directors: J. S. Moss-Solomon, S. C. Alexander, S. Soutar, G. E. Dodd, and

---

17   Ibid.

18   George and Branday Ltd. was formed in 1879 by Mr. Arthur George, who died in 1922 and Mr. J. W. Branday, who died in 1909.

19   I have learned from many sources that if a bank ever requested a guarantee from Grace, Kennedy to secure a loan, Mr. Kennedy would flatly refuse. If the bank wanted to do business, it would have to be done on his terms.

20   Grace, Kennedy & Co. Ltd., October 25, 1967.

J. A. Kennedy; and one external: James L. R. Bovell.[21] Mr. Harold Aarons[22] served as accountant and company secretary.

Harold H. Dunn (whom James Bovell of Dunn Cox and Orrett replaced), died at the age of eighty-five in 1967. The *Daily Gleaner* carried a full editorial on H. H. Dunn, July 19, praising him as "one of the finest legal brains"; he was regarded as the patriarch of the legal profession in Jamaica, a man of "impeccable integrity ... with an uncompromising adherence to the highest ideals of his profession."[23] Daddy Dunn had been a founding director and shareholder of Grace, Kennedy and contemporary and good friend of Fred William Kennedy.

Board meetings became lengthier and more frequent.[24] Luis Fred established sub-committees in management to deal with restructuring departments. In finance, he created an advisory body to the financial director to reduce borrowings and trade debts. He adopted a more consultative approach to governance, through which he invited regular reports to the board on the group's performance and sought the opinions of directors to arrive at consensus to approve new strategies, expenditures, and initiatives.[25]

Mr. Kennedy also took steps towards succession. James Moss-Solomon remained in the position of managing director but was slowly withdrawing from active management. In his place, Carlton Alexander assumed greater responsibility. In addition to serving on the main board, both he and Luis Fred were chairmen of subsidiaries: Carlton of George and Branday Ltd., Cecil deCordova, and Sheffield and Co.; Luis Fred of Grace, Kennedy Shipping, Jamaica Rums and National Processors.[26]

They worked jointly on oversight of the newly established divisions of the group: trading companies, insurance and real estate, shipping, manufacturing, overseas (Montreal, UK and Rotterdam), and financial operations. Managers of each of these corporate entities reported directly to Luis Fred Kennedy and to a committee appointed in 1969 to manage the operations of the Grace Group. The committee was composed of Luis Fred Kennedy, S. Carlton Alexander, Harold M. Aarons, L. Paul Scott, Francis (Paco) X. Kennedy, Edward G. Muschett, Ian M. Mordecai, and Bruce Rickards.

---

21 James Lytcott Reece Bovell, Solicitor, Partner, 1944, Dunn, Cox & Orrett Solicitors. Born June 1, 1915, educated at Jamaica College and Cambridge University, called to the Jamaica Bar, 1940, and married, 1942, to Marion Isabel Davidson (*Who's Who Jamaica* 1963, 57).

22 Gladstone Kamicka retired as company secretary in 1960 and remained a director of the board until the year of his death in 1966. Mr. Kamicka had joined the company in 1942.

23 *The Daily Gleaner*, Wednesday, July 19, 1967, 12.

24 Information about Grace, Kennedy & Co. Ltd. in this chapter is sourced from minutes of board of directors' meetings, 1962–1971.

25 See Bill Valliant's tribute in Appendices.

26 Grace, Kennedy established National Processors Ltd. in the mid-1960s for the production and packaging of non-food items for Chesebrough-Ponds.

Net profits across the group increased over the period, which kept the debt ratio stable and provided adequate cash reserves to be converted into assets of value. In December 1966, Luis Fred proposed a bonus share issue of one share for every five held, being part of the amount standing to the credit of the General Reserve Account to be capitalized. The authorized share capital of the company was increased from £800,000 to £1,000,000 by the creation of 1,800 ordinary shares and 200 employee shares of £100 each.[27]

In 1967 the chairman proposed that a trust be established for acquiring employee shares, which it would hold until such time as shares were to be offered to employees. The trust would borrow money from the company to acquire shares and pay interest only to the extent of dividends received. It would purchase shares from the estates of deceased members or from those leaving the company at the auditors' valuation or at par, whichever was higher. It would also be authorized to purchase shares held by employees.[28] It was named the Grace, Kennedy & Company Limited Employee Investment Trust, established October 1967. The trustees appointed were L. F. Kennedy, J. S. Moss-Solomon, S. C. Alexander, J. A. Kennedy, and Eric Chin.

Mentorship and Training

Ed Muschett[29] recalled his first meeting with Luis Fred Kennedy:

> When Fred Kennedy came back from England, it was 1964, I went for an interview with him and he talked to me for over an hour learning a lot about my history and eventually he said that he would like me to work for him and he said I would be getting sixteen hundred pounds per year. I worked directly under Fred Kennedy, not Alexander.[30]

Together they built the diversification of Grace's insurance interests with Insurance Holdings Ltd., which became the holding company for this expansion. Ed Muschett became managing director of George & Branday Ltd., which assumed Grace, Kennedy's pre-existing insurance business. In 1967, Neville Bolton was appointed managing director of United Reliance Insurance Ltd., General Agents

---

27   Grace, Kennedy & Co. Ltd., December 10, 1966.

28   Ibid., February 22, 1967.

29   Muschett reported directly to Luis Fred, who took a personal interest in the expansion of the insurance business. Mr. Muschett rose quickly through the ranks, building a network of insurance and brokerage companies for the group. He was appointed director to the board in 1970 and served as general manager for a couple of periods during the 1980s. He was also highly successful in operating his own business, B. L. Williams International, a freight forwarder and an international shipper of household effects. His son, Edward Muschett, is present chairman of the B. L. Williams Group.

30   Ed Muschett, interview with author, February 27, 2019.

for Jamaica Cooperative Fire and General Insurance Co. Ltd., and in 1969, Paul Bitter was placed in charge of Allied Insurance Brokers Ltd., one of three registered independent brokerages that provided access to both local and overseas underwriting capacities.[31]

Muschett became Luis Fred's trainee. In addition to overseeing all the insurance companies of the group, he managed the Lufthansa Agency, was assigned to special projects to investigate net costs of acquiring real estate for industrial and commercial purposes, and he prepared routine reports on the superannuation scheme. In 1971, they travelled together to New York to meet with managers of American International Underwriters. In an interview held at his home in St. Andrew, Ed shared his memories:

> My own relationship with your dad, he treated me almost like a son. He was interested in my life in general as well as in what I was trying to accomplish ... Alexander treated me like a brother and Fred Kennedy treated me like a son ... I had a party at my house and my mother was there and he was sitting and chatting with her, and he said to her, "Ed Muschett, Ed's one of my sons." That's the way he put it.
>
> He would discuss anything with you, we would chat for hours, and sometimes I would come up with a solution that didn't quite satisfy him. For instance, I would take to him some decisions and he'd tell me that he didn't agree with them, but he'd say, "You are not to come back and tell me when they go bad." He would just leave the matter alone if it was the wrong decision. He allowed me to make mistakes, and I learned from that.
>
> I came to learn how to discuss anything with your father even if the conclusion didn't quite satisfy him.[32]

One of Luis Fred's top priorities was employee training. In March 1967, Peter Moss-Solomon was articled at Messrs. Price Waterhouse to certify as an accountant, and Paco was slated for a six-month secondment to Canadian Imperial Bank of Commerce.[33] The company also provided study leave for employees to take

---

31   In the early 1970s, Richard Davidson was appointed managing director of Gamble & Davidson Insurance Brokers Ltd. Information provided by Paul Bitter. See Paul Bitter's tribute to Luis Fred Kennedy in Appendices and a bio and interview: http://www.jaweb2.com/jaalumni/stgc/profile/pfile07.html.

32   Ed Muschett (2019).

33   Plans for Paco's secondment to CIBC were cancelled because of his appointment as manager of Harbour Cold Stores in May 1967.

up scholarships abroad and offered to pay for university business training.[34] In September 1967, the chairman recommended to the board employment of a training officer to be recruited in Canada and of a Jamaican understudy to take over the position at the end of a three-year term.

Luis Fred continued to invest time and resources in the promotion of his son, Francis X. Kennedy. In 1967, Paco was appointed a director of three subsidiaries: Grace, Kennedy & Co. (Shipping) Ltd., Cecil deCordova & Co. Ltd., and National Processors Ltd. In May 1967, Luis Fred sought approval of the directors to appoint Francis X. Kennedy manager of Harbour Cold Stores Ltd. and of Dairy Industries (Ja) Ltd. at an annual salary of £2,500.[35] At twenty-six years of age, this was Paco's first managerial position. In March of that year, the chairman announced that Grace, Kennedy was entrusted with the management of both companies, "a joint enterprise in which they were associated with the New Zealand Dairy Board, Adolph Levy & Bros., Bryden and Evelyn Ltd., and T. Geddes Grant Ltd. for the packaging and marketing of dairy products in Jamaica."[36]

---

34   In 1969, Grace, Kennedy provided leave to Raymond Evans to take up a scholarship at Bowling Green State University and paid for Mr. Winston Ho Fatt to study business at the University of Waterloo.

35   Grace, Kennedy & Co. Ltd., October 25, 1967. On September 1, 1967, a full-page ad appeared in *The Daily Gleaner* advertising "The Largest and the Newest Cold Storage Warehouse in Jamaica, Harbour Cold Stores Ltd.," with 60,000 cu. feet of storage space and pictures of Grace Line Ship, the *Santa Rosa*, discharging forty-two tons of refrigerated cargo.

36   Douglas Hall (1992), 59.

OFFICERS AND MANAGING COMMITTEE OF ST. GEORGE'S OLD BOYS' ASSOCIATION.

Back Row: Messrs. B. Bennett, K. Campbell, C. Knight.
Middle Row: Messrs. F. X. Rankine, L. F. Kennedy, H. Brownlow, G. deLeon, V. Sasso, L. Fogarty, Jnr.
Front Row: Mr. R. E. Taylor, Rev. D. Cruchley, Hon. D. Judah, Rev. D. Tobin, Mr. A. Wynter, Mr. G. Desnoes.
Committee members not in picture are: Messrs. W. Meeks, Allan Taylor, J. C. Breakspeare and G. Bowen.

(Photo, courtesy of © St. George's College)
Ad Majorem Dei Gloriam (1850–1950)

Luis Fred Kennedy, flanked by His Lordship, Bishop John J. McEleney and Rev. Fr. Charles L. Judah S.J., addresses volunteers at Sts. Peter and Paul fundraiser dinner.

The Catholic Opinion, Vol. 65, No. 20, May 19, 1961.

(Photo, courtesy of © The Society of Jesuits in Jamaica)

Celia clothed as a Servite nun, Sr. Martine OSM.
1961, 45 Lady Musgrave Road.
(Photo from family collection, original by Wally Allen)

Runaway Bay Hotel, 1967. Luis Fred and Tillie Kennedy with family:
L–R: Fred (age 16), Paco (26), Elizabeth (13), Celia (Sr. Martine OSM) (24),
and Mary (Sr. Francis Xavier OSM) (21). (Photo from family collection)

Gracie and José Kennedy with their children:
L–R: Patrick Joseph, Michèle Marie, Anthony Charles,
and Philip Thomas, circa 1964.
(Photo, courtesy of Kennedy family, Philip, Tony, Michèle and Patrick)

32

HOLY MOTHER CHURCH

(1960s)

Dad followed the devout practice of his mother in attending daily Mass. He would wake me early in the mornings to accompany him to either the 6:00 or 6:30 Mass. The liturgy at Sts. Peter and Paul Church lasted eighteen minutes—spoken in ecclesiastical Latin and stripped of the trappings of Sunday service: no hymns, sermons, or prayers of the faithful. Devotion was long enough for him to seek spiritual guidance and to collect his thoughts for the day without having to listen to the priest sermonize.

Influenced by a strict Catholic upbringing and by her training at boarding school, my sister Celia expressed an interest in becoming a nun. As dutiful Catholic parents, Mom and Dad were expected to encourage their daughter's vocation. Yet, they had their reservations. "I didn't like it at all. We tried to dissuade Celia, to show her what the world outside was like, and maybe entice her to change her mind,"[1] Mom said in her memoirs.

Dad booked passages on board a Banana Producers ship from Port Antonio to England in early 1960 so that Celia could share in one of his most thrilling experiences—a North Atlantic crossing on a banana boat. As luck would have it,

---

1 Lydia M. Kennedy, interview with Georgianne Thompson Kennedy, 2002.

they met upon a treacherous storm at sea that lasted for days—the boat pitching, and pounding the waves:

> Fred loved the sea and prided himself on being a good sailor, which he was, except for once when crossing the Atlantic on one of the Banana Boats we had hit a bad storm. My daughter Celia and I were the first to succumb, followed shortly by Fred who claimed he had eaten a bad lobster.[2]

Mom, Dad, and Paco took Celia on a whirlwind tour of Europe before bidding her farewell at St. Mary's Priory in London. The intention was to expose the impressionable seventeen-year-old to the ways of the world. Yet none of the lure of the great sites of Europe, the wine, fine dining, or dance, changed Celia's mind.

On Saturday, February 25, 1961, Kingston, Jamaica, Miss Celia Kennedy knelt at the high altar of Sts. Peter and Paul Church, before his Lordship Bishop McEleney. She was attired in a white wedding gown and veil, a bride of Christ to be clothed in the habit of a Servite nun.[3] As a symbol of renouncing the yoke of the world, the young postulant (aged eighteen) removed her gold jewellery, which she handed to her attendant, her youngest sister, Elizabeth (aged seven). The bishop cut her hair, after which he placed sections of the holy tunic on her outstretched arms. The bride retired to the sacristy, where sisters of her religious community clothed her as a nun. When she returned to the altar, the bishop handed her a crucifix, placed a wreath of white flowers on her head, and proclaimed, "Come, spouse of Christ, receive the crown which the Lord hath prepared for thee forever."

Celia adopted a new name: Sr. Martine OSM (Order of the Servants of Mary). Martine was a derivative of San Martín (de Loinaz), one of twenty-six Franciscan friars martyred in Japan in the sixteenth century. He was born in Beasaín, Spain, of the Loinaz family, from whom my mom was descended.

In 1963 my sister Mary also entered the Servites.[4] The family sailed from Port Antonio to UK on board the eleven-passenger SS *Northern Lights* of the Jamaica Banana Producers Steamship Co. We encountered no storms, but it did not take much of a rough sea to cause the banana boat to rock with the waves. My parents' attempts to dissuade Mary also failed. She returned to Jamaica a year later for a clothing ceremony like Celia's at Sts. Peter and Paul. She received the Servite habit and chose the religious name, Sr. Francis Xavier O.S.M, in honour of Mom and Dad's special devotion to the saint.

---

2   Lydia M. Kennedy, "A Giant of a Man" (1997).

3   *The Catholic Opinion*, Friday, March 3, 1961, 1.

4   Earlier that year, on March 11, Mom's mother, Carmen Celia Arzeno Loinaz, died at the age of eighty-six. Mom was devastated at the news, even though her mother had lived a long, healthy life.

Mom commented about my sisters' decision to enter the convent:

> I don't think their time in the convent was time wasted, they may be better people because of it. We went visiting them in England. When we had a meal there at the convent, they weren't allowed to sit with us and eat, we had to eat by ourselves. They could stand there and watch us eat. They had to eat with their community. Imagine, their own parents, you know … It wasn't all nice for them, they had to kneel in front of the mother superior and all that kind of thing.[5]

When Celia and Mary returned from England, they were assigned to teach at the Servite boarding school in Brown's Town.

Fr. Charles Judah S. J.[6] had an interest in expanding the Servite mission in Jamaica by establishing a new school and convent in Kingston. Dad assisted him by chairing a fundraiser campaign that canvassed over 200 volunteers to raise money from parishioners of Sts. Peter and Paul Church. Two years later, February 23, 1963, the Catholic bishop of Kingston, the Rt. Reverend John J. McEleney S.J., blessed and officially opened Sts. Peter and Paul Preparatory School, Old Hope Road in Kingston. Dad never divulged the amount of his personal donations nor spoke of other charities and endowments he supported.[7]

On February 27, 1965, the family received unexpected news of the death of Dad's youngest brother, Francis, at age forty. He was on a business trip to Port-of-Spain, Trinidad, where he suffered a fatal heart attack while swimming with friends. Two years earlier, he had migrated from Jamaica to settle in England with his family.

His elder son, Stephen Francis, who was six weeks shy of his fifteenth birthday when his father died, recalls, "The last thing we did together before he left on his trip was to go see *Lawrence of Arabia* at the Cinema in Horley, England. He loved the movies."[8] While they lived in Jamaica, Steve can remember what a treat it was going to the Carib Cinema with his father.

---

5   Lydia Kennedy (2002).

6   Rev. Charles L. Judah S.J., (1903–1988), older brother of Douglas Judah. Charles Judah was the pastor of St. Ignatius Church in Brown's Town (1948–1960) and was responsible for inviting the Servite sisters from England to establish a boarding school in Jamaica. Dad worked with Fr. Judah when he became pastor of Sts. Peter and Paul Church (1961–1975) to fund the building of a new church and school. Fr. Judah's obituary appears in *The Daily Gleaner,* June 2, 1988, 19.

7   He was an active member of the Old Boys' Association of his *alma mater*, St. George's College, and director and chairman of the school board.

8   Stephen Francis Kennedy, personal correspondence, November 2021.

Holy Mother Church (1960s)

Francis' brother, José, travelled to Trinidad to identify the body and arrange for it to be transported to Jamaica for burial. His wife, Anne, and children, Lareine (aged sixteen) and Stephen (fourteen), came out from Surrey, England to attend the funeral. He was survived also by children Anne-Marie (aged thirteen), Kevin (ten), Mary Lou (five), and stepdaughter, Hiltrud Angerhausen (twenty).

Members of the staff of Grace, Kennedy, friends, and relatives attended the service held at Sts. Peter and Paul Church, Sunday, March 7.[9] Pallbearers to the church included James Moss-Solomon, Carlton Alexander, Ferdie Figueroa, Bruce Rickards, Stuart DaCosta, and Geoffrey George. Pallbearers to the grave were his brothers, Luis Fred and José Kennedy, his son, Stephen Kennedy, his brothers-in-law Simon Soutar and Owen Plant, and nephew, Francis (Paco) Kennedy. Service was conducted by Bishop McEleney, Rev. Frs. Charles Judah S.J., and William Connolly S.J.

All Dad's siblings were fervent Catholics, attending church services, supporting fundraisers, and sending their children to Catholic schools and universities. (Seven of Luisa Matilde's grandchildren entered religious life.)

My Uncle Che's family and ours were close; two brothers had married two sisters. The children attended the same high schools (St. George's College and Sts. Peter and Paul), spent time together on weekends and shared summer vacations on the north coast. Tony, José's and Gracie's second son, and I were a pair—sharing hobbies, cycling, and playing tennis. Michèle, their daughter, and my sister Elizabeth were the same age and went through school together at Alvernia Prep, Sts. Peter and Paul, and Campion College.[10]

"My recollection overall is that it was a pretty sheltered upbringing,"[11] Tony comments about his childhood. "It was a structured society. My cousin Fred and I were altar boys, diligent students, and close to our twenty-two cousins." Tony admired his dad for the priority he placed on education: "He was definitely the driving force behind us doing well at school." (Tony had one of the top academic records at St. George's College, first in his class and in his form, throughout his five years of high school; his sister Michèle was equally smart.) Tony recalled a conversation with his dad:

---

9 Article, "Mr. F. X. Kennedy Buried," *The Daily Gleaner*, Monday, March 8, 1965.

10 Philip is the eldest (b. 1947) and Patrick, the youngest (b. 1958). Phil also attended Campion Hall and St. Georges College, where he excelled in track and field. In later years, Grace, Kennedy paid for him to go to Franklin Institute in 1974. He got a certificate in Mechanical Engineering Technology in 1976. He transferred to Boston University and received a Bachelor of Science degree in mechanical engineering in 1978. Patrick attended both Campion Hall and Campion College and migrated with his family to Boston in January 1972.

11 Interview with Anthony (Tony) Charles Kennedy, José's and Gracie's second born.

> One day I came home, I was all excited, there was an exam in third form, I said to him, "I got a 100% in this big exam." My Dad's response was, "Next time, you should be getting 120%." Wouldn't you know, a few months later, I remember, it was a difficult class, and the teacher said he wanted to give students additional points at the end of the term, and so, I came home with the report card and showed him the score, 120%.[12]

Tony was also influenced by a strict Catholic upbringing. He spoke in the interview about his father and the kind of man he was:

> My Dad was disciplined in his own way, always had to be at church right on time, angry if we left the house two minutes later than we should have. He was a devoted Catholic, more so than my mother. My father was over the top sometimes, going to Mass every morning at 6:00 AM. My mother never went to those early masses.[13]

Tony entered the Society of Jesus in September 1965, days after his sixteenth birthday.

I entered the Jesuits in 1968, by which time our parents accepted the choices we had made. Dad and I corresponded frequently when I was in the seminary.

> My dear Fred
>
> Just a few lines on the last day of this old year to send you all my good wishes for 1969. I pray that you will continue happy in your vocation and am sure that you are doing what you really want and love. It is wonderfully generous of you and Celia and Mary. May you have many, many years of happiness, peace of mind and good health![14]

His letters expressed a fondness for church practices:

> My dear Fred
>
> This note goes to you with all my best wishes and prayers for a very happy and holy Easter. It is a wonderful feast and I have often wondered if it is not even greater than Christmas. Even now I get a great

---

12   Ibid.

13   Ibid.

14   Luis Fred Kennedy. Letter to his son Fred W. Kennedy, December 31, 1968. Dad wrote letters regularly while I was at the Jesuit Novitiate in Lennox, Massachusetts, 1968–1970. They were filled with news of family, politics, business, and fatherly advice.

thrill from the *Gloria in Excelsis Deo* suddenly bursting forth after the days of sad and solemn ritual without music.[15]

Despite the influence of organized religion, Dad was an independent thinker. He struggled with dogmatism and with the hierarchical structure of the Church. He paid no attention to the Church's stamp of approval, the *Imprimatur,* which he said did not "*mean a damn thing.*" He found presumptuous and arrogant the idea that the Vatican could decide which books he was allowed to read. He welcomed the changes that Vatican II (1962–1965) brought: use of the vernacular in celebration of the liturgy, accepting people's freedom of choice of religion, and including other Christian denominations in a spirit of ecumenism.

He was critical of the Church's mission in Jamaica, which he thought represented the establishment, not relevant to the poor in society. In December 1968, he wrote to me about a Catholic blessing at his sister Gloria's new home in Barbican:

My Dear Fred,

Archbishop[16] McEleney, Fr. Whelan and Fr. Ballou concelebrated Mass … I don't understand why the priests do not visit the poor more, communities in August Town and elsewhere. I believe most think of our Church as one for the privileged classes and the "white" people. Why should the Church here not become more aggressive and take positive action in the areas where most of the people are and where there is the most need? Why limit *Cursillos* and Masses in houses to the so-called "better" classes? Instead of preaching silly, unprepared sermons to people who don't really need or want them, they should be visiting the poor. Our Church started in the hovels to give hope and love to the people who lived in them.[17]

The letters were lengthy and frequent, often six or seven pages, handwritten on Grace, Kennedy & Co. Ltd. stationery. He concluded his letter of December 31, 1968:

The Company's lunch for the staff last Saturday was a big affair and went over very well. I flew up from Montego Bay at about noon and returned by a 6 pm flight—this latter was a 727 and took only 17 minutes![18]

---

15   Ibid., March 27, 1969.
16   The Roman Catholic Diocese of Kingston was elevated to an Archdiocese in September 1967.
17   Luis Fred Kennedy (December 31, 1968).
18   The family spent Christmas at Discovery Bay.

We were all so glad that you had a wonderful Christmas. I, and I know that all of us, remember you constantly in our prayers! You will be glad to know that Mummy's birthday letter eventually arrived on Saturday. We all read it together. It was a lovely letter, son. Today, Uncle Che handed me what must be your Xmas letter to us, but I shall not open it till we are together again in Discovery Bay this afternoon!

God bless you my son & keep you happy and well always.

Your old Dad.

A proud moment for Luis Fred Kennedy (Managing Director and Chairman of Kingston Wharves Ltd.), presenting plaque to Captain of SS *Santa Rosa* of Grace Lines to commemorate the docking of the first ship, Newport West, July 2, 1966.
(Photo by Wally Allen, courtesy of © GraceKennedy Ltd.)

The Daily Gleaner, Monday, July 4, 1966, 8.
(Photo, courtesy of the © Gleaner Company (Media) Ltd.)

33

(1960s)[1]

In 1963, when Kingston Wharves incorporated the Royal Mail Line's pier (Church Street) into the group, the company operated a total of twelve berths.[2] The size and design of the finger piers were grossly inadequate. Raf Diaz comments:

> The old finger piers downtown had outgrown their usefulness, and coupled with growth of commerce in the city, were also extremely congested for getting cargo in and out of the wharves. It became clear that action had to be taken and it was one of Jamaica's visionaries, Moses Matalon,[3] who conceived the idea of developing a new and modern port on the outskirts of the city now known as Newport West. Fred

---

    1   KW company logo, used as title of chapter, courtesy of Kingston Wharves Ltd. © 2022 Kingston Wharves Limited. All Rights Reserved.

    2   See note, Chapter 25, about Grace, Kennedy & Company Shipping Ltd.

    3   Moses Matalon was business executive engineer and company director, managing director of Kingston Engineering Works Ltd., chairman of United Motors Ltd., and director of numerous companies. Born, Jamaica, January 10, 1921, son of the late Joseph Matalon, merchant of Kingston and Florizel, née Henriques. (*Who's Who Jamaica 1963,* 264) He died in 1992.

Kennedy, Charles DaCosta and Moses Matalon then spearheaded negotiations with the government for the move that was to create what became Port Bustamante in the 1970s.[4]

Commerce had grown dramatically since 1945 when the company was founded. The tonnage handled doubled in the first ten years and quadrupled by the end of the following decade, reaching over 600,000 tons in 1965 compared with 156,000 tons in 1946.[5] In 1964 the company made the decision to relocate its port facilities. The proposal was to purchase fifty-six acres of reclaimed land at a projected cost of £5,000,000.

In July 1965, Luis Fred held a luncheon to celebrate the driving of the first piles in the quay wall. Present were board directors of Kingston Wharves: Douglas Judah, chairman; Luis Fred Kennedy, managing director; E. E. Cox and Brian Challis, managers; Robert McConnell, secretary; and James Moss-Solomon, F. W. Harris, T. D. Bentley, H. D. Macaulay Orrett, Karl Gaynair, and H. T. Hart. In attendance also were the company's accountant, the architect, and the quantity surveyor of the project.

Luis Fred's dream of improved conditions for port workers had materialized. Chairman of the Shipping Association of Jamaica, Paul Scott, helped negotiate the Port of Kingston Accord that included three waterfront unions: the BITU, NWU, and TUC. The accord came into effect, January 1, 1966. Paul Scott commented:

> The Accord changed significantly the way the Port, ships, cargo, management and labour worked. It resulted in improved productivity and enabled the introduction of containers without industrial unrest or any job loss by port workers. Guaranteed weekly wages were introduced for port workers instead of the old system which meant pay only when there was work. New and improved fringe benefits were provided including vacation with pays, pension scheme for retiring port workers...and better organized work roster with more equitably allocated work.[6]

After a long, hard-fought battle of thirty years, Luis Fred lived to see industrial relations normalized on the waterfront.[7]

---

4 A. Rafael Diaz, "The Origins of Kingston Wharves Ltd." (1999), 1. See bio, note 6, Ch. 35.

5 *The Daily Gleaner,* Monday, July 4, 1966, 8.

6 Laurence Paul Scott. "Memories of Grace" (1997).

7 The Moody Commission (Commission of Enquiry for the Port of Kingston to enquire into all aspects of the operations of public wharves coming within the ambit of the Wharfage Law) was appointed by Cabinet on October 20, 1961. It set the framework for the development of employment and compensation policies and for binding arbitration. Luis Fred made numerous submissions to the Commission of Enquiry on behalf of wharf owners.

I remember the day the first three berths (#5, 6 and 7) opened on July 2, 1966. It was five o'clock, the faint rays of light just showing at second cock crow in the early dawn.[8] *"Time to wake up, son, or we'll be late. It's a historic day."*

Dad floored the accelerator of his Plymouth V8 as he raced out of the driveway. The whole way to the pier, he was animated and chatty, gleeful like a child in anticipation of a grand surprise.

The 15,000-ton SS *Santa Rosa* rose high out of the water, 600 feet in length and ninety feet wide, docked alongside the No. 5 Newport West Kingston Wharves berth. She and her sister ship, the SS *Santa Paula*, were built in 1958 to replace two earlier namesakes first launched in 1932. (Dad always referred to a ship as "she," as sailors do. *"It's a sign of respect; she protects you like a mother does,"* he used to say.)

She had four holds with fully equipped cargo-handling-gear and automatic conveyors to load pallets on and off. The structure was made of fireproofed aluminum alloys, and she had stabilizers that reduced rolling in heavy seas. The interior contained spacious cabins with their own bathrooms and balconies, and the public areas and staterooms were air-conditioned, luxuriously furnished, and lit with crystal chandeliers.

In port that day, she carried 234 cruise passengers and 250 crew. Dad and I climbed the gangplank to be greeted by the captain, who escorted us up through the ship to the officers' deck. Mr. Kennedy welcomed the captain of the SS *Santa Rosa*, the first vessel to dock at the new berths of Kingston Wharves, Newport West.

Later that Saturday morning, Kingston Wharves Ltd. held a luncheon on site at the new terminal to mark the twenty-first anniversary of the company. At the start of the ceremony, Luis Fred called for a moment of silence as a mark of respect for Mr. Vincent Louis George, his friend and business associate, who had died just days earlier on June 30. Uncle Vin, age seventy-eight, had suffered a fatal stroke at his office on Port Royal Street. He had been a founding member and co-managing director of Kingston Wharves Ltd.[9]

General Manager Brian Challis assisted Luis Fred in presenting long service awards to employees. A certificate and a gold watch were handed out to staffers with twenty and more years of service. Awardees included Everard Cox, manager, Miss Arthurs, office helper, Douglas Lewis, accountant, Vincent Melbourne, wharfinger, and fourteen others, including foremen, clerks, accountants, and wharfingers.

> This is a proud moment in my life, and in a very large measure, I owe a lot to every one of you here. I remember clearly twenty-one years ago, when Mr. Douglas Judah, chairman, the other managing director, Mr. Vin George and I went to Spanish Town to register the company. I

---

8   We had a mean-spirited rooster and a large chicken coop of hens in the back yard.

9   An obituary for Vincent Louis George appeared in *The Daily Gleaner*, Saturday, July 2, 1966, 2.

am sure we are all proud today of the progress of our company, of the tremendous new facilities which we have managed to erect here.[10]

He had everything to be proud of—a modernized port that would become the envy of the Caribbean. The initial authorized capital of £249,000 had grown over a twenty-one-year period to £1,159, 100.

In 1967, the government approved the construction of additional piers and negotiated with Kingston Wharves for the purchase of the Princess Street, Orange Street, and South Street Wharves.

In a general election held on February 21, 1967, Donald Sangster, as head of government,[11] won for the Labour Party thirty-three of fifty-five seats, with a popular vote of 50.65%. Less than two months later, he succumbed to a fatal brain hemorrhage. He died at age fifty-five in Canada at the Montreal Neurological Institute. On April 11 the Hon. Hugh Lawson Shearer was sworn in as the third prime minister of independent Jamaica.

During Shearer's tenure as prime minster (1967–72), Jamaica experienced steady growth in agriculture, mining, and tourism, with steep percentage increases in Gross Domestic Product. He secured loan agreements to build primary schools, fifty new junior secondary schools, and to invest in teacher-training colleges, a College of the Arts, Science and Technology, and the Jamaica School of Agriculture.

Despite economic prosperity, societal tensions were building. After a decade of independence, many felt that the dreams of self-determination and a break from colonialism had not come fast enough. The gap between the rich and poor widened, not lessened, with those at the bottom of a pyramidal structure of society feeling trapped and disenfranchised. Unemployed youth were disillusioned.[12]

The resurgence of Rastafarianism and the Black Power movement also posed challenges for the establishment. These groups viewed government, professional and moneyed classes, and established Churches as instruments of Babylon, agents of

---

10   Luis Fred Kennedy, July 2, 1966, on occasion of twenty-first anniversary of Kingston Wharves Ltd.

11   In 1965, at the age of 81, Bustamante suffered a stroke, which forced him to withdraw from active, public duty. He appointed Donald Sangster as Acting Prime Minister. Born in Black River, St. Elizabeth, October 1911, Donald Sangster was the son of a land surveyor, W. B. Sangster and his wife Cassandra Plummer. He was educated at Munro College and admitted as a solicitor of the Supreme Court in 1937. He entered politics in 1949 as an elected member of Parliament. In 1953 he was Minister of Finance and Leader of the House and became Deputy Prime Minister in 1962.

12   Unemployment was estimated at 30% of the labour force, and 67% of those employed earned less than $10 per week and 80% under $20. (Ralph Gonsalves, "The Rodney Affair and its Aftermath" September 1979, 1.)

corruption that oppressed the innocent. Conflicts with the authorities increased "with a hardening of positions on both sides."[13] It proved difficult for the Labour government to cope.[14]

My dad was concerned about changes of life in Jamaica. He wrote to me in 1969:

> My dear Fred
>
> Kingston and Montego Bay are engulfed by a terrible crime wave. Houses are not only burgled, but the crime is accompanied by violence-shooting and rape. There have been many cases reported, and I believe, just as many that are never publicized. The result is that people now keep their houses locked all the time, day and night. We still eat out on the back terrace, but after dinner we lock up the house and either watch TV in the locked drawing room, or sit on the balcony upstairs, which is much cooler. Things have changed in the short time you have been away.[15]

A new era dawned for the People's National Party. Norman Manley announced his intention at a banquet held on his seventy-fifth birthday[16] not to stand for re-election as party leader or president. His words were among the most poignant of his political career:

> Looking back ... over the years, may I declare that they have been great years. I have known all things in politics the hard way ... I would not have chosen my road in life any other way. I affirm of Jamaica that we are a great people ... patient and strong, quick to anger, quick to forgive, lusty and vigorous, but with deep reserves of loyalty and love, and a deep capacity for steadiness under stress, and for joy in all things that make life good and blessed. Bless this dear land and bless our people now and forever more.[17]

---

13 Ibid., 45.

14 One setback for the labour government was the handling of the Rodney Affair. On October 15, 1968, the Jamaican government banned Dr. Walter Rodney, lecturer in the Department of History at UWI, Mona, from re-entry to Jamaica. The following day, university students led a march in opposition, which triggered a series of events that came to be called "The Rodney Affair." http://www.jstor.org/stable/40653384.

15 Luis Fred Kennedy, letter dated March 27, 1969.

16 Norman Manley's seventy-fifth birthday, July 4, 1968. Carlton Alexander announced at the Grace, Kennedy director's board meeting the day before that the sum of £250 had been contributed to Norman Manley's birthday fund.

17 Norman Manley's speech, July 4, 1968, at his seventy-fifth birthday party. Quoted by Arnold Bertram, *N.W. Manley and the Making of Modern Jamaica* (2016), 359–60.

Seven months later in February 1969, Michael Manley was elected president of the People's National Party, defeating Vivian Blake 376 to 155 votes. At age forty-four, he became the youngest elected leader of a major political party in Jamaica.[18]

In a packed National Arena, his father welcomed him as his successor: "Comrades, I welcome you all to this the greatest in size and one of themes important in purpose, of all the conferences of the People's National Party": a change of leadership and a nation in crisis. He stated that Jamaica was at a "cross-roads of history," and spoke of the role his party should play:

> What I want to do as leader of the party, as a final contribution to party thinking, is to express what I hope to be accepted by the party. We have to understand ourselves and we dare not ignore Black Power ...
>
> You may ask, some of you who are here and certainly many in Jamaica outside this arena, to whom my words are also spoken, you may ask what does Black Power mean in a country where the vast majority are black?
>
> Black power means the acceptance with joy and pride, the fact of blackness, of black dignity and black beauty. It means the acceptance by the black man of his own place in the brotherhood of man. It is true that although we have achieved the power of self-government and the dignity of nationhood, there still lurks beneath the surface of many minds and consciences a feeling that the white man can do more and can achieve more than we can do ...
>
> But Black Power, Black Dignity, Black Pride in Black Beauty gain ground the world over, and will be accepted by ourselves and by all the world.[19]

September 2, 1969. **MANLEY IS DEAD.** "The famed Jamaican leader, patriot, statesman, Queen's Counsel and politician, Norman Manley died at his 4 Washington Drive residence yesterday afternoon shortly before 1:00 o'clock and plunged the nation into mourning."[20]

In October, Michael Manley addressed 3,000 delegates of the Annual Conference for the first time as President of the People's National Party. He announced that

---

18   Godfrey Smith, *Michael Manley* (2016), 93.

19   Rex Nettleford, *Manley & the New Jamaica: Selected speeches and writings: 1938–1968*, 378–379. Norman Manley's speech followed in the wake of The Rodney Affair (October 1968).

20   *The Daily Gleaner*, Wednesday, September 3, 1969, 1.

the policy of the party would be one of "economic nationalism" for Jamaica. Its economy will "be fundamentally under national control and responsive to national needs."[21]

Michael Manley hailed his constituents as "fellow crusaders for a new Jamaica, sufferers." He roused the passions of a nation with a message of liberation, of "giving power to the people." He would bring legitimacy to the Black Power movement and Rastafarianism; he would lure academics who had leftist leanings and persuade many in private business that better days lay ahead.

In November 1969, Kingston Wharves expanded with the addition of three new berths that provided 1,200 feet more of berthing space. They would become equipped for containerized cargo and large gantry cranes for more efficient embarking and unloading of freight. It was revolutionary technology that mitigated damage to goods and reduced pilferage.

The Daily Gleaner reported that the large shipping complex ranked Kingston among the most modern ports in the hemisphere. It accommodated the world's largest and fastest vessels, provided quayside bunkering and telephone facilities, and ensured a fast turn-around for vessels.[22]

Yet not all was smooth sailing. Although Grace, Kennedy was in an enviable position to finance large loans through insurance companies and banks, the outlay of capital was substantial, in the millions of pounds sterling. Expansion included not only new berths but warehouses and large tracts of land for stacking containers, office buildings, and cold storage. New installations would prove to be a considerable financial burden.

At age sixty-one, Dad was admitting for the first time that pressures were taking a toll on his health. He wrote to me on December 13, 1969:

> Business here becomes more difficult and complicated every day. Union problems, government interference, high costs, and other problems make it necessary for longer hours at the office and I confess that sometimes I begin to feel the strain. We bought out Western Meat Packers in Sav-la-Mar that makes bacon, hams sausages etc. I spent a few days there last month and it is an interesting operation which I believe will turn out to be profitable.

---

21   Ibid., Monday, October 27, 1969, 1.
22   *The Sunday Gleaner,* January 22, 1968, 13.

All my prayers, my fine son, for a very, very Happy Christmas and for many years of good health, success, and great happiness.

Your old Dad.[23]

Uncertain times lay ahead. Jamaican society was being transformed, the country adopting an economic system that depended on public rather than private ownership of the means of production. Few could have foreseen the effects on the lives of Jamaicans.

---

23 Luis Fred Kennedy, letter to Fred W. Kennedy, December 13, 1969.

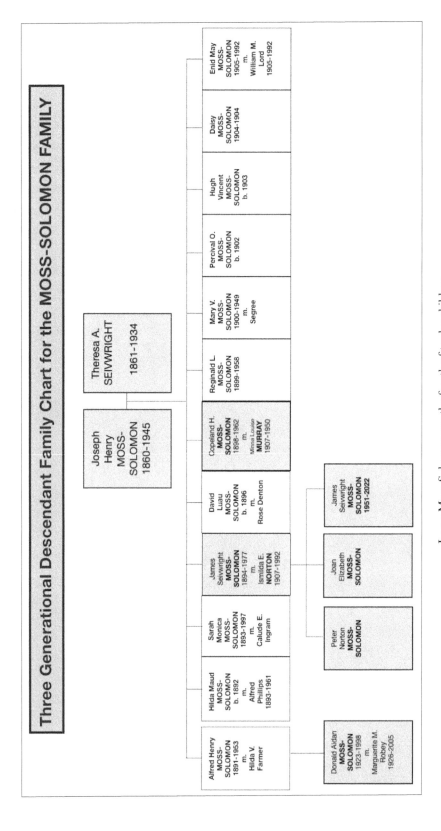

James Moss-Solomon was the fourth of twelve children.

His nephew, Donald Aidan Moss-Solomon, son of Alfred Henry, worked for Grace, Kennedy, as did his sons, Peter and James Jr. The Moss-Solomons are an extensive family. The chart does not show James Moss-Solomon's nephews, nieces, and their families.

Jimmy, Joan, and Peter with their parents, James Seivwright and Ismilda (Flossy) Moss-Solomon, circa 1954.
(Photo, courtesy of Joan Belcher, Peter and James Moss-Solomon)

MOSS-SOLOMON, James Seivright, Businessman; Manager & Director G r a c e, Kennedy & Co., Ltd.; Director Kgn. Wharves Ltd., Ja. Rums Ltd. & Grace Shipping Co., Ltd.; b. Dry Harbour, St. Ann, Dec. 11, 1894, s. late J o s e p h H e n·r y Moss - Solomon, merchant, & Theresa Seivright, his wife; Ed. Brown's Tn. Elem. Sch. & private tuition; clerical staff J. A. L e v y, Brown's Town, 1908-14; Asst. Secretary Brown's Town Benefit Building Society 1914-16; staff United Fruit Co., Montego Bay, 1916-18; Accountant Grace Ltd., Montego Bay, 1918-22; Religion: Anglican; twice m.; 1st: 1928, Olive Harrison (dec'd 1928); 2nd: June 10, 1936, Ismilda, d. late Benjamin Wallace Norton; 2 s., 1 d.; Recreation: Bridge & Reading; Clubs: Kingston, Wembley Athletic; Address: 64 Harbour St., Kingston, & 3 Musgrave Ave., Half Way Tree P.O., St. Andrew.

From *Who's Who Jamaica West Indies 1957*, 333.
(Photo, courtesy of © The Gleaner Company (Media) Ltd.)

34

UNCLE JIM

Chief among Fred's friends was Jim Moss-Solomon whom he loved and admired. He consulted him constantly about business and personal matters. Mr. Moss-Solomon was a wonderful person; benevolence just oozed from him, kind, wise, witty and very, very able.[1]

His son Peter remarked in an interview, "The Kennedys and the Moss-Solomons were like family. Our dads were good friends and worked side by side through thick and thin."[2]

Yet they differed in temperament. "My dad was quieter than Mr. Kennedy and quite a bit older, twelve years I believe, like a mentor, and he also had a long working relationship and friendship with your grandfather" (JSM).

His daughter Joan Belcher recalls a family ritual that always reminded her of the relationship between the families:

> My mother would go out into the garden with some mosquito netting bags and a pair of scissors. I remember very well that the first batch, the biggest bunch of grapes to come off the tree were put in a brown paper bag for your mother and father. The bag was labelled, Kennedy, and placed on the dining room table. No one was to touch them. (JEB)

---

1 Lydia M. Kennedy, "A Giant of a Man" (1997).

2 This chapter is based on excerpts from interviews in 2021 with James Moss-Solomon's three children: Peter Norton, Joan Elizabeth, and James Seivwright Jr. Their initials (PNM) (JEB) (JSM) are used after some quotations.

Uncle Jim

The association between the Moss-Solomons and Kennedys goes back to the nineteenth century. Like Dad's father and grandfather, James Moss-Solomon's father, Joseph Henry Moss-Solomon, and grandfather, Henry Solomon,[3] were both residents of the Town of Falmouth, Trelawny. Joseph Henry Moss-Solomon and Theresa Adelaide Seivwright were married in 1890 at the old Methodist chapel in the same town, Stewart Town, where Dad's grandfather, William Kennedy, operated a wholesale business.

James Mos-Solomon and his siblings grew up in Brown's Town, and after the passing of their mother, Theresa Adelaide, in 1934, they moved to Kingston. The families would visit Grandpa Joseph Henry Moss-Solomon at Bournemouth Baths every Sunday. Joan recalls those times with fondness; she would only have been five or six years of age:

> I remember my grandfather; I was very young. I remember going to visit him—a little, short Indianish-looking man. I remember seeing him sitting on one of those deck-type chairs on his verandah, you know, the type with the red polished floor, the kind you polish with the coconut husks. He'd always give me the most delicious biscuits. I loved them so much, I used to call them Grandpa Biscuits. They were graham crackers.

Despite Joseph Henry's ailments in middle age, he lived to the age of eighty-five. He died on April 16, 1945, at his residence, Bournemouth Gardens. In an obituary of April 21, 1945, *The Daily Gleaner* described him as a "well-known and respected resident of St. Ann—whose popularity that he enjoyed was responsible for the large and representative gathering that turned out to pay their last respects."[4]

James Moss-Solomon lost three siblings in the years that followed: his younger sister, Mary Veronica (1900–1949), who died in New York; his eldest sibling, Alfred Henry (1891–1953); and his younger brother, Reginald Lloyd Moss-Solomon (1899–1958). *The Daily Gleaner* of September 9, 1953, carried a lengthy obituary and photo of his brother, Alfred, who died at the age of sixty-two. His son, Donald Aidan Moss-Solomon, worked for Grace, Kennedy as an accountant with Kingston Wharves Ltd.

James Moss-Solomon's generosity became his legacy. He extended good will beyond family to countless numbers of people in acts of charity, some of which his children and wife were not even aware of. "He was a man of few words, yes very few words, and you would never know about all those he helped unless somebody else told you. 'You know it was your father who sent me to university, or it was your father who sent me to school, or I wouldn't have been a dentist if hadn't been for your father,' people would

---

3   See Chapter 11. Joseph Henry Moss-Solomon was born August 17, 1860, in Falmouth, Trelawny, the same year that his father Henry Solomon died three months earlier, May 3, 1860.

4   *The Daily Gleaner,* Saturday, April 21, 1945, 6.

tell me" (JEB). Jimmy concurred: "The most significant event of his life was leaving home at the age of thirteen to earn money to take care of his younger siblings. In later years, he continued supporting them, repatriated nieces and nephews. At one time, he was the one earning the most money, and in a position to help others."

He was dashing in his youth. Joan reminisces, "Lord knows, he had so many girlfriends. Even till this day people talk about him. 'After all,' his wife Flossie used to say, 'he was a good-looker, so handsome as a young man. For that matter, I wasn't so bad myself. I was a good looker too.' "

Although Uncle Jim (a term of endearment used by both his friends and casual acquaintances) identified with Grace, Kennedy as a family company, he never overtly pressured his children to join the firm. "Strangely enough I was never induced to work at Grace. Most of my mother's family were nurses, so my mother had the ambition that I should be a doctor" (PNM). Peter started in medicine at McGill, and with the encouragement of Luis Fred, later joined the company and trained as an accountant.[5]

Likewise, James Moss-Solomon did not pressure his daughter Joan to join Grace, Kennedy. He tried to persuade her to return to Jamaica after she had completed her studies in England, but this had more to do with wanting her close to home. Joan remembers her dad saying, "You sure you want to stay on in England? When girls leave school, they typically go to work for the bank or study typing and shorthand at Alpha." When she told him that her one and only career choice was nursing, he gladly afforded her the opportunity. "When you're young, you don't think that much about those things, about how much stuff costs. He would take us on holidays, travel, but you don't think about money when you're a child. He would never mention how much anything cost."

"I had the most wonderful father. I mean everybody thinks their father is wonderful, but I can tell you from all the goodness that he did for others and for the love he shared, I can tell you, he was very special. I had the best childhood anyone would ever want to have" (JEB).

Jimmy seemed to have received an early boyhood initiation into Grace, Kennedy. Peter had left for McGill University in Montreal in 1955, and Joan for finishing school at Queen Anne's School in Reading, England in 1956.[6] "He would take me to 64 Harbour Street from the time I was four years old. I've seen everything. I've seen every type of vessel offload, I got to know all the employees at Grace. He would take me to get a haircut at the barbershop on Church Street, where all the big men went—Charlie Johnston, Cecil Facey,[7] Abe Issa, Douglas Judah and the Matalons" (JSM). Jimmy

---

5   See Chapters 24 and 31.

6   Joan Moss-Solomon's parents accompanied her on her trip to England in September 1956 on board the United Fruit Company's *Jamaica Producer*.

7   Cecil Boswell Facey, retired by the late 1950s, was founder and chairman of Cecil B. Facey Ltd., which was founded in 1930, commission agents representing DuPont, GM (parts division) and other multinationals.

Uncle Jim

remembered his father walking him from 64 Harbour Street to the "Gentleman's Barbershop," which was owned by Joe Poveda. (The barber's assistant was Mr. King.)

"My dad's office door was always open. The staff knew they were welcome—especially on Saturday mornings, when anyone could drop by to chat about anyone or anything" (JSM). No appointment was necessary. "I remember there were people who had problems in their marriage or wanted some advice on something and Uncle Jim was the only person they would come to" (JEB).

Carlton Alexander shared similar impressions of Mr. Moss-Solomon when he first met him in 1933:

> Mr. Moss-Solomon, of course, I knew before and he knew me, because it was his brother who got me the job here—my uncle-in-law.[8] Mr. Moss-Solomon influenced me a lot because he was the main man in the trading section, and also he was a very fine person, very principled, very disciplined, very understanding, and a person to whom you could talk and one who was always ready to teach, and show and explain.[9]

Uncle Jim had a way of listening first and speaking second, of putting the needs of others above himself. "He would say to me, 'Darling, what's wrong? Your eyes give you away' " (JEB). And although reticent at first, before she knew what was happening, his daughter would find herself confiding in him and telling him her problems.

Family occasions were special: the day Jimmy was born in November 1951; the time Joan completed her nursing program at St. Mary Abbot's Hospital (top of her class) in London; and the births of his grandchildren. Only recently did Joan learn from her brother Jimmy of their father's burst of emotion at the news of her marriage in 1961. He had told her, "Do you know that when you got married and went off on your honeymoon Daddy cried like a baby?" Joan was in England then, just having completed her nursing degree.

But he could also be tough as nails. He stood firm in the face of adversity and refused to run from challenges when they presented themselves. His business persona showed steadfastness of purpose. Peter comments:

> In the early days, the company began trade in saltfish, and in order to do so, we had a share along with other importers and distributors in a local company called Fish Importers Ltd. Turned out that Grace, Kennedy's was quite small in comparison to the competition. License to distribute was pro rata to the percentage of holdings within the company. The company was growing fast, and the arrangement was

---

8   Copeland Moss-Solomon

9   Interview with Carlton Alexander, courtesy of Anna Jarvis, who received the transcript from Mavis Alexander. See Chapter 37.

not satisfactory. We needed more fish. As a senior person within the company at the time, and in charge of merchandising, he was able to call the shots. He made a request for a bigger share. It was turned down and so he immediately disassociated himself from the agreement. He sought out a group of independent Nova Scotian Fishers who saw the opportunity to compete in the market. As a result, Grace, Kennedy got full command of the supply chain, the supplies increased, and they eventually outstripped the competition.

One of many important lessons he taught his children was to stand up for their beliefs. Jimmy told the story of his father breaking away from the Baptist Church. The congressional leaders objected to his father's attendance at gymkhana, a regularly held fair with festivities, equestrian competitions, and occasional gambling. After a formal reprimand by the Baptists, he parted ways with the Church and never returned. His brothers tried to persuade him to reconsider, but he refused; he saw nothing wrong or immoral about his behaviour. Jimmy noted, "Look, my father was a person who taught me something, 'If you're wrong, say you're wrong and apologize, but if you know you're right, hold on to that belief, and don't ever apologize.' "

It would not be an exaggeration to say that the effect of Uncle Jim's life on those around him was immeasurably wholesome, and that his legacy of kindness lives on through his children, grandchildren, and great-grands. His beliefs and practices helped to establish the core purpose of the company: We Care, the cultural trademark of Grace, Kennedy. Without question, the professional ethics of the company today is rooted in the shared values of the Moss-Solomon and Kennedy families.

Lydia Mathilde and Luis Fred Kennedy
Portrait photograph, gift for my twentieth birthday, July 1970
(Photo from family collection, original by Amador Packer, Port Royal St., Kingston)

35

BOUNTY

(1970-71)

A young recruit by the name of Errol Donovan Anderson[1] was hired in 1970, a day in April that he considers to be "the most important date" of his life. He was soon promoted to company secretary and worked directly with his boss, Luis Fred, to become his friend and confidante. He shared with me reflections of the man he respected and revered:

> He loved people. When he graduated from university, he told me he could have gotten a job up there with any one of America's top companies, but he wanted to come back to Jamaica. He wanted to employ a thousand individuals—that was the goal he wanted to achieve before he died. And he told me that the people who worked for him, specifically in the warehouse, they were more important to him than the bags of flour that they moved around. That is why he brought into the company free medical care, for it was completely free for employees of Grace, Kennedy and Kingston Wharves, and Group Life Insurance, and that is why he brought in employee shares. When Michael Manley

---

[1] At the young age of twenty-four, Errol Donovan Anderson was employed by Grace, Kennedy & Co. Ltd., April 20, 1970. He was promoted from assistant to group corporate secretary in 1974. He became a close confidante of Luis Fred Kennedy, who depended on "young Anderson" for advice on group tax planning and legal entity management. Always vigilant to protect the good name of the company, Donovan became a key figure at Grace, Kennedy in managing legal liability risks. He was appointed a director of Grace, Kennedy in 1988.

came into power, he laughed, ha, ha, ha, he said to me, "Not Manley, young Anderson, I'm the original socialist, not Michael Manley."

One thing that concerned him deeply was unemployment, and we spoke of it almost every day. He'd call me into his office and say, "Young Anderson (that is what he called me), all these young people—when I go out today, I'm going to drive past George's, Kingston College, Alpha Academy and Alpha Commercial, and I will see all these young people on the road, tell me young Anderson, where do you think they are all going to find work? Tell me, where are they going to find work?!"[2]

Donovan joined Grace, Kennedy at a time when Luis Fred Kennedy, at age sixty, was officially beginning the transfer of power.

In April 1970, Mr. Kennedy took steps to change the Articles of Association of Kingston Wharves Ltd., to alter the rights of powers vested in preference shareholders. He obtained permission of the board of Kingston Wharves Ltd. that, in lieu of converting the preference shares to ordinary shares, the rate of dividend payable on the preference shares be increased as a form of compensation for losing the right of veto.

At a directors' meeting in May, the chairman proposed that the following be invited to serve as directors of Grace, Kennedy: L. P. Scott, E. G. Muschett, F. X. Kennedy, and B.H. Terfloth.[3] In addition, he proposed that Mr. S. C. Alexander be appointed his deputy chair; he had officially chosen his successor and was positioning young directors who were managers of the company.

On July 29, 1970, Mr. Kennedy advised the board of directors that he had resigned as chairman of Port Services Ltd. and appointed Mr. L. Paul Scott to take his place, and in December he announced that Kingston Wharves Ltd. had accepted his resignation as managing director. Paul Scott was appointed to assume the position.

Before full retirement, Luis Fred worked to improve the terms of free medical care, group life insurance, and superannuation schemes for all employees. He recommended that the medical scheme be revised with a view to extending benefits to members of the family by way of a staff contributory plan. He proposed to the board in 1970 that the company take out a new type of universal life insurance policy, which would cover all employees of the group, both local and foreign, irrespective of salary, to the extent of $200,000. Increased coverage became effective October 1, 1970.[4] With changes to regulations re superannuation payments, he was able to negotiate with the minister

---

2 Donovan Anderson, interview, May 30, 2019.
3 Although invited, Mr. Boerries Terfloth turned down the offer for personal reasons.
4 Grace, Kennedy & Co. Ltd., minutes of board of directors, October 28, 1970.

of finance that payments made to retired employees under the previous scheme be exempted from tax. Because new employees joining the fund would not enjoy full tax exemption, he proposed that the board investigate ways of providing adequate benefits to them on retirement.[5]

In January 1971, he appointed Paco manager of Western Meat Packers in Westmoreland, one year after Grace, Kennedy purchased the business. The board of directors included L. F. Kennedy, S. C. Alexander, L. P. Scott, F. X. Kennedy, E. Nelson, and Rafael Diaz[6] as secretary. Luis Fred took a personal interest in the project by visiting the plant in Westmoreland and holding directors' meetings on site. By March 1971, the chairman reported significant improvements in all aspects of the operation, including proper accounting and costing systems being put in place, and that he expected the company to be showing profits by the month of May.[7]

Paco faced enormous challenges, inheriting company debts of $400,000, supervising the upgrade of machinery, updating accounting systems, and dealing with labour strikes that had resulted in spoilage of meat products. Ten months after assuming the post, he suffered a heart attack at age thirty-one. In November the chairman reported that "due to the illness of Mr. F. X. Kennedy, it would be necessary to appoint a new general manager"[8] for the factory to be re-opened in the new year.

Luis Fred continued to expand the company's interests overseas with the objective of unlocking new supply sources. Instrumental in this expansion was Mr. Boerries Terfloth whom Mr. Kennedy thought to be a shrewd businessman, *"a born trader, who came by it naturally from his father."*

The Canadian company had begun its global exports with the introduction of chicken necks and backs to West Germany, followed by hundreds of thousands of Canadian GRACE wax beans to Europe. For the London and Rotterdam offices, business was also brisk from the start. The first year of operation in London, a harsh, cold winter in England, proved fortuitous. The severe frost triggered a sudden demand for the sale of canned carrots and potatoes.[9]

---

5   Ibid., June 9, 1971.

6   Rafael Diaz's name appears in the minutes of the board of directors of Grace, Kennedy & Co. Ltd., January 28, 1970, when he was appointed secretary of Western Meat Packers Co. Ltd. Carlton Alexander had hired him in 1969 to work as assistant accountant at Grace, Kennedy. Mr. Diaz's previous employment was with Myrtle Bank Hotel. At Grace, Kennedy, he was promoted during the 1970s to become financial controller in 1974; he was called to the board in 1975 and appointed finance director. He assumed the role of deputy chairman in 1980 and was appointed chairman and chief executive officer in 1989. He retired in 1998. Raf Diaz was born in Belize 1930, son of Rafael and Leonor Diaz. Educated at St. John's College, Belize, and before joining Myrtle Bank, worked as accountant (1950–1966) with Commonwealth Development Corporation..

7   Grace, Kennedy & Co. Ltd., March 31, 1971, April 28, 1971.

8   Ibid. (November 25, 1971).

9   B. H. Terfloth in *Memories of What Made Grace*. (1997), xiv.

Bounty (1970-71)

Boerries showed an interest in acquiring equity in the company. In May 1970, Luis Fred denied his first request to purchase shares in Grace, Kennedy & Co. (Canada) Ltd. Even though he had started to explore options to offer shares on the public market,[10] the chairman remained cautious, viewing himself as a steward of family assets. He struggled with the prospect that at some point he would need to let go of family control of the business.

In January 1971, Luis Fred conceded that Boerries would be allowed to own ten percent of the shareholding of Grace, Kennedy & Co. (Canada) Ltd., and in addition, offered him a position as director of the board of the parent company.

In July, the chairman reported that Mr. Terfloth had not replied to his letter setting out the terms under which he would be permitted to purchase shares. In September, Mr. Kennedy visited Montreal to meet with Boerries, who informed him that he would no longer be interested in acquiring shares in the Canadian Company.

Circumstances changed. Boerries agreed to remain with Grace, Kennedy:

> My decision to stay with Grace was fundamentally influenced by my special relationship with Messrs. Luis Fred Kennedy, James Moss-Solomon and Carlton Alexander, who had then come to mean a great deal to me.[11]

The following month, Carlton Alexander and Jim Bovell met with Terfloth and his lawyer in London to draw up a shareholders and employment agreement that outlined terms for equal participation by Grace, Kennedy & Co. (Canada) Ltd. and Mr. B. Terfloth. By-laws of the new company were drafted along with a letter of agreement re trading with Grace, Kennedy & Co. (Canada) Ltd. (which would continue to invoice goods to the parent company in Jamaica).

This was the start of a twenty-five-year-long relationship between Grace, Kennedy and Boerries Terfloth group of companies.

December 1971

Dad and I were on the pool deck in the backyard at 45 Lady Musgrave Road. I had left the seminary and begun studies at the University of the West Indies. Paco was also home; he was convalescing from his heart attack.

"I think the time has come, son."

---

10 Mr. Kennedy reported to the board, October 30, 1968, that on a recent trip to London, England, he had discussed with bankers options for Grace, Kennedy to offer its shares to be listed on the London Stock Exchange. The Jamaica Stock Exchange, incorporated in 1968, opened for business the following year.

11 B. H. Terfloth in *Memories of What Made Grace*. (1997), xiv.

"For what, Dad?"

"Your ol' man hasn't been well. I will soon be handing over full management to Carlton.[12] If anything should happen to me, the company is in good hands. I've talked it over with your mother."

"But you're still active and strong."

"Not fully, and besides, there comes a time when you must let others take the reins. I've spoken to your brother. I want you both to be executors and trustees of my estate. It's quite straight forward. In the event of my death, everything passes to your mother as sole beneficiary, and when she passes, the residue of the estate is divided into five equal parts."

"This stuff is hard to think about."

"I learned the lesson from my father who died intestate. It was a bloody mess, taking years to settle. I have already transferred shares to your mother and intend also to transfer the other half of my holdings to you, Paco, and your sisters, in five equal shares."

"Why would you do that?"

"It makes sense. It's what I have worked for, and there's no point in you waiting for your full inheritance until after I die. You turned twenty-one this year, and Elizabeth is eighteen; you're all adults, and it will help you get set up in life. And family trusts can be a damn mess, so they're out of the question."

"How do you mean?"

"A family trust is created by transferring assets into a company in the interest of beneficiaries; its purpose is to guarantee that ownership will remain within the family. But it seldom works, as family members always end up bickering."

"Yes. I can see how that would happen."

"Only one bloody thing I worry about."

"What's that?"

"That you'll turn wotless if I give you your inheritance too early." He let out one of his hearty laughs. "And then, of course, there's the god-damn taxman."

"What about him?"

"It's better for your mother and me to pay an inter-vivos tax than burden the estate with expenses. Remember one thing, though, about your inheritance.

"What's that?"

"Try never to spend capital; let it grow, and there's no better place than having it invested in a well-managed company."

Wednesday, December 22, 1971. At an Extraordinary General Meeting of the shareholders of Grace, Kennedy, it was resolved that the authorized share capital of the company be increased from $2,000,000 to $2,500,000 by the creation of 2,000 ordinary shares of $200 each and 500 employee shares of $200 each. Part of the amount

---

12 See Chapter 39. Occurred one year later.

standing to the credit of the Profit and Loss Account would be capitalized and applied to the issuing of fully paid bonus shares to shareholders on record in the ratio of one (1) share for every three (3) held by them on December 22.[13]

At a director's meeting held three days later, the issue was allotted to twenty-five holders of ordinary shares and the company's seal affixed to the relevant share certificates. The meeting also approved the transfer of employee shares from the Grace, Kennedy Employee Investment Trust to twenty-four employees, and the directors authorized the transfer of a total of 1200 shares from Luis Fred Kennedy and Lydia M. Kennedy to their five children.

Preparations were well underway for the celebration of Grace, Kennedy's Golden Jubilee. At the meeting, the chairman announced that a function would be held at the National Arena, Saturday, February 12, to which all members of staff, their wives and husbands would be invited. Special overseas guests, family, and business associates were invited. Transportation and hotel accommodation were to be provided.

At the January 26 board meeting, the chairman advised directors that Mr. James S. Moss-Solomon (aged seventy-eight) would be retiring from the company at the end of February after fifty years of service. Geoffrey Dodd and Aubrey Grant would also be retiring. To recognize their years of long service, the chairman recommended that they receive supplementary pensions from the superannuation fund to provide them with two-thirds of their salaries.

---

13   Grace, Kennedy & Co. Ltd., minutes of the Extraordinary General Meeting of shareholders, December 22, 1971.

PART SIX

Transitions (1972–1976)

"This is the Jamaica we did not often want to talk about, but it was real, and posed the greatest threat to business and to civil order in our society" (**LFK**).

Luis Fred Kennedy with James L. R. Bovell and His Excellency, Sir Clifford Campbell, Governor General of Jamaica, at the Fiftieth Anniversary celebrations of Grace, Kennedy & Co. Ltd, held at the National Arena, Saturday, February 12, 1972.

(Photographs of Fiftieth Anniversary celebrations, courtesy of © GraceKennedy Ltd.)

Mrs. Lydia M. Kennedy receiving bouquet from Mrs. Faustine Sharp at Grace, Kennedy's Fiftieth Anniversary celebrations, February 1972. Luis Fred in background, and Mrs. Flossie Moss-Solomon, partly hidden.

Gathering of Grace, Kennedy employees and guests at the National Arena, February 1972. A dream come true!

At head-table, L–R: Mrs. Moy Alexander, James S. Moss-Solomon, Lydia M. Kennedy, Luis Fred Kennedy, Flossie Moss-Solomon, Carlton Alexander, Paco Kennedy, and Paul Scott.

L–R: Lydia M. Kennedy, Anne de Brosses, Luis Fred Kennedy, Cynthia Muir, John Grace, Countess Alicia de Brosses, and Carlton Alexander.

Carlton Alexander, Hon. Edward Seaga, Minister of Finance and Planning, and Lydia M. Kennedy.

Geoffrey Dodd, Boerries Terfloth, Owen Plant, and Mrs. Arnold Foote.

Luis Fred Kennedy, Mary Kennedy, Elizabeth Kennedy (partly hidden), G. Arthur Brown, Governor of Bank of Jamaica, Sir Clifford Campbell, Governor General.

Junior Foote, Countess Alicia de Brosses, Paul Scott, and Mrs. Carmen Valliant.

L–R: Mrs. Cynthia Muir, José A. Kennedy, Mrs. Louise Soutar, Walter Fletcher (aged 91, original shareholder), Mrs. Alta Gracia Kennedy.

Mrs. G.E. Dodd, Simon Soutar, John Grace, and Gloria Plant.

36

GOLDEN JUBILEE
(1972)

The company was founded by two men who had complete confidence in and mutual respect for each other. It was founded on the premise of full cooperation of the people involved who were to establish a tradition of complete integrity (LFK).[1]

Luis Fred was the lead speaker at the Fiftieth Anniversary celebrations of Grace, Kennedy, held at the National Arena on Saturday, February 12, 1972. He stood before a crowd of hundreds to thank family, managers, employees, stakeholders, and customers for their loyalty to a great enterprise. He boasted about a bumper year in which sales reached what the founding fathers would have thought inconceivable, a total of $85 million. He praised the recipients of long service awards and paid special tribute to Carlton Alexander and James Moss-Solomon "for their continuous and tremendous contributions." The rhythm and tempo of his voice captivated the audience.

Carlton Alexander spoke next: "The fifty-year history of Grace, Kennedy is a history of which every single member can be proud." He said that that his own personal development was linked with Grace, Kennedy—he had joined at seventeen years of age, eleven years and three months after it had been formed. He grew with a company that provided "tremendous opportunities for young people with the will to succeed." It was important to look to the future, to be mindful of succession: "It is my sincere hope,

---

1 *The Daily Gleaner,* February 19, 1972, 40.

Golden Jubilee (1972)

and I am sure of Mr. Kennedy, that those destined in time to take over the reins of the company will do so in the same spirit of faith and love which has contributed to our success in the past."[2]

It was a grand affair, the largest the company had ever hosted. Invited guests included employees and spouses of the parent company and its local subsidiaries; managers and representatives of the overseas offices, London, Rotterdam and Montreal; children of founder Dr. John J. Grace, John Grace, Countess Alicia de Brossses, and Mrs. Cynthia Muir; members of the Kennedy family living abroad, Carmen Valliant, José and Alta Gracia Kennedy, and Anne Kennedy, widow of Francis Kennedy; and government officials, Governor General Sir Clifford Campbell, Governor of the Bank of Jamaica, G. Arthur Brown, and Minister of Finance, Hon. Edward Seaga.

Special guests were forty-six employees who received long-service awards (fifteen to fifty years of service). The governing director presented gold platters to James Moss-Solomon for fifty years and to Carlton Alexander for thirty-nine. Lydia M. Kennedy assumed the role of her late mother-in-law to present staff awards: gold watches to men and gold bracelets to women who had worked twenty-five years or more, gold cuff links to men, and gold earrings to women who had served fifteen years or more.

Among those recognized for twenty-five and more years were Geoffrey Dodd (thirty-eight), Aubrey Grant (thirty-seven), Simon Soutar (thirty-one), José Kennedy and Garfield Bourke (thirty), Faustine Sharp (Mrs.) (twenty-seven), Arthur Spence, Hezekiah Pratt, Vincent Girod and Alfred Williams (twenty-six), and Ronnie Rickards, Rosalie West (Mrs.), Donaldson Hay, Eric Chin, Clarabelle James (Mrs.), and Louise Law (Mrs.) (twenty-five).[3]

Grace, Kennedy hosted a cocktail party at the Terra Nova Hotel the following Monday, February 14, the birthdate of the company. Over two hundred guests attended, including government officials, local and overseas managers and their spouses, family, and members of the business community. The Jamaica Military Band, dressed in red Zouave uniforms, provided musical entertainment with its repertoire of classical and contemporary tunes.

MANLEY LEADS PARTY TO LANDSLIDE WIN.[4]

The People's National Party defeated the Jamaica Labour Party in the general election, February 29, 1972. They won thirty-seven of fifty-three seats with fifty-six percent

---

2 Ibid.

3 *The Daily Gleaner,* Saturday, February 19, 1972, page 40, contains a complete list of staff awardees at Grace, Kennedy's Fiftieth Anniversary function held at the National Arena, Saturday, February 12.

4 Ibid., Wednesday, March 1, 1972, 1.

of the popular vote (seventy-nine percent voter turnout). Jamaicans gave the PNP their first victory since Independence.

Manley emerged as the more populist leader of choice. For the "have-nots" of society, he became their messiah who promised to bring social equality and rid the country of corruption. He personified himself as Joshua, the prophet anointed by God to deliver his people, the Israelites, out of the wilderness into the Promised Land. Religious symbolism of Joshua's "rod of correction," a cane which Michael Manley claimed he had received from Emperor Haile Selassie of Ethiopia, captured the imagination of the Jamaican people. Delroy Wilson's "Better Must Come," Max Romeo's "Let the Power Fall on I," and other tunes became part of the political bandwagon, a euphoria never seen in the nation's history.

"To heal some of the bitter divisions, to restore confidence in public integrity" were Michael Manley's words to supporters at the PNP Headquarters, 25 South Camp Road[5] on election night. The crowds jammed the building and overflowed into the street. "I would like to carry out a policy of government by participation in which every Jamaican will feel that they can share in the government of their country and work together for the betterment of all."[6]

His was a message of transformation. The first task of a post-colonial society was to develop "a strategy designed to replace the psychology of dependence with the spirit of individual and collective self-reliance."[7] His was the dream of a classless society, a call for egalitarianism that was "the only enduring moral basis for social organization."[8] Jamaica was suffering from unemployment, from crime, from the marginalization of youth. Manley's promise was to tackle the problems, to "make an assault" on them for a better Jamaica.

His "politics of change" affected every aspect of Jamaican society. As his social agenda became more defined in practice, the doctrine of democratic socialism shook the foundations of capitalism in ways no one could have anticipated. With an ideological shift away from an imperialistic past, Michael Manley created a cultural revolution that influenced the way people dressed, the language they spoke, the literature they read, and the music they listened to. A new nationalism of social and racial equality became part of a new consciousness.

At the beginning of 1972, Dad's brother José and his wife established residence in Boston. Motivated by personal, family reasons, not by economic necessity, they migrated to seek specialized medical treatment for their youngest son, Patrick. José's migration was significant: Dad's only remaining brother, a director and major shareholder of the

---

5 PNP Headquarters was located at 25 South Camp Road, the property that Fred William and Luisa Kennedy purchased in 1917.
6 *The Daily Gleaner,* Wednesday, March 1, 1972, 1.
7 Michael Manley, *The Politics of Change* (1974), 23.
8 Ibid., 51.

company, was departing after thirty years of service. José had worked in a managerial position within the Shipping Division, as a director of subsidiaries, and as legal counsel to the board of directors. Dad admired and depended on his legal expertise.

José did not exhibit the aggression and Type-A personality traits of his brother. Tony Kennedy, José's second son, commented in an interview, "As adults we have benchmarks against which to judge people. For example, Bill Gates. You can see that your dad had the characteristics of an entrepreneur like Bill Gates; he took risks and found ways to grow the company."[9] He said of his own father, "I think my dad was a very conservative person who acted cautiously and only after much deliberation. His lifestyle reflected his legal training. He had a sentimental attachment to Grace Kennedy,"[10]

Tony's sister Michèle elaborates:

> My father was humble, yet so confident and self-assured, so clear on what his aspirations were, what his beliefs were, and where his love lay. His love of family was constant, steady, and very deep. He was an interesting combination of character traits. He was remarkably reserved in his opinions, always conforming to societal norms, always toeing the line. His thinking was logical and difficult to challenge—that was the lawyer in him, I suppose—and his approach to everything he did was methodical, precise, and exact. Yet at the same time, he was out-going, had a sharp wit, and was a charmer with the ladies.[11]

Tony reminisced, "My father left [Jamaica] in 1972, right after his birthday. My parents called me, 'Can you get us an apartment? We're coming to Boston.' " (Tony had left the seminary and recently graduated from Boston College.) They dropped everything and migrated. "It was absolutely difficult for them to have moved. It was torture for them, because of the uncertainty, you don't know how it's going to play out." But his parents were steadfast; his mother went to work, and they made ends meet. Living on a tight budget meant a change of lifestyle.

José chose not to sell his Grace, Kennedy shares. He had the reputation of a shrewd businessman and possessed an extreme loyalty to the company, a "sentimental attachment," as Tony called it in his interview. "In hindsight," Tony reflected on his father's decision, "it may have been the right thing financially not to have sold." In 1972, his dad judged that the share price being offered at par to family members was too low. His decision not to sell paid off: the value of Grace, Kennedy shares appreciated exponentially during his retirement. The difficulty for him and for others of the Kennedy family who migrated was access to dividend income and proceeds from the sale of shares.

---

9   Interview with Anthony (Tony) Charles Kennedy, June 14, 2021.
10 Ibid.
11 Michèle Kennedy, personal correspondence (November 2021).

Ancestry Chart showing Family Relation: Alexanders and Moss-Solomons

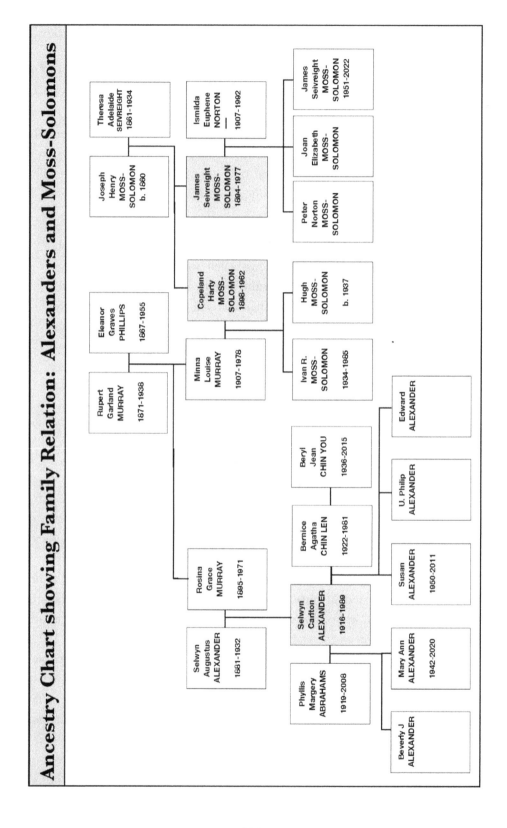

Carlton Alexander receiving gift from Luis Fred Kennedy at Fiftieth Anniversary celebrations of Grace, Kennedy, February 1972.
(Photo, courtesy of © GraceKennedy Ltd.)

37

CARLTON

Fred's health started to deteriorate in the early '70s and he turned more and more to his right hand, Carlton Alexander, for assistance. Carlton had come straight from Jamaica College to Grace, Kennedy, and Fred saw his ability and drive from the start. Carlton learned the business from the ground up as Fred had, and soon rose to the top. Aggressive, hard driving, intelligent, and an accomplished speaker, he became indispensable, and Fred primed him for the top job, knowing soon he would have to give up the reins.[1]

Carlton was born in Cambridge, St. James, on May 9, 1916, the first of nine children to Selwyn and Rosina Alexander. He attended Orange Hill Preparatory School in St. James, where he had fond memories of "playing backyard cricket and backyard football" (SCA)[2] with his cousins and the older boys, who attended Cornwall College. They used to call him "Lilliputian" because of his size.

Carlton recalled that on one occasion his father had bought a new gramophone. "He thought it was a wonderful thing. He thought we should listen to music, whereas I thought it was far more important to go and play cricket. I have never forgotten that. And he wouldn't allow me to go, and I was extremely upset" (SCA).

---

1   Lydia M. Kennedy, "A Giant of a Man" (1997).

2   Transcript of interview with Selwyn Carlton Alexander (SCA), with permission and courtesy of Anna Jarvis, who received the text and handwritten notes from Carlton's sister, Mavis Alexander. Attempts have been unsuccessful in identifying the interviewer. The initial "E" is the only identifier used in the original text. Excerpts quoted from the interview in this chapter are followed by Carlton's initials, SCA.

Up until the time Carlton was ten years of age, his father, Selwyn Alexander, worked as an accountant, employed with Browne Bros., a dry goods and retail business in Montego Bay. When the business folded, his father-in-law, Rupert Garland Murray (Carlton's grandfather, who was a travelling salesman with the Hannas), invited the family to reside with them in Kingston. They moved in with the Murrays at 26 1/2 East Queen Street. Carlton and his sister Dorothy (Miss D) remained close to their Murray grandparents, who invited them to spend time visiting, even after the Alexanders had moved out to establish their own home.

Carlton attended Jamaica College, where he excelled in both sports and academics. He was a star athlete, performing as a member of the track and field, Manning Cup, and Sunlight Cup teams at JC, and he was an academic, who in his senior years was groomed to take the Jamaica Scholarship examinations. People said he had a photographic memory, which he inherited from his father. His niece Anna Jarvis commented in an interview, "If he decided to call someone, he did not have to look in a directory or anything like that; he had the phone numbers in his head."[3] His son Philip remarked about his father's genius, "He also liked to 'race' the other clerks to add the figures in the ledgers; he did it in his head while they used calculators. He never lost!"[4]

Carlton was a high achiever from an early age. "There was a thing that always influenced me. I loved to win, not so much to win, but I liked to come first" (SCA). He recalled the time he was promoted at Jamaica College:

> When I was in school, and I was in 3B, old man Cooper brought some boys up to 3A. He said, I'll never forget it, "Lead will sink, and cork will float." So I said to myself, "All right, you son of a gun, I'm going to show you who is the cork and who is the lead." Right away, I came first in 3A and I never looked back." (SCA)

His father died twenty days after Carlton's sixteenth birthday, leaving the adolescent boy with the responsibility of supporting his mother and eight younger siblings (four boys and four girls). His Murray grandparents "took Carlton over as a project, seeing that his father had died when he was so young" (AJ). His mother too was his strength:

> When my father died, I thought the world had ended, but my mother had nine of us. I was the first, so we had to adjust our plans.

---

3 Anna Jarvis (née Figueroa), interview with author (2021). Anna stayed in touch with her Uncle Carlton's family all her adult life. Initials AJ are used after some of her quotations.

4 Philip Alexander, interview with author (February 15, 2021). Philip is the second born of Carlton's second marriage. Philip joined Grace, Kennedy in 1980 as a project engineer, and retired in 2012. During his time at Grace, he held positions of chief risk officer, divisional projects manager, engineering manager of Kingston Container Terminal, quality director, manufacturing director, and member of Grace, Kennedy board of directors. He has a BSC in mechanical engineering. Initials PA are used after some of his quotations.

> Circumstances had changed. My mother was a strong woman; she helped us to understand that there is no use going bitter through life, but get on with your life, don't dwell on disappointment and disaster.
>
> I learned a lot from my mother: not to hold a grudge, not to have animosity, try to understand what happens. Take the crisis brought on by our father's death, accept that, and move on. She said, "Maybe that will open up doors for us, God moves in mysterious ways." (SCA)

Carlton's ambition was to train as a lawyer, but he was forced to choose employment instead after graduation from JC. "I couldn't afford it, but every time I got holidays from Grace in the early days, I used to go to the courts and listen to Norman Manley" (SCA).

In May 1933, Carlton joined Grace, Kennedy (which then had a staff of fifteen employees) as a book clerk. Anna tells the story: "Uncle Jim had spoken to Mr. Kennedy. 'Fred, my sister-in-law, you must have heard her husband died. She's in hard times, and she has a very bright boy there by the name of Carlton.' Mr. Kennedy agreed to give him a chance. 'Well, bring him down. Let's try him out and if he shows he can do the work and he's serious about working, we'll keep him.' "

He turned out to be a keeper!

Carlton commented about his early days with the company:

> And then, of course, after I got to know Mr. Kennedy, I got interested in him. I found he was completely different. He was much younger than Mr. Moss-Solomon, but I also found that Mr. Kennedy was more of an outgoing person, and if you didn't understand him, you would try to avoid him because you might be afraid of him, but you would find really that he was a very shy person and very understanding, and I remember him as a person with a very human feeling—he was the entrepreneur. (SCA)

Carlton attributed his early successes to the relationships he established with the Chinese community. "Just before I joined Grace, Kennedy in 1933, the Company had employed its first Chinese salesman, Albert Chang A Loy. He was the new boy, and I was the newer boy." They made friends. "He took me under his wing; he introduced me to the entire Chinese community, and I learned how that all operated … I got totally involved in the whole thing, which was the life of the business prior to the War, during the War, and after" (SCA).

Connections to the Chinese community became more than business for Carlton. He gained for them what he said was a "deep understanding and affection." He would attend weddings, funerals, birthdays parties. "On Christmas Eve," his son Philip Alexander observed, "he would go visiting, he'd see every major customer to wish them well and ensure everything was okay."

He expected his sales staff to adopt the same habits. Anna Jarvis recalled a conversation between Carlton and Wallace Campbell, one of his top salesmen:

> CARLTON (to his salesmen): You remember my people's birthdays and make sure you send them a basket of flowers.
>
> Wallace Campbell: But Mr. Alexander, we can't remember the birthdays of all our clients or of all the Chinese people.
>
> CARLTON: Don't tell me any @#&! nonsense. Just bloody well do as I say.

The relationship was reciprocal. His Chinese customers came to regard Carlton as one of them, granting him special privileges as an honorary member of the Chinese Club (where he met his second wife, Moy).[5] His son, Teddy Alexander, observed in an interview, "He merged his family and business life; they almost became one and the same thing."[6]

Carlton was ambitious. "I just automatically went up. I just went from billing clerk, then to chief clerk—chief clerk was everything. Then I just took over sales. As Mr. Kennedy said, nobody promoted me. I took over people's work. He didn't have to promote me" (SCA).

Four years after he joined Grace, Kennedy, Carlton married Phyllis Margery Abrahams, October 1937 (the same month and year as his boss's wedding). He was twenty-one, and Phyllis, eighteen. In 1938,[7] Carlton and Phyllis welcomed their first born, Beverly Alexander. Two children followed: son, Heifetz, (who died shortly after birth, named after the famous Russian violinist, Jascha Heifetz),[8] and daughter Mary Ann Alexander.

---

5   Bernice Agatha Chin Lenn (1922–1981), Carlton's second wife; they were married in 1949.

6   Interview with Edward Alexander (February 18, 2021). Teddy is the youngest of Carlton's children. He is the founder and CEO of tTech Ltd. He had an eighteen-year career (1988–2006) with Grace, Kennedy, where he served as chief information officer and member of the board of directors. Initials EA are used after some of his quotations.

7   Carlton's grandfather, Rupert Garland Murray, died in 1938. A service was held at the All Saints Anglican Church in Kingston to celebrate the life of the devoted choir master, "a man with a big heart, with every zeal to service his God and his Church" (*The Daily Gleaner*, Wednesday, March 9, 1938, 11).

8   Named to honour Carlton's grandfather, Rupert Garland Murray, who was a gifted musician. Mr. Murray always wanted Carlton to study music. "His great disappointment in life was that I couldn't turn a tune" (SCA).

"It was about 1947," Beverly Valentine[9] recalled. "My parents divorced when I was nine or ten ... my mother and father were very civil to each other both during and after the marriage. My mother always made me respect my father, and I was never bitter." Beverly's dad was a role model for her, "a beacon in life," someone who "had a passion for excellence" and wanted the best for others. "I remember vividly, whenever I got my report card, my mother would tell me to take it down to Grace, Kennedy and show it to my father, and you know what he'd say: 'A little more effort, you can always do better.' And I thought to myself, you know, he's absolutely right."

Beverly left Jamaica at an early age, but her dad always stayed in touch. "I was in Massachusetts, you know, he would be sure to see me, he came for my graduation, and even as a teenager when they were divorced, he would always come for my birthday parties." When she returned to Jamaica in the early sixties to teach, her father told her, "You should be staying and contributing to the development of the country. My dad had a complete commitment to community, to Jamaica, and wanted his family to have the same passion." But teaching was Beverly's calling, and teachers were not being paid a living wage. Although she migrated to Canada shortly after, her father never once held it against her.

Her father remained connected as best he could. She commented:

> He always wanted to celebrate moments. He loved to take you out to dinner. I remember when my daughter was in boarding school in England, whenever he came from Jamaica, he'd always contact her and say, "Where are we meeting?" He would take her to these fancy places. I remember also, ten years after I was married, my husband and I were on our way from Germany to South America. We stopped in Jamaica, of course, and my father held a reception for us at Terra Nova to make sure we met all his friends. He regarded these as ceremonial moments. (BV)

Carlton had space in his heart for members of his immediate family. Beverly recalls, "My father had this loopy handwriting, and when we were living in Texas, I remember vividly, I think it was 1977, he wrote me a letter saying that he was putting 3,000 shares in my name, and that he wished it could be more." Years later, with her generous inheritance, Beverly was able to realize one of her dreams—to purchase an

---

9 Beverly Valentine (nee Alexander) left Jamaica at the young age of seventeen in 1955 to study in Massachusetts. Having trained as a teacher, she returned to Jamaica to work at Immaculate Conception High School but decided shortly after to migrate to Canada in 1962. She married a Canadian diplomat, Douglas Valentine, and spent thirty-five years of her adult life travelling to countries where he was posted. They have three children and eight grandchildren, one of whom is named after her great grandfather—Carly (Carlton's nickname), who is studying law at Osgoode Hall in Toronto. Initials BV are used after some of her quotations.

apartment in Stratford, Ontario, where she and her husband, Douglas Valentine, enjoy Shakespearean theatre. Her father's legacy lives on, she realizes, for her children and grandchildren to cherish.

During the war years, Luis Fred mentored and trained Carlton in the importance of civic duty and of defending the rights of the private sector. "Mr. Kennedy had set up a Food Trade Committee with the government and was chairman of it. He used to allow me to sit in on the meetings and do a lot of the basic work for him" (SCA). In 1945 Carlton was appointed assistant manager of Grace, Kennedy.

Carlton re-married in 1949—Bernice Agatha (Moy) Chin Lenn.[10] "While my time was tied up here [Grace, Kennedy], she took complete charge of home, fully understanding the seven-day work-week. She did a marvelous job with the children, looking after them, school, and things" (SCA). She had done brilliantly at school, Titchfield in Port Antonio, where she completed her Senior Cambridge. "If at the time she left school, had the university been there, I have no doubt she would have done medicine" (SCA).

In later years, after the responsibilities of childrearing had lessened, Moy became involved in community affairs. She was president of the Inner Wheel Club, and in her capacity as an executive of the Jamaica Federation of Women was committed to urban renewal projects. "She was an inspiration to her family and friends."[11] Her son Philip recalls accompanying her to the old age home at High Holborn Street and once a month taking lunch and food packages for the women who lived there.

They had three children: Susan (1950–2011), Philip, and Edward (Teddy). Both sons understood from a young age that their dad "gave his total commitment to the company; he never wavered from that" (PA). Yet despite the long workdays and nights, he found time for family. "He and I were reasonably close, for the time, yes, I think we were as close as a normal father would be to his son. You must get a decent education; he didn't care what you got, as long you studied for a profession. You'd always have an education and training to fall back on, even if you didn't end up choosing that profession" (PA). Philip's brother Teddy concurs, "He was quite involved in my life; he provided us with guidance in terms of choosing a career and becoming professionally qualified." They saw their dad as forward thinking. In the 1970s they remember him saying, " 'You boys should get into computers. That is the future. And software is this thing.' So looking back fifty years, yes, he was spot on, as it is the software side of technology that is driving the changes we see in society" (EA).

---

10   Bernice Agatha Chin Lenn, Carlton Alexander's second wife, was born in 1922, Nonesuch, Port Antonio, Portland and died suddenly of a heart attack at her residence in Kingston, January 18, 1981.

11   *The Daily Gleaner,* Thursday, January 22, 1981, page 10.

"One thing he would not tolerate was dishonesty. If you were a thief or a liar, you would be fired, no questions asked. You know, they have a saying, my word is my bond" (PA). Carlton recognized this in himself:

> I believe that the greatest asset a man has is his word, and in business, if a person cannot accept your word, you are in trouble. I hope that all the people I deal with know that I am a man of my word. This is very important to me; once you have given your word, that's it, that's binding, you don't have to sign documents.[12]

That was his creed and Grace, Kennedy's creed. "I never ever got the impression he was driven by the need to have more cash. He had a very big heart to help others, to help them succeed. You hear today of numerous examples of people he helped, the many he drew cheques for" (EA). He extended credit to customers who could not afford to pay their bills, and he assisted employees in starting up their own businesses as a way of advancing themselves. His acts of generosity engendered trust, laying the foundation for loyalty that became the hallmark of employee and customer relations at Grace, Kennedy.

By 1960, "around then, I'll never forget, Mr. Kennedy started to say, 'Well, damnit all, you should be manager.' He looked at me and he said, 'Well I suppose you know we are going to have to call you, Managing Director.' So I just became a managing director along with Mr. Solomon" (SCA).

Carlton entered the public arena when he was elected president of the Jamaica Chamber of Commerce in 1968, which gave him a platform for leadership within the private sector. Eight years later, he became a founding member of the Private Sector Organization of Jamaica (PSOJ). His community involvement included chairman of his *alma mater*, Jamaica College, of the Jamaica National Export Corporation, and of the Council of Voluntary Social Services. He would play a critical role in charting a way for the company through the turbulent years, 1972-1980.

Carlton had a fondness and respect for his boss. They were alike in character: exacting and dauntless, quick to anger, but blessed with a heart of gold. They both worked hard and played hard too. They enjoyed their liquor and the challenge of a poker game yet were never on the same team. (My dad had given up poker in the 1940s.)

Carlton spoke of him endearingly, "I remember him as a friend, and he considered me a friend and partner in the business. We did many things together, and I have many fond memories of our travels over the years."

---

12 Carlton Alexander, interviewed by Elaine Ferguson, "Carlton Alexander, the modest, caring giant of integrity" (1989).
http://www.nlj.gov.jm/BN/Alexander_Selwyn_Carlton/bn_alexander_scm_0031.pdf.

> In our travels, we talked for long periods—some in jest and at other times we discussed the big love of our lives, which was Grace. At times we felt that we were solving some of the major problems of the world, and too often we felt that we had done just that after we had imbibed sufficient slugs of good brandy, aided and abetted by the culinary skills of the hotels' chefs.
>
> He was a generous person. I remember my twenty-first birthday, which was celebrated at the Myrtle Bank Hotel. He cracked open a bottle of champagne, and with profound and magnanimous gestures, he welcomed me into the folds of adulthood. Those small expressions of kindness and the generosity of his nature helped to cement our relationship. A relationship which transcended political, social, and business differences and enabled us to work together for nearly half a century.[13]

Carlton spoke of his boss as a mentor:

> He was a free thinker. I learned a lot from Fred Kennedy. He was unselfish in the sharing of his business knowledge with me and other employees of the company. He encouraged discussions on the business, invited questions, and was patient enough to listen and to give answers to those questions. He was a man of strong character, deeply religious and with a strong commitment to the family.[14]

The feelings were mutual. Dad felt that Carlton was worth more than what the company could ever compensate him for. He was fiercely competitive and coupled with this drive was his ability to relate to customers, employees, shareholders, local politicians, and overseas principals. Carlton said of himself, "My philosophy is to have a strong belief in people, let your conscience be your guide. My basic philosophy is doing what I would like done unto myself. My bark is worse than my bite … If I have succeeded and had a better opportunity than others, then it is my duty to give back by helping somebody else" (SCA).

Carlton saw business not solely as a means of earning profit but as a way of giving back to employees, to family, and to the wider community, Jamaica. Through a deep conviction of service and caring, he built on the values of Grace, Kennedy's founding fathers to create prosperity and an enduring legacy.

---

13   Carlton Alexander, "My memories of Luis Fred Kennedy," *The Grace News*. Vol. 7, No. 2, August 1982.

14   Ibid.

"Carlton lived by a simple edict: 'My word is my bond.' Once a commitment was made, it was honoured no matter what. He never signed a contract with a union; the contract was sealed with a handshake. This was the essence of the man" (PA).

Philosopher, reciting verses of Cicero's orations in Latin on the front steps of Fenwick Hall, Holy Cross College, Worcester, Massachusetts, August 1972.
(Photo by author)

Francis (Paco) Xavier Kennedy and
Marjory Lesa Johnston, May 19, 1973.
(Photo, courtesy of Marjory Kennedy)

Mary Teresa Kennedy and Donald Ian Cameron,
July 27, 1974.
(Photo, courtesy of Mary and Donald Cameron)

38

PHILOSOPHER

(1970s)

Mom, Dad and I made plans to meet up in Mexico City in 1972, when I enrolled at UNAM (*Universidad Nacional Autónoma de México*) for summer courses in Latin American History. However, doctors advised that the high altitude of Mexico City was too risky for Dad, given his recently diagnosed atherosclerosis. He thought the advice *"damn foolishness,"* but Mom insisted he listen to the medics. My sisters Mary and Elizabeth joined me instead, and later that summer, I met my parents in Boston to be with family and friends.[1]

The highlight of the vacation was a trip to Holy Cross College. Dad had only returned once (and swore never again) for a ten-year reunion during which his former classmates spent time playing pranks and throwing toilet paper rolls out the dormitory windows.

This time it was different. While we walked the campus, he reminisced about his college days—philosophy, religion, and science classes, Prohibition, and his best friends, Jimmy Quinn and Bill O'Connor. He climbed the steps of Fenwick Hall, whose gothic-like columns towered over the campus, and before he reached the top landing, he turned to survey the expanse and beauty of the campus. Placing his right hand on his cheek, he paused for a moment in awe of what he saw. Then with a sudden show of emotion, he recited, in a theatrical, resounding voice, verses in Latin from one of Cicero's orations.

---

1 My Uncle Che and family had settled in Boston. I also stayed in touch with friends, Jack Allen and Richard Iandoli, and others whom I had met in the seminary. Rich Iandoli accompanied us for the tour of the Holy Cross College campus in 1972. His family was originally from Worcester, Massachusetts, and he was an alumnus of Holy Cross.

Dad returned to Jamaica in need of rest. He ignored doctors' orders, whereas Mom took his medical condition seriously. She terminated the cook's employment and assumed responsibility for preparing meals. "Housewives in Upper St. Andrew were not expected to do their own cooking."[2] But criticism from family and friends did not faze her; she was determined to save his life.

Dad did not make it easy for her; his habits were hard to change. He would say, *"I know what's good for me—all the foods I dislike, such as salads, porridge, and lean meat. And I know what's bad for me—all the foods I like, such as coconut oil, beef stews, and eggs."*

Mom was a perfectionist. Any project she took on, she embraced whole-heartedly, refusing to accept any standard short of excellence. She studied the culinary science of reducing trans fats and sodium in his diet. She ordered magazines, cut out recipes from newspapers, and resurrected old cookbooks that her mother had given her. The result was a gourmet cuisine, fat free and healthy for the heart. She discovered creative ways to serve dishes he relished (oxtail, pig trotters, and beef stew) by cooking them days ahead, refrigerating, and skimming off animal fats before re-heating and serving. Doctors were amazed; they claimed she extended his life by years.

Schedules were not rigid when we lived at home as young adults. We had the freedom to come and go as we wished, subject to a degree of parental monitoring, but without curfews or pressure to conform to rules. There was one exception—our parents did like us home for meals. It was an open house; our friends[3] felt free to drop in and often joined family chats on the front verandah.

Sometimes I would find Dad alone in the evenings while the rest of the family was watching television. He thought TV robbed families of the opportunity to engage in meaningful conversation. He did like a few shows, though, and on occasion would watch *Perry Mason* or *Alfred Hitchcock*.

He resumed drinking alcohol in the early seventies, although not to the same extent. After a few whiskeys, he became loquacious, wanting to talk philosophy.

"You studied philosophy when you were in the Jesuits."

"Yes, I did, but we didn't debate in Latin." We both chuckled.

"I'm sure you've heard the age-old question: If a tree falls in the forest, and no one is there to hear it, does it make a sound?"

"I think it would make a sound."

---

2 Lydia M. Kennedy, interview with Georgianne Thompson Kennedy (2002).

3 Ángel Alomá, former classmate at STGC, enjoyed conversing with my parents in Spanish. He was born in Cuba. And Swithin Wilmot spent many evenings chatting with Dad on every topic imaginable—philosophy, politics, religion.

Philosopher (1970s)

"Depends on how you define sound. The falling tree creates vibrations in the air, but does it produce sound if no one is there to hear it?"

"Hmm! But someone would have heard it if he had been there, no?"

"Well, that's the point."

"I suppose."

"And the same goes for sight. This ashtray on the table next to me exists, does it not?"

"For sure it does. That's not supposition. We both see it."

"Yes, but if there were no one here to see it, would it exist?"

"I definitely think so."

"I must be boring you."

"No, not at all."

Paco had fully recovered from his heart attack and was back in full swing at Grace, Kennedy, where he was assigned to work with Carlton Alexander. In May 1973 he and Marjory Lesa Johnston[4] were married at St. Andrew Parish Church, Half Way Tree, with a reception held afterwards for 500 guests on the lawns of Liguanea Club. It was a joyous occasion, celebrated by extended family and friends on both sides. The significance of their marriage did not go unnoticed: a union between two families who had been in competition in the shipping business for over fifty years.[5] In August 1975, Marjory gave birth to Cathrine Lesa, my parents' first grandchild.

Celia never married. While she was in the convent, Dad had persuaded her Mother Superior that Celia be given the opportunity to attend university by offering to pay her tuition. She obtained a BA in Spanish and French in 1968, after which she taught high school at the Servite Convent in Brown's Town. She left the convent in 1971.

Her first job was with Long Pond Estate in Trelawny, where she was hired as a social worker to assist families of cane workers. She lived in a small cottage in the middle of the sugar estate outside Clarke's Town. To visit the factory and employees' families, she drove about the hills of Trelawny in a Volkswagen bug. Every morning, she would give a policeman who lived nearby a ride into town. Little did he know that she did not have a driver's license. The Long Pond social worker became famous as the policeman's chauffeur who scooted around the district in her Beetle. (Like her father, Celia had a heavy foot on the accelerator.) My sister Mary remembers the VW Bug: "It was grayish-beige; I bought it from her when she left for Boston College and had it painted orange."

In 1973 Celia enrolled at Boston College to read for a Master of Social Work. When she graduated two years later, Dad employed her at Grace, Kennedy to run a program to provide social services for factory workers.

---

4   Marjory is the daughter of Ernest Michael Johnston and Marjory Lisa Williamson, and granddaughter of Edward Charles Johnston.

5   See Chapter 44.

Celia was a free spirit, a maverick, a generous and an affable soul, beloved by those whom she served.

When Mary returned from England in the 1960s, she applied for and earned a full scholarship to the University of the West Indies, but her Mother Superior said that her services were required to teach prep school at the convent in Brown's Town. Two years later in 1968, Dad persuaded the nuns to allow Mary to attend the University of the West Indies. She spent the summers of '69 and '70 perfecting her French at the Sorbonne and Alliance Française in Paris. She left the convent in February of 1971 in her final year at UWI and graduated a few months later with a BA (Hons) in French and Spanish. In September she entered a two-year program at the University of Toronto to read for a Master of Arts in French literature, which she completed in half the allotted time. Mary returned to Jamaica to teach French at Holy Childhood High School and at the University of the West Indies before marrying Donald Ian Cameron,[6] a medical student at UWI, in July 1974.

I graduated from the University of the West Indies in 1973, after which I taught at York Castle High School in Brown's Town, St. Ann, where I met my wife-to-be, Georgianne Ruth Thompson. She had travelled to Jamaica in 1972 to work as a teacher after completing her BA in English at the University of Western Ontario in London.[7] After our stint at YC, Georgie taught for a year at St. George's College, and I at Trench Town Comprehensive.

Elizabeth left sixth form at Campion College in 1971 to study at the College of St. Benedict's, a small Roman Catholic women's college in St. Joseph, Minnesota. She never understood why Mom and Dad sent her to the "god-forsaken" regions of northern Minnesota. Liz survived, but not without stories of freezing temperatures and rustic life in the "boonies." One saving grace was that the college in the 1960s had started to twin with St. John's University by holding co-educational classes. She returned to Jamaica in 1975 to take up a teaching position in Spanish at her *alma mater*, Campion College.

---

6   Donald Ian Cameron, born in Trinidad, is the son of Hugh Elliott Cameron (1915–2008) and Morita Casoetto (1921–2017). He studied and practised as a pediatric neurologist in the United States.

7   Georgianne Ruth Thompson, born in Tillsonburg, Ontario, Canada, is the daughter of Kenneth Sterling Thompson (1925–2015) and Abbie Herberta Barnes (1924–2019).

"I most clearly remember Carlton Alexander greeting his people on the stage at the National Arena. It was the best of times, it was the worst of times. This was the 1975 staff Christmas party. The country was in a state of fear and uncertainty. Some people were leaving their homes and disappearing to USA and Canada, others just riding it out. Carlton was lifting the spirits of his staff. He reassured all of us that we would overcome the hard times." (A. Rafael Diaz, 2020) (Photo, courtesy of © GraceKennedy Ltd.)

39

GRACE, KENNEDY SURVIVED ITSELF

(1972–1976)

Michael Manley took office in a time when foreign ownership controlled the "commanding heights" of post-colonial Jamaica.[1] Bauxite and alumina, the sugar industry (more than fifty percent), and segments of public utilities, banking, insurance, and tourism were in the hands of foreigners. He viewed the colonial period as one that bred a psychology of dependence, "the most insidious, elusive and intractable pattern"[2] that Jamaica had inherited. In his view, Jamaicans needed to be liberated, to assume a sense of confidence and self-reliance. He claimed that local control of the strategic sectors of the economy was paramount to attaining national sovereignty. His objective was the pursuit of equity and the re-distribution of wealth.[3] Manley believed that managers of private corporations and owners of capital felt entitled to the lion's share of economic activity.

---

1 The phrase, "commanding heights" was coined by Aneurin Bevan (1897–1960), deputy Leader of the Labour Party (1959–60) in the UK. He formed part of the left-wing of the party that subscribed to principles of democratic socialism.

2 Michael Manley, *Politics of Change* (1973), 2. His book is a bold manifesto and testament of his socialist ideology.

3 Damien King, "The Evolution of Structural Adjustment" (March 2001), 8.

To achieve his goals, state intervention was necessary. Manley deemed government responsible for the purchase and distribution of raw materials, especially those on which the staple diet depended, commodities like wheat, corn, soya beans, and rice.[4]

> Broadly speaking, we identified "the commanding heights" as those sections of the economy which occupy a strategic position. Hence, the public utilities, the banking system, the bauxite industry, and the sugar industry can be said to constitute the principal elements in the "commanding heights" of the Jamaican economy. These quite clearly belong in public ownership and control. As a matter of common sense and reality, public ownership will have to work tougher with foreign and local private capital in some areas in the foreseeable future.[5]

His plans shook the confidence of the private sector, and when widened to include nationalization of local businesses, they induced panic, resulting in the flight of capital. The Jamaican economy contracted. Gross Domestic Product declined from an annual growth rate of 7.9% in 1972 to -6.3% in 1976.[6] 1973 was the last year of positive economic growth for the next seven years (1974–1980).[7]

He invested heavily in reform programs: Special Employment Program (SEP), Adult Literacy (JAMAL), Operation GROW, co-operatives, and free tuition for secondary and tertiary level students. To offset expenses, he aimed to boost agricultural production, expand exports, and restrict imports.[8] The volume of imports fell by twenty percent in 1973, but measures were not sufficient to stem the growing deficit in the Balance of Payments—values kept exceeding those of exports. The public sector debt grew exponentially, as did the foreign component of the debt, which rose by 75.6%,[9] creating a significant drop in net foreign reserves.[10] Global economic conditions were partly to blame. The price of oil (Jamaica's oil import bill more than doubled from \$65M to \$177M between 1973 and 1974)[11] caused the price of goods and services to rise by 17.6%. Scores of people were unemployed and unable to afford basic foods, which were short in supply. It was not long before civil unrest erupted to further de-stabilize Jamaica's economy and society.

---

4   Manley (1974), 117.

5   Ibid., 118.

6   Damien King (March 2001), 7.

7   Michael Witter, "Exchange Rate Policy in Jamaica: A Critical Assessment," December 1983), 14.

8   Godfrey Smith, *Michael Manley: The Biography* (2016), 126.

9   David Panton, *Jamaica's Michael Manley: The Great Transformation: 1972–1992* (1993), 37. Between 1972 and 1974, Jamaica's public sector debt rose by 56.7% from J\$332.6M to J\$520.8M and the foreign component of the debt rose from J\$117.3M to J\$206.3M.

10   Ibid., 58. Net Foreign Reserves fell from J\$87.2M in 1972 to -J\$969.8 in 1980.

11   Ibid., 37.

GDP contractions affected profit margins, and a new political ideology constrained traditional ways of conducting business. Increased government intervention meant more foreign exchange controls, quotas, and restrictions on imports. Private enterprise needed to adjust to a new ethos: "under socialism, the private businessman is expected to work within the bounds of the national interest and the rights of the people."[12] At the PNP Annual Conference, September 15, 1974, Manley officially adopted "democratic socialism" as the model for economic and social change.

In December 1972,[13] Luis Fred advised the directors that he was resigning as chairman of the Executive Committee owing to health reasons. He would also relinquish his position as chairman of any subsidiary or managed company on whose boards he served. Mr. Alexander would take over as chairman of the Executive Committee and report directly to the Governing Director and Chairman of Grace, Kennedy. Together they would revamp the committee structure based on the advice of management consultants who arrived on the island in January 1973.

Grace, Kennedy fought for survival and emerged stronger. How was this possible in a climate of economic turmoil? How possible, when other businesses were failing, and capital was fleeing the country? To outsiders, its performance seemed miraculous, given the odds.

The company reinforced its managerial team, pursued mergers and acquisitions, and held fast to traditions of caring for employees, customers, and stockholders. But this would not have been enough had it not been for the reputation Grace, Kennedy had built for itself; it held the public trust for ethical business practices.

When asked how Grace, Kennedy managed to survive the Manley years, Mr. Rafael Diaz said:

> Grace, Kennedy survived itself. With the name of Grace, Kennedy behind us, we were able to keep our terms of credit quite well. So it was a matter of having a good name. That was No. 1. Our suppliers trusted Grace, Kennedy. Some would even come to Jamaica to see what was going on, to see how they could help.[14]

---

12   Michael Kaufman, "Democracy and Social Transformation in Jamaica" (Sept. 1988), 50, quoting Michael Manley, *The Search for Solutions* (1976), 160.

13   Grace, Kennedy & Co. Ltd., minutes of the board of directors, December 6, 1972. At this meeting, Luis Fred also announced that his brother's official retirement from the company would be January 31, 1973, at age sixty. José had remained a director and made periodic trips from Boston to attend meetings during the year.

14   A. Rafael Diaz, telephone interview, April 16, 2021

Often referred to as Mr. Jamaica, Carlton became a leader of the private sector at a time when government policy threatened its very existence. Ministers recognized Mr. Alexander's stature as a business leader, whose image was enhanced by the conciliatory approach he took with government. In August 1972, at the request of Mr. Wills Isaacs, minister of Commerce and Consumer Protection, he led a trade mission aimed at opening markets in the Far East and Eastern Europe. As chairman of the Jamaica National Export Corporation (JNEC), he promoted trade exhibitions in both Toronto and London, and with the support of Mr. P. J. Patterson, Minister of Industry and Tourism, launched island-wide activities for Export Month, May 1973.

In reflecting on the 1970s, Mr. Boerries Terfloth commented:

> Throughout these years, Mr. Carlton Alexander was our bulwark and guiding light. His confidence never failed and his optimism was always there to inspire us.
>
> Mr. Alexander was in touch with everyone in the government, from permanent secretaries and ministers right up to the prime minister. There were others like Miss Mabel Tenn and myself, who also became involved in this new way of doing business.
>
> It was not so much what you knew but who you knew. While this might just be an advantage in a capitalist country, it is a real necessity in a socialist one. So we survived.[15]

Problems loomed large. In March 1973, Carlton Alexander reported to the board of directors that the Trading Group of Grace, Kennedy was experiencing high costs of operation that were shrinking profit margins. The Jamaican dollar had devalued,[16] interest rates were climbing, and government was legislating price controls on basic food items and restricting imports on luxury items. Removal of subsidies on items considered non-essential caused increases in the cost of raw materials. The price of pig feed, for example, increased over thirty percent. Grace Food Processors Ltd. in Savannah-la-Mar was expected to post a loss of $300,000 for 1972.[17] Grace, Kennedy also experienced shortage of local agricultural produce for canning of vegetables, imported glass bottles for bottling ketchup and syrup, and cans for other Grace products.

---

15   Boerries Terfloth, "Memories of What Made Grace" (1981), liv.

16   In January 1973, the Jamaican dollar was devalued by 5.6% as one of the measures to deal with adverse balance of payments.

17   In May 1973, after a drawn-out series of discussions with Hon. Keble Munn, Minister of Agriculture and National Security, Mr. Alexander was able to dissuade government from a take-over of the company's pig factory in Savannah-la-Mar.

Dad asked me to attend the Annual General Meeting, September 11, 1973. Fifteen shareholders were gathered in the board room at 64 Harbour Street. The directors' report and company's balance sheet were presented, showing profits for the fiscal year 1972, an exceptional performance considering the early economic setbacks. What I was not expecting to hear was Dad's declaration that with the consent of the board, he had officially retired as Governing Director, effective August 30, 1973. Even though he made the announcement *with a heavy heart*, he told me he knew the time was right. Raf Diaz commented, "Mr. Kennedy groomed Mr. Alexander, and he appointed him. What Mr. Moss-Solomon and Mr. Kennedy had done had worked. Now it was time; he made Carlton the front man."

According to the Articles of Association of the company, in the event of the governing director leaving office, all directors of the board, excepting the managing director, were required to resign and to elect a new board of no more than ten directors, five by the preference shareholders and five by the ordinary and employee shareholders. That day they elected a board that consisted of Messrs. L. F. Kennedy (Chairman), S. C. Alexander (Group Managing Director), C. Heron, J. S. Moss-Solomon Sr.,[18] and F. X. Kennedy, chosen by the preference shareholders; and Messrs. L. P. Scott, S. Soutar, E. G. Muschett, J. L. R. Bovell, and C. Tame by ordinary and employee shareholders.

The office of governing director, established by Dr. J. J. Grace in 1922, now ceased to exist. Luis Fred would remain as chairman—a motion tabled by James L. R. Bovell and passed unanimously at the board meeting of August 30. Carlton Alexander would serve as group managing director and deputy chairman, but not as governing director.

Increases in interest rates on loans and overdrafts in 1973 and stringent exchange controls implemented in 1974 were serious issues for companies needing credit to stay alive. Carlton obtained approval from the Bank of Jamaica for extension of credit from commercial banks to the manufacturing division, which in 1973 owed $6,000,000 in USD and CAD currency for goods supplied.[19] And Mr. Diaz, who was appointed finance director[20] after the retirement of Mr. Colin Heron in 1974, negotiated loans from Barclays Bank and Canadian Imperial Bank of Commerce. The company also signed line of credit agreements with the Jamaica Citizens Bank and Chase Merchant

---

18   James Moss-Solomon, although officially retired, remained as a director of the board.

19   Grace, Kennedy & Co. Ltd., December 11, 1973.

20   March 28, 1974, board meeting, chairman announced the retirement of Mr. Colin Heron as finance director and the appointment of Mr. Donovan Anderson as secretary on retirement of Mr. Harold Aarons. Mr. A. Rafael Diaz was invited to the board to act in position of finance director and at recommendation of the chairman was appointed a preference shareholder director at the AGM, July 17, 1975.

Bankers Jamaica Ltd. With these loans, the company financed trade, kept subsidiaries afloat, and purchased new ones.[21]

Raf Diaz commented in an interview:

> The problem we had was paying our bills. That was the most difficult problem we had with the Bank of Jamaica; they took very long to approve payments. The suppliers abroad became very angry with us. The Bank of Jamaica finally told us that if we were able to extend our credit from three months to six months, they would find the necessary foreign exchange for us to pay our bills. "What you need to do is to get foreign credit from your faithful suppliers," they told us. Well, we went to people in Australia, for example, who sold us mutton. They were very good to Jamaica, and very good to Grace, Kennedy because we always had a good reputation with them. They extended credit for six months to allow us to pay our bills. Same thing with New Zealand. We packaged their cheese in Jamaica; it suited them to extend credit. We were also able to get lines of credit through our foreign subsidiaries, from Terfloth & Kennedy in Montreal, and from our offices in London and Rotterdam.[22]

An aggressive policy of acquiring new subsidiaries proved to be a key growth strategy during the recession of the 1970s. At an Extraordinary General Meeting (EGM) held September 26, 1974, Chairman Luis Fred Kennedy informed shareholders of the new companies the group had acquired. Grace, Kennedy was also investing in agriculture as a way of securing raw materials for factories and growing crops for export.[23]

Directors met more frequently, the first Tuesday and the last Thursday of every month. Meetings were longer, presentations more detailed, and deliberations more technical in nature. Chairman Luis Fred Kennedy took a less directive role, allowing Mr. Alexander to take charge of the management team.

---

21   Grace, Kennedy continued its aggressive policy of acquisitions and mergers during the 1970s: mergers of Grace Food Processors (Canning) with DaCosta Bros. Ltd., and R S Gamble & Sons with Davidson Insurance Brokers Ltd. Acquisitions included Cameo Products, United Merchants Ltd., General Printing Equipment and Supplies Ltd., Security Wire Woven Products Ltd, Metal Fencing Ltd., Rapid Vulcanizing Co. Ltd., Caribbean Greetings Corporation.

22   A. Rafael Diaz (April 16, 2021).

23   In 1974, Luis Fred Kennedy officially appointed Mable Tenn (who had been acting for Paco Kennedy while he was assigned to the Montreal office) as director in her own right. She was placed in charge of manufacturing and agricultural operations and spearheaded the Halse Hall Project—lease of acreage from Alcoa to grow fresh fruit and vegetables for export.

It was not business as usual. On November 28, 1974, Mr. Alexander recommended that the Articles of Association be changed to increase the number of directors. A new management structure was implemented to assign directors to major divisions of the company.[24] Carlton viewed the expansion as a way of ensuring the retention of managers in a time of national crisis when skilled workers were beginning to migrate.

The following month, at the board meeting of December 10, 1974, James S. Moss-Solomon Sr. tendered his resignation after thirty-nine years as director and fifty-two years of service.[25]

A further increase in the number of directors to a maximum of sixteen and no fewer than six was voted on at an EGM, July 17, 1975. The new board consisted of L. F. Kennedy (Chairman), S. C. Alexander (Group Managing Director), A. R. Diaz (Finance Director), F. X. Kennedy, P. N. Moss-Solomon, E. G. Muschett, Mabel Tenn, C. Tame, L. P. Scott, J. L. R. Bovell, S. Soutar, N. C. Bolton, W. Campbell, J. Lee, B. Rickards, R. McConnell; and Secretary, E. D. Anderson. Messrs. Peter N. Moss-Solomon and A Rafael Diaz were appointed as two of seven preference shareholders.[26] The following month, the company held another EGM, at which the total number of directors was increased to eighteen to accommodate the appointments of James Chin and Leroy Lukong.

Corporate generosity, which had been part of the culture of Grace, Kennedy from its inception, now proved critical to employee engagement. The need existed, now more than ever, to promote loyalty and increase job satisfaction and productivity.

On October 25, 1973, a committee had reported to the board on an employee share purchasing plan to make it easier for all employees to purchase shares. The company made an offer to all permanent employees on July 18, 1975, with an acceptance date of August 14; the units approved for issue would be sold at par and allotted in full, held by the Employee Investment Trust until they had been paid in full. Payment would be spread over five years in sixty equal monthly deductions from salary. Employees took up offers of purchase of a total of 626,975 units.

---

24 Board Structure of Management in 1974: Neville Bolton (Insurance and Properties); Ed Muschett (retired from management, December 31, 1974 but remained as external director); L. P. Scott (Wharves, Stevedoring and Shipping Agencies); S. C. Alexander, assisted by F. X. Kennedy (Industrial); Wallace Campbell (Food Distribution), A. R. Diaz (Finance); Mabel Tenn (Manufacturing and Agriculture); F. X. Kennedy (Special Projects and Corporate Planning).

25 James S. Moss-Solomon had been appointed a director by Dr. John Grace, January 24, 1936. See Chapter 14.

26 Directors with preference shares appointed on July 17, 1975 included L. F. Kennedy, E. G. Muschett, F. X. Kennedy, P. N. Moss-Solomon, J. L. R. Bovell, Cyril Tame and A. R. Diaz. S. C. Alexander, Group Managing Director, was also a preference shareholder.

Grace, Kennedy continued to provide salary bonuses during lean times. In December 1973, the board agreed to increase the salaries of managers and executive directors and approved bonuses to be paid by each subsidiary in the amount of $580,000.²⁷

In 1975 Carlton Alexander suggested to the board that instead of each subsidiary having separate Christmas parties as had been the tradition, a large event be planned for the National Arena to include all staff. It was an occasion to rally the troops in times of hardship, to show appreciation for their work, and to give hope that better was yet to come—a victory message all needed to hear.

Luis Fred Kennedy served on advisory committees, including the establishment of a new management structure and a new pension scheme, and held regular meetings with ministers and Mr. G. Arthur Brown, Governor of Bank of Jamaica.²⁸

He assisted in negotiations with government. With its increased monopoly over the food trade, the PNP government encroached directly on the operations of Grace, Kennedy. It established Jamaica Nutrition Holdings Ltd., a fully owned public enterprise under the authority of the Ministry of Industry, Commerce, and Tourism, with the objective to buy flour, grains, and other commodities in bulk to protect the consumer against price increases. Grace, Kennedy was placed in a position to negotiate with government to establish quotas and import licenses.

The chairman resented the government's threat to take over wharfage operations. After opening berths 5–9, his dream was to develop a transshipment port using an additional parcel of land containing berths 10–11. As early as December 1972, Mr. Paul Scott advised the board of directors that government had decided to establish its own facilities on the land owned by Kingston Wharves Ltd. Shipping was also in trouble because of a decline in imports. In April 1973, the division had fallen behind budget; Mr. Scott reported that Kingston Wharves faced a $2.2M loss in 1974.²⁹

Luis Fred Kennedy negotiated with Mr. Alfred Rattray, chairman of the Port Authority, for the sale of Berths 10 and 11 with the proviso that Kingston Wharves Ltd. be appointed operators of the facility at a per ton rate to be agreed on. Initial land sale agreements he found unsatisfactory and directed James Bovell to revise the contracts. An interim agreement was arrived at between the Port Authority of Jamaica, Kingston Wharves, and Western Terminals for container operations at Newport West and for

---

27 Grace, Kennedy & Co. Ltd. (November 13, 1973 and December 11, 1973).

28 The government had amended the law relating to superannuation schemes, effective October 21, 1971, with the result that staff employed after that date could not become part of the scheme. Grace, Kennedy introduced a new pension scheme to accommodate these new employees.

29 *The Daily Gleaner,* Thursday, April 11, 1974, 25.

a new joint organization known as Kingston Terminal Operations to be formed. The management agreement was signed April 23, 1975.[30]

Luis Fred Kennedy set a timetable for 1976 for Grace, Kennedy to go public—a necessary step he believed for its survival and future prosperity. At the August 1973 board meeting, the chairman had invited directors to consider registering the company on the Jamaica Stock Exchange. He suggested that shares then valued at $200 be converted to stock units of one dollar each, which would facilitate transfers on the exchange.[31] In April 1975, a committee consisting of the chairman, the managing director, the finance director, the company secretary, and Mr. Bob Humphries of Price Waterhouse commenced preparatory work.

November 25, 1975, 6:00 PM, Jamaica Pegasus Hotel, Kingston, Luis Fred Kennedy chaired an Extraordinary General Meeting. He was pleased with the turnout—ninety-eight shareholders, the majority of whom were employees of the company. The purpose of the meeting was to pass three resolutions:

1. That each of the existing Employee Stock Units of $1.00 each in the capital of the Company be converted into Ordinary Stock Units of $1.00 each.
2. That each of the 8% Cumulative Preference Stock Units of $1.00 each … be converted into Ordinary Stock Units.
3. That in consideration of the former holders of the 8% Cumulative Preference Stock Units having given up their rights attached to the said Preference Stock Units a capital distribution of $378,114 be made to those persons … in proportion to their respective holdings.[32]

With these amendments passed, Luis Fred Kennedy relinquished the privilege of his majority holding as a preference shareholder, an agreement established between himself and James S. Moss-Solomon in 1947. The arrangement had granted him veto power over decisions made by the board of directors, one which, according to Donovan Anderson, he never used. "Although he held the majority of preference shares and could exercise veto power, he never used it, not even for his own salary, which was modest, to say the least."[33]

---

30   Ibid., May 2, 1975, 25. The management agreement was signed by Mr. Alfred Rattray, chairman of the Port Authority; Mr. Paul Scott, managing director of Kingston Wharves Ltd.; and Mr. Peter Harty, managing director of Western Terminals. Berths 8–9, owned by Kingston Wharves, and Berths 10 and 11 of the new facilities would be used for handling of containment shipping.

31   The JSE required companies wishing to place shares on the exchange to have one hundred shareholders holding 20% of the stock.

32   Grace, Kennedy & Co. Ltd., Extraordinary General Meeting, Tuesday, November 25, 1975.

33   Donovan Anderson, interview with author, May 30, 2019.

Rafael Diaz concurred, "He didn't need to, he was the boss; he managed the place, he was fully aware, fully informed. Everybody knew he was head of the company."[34]

Disposing of the right to own preference shares was necessary for the company to go public; the chairman understood that the days of *"playing god,"* as he called it, were over, and that the privilege was too great for any one person to hold.

Mr. Alexander was Deputy Chairman and Group Managing Director. Finance Director Rafael Diaz was rapidly gaining the experience necessary to succeed Mr. Alexander, and Paco was Carlton's deputy. Paco had been appointed a preference shareholder and Deputy Group Managing Director at the AGM, August 4, 1974, and Assistant-deputy to Carlton Alexander, four months later, on December 10.

Threats to the private sector grew more menacing as the 1976 elections neared. The country was unstable, with violence escalating and the economy slipping further into recession. Mr. Manley was desperate.

The PM advised the governor general to declare a Public State of Emergency, which went into effect June 19, 1976. Security forces were given instructions to target "gunmen and terrorists," to use powers to "lock up and keep locked up all persons whose activities are likely to endanger the public safety."[35] The prime minister's speeches became more radical, and the leftist wing in his party, more vocal. He shrugged off the need for help from the United States and the International Monetary Fund and adopted the stance that "We are masters in our own house."[36]

At the board of directors meeting of September 11, 1976, Mr. Alexander reported that private companies in Jamaica were in dire straits, given the country's negative balance of payments and declining revenues. Many were being forced to lay off staff, and others were going bankrupt. "We must consolidate our position at this time and preserve our present assets until after the elections … Our planning should now be done so that after the elections if these plans are still relevant, then they would be put into effect immediately."[37] Grace, Kennedy had to make itself recession-proof. It was a time to take stock of outstanding accounts receivable, to reduce foreign debts and halt further capital expenditure. Mr. Diaz planned to mobilize funds to liquidate foreign debts and minimize losses caused by future devaluations of the Jamaican dollar.

Directors and managers hoped the period of wait-and-see would not last long. Grace, Kennedy, Jamaica, and the world kept a watchful eye on conditions in anticipation of the outcome of the next general election.

---

34   A. Rafael Diaz (April 16, 2021).

35   *The Sunday Gleaner,* June 20, 1976, 1. Headlines read, "Governor-General on advice of Cabinet declares … STATE OF PUBLIC EMERGENCY."

36   Panton, 52.

37   Grace, Kennedy & Co. Ltd. (September 11, 1976).

SEASONS GREETINGS

As I expect that the coming issue of the "Grace" News will be the last to be published before Christmas and the New Year, I should very much like to have you publish on my behalf, my greetings and all best wishes to all the Group's management, staff, your other readers and the families of everyone, for much happiness at Christmas and through- out 1976.

In wishing happiness, I include good health, peace of mind, financial security and an individual personal sense of accomplishment – whether it be a job well and conscientiously done in formal employment or even the achievement of success in outside activities be they hobbies or anything else.

I also want to say how grateful I am for the co-operation and immense dedication to our Group by all the members of management and staff. Given a continuation of these, we can all end the year and commence a new one with full confidence in the future, despite the pessimistic predictions that one so frequently hears in these confused days.

Luis Kennedy

Grace News, December 1975. A Christmas message of hope from Chairman Luis Fred Kennedy. (Photo, courtesy of © GraceKennedy Ltd.)

40

AFTER FORTY-EIGHT YEARS OF SERVICE (August-December 1976)

August 17, 1976

Dr. Shoucair[1] recommended that Dad be flown to Miami for treatment at the Cedars of Lebanon Hospital.[2] The risk was that a suspected blood clot in his right leg could break free and travel through the heart to his lungs.

For the month of September, Mom rented an apartment spacious enough to accommodate family members who travelled back and forth from Jamaica. With daily visits to the hospital, we kept Dad in good spirits—a difficult task as he was not a good patient at the best of times. He underwent angioplasty, a procedure that, although less life-threatening than bypass surgery, did not guarantee a lasting cure. It involved inserting a catheter and stents to unblock the blood vessel. The operation was successful, resulting in some swelling and discomfort and a recovery period of six to eight weeks. He missed his coconut water and salt fish and ackee and could not wait to return to smoking cigarettes. (He had given up cigars by then.)

Early October.

Dad was seated in a wheelchair on the front verandah at home. With the help of regular visits from physiotherapists, he was responding well to treatment from the operation and was

---

1 Dr. Rudolph "Ruddy" Shoucair, family doctor during the 1970s.
2 See Introduction.

occasionally moving about with the aid of a walking stick. We had converted the dining room into a bedroom for easy access to the ground floor and to the garden, which allowed for greater mobility and daily exercise. [3]

On the table next to him was Jaws *by Peter Benchley.*

"Are you enjoying the book?"

"Yes, it's a gripping, fascinating tale. An easy read, just what I need to distract me from all that's going on. Thank you for recommending it. Maybe when I'm done, everyone else in the family can have a chance to read it, and that way, we can all have a conversation about it."

"That would be great."

"Your adventures abroad were not successful, I gather."

"I was able to see Georgie,[4] *so that was good. But the other part, visiting family in Montreal, was not that successful. They were non-committal, saying they needed more details."*

"It's just as well. I don't think I could have gotten permission from The Bank of Jamaica to move money out. They're now allowing only $10,000 when you migrate. We will have to find another way. Jamaica is broke as hell."

"We'll manage somehow. I know people who are buying dollars on the street."

"You mean on the black market?"

"Yes, people are smuggling cash out that way."

"No, I don't want us to be doing that, Fred. We're in one hell of a situation; it's hard to believe that we've gotten ourselves into this mess. It goes to show you that market forces will always outsmart controls. You know what Manley's big mistake is?"

"What's that?"

"He's spent the country's resources on a scale that he is incapable of repaying. He's run out of money and scared everyone away at the same time."

"Don't you think the US is partly to blame? Rumour has it that the CIA is working behind the scenes."

"Maybe this is true, but he has only himself to blame, son. He wants it both ways."

"How do you mean?"

"He wants two systems of government to work in harmony, a two-party system of British Parliamentary democracy and a socialist state with a planned economy. He's tried to build a mixed economy, but he's killed the markets. He's in trouble and doesn't know how the rass to get out of it."

"People want a change from the past. They're saying that despite everything he has a chance of winning a second term."

---

3   Alva Russell, Luis Fred's personal executive secretary, took charge of preparing for her boss's return from Miami. See Chapter 41.

4   Georgie had returned to Canada in 1975, and in 1976 enrolled at University of Western Ontario for her Bachelor of Education degree to qualify as a teacher. We were engaged to be married.

"I'm afraid so. He can work a crowd in a way his father could not. Norman Manley was more serious minded, a brilliant criminal lawyer. His son is smart too and popular as hell, but he has not been able to deliver."

"He's done a lot, Dad ... established cooperatives for the farmers, minimum national wage, jobs through the Impact Program."

"Look, son, you don't have to look much further than ninety miles north of us to realize that socialism would be unworkable in Jamaica. When you have the state making decisions for you, that's a recipe for disaster. He's taken away the incentive to earn money. For every dollar I earn, I give seventy-five cents to the government. The pressing issue right now, though, is for us to find a way to migrate."

"You know, Georgie and I have decided to settle in Canada after we get married. We could sponsor you once we get settled."

"That may take too long; you will first have to establish residence before applying for us. So damn cold in that country. I have always said that Montreal was the place where God invented snow." He laughed. He had a habit of laughing at his own jokes, which made them even more amusing. "Do you know where you will be settling?"

"Georgie is doing her teacher training at the University of Western Ontario. So I expect we will make London our first home."

"I'm happy for you both. I know she loves you."

"How do you know that?"

"I've seen her face light up when your mother shows her family albums. It's always a good sign when your girl enjoys seeing pictures of you when you were a child."

"That's funny." We both laughed.

"The other test of whether you're marrying the right girl is to check out her mother."

"Why's that?"

"Because she's going to end up just like her ol' lady one day."

"Well, Abbie is a wonderful lady."[5]

"Yes, she is. If things work out for you in London, perhaps Paco could travel ahead and see about a suitable place for us to live. I expect we will not be allowed to take out enough money to buy a place. If you don't have enough cash, I will advance you some money. You can pay us back after you get yourself a job up there. I was thinking too that before we leave, we will need to appoint someone here to look after our affairs."

"Who do you suggest?"

---

5   Abbie Herberta Barnes Thompson (1924–2019), my mother-in-law, visited Jamaica in 1974.

"I have full trust in Chris Bovell [6] and Donovan Anderson. We could appoint either or both as power of attorney. Both have given me sound advice on the tax situation. Alva Russell will look after my personal and business affairs at the office."

"I hope we're all making the right decision."

"Obtaining a green card for the US is impossible for us at this juncture. It's time to leave. Doctors are saying I should go but it's also not safe for your mother. God knows."

"What about Grace, Kennedy?"

"We will do our damndest to ensure continued prosperity. Carlton is in charge and will soon be appointed chairman of the board. There is the problem of what to do with the two houses."

"What about them?"

"If we migrate, we will have to decide about Lady Musgrave Road. Real estate prices have plummeted."

"Does it really make sense selling at this time?"

"Who the hell will look after them? You know what's happened to houses that people have left behind."

"You'd get little or nothing for them though."

"I received an offer of $30,000 for Discovery Bay."

"Don't do it, Dad; that makes no sense. You'll be throwing it away."

"Well, we'll have to figure out something."

October 28

Unassisted by a walking stick, Luis Fred Kennedy climbed the stairs to the board room on the second floor at 64 Harbour Street. It would be his last directors' meeting as chairman of the board.

Days before his sixty-eighth birthday, he made the announcement to his fellow directors: "It is with the deepest regrets that I resign as Chairman of Grace, Kennedy & Co. Ltd., I am not able to continue anymore."[7] The board of directors passed a unanimous motion to appoint Mr. S. Carlton Alexander as chairman. Mr. Alexander thanked Mr. Kennedy for his many years of service and said that he was sorry that he had to resign but was happy that he would continue as a director of the company.[8] Luis Fred had joined the company in 1928, appointed director and assistant manager in 1930, and governing director and chairman in 1947.

---

6 Christopher D. R. Bovell, attorney-at-law, was appointed to the Grace, Kennedy board as alternate director to his uncle, James L. R. Bovell, September 1975. Dad thought Chris to be "one of the most brilliant legal minds" he had ever encountered. Chris gained a reputation for excellence in Jamaica in the areas of banking and corporate and commercial law. He retired as a director of Grace, Kennedy in 2010.

7 Grace, Kennedy & Co. Ltd., minutes of the board of directors' meeting, October 28. 1976.

8 Ibid.

Carlton Alexander was now fully in charge. That day, the Private Sector Organization of Jamaica,[9] of which Carlton was a founding member and first president, held its Annual General Meeting at the Jamaica Pegasus. Mr. Alexander outlined a three-point economic policy[10] aimed to reduce unemployment, support local industry, and re-assess government policies on price controls and taxation. As a leader of the private sector, he threw his full support behind helping government succeed. He re-affirmed his position that the PSOJ was not part of any "clique" threatening to ruin the country. "We propose to work hand-in-hand with our government. We propose to develop plans and projects that can motivate and move our country."[11]

October 30

A sprinkling of rain at home and heavier showers in the vicinity of Kingston blessed the last Saturday morning in October, the day when Georgianne Ruth Thompson and Fred William Kennedy were to be wed. The ceremony was held that afternoon at the Church of St. Thomas Aquinas, and my parents held a reception afterward at Lady Musgrave Road. Georgie's sister Shirley and my sister Liz were bridesmaids. My friend Jack Allen from Boston was best man, and my brother Paco, groomsman and master of ceremonies.

November 9

Mr. Luis Fred Kennedy attended his first directors' meeting after retirement as chairman. His brother-in-law, Simon Soutar, in the absence of Mr. Alexander, was unanimously appointed chairman of the meeting. Messrs. Carlton Alexander, Francis Kennedy, Peter Moss-Solomon and A. Rafael Diaz were visiting Grace, Kennedy's foreign subsidiaries in London, Rotterdam and Montreal.

November 21

At a political rally in Sam Sharpe Square, Montego Bay, Michael Manley announced that the general election would be held on December 15, 1976. Parliament was dissolved on November 23. Edward Seaga was Leader of the Opposition, having been

---

9   The Private Sector Organization of Jamaica was founded in 1976. "It is a national organization of private sector associations, companies and individuals working together to promote a competitive and productive private sector." https://www.psoj.org. It was formed at a time when the private sector was fighting for its survival. As its first president, Mr. Alexander led the charge for economic reconstruction as a national priority.

10 *The Daily Gleaner*, Friday, October 29, 1976, 1.

11 Ibid., Wednesday, November 3, 1976, 21.

elected in November 1974 at the Annual General Conference of the JLP, winning the support of 2,052 votes out of a total of 2,400.[12]

The foreign press carried news stories of the upcoming general election:

> KINGSTON, Jamaica, Nov. 22—This island, still under a state of emergency declared in June, will hold national elections on Dec. 15, just before the start of the winter tourist season.
>
> Prime Minister Michael Manley announced the date at a People's National Party rally last night. Edward Seaga, the Leader of the Opposition Jamaica Labor Party, said at a news conference this afternoon, that "we cannot expect that free elections are going to be held in Jamaica."
>
> Mr. Seaga said that security forces searching for guns raided a Labor Party office this afternoon in the northern part of St. James Parish, not far from the resort of Montego Bay.
>
> Combined police and army units have been raiding houses and searching cars all over the island in an effort to seize firearms and prevent repetition of recent brief exchanges of gunfire between political factions.[13]

The island was thrown into a pre-election frenzy.[14] Citizens of Rose Town and other areas of Kingston reported homes being destroyed and people killed, victims of political violence.[15] Calls for police protection went unheeded, with the public left vulnerable to intimidation by gangs armed with guns, bombs, stones, and bottles. Each party blamed the other for the upsurge in violence.

---

   12   Edward Seaga had been undecided about running for leadership of the JLP. One of the deciding factors was a conversation he had with Bustamante at his home in Irish Town in 1974:
"Can you handle the leadership, son?" he asked.
"I would like some more time to enjoy life," I said, "but who would want a future in which Manley is not kept in check; if I have your support, I will do my best."
"My support I can give, but you must go out and win the support of the people. Don't forget the poor. They are suffering hell." Edward Seaga in *Edward Seaga: My Life and Leadership* (2009), 215.

   13   *New York Times*, November 23, 1976, 12

   14   The number of murders per year increased from 145 in 1971 to 943 in 1980. *Statistical Yearbook of Jamaica* published by STATIN, quoted by Edward Seaga (2009), 284. During the period leading up the general election, over one hundred persons were shot and injured (Carl Stone, 1979, 7).

   15   *The Daily Gleaner*, December 6, 1976, 12.

November 25

Mom and Dad arrived in Toronto as landed immigrants, naming me as their next of kin and CIBC, Wonderland Road South, London, Ontario, as their official Canadian address. By coincidence, Mr. Greg Morris, who had been general manager of CIBC in Jamaica, was the newly assigned branch manager in Westmount in London, Ontario.

Before arriving in London, they spent a few days in Toronto, where they entertained my in-laws, Abbie and Ken Thompson, who lived in the town of Galt, midway between London and Toronto. They got along famously. Dad's eldest niece, Mary Lou, who was living in Toronto, toured my parents around the city and arranged for excursions to see the farms and forests north of Toronto. Dad was fond of autumn colours, but it was too late in the year, and the leaves had already fallen.

On doctor's orders, Dad was told not to remain in Canada long, for the cold would be detrimental to his health. He welcomed the suggestion.

December 3

The nation was rattled by news of the attempted murder of Bob Marley at his Island House at 56 Hope Road, where he and his band were rehearsing for a concert scheduled to be held two days later. The Smile Jamaica Concert at National Heroes Park went ahead on Sunday as planned, despite Marley having suffered a gunshot wound to his forearm;[16] Bob Marley and the Third World Band gave a free performance before a crowd of 50,000 to share their message of peace and love. In his closing song, "So Jah S'eh," he repeated the verses, "Puss and dog they get together. What's wrong with you, my brother?" Bob unbuttoned his shirt for all to see where he had been shot.

December 9

In the wake of political disturbances, curfews were enforced as part of the National State of Emergency, and in early December, a ban was imposed on public political gatherings, marches, processions, and motorcades in the Corporate Area. "We want to make it abundantly clear that to everyone, whatever their political affiliation, that anyone found committing a breach of the peace during the election campaign and on election day will be detained under the emergency regulations" (The Competent Authority). [17]

---

16   Rita Marley was injured with a shot that grazed her head; Don Taylor, manager, and another man from the Twelve Tribes Band were hospitalized. In subsequent weeks leading up to the election, the incident became the subject of commentaries in the press, which speculated about political motivations behind the shooting.

17   *The Daily Gleaner,* Thursday, December 9, 1976, 1–2.

December 15

PNP RETURNED IN LANDSLIDE. From early hours, electors were lined up in heavy numbers awaiting the opening of polling stations at 7:00 AM. The People's National Party made a clean sweep, taking JLP strongholds like those in St. Thomas. Michael Manley's party won forty-seven of sixty seats with fifty-seven percent of the popular vote and eighty-five percent voter turnout. Manley viewed the election as a clear mandate for democratic socialism. He saw it as "total rejection by the people of the suggestion that Jamaica was going Communist"[18] and as an acceptance of the government's policies of land reform and social development. He urged all Jamaicans to understand that they must unite and pleaded with his supporters to "take this victory with tremendous humility, modesty and with a sense of responsibility."[19]

Edward Seaga offered the PNP his congratulations, acknowledging that they had "scored a very clear and decisive victory" and the people had chosen one ideology over another.[20]

Manley maintained party support among most groups of voters except for white collar workers, whose votes declined from seventy-five percent in 1972 to fifty-seven percent in 1976, and for business, management, and high-income professionals, whose support fell from sixty percent to twenty percent.[21] Manley secured the patronage of wealthy families such as the Hendricksons, Matalons, Ashenheims, and Rousseaus,[22] but this was not enough to stem the migration of thousands who feared economic collapse and a communist takeover.

Manley was now faced with a critical choice: to follow an Emergency Production Plan proposed by the left-wing faction in his party or sign an agreement with the IMF, proposed by the moderates.[23] Either way, business was nervous, anxious about his next steps.

---

18   Ibid., December 16, 1976, 1.
19   Ibid.
20   Ibid.
21   David Panton, *Jamaica's Michael Manley: The Great Transformation 1972–92,* (1993), 51.
22   Ibid., 50.
23   Ibid., 51–52. The left consisted of political activists (D. K. Duncan, Hugh Small, and Anthony Spaulding) and UWI social scientists (Norman Girvan, Louis Lindsay, and George Beckford). Moderates included some Cabinet members (P.J. Patterson, David Coore, Eric Bell) and finance technocrats (G. Arthur Brown, Governor of Bank of Jamaica; Horace Baker, Permanent Secretary, Ministry of Finance; and Gladstone Bonnick, director of National Planning Agency).

Fred William Kennedy and Georgianne Ruth
Thompson were married, Kingston, Jamaica,
October 30, 1976.
(Photo from family collection)

Recently migrated to Canada, November 1976, Tillie and Luis Fred seen here with
Abbie Thompson in Toronto.

(Photo from family collection)

41

THE

DEBACLE

(January–May 1977)

Manley declared 1977 as the Year of Economic Emergency. Jamaica was in trouble. Since 1972, GDP per capita had dipped by a cumulative decline of twenty percent,[1] with no signs of improvement in sight. In desperation, he embraced an IMF contract (against the advice of the left wing of the party), the only option he thought would bring recovery. But he was unable to restore confidence.

USAID intervened. It increased aid with a $65M package for 1977–1978. The objective was to promote long-term growth for Jamaica by reducing balance-of-payment deficits of $250M. Yet even with the stimulus, economic conditions worsened. Fifty percent of people living in Kingston were out of jobs, and the rate of inflation was astronomical—the Jamaica Consumer Price Index rose by forty-nine percent in 1978–1979. In 1978, approximately seventy percent of urban households had income below the poverty line: $2,011 per annum.[2] Austerity had set in, supermarket shelves were empty, and the crime rate was on the rise.

The private sector blamed the government for mismanagement, and the government blamed the private sector for not re-investing in the country. They were at loggerheads. The powerful voices of business, represented by the Private

---

1 Richard S. Schweiker, "U.S. Response to Jamaica's Economic Crisis" (1980), 1.
2 Ibid., Statistics taken from Schweiker (1980).

Sector Organization of Jamaica,[3] the Chamber of Commerce, and the Jamaica Manufacturers' Association, became more oppositional to the Manley government as they saw company profits shrinking and the country's economy collapsing. Government, in turn, became more disenchanted with private sector; they took the view that the corporations were part of a monopoly,[4] at odds with the socialist agenda of the People's National Party.

THE ECONOMIC CRUNCH IS HERE was the lead headline of *the Daily Gleaner,* January 20, 1977. Michael Manley outlined in a speech to Parliament the prime strategy for recovery, which he described as a "battle of production."[5] He declared the priorities to be employment, goods, clothing, medical care, and education. Government intended to impose further restrictions on imports, foreign exchange spending, and on remittances abroad. He proposed to increase taxation on commodities and impose a surtax on incomes in excess of $30,000. Local expenditure would be reduced by scaling back on social programs and by implementing a wage freeze for civil servants. He aimed to set an example by reducing salaries of ministers of government and parliamentary secretaries. He reduced his own by $6000, to $16,000.

Manley took a radical step forward, one that shook the foundations of private enterprise:

> I want to announce tonight that the government will, in concert with its existing financial institutions, enter into negotiations with a view to acquisition of one of the major foreign-controlled banks and two of the smaller banks …
>
> The government proposes as rapidly as possible to take control of all imports which are deemed to be essential for national development and in the public interest … Government intends to enter into negotiations with certain organizations that are now engaged in the importation and distribution of essential commodities …[6]

---

[3] Carlton Alexander initially accepted the austerity Economic Package. He took a more conciliatory approach to PNP policies than more conservative members of the PSOJ would have liked. But his was not a socialist stance; it was his way of wooing the radical elements within the PNP to forge a better working relationship between private sector and government. He also thought it reasonable to expect people in higher income brackets to make sacrifices but warned against "confiscatory income taxes which so reduce earnings that the incentive to produce is removed."

[4] Schweiker (1980), 12.

[5] *The Daily Gleaner,* Thursday, January 20, 1977, 1.

[6] *The Daily Gleaner,* Thursday, January 20, 13.

He promised not to repress private businesses if they were willing to go along with the government's philosophy and objectives: "We seek nothing more than cooperation, understanding, and commitment." Yet, he warned them: "Let them be reminded that under the Emergency Powers Regulations, the Competent Authority has the right to take over plants and business in the public interest."[7]

Government intended to take control of the importation and distribution of foods. In Manley's view, a rationalized approach to food distribution was necessary for shopkeepers to benefit from centralized bulk purchasing. Jamaica Nutrition Holdings Ltd. became an umbrella organization under which the importation of all commodities would be developed.[8]

Nationalization was no longer an idle promise; it was an intrinsic ingredient of "democratic socialism":

> This is the way, Mr. Speaker, in which Democratic Socialism approaches the question of nationalization. We use it as a tool which will provide us with greater discretional powers to redistribute incomes and stimulate rapid development.[9]

In his speech to Parliament, January 1977, Manley also announced government's intention to acquire two additional private companies: Radio Jamaica (RJR) (Douglas Judah subsequently resigned as chairman) and the Caribbean Cement Company. They were both in public hands by the end of 1977. In March, the government released information that the "major bank" referred to in the prime minister's speech was Barclay's Bank, and the two "minor banks" were Jamaica Citizens Bank and Bank of Montreal Jamaica Ltd.

"Too much too late" was the *Gleaner's* editorial comment of January 20, 1977. Manley's economic reforms could not rescue Jamaica.

Instead of unifying the country, the prime minister's announcements caused further discord. The Private Sector Organization of Jamaica, the Chamber of Commerce, and Jamaica Manufacturers Association became more vocal in their attacks, and *The Daily Gleaner* intensified its anti-PNP campaign. Migration of capital and of skilled labour escalated.[10]

---

7   Ibid.

8   Ibid.

9   Ibid.

10 "Over the period 1974–6 and 1977–9, migration of those categories (managers, officials and proprietors) jumped 300%, from 2,140 to 6,610." Edward Seaga in *Edward Seaga: My Life and Leadership* (2009), 267.

The Debacle (January–May 1977)

A story appeared on the front page of *The Daily Gleaner*, February 16, 1977:

Govt. takeover: distributors deny report.

Rumours and speculations developed as a result of an announcement made by Prime Minister Michael Manley in his economic package of January 19 that government intends to enter into negotiations with certain major organizations in the food distribution trade in order to rationalize the distribution system and to realize the attendant benefit to the public at large ...

Hon. David Coore told the GLEANER yesterday that his Ministry was not involved in any negotiations with these companies (Industrial and Commercial Developments, Bryden and Evelyn, Grace, Kennedy and National Continental Corporation) and the Minister of Industry and Commerce was not available, and the Permanent Secretary refused to comment on the matter, but instead referred the GLEANER to the unavailable Minister.[11]

One week later, Carlton Alexander made the solemn announcement at the February 24 meeting of the board of directors that he "had been officially informed by the Minister of Marketing, Mr. Vivian Blake, that the government intended to commence negotiations for the acquisition of Grace, Kennedy & Co. Ltd."[12] He hoped that plans to gain full control might be thwarted by government's inability to pay.

Luis Fred Kennedy was present for the meeting. When he had first heard the news from Carlton, he was devastated. This was his worst nightmare possible—government ownership of Grace, Kennedy. The announcement came also on the heels of forecasts that, given the government's intended reduction in import quotas, the company's sales of imported goods for 1977 would be reduced by $12 million.[13]

In Carlton's public speeches, he acknowledged the threat: "It is the professed desire of our government to acquire greater interest in the private sector."[14] And he was emphatic about the dire prospects: "We must not allow ourselves to be victims of excessive state control, nor should we unwittingly submit to any ideological system that

---

11 *The Daily Gleaner*, Wednesday, February 16, 1977, 1, 16.

12 Grace, Kennedy & Co. Ltd., minutes of meeting of board of directors, February 24, 1977.

13 Ibid.

14 Grace, Kennedy & Co. Ltd., *Nation Building: A Businessman's Perspective. Excerpts from Selected Speeches by Carlton Alexander* (1983), 43.

might rob us of our basic freedoms."[15] He was determined to rescue the company from the clutches of state control.

Boerries Terfloth, who held fifty percent holdings in Terfloth & Kennedy (Montreal, Rotterdam, UK, and USA), was nervous about recent developments. Yet, despite viewing Jamaica as a failed socialist state, he remained optimistic, holding fast to a long-term vision of continued trade with the Caribbean. A business association with Grace, Kennedy was profitable for Terfloth & Kennedy, and he did not want to surrender this. What stood in the way was the Jamaican government's stated intention to nationalize Grace, Kennedy. The stakes were high for both Grace, Kennedy & Co. Ltd. and Terfloth & Kennedy Ltd.

The government's announcement triggered an immediate reaction. The two companies needed to find a way out of the impasse. If nationalization was *a fait accompli*, Luis Fred Kennedy would not, *"over his dead body,"* as he said, have the state assume total ownership of the Grace brand. And Boerries Terfloth, who had been in business with Grace, Kennedy since 1952, was not prepared to partner with a government-owned business.

Informal discussions followed.

> It was at his home (February 1977) at Lady Musgrave Road, Luis Fred Kennedy called a meeting with Carlton Alexander, Boerries Terfloth, Rafael Diaz, and Donovan Anderson. The concern was the great possibility that the government's takeover of the "Commanding Heights" appeared to be a forgone conclusion. One thing was sure: the government should never ever be allowed to own the GRACE brand.[16]

If the company sold its interests in the overseas subsidiaries, it would prevent government control of Grace, Kennedy's fifty percent share of Terfloth & Kennedy. This suited Terfloth. However, the ownership of the Grace brand remained a contentious issue, the subject of heated arguments. Luis Fred Kennedy left the final decision to be ratified by the chairman, directors, and shareholders of the company.

On March 8, 1977, Chairman Carlton Alexander announced to the board that discussions with Mr. Vivian Blake were ongoing, talks that detailed the history and nature of Grace, Kennedy's operations.

On March 14, Mr. Alexander reported on negotiations with the Terfloth group of companies:

> There was a dispute between Grace, Kennedy & Co. (Canada) Ltd., Boerries H. Terfloth and Greven Holdings Limited as to their ongoing

---

15   Ibid. Excerpt from speech made to the Jaycees 11th State Convention Senators' Luncheon, August 28, 1976.
16   A. Rafael Diaz, "I Remember Luis Fred Kennedy" (2019).

relationships as Shareholders of Terfloth & Kennedy Limited, Terfloth & Kennedy B.V., Terfloth & Kennedy (UK) Ltd. and Terfloth & Kennedy (U.S.A.) Inc. After discussion, it was decided that the only course of action open to the company was for it to dispose of its interests in these companies and in the trademark "Grace" to Boerries H. Terfloth and Greven Holdings Limited for a consideration of not less than $700,000 Cdn, and it was agreed that this course of action should be recommended to the shareholders of the company for their agreement.[17]

He announced that a special meeting of the shareholders of Grace, Kennedy & Company (Canada) Ltd. would take place at the head office of the company in Montreal, March 18, 1977, to consider the proposed agreements.

The deal went through. Terfloth & Co. Ltd. was formed. It acquired Grace, Kennedy's fifty percent share of its foreign subsidiaries and international rights to the Grace brand. Grace, Kennedy & Co. Ltd. had hedged its bets. With the threat of an imminent government take-over, the company would depend on the long-standing business relationship with the Terfloth Group to re-establish its partnership and re-acquire the Grace brand abroad in the event of conditions normalizing in Jamaica.

Chis Bovell, commented in an interview:

> I think that your dad's motive frankly at the time of giving it was to rescue it so that there would still be a Grace, Kennedy overseas and it could've been in conjunction with Terfloth. In other words, it was not meant as an outright gift once and for all. It could've been rather, "Look, I'm giving it to you in safe keeping. Hold it for me and then I expect to get it back." But that is mere supposition on my part.[18]

A trading relationship continued with the Terfloth Group of Companies that, given their experience in marketing and distribution, proved mutually beneficial to both Grace, Kennedy and themselves. Twenty years passed before Grace re-acquired international ownership rights of the Grace brand.[19]

April 1977

Government had issued new quotas and licenses for imports, and the company was expecting improvement for the second quarter. Subsidiaries were controlling expenses: a reduction of $100,000 in the Food Division and Holding Company for the three

---

17   Grace, Kennedy & Co. Ltd., minutes of meeting of board of directors, March 8, 1977.
18   Interview with Christopher Bovell, October 18, 2019.
19   See details in Postscript.

months ending March 31, 1977, as compared to the same period a year earlier. And export sales were earning more due to the devaluation of the Jamaican dollar.

The political climate remained volatile. At Mr. Kennedy's request, consideration of Grace, Kennedy going public was put on hold. On April 28, at the last board meeting he attended before returning to Canada, he received disconcerting news: Chairman Carlton Alexander reported that the company had begun to restructure the Food Division based on government's announcement to become the sole purchasers of imported basic foods within the next twelve to fifteen months.

May 1977

Mom and Dad made plans to return to Canada. Dad's secretary, Alva Russell, was a champion, looking after auctions, the sale of 45 Lady Musgrave Road, the shipment of personal effects, and clearing office files out at 64 Harbour Street.[20] (Her boss's office was adjacent to the board room on the second floor in the north-east corner of the building.) The only piece of furniture Dad kept was his father's mahogany desk with its glass top, a vintage piece from the 1920s.

Alva commented in an interview:

> When he migrated, he would come during the winter months. That is when I knew Mr. Kennedy trusted me. I was in charge of payments of all his bills. When he moved to Canada (he was a gentleman who did things legally), he advised them that he was leaving. I said to him, "Why did you do that, Mr. Kennedy?" He said to me, "No, no, I like to do things the right way." Then they froze his money. I had to go to the Bank of Jamaica every time the money finished. He opened a chequing account, and I alone could sign it or Paco could sign it or he could sign it. I said to myself, "Mr. Kennedy really trusts me."[21]

Alva Russell left Grace, Kennedy in 1978 when Dad secured a job for her with CIBC in Toronto.

---

20 Alva was hired by Carlton Alexander in 1959. She began her position as Mr. Kennedy's secretary in 1967 after Mavis Chang migrated to Canada. She had succeeded Dolly Chevolleau. "When Miss Chevolleau retired, young, at age fifty-five, Mavis was working for Mr. Moss-Solomon. She moved from Mr. Moss to Mr. Kennedy, and I moved from Mr. Alexander to Mr. Moss. When Mavis migrated to Canada in 1967, I moved from Mr. Moss to your father—that is how they operated in those days." (Conversation with Alva Russell, November 2021).

21 Alva Russell, November 2021.

PART SEVEN

Exile (1977–1982)

"All the time he was away,
he pined for Jamaica" (LMK).[1]

---

[1] Lydia M. Kennedy, "A Giant of a Man" (1997).

Celebrating fortieth wedding anniversary, October 1977, London, Ontario, Canada.

Mother's Day, 1977. Recently arrived in London, Ontario, meeting their granddaughter Amanda Mary, born Thursday, May 5, 1977.
(Photo from family collection)

Celebrating fortieth wedding anniversary, London, Ontario, October 2, 1977. Standing, L–R: Celia, Donald and Mary, Liz, Fred and Georgie; seated with Dad and Mom: Marjory and Paco.
(Photo from family collection)

42

LONDON, ONTARIO
(May–December 1977)

Mother's Day, May 8, 1977. Mom and Dad arrived in London, Ontario, three days after the birth of Amanda Mary, their second grandchild, born to Fred and Georgianne Kennedy. It was a joyous time, but one filled with apprehensions about adapting to a new life in Canada.

Paco and Marjory had travelled ahead to London months earlier to find Mom and Dad a two-bedroom, ground-floor rental unit in the newly developed Westmount neighbourhood. It was an upscale residential area with a short commute to the city centre, easy access to shopping and nearby parklands, and ideal for walks and picnics by the Thames River. Georgie and I received shipments of their personal effects in Canada and arranged for unpacking. Mom and Dad arrived to find a fully furnished home, ready to move in.

A first order of business as new immigrants to Canada was to obtain their Social Insurance Number (SIN) card, needed for conducting official transactions: filing taxes, applying for health care and Canada pension, and obtaining a driver's licence.

Mom and Dad wasted no time in applying.

Within a few weeks, they received an appointment to meet a federal government representative at Service Centre Canada. I accompanied them to the Dominion Public Building in downtown London, only to discover that there had been a power cut. The elevators were not operating, and the meeting was on the sixth floor.

"We shall have to take the stairs."

"I'm not sure that's the best idea, Dad," I told him.

"We have no choice."

London, Ontario   (May–December 1977)

Once Dad's mind was made up, there was no point arguing. Despite his heart condition, he climbed with stubborn determination. Mom, worried because of the steep ascent, kept telling him to slow down.

"I apologize for being late."

"Fully understandable under the circumstances, Mr. Kennedy," the officer replied.

"You really make us work for our SINs, don't you?" Dad laughed aloud, and Mom and I joined in.

The interviewer gave him a bemused look, which did not seem to faze Dad in the least. He continued laughing at his own joke, ignoring the man's unfriendly reaction. Conversation was awkward at first, but by the end of the interview, the gentleman was charmed by Dad's wit and good humour. Dad remarked after the meeting that it must have been his Jamaican accent that was off-putting. The SIN cards arrived a few weeks later in the mail.

As the warm weather set in, my parents enjoyed strolls in the park, picnics, and drives in the countryside. They became members of St. John the Divine Catholic Church, made acquaintance with tenants in the apartment complex, and contacted Jamaicans they knew who lived in London. Mom applied for an Ontario driver's license and went shopping for a new car. She surprised everyone with the purchase of a new 1977 Datsun Z280, a flashy, two-seater sports car. They used it for short trips around London but were reticent at first to take it on the highway. For longer trips, they took the train, for example, to meet with my in-laws, who lived a hundred kilometres from London.

Mom adjusted to life in Canada more easily than Dad. Having grown up in New York in the days of Lou Gehrig and Babe Ruth of the Yankees, she was an avid baseball fan. A month before Mom and Dad arrived in Canada, the Blue Jays had played their first Major League game on April 7, 1977, against the Chicago White Sox, winning 9–5. They ended up placing last in their opening season but improved steadily after that. Mom became a Blue Jays afficionado. She also felt at home with the seasons. "When I moved to Jamaica, I missed changes in seasons; I loved to go shopping for different fashions, especially in the fall and winter months."[1] She did not mind the cold and the snow, whereas Dad preferred the Caribbean climate.

Finances were complicated. Under the Exchange Control Act of Jamaica, the government restricted remittances abroad. As Canadian residents, they were unable to receive capital or income from Jamaica. When they returned to Jamaica, they were not permitted to use their money without application to the Bank of Jamaica because assets and income were frozen. They were required also to register their Grace, Kennedy shares with an authorized depository to prevent any disposition of stocks without permission of the Exchange Control.

Donovan Anderson commented in an interview:

---

1   Lydia M. Kennedy, interview with Georgianne Thompson Kennedy (2002).

Mr. Kennedy hated the idea of going to Canada, totally one thousand percent. He didn't want to leave. He also thought that the Canadian government would harass him for taxation, you know, because he wasn't taking any money there. They were not ready to believe him, and they would want to tax the dividends that were blocked in Jamaica.[2]

Before migrating, Dad bought an annuity and life insurance policy through the Canadian office. The joint and survivor annuity guaranteed an annual income of CAD 30,000 while he was alive and a payment of $15,000 per annum to my mom, who was the secondary annuitant. To pay for these, he forfeited interest earned on deposits at Grace, Kennedy and declined the offer of a pension. Although these arrangements covered living expenses and rent on the apartment, they did not resolve issues with Revenue Canada.

Friends advised him, "Mr. Kennedy, you think when Jamaicans go abroad, they tell anybody about their business?" But he would not have it any other way; everything had to be transparent and legal. He declared all his assets. Battles with Revenue Canada ensued for years. He hired lawyers in Toronto to argue that money not received should not be classified as income and therefore not subject to taxation. Lawyers quoted precedents established with the Supreme Court of Canada: "A taxpayer must have a right which is absolute and under no restriction, directly or otherwise, as to its disposition, use or enjoyment." But none of this convinced the Department of National Revenue to make exceptions.

Mom became his money manager. She invested, saved, and spent judiciously. She created annual budgets to ensure they met expenses, and she consulted with bankers at CIBC to negotiate optimal returns on investments. Mom enjoyed certain luxuries, but having lived through the Great Depression, she was cautious, wary of extravagance. She took calculated financial risks and only spent what she could afford.

Dad received news of the passing of his long-time friend, Alexander Bustamante, who died on the fifteenth anniversary of Jamaica's Independence, August 6, 1977, at the age of ninety-three. Bustamante died at his home in Irish Town, *Bellencita,* with his wife, Lady Gladys, at his side.[3] Busta had led a heroic life as political activist (champion of the labour movement) and as the first prime minister of Independent Jamaica. Thousands from every parish crowded the Holy Trinity Cathedral to pay

---

2 Donovan Anderson, interview with author, May 30, 2019.

3 Bustamante and Lady B had bought "The Hill" from the Kennedys, built a retirement home, and named it *Bellencita*. Mr. Seaga claimed that the name had a double *entendre:* bell, symbol of the Jamaica Labour Party, and the Spanish suffix, *encita*. It may also be a derivative of the Spanish, *belleza,* meaning beauty.

London, Ontario (May–December 1977)

their last respects, and world leaders sent messages of condolences to the family and to the nation. Sir Alexander Bustamante, Founding Father of the nation, died a National Hero, beloved by the Jamaican people.

In October, Mom and Dad celebrated their fortieth anniversary with family: Paco, Marjory, and Cathrine Lesa (aged two), and Celia and Elizabeth from Jamaica; Mary, Donald, and Jan Michael (three months) from Cleveland; and Georgie, me, and Amanda Mary (five months) from Toronto. Mom prepared a lavish meal and ordered champagne.

My friend Swithin Wilmot commented in an interview:

> On constant display was his loyalty to family, to his children and more than to anyone else, his wife. He was very tender to her, because sometimes your father could be grumpy. Loyalty to family, I guess he passed that on. Loyalty to Jamaica too. He really had a passion for Jamaica, and he spoke often about that.[4]

The anniversary marked a new phase in their lives, the beginning of what Mom had looked forward to as "the golden years" of retirement.

It was also a change for us children. As new immigrants, Mary and I were in search of employment[5] and were looking to further our education. Dad sent letters to Canadian Manpower and Immigration in Kingston soon after arriving in Canada to sponsor Celia and Elizabeth for permanent residence in Canada. At his request, Raf Diaz explored opportunities for Paco to purchase a business in Toronto. The original plan was for everyone to reside in Canada.

While we were gathered in London for anniversary celebrations, Dad received word of the passing of his good friend, mentor, and business associate of forty-seven years, James Seivwright Moss-Solomon. "It was a great shock to Fred to learn one morning that this great friend had died suddenly of a heart attack. I don't think that he ever got over his loss" (LMK).[6]

In an interview with Joan Belcher, she reminisced about her father's passing. It was October 4, 1977. He was lying down, and she was handed an envelope. "Inside the envelope was a cheque for £400, and I said to him, 'What is this for? My birthday is not till next week.' He said to me, 'You ask too many questions. I just want you to have it.' He died two days later."[7]

An obituary appeared in *The Daily Gleaner*, Tuesday, October 11, 1977:

---

[4] Swithin Wilmot, interview with author, May 12, 2019.
[5] With a glut of qualified teachers on the market, jobs were scarce. To make ends meet, I took a job as a fast-food delivery driver.
[6] Lydia M. Kennedy, "A Giant of a Man" (1997).
[7] Joan Belcher, interview with author, January 18, 2021.

J. Moss-Solomon dies at 82

Mr. James Seivright Moss-Solomon, retired businessman who up to his retirement in 1972, was manager and director of Grace Kennedy & Company Ltd., Director of Kingston Wharves, Ltd., and Sheffield & Company Ltd., died at his residence in St. Andrew on Thursday last, aged 82 years.

Born in Dry Harbour, St. Ann, on December 11, 1894 he was a son of late Joseph H. Moss-Solomon and Theresa Seivright, his wife.

Educated at Brown's Town Primary School and by private tuition he joined the staff of J. H. Levy, Brown's Town and served from 1908 to 1914. He was assistant secretary of the Brown's Town Building Society from 1914-16 and for the next two years served on the staff of the United Fruit Company; Montego Bay. From 1918 to 1922 he was accountant at Grace Ltd.

Twice married, he is survived by his second wife Flossie, sons Peter, Deputy Group Finance Director and Director of Grace Kennedy and Company Ltd., and James S. Jnr., executive in the food distribution division of Grace, Kennedy and Company Ltd., daughter Joan, S.R.N., wife of Mr. Michael Belcher, Manager of Kingston Wharves, Ltd., sister, Mrs. Enid Lord and Mrs. Hilda Phillips (Canada), seven grandchildren, nieces and nephews.

An Anglican, after a service in the St. Andrew Parish Church, Half-Way-Tree (tomorrow) at 4.30 p.m., interment will follow in the church cemetery.

The Daily Gleaner, Tuesday, October 11, 1977, 13.

(Photo, courtesy of © The Gleaner Company (Media) Ltd.)

Carlton Alexander and Raf Diaz often came to London to seek business advice and keep Dad abreast of the political situation in Jamaica. Raf Diaz described his relationship: "In my mind, he was still the boss."[8]

Carlton and Raf were not always the bearers of bad news. Although the government had instructed Nutrition Holdings Ltd. to take over the importation and distribution of basic foods in July, Carlton reported that by October, Grace, Kennedy was able to negotiate a forty percent share of importation of counter flour and that their portion for bulk rice exceeded that percentage.[9] The company received quotas for the third quarter of 1977 for the importation of chicken necks and backs, pickled beef, and frozen meats. Mr. Diaz, Mr. Rickards, other managers from Grace, Kennedy, and representatives from eight of the largest food importers on the island had successfully negotiated concessions.

Dad was also pleased to learn that Paco was assuming increased managerial responsibilities: meeting with ministers of government, working with Peter Moss-Solomon

---

8   A. Rafael Diaz, "I Remember Luis Fred Kennedy" (2019).

9   Grace, Kennedy & Co. Ltd., minutes of meeting of the board of directors, September 1977.

London, Ontario (May–December 1977)

on restructuring the Employee Investment Trust, reporting on operations of Cash and Carry,[10] and evaluating the retail market with the prospect of Grace, Kennedy's purchase of supermarkets. Paco also chaired board meetings when Carlton and Raf were off the island (October and November 1977).

To escape the winter, my parents returned to Jamaica at the end of November, in time for Dad to attend the December meeting of the board of directors.

James S. Moss-Solomon died Tuesday, October 11, 1977.
(Photo, courtesy of © GraceKennedy Ltd.)

---

10   During the 1970s, the company made a commitment to customers to distribute products island wide through Grace Cash and Carry by providing direct access of imported and locally manufactured food products to shop-keepers, higglers, and consumers. However, owing to shortage of foreign exchange and of goods for sale, and restrictions imposed by price controls, the establishment encountered difficulties. By the start of 1978, there were five centres across the island; in April 1982, the last remaining Cash and Carry was closed.

Christmas 1977, Puerto Plata, Discovery Bay, with granddaughter Amanda.
(Photo from family collection)

43

BACK HOME

(1978)

Puerto Plata
January 1978

I joined Dad at the gazebo. He was relaxed, smoking a Kent cigarette, too mild for his taste, but what the doctors had recommended as a substitute for the stronger Camels or Jamaican cigars. He convinced them that smoking calmed his nerves and was better for him than Valium.

"It's good to be back home," Dad said.

"It's amazing to be here. How has your health been?"

"I feel surprisingly well and in good spirits. Your mother is taking good care of me. Thank God we are still well enough to travel to spend the winter here." He flicked the half-finished cigarette into the sea. "I met with Bovell when I was in town."

"What did he have to say?"

"We talked about tax arrears in Canada. Price Waterhouse in London has filed for a ninety-day objection to the notice of assessment to give your mother and me time to figure out what to do."

"What does Mr. Bovell suggest?"

"He's mulling over some options. One of them is to form a private company in Jamaica to hold the shares that I've given to your mother and to you children. Lawyers in Toronto have advised that if the holding company owns a given percentage, I believe if it is ten percent of the shares of Grace, Kennedy, then we may be eligible for a tax exemption in Canada."

"You think it's a good idea to have all our family holdings held jointly in one company?"

"Better than having outstanding balances and interest accruing year after year. It's becoming expensive, Fred."

"I suppose."

"We also discussed your mother's and my wills. I'm concerned about the estate's liability in the face of succession duties in Canada. Assets are deemed to have been sold on the death of the *second spouse. Whatever happens, you and Paco must ensure that everything is divided equally among the five of you after your mother passes. Nothing is tied up, so everything can be used by the beneficiaries at their discretion. One blessing is that we're here now so we can forget about all these problems, even for a short while. What a gorgeous day it's turned out to be!"*

"It's beautiful."

"You know, son, it may sound strange to say, but I've been meaning to tell you that if you ever find yourself in trouble, prayer is a powerful thing. I have always had a special devotion to Mary; she has interceded for me through challenging periods in my life. If there's no one there to help, you can always turn to prayer, and miracles can and do happen."

"Those are comforting words, Dad."

"Make sure you look after your mother when I'm gone. And don't listen to gossip."

"How do you mean?"

"I mean that people will talk things that they won't say to my face."

"Why are you telling me this?"

"I'm sure I've made some enemies along the way. It's human nature, that's how it is–people will gossip. We faced that when my father died." He paused and looked out over the bay. *"How have you been getting along with your job in Canada?"*

"I have applied for teaching positions, but there's a real shortage right now."

"I want you to know that your ol' man is proud of you for taking up that delivery job in the meantime."

"Thanks, Dad. That means a lot."

"Why don't you get your bathing suit on? We should take advantage of this day. If you're going up to the house, ask Douglas to cut me down a jelly coconut."

Carlton remained optimistic. In his 1978 New Year message, he stated, "The year 1977 has been one of the most trying for business in the history of our country, but we look forward to 1978 as the year of expected turn-around in the Jamaican economy, and I hope that all Jamaicans of goodwill will join together in bringing this to fruition."[1]

A new management structure was presented at the January 10 directors' board meeting:

>   S. C. Alexander: Chairman
>   A. R. Diaz: Finance Director
>   P. N. Moss-Solomon: Deputy Finance Director
>   E. D. Anderson: Company Secretary
>   L. E. Lukong: Group Chief Accountant

---

1 *The Sunday Gleaner,* January 1, 1978.

Back Home (1978)

Trading Division (housed at 64 Harbour Street)
F. X. Kennedy: Divisional Director of Trading
Members of the Management Team for the Trading Division:
Bruce Rickards: General Manager
Joe Lee and Orville Garrick: Assistant General Managers
Basil Morris: Accountant
Ivanhoe Yee: General Sales Manager
Tony Wright: Brand Manager
Mrs. Faustine Sharp: Credit Manager.

Paul Bitter: Divisional Director, Insurance Group of Companies.[2]

The board consisted of eighteen directors and Donovan Anderson as company secretary. Of these, the newly appointed directors were Messrs. Ernest Girod, Michael Belcher, Paul Bitter, and Gordon Sharp. Mr. Alexander also invited key personnel like Orville Garrick[3] as guests to present reports at board meetings.

The board had announced Simon Soutar's retirement at age sixty-five at the board of directors' meeting, October 27, 1977. He departed the island with his wife, Louise, to take up residence in Toronto. Mr. Soutar joined Grace, Kennedy in 1941, served as director of the company and assistant manager to Luis Fred Kennedy and James Moss-Solomon in the 1950s. He was a director of subsidiaries, George & Branday Ltd., Cecil deCordova & Co. Ltd., and National Processors Ltd. in the 1960s, and chairman of the new merger of Cecil deCordova Ltd. and Grace Agricultural and Industrial Company (GAICO) in 1975. Mr. Soutar served thirty-six years of a productive business career at Grace, Kennedy.

Luis Fred was saddened by the resignation of Mr. Paul Scott, whom he had groomed over a twenty-year period. Paul, who had started work at the age of sixteen,[4] reminisced about his experiences at the "Grace, Kennedy University":[5]

> In Mr. Luis Fred Kennedy, Jim Moss-Solomon, and Carlton Alexander we had the best teachers and the best of businessmen. Each was quite different, and each complemented the other. Together they provided the company with a vision, dedicated leadership, and

---

2   Grace, Kennedy & Co. Ltd., minutes of meeting of board of directors, January 10, 1978.

3   Orville Garrick was appointed in charge of the Distribution Centre Inventory Control systems and made routine reports to the board.

4   He graduated from Priory School in 1951 and first worked for the Bank of Nova Scotia (1952–1954).

5   Many managers, who were either formally trained by the company and/or mentored by Luis Fred Kennedy, James Moss-Solomon, and Carlton Alexander, often referred to Grace, Kennedy as the University of Harbour Street.

results oriented drive that was also reflected in the employees. This is what I believe separated GKCO from the rest.

When I left Jamaica in 1978, I was privileged to be a shareholder, member of the board and executive committee of the parent company and chaired or sat on the boards of many of GKCO's subsidiaries with divisional responsibility for shipping, wharf, and services groups.

It was with sadness that I decided to leave Jamaica and therefore GKCO and move to Canada.[6]

February

Group management reports for the month of February 1978 showed positive results despite shortage of basic food items for the Merchandise Division and of raw materials for the factories. Showing profits were Grace Food Processors (Canning), Grace Agricultural and Industrial Co. Ltd., Metal Fencing/Security Woven Wire Products Ltd., Rapid & Sheffield Co. Ltd. (recording highest monthly sales to date), Pilkington Glass (Jamaica) Ltd., the Insurance Division, and the Shipping Group of Companies (Kingston Wharves Ltd. reporting increases in revenue per ton).

At the February 23 board meeting, Carlton Alexander paid his respects to one of the company's original shareholders, the Hon. Walter Fletcher OBE JP, Fred William Kennedy's friend and contemporary. A resident of Reading, St. James, Mr. Fletcher died in his ninety-eighth year on the fifty-sixth anniversary of Grace, Kennedy, February 14, 1978. At a memorial service held in Montego Bay, Governor General Florizel Glasspole praised him for his sterling character and outstanding community service.[7] Mr. Fletcher became custos of St. James Parish at the age of ninety and was president of the Montego Bay Chamber of Commerce and director of the Tourist Trade Development Board for over twenty-five years.[8] He was buried at sea at a private ceremony, Friday, February 24.[9]

---

6 Paul Laurence Scott. "Memories of Grace" in "Memories of What Made Grace" (1997), xxxii. Paul Scott migrated with his family to Canada in 1978 and subsequently moved to Charlotte, NC, where he worked at the World Headquarters of Sea-Land Service Inc.

7 *The Daily Gleaner,* Monday, February 27, 1978, 2.

8 Ibid., Friday, February 24, 1978, 10. An editorial entitled, "A Long and Useful Life" was dedicated to the Hon. Walter Fletcher OBE JP.

9 Mr. Fletcher was an avid yachtsman and one of the initiators of the Miami-Montego Bay Yacht Race. His son, Robert Fletcher, was commodore of the Montego Bay Yacht Club.

Back Home (1978)

By the end of 1977, Jamaica had failed the initial Stand-by Agreement signed with the IMF. The most stringent demand was that "all outstanding arrears in foreign payments were to be eliminated in three months"[10] and net foreign assets with the Bank of Jamaica, stabilized. The IMF loan had been too small, and despite Manley's political friendships with Pierre Elliott Trudeau of Canada and with newly elected President Jimmy Carter, Jamaica only received a fraction of capital inflows necessary to meet IMF demands. In 1978 a new, more austere plan was negotiated. The new Extended Fund Facility involved a larger IMF loan ($235M) and a more direct role in Jamaica's internal affairs, but funding was still too short to create an economic recovery.

Michael Manley was up against a force greater than himself: a world economic order of "big power imperialism"[11] that discriminated against poorer countries. On the one hand, sympathizers claimed that he was being forced to "put the country's economy through the wringer, precisely to satisfy the demands of these very industrial nations;"[12] and on the other, moderates within his party stepped up their criticism of Dr. D.K. Duncan[13] and others for refusing to admit that the socialist policies of the government were bankrupting Jamaica. Without the means to resolve the conflicts, Manley was trapped. The austerity measures were destroying an already ailing economy: devaluation of the dollar,[14] further controls on imports, increases in taxes and the interest rate, and restrictions on credit and wages. The balance of payments worsened in 1978 and 1979, higher import prices fueled domestic inflation (thirty-five percent in 1978)[15] and the real GDP continued to decline.

Mom and Dad returned to Canada, April 1978. Towards the end of that year, patient prognosis was not good. Dad's doctor in London wrote to me: "All things considered the long-term outlook is rather gloomy and the short-term outlook is

---

10   David Panton. *Jamaica's Michael Manley: The Great Transformation (1972–92)*, (1993), 64.

11   *The Sunday Gleaner*, October 8, 1978, 11.

12   Ibid.

13   Donald Keith Duncan (1940–2020), dental surgeon and politician, general secretary of the PNP (1974), appointed minister of National Mobilization and Human Resource Development (1977–1980).

14   Michael Witter, "Exchange Rate Policy in Jamaica: A Critical Assessment" (1983), 2, 13–18. By May 1979, the exchange rate was J$1.78: US$1.00 or a total devaluation of 97% in two years. The result was that "Devaluations failed to stimulate exports, to contain the growth of imports and to eliminate the balance of payments deficits."

15   Ibid., 27.

unpredictable."[16] Dad's cardiac status was stable; he was not experiencing angina attacks, and the vascular supply had improved in his legs. But the concern was a large abdominal aneurism (swelling in the abdominal aorta), inoperable and potentially lethal. He would not be able to return to Jamaica to spend the winter months. The golden years would be cut short.

In the 1970s, my sister Liz met John Patrick Riley who was living in Jamaica and working for the United States Foreign Service. John was born in New York, New York. When they became engaged, the US Department of State required that Liz become a US citizen as a prerequisite for marriage. She underwent a cumbersome and rigorous process of "expeditious naturalization," which lasted six to eight months. They were married in London, Ontario, July 28, 1979, at my parents' parish church, St. John the Divine.

Elizabeth Louise Kennedy and John Patrick Riley,
London, Ontario, July 28, 1979.

---

16   R. Stuart Eberhard MD (internal medicine, London, Ontario), personal letter to Fred W. Kennedy, dated November 1, 1978. Georgie introduced Dr. Eberhard to Mom and Dad. When she had returned to Canada in 1975, she worked for Dr. Eberhard as a live-in nanny for his children.

Francis Xavier Kennedy (aged 17), Georgetown Prep, Maryland, June 1958.

(Photo, courtesy of © Georgetown Preparatory School)

44

PACO

Gentle Giant[1]

> Our eldest son, Paco, joined the company in 1959. Like his father he started from the ground up, and over the years worked in every department of the firm. He was even attached to the Montreal office for a while, and in the early seventies attended Harvard Business School … He has a keen mind and business sense and on many occasions he has been called upon to turn around a floundering department or branch of the company.[2]

Francis Xavier Kennedy was born September 10, 1940. "He was the most beautiful baby, perfect in every way," Mom said of her first born. As the eldest of his siblings and of the entire Kennedy clan of first cousins, Paco was destined to hold a position of being first. Dad, having himself held pride of place within his own family, wished the same for his son. As a result, Paco found himself compelled to conform to extraordinary expectations. He charted his way, but it was difficult. When tensions flared between him and Dad, Mom tried to put out the flames. She had a special place in her heart

---

1 Milton Samuda (past-president, Jamaica Chamber of Commerce) referred to Paco as "The Gentle Giant of the Private Sector" in a Tribute published in *Who's Who in Jamaica Business,* October 2014.

2 Lydia M. Kennedy, "A Giant of a Man" (1997).

for Paco; she nurtured him through childhood illnesses and became his best friend and advocate in his adult years.

Paco was brilliant. (I remember the day Dad told me of his IQ scores; they were off the charts.)[3] He had all the characteristics of giftedness: top of his class, a problem solver and keen learner who had a wide array of interests—mathematics, sports, music, economics, and politics. He was high-strung: energetic, impatient, and driven, motivated by a strong sense of fairness and equity. Teachers at both Alpha Prep and St. George's College wanted Paco to skip grades, but Dad rejected the offers. He did not want Paco to experience the emotional challenges he himself had known as the youngest in his class.

Part of the decision to enroll Paco in boarding school in the United States was the opportunity for him to complete four years of high school at an age-appropriate level. He was pulled out of third form at St. George's to enroll at Georgetown Prep, September 1954, at the age of fourteen.

The Jesuit all-boys school, founded in 1789 (the oldest Catholic boarding and day school in the US for young men in grades 9–12) is located on a campus of ninety-three acres in Bethesda, Maryland. Paco's class consisted of fifty-eight boys, the majority Catholic, of American Irish, Italian, and Polish background. Of the few foreign students (mostly Latin Americans), Paco was the only boy from the British West Indies. The total enrolment was 250.

Since the 1920s, not much had changed in Jesuit curriculum with its focus on the humanities, language, rhetoric, and philosophy. As a Catholic school, Georgetown Prep aimed to educate young men "in their faith and understanding of the teachings of the Church and to learn to put their faith into action in the service of others."[4] Paco's subjects included religion, Greek, Latin, English, history, civics, sciences, and mathematics, at which he excelled. He gained the reputation as the best math tutor at Georgetown. Classes were small, not exceeding fifteen.

Paco joined the Sodality, which encouraged attendance at daily Mass, devotion to prayer, and dedication to works of the apostolate, and St. John Berchmans' Society, an assemblage of almost half the student body that assisted with daily Masses and religious functions. He was a member of the Library Club and of the editorial staff of two school publications, *Little Hoya* and *The Blue and Gray*. As a campus reporter, he wrote news and literary articles for publication on topics of labour, racial integration, Catholic education, Sputnik, and communism. Some of his *Blue and Gray* student staff had the opportunity to interview US labour leader George Meany, Senator John F. Kennedy, and former President of the United States Harry S. Truman. His literary interests drew him to the Homeric Club, a small group that read and studied the *Iliad and the Odyssey*

---

3   Dad arranged for Paco to take a management profile evaluation conducted by LeBlanc, Mahoney, Howard Inc. in Montreal, 1975.

4   https://www.gprep.org/about/mission.

in its original Greek and held discussions about the ancient civilizations of the Aegean people. When he was not busy in intellectual pursuits, he enjoyed competing in matches with members of the tennis team. Paco became a skilled and avid tennis player.

After graduating in 1958, he was the right age, eighteen, to enter freshman year at Holy Cross College to read for a bachelor's degree in philosophy and the humanities. During his first year, the dean called Dad to say that Paco was not attending classes regularly and would be found playing bridge in the cafeteria. Dad's trips to the college did not rectify the situation. He spoke with the priests, with Paco, and with his nephew, Bill Valliant, who had also started at the college that same year. Paco confided in Bill that "he never felt at home at the Cross and yearned to return to Jamaica."[5]

Paco started his career at Grace, Kennedy at the age of nineteen in the summer of 1959. Young Kennedy stood every morning at the #14 Jolly Joseph[6] stop on Lady Musgrave Road to catch the 7:30 bus to Crossroads. There he changed buses to arrive in time to start work at Harbour Street at 8:00 AM sharp. He was too young to be given the use of a company car.

He began without privilege or status in a junior clerk position with the Shipping Division, and later with Merchandise. From the very start, he wholeheartedly embraced his work at Grace, Kennedy. Yet he found time for recreation. Every Thursday night, he and his bridge partners, Peter Moss-Solomon, Trevor Jones, and Peter's cousin, William Lord, took turns meeting at each other's homes. Once a month, it was bridge night at #45 Lady Musgrave Road. The felt-covered mahogany card table was set out on the front verandah, and drinks and snacks were served. The rest of the family found other places in the home to spend the evening without disturbing the players. Paco was a card sharp, becoming an expert contract bridge player.

Even though we were ten years apart, Paco included me in recreational activities. He taught me how to use his BB gun; we went to the movies, matinees at Carib, Palace, and Rialto theatres; every time there was a boxing match in town, we went to the Race Course to see the fight—ring-side seats. I learned about table saws and carpentry while sharing woodwork projects; when I was in high school, he invited me to Liguanea Club for tennis matches, and at home, we spent hours playing table tennis matches on the back verandah.

Paco also learned to fly. He and Paul Scott took flying lessons out of Tinson Pen with the idea that they could travel more easily to Westmoreland and other parishes, as needed, for business, but Dad never agreed to purchase or lease a company plane.

Paco was compelled to work doubly hard, and within a short space of time, his efforts paid off. He demonstrated that he had the makings of a business leader, capable of succeeding on his own merits. He inspired the confidence of his fellow-workers and

---

5   Personal correspondence with Bill Valliant, December 7, 2020.

6   Jolly Joseph, colloquial name for company and its buses, Jamaica Omnibus Service, founded in 1953.

customers and was by nature and inclination a Grace, Kennedy man, determined to excel. Dad rewarded him with transfers of bonus shares (1967) and managerial appointments. He moved Paco around a lot within the organization so that he would become familiar with every aspect of the business.

Dad wanted what he thought was best for Paco—success in the family business. It was a situation of "tough love," and conflicts did arise. Paco was not completely free to be his own self; he felt constrained by normative obligations of his duties as the boss's eldest child. Yet despite the complex relationship, they shared a mutual admiration: Dad thought Paco capable of any task, and Paco, beyond the shadow of a doubt, loved his father, worshipped him as a hero. Both were goal-oriented, driven to succeed for the good of Grace, Kennedy, but they remained essentially different personalities.

Mom was concerned about the pace at which Paco pushed himself. "I kept telling him he burned the candle at both ends, but like his father, he did not listen." Her worries were not ill-founded. He suffered a heart attack in 1971 while serving as manager of Western Meat Packers in Westmoreland. With a strong constitution and a strict convalescence routine that included rest and daily exercises, he recovered quickly.

It was during this period of recuperation that he first met Marjory Johnston. On doctor's orders, he worked half days at the office. Marjory recalls:

> We used to meet at a place called Paul's 104 on Harbour Street. Sometimes he would have his Grace, Kennedy colleagues join him, but other times, it would be just the two of us. He'd buy me lunch and, you know, we would just chat. And we did that for over a year. Never took the relationship further than lunch. But I mean we talked about every subject that you could imagine. And then one day, out of the blue, he asked me on a date. He said, "Well, can I buy you a nice dinner somewhere?" So I said, "All right, fine." So it was dinner at Swiss Chalet.[7]

Paco had only ever seen Marjory either in casual wear or in her Pan Am uniform. For the special evening out, she bought a designer dress. "He was completely knocked off his feet when he saw me; he literally rolled backwards" (MJ). There was no turning back after that. "We started to date, and we must have gone out on I don't know how many dates before he even attempted to kiss me" (MJ). Before she knew it, he was pressuring her to get married. "I said, What? I mean that was something I had not even thought of. My brain hadn't gone there" (MJ).

Marjory needed space, and he respected this. She travelled to South America with her mother to clear her head.

---

7   Interview with Marjory Kennedy, April 5, 2021. Paco often called her MJ, initials for Marjory Johnston. Quotations from the interview are labelled, MJ.

Paco never said anything more to me when I got back. I'd call and we went out, but certainly he never brought up the topic. I was getting nervous because by this time I'm thinking we had a future together, and so one night I said to him, "Remember that question that you asked me?"

And he said, "What question?"

I was flabbergasted. I lost my nerve. "When you asked me to marry you."

So he said, "What's your answer?

I said, "Yes." And that was that. (MJ)

Paco made a formal visit to the Johnston family home to ask for her hand in marriage. Ernest Johnston gave his blessing. "He was thrilled. He had not known Paco personally but had worked with your dad in business, Kingston Wharves; he knew of the historical relationship between the two families and knew of the relationship that Paco had with his father" (MJ).

Shy as he was, Paco found his soul mate in life. "Our values were the same, we were good friends, and we both came from a similar business background—shipping. I could not have found a better husband. I was very, very lucky. It wasn't all smooth sailing, but he sure was good to me, you know" (MJ).

In the months following their wedding, Dad arranged for Paco to be assigned to the Montreal office so that he would gain firsthand experience with the international side of the business. During his one-year secondment, he enrolled in a fourteen-week course in business administration at Harvard University. A year after his return to Jamaica, Carlton Alexander recommended to the board of directors that Paco be appointed deputy group managing director "with immediate effect."[8] The motion was passed with unanimous approval.

For the following four years, Paco became Carlton's understudy. He worked closely with Raf Diaz, Mable Tenn, and Bruce Rickards on management projects, and after Dad retired in 1976, assumed the role of chair of directors' board meetings and of Extraordinary General Meetings of shareholders when Mr. Diaz and Mr. Alexander were absent. When Mr. Alexander went on leave in June 1978, he placed Mable Tenn and Paco in charge as co-managers of Grace, Kennedy, to report to Mr. Raf Diaz who was acting chairman of the company.[9]

Paco's public profile became more prominent. As members of the Jamaica Export Association, he and Mable Tenn (director of Products Division) were spokespersons

---

8   Grace, Kennedy & Co. Ltd. (April 24, 1975).

9   Francis Kennedy, appointed chair of directors' board meetings, October 27 and November 8, 1977, March 1978 and of EGM, December 29, 1978.

at Export Awards functions sponsored by the company. "Grace, Kennedy believes in exports and Jamaica has tremendous potential, but much more needs to be done to get out and put the Jamaican name on the market."[10] The company explored new markets in the Middle East, and with the help of the World Trade Institute, embarked on exportation of canned juices to the USA market. As deputy managing director, Paco was invited to speak at civic functions, at which he made public announcements about the current state of affairs in Jamaica. "There is a breakdown in discipline that is affecting every aspect of life in Jamaica; change must start in the home. One of the by-products of good discipline is high productivity ... as a way of creating economic wealth and development for the country."[11]

His business career became his identity. There were no substitutes for Grace; nothing could compare. His son Charles recalls an incident from his own childhood:

> One time I remember there was no Grace Ketchup in the supermarket. So I bought what I thought was the next best brand, _____. I needed ketchup. I think it was for cooking some jerk chicken. So seeing no Grace, I bought the other brand. When I got home, I put the bottle on the kitchen counter, and when I came back the next day, I am looking for this ketchup, and my mommy said to me, "Your father must have found it and thrown it in the garbage." He later cussed some bad word and told me I should never buy any other brand of ketchup, no matter what.[12]

His daughter Cathrine observed in an interview: "He was very humble. Everyone knew he had the best interests of the business at heart, that he wanted to preserve the harmony of the group. 'My individual interests are not what are important,' he would say. He didn't feel entitled, not entitled to the accolades, not entitled to the recognition, or to holding a special position within the company" (CLK).[13]

Paco and Marjory decided to migrate in 1979. He held a top managerial position as divisional director of the Merchandise Division, and they had a young family: Cathrine was turning four, and another baby was on the way. He hoped the move to Miami would be temporary, but for the sake of his own safety and the well-being of his young family, it seemed the right choice at the time. Marjory recalls:

---

10   *The Sunday Gleaner*, March 7, 1976, 19.

11   Address to Boy Scouts and Girl Guides Association, *The Daily Gleaner,* March 5, 1976, 19.

12   Interview with Charles David Kennedy, May 6, 2021. Initials CDK are used to cite quotations.

13   Interview with Cathrine Lesa Kennedy April 27, 2021. CLK initials are used for some of her direct quotations.

> I remember, it was '79. I was pregnant with Charles. Bad times came; Paco was caught downtown hiding under a desk in a crossfire. It was the time of the Estrada riots. Estrada[14] was the Cuban Ambassador, you remember. Things weren't quite right then in Jamaica. He was receiving threats, "White man" this, and "white man" that; you know it was very difficult, very, very difficult. But he never settled in the States. He never liked it. He always hankered to come home. He regretted leaving Jamaica. (MJ)

Cathrine recalls those years:

> One of my earliest memories of Dad was when we were living in Florida, and interestingly, they owned a supermarket,[15] and I had lovely memories of him because Mommy and Daddy would often trade off. I remember Daddy reading to me and him cooking for me.
>
> He was reading to me in bed one night, I was four maybe. "Daddy, Daddy, read the story to me." And he said to me, "No, no, you read to me." And actually, I came out and started reading aloud, just like that. For me it was a shock, but the look of surprise on his face said it all. You know, as a child you spend your whole life wanting to make your parents proud, and that was my first moment of feeling, wow, I made my dad proud. (CLK)

Paco had a gift for bringing out the best in others—his colleagues, his wife, his children. "He was the kindest person you'd ever want to meet. He was generous, extremely caring, and respectful as a husband. You know, he encouraged me to do whatever I wanted. And so too with the children" (MJ). "He helped a lot of people very quietly; he helped a lot of people financially, emotionally. A lot of people went to him for help in many ways and he mentored them, within the Group and out" (CLK).

As eldest child and big brother, Paco was in a position to take charge, to look after family matters, financial and otherwise. He managed Mom's and Dad's affairs when they migrated, and when matters arose that needed attention in Jamaica, Paco never refused to help. He encouraged us all, my sisters and me alike, to stay close to Grace, Kennedy, to safeguard our shareholdings and show regular attendance at Annual

---

14   Ulises Estrada served as Cuban ambassador to Jamaica (August 1978–November 1980). The Jamaica Labour Party accused Estrada of interference in Jamaica's internal affairs after he publicly criticized the party. He was declared *persona non grata* and expelled by the JLP in 1980 when Edward Seaga became prime minister. He died at the age of seventy-nine in 2014.

15   Cathrine Kennedy is General Manager of Hi-Lo Food Stores.

General Meetings. He understood the significance of the legacy of a family business and of the need for participation.

Paco was in a hurry to return to Jamaica. As soon as the government changed in 1980,[16] the family packed up and sold the supermarket. When he returned to Grace, Kennedy, he worked harder than ever. His son Charles recalls: "That was the number one thing I knew about him, that he was a man who never slept. You know, he would do his Grace, Kennedy work, and then in later years, work for the government." And his daughter Cathrine remembers: "When I was thirteen or fourteen, I backed Daddy up one day and said, 'You know, I miss having you in my life.' And I think that really shook him up. There was a big change in Daddy's engagement with Charles and me because of that conversation, and he apologized."

Despite the competing business interests of their parents, their children were raised in a household in which they felt secure. "My mom and dad, they were like best friends, always in love, always respectful. They supported one another. 'Stop the smoking, stop the drinking, start eating more healthy.' My mom was always working with him. She cared for him. And he was always encouraging her to do her best in business" (CDK).

Paco's mind never rested. He internalized pressures but found creative ways to express his genius. As he turned his attention to family to regain balance, he found pleasure in building a library of thousands of volumes (he was a voracious reader) and recording movies and collections of every imaginable genre, so large that he set up a card catalogue system by which friends and family could borrow books and tapes. Paco was tone deaf like his father but shared with Mom a deep appreciation for classical music. He spent hours listening to Luciano Pavarotti and made special recordings for Mom.

Paco was proficient in data-analysis and the use of computers—equipping his study with numerous monitors and hard-drives. Mom often said, "He was considered one of the best computer experts on the island." He was a multitasker, a master at performing myriad functions simultaneously: writing emails, messaging and phoning, viewing news and sports reports, and recording his favourite media. Paco had a unique way of using technology to switch from business to personal pursuits, moving seamlessly from one to the other.

Paco's style of management was different than Dad's but no less effective. Charles observes: "He loved dealing with people; he loved being the person behind the scenes. Once you take that high seat, you know, you have to keep your distance from people; you don't get to be on the ground." Paco embodied the best of what he learned from his father: a deep love for the company and a strict adherence to ethical values. But he carved out his own destiny, his own legacy, no less significant than that of his father's or grandfather's.

---

16   See Chapter 45

Paco was popular both within Grace, Kennedy and in wider circles of commerce and government. People were fond of him, and they trusted him. Cathrine speaks of her dad as someone who was an egalitarian: "He was completely accepting of who you were. He treated everyone the same. Whether you were the office bearer or the chairman, he treated every person with the same level of respect. As I became an adult, he was my mentor and very close friend."

Marjory Lesa Johnston with her new father-in-law, May 19, 1973.

(Photo, courtesy of Marjory Kennedy)

Paco (aged 40) with Charles and Cathrine visiting Mom and Dad in London, Ontario, April 1981.
London, Ontario, 1981.

(Photo from family collection)

45

BLOODBATH

(1980)

Gang violence intensified after Paco and Marjory migrated. Civil unrest spread throughout Jamaica at an alarming rate, with garrison lords gaining control.

Warring political posses drew unofficial boundaries within constituencies such as Gold Street,[1] a line that no one dared cross. They fought bloody battles with M16s on the streets of Kingston, names of the dead too numerous for anyone to remember.[2] Scores of party supporters on both sides were killed and others left homeless by the burning of homes in the shantytowns, which became the battlegrounds of the capital city. Security forces were ill-equipped and private citizens, terrified.

The city gangs consisted of youths who had flocked to Kingston to form fraternities of hoodlums. Jobs were unavailable; guns and drugs instead became the sources of livelihood. *"This is the Jamaica we did not often want to talk about, but it was real, and posed the greatest threat to business and to civil order in our society,"* Dad used to say.

By the end of 1979, it was impossible for the Manley government to meet conditions of the International Monetary Fund. Another round of negotiations began in September to prepare a package to stimulate economic growth and strengthen balance

---

1 Gold Street runs north to south from East Queen Street to Port Royal Street in the heart of downtown Kingston, a few blocks from Grace, Kennedy & Co. Ltd. on Harbour Street. It is located within the Kingston Central Constituency, which seat was held by Michael Manley during the 1970s. Within its borders, just east of Gold Street, are JLP strongholds, Rae Town, ruled in 1980 by Chubby and the Southside gang.

2 Mark Kurlansky, "Showdown in Jamaica" in *The New York Times,* November 27, 1988, Section 6, 48.

of payments. With political and social unrest mounting, by January 1980, Jamaica faced stark realities: for economic survival, it needed assistance to reduce the budget deficit by JMD 150 million. The IMF would grant the government a waiver, but the PNP felt trapped. The strategy of cuts in spending and increases in taxes would further destabilize the country, put more Jamaicans out of work, and force the closure of social welfare programs. In March, negotiations broke down. Jamaica's economy kept sliding in a downward spiral, incapable of meeting its foreign debt obligations, which totaled a billion dollars. Government scrambled for funds outside of traditional sources, but they were insufficient.

For Manley, the IMF was largely to blame. The international body used methods and philosophies diametrically opposed to democratic socialism. Its political agenda was to align to the West: reduce the role of the state, establish free markets, and encourage foreign capital ownership.[3] The conflict brought Jamaica to the brink of a social explosion. "The whole IMF approach is a disaster" because it tightened money at the bottom end of society rather than at the top.[4]

JLP supporters blamed Manley, not the IMF, for mismanagement of the economy, and PNP supporters accused Seaga of soliciting the aid of the CIA to destabilize the nation. Graffiti in downtown Kingston cast blame on both sides:

Is

Manley

Fault

CIAGA

Regardless of whose fault it was, the government was backed into a corner with the threat of mob rule intensifying daily—a political climate in which levels of terrorism proved uncontrollable.[5]

Despite periodic attempts by both political parties to reach a peace accord,[6] bloody feuds[7] prevailed and dominated daily life during the days leading up to

---

3   Richard L. Bernal, "The IMF and Class Struggle in Jamaica, 1977-1980" (1984), 65–66.

4   Michael Manley as quoted by Karen DeYoung, "Manley's Rift with IMF Dominates Jamaican Economics" in *The Washington Post*, September 6, 1980.

5   Kareen Felicia Williams, *The Evolution of Political Violence in Jamaica 1940–1980* (2011), 244.

6   *The Daily Gleaner,* Friday, August 1, 1980, front page headlines: "PNP and JLP re-affirm peace pledge."

7   Mr. Roy McGann, deputy national security minister, and his bodyguard were killed in a shootout with the police, October 13, 1980. The PNP accused the JLP of assassinating a minister of government, and the JLP defended its position by saying that Mr. McGann had opened fire on the police. *The Daily Gleaner,* Saturday, October 18, 1980.

the general election of October 30, 1980, the bloodiest in Jamaica's history. Seven hundred murders were reported during the period between February, when Manley announced the election, and October. Even in the face of extreme conditions of social unrest and financial ruin, Manley remained hopeful. He announced before a mammoth crowd at a rally in Sam Sharpe Square, October 5, 1980, "One hundred and fifty thousand strong cannot be wrong." But the people of Jamaica spoke differently through the ballot box.

On October 30, 1980, the People's National Party was defeated in a landslide victory by the Jamaica Labour Party, which won fifty-one of the sixty parliamentary seats. PNP support declined among all groups in Jamaican society: the unemployed, skilled labour, farm labour, and professional and business classes.[8] Key members, P. J. Patterson, Howard Cooke, and Arnold Bertram, lost their seats.

Seaga's campaign messages of the government's mismanagement of the economy and the threat of communism also contributed to Manley's defeat: "The country has come to believe our view … that Manley's ultimate aim is to take Jamaica into the Cuban orbit."[9] Manley had tried, but to no avail, to counter the accusations: "We [Fidel Castro and I] share causes. I don't support his causes … but the relationship has contributed to the cause of strains with the United States of America—there is no question about that."[10]

The JLP victory on October 30 was decisive, with eighty-seven percent voter turnout, largest to date under adult suffrage. Papa Eddie[11] won the general election with over 500,000, or 58.88% of votes cast for the Jamaica Labour Party. He was sworn in as the fifth prime minister of Independent Jamaica on November 1. A decisive first action as PM was to break diplomatic ties with Cuba by expelling Cuban Ambassador Señor Ulises Estrada.[12] A strong signal to Jamaica and the world!

Seaga's claim to pragmatism signaled a new period in Jamaica's development as an independent nation. It eliminated the perception of a communist threat both for Jamaicans and the Western Alliance with aims to attract foreign investment and relax foreign exchange and import controls. Jamaica was open for business. The new prime minister realized, though, that it would be folly to dismantle all programs initiated under a democratic socialist government. "Concessions such as pay equity for women, the 1975 Labour Relations and Industrial Disputes Act,

---

8   Richard Bernal, page 79.

9   BBC, *A Look at Michael Manley's Jamaica in 1980*: https://www.youtube.com/watch?v=6GbOcq4GwHA.

10   Ibid.

11   Edward Seaga was affectionately called Papa Eddie by his supporters.

12   *The Daily Gleaner,* "Flashback: 1980 General Election, Ballot, Blood and Bullets," October 30, 2010.

which had helped to stabilize labour relations, free secondary education, and the National Housing Trust could not simply be withdrawn."[13]

Troubles were not over. The IMF reported serious recessionary pressures[14] during the 1980s, which meant shrinking export markets for developing countries. Supplies of foreign exchange remained tight. In the years ahead, Carlton Alexander and Raf Diaz struggled with government for import licenses and for purchase of foreign exchange. When restrictions were finally lifted by the end of the decade, the Jamaican currency devalued further, and shortages were again the order of the day. Mr. Seaga found it difficult to deliver on the promise to make "money jingle" in the pockets of the people.

The general election of October 30, 1980 re-affirmed the strength of democratic institutions in Jamaica. In the words of Dr. Swithin Wilmot, "We had reached the precipice and halted."[15] The people had voted. For Grace, Kennedy, it meant that "it had survived itself." The threat of nationalization was now over.

With a change of government, many Jamaicans who had migrated during the Manley period returned home. It was too late for Dad, but for those who were younger, like Paco, selling businesses they had acquired abroad and uprooting their young families were risks they were willing to take to repatriate.

---

13 Patrick E. Bryan (2009), 209–210.

14 Norris McDonald, "Machiavelli. 'Puppa Eddie' and an Analysis of JLP's 1980-1989 Mandate." *The Jamaica Gleaner*, June 16, 2019.

15 Interview with Swithin Wilmot (2019).

Dad with me and Amanda at Georgie's and my home in Scarborough, Ontario.

Easter 1980. Amanda, almost three years, holding her favourite toy, Fluffy.

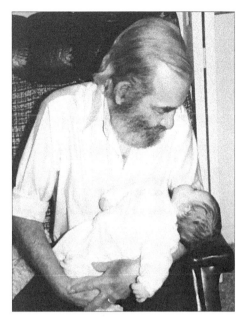

Welcoming Sarah Elise, born March 8, 1981.

Grandkids visiting, London, Ontario, April 1981, for Sarah's christening.
Standing, L–R: Andrew Marc and Ian Michael Cameron; Seated,
L– R: Charles David, Cathrine Lesa and Amanda Kennedy.
Paco in background, enjoying their kids' show on TV.
(Photos from family collection)

46

TWILIGHT YEARS

Grandkids and Diamond Jubilee

(1981–1982)

Dad's third winter in Canada was in 1981. He blamed the cold for arthritic pains, which the doctors claimed were minor compared to his more serious ailments. Yet they caused him extreme discomfort. Amidst all the gloom, grandkids brightened his days and brought him solace. He welcomed new life with a sense of awe, especially his own progeny, who were a beacon of hope.

Mom too was thrilled. She would often joke, *"I was getting worried that I wouldn't get to see any grandchildren. Your father and I are not getting any younger, you know. You children waited too long to start."* They now had six: Cathrine Kennedy (born August 1975); Amanda Kennedy (May 1977); Jan Michael Cameron (July 1977); Andrew Marc Cameron (February 1979); Charles David Kennedy (August 1979); and Sarah Elise Kennedy (March 1981).

When we came to visit, we would find Dad ensconced in his La-Z-Boy. His beard was bushy and hair disheveled, his eyes fatigued from lack of sleep. As difficult as it was for him, he made an effort to raise himself out of the chair to greet us and the kids with warm hugs and kisses. (Many found it odd that Dad kissed family and friends, whether male or female, it did not matter, on both cheeks; he said it was the custom in Spanish America.) He greeted his grandchildren by sitting them up on his lap and making shapes with his fingers:

> Here is the church
> Here is the steeple
> Open the doors and see all the people.

Or he played games by placing grains of rice on his fingers and making them magically disappear and re-appear as he sang the nursery rhyme:

> Two little blackbirds sitting on a wall
> One named Peter, one named Paul.
> Fly away Peter! Fly away Paul!
> Come back Peter! Come back Paul!

Although afraid of his Santa Claus beard, the grandkids loved the ol' man.

Mary and Donald often visited from Ohio with their two children, Jan Michael and Andrew Marc. Georgie and I would drive weekly with Amanda and Sarah from Toronto. (I found employment teaching English as a Second Dialect programs for West Indian immigrants in Scarborough.) Paco and Marjory had returned to Jamaica after their stint in Miami and would come up with their two, Cathrine and Charles, for special family events. One such was Sarah's christening in April 1981, when the three families and Celia and Liz gathered in London. It was a joyous time for Mom and Dad.

Dad's mind was alert enough to converse with adults about business and current affairs, but his concentration span was short. He passed the time listening to world news on the radio and scanning headlines of *The London Free Press* and *The Jamaican Weekly Gleaner: Highlights of the News from Home*.[1] Despite his best efforts, his health declined. Medicines were not curing his ailments or slowing their progression. His sleepless nights took their toll on Mom, who attended to his every need.

Dad was not well enough to conduct family business meetings. He realized that with the change of government, family members would be wanting to liquidate assets in Grace, Kennedy, to move capital abroad. By the end of the 1970s, all Kennedy families, without exception, had migrated either to Canada or the USA. He wanted to facilitate their needs. Unable to manage the negotiations, he asked Carlton Alexander to host a meeting in London, Ontario in 1981. He met with Paco the morning of the family conference. *"I can no longer carry on, son,"* Dad told him. *"You are head of the family now. I want you to represent me at the meeting."*

---

1   *The Jamaican Weekly Gleaner: Highlights from the News at Home*, first published in 1951 as *The Overseas Gleaner*, began out of efforts to keep the public informed after "the worst modern-day strike by a hurricane [Charlie] on the shores of Jamaica."

The transactions were not easy for those who wanted to sell. Foreign exchange for remittance of funds was still difficult to obtain despite Seaga's intention to open the markets. And the amount of capital the Kennedys wished to move was substantial; in December 1979, the family owned 47.5% of the company. Grace, Kennedy fixed a price and made offers of sale through the Employee Investment Trust, which owned twenty percent of shares. José Kennedy declined and retained his holdings, as did Mom and most of Luis Fred Kennedy's family.

DIAMOND JUBILEE [2]

In February 1982, Grace, Kennedy reached another milestone: sixty years of business success and contributions to the welfare of Jamaicans.

> When Grace, Kennedy and Company Limited holds its Long Service Awards Ceremony at the National Arena on Friday of this week, it will be the biggest such event in the country's history.
>
> Over 600 Grace employees will be honoured for service ten years and over. This represents about a third of the Grace staff of 2,000. About 15 employees will also be honoured for service to the company of 30 years or more.[3]

Peter Moss-Solomon chaired the committee for the Sixtieth Anniversary celebrations. Two thanksgiving services were held on the birthday of the company: one in Savnnah-la-Mar for Grace staff residing in the western area of the island, and the other at the National Arena, attended by over 3,000 people.

The board of directors agreed to invite members of the Grace and Kennedy families living abroad and to provide them with complimentary travel and hotel

---

2 Image of Grace logo, courtesy of © GraceKennedy Ltd.

3 *The Daily Gleaner*, February 14, 1982, 16.

accommodations. Dad's children and their families, his siblings, in-laws, nephews, and nieces were all invited.

Paul Bitter was MC of the staff awards ceremony held on Friday, February 19. Mrs. Seaga handed out some of the long-service awards, and Carlton Alexander received a citation in honour of his forty-nine years with the company. Mrs. Carmen Valliant spoke on behalf of the Kennedy family; the vote of thanks was moved by Mr. Bruce Rickards.

Carlton Alexander announced the launch of the Grace, Kennedy Foundation. He said that "it would participate in and support national projects aimed at the development of the human personality, improvement in social and welfare conditions and the furtherance of goals synonymous with the company's basic objectives."[4] Prime Minister Seaga proposed to "examine the possibility of offering the Grace, Kennedy Foundation tax exemption under the amendment to the Income Tax Law."

Carlton also praised Luis Fred Kennedy and the three basic principles of business management that had guided him over the years and "which have been the cornerstone of the company's development":

> Helping people—customers and/or staff in establishing their own businesses.
>
> Benefit schemes—which ensure effective medical, insurance and pension coverage for staff.
>
> Investment—of the major portion of the company's profits into the business to facilitate continued growth and development.[5]

Grace also celebrated the anniversary with staff parties in Kingston, and cocktail parties in different parts of the island to include customers and their communities. Radio commercials and broadcasts featured testimonials of consumers and employees, and local musicians produced songs to commemorate sixty years. The company created a fifteen-minute documentary film and printed brochures and anniversary souvenirs for all staff, and the Grace Sports Club planned a festival for the month of May. It was a Diamond Jubilee to be remembered.

Congratulations came in from local politicians, customers, staff, shareholders, competitors, and foreign suppliers. As part of a thirty-two-page *Sunday Gleaner Supplement*, published February 14, 1982, political leaders offered praise for the company's reputation, the quality of its brand, and its service to the nation:

---

4   Ibid., February 24, 1982, 1.

5   Grace, Kennedy & Co. Ltd., "*The Grace News*, 60th Anniversary: The Chairman pays tribute," April 1982, Vol. 7. No.1, 1.

> The name of Grace is a household name in Jamaica's shopping. Over all these years the company brought a quality of service to their customers that has earned them unqualified respect. (Governor General the Most Hon. Florizel Glasspole)

> Grace has become well known for its corporate citizenship. Its sponsorship of cultural activities, its willingness to spare senior executives to assist the national effort, its involvement in programs to assist to the young and handicapped. These are all characteristics of the way the company has discharged its social responsibility and are the reason it has gained a unique national stature and recognition. (Prime Minister Rt. Hon. Edward P.G. Seaga)

> The company and all associated with it can derive even greater satisfaction from the fact that its name is associated with quality, service to customers and service to the community; for in the end, no enterprise, however clever its business dealings, however vigorous its salesmanship, will survive and grow if it does not take these factors into account. (Leader of the Opposition, Mr. Michael Manley)[6]

Joining in extending best wishes were Mrs. Aves Henriques, President of the Private Organization of Jamaica; Mr. Leslie E. Ashenheim, President of the Jamaican Employers Federation; The ICD Group of Companies; Jamaica National Corporation Group of Companies; National Commercial Bank; Chas E. Ramson Ltd.; Appliance Traders; Lenn Happ Supermarket; the Wray and Nephew Group of Companies; and the Jamaica Biscuit Company—manufacturers of the Excelsior range of biscuits and crackers, for which Grace, Kennedy had been a major distributor since 1920s. A full-page advertisement was published in the supplement with pictures and names of the twenty directors and company secretary, Donovan Anderson.

The Gleaner editorial of March 14, 1982, congratulated Grace, Kennedy & Co. Ltd. for its record of leadership, for helping to "make Jamaica a better place for all of us Jamaicans to live—economically, socially, and with grace."

> There can be no doubt that this firm is one of the business giants of Jamaica. As a commercial and industrial enterprise, it takes second place to none. In merchandising, in manufacturing, in insurance, in travel, in shipping, in multiple operations, it takes its place in the

---

6   *The Daily Gleaner*, February 14, 1982, 2.

forefront of Jamaican economic activity, contributing in large measure to the prosperity of the nation.⁷

"Each of the company's factories has its own laboratory which is concerned not only with the formulation of new products but with improvement to existing product lines."⁸ With a Quality Control Division, "the company ensures that the buyer gets what he pays for—a Grace product of the highest quality." In both 1980 and 1981, the company had won gold and silver medals in the prestigious Monde (World) Selection of Canned Food Products for Calypso Punch, Seville Orange Marmalade, Tomato Ketchup, Quench Aid, and Ortanique Juice.

The Daily Gleaner cited excerpts from the Grace corporate philosophy: "The company believes in creating ideal working conditions, training, and development for its staff."⁹ And "The company believes that the customer is always right, that once a commitment is made to a customer, it must be fulfilled."¹⁰

The company was proud of its community work, its donations to charities and sports, and its summer employment programs. Between 1975 and 1980, Grace, Kennedy had invested $200,000 to provide over 500 summer jobs for students from secondary and technical schools.¹¹ On July 24, 1979, the Grace and Staff Community Development Foundation had been founded to identify and finance community projects with the objective to "Build a Bridge of Human Care and Understanding." It was designed so that financial contributions made by staff in support of programs would be doubled by the company. Care for communities was part of the corporate philosophy:

> We believe that our development programs must fit within the framework of national aspirations and objectives. The community is the source of our skills, raw materials and capital … and ultimately the market of our goods. We accept the responsibilities of citizenship, and as a private enterprise Grace, Kennedy makes contributions in the form of money and gifts and employees' time.¹²

Carlton Alexander called Dad to wish him congratulations on the sixtieth birthday of the company. Dad was not his loquacious and animated self but knew enough to

---

7   Ibid., March 14, 1982, 10.

8   *A Sunday Gleaner Supplement,* February 14, 1982, 14.

9   Ibid.,13–15.

10 Ibid.,14.

11 Interview with Carlton Alexander, "Sixty Years of Grace," *The Daily Gleaner,* March 21, 1982, 28.

12 *A Sunday Gleaner Supplement,* February 14, 1982, 15.

understand the significance of sixty years and was comforted by Carlton's assurances that all would be well.

I went to London the following month. Dad's condition had worsened to the point where his speech was not cogent. I asked Mom if she did not think it was time to get some help, even a nurse.

"He doesn't want anyone else tending to him," Mom said.

"As a family, we've talked about a specialized care facility."

"It may be too late for that, Fred."

"We're worried about you, though, and the toll it's taking on your own health."

"Your dad and I made a promise to each other if it came to this. I know some people think differently, but it's never been the custom in our families. Probably the way we were brought up."

A couple months passed with sleepless nights for them both.

In the second week of May, Mom phoned. "Fred, I needed to call an ambulance. He's at Joseph's Hospital; I think you should come over to London."

Celebrating Sixtieth Anniversary, Grace, Kennedy & Co. Ltd.

S. Carlton Alexander (Chairman), A. Rafael Diaz (Finance Director), Edward Muschett (Managing Director), February 1982.

The Grace News Vol 7, No. 1, page 5, April 1982. (Photos, courtesy of © GraceKennedy Ltd.)

Mrs. Carmen Valliant bringing greetings on behalf of her brother and of the Kennedy family.

47

Adiós

(May 1982)

St. Joseph's Catholic Hospital, London, Ontario. Mom, Celia, and I were at Dad's bedside—comforting him but now with little effect. He had reached the stage where he was slipping in and out of consciousness. We called Paco and Liz in Jamaica and Mary in Ohio to let them know that we had received the fateful doctor conference: "I know this must be a very difficult time for you all, but I feel I must tell you that we have done all that we can medically."

Dad's eyes told me that he wanted to speak, to say that he was sorry for having put Mom through months of hardship. He never wanted to be a burden or nuisance to anyone. He vocalized one wish: he asked to see Mary Lou, his sister Louise's eldest daughter, one last time. She recalls meeting her uncle in May 1982:

> I do remember having a brief visit with your dad in his hospital room in London before he died. He was very weak, had trouble speaking, was getting oxygen ... I remember your mom was also in the room ... I was very touched that he had wanted to see me before he died; I had driven down to London from Toronto ...
>
> It's been a long time, but I do know it was a loving exchange on both our parts. I remember leaning over and hugging him as he lay in his hospital bed, wanting to assure him that I loved him, that he was very special to me, that I was so glad he was my uncle and godfather, and

proud to be his niece. Though he was having trouble speaking, I left assured that the feelings were mutual.

I have always treasured being able to see him that one last time. And especially knowing that he had wanted to see me. Remembering those moments still brings me to tears.[1]

Sunday afternoon, May 16.

Mom, Celia, Paco, and I were with him most of the day and had just retired to the hospital waiting room when the nurse came with the news.

Dad lay quiet, peaceful at last, as if asleep. He had died alone.

I put my arm around Mom's shoulder as we stood by the bedside. She lowered her head as if in prayer and reached for his hand. She breathed a deep sigh but did not cry—*Adiós, mi amorcito*—she whispered. Mom seldom cried, not that she was stoical or that her heart was hardened. On the contrary, she was loving and compassionate but private and guarded with her emotions. "He loved you children; he would have done anything for you. I was even jealous sometimes," she said.[2]

Paco was standing in the corridor when I left the room. He gave me a spontaneous bear hug that opened for me a floodgate of emotions. He remained silent—like Mom, Paco felt deeply but seldom shed a tear.

The hospital agreed to have the body remain in the room for a couple of hours until my sisters Mary and Liz arrived. Dad's eldest sibling, Mim, who was living in Toronto, also had a chance to pay her last respects.

Dad was buried at St. Peter's Catholic Cemetery, May 20, 1982, following a funeral service at the Church of St. John the Divine. Our immediate family was present as well as Dad's brother José and sisters Mim and Louise; sister-in-law Anne Kennedy; niece Mary Lou and nephews Bill and Jim Valliant, Steve Kennedy, and Philip Kennedy. Ten representatives of Grace, Kennedy attended from Jamaica: Carlton Alexander, Mable Tenn, Faustine Sharp,[3] Ed Muschett, Raf Diaz, Everard Cox, Chris Bovell, Peter Moss-Solomon, Donovan Anderson, and Ernest Girod. At the funeral, "young Anderson" greeted my mom with his condolences, "Mrs. Kennedy, your husband was a great, great man, a great human being." To which she replied, "I know, I know, Donovan, thank you."

---

1   Mary Lou Soutar-Hynes, excerpts from an email, June 9, 2021.

2   Lydia M. Kennedy, interview with Georgianne Thompson Kennedy (2002).

3   Faustine Sharp joined the company in 1945, two years before Dr. John Grace retired. One of two female directors of Grace, Kennedy, Faustine Sharp, was appointed credit manager of the Merchandise Division in 1977, a position she held until her retirement in May 1993. She passed in 2007.

Adiós (May 1982)

Business associates from the Montreal office, Consul General of Jamaica for Toronto, Oswald Murray, and Jamaica's trade commissioner to Canada also attended. Former secretaries Alva Russell and Mavis Chang were in attendance, along with CIBC Manager Mr. Jim Bickford, who escorted them in a limousine from Toronto to London.[4]

On June 3, 1982, Paco organized a memorial service that was held at Sts. Peter and Paul Church in Kingston. Mass was officiated by Very Rev. Alwyn Harry S.J., Rev. Kenneth J. Mock Yen, and the Rev. John J. Sullivan S.J. Lessons were read by Paco and Uncle Simon Soutar (who had travelled specially from Toronto), and Carlton Alexander delivered a tribute in remembrance of his friend and associate. Singing of the "Ave Maria" by Miss Theresa Mendes and playing of recessional hymn, "The Strife is O'er," followed. The church was packed.

For several months, condolences poured in from friends and colleagues as well as politicians, such as the Prime Minister of Jamaica, Hon. Edward Seaga, and Leader of the Opposition, Hon. Michael Manley. Mom was heartened by Michael Manley's praises for Luis Fred.

Michael Manley's letter, in a curious way, seemed to Mom a vindication of her husband's life's work.

FRED KENNEDY DIES IN CANADA

MR. LUIS FRED KENNEDY, retired Chairman of Grace, Kennedy & Company Limited, died on Sunday, in London, Ontario, Canada. He was 75 years old.

Mr. Kennedy became Joint Manager of Grace, Kennedy & Company Limited along with the late Mr. James S. Moss-Solomon in 1950. He then succeeded Dr. John Grace as Governing Director, and became Chairman of the Board of Directors.

Mr. Kennedy retired from active service at the end of 1972 but remained a member of the Board until his death.

HIS CONTRIBUTION to the company was the major expansion of the activities of Grace, Kennedy in Commerce, Manufacturing, Insurance and Shipping and particularly the expansion and development of the Kingston Waterfront.

Mr. Kennedy was a pioneer in progressive employment practices. From as early as the 1940s he was convinced that employers should provide as a right to employees, pension, health scheme, and group life insurance, and that these benefits should apply to all.

FRED KENNEDY served on several Government bodies, particularly during World War II, and once acted as a Nominated Member of the Legislative Council.

A passionate advocate the free enterprise system, he was active in the Shipping Association of Jamaica, the Jamaica Chamber of Commerce and other trade associations. He believed that companies should be involved in the organizations which represent them.

He also gave outstanding service to the Roman Catholic Church and educational and charitable causes associated with it.

Mr. Kennedy is survived by his wife Lydia (Tillie), sons Francis (Paco) and Fred, and daughters Celia, Mary and Elizabeth, brother Jose (Joe), sisters Carmen, Louise, Gloria, grandchildren, nieces, nephews, grand nieces, grand nephews, and other relatives.

The funeral service will take place in London, Ontario.

The Daily Gleaner, Tuesday, May 18, 1982, 1.
(Photo, courtesy of The Gleaner Company (Media) Ltd.)

---

4   James Gordon Bickford (1928–2011) had been manager of CIBC in Jamaica. He returned to Canada to head up the bank's International Division, which encompassed countries outside Canada. Dad secured positions with the bank for Mavis Chang and Alva Russell when they migrated to Canada.

Michael Manley
2 Washington Close
Kingston 8, Jamaica

May 18, 1982

Mrs Lydia Kennedy
577 Cranbrook Road
London
Ontario
CANADA   N6K 2Y4

Dear Mrs Kennedy,

My wife and I offer you our sincerest condolences on the sad passing of your husband.

Mr Kennedy made a great contribution to the development of Jamaica, a contribution that can be seen in many aspects of our national life.

I particularly remember his sympathetic and progressive outlook towards the many categories of workers who came within his responsibility. Of course, his service to the Church, to education and to charitable causes was legendary.

I hope you will find it possible to convey these sincere feelings of loss and sympathy to your sons, Francis and Fred and daughters, Celia, Mary and Elizabeth. We know that you will be supported at this sad time by the prayers of your many friends and the great number of admirers which Mr Kennedy won by his quality of sincerity no less than his great personal achievements.

Yours sincerely,

MICHAEL   MANLEY

POSTSCRIPT
(1982—2022)

Lydia Kennedy is seated with her nine grandchildren at her eightieth birthday celebration in Markham, Ontario, 1990.

Standing, back row, L–R: Charles David, Cathrine Lesa, Andrew Marc, Amanda Mary and Jan Michael.

Standing in front, L–R: Julia Claire, Laura Catherine, Sarah Elise, and Claire Marie with her mother, Mary.

(Photo from family collection)

48

PARADOX

Mom spent the first while after Dad's death sorting through and clearing out his personal belongings. She gave me the leather-bound booklet of *A Question* and a few gold accessories: cuff links, belt buckle, and tie pin, which contained my grandfather's initials, FWK.[1] "He wanted you to have these," she said.

Mom devoted much of her leisure time to honing her skills of crochet to make intricately designed gifts of tablecloths for her children, and Christmas stockings for her grandkids. She challenged herself mentally to the most difficult crossword puzzles and kept abreast of business, news, and sports, especially baseball. She continued to show a keen interest in Grace, Kennedy, and Raf Diaz, whom she grew to know and trust like family, kept her fully informed. She increased her shareholding at every opportunity that shares became available for purchase.

Before long, she travelled to spend time with families in Toronto, Toledo, Kingston, and even Paris, where Elizabeth had moved with her daughter, Laura (born October 1982), and husband, John, who had a new assignment at the US Embassy there.

In 1990, when Mom was approaching her eightieth birthday, she decided it would be best to return to Jamaica. After having lived there for forty years, she considered herself a Jamaican. Paco and Marjory assisted with her move, which included a stopover in Miami where she bought new sets of clothes and furniture. She purchased a townhouse in Kingston and a new Toyota Camry, four door sedan, deluxe. She reconnected with friends and with business associates of Dad's and enjoyed ten more years of good health and prosperity.

---

1 Dad had also wished for me to have his father's wedding band, but a couple of years before he died, he lost it down the sink drain when he was washing his hands. He used to wear the ring on his little finger; his father's hands were far slenderer than his.

Although Mom became legally blind due to macular degeneration in her last couple years, her mental faculties were intact. The disadvantage to living to an old age, she used to say, was to see her friends pass before her. (Douglas Judah died at the age of eighty in 1986 and his wife, Mary Horn, eighty-nine, in 1999.)[2]

Lucid to the end, she bade loving farewells to each of her children and grandchildren. Her parting words to me were, "Fred, please tell your girls that I love each of them very much." The same was true of her feelings for all her grandkids. She used to say, *"Love is a curious thing, the more you give of it, the more you have."* Mom's spirituality ran deep. She once said to me, *"I always felt that those who are afraid of death have neither faith in God nor a belief in the afterlife."* She knew she was going to a better place. She said she wanted to come back to visit me, but I told her I did not like duppies. (I regret now that I declined her offer.)

One of her last acts was to summon Raf Diaz to her bedside to ask him to witness the transfer of her shares to her five children. She died peacefully in her ninety-second year, October 4, 2002.

Celia, who had returned from Canada to live with Mom, died six years later from complications related to adult epilepsy. She found it hard to cope without Mom. Celia had lived a life of extreme generosity, continuing to devote much of her time to social work as she had done during her years in the convent.

After returning from Miami, Paco rejoined Grace, Kennedy and worked as head of the Shipping Division and later as project manager for the group. In his retirement (2005), he embraced the mission to restore cultural sites in Kingston and to launch renovation projects for the downtown core by creating private-public sector networks. As chairman of the Kingston City Centre Improvement Company, he worked with the Urban Development Corporation to host New Year's Eve celebrations on the Kingston Waterfront; the events attracted crowds of over 100,000 spectators.[3] Cathrine commented, "After retirement, he was at peace with himself. He was free to do what he wanted and to challenge his genius in the ways that he saw fit."[4]

Heroes Day, October 20, 2014, I accompanied Cathrine, Charles and their mom to King's House for the National Honours and Awards Investiture Ceremony. On behalf of her husband, Marjory Kennedy accepted from Governor General Sir Patrick Allen, the insignia of the Order of Distinction (Commander). Paco was awarded the National Honour for his contribution in the field of business and voluntary service. He was unable to attend for he had been admitted to ICU at University Hospital of the West

---

2   Douglas Judah's obituary appeared in *The Daily Gleaner* of November 1, 1986. The family had migrated to Virginia, USA.

3   New Year's Eve celebrations on the waterfront started as early as 1999/2000 to herald the start of the new century.

4   Interview with Cathrine Kennedy, April 27, 2021. Both she and her brother Charles have two sons each. See Descendant Family Chart in Appendices.

Indies. Marjory shared the joy with Paco by showing him the badge of honour that contained the heraldic Coat of Arms of Jamaica—a magnificent tribute to the work of a true patriot. Paco died six days later. A funeral service was held at Holy Trinity Cathedral with standing room only.

Mary and Donald made successful careers in education and medicine respectfully. Mary received her PhD in education and worked as professor at the University of Findlay in Ohio. Donald specialized in pediatric neurology, opened a medical practice, and was affiliated with multiple hospitals in the Toledo area. Their third child, Claire Marie, was born in 1989. They both enjoy their retirement, yachting, and spending time with their families and two grandchildren.

Liz remarried in 2003 to Trinidadian Dr. Francis Kenneth (Ken) Sealey who, after studying psychiatry at the University of the West Indies, had migrated to Toronto in the 1970s. They enjoyed sixteen years of marriage until, in May 2019, Liz was diagnosed with lung cancer, which quickly metastasized to other parts of her body. She accepted her illness with courage, and remained positive and serene to the very end. Liz died four months later at the age of sixty-five, a passing that was a shock to everyone who knew her. Liz was a vivacious, loving, and caring soul.

> Ya tú sabes, Liz,
> You now know, Liz,
>
> cuánto te queremos.
> how much we love you.
>
> Gracias a tí por todo,
> Thank you for everything,
>
> el amor, los tiempos felices y las risas.
> your love, the happy times and laughter.
>
> Qué viva tu espíritu por toda la eternidad,
> May your spirit live for all eternity,
>
> qué dios te bendiga, mi hermanita,
> may God bless you, my sister,
>
> descansa tú en paz para siempre.
> may you forever Rest in Peace.[5]

Liz lived to see her first grandchild, who arrived in March 2018. Her daughter Laura and husband, Stan, have had a second child since her passing.

---

5   Excerpt from Fred W. Kennedy's tribute to his sister, Elizabeth Sealey, Pasadena, California, September 14, 2019.

Georgie and I remained in Canada and pursued careers in education. We were blessed with our third child, Julia Claire, our home-birth baby, who was born February 03, 1983. Our three girls have given us the joy of three wonderful sons-in-law[6] and seven grandchildren, who all live in the Toronto area.

Of Dad's siblings, only José and his wife returned to Jamaica; they repatriated shortly after Mom took up residence. Louise and Simon remained in Toronto, Gloria and Owen in the US, and Mim joined her daughter in Italy. They have all since passed. Three of my first cousins have died: Jeanne Valliant (2009), Anne Marie Kennedy (2015), and William Plant (2018). The families, except for Dad's and his brother José's, have sold their GraceKennedy shares.

Both homes, Puerto Plata and 45 Lady Musgrave Road, are still in the family. Georgie and I spend time with our children and grandchildren in Discovery Bay, where we enjoy fishing, swimming, gardening, and pure relaxation; it is a place brimming with memories of friends and family. Paco and Marjory purchased the Kingston home shortly after returning to Jamaica in the 1980s, and now, forty years later, Dad's and Mom's great grandsons enjoy the family home.

Dad was not meek in any sense of the word, but he was humble enough to recognize that there was a loftier purpose in life than the pursuit of his own interests. Influenced by his Jesuit education, he pursued social justice and made as priorities in life actions of service and love to others. Yet he was driven to succeed, and this meant putting himself first. Barack Obama, in *A Promised Land,* tells of his own struggles in wanting to be first:

> I recalled a sermon by Dr. Martin Luther King Jr., called "The Drum Instinct." In it, he talks about how, deep down, we all want to be first, celebrated for our greatness; we all "want to lead the parade." He goes on to point out that such selfish impulses can be reconciled by aligning that quest for greatness with more selfless aims. You can be first in service, first in love. For me, it seemed a satisfying way to square the circle when it came to one's baser and higher instincts.[7]

Dad's life was not free of mistakes or of regrets for times of arrogance and self-indulgence, but he rose above these baser instincts. He aspired to be first, knowing it was his calling to be head of the business and of his family and an opportunity for him to do good in the world.

---

6   Sarah married James Patrick Near in 2006, Amanda married Wayne Anthony Williams in 2011, and Julia married Martin Joseph Vaz-Jones in 2012.

7   Barak Obama, *A Promised Land* (2020), 71.

Paradox

He was a paradox, a man of contradictions—conservative because of his Catholic upbringing yet ambitious with a propensity for taking risks. His life was a story of keeping the two in balance. Although fiercely competitive in his drive to succeed, he acknowledged the need to depend on others. At home, Mom tempered his wild side; in business, Uncle Jim had a calming influence on him.

He did not always listen to advice, but if he was wrong, he would be the first to admit it. He could laugh at himself. *"I did some dumb things, like invest in growing rice. I should have realized that Jamaicans hate getting their feet wet; they will not wade in water even if they have their boots on."* The important lesson was that he learned from mistakes—the next step would be bolder and more successful. Allowing himself to be vulnerable, not afraid to admit wrong, not afraid to cry, to hug, made him quintessentially human, a courageous leader and compassionate father.

He grew up in Caribbean society with all its complexities of race, class, and culture. Born into privilege, he benefited from opportunities afforded to "brown" men of that period. But he rejected the colonial values inherent in this status; he aspired instead to create for himself his own identity and for his country a new nationalism. Although he held a position of primacy, he did not abuse it. He was never apologetic for the material wealth he acquired but rather gave thanks that he was able to use it for the benefit of others. He was an egalitarian, who believed in and practised social and racial equality.

His actions were driven by a set of principles. As a philosopher, he was guided by the precepts of classical liberalism, a construct of ideas that advocates civil liberties and economic freedom.[8] He protected the rights and welfare of individuals and fought to establish the place of free enterprise within a mixed economy. He showed no tolerance, absolutely none, for those who violated ethics. The most serious offence was dishonesty, for which he would fire employees.

Did he have regrets? Complicated as the political situation had been in the 1970s, Mom recognized that one of Dad's reasons for migrating was that his "health had broken. The medical specialists who were treating him had left the island and the drugs he was taking were unavailable."[9] Yet his migration worsened his condition. "The hardest decision of his life was to leave Jamaica and his company. We went to live in Canada. All the time he was away, he pined for Jamaica and wanted to return but we were only able to make it down once for the first winter we were away. After that he was unable to travel."[10]

Dad believed that goodness in the world emanated as much from the countless numbers whom history failed to recognize as from the select few it has decorated with honours. He was fond of the closing sentence of George Elliot's *Middlemarch*:

---

8   His philosophy was most akin to that of John Stuart Mill (1806–1873), a nineteenth-century British philosopher.

9   Lydia M. Kennedy, "A Giant of a Man" (1997).

10   Ibid.

> But the effect of her being on those around her was incalculably diffusive: for the growing good of the world is partly dependent on unhistoric acts; and that things are not so ill with you and me as they might have been is half owing to the number who lived faithfully a hidden life, and rest in unvisited tombs.[11]

His own life was anything but hidden, anything but unhistoric. But he recognized the goodness and heroism of those who might have lived quieter, more private lives, and he strove always to empower them.

---

11   George Eliot, *Middlemarch* (1871), 720.

Resting place: St. Peter's Catholic Cemetery, London, Ontario, Canada.

Mom chose the inscription, *Qué amante siempre te adoraré*, a verse from their courtship song, *"Quiéreme Mucho."*

She also had sculptured a marble statue of Mary in honour of Dad's Marian devotion.

She survived Dad by twenty years.

(Photo by author)

A. Rafael Diaz, Chairman/CEO, Grace, Kennedy, 1989-1998.
(Photo, 1986, courtesy of Raf Diaz.)

Douglas Orane succeeded Raf Diaz as Chairman/CEO, Grace, Kennedy & Co. Ltd. in (1998-2011).
(Photo, courtesy of Douglas Orane)

Prime Minister Hon. Andrew Holness opens GraceKennedy's new Corporate Headquarters at 42–56 Harbour Street, May 29, 2019. Greeting him are Group CEO, Hon. Don Wehby, and Chairman Professor Gordon Shirley.
(Photo, courtesy of © Ricardo Makyn)

49

SUCCESSION

> As managers and leaders, we are like runners in a relay race. We took the baton from the previous generation, who themselves saw notable records on their leg. We must not drop it. We must set distinguished records of our own, so that we can pass these on, along with the baton, to the generation that will succeed us.[1]
>
> Douglas R. Orane

Post-1982, the world order changed. The economics of Ronald Reagan and Margaret Thatcher moved countries away from state control of the commanding heights. The fall of the Soviet Union and China's embrace of capitalism further pushed the world towards the increased role of private enterprise. With the twenty-first century came new technologies that drew nations closer together, freed markets, and diminished collectivism, setting the world on a course of globalization.[2]

---

1 Douglas Orane, *The Business of Nation Building* (2016), 13. In 1998, Douglas Orane CD, Hon. LLD became the fifth to hold the position as head of GraceKennedy, appointed CEO in 1995 and chairman/CEO in 1998. He joined the company in 1981. He holds a degree in Mechanical Engineering, Glasgow University, and a master's degree in Business Administration, Harvard University. As President of the PSOJ (1992–94), he achieved national recognition as a leader within the private sector. In 1998 he was appointed an Independent Senator, a position he held for four and a half years. In 2002 he was named Commander of the Order of Distinction (CD) for his contributions to commerce. In 2009 he was appointed to Jamaica's Privy Council. He now devotes his time mentoring the younger generation, in particular, students of Wolmer's Schools, his *alma mater,* in Kingston, Jamaica.

2 Daniel Yergin and Joseph Stanislaw, *Commanding Heights* (2002).

Jamaica kept in step by becoming a leading "globalizer"[3] among developing countries—with successive governments committed to the development of mixed economies, the reduction of tariffs, and opening markets to international trade. In recent years, the country has launched ambitious programs "to stabilize the economy, reduce debt, and fuel growth"[4]—steps that have attracted both local and international investment.

Liberalization of the economy drove Grace, Kennedy to restructure business operations and governance, paving the way to becoming a global consumer group.

Carlton Alexander

A first step towards expansion was going public. In a message printed in *The Daily Gleaner*, September 5, 1986, Carlton Alexander expressed gratitude to the 600 shareholders of the company, many of whom were employees. He thanked his customers, acknowledged the contributions of the founders, Dr. John J. Grace and Fred W. Kennedy, and Luis Fred Kennedy and James S. Moss-Solomon Sr, who were joint managing directors when he joined Grace, Kennedy in May 1933:

> Today marks a milestone in the history of the Company. As Chairman, I write to express on behalf of my fellow directors and members of staff our thanks to all our customers and the public at large who have supported the company over the last 64 years. This has made it possible for the Company to have grown from its small beginnings to the size it is today.[5]
>
> S. Carlton Alexander.

It was a momentous step for the company, one that would unlock the control of ownership and create visibility for the corporation. The public would now have a stake in a Jamaican enterprise known for its service to employees, customers, and the community. Carlton Alexander, Rafael Diaz, and senior management successfully launched Grace, Kennedy as a publicly listed company, thirteen years after Chairman Luis Fred Kennedy had first put forward the motion in August 1973 to the board of directors. "Take my company public," Mr. Kennedy used to tell Carlton Alexander and Raf Diaz.

Carlton Alexander's public profile increased during the 1980s. He was recognized for his contributions to Jamaica, not only as chairman and CEO of Grace, Kennedy

---

3   Jamaica ranked in the top one-third of a group of seventy-two developing countries in terms of the increase in trade relative to GDP. World Bank, *The Road to Sustained Growth in Jamaica* (2004), 34. https://elibrary.worldbank.org/doi/abs/10.1596/0-8213-5826-X.

4   The World Bank: https://www.worldbank.org/en/country/jamaica/overview.

5   *The Daily Gleaner*, Friday, September 5, 1986, 19. A four-page spread was printed to include a message by the chairman, the forty subsidiaries, and associated companies of Grace, Kennedy, pictures and names of directors and senior management, and the Profit and Loss Account and Group Balance Sheet for 1985.

Succession

but for his directorship/chairmanship of private/public sector organizations: Jamaica Chamber of Commerce, Private Sector Organization of Jamaica, Council of Voluntary Social Services, Jamaica College, Jamaica Promotions Corporation (JAMPRO), and Jamaica National Export Corporation.

Carlton received Jamaica's highest honour, the Order of Jamaica, in 1983 and an honorary Doctor of Laws from the University of the West Indies in 1986, on which occasion he was described as "an exemplary exponent of the virtues of free enterprise, tireless worker for the general good, patriot as in deed as in word."[6]

With his growing stature came the enhanced role of Grace, Kennedy. Carlton understood the vital importance of private capital to national development and had likewise re-oriented the operations of the company towards manufacturing and exports. He sponsored sports and cultural events, involved the staff in community development, and granted scholarships through the Grace, Kennedy Foundation.

In the early months of 1989, he knew that he had only a short time left. "I would like you to take over until such time as you believe your successor is ready,"[7] he told Raf Diaz. Raf had been his deputy for nine years—the person most knowledgeable and capable to assume the position of chairman/CEO. After a battle with cancer, Carlton Alexander died at age seventy-three, May 23, 1989, in the same month and at the same age as his predecessor, Luis Fred Kennedy.

At the board meeting of May 25, 1989, the directors stood in silence to pay their respects to the Hon. S. Carlton Alexander. Mr. A. Rafael Diaz was unanimously appointed Chairman and Chief Executive Officer of Grace, Kennedy. Peter Moss-Solomon[8] was chosen to take Raf's place as finance director, and Edward G. Muschett and Douglas R. Orane became co-managing directors.

Carlton's name is memorialized through student awards offered by his *alma mater*, Jamaica College, through scholarships by Grace, Kennedy Foundation, and through the Carlton Alexander Endowment Chair at the University of the West Indies.

Rafael Diaz

It was a cool October morning in 2019. Relaxed and composed, Raf Diaz was seated in his favourite spot on his verandah, a spacious, high-roofed gazebo facing a lush tropical garden.[9] Laid out on the table before us was an extensive dossier of

---

6   http://www.nlj.gov.jm/BN/Alexander_Selwyn_Carlton/bn_alexander_scm_0007.pdf.

7   Interview with Raf Diaz, July 20, 2021.

8   Peter Moss-Solomon served as chief financial officer of Grace, Kennedy from May 1989 until December 2006, when Fay McIntosh succeeded him. He remained on the board of directors until May 2008.

9   I interviewed Amauri Rafael Diaz at his home in Kingston, October 2, 2019, and had numerous telephone conversations after this in 2020–2022. Excerpts are quoted in this chapter using his initials, ARD.

pictures and remembrances of my dad. He welcomed me and my wife and offered us a glass of homemade lemonade.

"Thank you so much for having us, Raf, and for going to all this work."

"It is my pleasure, Fred. I always thought of myself as part of the family. Your mother and others of the Kennedy family always made me feel welcome."

He spoke with us for hours, sharing a lifetime of memories of his thirty-six years of service with the company.

In 1989 Mr. Diaz inherited the chairmanship of a conglomerate of 1,600 shareholders, 76 subsidiaries and associated companies, 2,000 full time employees, and an annual turnover of JMD 1.5 billion. He brought his unique style of management, distinct from that of Carlton Alexander and Luis Fred Kennedy; his manner was calm and composed, reminiscent of James Moss-Solomon Sr. Many friends and family affectionately call him Uncle Raf.

Raf faced enormous challenges. Jamaica had experienced small margins of growth since 1985 under the trade liberalization policies of the Seaga government.[10] In February 1989, Michael Manley was elected to serve his third term but was unable to rescue Jamaica from the downward trend set in motion by a devaluation of the dollar and an increase in trade deficits. (Unlike in his two previous terms, Manley did not reverse the JLP's pro-business policies.)

Conditions worsened. In the early 1990s, "the economy of Jamaica was collapsing—some banks shut down, the insurance sector crashed, and the hotels went under" (ARD). The government could no longer guarantee Grace, Kennedy a fixed rate of foreign exchange to purchase its goods from abroad. "I think in my first year, 1990, we lost a million dollars due to changes in foreign exchange rates" (ARD).

When Michael Manley lifted foreign exchange controls in 1992, the natural consequence occurred: the flight of capital, exacerbated by the fact that the Bank of Jamaica lacked the net international reserves to cope with demand.[11] Manley resigned later in 1992 because of ill-health,[12] and P. J. Patterson succeeded him as prime minister, an office he held for fourteen years.

"A prolonged period of economic stagnation became entrenched for the entire decade and beyond."[13] Although the Patterson years led to increased investment in tourism, mining, technology, education, and energy, and ended Jamaica's eighteen-year-long borrowing relationship with the IMF, adverse conditions prevailed. Depreciation

---

10   Kim Namsuk and Marta Serra-Garcia, *Economic Crises, Health and Education in Jamaica* (2010), 108.

11   Ibid. The reserve balance was in deficit by USD 372M.

12   Michael Manley died, March 6, 1997, and is interred at National Heroes Park, where his father, Norman Manley is also buried.

13   Namsuk and Serra-Garcia (2010).

in the rate of exchange led to unemployment rates of 16%, hyper-inflation peaking at 68.8%, and average loan rates at 49.6% in 1993.[14]

Grace, Kennedy scrambled for foreign exchange: "We went to our suppliers, we went to New Zealand, we went to Australia, and we went to Terfloth, who was able to borrow on the strength of the products that he was selling for us" (ARD). The good name and credit rating of the company rescued it once more from demise. "Our suppliers trusted Grace, Kennedy; they knew we would pay our bills" (ARD).

Yet freeing of the markets was a blessing in disguise. "It meant that we could buy what we wanted, we could import what we wanted, the government would no longer issue quotas, they would no longer have control over foreign exchange—a time that your father, Luis Fred, always dreamed would happen" (ARD).

Raf Diaz seized opportunities to strengthen Grace, Kennedy's position. "Our goal must be to become net earners of foreign exchange, to transform Grace from a net user to a net earner. We will move the company into global marketing and international business development" (ARD).

Under his watch came an internal re-organization of management,[15] the purchase of fifty percent ownership of Terfloth Group of Companies operating in North America, Britain, and Europe;[16] ventures into the remittance business with the formation of Grace, Kennedy Remittance Services Ltd. (1990); and the establishment of new international subsidiaries: Grace, Kennedy Belize Ltd., Grace, Kennedy Remittance Guyana Ltd. and Grace, Kennedy Ontario Inc.

Investment in remittance services changed the direction of Grace, Kennedy by creating a conduit for the direct transfer of foreign exchange into Jamaica.

> It must have been a guardian angel. The agent for Western Union had been NCB. It was Paul Bitter; I must give him credit. He had heard that they were looking for a new agent. Here was an opportunity for foreign exchange to be earned. I told him to go about it quickly. So we

---

14   https://boj.org.jm/uploads/pdf/papers_pamphlets/papers_pamphlets_commercial_bank_interest_rate_spreads_in_jamaica.pdf.

15   Chairman and CEO, A. Rafael Diaz oversaw all operations with a special watch on international trading. Robert McDonald was named Manager of Overseas Operations; Cecil Ho, Manager of Export Trading Co.; and Douglas Orane, Group Managing Director responsible for the Trading Division comprised of twenty-eight subsidiaries. Mike Belcher, Francis Kennedy, Ernest Girod and George Phillip, directors in charge of Transportation Division; Peter Moss-Solomon, Group Finances; Paul Bitter, Financial Services Division; and Anthony Barnes, director, Information Division.

16   Raf Diaz and Douglas Orane in 1989 successfully negotiated with Boerries Terfloth to have Terfloth and Kennedy (Bermuda), which was owned 50/50 by Grace, Kennedy and B. Terfloth & Co. Ltd., acquire 100% shareholding in the Terfloth Group of Companies. This fully restored 50% ownership of the original Terfloth and Kennedy companies of the early 1970s. Douglas Hall (1992), 153.

became their agent, and we gradually developed remittance services. We would fund transactions from monies in the bank, and at the end of the month, they would send us a cheque in US dollars. We became traders in foreign exchange. We were then able to sell to Carreras, to Musson, and to Matalon because we had surpluses. (ARD)

Raf achieved what he wanted for the company: "We became net earners of foreign exchange, and we still are today. If that had not happened, I don't know where Grace, Kennedy would be today" (ARD).

As a means of strengthening the Finance Division, Raf expanded the business into banking. Grace, Kennedy acquired shares in Trafalgar Commercial Bank. In 1997 it owned forty-nine percent and took controlling interests five years later under the new name, First Global Bank. In his modest, matter-of-fact way, Raf reminded me that he was following through with the wishes of Luis Fred Kennedy: "Your father always wanted to own a bank. I remember Mr. Kennedy saying, 'My ambition is to own a bank.'"

Raf is credited with publishing, *Philosophies & Policies of Grace, Kennedy & Company Ltd.* (1993). "When I was at Grace, I got all the philosophies and values of the Company that Mr. Kennedy had established. I posted them on the walls of my office; I produced a booklet and expanded on his ideas" (ARD). His work evolved into the current *Code of Ethics and Business Conduct*. And in keeping with the "We Care" mantra of the company, to commemorate its seventieth anniversary, he established the James S. Moss-Solomon Sr. Chair in Environmental Management and the Luis Fred Kennedy Foundation at the University of the West Indies, Mona.

Raf's contribution to nation building did not go unnoticed.[17] In recognition of his service to business, he was inducted into the Private Sector Hall of Fame in 1995,[18] and the Government of Jamaica honoured Mr. Diaz in 2005 with the Order of Distinction (Commander Class) for outstanding service in the field of commerce.

GraceKennedy

---

[17] Like former leaders, Raf Diaz set a personal example by volunteering with organizations: Addiction Alert Organization (AAO) and The University of the West Indies Development and Endowment Fund. He was honorary vice-president of the Young Men's Christian Association (YMCA) and chairman of the Jamaica Maritime Institute Trust Fund.

[18] "Mr. Amauri Rafael Diaz, CD, is a distinguished leader, who has established himself within the Jamaican business community as a man of great vision, fortitude, integrity, and loyalty. He is a celebrated expert in the field of finance with a proven track record for excellence." *Wealth Magazine*. Issue # 33. *25th Anniversary Hall of Fame*. Raf Diaz was inducted into the PSOJ Hall of Fame in 1995.

Douglas Orane

> I owe a big debt of gratitude to Raf because he was very supportive in helping to guide me in terms of my development as a businessperson. I have a lot of respect for him, he helped me tremendously. He recognized also how important succession planning was, the continuity of the organization. I am a huge fan of succession planning.[19]

As I sat at a small, wrought-iron table in a garden patio at Doug Orane's home in Acadia, St. Andrew, I could not help but reflect on the immense responsibility this man, like others before him, had shouldered as CEO of GraceKennedy. Before us on the table were two books: *Grace, Kennedy: A Story of a Jamaican Enterprise* by Douglas Hall and *The Business of Nation Building,* a compilation Doug had published of excerpts from his own speeches. Relevant pages were bookmarked and careful notations made. "I never met your father, but I have read so much about the history of the company, I almost feel as if I had known him" (DRO). For the interview, and for most other occasions in life, Doug is known for his preparedness. He is deeply reflective and astute at painting the big picture and articulating his vision with clarity and feeling.

The board appointed Douglas Orane as CEO in 1995, and Raf Diaz remained on as executive chairman for three more years, at which time, Mr. Orane became Chairman and CEO of Grace, Kennedy. Raf continued to serve as director until 2005, when at the age of seventy-five, he terminated his contract with Grace, Kennedy, thirty-six years after Carlton Alexander had asked him to move next door from Myrtle Bank Hotel to be Peter Moss-Solomon's assistant.

Passing of the baton coincided with the opening of new corporate headquarters at 73 Harbour Street.[20] "The ground had started to shift under our feet" (DRO). "It was time," said Raf, "to hand over the leadership, to have young people run with things that were coming; Grace, Kennedy had already started investing in technology" (ARD).[21]

Doug's entrepreneurial nature spurred him on to immediate action. "Severe adversity in Jamaica caused us to rethink the entire future of our company."[22] When

---

19   Interview with Douglas Orane took place at his home, February 27, 2019. After the onset of COVID-19, we continued conversations by telephone (2020-2022). Excerpts are quoted in this chapter, marked by his initials, DRO.

20   Paco Kennedy, then Project Manager, oversaw the planning, design, and construction of the new corporate headquarters at 73 Harbour Street. Paco also had a hand in the design of the new company logo (2005) with the name GraceKennedy and crown insignia, which is used as a section divider in this chapter. Courtesy of © GraceKennedy Ltd.

21   In 1990, Grace, Kennedy invested in Webb Terrelonge Gibb (WTG), which acquired APTEC from the ICD Group, re-named WTG-APTEC Systems Ltd. It also entered into a joint venture with Unisys Corporation forming Grace Unisys (Ja.) Ltd. to sell computer hardware.

22   Douglas Orane, *The Business of Nation Building* (2016), 25.

asked what characterized his tenure as chairman/CEO of GraceKennedy, Douglas Orane replied:

> I would say the conceptualization, the launch, and the implementation of the 2020 vision. The 2020 vision we launched in January 1996 at our annual business conference was a vision of what we wanted Grace, Kennedy to be in 25 years: to convert Grace, Kennedy from being a domestic conglomerate to being a global consumer group.

It was a vision born out of necessity, a hedge against extreme dangers the company faced in the '70s, '80s, and '90s. "We must never find ourselves in that position again. The aim was to own more companies outside of Jamaica, ones that make profits and pay dividends with the objective of generating fifty percent of profits outside of Jamaica by 2020. Your father laid the foundation for this" (DRO). (Doug was referring to the reputation of the Grace brand and the foreign subsidiaries established in Montreal, Rotterdam, and UK in the 1950s and '60s.)

The 2020 vision was based on the value of a diaspora market of millions of Jamaicans worldwide and on the recognition that Grace, Kennedy needed to become globally competitive. To double productivity, management needed to be innovative and entrepreneurial. "Like Carlton Alexander, I became obsessed with customer service" (DRO).

To go global, it was necessary to re-acquire full ownership of the Grace brand overseas. "In some of the toughest bargaining sessions in our experience we reached agreement at 11:45 p.m. on October 31, 1997."[23] Orane and team sealed a deal for US$3 million to regain full ownership of the international trademark rights of the Grace brand. The company became free to position itself as the premier brand of choice for the Caribbean diaspora wherever they lived in the world. January 1998 marked the re-launch of the Grace brand, "the true beginning of international expansion" (DRO). It was the fulfilment of a promise made to Luis Fred Kennedy and Carlton Alexander "to continue negotiations with Terfloth to buy back the brand" (ARD).

The company changed from being a conglomerate that held controlling stakes in smaller, multi-industry entities, to one that strengthened its core business. The shift was a function of a changed economic climate:

> Your father was absolutely correct for the era he lived in, when Jamaica was a closed economy. You could not import what you wanted, you needed licenses, you needed foreign exchange approval to pay for your goods abroad. If you created a business locally, you could not readily replicate it elsewhere in the world. So you diversified. But once Jamaica opened up as we are now, we needed to do things differently. We had to adjust. But your father was a visionary, he had created the

---

23   Ibid. 29.

foundation and methodology for Grace, Kennedy to become a multi-national corporation. (DRO)

Transitioning meant divestiture, the sale of subsidiaries and associated companies (even if they were profitable), to focus on building food brands globally and financial services regionally.[24] Divesting of interests in Kingston Wharves Ltd. in January 2004 was one step in this direction.

With Kingston Wharves' relocation to Newport West in the 1960s, the company expanded its operations to become a major container port for transshipment of goods in the Caribbean. To increase its efficiency, Mr. Diaz negotiated a merger with Western Terminals in 1994 to assume operation of its four berths by occupying all nine deep-water berths. As a result of the merger, Grace, Kennedy's shareholding of fifty-five percent of the issued ordinary shares was reduced to forty-four percent; it remained the largest shareholder and holder of the management contract of the wharves. The following year, KWL was listed on the Jamaica Stock Exchange—showing profits of JMD 57M with assets of 700.3M.

Grace, Kennedy acquired an additional seven percent of stock but was forced by a court decision to make an offer for purchase of the entire shareholding of the company or reduce its share to <50% by selling seven percent of its stock. It decided on the latter. Minority shareholders bought up publicly listed shares to gain more seats on the board. With a prolonged legal battle and the loss of its management contract, Grace, Kennedy decided on full divestment of GK stock. January 21, 2004, the company sold its remaining 43.9% to National Commercial Bank. Following the sale, Mr. Orane made the announcement: "We are pleased that the sale has now been completed and happy that the new shareholder is a strong, well-capitalised and well-managed Jamaican Company."[25]

Raf Diaz commented:

> It was a blessing in disguise. It was no longer feasible for Grace to build a new port, renovate it, to invest in new machinery, to buy lands, build buildings. It would have been such a burden on Grace, Kennedy that it would have crashed Grace, Kennedy. We sold our shares to NCB and we got a fair price. We got back our money.[26]

---

24 In 2006, GraceKennedy introduced a new corporate structure, reducing the number of divisions to two business segments: GK Foods and GK Investments.

25 JSE Posted January 21, 2004. "The transaction was completed across the Jamaica Stock Exchange. Some 470 million shares were sold at $1.30 each for a transaction totalling approximately J$610 million."

26 Interview with Raf Diaz, July 20, 2021.

A break with tradition? Most certainly! Wharves and shipping had been an integral part of the operations of Grace, Kennedy since its formation. Founders Fred William Kennedy and Dr. John Grace acquired the Grace Wharf in the 1920s, Luis Fred Kennedy formed Kingston Wharves Ltd. in the 1940s, and Carlton Alexander and Raf Diaz succeeded him as chairmen. Was it fortuitous? Yes! Full disposal of shares in Kingston Wharves gave Grace, Kennedy the opportunity to maximize its own value.

As part of his vision to modernize Grace, Kennedy, Douglas re-designed the structure of the board of directors in the early years of 2000.

> I give credit to Chris Bovell who headed the governance committee; he studied models in the UK, Australia and New Zealand and made recommendations. We streamlined the Board by reducing the number of directors from twenty-four to twelve—we retained six external directors and made internal appointments based on managerial positions of the major divisions of the company … Decisions were not easy; I know many were upset. (DRO)

The face of the company changed. The only remaining member of the Kennedy, Moss-Solomon, and Alexander families on the board of directors in January 2002 was CFO Peter N. Moss-Solomon.[27] It was a difficult transition for many who wanted to see the continuity of family names, yet for others, it represented an inevitable step in the evolution of Grace, Kennedy.

Under Doug Orane's chairmanship, board appointments of non-executive directors also supported a vision of international diversity: Tom Craig of USA (2002),[28] G. Raymond Chang[29] of Canada (2004), LeRoy Bookal of USA (2005),[30] Joe Esau of Trinidad and Tobago (2006),[31] and Mary Anne Chambers[32] of Canada (2011).

---

27   Grace, Kennedy & Co. Ltd., minutes of meeting of board of directors, January 31, 2002.

28   Tom Craig was a founding partner of Monitor Group (now Monitor Deloitte). He is active with organizations focused on education and economic opportunities in disadvantaged communities.

29   G. Raymond Chang (1948–2014) was then chairman of the board of trustees of C.I. Financial, Canada and a resident of Canada. Native son of Jamaica, graduate of STGC, he migrated to Toronto in 1967 to pursue studies in engineering; renowned businessman and philanthropist, third chancellor of the Ryerson University (2006–2012), recipient of the Order of Canada and Order of Jamaica. (Ryerson changed its name to Toronto Metropolitan University in 2022.)

30   LeRoy Bookal is a retired Auditor General of the World Bank and retired General Auditor of Texaco Inc.

31   Joe Esau is a former Group CEO of McEnearny Alstons Ltd in Trinidad and Tobago (now the ANSA McAL Group).

32   See note later in chapter.

The embrace of corporate governance resulted in eliminating layers of bureaucracy and establishing more effective internal audit service. The results were increased levels of productivity and transparency as a public corporation for customers and investors. The transformation has been ongoing, begun in the late 1990s and continuing today, a look now different from that of the Alexander era and from that of the earlier days of governance.

"And it worked," as Doug Orane is wont to say. The company weathered the Global Financial Crisis (GFC), the most serious since the Great Depression. In 2011, the net profit attributable to shareholders of GraceKennedy increased by 22.2% over the previous year;[33] other indicators of financial success showed improvement: turnover, return on equity, profit before taxation, productivity per employee, and debt to equity ratios.

GraceKennedy

Donald Wehby

On July 1, 2011, the board appointed Donald George Wehby[34] as CEO to succeed Douglas Orane, who retired as an employee of GraceKennedy after thirty years. (Doug was re-hired to serve as executive chairman for three more years.) "I must say that Doug Orane groomed me extremely well to take that baton, and it's one of the smoothest transitions that I've seen, but it wasn't overnight. It was in the making a long time. Peter Moss-Solomon was my first boss, interviewed me and offered me the position of financial accountant in 1995" (DGW).[35] Don was appointed to the board within two years, and one year later, was promoted to group chief financial officer at the age of thirty-four.

---

33   GraceKennedy Annual Report 2011 https://www.gracekennedy.com/wp-content/uploads/2017/1/GK_Annual_Report_2011.pdf.

34   Donald George Wehby serves as a government senator under the Andrew Holness Administration (reappointed in April 2016 and June 2020). A St. George's College graduate (1980), he later attended the University of the West Indies where he obtained a Bachelor of Science (Hons.) and a Master of Science in Accounting, and Stanford University where he completed an Advanced Management College Certificate. He is a Fellow Chartered Accountant. In 2017 he was conferred with the National Honour of the Order of Distinction (Commander Class) and is a recipient of many other awards for his distinguished service to Jamaica. He served as chairman of JAMPRO and is Honorary Consul to New Zealand. Don is married to Hilary Wehby (née Moss-Solomon) with three children. He is an avid cricket fan and member of the Kingston Cricket Club.

35   Excerpts are quoted, using the initials DGW, from an interview with Don Wehby (March 12, 2021).

Don is a St. George's College graduate, a product of Jesuit education, which is known for its focus on ethical values and service to others. "Like your dad," he told me, "money has never been my driver. What motivates me as a CEO is realizing the vision that we've created, implementing strategies, and developing people to help me implement them. People are our #1 resource" (DGW).

Don Wehby is proud of his achievements as CEO, yet he is not pretentious or boastful. "GraceKennedy is a great company," says Don Wehby, "but remember, Fred, humility is important. Talking about humility, I have great memories of Mr. Paco Kennedy; he was my good friend" (DGW). "He always put the company first and never used his Kennedy name for any personal gain."[36] "He set the tone for young executives like me, and I think that has carried through the culture of GraceKennedy. Family values are most important, and Fred, also remember that even though I'm not classified as such, I'm a family member too" (DGW).[37]

Don has taken a personal interest in the Grace and Staff Community Development Foundation by assuming the position of deputy chairman. His wish is for the staff "to fully appreciate what GraceKennedy is about in terms of giving back to our communities" (DGW). Establishing a first-rate education system is the answer to reducing crime, which, in his view, is the biggest problem facing Jamaica today. "I have set myself a target. I want to be sending 1,000 children to school. These are underprivileged children in vulnerable communities; now we are at 850" (DGW). "My heart," he calls the work of the foundations, the embodiment of the *We Care* ethos of the company.

Don Wehby charged forward like a champion in the face of continuing economic hardships. In 2011 Jamaica "was on the verge of an economic meltdown with no access to international capital markets,"[38] despite a National Debt Exchange (NDX) held the previous year. To stabilize the economy, government was forced to launch a second NDX in 2013.[39] Through these crises, GraceKennedy reported improved performances in all business segments. In 2013 it showed a 23.7% increase over the previous year in profits before taxation.[40]

In keeping with the company's 2020 vision, Don Wehby's team took steps to grow the Grace brand by expanding its international footprint. "We have been aggressive in terms of being a Global Consumer Group, with the acquisition of food companies

---

36   Excerpt from Don Wehby's eulogy at Paco Kennedy's funeral, November 8, 2014.

37   Don and Hilary Wehby's three children are the great grandchildren of James S. Moss-Solomon Sr.

38   "Jamaica and the IMF: The Power of Partnership and Ownership," https://www.imf.org/en/Countries/JAM/jamaica-lending-case-study.

39   Participation by GraceKennedy and other private companies/bondholders helped the government reduce its debt-GDP ratio by 8.5% or JDM 17 billion between 2013 and 2020. (*The Daily Gleaner*, Tuesday February 12, 2013.)

40   GraceKennedy Annual Report 2013: https://www.gracekennedy.com/wp-content/uploads/2017/1/GK-Annual-Report_2013.pdf.

in the USA,[41] and we have also grown successfully in Canada and the U.K" (DGW). GraceKennedy Foods distributes the Grace brand in over forty countries through its wholly owned subsidiaries in Jamaica, Canada, Belize, Ghana, the United States, and the United Kingdom.

Don has developed the Financial Services Division locally and internationally by expanding GK Money Services, which has a network of over 300 locations in Jamaica; banking and investments, which include commercial banking, stock brokerage, corporate finance, and advisory services; and insurance with subsidiaries located in Jamaica and the Eastern Caribbean.[42] He has also set as a priority the development of the IT infrastructure of the group, with an investment over the past five years of USD 25 million.[43]

A proud moment for Mr. Wehby was the opening of the new corporate headquarters on May 29, 2019, at 42-56 Harbour Street, the site of the former Myrtle Bank Hotel:

> This building is so much more than a physical structure. It is really a manifestation of the vision of generations of GraceKennedy leaders, many of whom are with us here today. It shows where GraceKennedy is coming from, and a demonstration of the commitment we made in 1922 that downtown Kingston would always be our home.[44]

Over 150 Jamaicans from communities in downtown Kingston worked on the construction project. "We teamed up … with the community to make this dream a reality … What you are seeing here is a product of Jamaican hands from this community, building the GraceKennedy building" (DGW). It shows that "a relatively small company, by global standards, can be the best in the world."[45]

In recognition of the company as a leader in commerce, the GK Group was awarded the Jamaica Stock Exchange Governor General's Award for Excellence in 2019. The company ranked highest in the areas of "investor relations, corporate disclosures, governance website design, annual reporting, and overall performance on the

---

41   In 2014, GraceKennedy acquired La Fe Foods in the United States.

42   GK Money services is the umbrella brand for Western Union, FX Traders, and Bill Express; banking and investment subsidiaries include First Global Bank, Globe Financial Group, GK Investments, and GK Capital Management. Insurance segments include subsidiaries, Allied Insurance Brokers; GK Insurance; GK Insurance (Eastern Caribbean) Ltd., GK Insurance Brokers; Canopy Insurance Ltd., and Key Insurance Company Ltd.

43   "GraceKennedy taps Tech talent for digital business growth" in *The Jamaica Gleaner*, Friday, May 14, 2021.

44   *The Gleaner*, Friday, June 7, 2019.

45   Don Wehby, GraceKennedy Annual Report 2018: https://www.gracekennedy.com/wp-content/uploads/media-center-reports/GKAR-2018.pdf.

JSE."[46] Extraordinary achievements? Yes! But Don qualifies his success by reminding everyone, "The first thing you need is a great team working with you directly—a great team of executives[47] and employees throughout the GraceKennedy Group who are fully engaged in terms of the vision we have" (DGW). He supports the notion that there exists an indisputable correlation between employee engagement and profit.[48]

"There is a clear vision of where we want to go," says Don. With group revenue exceeding JMD 100 billion and profits, JMD 10 billion, GraceKennedy surpassed performance targets for 2020, fast approaching its goals of revenue earnings of sixty percent and profits of fifty percent to be derived from countries outside of Jamaica. "I am very proud to be leading this great Jamaican company towards our vision of becoming a global consumer group," and by 2030, being "the number one Caribbean brand in the world."[49]

"Jamaica, the greatest country in the world" is one of Don's favourite quotes on social media. He is a Jamaican ambassador, *par excellence*.

Going forward, Don and his team have identified a ten-point strategic plan[50] to drive profitability and to create a "performance driven culture" (DGW). He is keen to concentrate on the big picture, choosing to delegate operations to a select management team. His goal is for young executives to "understand very well and embrace the culture of GraceKennedy so that there is continuity" (DGW). "I am mandated to have a robust succession plan for my position as CEO. I need to present at least three senior executives who are able to succeed me. Douglas Orane took me under his wing and mentored me, and I want to do the same (for others)."[51]

---

46   GraceKennedy Annual Report 2019 https://www.gracekennedy.com/wp-content/uploads/media-center-reports/GK-2019-AnnualReport.pdf.

47   Members of the Executive Committee of GraceKennedy 2021: Don Wehby (Group Chief Executive Officer); Grace Burnett CEO, GK Financial Group); Andrea Coy (CEO, GK International Business); Andrew Messado (Group Chief Financial Officer); Frank James (CEO, GK Foods Domestic Business); Gail Moss-Solomon (General Counsel and Chief Corporate Secretary); Mariame McIntosh Robinson (President and CEO, First Global Bank); Steven Whittingham (Chief Operating Officer, GK Financial Group); and Naomi Holness (Chief Resources Officer).

48   In 2020, GraceKennedy recorded its highest level of employee engagement—a score of 71%, 6% above the global benchmark (GK Annual Report 2020).

49   GraceKennedy Annual Report 2020

50   Ten Point Strategic Plan: Digital Transformation; Mergers and Acquisitions; Invest in our Brands; Working Capital Management; CSR; Operational Efficiency; Margin Management; Compliance and Risk Management; Strong Talent Management; and Market Share. (GraceKennedy Annual General Meeting, May 26, 2021).

51   Al Edwards, "How Don Wehby is positioning GraceKennedy for evolutionary change" In *Jamaica Observer*, Sunday, February 3, 2019.

GraceKennedy

Gordon Shirley

Professor Gordon Shirley's[52] appointment, January 1, 2014, as an independent non-executive chairman of the board, marked a radical step in the modernization of GraceKennedy. The governance and management roles of the company were separated; the CEO would no longer be chairman of the board. The composition of the board has also evolved to become smaller and to represent a larger percentage of independent directors. It now comprises nine members,[53] seven of whom are non-executive directors. "GK was a pioneer in embracing corporate governance practices, which were being adopted in North America and Europe"[54] (GVS).

Gordon Shirley's association with GraceKennedy began when he returned to Jamaica in 1991 after completing his doctorate in business administration at Harvard University and teaching at UCLA. Shortly after taking up positions as head of department of Management Studies at UWI and the Carlton Alexander Chair in Business, he was asked by Rafael Diaz and Mr. Douglas Orane to serve on some of the Grace subsidiary boards. "In 1996, Mr. Diaz invited me to join the main board. It was just something very special, a huge board, all managers and directors. Never before had there been an academic as director" (GVS). When he returned from Washington, where he served as Jamaica's ambassador to the United States, he found a changed governance structure in place. GraceKennedy was looking for a non-executive chairman. In 2014, "my fellow Board members elected me as Board Chair from among the group. It was a real and deep honour" (GVS).

As a former diplomat and academic with international experience, Gordon Shirley was fully qualified. "As the chair of a company like this, you must be open to different

---

52   Gordon Shirley is a graduate of UWI, St. Augustine (BSc, engineering) and Harvard University (MBA, Operations and Finance, and Doctorate in Business Administration and Operations Management). The Order of Jamaica was conferred on him August 6, 2009. He was appointed pro vice chancellor and principal of UWI (2007–2013) and served as chairman of The Port Authority of Jamaica Board of Directors (2013–2016). He is currently president and CEO of the Port Authority of Jamaica and Chairman, GraceKennedy Ltd.

53   In 2021, directors of the Board of GraceKennedy: Gordon Valentine Shirley O.J. (Chairman), Donald George Wehby C.D. (Group Chief Executive Officer), Mary Anne Chambers, Andrew Messado (Group Chief Financial Officer), Peter Moses, Dr. Parris Lyew-Ayee Jr., Gina Phillipps Black; Indianna Minto-Coy, and Peter E. Williams. Gail Moss-Solomon is General Counsel and Chief Corporate Secretary. Initials, GVS, stand for Gordon Valentine Shirley.

54   In 2012, 44% of S&P 500 companies had adopted governance model separating roles of CEO and board chair. https://www.russellreynolds.com/newsroom/splitting-the-ceo-and-chairman-roles-yes-or-no.

perspectives to arrive at decisions, you have to be open to listen to all perspectives because the way people think about things in Canada may be different from United States, may be different from here. We are now a global company" (GVS).

Mary Anne Chambers, Gordon contends, provides this international perspective as a director of the board. She resides in Toronto, Canada and serves as chair of Grace Foods Canada Inc. Mary Anne comments about her association with GraceKennedy:

> It has been very satisfying to be associated with such a highly skilled, innovative, and dedicated team, wherever they happen to be in the organization. Their commitment to the success of the business is anchored by the values of honesty, integrity and trust and matched by an equally strong commitment to serving and strengthening the communities in which they operate. It is those factors that make me so proud of my relationship with the GraceKennedy Group.[55]

Gordon identifies strongly with the values of the corporation. "The Grace *We Care* tradition remains strong. It has evolved to mean Grace, *We Care* for the country, Jamaica; Grace, *We Care* for all communities and for the people we serve. It's care for customers, it's care for our employees, for our shareholders, and it's care more broadly for our suppliers, and for the products we distribute" (GVS).

As chairman, he sees one of his prime responsibilities as guiding the vision. "We see GraceKennedy becoming the number one Caribbean brand in the world; we also see our company being listed on an international stock exchange; we see our company becoming industry leaders for the digital era. We see our company continuing to develop a highly skilled and motivated team. We see ourselves returning a high return on investment to our shareholders" (GVS).

The pandemic has tested the resilience of GraceKennedy. "I have seen through this pandemic the notion that 'we are stronger together' (GVS). It has allowed us to give back to the country in many ways. Last year (2020), we enjoyed the greatest performance, the greatest profits that have been generated. It's absolutely remarkable. You see the magic. You see the essence of your dad and of that generation. The best is yet to come" (GVS).

---

55   Mary Anne Chambers, personal correspondence, 2021. Mary Anne Chambers (Order of Ontario) is recipient of the Governor General of Canada's Meritorious Service Medal for her outstanding community service. She is a former Scotiabank Senior Vice-President, Ontario Minister of Training Colleges and Universities and Minister of Children and Youth Services. She has written her autobiography, *From the Heart* (2022) and has recently been named Chancellor of the University of Guelph, Ontario. (Mary Anne kindly endorsed *Firstborn*.)

GraceKennedy

As GraceKennedy celebrates its centenary, it enters a new phase by embracing a 2030 vision. In an interview with business reporter of *The Jamaica Observer*, Don commented:

> We have created a 2030 vision which sees our revenues growing by about four times – this currently stands at about US$800 million, and our projection is to move it to US$2.5 billion by 2030.
>
> The growth is going to come from further geographic expansions in our major food markets across USA, Canada, and the United Kingdom —and, of course, acquisitions will also play a key strategic role in helping us to get to this very ambitious number.[56]

GraceKennedy has defied the odds. The average age of the 500 largest companies listed on the S&P Index has fallen by 80% in the last eighty years from sixty-seven to fifteen.[57] It survived its hundred years. And prospered. And will likely prosper for another hundred years.

How did the company do it? A stable core has existed at its heart, "an unchanging organizational purpose"[58] —*We Care*. Centenarian companies generally "stabilize their core to safeguard what they stand for. They are "radically traditional … but they also stay fresh and create a continuous flow of new ideas."[59] The balance between tradition and innovation has been a distinguishing feature of GraceKennedy governance.

When Luis Fred Kennedy requested the board to have the company publicly listed, he knew full well that the Kennedy family share of ownership would be reduced, that its majority holdings would likely not be passed down to a third generation.[60] "There are many who made the opposite decision and said, 'I'm going to keep everything tight to my chest and I'm going to remain a private company. I won't open up to professional management.' Many of these companies have disappeared" (DRO).

---

56 *Jamaica Observer,* February 16, 2022. https://www.jamaicaobserver.com/business/ready-for-next-century-gk-anticipates-fourfold-growth-in-coming-years/

57 Alex Hill et al., "How Winning Organizations Last 100 Years," in *Harvard Business Review:* https://hbr.org/2018/9/how-winning-organizations-last-100-years September 27, 2018.

58 Ibid. The authors identified seven celebrated centennials who have outperformed their peers over the last one hundred years.

59 Ibid.

60 Only 40% of family-owned companies in US turn into second-generation businesses, and approximately 13% are passed down successfully to a third generation. The average life span is twenty-four years (familybusinesscenter.com).

Douglas Orane recalls a meeting with Carlton Alexander soon after he joined the Company in 1981.

> Early on, when I joined Grace, Carlton Alexander would get a group of us young people together and say, "Have you ever heard of the names of these companies?" And most of us said no. "When I joined Grace, Kennedy in the 1930s, every one of these was bigger than Grace, Kennedy, and now, they no longer exist." (DRO)

"The world does not owe us a living; the world does not owe GraceKennedy the right to exist. We need to be constantly on the watch to ensure that we keep relevant to our customers and to society" (DRO).

The centre has held strong. The company has preserved a culture rooted in a shared set of values—"honesty, integrity, a belief in free enterprise, and a conviction that all people are created equal" (DRO). Succession, which began with Dr. Grace's appointment of Luis Fred Kennedy as Governing Director in 1947, through to the board's appointment of Don Wehby as Group CEO in 2011, has seen the appointment of leaders who have identified themselves with the core values, placed the interests of the company first, and ensured the preservation of its culture. "I was really building on the shoulders of those who went before me, and now Don and his team are doing the same and that's why we are where we are today" (DRO).

Like competitive athletes, each with his own style, successive leaders of GraceKennedy remained focused, with grit and determination, to pass the baton[61] to the next generation at the end of his leg run.

AD MULTOS ANNOS
A Latinate saying my dad used for sending
congratulations and/or best wishes for a long life.

---

61   Metaphor used by Douglas Orane in *The Business of Nation Building* (2016).

APPENDICES

I
Family Trees

II
Tributes

III
Acknowledgements

IV
Bibliography

V
Index

FIRSTBORN The Life of Luis Fred Kennedy 1908-1982

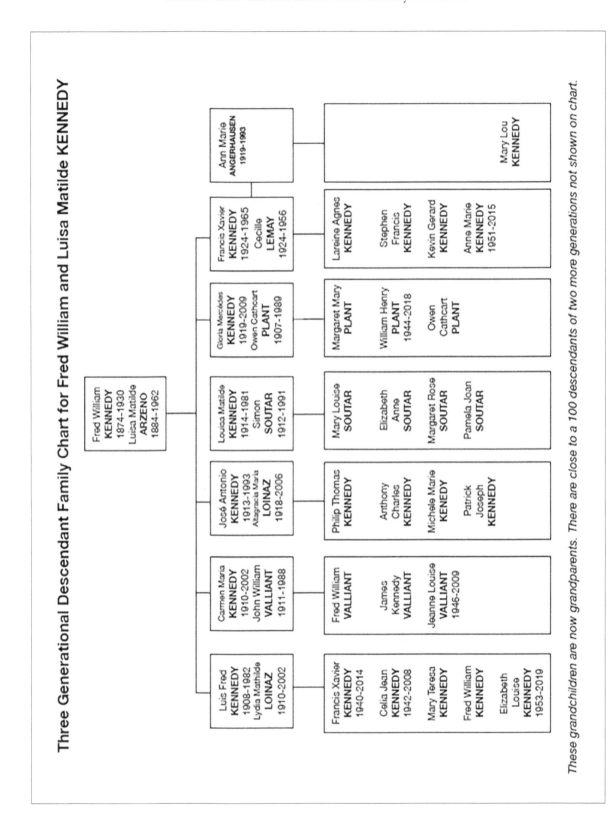

Family Trees

Ancestry Chart showing Common Maternal Grandparents of Luis Fred Kennedy and Lydia Mathilde Loinaz.

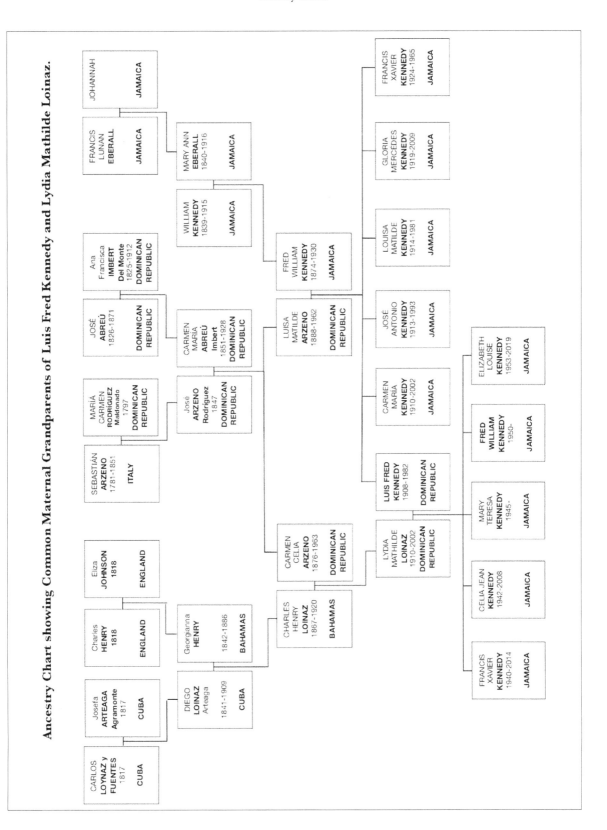

445

DNA Results Summary
Fred William Kennedy
son of Luis Fred Kennedy and Lydia Mathilde Loinaz

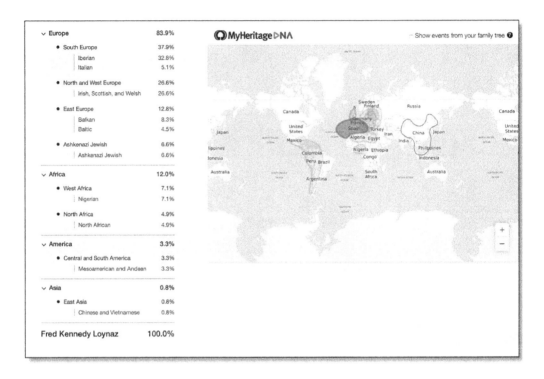

A Caribbean mix: Spanish, English, African and Taíno. DNA results show predominantly Spanish and Italian ancestry (37.9%), genes inherited from both parents. The second dominant grouping is Irish, Scottish, and Welsh (26.6%) from my dad's paternal grandparents and from my mother's paternal grandmother, who was English. Eastern European is 12.8%, and Jewish ancestry, 6.6%, is from the Arzeno line on both sides. African ancestry (12.0%) is exclusively from my paternal grandparents, Kennedy and Eberall, and Taíno (3.3%) from my mother's father's line, Loinaz, which traces back to Cuba.

Information and image, © MyHeritage.

Family Trees

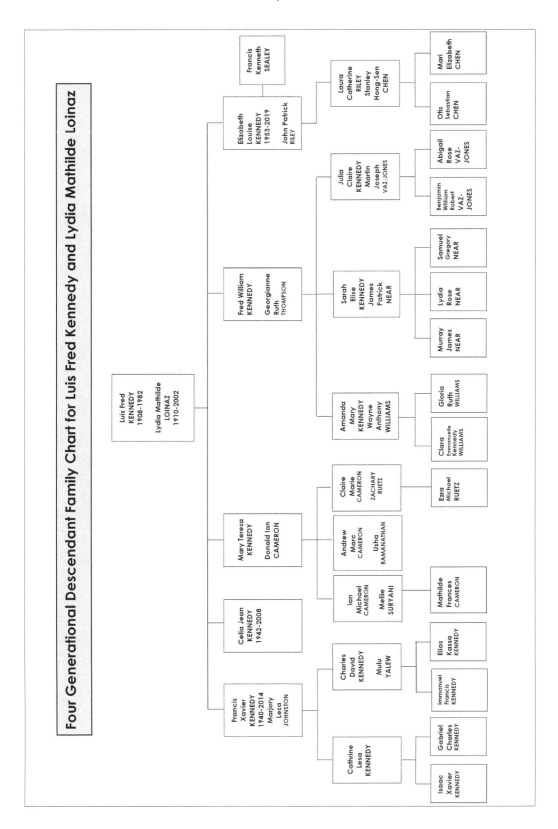

II
Tributes to Luis Fred Kennedy

Thank you to those who shared their tributes that appear in alphabetical order by contributor.

Paul Bitter (former Manager/Director of GraceKennedy)

I was recruited (1969) by Ed Muschett to be a founder (and minority shareholder) of an insurance brokerage to complement their then existing Underwriting Agency for the Jamaica Cooperative Fire & General Ins. Co. through George & Branday Ltd.

On the first week that I joined, I was advised that Mr. Fred Kennedy, Chairman of Grace, wished to interview me in his offices at the then revered offices of Grace, Kennedy on 64 Harbour St. Kingston at 10:00am.

With great trepidation, I attended and was duly ushered into his private office. As expected, he welcomed me, and for the majority of the twenty-minute interview he explored with me the business nature of this new venture and its opportunities as I saw it.

Presumably, when he was satisfied with my responses, he sat back in his chair and handed me a copy of an audited Balance Sheet he had been mulling over when I had first entered and said, "Young Bitter, take a look at this balance sheet, and you will observe that apart from figures, there are NO columns for excuses. I expect you to be guided by this as you grow this new venture into a profitable business."

In that one statement from your dad, he emblazoned in my mind how I was to fashion the running a successful business.

I enjoyed subsequently a few informal meetings with your dad on the verandah at 45 Lady Musgrave Rd. on the occasional Saturday after closing the office at mid-day (all businesses were open half day in those days) when he took an interest in asking me to visit to keep him abreast of how life was treating me.

Donald Cameron (son-in-law)

Dear Dad:

Well, no, this is not a letter from either of your biological sons but from your first son-in-law, Mary's husband. It has been forty-eight years since we first met. It does feel like just yesterday since our afternoon chats on the verandah. I enjoyed the clarity of your thinking, the breadth of your interests, and the uniqueness of your life experiences. As I replay these memories, I think of the generosity you displayed in sharing your time with the young student I was at the time. Those times were formative ones for me, having grown up without a father figure in my life, and I have been grateful for the warmth, the confidence building, and the wisdom that you contributed to my life. I am certain that tributes to you will have ample descriptions of your skills as a leader in business, your hard work, your solid values, your loyalty to your employees, your vision for the future of your business, your country, and your family. For me, however, it is the hugs and consoling that you gave me when I needed to face Mary's emergency surgery that fill me with gratitude.

Indeed, beyond the talents that L. F. Kennedy displayed, it is the caring and love that were at the core of what made him a great man.

Thank you, Mom and Dad, for all the caring and love.

Donald

Mary Cameron (daughter)

One of my earliest memories, I must have been about three or four, was of Dad holding me; I was very sick, scared, and on my way to the hospital. He told me I could have anything I wanted to take with me. Celia, my older sister, had a special book about horses that she would not let me look at, much less touch, and yes, you guessed it, I asked to take that book with me to the hospital. True to his promise, Dad let me have it!

Dad was warm, caring, sentimental, fun, and devoted to us, never concealing his love and adoration of Mom; he was the one the whole family leaned on and looked to for organization, planning fun trips, eating out, playing "Tom Tiddler" with us after dinner, reading to us from the classics in the evening.

We had many wonderful trips to New York to visit Grandma, Chela, Mecha, Auntie Georgie, and Uncle Charlie—which were some of the most exciting times that I remember from my childhood. Among other good memories, too numerous to recount, were the week or two during the summer holidays spent at Tower Isle Hotel—swimming all day, playing crab races, listening to calypso, dancing in the evening, and where we made friends with the bartender who loaded up our Cokes with maraschino cherries. Dad loved to have fun and relished seeing us enjoy ourselves. Family was everything to him.

Despite the overwhelming responsibilities of work, he came home from work every day by 4:00, at which time we all piled into the car to visit Grandma. We always ate dinner together followed by together-time on the verandah, where, smoking his cigar, he sat in his rocker, one leg draped over one arm. It was a peaceful time; sometimes he was quiet and pensive, reading his *Time* magazine, and Mom would be crocheting; sometimes we would get an impromptu visit from my Aunt Louise. I can still hear his chuckle as he enjoyed her jokes.

All that being said, he was a strict disciplinarian, intolerant of noncompliant behaviour, and like his father, believed strongly in the age old saying: "spare the rod, spoil the child." I tried very hard to toe the line and fly under the radar, unlike Paco and Celia, so I rarely was the brunt of his hot temper. Mom called it a healthy fear!

Although he had traditional ideas of the roles of men and women in society, he insisted that all of us pursue a high level of education, become independent, and be successful in whatever we chose to do. I know now that his ideals of honesty, moral integrity, and forthrightness have influenced me in my life's journey and contributed to me embracing these same values in my life.

"The quality of a father can be seen in the goals, dreams and aspirations he sets not only for himself, but for his family" Reed Markham.

Rafael Diaz (friend, former CEO/Chairman GraceKennedy)

I worked very closely with Mr. Kennedy and found him to have a terrific brain; he was a master of business and finance. We would review the financial results of the company and its subsidiaries monthly. He had the ability to interpret the balance sheet at first sight.

Laughingly he would say to me, "Raf, I am not an accountant, you know," and laugh at this. These times would bring out his tremendous sense of humour.

In our leisure, we would sometimes chat and exchange experiences we both had with connections in my native land Belize, and with mutual friends and acquaintances he made whilst on business trips.

I deem myself so fortunate that Mr. Kennedy took me into his confidences, both in business and in his private life. I became like family. Mrs. Kennedy was such a lovely person; she welcomed me into her home and always beamed when she saw me.

I also had my good and sad moments with Mr. Kennedy. I could never travel to Toronto and Montreal that I didn't branch off to visit them at their retirement home in London, Ontario. There I had to give him the "full bill and receipt" of his beloved Grace, Kennedy.[1]

Anna Jarvis (family friend, and niece to Carlton Alexander)

Carlton Alexander often held social events at his home where colleagues from work mingled with business associates, clients, friends, and family. When I was old enough to attend, I met Mr. and Mrs. Fred (often with Liz in tow) at some of these events. The whole family held Mr. Fred in high esteem. I also learned in later years that my father (typical academic with no financial talents) received a generous donation to publish his first book of poetry, *Blue Mountain Peak*. Mr. Fred was known to be stern and demanded high standards. He gave generously to many charitable causes.

During the years I dated Paco, I discovered other human sides to your father. He was a family man who adored his wife.

When his beloved Tillie went into labour the first time, she was very reluctant to take the routine dose of castor oil "to help move things along." The nuns were adamant. Mr. Fred said he was "such a damn fool in love" that he offered to take a dose of castor oil with her. Two generous doses were poured and on the count of three he swallowed his dose completely. She never took more than a whiff of her medicine cup. He reported that although he still loved her, he never got caught taking castor oil again during subsequent labours. He claimed the first labour was harder on him than her!

Mr. Kennedy took a keen interest in my studies at medical school and always asked about my current rotations. He was curious about changes in therapies and the advice of doctors over the years. He was generally well informed. He used to chuckle about politicians and how Bustamante got him to give up his first retreat in the hills, as "the Chief" loved the view from Mr. Fred's land.

The first time I offered to get him a drink refill I made a terrible error. In my family, many people liked a foamy head on their beer; he did not. He growled, "Young lady, never pour a man's beer without knowing what he wants." Anyway, he forgave quickly and allowed me to pour future drinks, luckily without any mishaps! He was very respectful of my time. On one occasion, he forgot his medicines at home and he apologised sincerely for interrupting Paco's and my Friday night plans. We drove them to Discovery Bay, then returned to Kingston in the middle of the night.

He liked having his extended family around and loved them all in spite of each person's foibles. He often shared with me stories about various family members. Wish I had written down his stories as they were always insightful, not judgemental, often

---

1   These excerpts by Raf Diaz are part of an extensive tribute entitled, "I Remember Luis Fred Kennedy," which is referenced in the book. Essentially, Raf viewed Dad as a humanitarian and forward-thinking entrepreneur, one who pioneered for his staff, death benefit insurance, pregnancy leave, a free medical scheme, and the establishment of a superannuation fund.

amusing though never mean. His wisdom to see people as they really were, appreciating different talents and faults, was an important part of my education. I believe he was orienting me to the Kennedy clan in a gentle way. I have always appreciated his caring and his trust and loved both your parents deeply.

Anthony (Tony) Charles Kennedy (nephew)

ONE word characterizes my Uncle Luli: ZEST! He always showed great enthusiasm for life, with family, at church, partying, and in his work ethic.

Our family spent countless hours with Uncle Luli, Aunt Tillie, and their five children. After all, they are our first cousins! Frequently, on Sundays we were invited over to Lady Musgrave Road to swim in the pool and eat out around the big almond tree. When we weren't there, we would often be invited to sleep over at our uncle's vacation home in Discovery Bay. He knew how to live!

Not only did he live with zest, but he carried out his caring through action. He always made you feel like you belonged and often arranged for the company to contribute to the education costs of his extended family.

Thanks for everything, Uncle Luli!

Georgianne Thompson Kennedy (daughter-in-law)

My new boyfriend took me to meet his parents at their weekend beach home in Discovery Bay, October 1973. There on the front porch sat his father, short and stocky, relaxed and welcoming in shorts and cotton shirt, catching a cool breeze on the green glider. In almost no time at all, he and Fred's mother, Tillie, accepted me as a regular guest, both in the country and in "town." Many evenings the family sat out under the stars in Discovery Bay and philosophized. All questions were open for honest debate and reflection.

As I came to know him, I realized Luis Fred Kennedy was a man of contradictions. One Friday evening, Fred and I made plans to stay in St Ann, but his father had a different idea. "Damn foolishness, son! You're coming into Kingston. I'm sending a plane for you and Georgie, and you'd better be on it."

He and Tillie invited a large group for a tropical evening meal by the pool at Club Caribbean in Runaway Bay. Steel drums, coconut palms swaying, laughter, and full bellies. Then the waiter asked how he was paying for our meal and Luis Fred said, "Don't you know who I am? Just send it to my office."

I wish I had listened more closely to the financial discussions held on the Lady Musgrave Road verandah. Broken bits of marble gravestones used for flooring and misty mountains in the distance framed the scene. Luis Fred, ensconced in his white wicker chair, legs up, presided over an intriguing parade of eccentric relatives arriving for after-dinner drinks and advice. Business bored me, but I realize now that historic events were transpiring.

Despite his quick temper and traditional views about women, he had a generous heart. When Fred and I became engaged, my own father raised a fuss about the costs of a wedding in Canada. Dad felt he had done his paternal duty by helping with my university expenses. Fred's parents kindly and simply offered to host the reception at #45.

Luis Fred had a brush with death in August 1976, but he was determined to be able to walk for our October date. Will power brought him out of a sick bed set up in the dining room. In the formal photos, we are standing in front of a green curtain left over from the makeshift sickroom he had just vacated. The wedding celebration was a cheerful affair with pink roses and a full Jamaican banquet, followed by speeches and dancing. Fred's dad gave a magnificent address and wrapped me in a warm hug. A week later, my parents gave us a small party at home in Galt (Cambridge), Ontario.

Although I met him just before he entered his declining years, I recognized his clear authority as oldest brother in his family and wise counsel to his business associates. It delighted him to live to see the legacy of grandchildren, as he and Tillie were starting to believe it might never happen. The only fear he ever expressed was of losing his mental faculties in old age. That very affliction came upon him so rapidly and so hard that he was terrified, but Fred's mother never abandoned him even when she found it difficult to cope. She stayed true and cared for him the entire way.

Stephen (Steve) Francis Kennedy (nephew)

I distinctly remember leaving a meeting with him with a deep sense that he generally cared about people and their "lot in life" on a more than superficial level. He particularly cared about family and education.

Family

Specifically, he would make a point of calling the house in Germany every New Year's Eve not only to speak to Mom but also to take the time to have a word with all of us on an individual basis. He always sincerely wanted to know how we were doing in our daily lives and wanted to know if there was anything he could help us with.

Education

It was on this annual call in 1970 that he asked me about what my plans were going forward regarding my education. He seemed to know/sense that I was at a bit of a crossroads in that regard, so he made a suggestion. He said he would be in Rotterdam in a couple of weeks (January 1971) and that I could take the train up there to meet with him for a few hours to discuss. Obviously, I went, and we talked about a plan to secure a university education for me in Canada if I were so inclined/interested. I then met him that March in Montreal for another discussion, and the rest is history, so to speak. We kept in touch via letters a couple of times each year, as his interest in my progression remained high.

I have always been very appreciative of his generosity, caring, and mentorship. Every once in a while, I reflect upon his act of kindness for family and quietly say thank you.

Owen Cathcart (OC) Plant (nephew)

For as long as I can remember in my childhood, I always heard from my mother that Uncle Fred really liked me. Why that might have been the case was a mystery to me. Yet his demonstrable pleasure whenever he was with me, and the help he gave me, were very clear.

I always enjoyed being in his presence. My earliest memories of my uncle were at his home. He would sit in a wicker chair on his verandah telling jokes and stories and playing the role of family patriarch during gatherings with the many relatives.

I remember him at work, in his very large office in the corner of an upper floor of the Grace, Kennedy building. Here Uncle Fred would hold court. Most of the time he would have a lit cigar filling the room with its aroma. Sometimes he would put a leg up over the arm of his chair while he met with employees and then provided his answers in a low growl. To me, as I watched him during my time working at Grace, Kennedy, it was clear he intended to intimidate and pressure all who approached. Even so, I never once felt intimidated. Rather, I felt I had a wise and supportive relative, and I appreciated it.

In business, I saw a brilliant and dynamic person driving himself and his company to great success. He always seemed to get to the most pertinent point of each issue and was able to make his decisions quickly. To indicate his level of brilliance, I vividly remember seeing him when I was about twelve years old as I accompanied my father to a meeting with union delegates threatening a strike of the employees at Kingston Wharves Ltd. The meeting was in the office of Kingston Wharves. After listening to their points, threats, and arguments, Uncle Fred stood, walked to one side of the room, and began talking. Without notes, he talked for what I remember to be over half-an-hour. He presented information that included tonnage, costs, timelines, competitive information, history, and projections. When he sat down, it was as if all the air had been removed from the room. No one said anything more and the meeting broke up. There was no strike.

Paul Scott (former Manager/Director, GraceKennedy)

Mr. Luis Fred was the finest man, business leader, organization planner, business planner, and developer, forward thinker, and my revered, admired, trusted, and respected boss.

Everything I learnt about business, forward thinking, business development, and human resources, was from Mr. LFK—watching and learning from him, consulting, and striving to emulate the example, extraordinary vision, and commitment to the growth of our organization that he demonstrated.

I was/am very proud to have been part of the Grace, Kennedy team under the leadership of my mentor, respected leader, and wonderful head of the finest organization in Jamaica at that time. "Amazing Grace" is one of my favorite tunes that I often listen to, and which also reminds me of working for my mentor, Mr. Luis Fred Kennedy, in my days at Grace (GraceKennedy Group of Companies).

Mary-Lou Soutar-Hynes (niece)

My memories of my uncle and godfather are impossible to understand without some insight into my mother's relationship with her eldest brother. My mother, Louisa Matilde Kennedy—fourth child of Fred William Kennedy—was only twelve years old when her adored father died. Luis Fred, then barely twenty-one, would become the *de facto* head of the family, shouldering the responsibilities of his father's fledgling company and ensuring that his mother and siblings were taken care of. While she never stopped grieving her father's loss, my mother trusted her brother implicitly.

When she became a mother, he would become my godfather and well-loved uncle. Like him, I was the first-born. I felt that kinship. Unlike my uncle, however, I was a girl, my Kennedy mother's first child, a first granddaughter, a first niece. I grew up serenely unaware that being a girl might entail any diminishment of worth or possibility. But there was never a day that I felt lesser-than because of my gender, affirmed by my parents, by both families, and, in particular, by my Uncle Luli.

I would come to know him as a multi-faceted and complex man, who seemed to stride through our lives as a colossus. The man on whom so many depended was the man who encouraged me to let go and swim across the water-tank-turned pool at their country home near Irish Town, the man who always seemed genuinely pleased to include his tom-boy niece on outings. The man who supported my decision to become a Sister of Mercy, and equally enthusiastically, my decision to leave the religious community twelve years later. The man for whom I would rush from Toronto to his hospital bed in London, Ontario, because he wanted to see me before he died.

If I were to try to pin-point one shining quality that epitomized my bond with my uncle, it would be a deep connection to family that would transcend time and space—a sense of belonging, and the desire to be of service in whatever context or place I would ultimately find myself.

Fred William (Bill) Valliant (nephew)

Shortly after I started at Grace, Kennedy (1962), there was a sharp downturn in the local economy. Most importers, including GKCO, found themselves strapped for funds. LFK simply called the foreign suppliers and personally requested an extension of the payment terms. They were more than willing to do so based on the company's track record. Several of the other local importers could not get these extended terms and found themselves without goods to sell.

In about 1963, LFK took the bold step of creating a Management Group of five employees. I was fortunate to have been chosen as one of the five. LFK called a meeting of all the department heads and informed them that any member of this group could ask any question of them. In addition, each Wednesday, all of us were required to attend a luncheon meeting in the board room—lunch catered by Clara. There was no set agenda. Any topic could be tabled. From time to time, LFK would invite a department head to attend, but again there would be no agenda. For example, if Boerries Terfloth was in town, we would be treated to his words of wisdom. Old man Solomon would also be in attendance.

LFK obviously realized that it was very important for each member of this management team, in addition to being responsible for his own department, to become familiar with the other operations of the company. The obligatory luncheon meetings allowed us to pause and appreciate problems faced by other managers. It was an attempt for us to see the big picture rather than being tied-up with the details of our specific departments.

III

Acknowledgements

My wife Georgianne is a pillar of strength. Her superior editing skills helped polish the narrative, and her extensive genealogical research into family history and DNA were critical to the success of the project. It was fun working with her to put together pieces of the puzzle. I treasure her loving and moral support.

My three daughters—Amanda, Sarah, and Julia—were always there to encourage their dad. I am indebted to Sarah for her contributions of artwork, original book cover design, and her painting of my dad's friend, Sullivan. She also participated in conferences with the publisher to contribute ideas about the interior design of the book.

Mary, my sole surviving sibling, with whom I feel a special bond, helped validate our shared memories of childhood. She read and critiqued the manuscript, contributed photos, family letters, her own narrative accounts, and a tribute to Dad. Thanks also to Mary for endorsing my work. I am grateful to her husband, Donald, for editing medical information and for his tribute to Dad. He had a special fondness for my parents.

My youngest sister, Liz, who passed away August 30, 2019, was more excited about the project than I was. She loved Dad deeply. Her wish was to have written her own tribute to Dad, but her illness advanced so quickly that she was not able to complete it. She was with me in spirit when I was composing our family story.

My long-time friend of over fifty years, Swithin Wilmot, deserves special mention. He first gave me the idea to embark on the project, and throughout the process, encouraged and inspired me to the finish. He reviewed chapters for historical accuracy and recommended additional resources for me to consult. I respect his scholarship, treasure his friendship, and am ever grateful for his contributions. "Your father showed so much kindness to me, I could not say no," he said when I asked for his help. He also kindly accepted the invitation to be one of those who endorsed the book. Swithin R. Wilmot (D. Phil. Oxon) is former Head, Department of History and Archaeology, and Dean of the Faculty of Humanities, University of West Indies, Mona.

I am thankful also to those who wrote tributes. Their personal testimonies enriched the biography: Paul Bitter, Mary and Donald Cameron, Raf Diaz, Anna Jarvis, Tony Kennedy, Georgie Kennedy, Steve Kennedy, OC Plant, Paul Scott, Mary Lou Soutar-Hynes, and Bill Valliant. Mary Lou took a special interest in the project by sharing memories of her childhood and contributing texts. Thanks to my cousins, Philip, Tony and Michèle for their input and reflections on their dad, my Uncle Che. Tony Kennedy agreed to be interviewed and to have his Y-DNA tested.

Raf Díaz was a major contributor. He invited me and my wife to his home in St. Andrew. We spent hours reminiscing. We had many subsequent conversations by phone and corresponded often by email. He did an immense amount of research on the company and was generous with sharing his written reflections with me. Raf brought a special perspective as former CEO and Chairman of Grace, Kennedy. He became a true and loyal friend of the Kennedy family. And thanks to him, for endorsing *Firstborn*.

Donovan Anderson, former company secretary, was Dad's business associate who became his good friend. "I was his disciple," Donovan told me. Donovan spent hours recounting his experiences with the company and his interactions with my dad. When I asked Donovan if he would review chapters, he said, "Your father meant so much to me, I would not hesitate for a moment." A big debt of gratitude to Donovan and respect for his expert knowledge of corporate law, and thanks to him too for endorsing the book.

I am grateful to Marjory, Cathrine, and Charles for sharing their sentiments of Paco. I dedicated the book to my brother and included a chapter devoted to him. Marjory's brother, Charlie Johnston, contributed a picture of his grandfather, Charles. E. Johnston, and information on the history of Johnston family businesses.

I spent many hours conversing with the Moss-Solomons, Peter, Joan, and Jimmy, who shared precious memories of Grace, Kennedy, of their father, James S. Moss-Solomon Sr., and his business and personal relationship with my dad. They contributed photographs and anecdotes that I used for writing a chapter about Uncle Jim. Relationship between the families is a central part of the narrative. Jimmy, who passed on January 4, 2022, had a keen interest (as do his siblings) in family and company history. He told me many times about how much he looked forward to reading the biography. RIP, Jimmy.

Thanks to Anna Jarvis, Beverley Valentine, Philip and Edward (Teddy) Alexander, with whom I had long conversations about the life of Carlton Alexander. They enlightened me about what a great man he was and provided resources that helped me write the chapter on Carlton.

Others whom I interviewed were Boerries Terfloth, Alva Russell, Arnold (Junior) Foote, Ed Muschett, Chris Bovell, Donovan Anderson, Mable Tenn, John Issa, George Munro, Alvin Lee, and Charlie Johnston. Both Junior Foote and Boerries Terfloth have since passed.

Acknowledgements

I am thankful also to Professor Gordon Shirley, Sen. Don Wehby, and Douglas Orane, who even though they did not know Dad, were able to shed light on how he shaped the culture of the company. Their conversations formed the basis of Chapter 49 of the postscript. Douglas took a special interest in the project and volunteered his time to review and edit chapters. Thanks to Doug also for agreeing to endorse the book.

Thanks to family members Georgianne Thompson Kennedy, Cathrine Kennedy, Mary Cameron, Lisa Valliant, and others for the use of photographs from family collections.

I am thankful to Gail Moss-Solomon, GraceKennedy's General Counsel and Chief Corporate Secretary, who granted permission for use of the minutes of Grace, Kennedy & Co. Ltd. (1922–1982) and for the right to publish photographs. Thanks go out also to IT personnel at tTech, Fitzroy Gray, who assisted with access to the Central Repository website, and Lurline Cummings, former GraceKennedy Information Resource Specialist who shared her time and expertise.

Thanks to Glen Pawelski, Project Manager of Mapping Specialists Ltd., Fitchburg, WI, for the creation of the maps of Kingston, Jamaica, 1920s and Falmouth, Trelawny, 1870s.

I am indebted to Glenn Haniboll of retouch.ca Photo Retouching Services for enhancing the quality of photographs.

I am grateful to Sheree Rhoden, Information Systems Department, of Gleaner Company (Media) Ltd. for facilitating license to publish numerous articles and photographs.

And finally, but by no means least, Cam Bradley, Publishing Specialist, and the team of editors at FriesenPress Publishers who collaborated with me to make this production possible.

IV
Bibliography

Altink, Henrice. *Public Secrets: Race and Colour in Colonial and Independent Jamaica.* Liverpool, UK: Liverpool University Press, 2019.

Arévalo, Manuel A. García. "Orígenes del Mesitjaze y de la Multización en Santo Domingo." https://revistas.unphu.edu.do/index.php/aula, 1995.

Ayre, Michael. *The Caribbean in Sepia: A History in Photographs 1840-1900.* Kingston, Jamaica: IanRandle Publishers, 2012.

Bell, Wendell. "Independent Jamaica Enters World Politics: Foreign Policy in a New State." In *Political Science Quarterly.* Vol. 92, No. 4, Winter 1977–1978, pages 683–703. Published by The Academy of Political Science.

Bennett, Louise. *Jamaica Labrish.* Jamaica: Sangsters Book Stores Ltd., 1966.

Bernal, Richard L. "The Great Depression, Colonial Policy and Industrialization in Jamaica." In *Social and Economic Studies,* Volume 37, Nos. 1 & 2, March-June 1988, pages 33-64.

_____. "The IMF and Class Struggle in Jamaica, 1977–1980." In *Latin American Perspectives,* Vol. 11, No. 3, Summer 1984, pages 53-82.

Bertram, Arnold. *N. W. Manley and The Making of Modern Jamaica.* Kingston Jamaica: Arawak Publications, 2016.

Blake, Duane. *Shower Posse: The Most Notorious Jamaican Criminal Organization.* New York, New York: Diamond Publishing, 2002.

Blake, J. Herman. "Black Nationalism" *The Annals of the American Academy of Political and Social Science* 382 (1969): 15–25. http://www.jstor.org/stable/1037110

Bibliography

Bradley, Paul C. "Mass Parties in Jamaica: Structure and Organization." In *Social and Economic Studies*. Vol. 9, NO. 4, pages 375–416. Published by Sir Arthur Lewis Institute of Social and Economic Studies, University of the West Indies, December 1960.

Brodber, Erna. *The Second Generation of Freemen in Jamaica, 1907–1944*. Gainesville, FL, USA: University of Florida, 2004.

Bryan, Patrick. *Inside Out, Outside In: Factors in the Creation of Contemporary Jamaica*. GraceKennedy Lecture. Kingston, Jamaica: GraceKennedy, 2000.

_____. *Jamaican People 1880–1902: Race, Class and Social Control*. Mona, Jamaica: The University of the West Indies, 2002.

Bryan, Patrick and Karl Watson, editors. *Not for Wages Alone: Eyewitness Summaries of the 1938 Labour Rebellion in Jamaica*. Department of History, University of West Indies, Mona, 2003.

Buddan, Robert. "Universal Adult Suffrage in Jamaica and the Caribbean since 1944." In *Social and Economic Studies* 53:4, pages 135-162, 2004.

Camarena, Germán. *Historia de la ciudad de Puerto Plata*. Santo Domingo, República Dominicana: Corripio, 2003.

Cassá, Roberto. *Historia Social y Económica de la República Dominicana. Tomo II*. Santo Domingo, República Dominicana: Editora Alfa y Omega, 1980.

Cassidy, F.G. and R.B. Le Page. *Dictionary of Jamaican English*. Cambridge at the University Press, 1967.

Clarke, Carl F. "A Concise History of St. George's College." Unpublished, 2004.

Cocks, Catherine. *Tropical Whites: The Rise of the Tourist South in the Americas*. Philadelphia: University of Pennsylvania Press, 2013.

College of the Holy Cross (John. T. MacPherson, editor). *The Holy Cross Purple Patcher Volume XXII*. Worcester, Massachusetts, 1928.

Colonial Office. *Annual Report on Jamaica for the Year 1948*. London: His Majesty's Stationery Office, 1950. Printed in Jamaica.

Cumper, G.E. "Population Movements in Jamaica, 1830–1950." In *Social and Economic Studies*, Vol. 5, No. 3, September 1956, pages 261–280.

Cundall, Frank. *The Handbook of Jamaica for 1902, Comprising Historical, Statistical and General Information Concerning the Island*. Kingston, Jamaica: Government Printing Office, 1902.

De Lisser, Herbert G. 1930-3. *Planters' Punch: A Jamaica Magazine.* © National Library of Jamaica. Journal originally printed by The Gleaner Co. Ltd. Source: Digital Library of the Caribbean, University of Florida, ufdc.ufl.edu.

_____. *Planters' Punch: A Jamaica Magazine.* Vol. 4. No. 1 © National Library of Jamaica. Journal originally printed by The Gleaner Co. Ltd, 1938-39.

Diaz, A. Rafael. "The Origins of Kingston Wharves Ltd." December 19, 1999.

_____. "I Remember Luis Fred Kennedy." 2019.

Dickens, Charles. *A Tale of Two Cities.* London: Chapman and Hall, 1906.

Eaton, George E. *Alexander Bustamante and Modern Jamaica.* Kingston, Jamaica: Kingston Publishers Limited, 1975.

Eccles, Karen E. and Debbie McCollin, editors. *World War II and the Caribbean.* Jamaica-Barbados-Trinidad and Tobago: The University of the West Indies, 2017.

Eisner, Gisela. *Jamaica 1830–1930: A Study of Economic Growth.* Manchester University Press, 1961.

Eliot, George. *Middlemarch* (first published 1871–2). UK: Alma Classics, 2010.

Foster, Henry Blaine. *Rise and Progress of Wesleyan-Methodism in Jamaica.* London, England: Wesleyan Conference Office, Hazeu, Watson, and Viney Printers, 1881.

Garvey, Marcus. *Message to the People.* Originally published 1937. Re-printed Bensenville, Illinois, USA: Lushena Books Inc., 2014.

Gilbert, Martin. *Churchill: A Life.* Great Britain: Pimlico Publishers Ltd., 2000.

Girling, Robert and Sherry Keith. "The Planning and Management of Jamaica's Special Employment Programme: Lessons and Limitations." In *Social and Economic Studies.* June/September 1980. Vol. 29, No. 2/3, pages 1-34. Published by Sir Arthur Lewis Institute of Social and Economic Studies: University of the West Indies.

Glassner, Martin Ira. "The Foreign Relations of Jamaica and Trinidad and Tobago 1960–1965." In *Caribbean Studies,* Vol. 10, No. 3, pages 116-153, October 1970.

Gleaner Company Limited. *Jamaica: 50 Golden Moments 1962–2012.* Kingston, Jamaica: IanRandle Publishers, 2012.

Goffe, Leslie Gordon. *When Banana Was King: A Jamaican Banana King in Jim Crow America.* Kingston Jamaica: LMH Publishing Ltd., 2007.

Bibliography

Gonsalves, Ralph. "The Rodney Affair and its Aftermath." In *Caribbean Quarterly,* Vol. 25, No. 3, pages 1–25, September 1979.

Grace, Kennedy & Co. Ltd. *Minutes of Meetings of the Board of Directors* (1922-1982).

_____. "Memorandum of Association and the Articles of Association." *The Companies Act*. Published by Ministry of Justice, Government of Jamaica, 1965.

_____. *The Grace News:* Vol. 7, No. 2., August 1982.

_____. *Nation Building: A Businessman's Perspective. Excerpts from Selected Speeches by Carlton Alexander*. Edited by George J. Phillip. Kingston, Jamaica: Media Advisory and Research Services Ltd., 1983.

_____. *S. Carlton Alexander: May 1933–May 1983. A Gift from the Employees of Grace, Kennedy & Co. Ltd.* Compiled by Lilleth C. Morris, April 1983.

Hall, Douglas. *Free Jamaica 1838–1865*. Portsmouth, NH, USA: Heinnemann Educational Books, 1959.

_____. *Grace, Kennedy & Company Limited: A Story of a Jamaican Enterprise*. Kingston, Jamaica: Grace, Kennedy & Company Limited, 1992.

Hart, Ansel. *Monthly Comments*. Volume 6, No. 1–24. Newport (Manchester), Jamaica, West Indies, Dec. 1967–Nov. 1969.

_____. *Old York Castle: 1876–1906* (unpublished material).

http://www.ycalumni.org/Historical%20Info_YCHS.pdf posted on York Castle High School Alumni website, https://yorkcastlealumni.org/history/.

Hart, Richard. *Rise and Organise: The Birth of The Workers and National Movements in Jamaica (1936–1939)*. London, UK: Karia Press, 1989.

_____. *Towards Decolonisation: Political, labour and economic development in Jamaica 1938–1945*. Kingston, Jamaica: Canoe Press University of the West Indies, 1999.

Henriques, Fernando. *Family and Colour in Jamaica*. Jamaica: Sangster's Book Stores Ltd. 1953.

Higman, B. W., editor. *The Jamaican censuses of 1844 and 1861*. Jamaica Archives, 1980.

Hill, Frank. *Bustamante and His Letters*. Kingston, Jamaica: Kingston Publishers Ltd., 1976.

Hill, Stephen A., editor. *Who's Who in Jamaica 1916. A Biennial Record containing careers of Principal Public Men and Women of Jamaica*. Kingston, Jamaica: The Gleaner Co. Ltd., 1916.

_____., editor. *Who's Who in Jamaica, 1919-1920.* Kingston, Jamaica: Gleaner Co. Ltd., 1920.

Horn, Michiel. *The Great Depression in the 1930s in Canada.* Ottawa: Canadian Historical Association Historical Booklet, No. 39, 1984.

Howson, Susan. "The Management of Sterling, 1932–1939." In *The Journal of Economic History*, vol. 40, no. 1, Cambridge University Press, 1980, pp. 53–60. http://www.jstor.org/stable/2120421.

Institute of Jamaica. *A Handbook of Information for Intending Travelers and Others.* Kingston, Jamaica: Aston Gardner & Publishers, 1896.

James, Marquis. *Merchant Adventurer: The Story of W. R. Grace.* © W. R. Grace & Co. Wilmington, Delaware, USA: Scholarly Resource Inc., 1993.

Jeffrey, Duncan James. "Education, Economy and Class in Jamaica, 1700–1944." Master of arts thesis, unpublished, McMaster University, 1980.

Jensen, Peta Gay. *The Last Colonials: The Story of Two European Families in Jamaica.* London: Radcliffe, 2005.

Kaluta-Crumpton, Anita. "The inclusion of the term 'colour' in any racial label is racist, is it not?" In *Ethnicities.* Vol. 20, No. 1, pages 115-135, 2020.

King, Damien. "The Evolution of Structural Adjustment and Stabilization Policy in Jamaica." In *Social and Economic Studies.* Vol. 50, No. 1, pages 1–53, March 1971.

King, Ruby. *Education in the British Caribbean: The Legacy of the Nineteenth Century.* Article published online, partly based on research grant from University of the West Indies, Mona, Jamaica, 1998.

Kramer, Mark and Wendy Call, editors. *Telling True Stories: A Nonfiction Writer's Guide from the Nieman Foundation at Harvard University.* USA: Plume Book, Penguin Publishers, 2007.

Lacey, Terry. *Violence and politics in Jamaica, 1960–1970: Internal security in a developing country.* UK: Manchester University Press, 1977.

Liston, Thomas S. *The Kingston Kingpins: how a powerful mobster brought the Jamaican government down.* USA: Page Publishing Inc., 2016.

Lowenthall, David, editor. *The West Indies Federation.* New York: Columbia University Press, 1961.

Lusaka, Paul Firmino. "The Dissolution of the West Indies Federation: A Study in Political Geography." unpublished master's thesis, Department of Geography, McGill University, Montreal, November 1963.

Bibliography

Macmillan, Allister, editor. *The Red Book of the West Indies: Historical and Descriptive, Commercial and Industrial Facts, Figures and Resources.* London: W.H. & L. Collinbridge, 1922. Google Books, Reprints from the Collection of the University of Michigan Library.

Manderson-Jones, R. B. *Jamaican Foreign Policy in the Caribbean: 1962–1988.* Kingston: CARICOM, 1990.

Manley, Beverley. *The Manley Memoirs.* Kingston, Jamaica: IanRandle Publishers, 2008.

Manley, Michael. *A Voice at the Workplace: Reflections on Colonialism and the Jamaican Worker.* London, England: André Deutsch Limited, 1975.

Manley, Rachel. *Drumblair: Memories of a Jamaican Childhood.* Kingston, Jamaica: IanRandle Publishers, 1996.

Martin, James W. *Banana Cowboys: The United Fruit Company and the Culture of Corporate Colonialism.* Albuquerque, USA: University of Mexico Press, 2018.

May, Stacy. *The United Fruit Company in Latin America.* Miami, FL: HardPress Publishing, 2004.

McCann, Thomas P. *An American Company: The Tragedy of United Fruit.* New York, New York: Crown Publishers Inc, 1976.

McLaughlin, Gerard Leo S.J. *Jesuitana Jamaica Historical Profiles 1837–1996.* Kingston, Jamaica: Arawak Publications, 2000.

McMahon, Benjamin. *Jamaica Plantership. Eighteen Years Employed in the Planting Life in that Island.* London: Effingham Wilson, 1839.

Morris, Lilleth C., editor. *S. Carlton Alexander, May 1933–May 1983: A Gift from the Employees of Grace, Kennedy & Company Limited.* Kingston, Jamaica: GraceKennedy Ltd., 1983.

Moss-Solomon, James. *Jamaica and GraceKennedy: Dreams Converging, Roads Diverging.* GraceKennedy Foundation Lecture. ©2012 GraceKennedy Foundation. Printed in Jamaica by The Phoenix Printery Limited, 2012.

Namsuk, Kim and Marta Serra-Garcia. "Economic Crises, Health and Education in Jamaica." In *Estudios Económicos.* Vol 25, No. 1 (49), Enero-Junio de 2010, pages 105–134.

Neita, Clifton, editor. *Who's Who in Jamaica 1963: An Illustrated Biographical Record of Outstanding People in Jamaica.* Jamaica: City Printery Ltd, 1963.

_____. *Who's Who Jamaica British West Indies, 1957: An Illustrated Biographical Record of Outstanding People in Jamaica.* Kingston, Jamaica: Who's Who Jamaica Ltd. (The Gleaner Co. Ltd.), 1957.

Nettleford, Rex, editor. *Manley & The New Jamaica: Selected Speeches and Writings 1938–1968.* Trinidad and Jamaica: Longman Caribbean, 1971.

Nettleford, Rex. *Mirror, Mirror: Identity, Race and Protest in Jamaica.* Kingston, Jamaica: LMH Publishing Ltd., 2001.

Newfoundland Fisheries Board. "Report of Fisheries Post-War Planning Committee: Comprising A Review of Newfoundland's Salt Codfish Trade and Recommendations on Post-War Marketing." February 1946.

Newman, Joanna. *Nearly the New World: The British West Indies and the Flight from Nazism, 1933–1945.* New York-Oxford: Berghahn Books, 2019.

Norton, Anne and Richard Symanski. "The Internal Marketing Systems of Jamaica." In *Geographical Review.* Vol. 65, No. 4, (Oct. 1975), pages 461–475.

Obama, Barack. *A Promised Land.* New York: Penguin Random House, 2020.

Orane, Douglas. *The Business of Nationbuilding: Excerpts from the Selected Speeches of Douglas Orane.* Kingston, Jamaica: IanRandle Publishers. 2016.

Padmore, George. "Labour Unrest in Jamaica." In *International African Opinion*, Vol. 1, No. 1, July 1938.

Padmore, Overand R. "Federation: The Denise of an Idea." In *Social and Economic Studies*, Vol. 48, No. 4, December 1999, pages 21–63.

Palmer, Colin A. *Freedom's Children: The 1938 Labour Rebellion and the Birth of Modern Jamaica.* The University of North Carolina Press, 2014.

Panton, David. *Jamaica's Michael Manley: The Great Transformation (1972–92).* Kingston, Jamaica: LMH Publishing Ltd., 1993.

Phelps, O. W. "Rise of the Labour Movement in Jamaica." In *Social and Economic Studies:* Vol 9, No. 4., December 1960, pages 417–468.

Post, Ken. *Arise Ye Starvelings: The Jamaican Labour Rebellion of 1938 and its Aftermath.* The Hague/Boston/London: Martinus Nijhoff, 1978.

Proctor, Jesse Harris, Jr. "British West Indian Society and Government in Transition: 1920–1960." In *Social and Economic Studies*, Vol. 11, No. 4, December 1962.

Ranston, Jack (researcher and compiler) and Ken Jones (editor). *The Best of Bustamante: Selected Quotations 1935–74.* Red Hills, Jamaica: Twin Guinep Ltd., 1977.

Reynolds, Julian. "A Genuine Jamaican Brown Man." *The Gleaner,* November 23, 2014.

Bibliography

Robinson, Carey. *The Rise and Fall of Falmouth Jamaica.* Kingston, Jamaica: LMH Publishing Ltd., 2007.

Roorda, Eric Paul. *Historical Dictionary of the Dominican Republic.* Latham, Maryland: Rowman and Littlefield, 2016.

Schroeder, Alice. *The Snowball: Warren Buffett and the Business of Life.* New York, New York: Random House Inc., 2009.

Schweiker, Richard S. (comptroller general of the United States). "U.S. Response to Jamaica's Economic Crisis." United States General Accounting Office, July 17, 1980.

Seaga, Edward. *Edward Seaga: My Life and Leadership. Volume 1: Clash of Ideologies 1930–1980.* Oxford: Macmillan Publishers Ltd., 2009.

Sewell, William G. *The Ordeal of Free Labour in the British West Indies.* New York: Harper and Bros. Publishers, 1861.

Shragai, Atalia. "Do Bananas Have a Culture? United Fruit Company Colonies in Central America: 1900–1960." *Iberoamericana (2001–) Nueva época, Año 11, No. 42 (Junio de 2011), pages 65–82.* Published by *Iberoamericana Editorial Vervuert.*

Sinclair, A.C. *The Handbook of Jamaica for 1886–1887.* Kingston, Jamaica, Government Printing Office.

Smith, Godfrey. *Michael Manley: The Biography.* Kingston, Jamaica: Ian Randle Publishers, 2016.

Smith, Richard. *Jamaican Volunteers in the First World War: Race, masculinity, and the development of national consciousness.* Manchester and New York: Manchester University Press, 2004.

Springer, Hugh Worrell. *Reflections on the Failure of the First West Indian Federation.* Boston, Mass. Center for International Affairs, Harvard University, 1962.

Spruill, Larry H. Dr. And Donna M. Jackson. *Images of America: Mount Vernon Revisited.* Charleston, South Carolina: Arcadia Publishing, 2014.

St. George's College. *St. George's College 1850–1950 Commemorates its 100th Anniversary.* Kingston, Jamaica, BWI: St. George's College, 1950.

Stone, Carl. "The 1976 Parliamentary Election in Jamaica." In *Caribbean Studies.* Apr–Jul 1979. Col. 19. No 1/2, pages 22–50. Published by The Institute of Caribbean Studies, UPR.

Terfloth & Kennedy (U.K.) Ltd. *Memories of What Made Grace.* Richmond, Surrey, England TW9 2RD, 1997. Compliments of Boerries Terfloth.

Underhill, Edward Bean. *The West Indies: Their Social and Religious Condition.* London: Jackson, Walford, and Hodder, 1862.

United Fruit Company. *The Story of the Banana.* Boston, Massachusetts, 1921. https://ia600909.us.archive.org/4/items/storyofbanana00unit/storyofbanana00unit_bw.pdf.

US Department of Commerce and Labour, Bureau of Foreign and Domestic Commerce. *Daily Consular and Trade Reports.* Vol 3, Nos. 154–230. Washington, USA: Washington Government Printing Office, 1912.

Valliant Fred William. *Jigga Foot Boy at Quisqueya – A Proud Jamaican Heritage.* January 2020.

Vega, Bernardo, editor. *Dominican Cultures: The Making of a Caribbean Society.* Translated by Christine Ayorinde. Princeton, N.J.: Markus Wiener, 2007.

Von der Porton, Arnold. *The Nine Lives of Arnold.* AuthorHouse, 2003.

Williams, Kareen Felicia. "The Evolution of Political Violence in Jamaica 1940–1980." PhD Dissertation, Columbia University, 2011.

Wilmot, Swithin R. "My Browning Experience." Unpublished paper delivered at University of the West Indies, October 8, 1999.

Wirtenberger, Kathryn. "The Jesuits in Jamaica." Unpublished Master of Arts thesis, Loyola University, Chicago, 1942.

Withers, Harry C. "Summary of World War II." In *Southwest Review.* Fall 1945, Vol. 31, No. 1, pages 101–107. Published by Southern Methodist University.

Witter, Michael. "Exchange Rate Policy in Jamaica: A Critical Assessment." In *Social and Economic Studies,* December 1983, pages 1–50. Published by Sir Arthur Lewis Institute of Social and Economic Studies, University of the West Indies, 1983.

Yergin, Daniel and Joseph Stanislaw. *The Commanding Heights: The Battle for the World Economy.* New York, New York: A Touchstone Book, Simon and Schuster, 2002.

Zeidenfelt, Alex. "Political and Constitutional Developments in Jamaica." *The Journal of Politics,* Vol. 14, No. 3, pp. 512–540. Published by The University of Chicago Press on behalf of the Southern Political Science Association, August 1952.

Zeller, Neici M. "Puerto Plata en el Siglo XIX." Eme Eme: Estudios Dominicanos, 5(28), pages 27–52. *Santiago del los Caballeros, Republica Dominicana: Universidad Carolina Madre y Maestra* (UCMM), 1977.

Zinsser, William. *On Writing Well.* New York: Harper Perennial, 2006.

V

Index

Note re abbreviations used:
LFK for Luis Fred Kennedy
GKCO for Grace, Kennedy & Co. Ltd.
LMK for Lydia Mathilde Kennedy

A

Aarons, Harold, 253, 346*n*20
Abrahams, Phyllis Margery (m. Selwyn Carlton Alexander), 329
Abreu, Imbert, Carmen María (LFK's and LMK's maternal grandmother, m. José Antonio Arzeno Rodriguez), 39, 43, 50, 139, 140
Adams, Sir Grantley, 237, 241
adult suffrage (1944). *See under* Jamaica, POLITICS AND SOCIETY
air travel. *See* Pan American World Airways
Alcoa Steamship Company, 170, 204, 232
Alexander, Beverly (m. Douglas Valentine), 329, 330–331, 330*n*9
Alexander, Edward (Teddy), 329, 329*n*6, 331 through 334
Alexander, Heifetz, 329
Alexander, Mary Ann, 329
Alexander, Mavis, 326
Alexander, Philip, 274, 274*n*9, 327*n*4, 328, 331
Alexander, Selwyn Carlton
  as assistant manager to James Moss-Solomon, GKCO, 331
  birth (1916), 326
  as chairman, executive committee (1972), GKCO, 344
  as chairman, group managing director (1976), GKCO, 356
  as chairman, subsidiaries, GKCO, 277
  as committee member (1950s), Jamaica Chamber of Commerce, 240, 242
  death (1989), 427
  death of father, 327-328
  as deputy chair (1970) to LFK, 309
  as director of the board, GKCO, 229
  elementary schooling, 326
  family relation to Moss-Solomons, 132*n*19
  first job at GKCO (1933), 131-132, 328
  first marriage to Phyllis Margery Abrahams, 329
  Grace brand (1960s) and, 271, 274
  as group managing director (1973), GKCO, 11, 346
  Jamaica College and, 327, 332, 427
  as manager (1950s), merchandise division, 233
  as mentee of Luis Fred Kennedy, 332, 333
  Order of Jamaica (1983) and, 427
  as president of Jamaica Chamber of Commerce, 332
  as president of PSOJ (Private Sector Organization of Jamaica), 332, 357, 363*n*3
  as public figure, 345, 426-427
  relationship with Jamaican Chinese community, 328-329
  as salesman, 273, 274, 329
  second marriage, to Bernice Agatha Chin Lenn, 331
  service awards (GKCO): (1962), 253, 255; (1972), 321, 325
  as trustee, Employee Investment Trust, GKCO, 278
Alexander, Susan, 331
Allen, Jack, 337*n*1, 357

Allied Insurance Brokers Ltd., 279
All-Island Banana Growers Association, 223
Amberly cottage, Irish Town. *See under* Kennedy, Luis Fred, MARRIED/FAMILY LIFE; Kennedy, Celia Jean; and Kennedy, Mary Teresa
American International Underwriters (AIU) Corporation, 92, 141, 141n20, 147, 150, 279
Anderson, Errol Donovan, 308-309, 308n1, 356, 410
Angerhausen, Anne Marie (m, Francis Xavier Kennedy), 256–257, 256n2, 287, 321, 410
Angerhausen, Hiltrud, 287
Anglo-American Destroyers for Bases Agreement (1940), 169-170
Arzeno, Carmen Celia (LMK's mother) (m. Charles Henry Loinaz), 39, 91, 91n12, 135, 137, 177n2, 285n4
Arzeno, Gloria Mercedes (Tía Llolla), 39, 82, 177, 177n1
Arzeno, José Antonio (Papa Che) (LFK'S and LMK'S maternal grandfather), 39, 54, 43, 138–139
Arzeno, Luisa Matilde (LFK's mother) (m. Fred W. Kennedy)
  birth (1888), 43n14
  as child in Dominican Republic, 39
  death, (1962), 256
  family photo (1926), 104
  fortieth anniversary of GKCO and, 252 through 255
  Kingston home, 8 Seaview Avenue, *Quisquea*, 175, 179
  Kingston home, 25 South Camp Road, *Quisqueya*, 77
  marriage to Fred W. Kennedy, 44-45
  as matriarch of Kennedy family, 179, 255
  migration to Port Antonio, Jamaica (1909), 50
  Montego Bay home, *Quisqueya*, 52
  relationship with her son, LFK, 255
  as shareholder, GKCO, 97
  visit to New York, 90-91
Arzeno, Rafael (Tío Rafael), 39, 138, 138n9
Arzeno, Sebastián (LFK's and LMK's maternal great-grandfather), 43
Atlantic Fruit Company, 51n8, 100

B
Baker, Lorenzo Dow, 33 through 37, 50, 51

Banana Cowboys: The United Fruit Company and the Culture of Corporate Colonialism (Martin), 36, 36n13–n14, 41n6, 42, 42n7–n9
banana industry
  banana cowboys and, 36, 42
  beginnings in Jamaica, 33–34
  culture in Caribbean, 41–42, 41n5
  decline of exports (1939), 168
  economic Jamaican nationalists, 35
  fierce competition and, 36
  Fred William Kennedy and, 33 through 37
  Goffe Bros. as threat to foreign control of, 35-36
  Gros Michel variety, 35, 35n7
  Panama disease (1950s) and, 35n7, 168
  plantation industry by late 1880s, 34–35
  slump in demand during Great Depression, 129
  tropical storms and, 59–60
  United Fruit Company (UFCO) and, 36 through 42 (*see also* United Fruit Company)
  during WW I, 60
  during WW II, 168
Bank of Jamaica, 261, 346, 349, 354, 375, 385, 397-398, 428
Barker, Robert Beacroft, 117
Barnes, Abbie Herberta (m. Kenneth Sterling Thompson), 340n7, 355, 359, 361
Bay of Pigs invasion (1961), 246n3
Belcher, Joan. *See* Moss-Solomon, Joan
Bennett, Louise, 263
"Better Must Come" (Wilson), 322
Bitter, Paul, 279, 383, 405, 429, 448
  tribute to LFK, 448-449
Blackburne, Sir Kenneth (governor general), 259, 260
Black power movement, 295, 297, 298
Blue Mountains, 10, 89, 179, 247
Bolton, Neville, 278
Bonnie View Hotel, Port Antonio, 220, 220n9
Bookal, LeRoy, 434, 434n30
Boston Fruit Company, 34 through 36
Bovell, Christopher D. R., 356, 356n6, 367, 367n18, 381, 410, 434, 458
Bovell, James Lytcott Reece, 277, 277n21, 311, 316, 346, 348, 348n26, 349, 356n6
Bradshaw, Thomas, 160
Britain. *See* Great Britain
British West Indies Regiment (WW I), 59

Index

Brown's Town, St. Ann, 24, 26, 31, 31n15, 53, 75, 98, 217, 268, 286, 303, 339, 340
Business of Nation Building, The (Orane), 425n1, 431, 442n61
Bustamante Industrial Trade Union (BITU), 160n25, 188, 212, 221, 293
Bustamante, William Alexander
    arrest and imprisonment, 158, 188
    as chief minister of Jamaica (1944), 193-194
    as chief minister of Jamaica, re-election (1949), 212
    conflict with wharfingers, and LFK, 157, 203
    death (1977), 376-377, 376n3
    friendship with LFK, 220, 376
    Frome Estate, Westmoreland, and, 155
    general elections, loss to Norman Manley: (1955), 238; (1959), 243
    Gladys Longbridge and, 259, 376
    illness (1965), 295n11
    Jamaica Labour Party, founder of, 187-189
    as National Hero, 376-377
    as populist leader (1937), 155
    as premier of Independent Jamaica, 259-260
    proposal of marriage to Luisa M. Kennedy, 220
    release from prison, 159, 212
    retirement home (Bellencita), 376n3
    as saviour of longshoremen, 159
    strikes on the waterfront and, 156-158
    trade unions and, 160

C

Cameron, Andrew Marc, 402, 403
Cameron, Claire Marie, 419
Cameron, Donald Ian, 340, 340n6, 377, 403, 419
    tribute to LFK, 449
Cameron, Jan Michael, 377, 402, 403
Campbell, Wallace, 329, 349n24
Canada-Jamaica Line, 232
Canada-West Indies Agreement, 242n22
Canadian Imperial Bank of Commerce (CIBC), 130, 133, 229n8, 279, 346, 359, 368, 376, 411
Canadian-West Indian Trade Agreement (1920), 130
Cargill, Cargill & Dunn, 89, 96, 98, 98n10
Cargill, John Henry, 89, 89n6, 90, 96, 118, 121
Cargill, Sidney Raynes, 96, 97, 118, 121
Caribbean Cement Company, 364
Caribbean Labour Congress, 239n

Carter, Jimmy (president), 385
Castro, Fidel, 262, 399
Cecil deCordova & Co. Ltd. *See under* Grace, Kennedy & Co. Ltd.
Cedars of Lebanon Hospital, Miami, 14, 353
C. E. Johnston & Co. Wholesale and Retail General Merchants and Commission Agents. *See under* Johnston, Charles Edward
census. *See under* Jamaica, POLITICS AND SOCIETY
Challis, Brian, 293, 294
Chambers, Mary Anne, 434, 439n53, 440, 440n55
Chang, G. Raymond, 434, 434n29
Charter for Women (JDP), 192
Chevolleau, Mary Dorothy (Dolly), 231, 231n23, 254, 368n20
Churchill, Sir Winston, 208
College of the Holy Cross, Massachusetts, 107n5. *See also under* Kennedy, Luis Fred
colonialism, 3, 27,33,154, 181, 189, 295, 342, 421
commanding heights, 342-343, 342n1, 366, 425
Competent Authority, 168, 171, 209
Constantine, Mitsy, 258, 261, 261n10
Coombs, Allan G. S., 155, 160, 160n26
Craig, Tom, 434, 434n28
Cripps, Sir Richard Stafford, 185, 185n2
Cronin S.J., Rev. Emmet, 84, 107, 107n4
Crown Colony government, 25,186
Cuban War of Independence (1895–1898), 41, 137

D

DaCosta, Charles (Charlie), 203, 211, 293
DaCosta, F.N., 59, 59n11,228
Daily Gleaner, 13n10, 26, 58, 61n22, 62, 63, 77–78, 84, 87,88, 89, 90, 91, 93, 98, 117, 120, 121, 122n15, 130, 133, 152, 155–156, 157, 158, 160, 161n31-n32, 166, 168, 170, 171n29, 172, 173, 179, 180, 184, 186, 189, 190, 192, 200–201, 204, 209, 210, 212, 221, 223, 224, 226, 232, 234, 235, 236, 238, 239n9, 240, 241, 242, 243, 256, 258, 260, 261n8, 270, 272, 273,274, 277, 280, 286n6, 287, 291, 293, 294n9, 297, 298, 301, 303, 321, 322, 329, 331, 357, 358, 363, 364, 365, 377–378, 384, 393, 398, 399, 400, 404, 406, 407, 411, 418n2, 426, 436, 437
*Defence of the Realm Ac*t (UK, 1914), 167–168
Diaz, Amauri Rafael

attendance at LFK's funeral (1982), 410
biography, 310n6
as chairman/CEO, GKCO (1989), 424, 427
challenges faced during 1990s and, 428-429
as deputy chairman, GKCO, 427
as director GKCO board, 346n20
expansion of GKCO financial division under, 429-430
as finance director of GKCO, 346n20
as friend of the Kennedy family, 417, 418
as inductee, Private Sector Hall of Fame (1995), 430
Philosophies & Policies of Grace, Kennedy & Company Ltd. (1993) and, 430
as preference shareholder, 348
retirement as executive chairman (1998), 431
as secretary of Western Meat Packers Ltd., 310
tribute to LFK, 450-451
visit to LFK, London, Ontario and, 37
Dinand S.J., Rev. Joseph N., 108, 108n7, 111
Do Bananas Have a Culture? United Fruit Company Colonies in Central America 1900–1960 (Shragai), 41n5, 43n10
Dodd, Geoffrey Evelyn, 229, 229n8, 276, 313, 321
Dominican Republic
  banana industry (UFCO) and, 40, 42
  caste system in, 44, 44n15
  Cibao region, 40
  "crisis phase" in political history (1902), 41
  grand reconstruction (end of 19th century), 137
  mestizaje, 44
  migration of Cubans to, 137
  Puerto Plata, 41, 42-43,138, 138n8, 137
  United States intervention and, 41, 91
Dunn, Harold Herbert (Daddy Dunn), 90, 97, 118, 121, 128, 129, 129n3, 131, 228, 277

E
earthquake (Jamaica, 1907), 76, 76n5
Eberall, Francis Lunan, 21, 22
Eberall, Gilbert, 61
Eberall, Mary Ann (LFK's maternal grandmother) (m. William Kennedy), 21, 22, 61, 95
Eberall, Rosabella, 61
Eberall, Sarah Louisa, 61
Eden, Ruth Agnes (m. Morgan Hatton Grace), 99
Edward Seaga: My Life and Leadership (Seaga), 358n12, 364n10
Elders and Fyffes, 60
Elizabeth II (queen), 216, 260

Esau, Joe, 434, 434n31

F
Fabian socialism. *See under* Manley, Norman Washington.
Falmouth
  birthplace of Fred W. Kennedy (LFK's father), 22, 95
  founding (1769), 23
  location of William Kennedy's (LFK's grandfather's) store, 24
  map (1870s), 18
  nineteenth century and, 23
  residence of Moss-Solomons, 303
Falmouth Post, 23–24, 24n19
Farquharson, Sir Arthur, 117, 239n9
Figueroa, Anna (m. Del Jarvis), 132, 132n19–n20, 233-234, 233n31, 234n32, 305, 305n9, 327 through 329, 327n3, 458
  tribute to LFK, 451-452
Figueroa, Ferdinand (Ferdie), 254, 275, 275n14, 287
FitzRitson, William, 60–61, 60n17
Fletcher, Walter, 90, 96, 97, 96n4, 191n36, 319, 384, 384n8–n9
Foot, Sir Hugh (governor), 221, 235
Foote, Arnold (Junior), 272-273, 272n3, 319, 458
Four Hs (Ken Hill, Frank Hill, Richard Hart & Arthur Henry), 187, 187n12
Free Jamaica (Hall), 21n10–n11, 23n18
Free Trade Act (UK, 1846), 25

G
gang violence. *See under* Jamaica, POLITICS AND SOCIETY
Garvey, Marcus, 79, 79n17–n18, 158n17
General Agreement on Tariffs and Trade (GATT), 241, 241n21, 242
general elections. *See under* Bustamante, William Alexander; Jamaica, POLITICS AND SOCIETY
George and Branday Company Ltd. 216n3, 276, 276n18, 277, 278, 283
George V (king), 57, 79
George, Vincent Louis (Uncle Vin), 216, 216n3, 294, 294n9
Georgetown Preparatory school, Maryland. *See under* Kennedy, Francis (Paco) Xavier
German U-boats, 167, 167n11, 168n20, 170

Index

Glasspole, Sir Florizel (governor general), 10n5, 384, 406
globalization, 425
Goffe. *See under* banana industry; *When Banana was King*
Grace, Agnes Maude (J. J. Grace's sister), 234
Grace, Alice May (J. J. Grace's sister) (m. William Mitchell Clarke), 206
Grace, Alicia Vere (Countess Charles De Brosses) (J. J. Grace's daughter), 102, 205–206, 205n20, 234, 318, 319, 321
Grace and Staff Community Development Foundation, 407, 436
Grace, Cleveland Raphael, 93, 93n22, 97, 99
Grace, Cynthia Beatrice (J. J. Grace's daughter), 90, 102, 205, 234, 310, 321
Grace, John Greenfield (J. J. Grace's son), 90, 205, 205n19, 234
Grace, John Johnston
  birth (1870), 63
  as co-founder (1922) of Grace, Kennedy & Co. Ltd., 95 through 97
  death (1956), 234
  divestment of stock in GKCO and, 204
  family tree, 86
  as governing director, GKCO, 96-97, 129, 129n4
  as majority preference shareholder, 118-119, 118n3, 129, 129n4, 205
  as managing director, Grace Ltd. BWI (1920), 89
  migration, 205
  as pallbearer at Fred W. Kennedy's funeral (1930), 121
  photo of, 87, 94, 102
  residence (Strawberry Hill), 89-90, 162
  retirement, GKCO, (1947), 204-205
  sale (1922) of Grace Ltd. BWI and, 93, 96
Grace, Kennedy & Company (Canada) in Montreal, 229, 275n10, 311, 366, 367
Grace, Kennedy & Company (Europa) in Rotterdam, 231, 277, 310, 321, 347, 357, 366, 432
Grace, Kennedy & Company Ltd.
  acquisitions and mergers: (1950s), 227; (1960s), 275; (1970s), 344, 347n21
  bonus shares, 228, 228n2, 313, 391
  capitalization of general reserve, 228, 278, 312-313
  Cecil deCordova & Co. Ltd., 76, 254, 256, 275–277, 275n15, 280, 383
  cigarette factory and, 146n3
  coconut industry and, 169
  conversion of preference shares to ordinary shares, 350-351
  co-ownership (1920s) of Grace Wharf with Jamaica Fruit & Shipping Co. Ltd., 100
  employee investment trust, 278, 313, 348, 379, 404
  employee shares (defined), 228n3
  employee shares, issuance of, 228, 278, 312
  first directors' meeting, 96
  fortieth anniversary, 252 through 255
  founding (1922), 93, 95 through 97
  free medical care for staff (1947), 207, 207n28, 308, 309, 405, 451
  general importers and commission agents (1930s), 130
  golden jubilee, 316 through 321
  governance (1960s), 276-277
  Grace brand, 271 through 274
  group life insurance, 228, 308, 309,
  Jamaica Stock Exchange (JSE) and, 5, 350, 350n31, 426 (*see also* Jamaica Stock Exchange)
  name change (GraceKennedy Ltd.), 4 (*see also under* GraceKennedy Ltd., new logo)
  nationalization proposal by PNP government, 365
  overseas subsidiaries, 229-230 (*see also* Grace, Kennedy & Co. (Canada); (Europa); and (UK)
  Philosophies & Policies of Grace, Kennedy & Company Ltd. (1993), 430 (*see also under* Diaz, Aumari Rafael)
  PNP government (1972-1976) and, 242 through 351
  post-war boom in shipping and, 170, 204
  preference shares and, 119, 205
  profits: (1922), 100; (1930s), 131; (1950s), 227; (1960s), 276; (1970s), 345; (1980s), 426n5)
  as public company (1989), 5, 426
  quotas during 1970s, 378
  quotas during WW II, 167n13
  reaction to proposed nationalization, 366-367
  registered offices, 64 Harbour Street, 94, 96, 97, 100, 130n10, 232, 304, 346, 356, 368, 383

473

restrictions on merchant shipping (1940s), effects on, 167
as shipping agents, 130, 131, 170, 232, 204, 224, 232
sixtieth anniversary, 404 through 407
strikes (1938) and, 154 through 159
superannuation scheme, 211, 279, 293, 309, 313, 349, 405, 451
training of staff and, 278-279
we care, as mantra, 306, 430, 436, 440, 441
Western Meat Packers Co. Ltd., 298, 310, 310n6, 391
See also Kingston Wharves Ltd.; Kennedy, Luis Fred.
Grace, Kennedy & Company Limited: Story of Jamaican Enterprise (Hall), 4, 21n10, 63, 96n2, 167n13, 168n20, 173n37, 275n11, 280n36, 429n16, 431
Grace, Kennedy & Company (Shipping) Ltd., 232, 232n26, 273,
Grace, Kennedy & Company (UK) in London, 231, 310, 321, 347, 357, 366, 432
Grace, Kennedy Foundation, 405, 427
GraceKennedy Ltd.
    acquisitions, 436-437
    as centenarian, 442
    corporate governance, 434
    financial services, 437, 437n42
    as global consumer group, 436
    group revenue (2020), 438, 441
    net profit increases (2011), 435
    new corporate headquarters, 437
    new logo, 431n20
    succession and, 425 through 442
    vision (2020), 432, 436
    vision (2030), 438, 441
Grace, Kennedy Remittance Services Ltd., 429
Grace Ltd. Jamaica (BWI), 62-63, 75, 89, 91-92, 93n21,
Grace, Michael Sheffield (J. J. Grace's brother), 62, 63, 87, 88, 93n23
Grace, Morgan Hatton (J. J. Grace's brother), 93, 93n23, 99–100, 206
Grace, Morgan Stanislaus, (J. J. Grace's father), 63, 99n13
Grace, Walter Henry (J. J. Grace's brother-in-law), 93, 97, 99, 206
Grace, William Russell (J. J. Grace's uncle), 62, 62n24, 99n12–n13, 206, 206n4, 207

Grant, William Wellington Wellwood, 158, 158n17, 159
Great Britain (UK)
    equalization of duties (19th century), 25
    general election (1945), 208
    preferential tariffs and, 262
    as principal trading partner of Jamaica, 130n13
    World War I and, 57 through 60
    World War II and, 168 through 171
Great Depression, 91, 129, 129n6, 142, 376, 435
Great War, The. *See* World War I
Greenfield, Eleanor Vere (m. John Johnston Grace), 89–90, 97, 205, 234
Greven Holdings Limited, 366, 367
Gun Court Act (1974), 12, 12n9

H

Haile Selassie, 322
Hart, Ansel Henry Lester. *See* "Memories and Reflections: More on Old York Castle" (Hart)
Hart, Richard, 4, 155m5–n6, 157n14, 161n32, 167n14, 168n20, 187, 188n18
Henry, Georgianna (LMK's paternal grandmother) (m. Diego Loinaz Arteaga), 137
Heron, Colin, 346, 346n20
Hitler, Adolf, 169
Holocaust, 201n2
Holy Trinity Cathedral, Kingston, 76, 77, 77n13, 89, 120, 376, 419
Horn, Mary (m. Douglas Judah), 179, 179n13, 418, 418n2
Huggins, Sir John (governor), 193, 210
hurricanes, 59-60, 59n12, 106, 106n2, 403n1

I

influenza epidemic, post-World War I, 79
International Monetary Fund (IMF). *See under* Jamaica, ECONOMY
Issa, Abe, 77, 84, 85, 190, 190n30, 191, 220, 304
Issa, John, 273n7

J

Jamaica
    ECONOMY:
        bauxite industry and, 11, 262, 342, 343
        budget deficit (1980) and, 398
        collapse (1990s), 428
        depression (1890s) and, 25

Index

dollar devaluation and, 345, 345*n*16, 351, 368, 385, 400, 428
GDP (1970s), decline in, 343
International Monetary Fund (IMF) and, 351, 360, 362, 385, 397–398, 398*n*3, 400, 428, 436*n*38
liberalization (1980s), 425-426
National Debt Exchanges (NDX) and, 436
post-independence policy, 262
prosperity (1960s), 295
sugar industry and, 23, 129
tourism, 50 through 52, 204, 220, 239, 242, 262, 295, 342
unemployment rates: (1944), 193; (1972), 295, 295*n*12; (1993), 429
Year of Economic Emergency (1977) and, 362
WW I and, 57-59, 79
WW II and, 166-167, 170
See also banana industry
POLITICS AND SOCIETY:
 adult suffrage (1944) and, 193
 census (1861), 20*n*3, 21; (1881), 76*n*6
 democratic socialism, 364 (*see also* Manley, Michael)
 gang violence: (1976), 358; (1980), 397
 general elections: (1944),193; (1955), 237-238; (1959) 243; (1962), 259; (1967), 295); (1972), 321-322; (1976), 360; (1980), 399; (1989), 428
 Independence, 259 through 261
 labour unrest, strikes (1938), 154-159
 migration (1838-1914), 129, 129*n*5
 nationalization of private companies (1976), 363-364
 National State of Public Emergency (1976), 10 through 13, 10*n*5, 160, 351, 358, 359
 post-emancipation, 20-21
 race and classism, 4, 20*n*3, 79, 91, 193, 421
 referendum (1961), 244 (*see also under* Manley, Norman Washington)
 retail trade and supermarkets (1960s), 274
 schooling (19th century), 27-28, 30
 social life (1940s), 181
 trade union movement (1930s), 160
 West Indies Federation, 239-244
Jamaica Banana Producers Association. *See under* Johnston, Charles E.

Jamaica Broadcasting Corporation, 10, 273, 273*n*6
Jamaica Chamber of Commerce 121, 171*n*29, 209, 240-242, 364
Jamaica Citizens Bank, 346, 364
Jamaica Coconut Producers Association Ltd., 169, 169*n*23
Jamaica Democratic Party, 184, 185, 190-192, 194
Jamaica Fruit and Shipping Co. Ltd., 100, 100*n*14, 129, 232. *See also under* Johnston, Charles Edward
Jamaica High School, The (Jamaica College), 28, 85*n*18
Jamaica Imperial Association (JIA), 239, 239*n*9, 241*n*2. *See also under* Kennedy, Luis Fred, PUBLIC LIFE
Jamaica Labour Party, 185, 185*n*1, 187-189, 221*n*3, 244. *See also under* Bustamante, William Alexander
Jamaica Manufacturer's Association, 242, 363
Jamaica National Export Corporation (JNEC), 332, 345, 427
Jamaica Nutrition Holdings Ltd., 349, 364
Jamaica Observer, 438 *n*51; 441, 441*n*56
Jamaica Promotions Corporation (JAMPRO), 427
Jamaica Public Service Ltd., 181*n*22
Jamaica Reserve Regiment 58–59, 63
Jamaica Scholarship, The (1881), 30, 30*n*13
Jamaica Stock Exchange (JSE), 5, 311*n*10, 350, 433, 433*n*25, 437–438
Jamaica War Relief Fund, 58
Jamaica Workers and Tradesmen's Union, 155
Jamaican Constitution (1944). *See under* Manley, Norman Washington
"Jamaicanization," 11
Jaws (Buckley), 354
Jensen, Peta Gay, 24*n*22, 53–54, 53*n*15, 54*n*17–*n*18, 54*n*20
Jesuits, 82–83, 84, 288, 338, 389, 420, 436
Jim Crow laws (US south), 110
Johnson, Lyndon (president), 260
Johnston, Agnes Mary (m. Morgan Stanislaus Grace), 63
Johnston, Charles Edward (Maas Charlie)
 Atlantic Fruit Co. (1904) and, 51*n*8
 banana industry and, 60, 100*n*15, 130
 birth (1871), 51*n*8
 as business mogul, 235
 C. E. Johnston & Co. Ltd. (1891) and, 5
 death, 234

as director of Kingston Wharves Ltd., 201
Grace wharf and, 129-130
Jamaica Banana Producers Association and, 117, 159, 235
Jamaica Co-operative Fruit and Trading Co. and, 51n8
Jamaica Fruit and Shipping (1919) and, 50n1
as Jamaican patriot, 100
Kennedy family and, 100
photo of, 48, 117
Shipping Association of Jamaica and, 160
Johnston, Ernest Michael, 235, 339n4, 392
Johnston, Lewis ("Daniel") Ewart, 30, 30n8
Johnston, Marjory Lesa (m. Francis (Paco) Xavier Kennedy), 339, 339n4, 374, 377, 390-392, 403, 409, 417, 418–419
Judah, Rev. Charles L., 286, 286n6, 287
Judah, Douglas, 84, 84n11, 179, 179n13, 190, 191, 222, 236, 239, 240, 241, 286n6, 293, 294, 304, 364, 418n2

K

Kamicka, Gladstone, 229, 230, 253, 255, 277
Keith, Minor Cooper, 33, 36
Kennedy, Alexander (overseer, 19th century), 19
Kennedy, Amanda Mary (m. Wayne Anthony Williams), 373, 374, 377, 380, 401, 402, 420n6, 457
Kennedy, Anne Marie, 234, 287, 420
Kennedy, Anthony (Tony) Charles, 283, 287–288, 287n11, 323
  tribute to LFK, 452
Kennedy, Carmen María (Mim) (LFK's sister) (m. John William Valliant), 50, 77, 77n12, 85, 103, 104, 111, 121, 122, 122n15–n16, 133, 133n23–n24, 150, 174, 176, 178, 183n28. 198, 257, 319, 321, 405, 408, 410, 420
Kennedy, Cathrine Lesa, 339, 377, 393 through 396, 393n13, 401, 402, 403, 416, 418, 458
Kennedy, Celia Jean (LFK's daughter)
  Amberly cottage and, 176, 180
  birth (1942), 178
  as boarder, Servite Convent of the Assumption, Brown's Town, 214, 217
  death (2008), 418
  as postgraduate student, Boston College, 339
  as resident of Canada, 377
  as Servite nun, 257, 282, 284-285
  as social worker, Jamaica, 19n1
  as student, University of the West Indies (UWI), 339
  as teacher at Servite Convent school, 286
Kennedy, Charles David, 393 through 396, 393n12, 401, 402, 416, 418, 418n4
Kennedy, Elizabeth Louise (LFK's daughter)
  birth (1953), 213, 216
  daughter, Laura Riley, 416, 419
  death (2019), 419
  first marriage (1979) to John Patrick Riley, 386
  grandchildren, 419
  as resident of Paris, 417
  second marriage (2003) to Francis Kenneth Sealey, 419
  as student, Campion College, 340
  tribute to, 419
  as student, St. Benedict's, Minnesota, 340
Kennedy, Fillan Mable Josephine (LFK's aunt), 22, 61, 82, 132n22, 161, 182-183, 183n28
Kennedy, Francis Xavier (LFK's brother)
  birth (1924), 108, 109
  childhood and schooling, 103, 104, 134, 178, 201
  children, 198, 207, 234, 287
  death, 286-287
  death of first wife, 234
  first marriage to Marie Margaret Cecile Lemay, 207
  Grace, Kennedy and, 207, 229 254, 275
  migration, 257
  second marriage to Anne Marie Angerhausen, 256, 256n2, 257
Kennedy, Francis Xavier (LFK's son, Paco)
  birth (1940), 176, 178, 388
  as college student at Holy Cross, 389-390
  courtship with Marjory Johnston, 391-392
  death (2014), 419
  as firstborn, 233, 388-389
  hobbies, 390, 395
  illness, 310, 311, 391, 418-419
  marriage (1973) to Marjory Lesa Johnston, 336, 392
  migration, 393
  nickname, 178
  Order of Distinction (Commander), 418
  relationship with children, 395-396
  as student, Georgetown Prep High School, 233, 389, 390
  GRACE, KENNEDY & CO. LTD. CAREER:

Index

deputy group managing director (1975), 351, 392
director of GKCO board (1970), 309
director of GKCO subsidiaries (1967), 280
divisional director of trading (1978), 383
first employment (1959), 233, 390
manager of Harbour Cold Stores Ltd. (1967), 279n33, 280
manager of Shipping Division, 418
manager of Western Meat Packers Ltd. (1971), 310
preference shareholder, 346
project manager, 431
public profile while at, 393
retirement (2005), 418
working relationship with LFK, 391

Kennedy, Fred William (LFK's father) (m. Luisa Matilde Arzeno Abreu)
as auditor and banana agent (1901-1909), UFCO, Sosúa, Dominican Republic, 37, 41
automobile (1915), 73, 78, 78n15
birth (1874), 22
as cabinet maker, 45, 45n2
children of, 104
convert to Catholicism, 44, 77, 84
as co-founder of GKCO, 95 through 97
death (1930), 119
employee of Boston Fruit Company, Port Antonio, 34
employee of J. H. Levy & Son, 26, 31
employee of UFCO, Port Antonio, 36, 50
letters to his son, LFK, 107, 111, 112
as manager of Grace Ltd. BWI, 88
as manager of Montego Bay Agency, UFCO, 51, 58
marriage to Luisa Arzeno Abreu, Puerto Plata, Dominican Republic (1908), 44-45
as Methodist (impediment to marriage), 44
obituary, *The Daily Gleaner*, March 6, 1930, 120
photo (1929) with C. E. Johnston, 117
poem, *A Question*, by, 64-72
purchase of shares from Grace family, 119
resident (1917), Kingston, 75
social life, Dominican Republic, 43
tributes to, 63, 121
York Castle High School student, 28-30

Kennedy, Fred William (LFK's son) (m. Georgianne Ruth Thompson)
birth (1950), 216
children, 373, 380, 416, 420
conversation with LFK re: Jamaican politics (1976), 10 through 13; migration (1976), 354-356; philosophy (1972), 338-339
engagement (1976) to Georgianne Ruth Thompson, 355
GKCO shares and, 313
as Jesuit, 288
letters from LFK, 268-269, 288-289
marriage to Georgianne Ruth Thompson (1976), 357, 361
migration to Canada (1976), 355
as student, St. George's College (STGC), 246, 246n3
as student, University of West Indies (UWI), 340
as teacher (Jamaica), Trench Town Comprehensive, 11n7
as teacher (Jamaica), York Castle High School, 340

Kennedy, Frederick Albert, 61

Kennedy, Gloria Mercedes (LFK's sister) (m. Owen Cathcart Plant), 82, 121, 122n15, 134, 150, 178, 182n24, 257, 420

Kennedy, Henrietta Eliza (LFK's aunt), 22

Kennedy, José Antonio (LFK's brother)
attributes, 323
birth (1913), 54
Catholicism and, 288
children, 283
courtship with Altagracia (Gracie) Loinaz, 201
death (1993), 420
as director of board of Grace, Kennedy, 228, 276
early education, 85, 121
first employment, GKCO (1942), 202
as law intern (1932), 122
as manager, shipping division, Grace, Kennedy, 228, 232
marriage to Alta Gracia (Gracie) Loinaz, 204n14, 207
migration (1972), 322
repatriation, 420
retirement (1973), 344n13
service awards (GKCO): (1962), 254-255; (1972), 321
as solicitor, called to the bar, 133

Kennedy, Julia Claire (m. Martin Joseph Vaz-Jones), 416, 420, 420n6, 457

Kennedy, Kevin, 234, 287
Kennedy, Lareine Agnes, 207, 234, 287
Kennedy, Lena Stewart (LFK's aunt), 22, 61
Kennedy, Long Andrew, 20, 20n4
Kennedy, Louisa (Louise) Matilde (LFK's sister)
    (m. Simon Soutar), 54–55, 85, 92n15, 121,
    122, 133-134, 178, 410, 420
Kennedy, Luis Fred
  as bachelor, 132
  birth (1908), 45
  childhood personality, 52-53
  courtship with Lydia Loinaz, 145-146
  death (1982), 410-411
  death (1930) of father, effects on, 119
  early childhood (Montego Bay), 52
  engagement (1937) to be married, 147
  as firstborn, 3, 121-122, 252, 420
  friendship with Carlton Alexander, 333
  friendship with James Moss-Solomon, 302
  honeymoon, 151-152
  marriage to Lydia M. Loinaz, 149
  relationship with his mother, 179
  as student, St. George's College
    (STGC), 82-85
  as teacher of Spanish, STGC (1924), 100-101
  upbringing in Kingston, 76-78
  visit to *alma mater* (1972), 335, 337
  COLLEGE OF THE HOLY
    CROSS EXPERIENCE:
    culture and race at Holy Cross, 110
    extra-curriculars, 112
    foundation of future success, 114
    freshman, enrolment (1924), 106-107
    friends, 105, 112, 150
    graduation, 113-114, 116
    Jesuit curriculum, 111
    pranks, 113
    roaring twenties, 109
    yearbook citation, 114-115, 115n20
  GRACE, KENNEDY & CO.
    LTD. CAREER:
    acquisitions and mergers/local subsidiaries:
      (1950s) 227; (1960s), 256, 275-276, 277,
      280; (1970s), 347; (19080s), 426n5, 428
    assistant manager (1930), 127 through 131
    award, 34 years of service (1962), 252, 255
    business trips (1930), 130-131
    director (1930), 128
    employee investment trust (1967), 278
    first employment, GKCO, (1928), 119

    formation of Kingston Wharves Ltd., 207
    formation of Shipping Association of
      Jamaica, 160
    fortieth anniversary, 255
    free medical care, 207
    governance structure (1960s), 276-277
    governing director, 205, 207
    Kingston Wharves Ltd., twenty-first
      anniversary, and, 294
    mentorship of Carlton Alexander, 331
    mentorship of family members, 233
    overseas subsidiaries (1950s), 229-231
    preference shares, 205, 309, 348n26,
      350- 351
    preparations for GKCO to go public
      (1976), 350-351
    promotion and training of young management, 231-232
    reaction to proposed nationalization, 366
    resignation as chairman of executive
      committee (1972), 344
    resignation as governing director
      (1973), 346
    retirement as chairman 1976), 356-357
    strikes on the waterfront (1938), 156
    succession, 277, 357
    superannuation scheme (*see under* Grace,
      Kennedy & Co. Ltd.)
    training of family/employees (1960s), 278
      through 280
    transfer of shares to Lydia M. Kennedy and
      five children, 313
    welcome speech at fiftieth anniversary, 320
  MARRIED/FAMILY LIFE:
    adjustment to life in Canada, 374-375
    Amberly cottage (1940s), 179-180
    Catholicism and, 77, 217, 266, 268, 284,
      288-289, 421
    family man (1940s), 177 through 179
    fortieth wedding anniversary, 373, 377
    grandchildren, 402-403
    hobbies: boating, 175, 180; farming, 245
      through 248; gardening, 266-267,
      267n2; reading, 10, 119, 218, 354, 450
    homelife (1950s), 215 through 218
    homelife (1960s), 284-285
    illness, 13-14, 337-338, 353, 385,
      386, 386n16
    letters to his son, Fred W. Kennedy, 268-
      269, 288-289

migration to Canada (1976), 359, 374 through 376
pressures, 220
Puerto Plata, Discovery Bay, beachside home and, 265 through 269
PHILOSOPHICAL BELIEFS:
   classical liberalism (John Stuart Mill), 421
   democratic ideals, 192
   egalitarianism, 421
   scholasticism, 114
   social justice, 212, 308-309, 420
PUBLIC LIFE:
   call for resignation of British trade commissioner, 171
   chair, fundraiser for Sts. Peter and Paul Church and school, 281
   co-chairman with Hugh Shearer, Port of Kingston Joint Industrial Council, 221-222
   defender of private enterprise, 211
   government mission to Cuba, 172
   member of Jamaica Imperial Association (JIA) in defence of federation, 240
   member of Legislative Council (1953), 221, 222-224
   member of St. George's College Old Boys' Association, 281, 286n7
   opponent of price controls and import quotas (1943), 173, 208
   opponent of socialism, 209
   organizer of JDP (Jamaica Democratic Party), 191
   philanthropy, 286
   public criticism of the governor (1939, 1948), 169, 210-211
   in support of Norman Manley's pitch for federation, 239, 240
   vice-president, Chamber of Commerce (1948), 209-211
Kennedy, Lunan Dexter (LFK's uncle), 22, 61–62
Kennedy, Mary Teresa (LFK's daughter), (m. Donald Ian Cameron)
   Amberly cottage and, 179-180
   birth (1945), 176, 178, 183
   as boarder at Servite Convent of the Assumption, 214, 217
   children, 377, 403, 416
   as graduate student, University of Toronto, 340
   grandchildren, 419
   marriage to Donald Ian Cameron, 336, 340
   as professor, University of Findlay, Ohio, 419
   as Servite nun, 282, 285, 286
   as student, University of West Indies (UWI), 340
   as teacher of French, 340
   tribute to LFK, 449-450
Kennedy, Michèle Marie, 283, 287, 323. 458
Kennedy, Patrick Joseph, 283, 287n10, 322
Kennedy, Philip Thomas, 204, 204n14, 283, 287n10, 294, 410. 458
Kennedy, Rose Eliza (LFK's aunt) (m. William Morris), 22, 45
Kennedy, Sarah Elise (m. James Patrick Near), 264, 401, 402, 403, 416, 420n6, 457
   as visual artist: front cover design, *Firstborn*; acrylic painting, 264
Kennedy, Stephen Francis, 234, 286, 286n8, 287, 410
   tribute to LFK, 453-454
Kennedy, Theodore (LFK's uncle), 22, 28-29, 30, 61
Kennedy, William (LFK's grandfather)
   ancestry, 19
   birth (1839), 19
   charitable works, 25
   children, 22
   death (1915), 60-61
   home, location in Falmouth, 18, 22, 267
   leg amputation, 53
   marriage (1867) to Mary Ann Eberall,
   as merchant, 21, 23-24, 25
   as Methodist, 21, 21n13, 44, 22, 25
   Methodist chapel, founded (1907) by, 48, 54
   obituary, 61
   store, location in Falmouth, 18, 24
King, Martin Luther, Jr., 420
Kingston
   Catholic population, 83
   hotels (early 1900s), 76
   labour protests (1938) and, 156 through 158
   map (1920s), 74
   original grid of city streets, 76
   population: (1861), 23; (1881), 76; (1914), 76
   railroad, 35
   tramcars, 76
   unemployment (1978) in, 362
   violence (1976) in: 11, 397, 397n1
Kingston Wharves Ltd.
   congestion at finger piers (1950s) and, 224

expansion (1969), 298
founding (1945), 173, 201
Grace, Kennedy's divestment of shares and, 433
medical plan for employees and, 207
Newport West, 292 through 295, 349, 433
Paul Scott and, 231, 232, 293, 309, 349, 350n30 (*see also* Scott, Laurence Paul
preference shares and, 309
Royal Mail Line's pier (1963) and, 292
S.S. Santa Rosa and, 291, 294
twenty-first anniversary (1966), 291
See also under Grace, Kennedy & Co. Ltd., and Kennedy, Luis Fred
Ku Klux Klan (KKK), 110

L

Lanasa, Antonio, 35
Last Colonials: The Story of Two European Families in Jamaica (Jensen), 24n22, 53–54, 53n15, 54n17–n18, 54n20
Lemay, Marie Margaret Cecille (m. Francis Xavier Kennedy), 198, 207, 207n29, 234, 256
Lenn, Bernice Agatha (Moy) Chin (m. Selwyn Carlton Alexander), 318, 329, 329n5, 331, 331n10
"Let the Power Fall on I" (Romeo), 322
Levy, Joseph Henry, 26, 31, 31n14
List, Sibrandt Duhn, 50, 50n1, 58, 100, 117, 121, 160, 160n29
Loinaz, Alta Gracia María (LMK's sister) (m. José Antonio Kennedy), 91, 138n6, 148n12, 150, 201, 201n4, 203, 283, 287, 420
Loinaz, Carmen Celia (Chela) (LMK's sister), 91, 113, 138n6, 148n12, 150, 150n21
Loinaz, Charles Eugenio (LMK's brother), 91, 138n6, 140, 150, 150n22, 150n23
Loinaz, Charles Henry (LMK's father), 91, 91n13, 137-138, 140, 150n22, 150n23
Loinaz, Diego (LMK's paternal grandfather), 137
Loinaz, Rev. Diego (LMK's uncle), 150
Loinaz, Ines Georgiana (LMK's sister, m. Emanuel Jacob Thomen), 91,138n6, 150n24
Loinaz, José (LMK's brother), 91, 135, 138n6, 139, 140
Loinaz, Lydia Mathilde (m. Luis Fred Kennedy)
adjustments to life, Jamaica (1940s), 180-182
adjustments to life, London, Ontario (1970s), 375
birth (1910), 137
as business executive, American International Underwriters (AIU), 140-141, 147
childhood and schooling (Mount Vernon, New York), 139-140
courtship with LFK, 145-149
death (2002), 418
as deed holder, Puerto Plata, Discovery Bay, 265
early childhood, Dominican Republic, 138
eightieth birthday, with grandchildren, 416
family life, 45 Lady Musgrave Road (1950s), 216-220
fiftieth anniversary, GKCO, and, 317-319, 321
fortieth wedding anniversary, 377
honeymoon, Bermuda, 151-152
illness, visits to doctors in New York, 165, 183, 201
LFK's death, London Ontario (1982) and, 410
marriage to LFK, 149-150
migration to Canada (1976), 359
repatriation to Jamaica (1990), 417
transfer of shares to, 313
wedding photo, 143
Loinaz y Amunabarro, San Martín, de, 137n1
Loinaz, Mercedes Amanda (LMK's sister), 91, 138n6
London Times, 238, 238n5
Lunan, John (LFK's ancestor), 22

M

Manchester Guardian, 166, 166n5
Manley and the New Jamaica: Selected Speeches and Writings 1938–1968 (ed. Rex Nettleford), 185n3–n4, 241n18, 241n27, 243n32–n33, 297n19
Manley, Michael
biography, 238n2
commanding heights and, 343
death (1997), 428n12
democratic socialism (1976), 363-364
as Joshua, 322
letter (1982) to Lydia M. Kennedy, 412
PNP campaign (1955) and, 237-238
in praise of GKCO, 406
as prime minister: (1972), 321-322; (1976), 360; (1989), 428
as president (1969) of PNP, 297-298
victory speech (1972), 322
world economy and, 385

Index

Manley, Norman Washington
    affidavit for release (1938) of Bustamante from prison and, 159
    as chief minister: (1955), 224; (1959)
    death (1969), 297
    Fabian socialism and, 185n2
    farewell speech (1969), PNP conference, 297
    father of the nation, 212
    Jamaican Constitution (1944) and, 186-187, 193
    as leader of the opposition: (1944), 194; (1949), 212; (1962), 260
    as nationalist, 243, 243n30
    negotiations (1938) with wharfingers, 159
    oratory, 186
    People's National Party (PNP), founder of, 185 through 187
    referendum and, 243-244
    self-government, proponent of, 186
    Trade Union Advisory Council and, 160
    West Indies Federation and, 237 through 244

Manley, Rachel, 238n2

Manning, Sir William Henry (governor, during World War I), 57, 57n3, 58

Marley, Bob, 359

Marley, Rita, 359n16

Matalon, Aaron, 173, 173n37

Matalon, Moses, 292, 292n3, 293

McKay, Claude, 79, 79n19

McLaughlin, Lucy Ellen (second wife of Dexter Lunan Kennedy), 62

Meaney, Charles (Chuck) Francis, 105, 112–113

"Memories and Reflections: More on Old York Castle" (Hart). 28–29, 28n3, 29n4–n6, 30n10–n11

Memories of What Made Grace (Terfloth), 232n24, 272n1, 311n11

Merchant Adventurer: Story of W. R. Grace (James), 9 3n20, 98n11, 99n12, 206n23–n24

Methodism. 25n24, 28-29, 44n17. *See also under* Kennedy, William, as Methodist; Kennedy, Fred William

Middlemarch (Elliot), 421

Mill, John Stuart, 421n8

"Minutes of Board of Directors" (Grace, Kennedy & Co. Ltd.), 14n13, 96n3, 118n2, 119n4, 127, 128n2, 130n8, 207n28, 228n2–n4, 275n13, 309n4, 310n6, 313n13, 344n13, 356n7, 365n12, 367n17, 378n9, 383n2, 434n27

Mocatta, Mary Barbara (m. John Greenfield Grace), 205, 205n19

Montego Bay
    census (1861), 23
    crime, 286, 358
    early twentieth century, 51-52
    UFCO agency, 59

Moody Commission of Enquiry for Port of Kingston, 293n7

Morant Bay Rebellion, 1865, 25

Morgan-Brown, Dorothy Ann (m. Michael Sheffield Grace), 63, 89

Morris, Katherine (LFK's first cousin), 45, 45n21

Morris, William, 45, 45n20

Mosley, John (Kennedy family doctor, 1940s), 10, 180, 180n16, 215, 216

Moss-Solomon, Alfred Henry, 300, 303

Moss-Solomon, Copeland, 132n19, 300, 305n8

Moss-Solomon, Donald Aidan, 229, 229n11, 253, 255, 300, 303

Moss-Solomon, Enid May, 98

Moss-Solomon, James (Jimmy) Seivwright, Jr., 300, 301, 302n2, 303 through 306, 458

Moss-Solomon, James (Uncle Jim) Seivwright, Sr.
    appointment as accountant (1922), GKCO, 96, 98
    as assistant manager (1930), GKCO, 127-128
    birth (1894), 97
    character, as described by Carlton Alexander, 305
    character, as described by his three children, Peter, Joan, and Jimmy, 303 through 306
    childhood years, 98
    as company secretary (1930), GKCO, 128
    death (1977), 377-378
    family chart, 300
    family relation to Carlton Alexander, 132n19
    as friend of Fred W. Kennedy, 97
    as general manager, GKCO, (1950s), 228
    preference shares and, 205
    relationship with Kennedys, 302-303
    retirement, board director, GKCO (1974), 348
    retirement, managing director GKCO (1972), 313
    service awards (GKCO): (1962), 252, 255; (1972), 321
    as shareholder, GKCO (1920s), 99
    as trustee of employee investment trust (1967), 278
    work experience prior to joining GKCO, 98

Moss-Solomon, Joan Elizabeth (m. Michael Belcher), 180, 229, 300, 301, 302 through 305, 302*n*2, 304*n*6, 377, 458
Moss-Solomon, Joseph Henry, 97, 300, 303, 303*n*3
Moss-Solomon, Peter Norton
    attendance at LFK's funeral, Canada, 410
    as bridge player, 390
    as chair (1982) of sixtieth celebrations, GKCO, 404
    family chart, 300
    as finance director, GKCO, 427, 417*n*8
    as firstborn, 190*n*29, 233
    first employment at GKCO (1958), 233
    as intern (1967), Price Waterhouse, 279
    LFK's invitation (1958) to join GKCO, 226, 233, 304
    as ordinary shareholder (1958), 229
    as preference shareholder (1967), 348
    retirement as chief financial officer (2006), 427*n*8
    as student, McGill University, 304
Mount Vernon, Washington. *See under* Loinaz, Lydia Mathilde, early childhood and schooling
Moyne Royal Commission, 186, 186*n*7, 239, 239*n*6
Murray, Dorothy Grace, 132*n*19, 275, 327
Murray, Minna Louise (m. Copeland Moss-Solomon), 132*n*19, 324
Murray, Rosina Grace, 132*n*19, 324, 326
Murray, Rupert Garland, 324, 327, 329*n*7–*n*8
Muschett, Edward (Ed), 276 through 279, 278*n*29–*n*30, 309, 346, 348, 348*n*24, 348*n*26, 408, 410, 427
Myrtle Bank Hotel, Kingston, 76, 132, 174, 182, 190*n*30, 333, 437

N
National Debt Exchange (NDX) *See under* Jamaica, ECONOMY
National Processors Ltd., 277, 277*n*26, 280, 383
National State of Public Emergency (1976). *See under* Jamaica, POLITICS AND SOCIETY
National Workers Union (NWU), 238, 238*n*2, 293
Near, James Patrick, 420*n*6
Nethersole, J. M., 128, 128*n*2
Nethersole, Noel Newton, 212, 212*n*16
Newport West, Kingston. *See* Kingston Wharves

New York Daily News, 150–151
New York Times, 358, 358*n*13, 397*n*12
Norton, Ismilda (Flossie) (m. James Seivwright Moss-Solomon), 190*n*29, 229, 301, 304, 317, 318
Nugent, Wilhelmina (m. Diego Loinaz Arteaga), 137*n*3
N.W. Manley and the Making of Modern Jamaica (Bertram), 79*n*16, 189*n*23, 193*n*48–*n*49, 194*n*51, 212*n*17, 238*n*2, 239*n*10, 296*n*17

O
Ocean Dominion Steamship Corporation, 170
O'Connor, William (Bill) Francis, 105, 112, 150, 337
Operation GROW, 343
Orane, Douglas, 424, 425, 425*n*1, 427, 429*n*15–*n*16, 431 through 435, 438, 439, 442
    appointment as CEO of GKCO, (1995)
    appointment as chairman/CEO (1998), 431
    retirement as CEO (2011)
    retirement as Executive Chairman (2014), 435
Order of Servants of Mary (OSM), 257, 285, 286, 286*n*6. *See also* Kennedy, Celia Jean; Kennedy, Mary Teresa
"Origins of Kingston Wharves Ltd." (Diaz), 293*n*4
Orrett, Marian Joan (m. Laurence Paul Scott), 231*n*22

P
Pan Africanism movement, 79
Pan American World Airways, 101*n*16, 131*n*16, 133*n*23, 151, 151*n*27, 165, 167, 170, 201*n*3
Patterson, P. J. (prime minister), 345, 360*n*23, 399, 428
Pearl Harbour bombing (December 1941), 172
People's National Movement (PNM). *See under* Williams, Eric
People's National Party (PNP). *See under* Manley, Norman Washington; Manley, Michael
Plant, Margaret, 257
Plant, Owen Cathcart, 51, 178, 178*n*7, 198, 319, 387, 420
Plant, Owen Cathcart (OC), (son of Owen Cathcart Plant and Gloria Kennedy)
    tribute to LFK, 454
Plant, William Henry, 178*n*7, 257, 420
Politics of Change (Michael Manley), 322, 322*n*7–*n*8, 342*n*2, 343*n*4

Port Antonio
  banana industry and, 34-35
  as birthplace of Jamaican tourism, 51
  as centre of operations for UFCO, 37
  extension of railroad from Kingston, 35
  Titchfield High School, 51, 51*n*7
  Tichfield Hotel, 51
Port Bustamante, 293
Port of Kingston Accord, 222, 293
Port of Kingston Joint Industrial Council (KJIC). 222, 232 *See also under* Kennedy, Luis Fred, PUBLIC LIFE
Port Services Ltd., 309
post-emancipation (1839–1908). *See under* Jamaica, POLITICS AND SOCIETY
Potsdam School (Munro College), 28, 28*n*2
Private Sector Organization of Jamaica (PSOJ). 332, 357, 357*n*9, 362–363, 427. *See also under* Alexander, Selwyn Carlton, as president of
Prohibition (US, 1920–1933). *See under* United States (US)
Promised Land, A (Obama), 420, 420*n*7
Public State of Emergency, Jamaica. *See under* Jamaica, POLITICS AND SOCIETY
Puerto Plata, Discovery Bay, Jamaica. *See under* Kennedy, Luis Fred, MARRIED/FAMILY LIFE
Puerto Plata, Dominican Republic. *See under* Dominican Republic
Puritanism, 110

Q
"Quiéreme Mucho" (song, by Roig), 145, 145*n*1, 165, 423
Quinn, James (Jimmy) F., 105, 112, 113, 113*n*15, 114, 337

R
Radio Jamaica (RJR), 364
Rastafarianism, 295, 298
Reagan, Ronald (president), 425
Red Book of the West Indies, The (Macmillan), 75*n*1–*n*2, 76*n*4, 88*n*3, 92*n*14, 261*n*9, 275*n*15
Red Scare (US, 1920s), 110
Revenue Canada, 376
Richards, Sir Arthur Frederick (governor, during World War II), 160, 160*n*27, 167, 167*n*16, 173, 187, 187*n*12, 188, 189

Rickards, Bruce Errol Anthony, 231, 231*n*21, 233*n*28, 272, 272*n*2, 277, 287, 348. 378, 383, 392, 405
Riley, John Patrick, 386, 417
Riley, Laura, 416, 417, 419
Rise and Fall of Falmouth, The (Robinson), 23*n*16, 24*n*23
Roblin, Edna (first wife of Dexter Lunan Kennedy), 61
Rodney Affair (1968), 295*n*12, 296*n*14, 297*n*19
Roosevelt, Franklin D. (president), 183
Roosevelt, Theodore (president), 98
Royal Bank of Canada, 130
Russell, Alva, 354, 354*n*3, 356, 368, 368*n*20-*n*21, 411, 411*n*4, 458

S
Sangster, Donald (prime minister), 295, 295*n*11
schooling in Jamaica. *See under* Jamaica, POLITICS AND SOCIETY
Scott, Laurence Paul, 231–232, 231*n*22, 232*n*24, 273, 277, 293, 293*n*6, 309, 310, 318, 346, 348 through 350, 383, 384*n*6, 390
  tribute to LFK, 454-455
Seaga, Edward, 357, 364*n*10, 398, 399-400, 411
  as fifth prime minister of Jamaica, 399
  as leader of opposition JLP, 357–358
Sealey, Francis Kenneth (Ken), 419
seaplanes (1930s). *See* Pan American World Airways
Second Industrial Revolution (1870–1950), 235
Seivwright, Theresa Adelaide (m. Joseph Henry Moss-Solomon), 97, 303
Servite. *See* Order of Servants of Mary (OSM)
Sharp, Faustine, 383, 410, 410*n*3
Sharp, Thomas Hicks, 190, 190*n*33
Shearer, Hugh Lawson, 221, 221*n*3, 222, 232*n*2
  as third prime minister of Jamaica, 295
Shipping Association of Jamaica (SAJ). *See under* Johnston, Charles Edward; and Kennedy, Luis Fred
Shirley, Gordon, 424, 439–440, 439*n*52–*n*53
  appointment, non-executive chairman, 2014, 439
Shoucair, Rudolph "Ruddy," (family doctor), 353, 353*n*1
Society of Jesus. *See* Jesuits
Sosúa, Dominican Republic. *See under* United Fruit Company (UFCO); Kennedy, Fred William

Soutar, Simon
   as board director, GKCO, 228-229, 276, 346, 348
   as chair, GKCO board meeting (1976), 357
   first employment at GKCO (1941), 229n7
   as manager, GKCO, assistant to James Moss-Solomon, (1959), 228-229
   marriage, 178
   parents, 178n5
   photo, 198, 319
   retirement (1977), GKCO, 383
   service award (1962), GKCO, 255
   service award (1972), GKCO, 321
Soutar, William Donald, 92n15, 178n5
Soutar-Hynes, Mary Louise (Mary Lou), (LFK's niece), 133, 134n27, 174, 178n8, 180, 183, 257, 359, 409-410
   tribute to LFK, 455
Special Employment Program (SEP), 343
Stewart Town, Trelawny, 19, 24, 48, 53, 54, 303
St. George's College (STGC). *See under* Kennedy, Luis Fred, as student; Kennedy, Fred William (LFK's son), as student
Stockhausen, John, 53, 53n15, 54, 54n18
Sts. Peter and Paul Church, 179n12, 217, 281, 284, 285, 286, 286n6
sugar industry in Jamaica. *See under* Jamaica, ECONOMY
Supreme Court of Canada, 376

T

Tale of Two Cities, A (Dickens), 218, 218n7
Taylor, Eliza, 50, 50n2, 52, 90
Tenn, Mable, 347n23, 392, 410, 458
Terfloth, Boerries, 230, 230n14, 233, 271-272, 275n10, 309, 309n3, 310, 311, 319, 345, 366, 367
Terfloth & Kennedy Ltd., 347, 366, 367, 429, ` 420n16, 432, 456
Thatcher, Margaret (prime minister), 425
Thiesmeier, Frances (m. Charles Eugenio Loinaz), 150n23
Thomen, Charles (LMK's nephew), 150n24
Thomen, Emanuel Jacob, 150n24
Thomen, Louis (LMK's nephew), 150n24
Thompson, Abbie Herberta Barnes, 340n7, 355, 355n5, 359, 361
Thompson, Georgianne (Georgie) Ruth (m. Fred W. Kennedy), 340, 340n7, 354, 355, 357, 361, 373, 374, 377, 386n16, 401, 403, 420, 457
   genealogical research, 4, 24n20, 45n21, 63n27, 86 44n18, 139n10, 147n7, 162n33, 170n24, 178n4, 202n5, 215n2, 284n1, 338n2, 375n1, 416n2
   tribute to LFK, 452-453
Thompson, Kenneth Sterling, 340n7, 359
Thompson, Shirley, 357
Titchfield Hotel. *See under* Port Antonio
tourism. *See under* Jamaica, ECONOMY
Tower Isle Hotel, 174, 179n12, 190n30, 220, 449
Trelawny, Sir William (governor), 23n14
Trudeau, Pierre Elliott (prime minister), 385

U

UK. *See* Great Britain
United Fruit Company (UFCO)
   formation, 1899, 36
   Great White Fleet, 40, 50, 90-91
   labour strikes (1938–1939) and, 156-157, 160n26
   Montego Bay and, 59
   Port Antonio and, 51
   racial discrimination and, 42
   Sosúa, Dominican Republic (1901) and, 40, 41, 41n4
   travel packages developed by, 51
   white macho elite and, 26-27
   See also under banana industry; Kennedy, Fred William (LFK's father)
United Negro Improvement Association (UNIA), 79, 79n18–n19, 158n17
United Port Workers Union, 222
United States (US)
   Anglo-American Destroyers for Bases Agreement (1940), 169
   anti-immigration laws (1920s), 110
   Bureau of Immigration, 91, 110
   employees of UFCO in Jamaica, 51
   expansionism in Caribbean (early 20th century), 41
   intervention in Cuba, 41
   Jamaica's alliance, post-Independence, with, 262
   Jamaica's relationship (1970s) and, 398- 388
   Jim Crow laws (1920s), 110
   occupation of Dominican Republic (1916), 91, 138
   Prohibition (1920-1933), 110

protection of convoy system during WW II, 172
strategic interests in the Caribbean WW II, 167
War Trade Board (1917), 60
University of West Indies (UWI), 34n5, 296, 311, 340, 419, 427, 430, 435n34, 360n23, 439n52
See also under Kennedy, Celia Jean, Kennedy, Fred W (LFK's son); Kennedy, Mary Teresa, as student of
Up Park Camp, Jamaica Defence Force, 166, 166n6, 187
USAID, 362

V

Valliant, Fred William (Bill), 77n12, 122n16, 133n24, 174, 178n8, 257, 390, 390n5, 410
tribute to LFK, 455-456
Valliant, James Kennedy (Jim), 257, 410
Valliant, Jeanne Louise, 420
Valliant, John William, 178, 178n6
Vander Starr, Cornelius, 141, 141n20, 147
Vatican II (1962–1965), 217, 289
Vaz-Jones, Martin Joseph, 420n6
Vendryes, Hilda Blanche (m. William Donald Soutar), 178n5
Vernon, Alexander Apfel, 30, 30n7, 78n14
Vernon, Robert (Robbie), 29n4, 30n7
Victoria (queen), 181

W

Wall Street Crash of 1929. See Great Depression
Wall Street Journal, 92, 92n17–n18
Walters, Norma, 31n15
War Contingent Committee (Jamaica, 1914), 57– 58
Wehby, Donald George, 424, 435 through 438, 435n34, 442
appointment as Group CEO, GraceKennedy Ltd. (2011), 435
Western Meat Packers Co. Ltd. See under Grace, Kennedy & Company Ltd; Kennedy, Francis Xavier (Paco)
West Indies Federal Labour Party (WIFLP), 241, 241n16
West Indies Federation. See under Jamaica, POLITICS AND SOCIETY; Manley, Norman Washington

When Banana Was King: A Jamaican Banana King in Jim Crow America (Goffe), 35n10–n11, 36n12, 60n13, 60n15
Who's Who Jamaica 1916 (Hill, ed.), 26, 31n14
Who's Who Jamaica 1919-1920 (Hill, ed.), 59n10
Who's Who Jamaica 1957 (Neita, ed.), 117, 212n16, 216n3, 226, 238n2, 240n13, 301
Who's Who Jamaica 1963 (Neita, ed.), 98n8, 179n13, 190n30, 231n22, 277n21, 292n3
Williams, Eric, 237, 237n1, 241
as president, People's National Movement (PNM) (1956), 244
Williams, Wayne Anthony, 420n6
Williamson, Marjory Lisa (m. Ernest Michael Johnston), 339n4
Wilmot, Swithin, 20, 20n6, 21n9, 25, 25n26, 30n7, 35n9, 76–77, 77n10, 239n8, 338n3, 377, 400, 457
Wolff, Jules, 132, 166, 166n4, 229
World Bank, 223, 426n3–n4
World Trade Organization (WTO), 241n21
World War I (The Great War), 57 through 60, 60n14, 78-79
World War II, 165 through 170, 166n7, 201n2
W. R. Grace & Co. of New York, 62, 62n24, 89, 89n7, 90, 92, 95, 98, 129, 131, 206-207, 206n23–n24, 275n10

Y

York Castle High School, Alderton, St. Ann. See under Kennedy, Fred William (LFK's father)

About the Author

"Happiest when boating in the Caribbean."
(Photo by Georgie Kennedy,
Ocho Rios, 2022)

Fred W. Kennedy was born and raised in Jamaica. He holds a Bachelor of Arts and Graduate Diploma in Education from the University of the West Indies (UWI) and a Master and Doctor of Education from the University of Toronto. After thirty years of serving as an educator and principal, Fred turned to writing Jamaican historical fiction. He is the author of *Daddy Sharpe* (2008) and *Huareo* (2015). *Firstborn* is his first published work of nonfiction.

He wrote his father's biography "to celebrate the relationship of love and trust that we shared" and as "a tribute to him in praise of his contributions to the national development of Jamaica." Fred remains connected to the company as Chair of the GraceKennedy Foundation, which funds educational, environmental and health initiatives in Jamaica.

He and his wife Georgianne share their time between his native Jamaica and adopted Canada, where their three daughters and families reside.